# Advances in Science, Technology & Innovation

## IEREK Interdisciplinary Series for Sustainable Development

**Advances in Science, Technology & Innovation (ASTI)** is a series of peer-reviewed books based on important emerging research that redefines the current disciplinary boundaries in science, technology and innovation (STI) in order to develop integrated concepts for sustainable development. It not only discusses the progress made towards securing more resources, allocating smarter solutions, and rebalancing the relationship between nature and people, but also provides in-depth insights from comprehensive research that addresses the **17 sustainable development goals (SDGs)** as set out by the UN for 2030.

The series draws on the best research papers from various IEREK and other international conferences to promote the creation and development of viable solutions for a **sustainable future and a positive societal** transformation with the help of integrated and innovative science-based approaches. Including interdisciplinary contributions, it presents innovative approaches and highlights how they can best support both economic and sustainable development, through better use of data, more effective institutions, and global, local and individual action, for the welfare of all societies.

The series particularly features conceptual and empirical contributions from various interrelated fields of science, technology and innovation, with an emphasis on digital transformation, that focus on providing practical solutions to **ensure food, water and energy security to achieve the SDGs.** It also presents new case studies offering concrete examples of how to resolve sustainable urbanization and environmental issues in different regions of the world.

The series is intended for professionals in research and teaching, consultancies and industry, and government and international organizations. Published in collaboration with IEREK, the Springer ASTI series will acquaint readers with essential new studies in STI for sustainable development.

**ASTI series has now been accepted for Scopus (September 2020). All content published in this series will start appearing on the Scopus site in early 2021.**

More information about this series at http://www.springer.com/series/15883

Manuel Abrunhosa · António Chambel ·
Silvia Peppoloni · Helder I. Chaminé
Editors

# Advances in Geoethics and Groundwater Management: Theory and Practice for a Sustainable Development

## Proceedings of the 1st Congress on Geoethics and Groundwater Management (GEOETH&GWM'20), Porto, Portugal 2020

Springer

*Editors*
Manuel Abrunhosa
Portuguese Chapter of International Association of Hydrogeologists
CITEUC, University of Coimbra
Porto, Portugal

Silvia Peppoloni
International Association for Promoting Geoethics (IAPG)
Istituto Nazionale di Geofisica e Vulcanologia (INVG)
Rome, Italy

António Chambel
International Association of Hydrogeologists (IAH), Department of Geosciences, Institute of Earth Sciences
University of Évora
Évora, Portugal

Helder I. Chaminé
Portuguese Chapter of International Association of Hydrogeologists, Laboratory of Cartography and Applied Geology, Department of Geotechnical Engineering, School of Engineering (ISEP)
Polytechnic of Porto
Porto, Portugal

ISSN 2522-8714 ISSN 2522-8722 (electronic)
Advances in Science, Technology & Innovation
IEREK Interdisciplinary Series for Sustainable Development
ISBN 978-3-030-59319-3 ISBN 978-3-030-59320-9 (eBook)
https://doi.org/10.1007/978-3-030-59320-9



This Springer imprint is published by the registered company Springer Nature Switzerland AG
The registered company address is: Gewerbestrasse 11, 6330 Cham, Switzerland

***In Memoriam***

***Professor Partha Sarathi Datta (1950–2018), India***

*"Instead of trying to prove the relevance of excellent fields let us develop excellence in relevant fields."*

***Professor Luís Ribeiro (1953–2020), Portugal***

*"Nature is full of music. Being the elements of nature, and especially water, a source of inspiration in the history of music, there is a sensation in this journey through the musical production of different eras that music can also be an inspiration and a means of reconnecting man with nature."*

# Committees

## Scientific Committee

### Chairs

Manuel Abrunhosa, IAH-PC and CITEUC, Portugal
António Chambel, IAH and UÉ, Portugal
Silvia Peppoloni, IAPG and INVG, Italy
Helder I. Chaminé, ISEP and IAH-PC, Portugal

### Board

Acacia Naves, UDC, Spain
Adilson Pinheiro, ABRH, Brazil
Ahmed Fekri, UHC, Morocco
Alberto Garrido, UPM, Spain
Alberto Gomes, FLUP, Portugal
Alcides Pereira, FCTUC, Portugal
Alexandra Aragão, FDUC, Portugal
Alexandre Tavares, FCTUC, Portugal
Alice Aureli, IHP-UNESCO, France
Ana Alencoão, UTAD, Portugal
Ana Isabel Andrade, CITEUC, Portugal
Ana Paula Pereira, LNEG, Portugal
Ana Raquel Moniz, FDUC, Portugal
Amélia Dill, UALG, Portugal
Andrés Sahuquillo, UPV, Spain
António Diniz Ferreira, ESAC, Portugal
Antonio Pulido-Bosch, UG, Spain
Axel Gosseries, UCL, Belgium
Bartolomé Andreo-Navarro, UMA, Spain
Broder Merkel, UTF, Germany
Bruce Misstear, TCD, Ireland
Carla Lourenço, DGEG, Portugal
Carlos Costa Almeida, FCUL, Portugal
Carlos Molano, ULA, Colombia
Cristina Delerue-Matos, ISEP, Portugal
Daniela Ducci, USN, Italy
Dave Kreamer, UNLV, USA
David Crookall, UNSA, France
Diana Allen, SFU, Canada
Eduardo Ferreira da Silva, UA, Portugal
Eduardo Vivas, APRH-NRN, Portugal

Emilia Bocanegra, UNMP, Argentina
Emilio Custodio, UPC, Spain
Estrella Santacruz Pérez, UNAICC, Cuba
Eva Hernández, MAVA|WWF, Spain
Ezzoura Errami, CDU, Morocco
Fernando Pacheco, UTAD, Portugal
Fernando Pedro Figueiredo, FCTUC, Portugal
Fernando Rocha, UA, Portugal
Francisco Javier Elorza, UPM, Spain
Francisco Javier Samper, UDC, Spain
Giuseppe Di Capua, INGV and IAPG, Italy
Giuseppe Sappa, ULR, Italy
Gopal krishan, NIHR, India
Helena Pereira de Melo, FDUNL, Portugal
Huaming Guo, CUG, China
Hycienth O. Nwankwoala, UPH, Nigeria
Jean-Christophe Comte, SG|UA, UK
James W. LaMoreaux, PELA, USA
João Carlos Nunes, UAç, Portugal
João Paulo Lobo-Ferreira, LNEC, Portugal
Joaquim Góis, FEUP, Portugal
Joaquim Poças Martins, FEUP, Portugal
John Cherry, G360|UG, Canada
Jorge Espinha Marques, FCUP, Portugal
Jorge Luis Loredo, UO, Spain
José António Simões Cortez, FEUP, Portugal
José Joel Carrillo-Rivera, UNAM, Mexico
José Manuel Azevedo, FCTUC, Portugal
José Manuel Marques, IST, Portugal
José Martins Carvalho, ISEP, Portugal
José Paulo Monteiro, UALG, Portugal
José Teixeira, FLUP, Portugal
José Virgílio Cruz, UAç, Portugal
Juan Carlos Santamarta, ULL, Spain
Juliana Freitas, UNIFESP, Brazil
Ken Howard, UTS, Canada
Liankai Zhang, IKG, China
Lucia De Stefano, UCM, Spain
Lúcia Guilhermino, CIIMAR and ICBAS|UP, Portugal
Luciana Cordeiro, UNICAMP, Brazil
Luís Aires-Barros, IST, Portugal
Luís Ferreira Gomes, UBI, Portugal
Luis González de Vallejo, UCM, Spain
Luís Neves, FCTUC, Portugal
Luís Ribeiro, IST, Portugal
Luís Zêzere, IGOT|UL, Portugal
Luiz Fernando Scheibe, UFSC, Brazil
Maite M. Aldaya, ISFOOD|UPN, Spain
Manuel Oliveira, LNEC, Portugal
Manuel Oliveira da Silva, FCUL, Portugal
Manuel Ramón Llamas, UCM, Spain
Manuela Simões, FCT|UNL, Portugal
Marcela Perez, UNL, Argentina

## Honour Committee

Under the High Patronage of the President of the Portuguese Republic—Prof. Marcelo Rebelo de Sousa
Adilson Pinheiro, ABRH, Brazil
Alice Aureli, IHP-UNESCO, France
Bartolomé Andreo-Navarro, CEHIUMA, Spain
Carlos Mineiro Aires, OE, Portugal
Fidel Ribera Urenda, FCIHS, Spain
João Oliveira, APPBG, Portugal
João Rocha, IPP, Portugal
José Ferreira Gomes, CdC, Portugal
José M. Sousa, OET, Portugal
José Romão, APG, Portugal
Juan Jose Durán Valsero, IAH-SC, Spain
Manuel Matos Fernandes, SPG, Portugal
Maria João Viamonte, ISEP, Portugal
Mário Rui Machado Leite, LNEG, Portugal
Susana Neto, APRH, Portugal

## Organizing Committee

### Chairman

Manuel Abrunhosa (IAH-PC)

### General Chairs

António Chambel (IAH)
Silvia Peppoloni (IAPG)
Bartolomé Andreo-Navarro (IAH-SC)
Helder I. Chaminé (IAH-PC and ISEP)

### Advisor Board

José Manuel Marques (IAH-PC)
Raquel Sousa (IAH-PC)
Rui Hugman (IAH-PC)
Maria José Afonso (ISEP)
Teodóra Szőcs (IAH)
Giuseppe Di Capua (IAPG)
Juan José Durán Valsero (IAH-SC)
Sergio Martos-Rosillo (IAH-SC)
Ana Isabel Andrade (IAH-PC)
Clara Vasconcelos (IAPG-PC)
Teresa Albuquerque (IPCB)

### Local Organizing Committee

Manuel Abrunhosa (IAH-PC)
Helder I. Chaminé (IAH-PC and LABCARGA|ISEP)
Liliana Freitas (LABCARGA|ISEP and FCTUC)
José Teixeira (FLUP and LABCARGA|ISEP)
João Paulo Meixedo (LABCARGA|ISEP)

## The Conference Logo: Creation, Design and Symbolic

A powerful logo was created by the designer Joel Vilas Boas (GDM|ISEP) for the Congress GEOETH&GWM'20. In its apparent simplicity, it represents a large amount of symbolism underlining the place and aims of the conference. The overall shape and colours are reminiscent of the square ceramic wall tiles in contrasting white and cobalt blue that are a recognized hallmark in Portugal's civilian, religious and military buildings, since the twenty-six century, in the Porto urban area. They were used domestically and exported worldwide, not only as a commodity hand made by the millions, but also as a fortunate association of its technological excellence and usefulness with the expression of new values, ideas and aesthetics proposed to other cultures around the world, not without setting a distinct enduring mark from its origins we, as Portuguese citizens, are proud of.

In the composition of the logo, only four simple glyphs in ten arrangements were used, harmoniously distributed inside a line bordered square. An undrawn but distinct central vertical axis of symmetry defines a path for a bottom-up reading of several symbols that set out by

**(Logo design: Joel Vilas Boas, 2019)**

**(Photo: Jaime Neto; Logo: Joel Vilas Boas, 2019)**

GEOETHICS & GROUNDWATER MANAGEMENT CONGRESS
PORTO · PORTUGAL
MAY 18-22 · 2020

**(Meeting related to the discussion of the logo design; Joel Vilas Botas and Manuel Abrunhosa)**

deconstruction and recomposing to define, in quite a few lines what, in our view, Geoethics and Groundwater Management is. A bottom line could represent the strong foundations of Geoethics and Groundwater Science, from which emerge recognized symbols of drawdown curves, and with them the consequences of the exploitation of aquifers underground. Above is what could be a water well head and the processes that may occur on the surface in interaction with the society and the underground. Then, the acronym of the Congress is brought to our minds through face-to-face double-G physiognomies of what might be geoethics and groundwater science/engineering, a wide M for management below. These opposing, often conflicting views, and the need of an informed dialogue, together with the underlying reflections on all values involved and on the consideration of foreseeable consequences of actions and omissions, are the correct supporting paths to foster the emergence of responsible management tools. This could be the readings underlining the perseverance and quality for the construction and maintenance of the sequence of arch bridges connecting the extremes on top of the logo. They evocate the at their times innovative six bridges of Porto over the Douro River, fed in summer by far away aquifers in its vast transboundary catchment in Iberian Peninsula. This supports a safe road to the future of the society and nature in resiliency and sustainability, through the combined efforts of geoscience, engineering, humanities, culture and arts, to cope with the challenges of a changing world.

## Side-Event Winning Video: *Before the Last Drop—Viva la Geoethics!*

A Side-Event to the GEOETH&GWM'20 Congress was organized in close cooperation with the section of IAPG–Portugal to run during the academic year 2018/19. It aims to promote geoethics and responsible groundwater management awareness among students and teachers in secondary schools in Portugal. This approach takes advantage of the recent legal framework that allows Secondary School free choice for a part of the syllabus. The outcome is a contest of 3 minute digital videos in free common layout produced by the students under the teacher's supervision. The content of videos may cover scientific, humanistic or artistic issues related to groundwater, in any combination, favouring transdisciplinary thought and emotional involvement on the themes of the GEOETH&GWM'20. Three best videos were selected by an independent jury and side-event chairs.

The 1st prize was awarded to the outstanding video called "*Before the Last Drop—Viva la Geoethics!*", authored by several students from class 11ºB (2018/19) of the Basic and Secondary School of Fontes Pereira de Melo (Porto, Portugal), and the advisers' teachers were Sandra Eustáquio Ferraz, Marta Paz and Maria de Lurdes Alves. The 2nd prize was given to the Colégio da Rainha Santa Isabel (Coimbra) and the 3rd prize ex-aequo to the schools: Agrupamento de Escolas Anselmo de Andrade (Almada), Colégio de Gaia (Vila Nova de Gaia) and Escola Básica e Secundária Caldas das Taipas (Guimarães).

Water is a vital asset for all terrestrial systems and has always been a determining factor in establishing life in general and human populations in particular. Historically, civilizations have always flourished in settlements with an abundance of water and collapsed when water no longer available. Groundwater, an invisible component of the water and hydrosphere cycle, accounts for more than 97% of all liquid freshwater on the planet and contributes a large percentage to all human uses of it, be it public or domestic supply, agricultural production, livestock or industry, also constituting the main source of supply in regions with water scarcity and acting as a buffer against extreme weather events. In a context marked by increasing environmental challenges imposed by societies that are increasingly overcrowded and marked by excessive consumerism, it is essential to involve the school in order to sensitize young people to these themes.

Currently, the curricular area of earth sciences is crucial for the exercise of responsible citizenship, given the need to understand problems and make informed decisions on issues that affect societies and subsystems on the planet Earth. Society in general, and the school in particular, must embrace the mission of training youth not only scientifically educated, but also ethically just and balanced, endowing them with the necessary skills to intervene in a well-founded manner in matters of a technical, scientific and ethical nature, in an increasingly challenging world and from the perspective of democratic citizenship.

The GEOETH&GWM'20 offers the unprecedented possibility for students of basic and or secondary education to participate in a scientific event of this nature, contributing to promote their motivation and enthusiasm regarding science. The work on a project whose theme is of indisputable importance in today's societies also allows for the scientific deepening of the theme and the development of several essential skills and attitudes in modern societies. In addition, the collaborative production of an audio-visual system favours students' motivation, their artistic sensitivity and the use of technology, which young people deal with on a daily basis, can be combined with pedagogy under the specialized guidance of the teacher.

Considering all these assumptions, the winning project ***Before the last drop—Viva la Geoethics!*** was developed with the class 11ºB of the Basic and Secondary School of Fontes Pereira de Melo (Porto, Portugal), within the course of biology and geology. The designed audio-visual consists of the adaptation of a lyrics suggesting to the theme of geoethics and sustainability of groundwater, starting from a song well known to the general public, called Viva La Vida, by the band Coldplay.

The class was divided into three groups, one being responsible for choosing the song and adapting the lyrics, another for choosing/drawing the images to be used and finally the third for collecting relevant data/keywords on the topic. Throughout the project, students showed interest and motivation, increasing their autonomy, responsibility and creativeness. Additionally, they all achieved the proposed goals. In addition, to increasing their scientific knowledge about groundwater, they also developed awareness of their importance and the key relationship to a sustainable development and geoethical issues.

Lastly, these types of initiatives, connecting the school community (students, educators and technicians), the scientific community and society, constitute real opportunities to simultaneously promote the building of awareness and knowledge, as well as values, attitudes and competences, from the perspective of active citizenship for a more sustainable and geoethical future.

***Winning Video Authors***

***Students from class 11ºB (2018/19) of the School of Fontes Pereira de Melo (Porto, Portugal)***: Alice Pereira; Ana Raquel Costa; Ana Rita Ferreira; António Ferreira; Beatriz Martins; Catarina Ribeiro; Henrique Macieira; Joana Oliveira; Leonor Ferreira; Luís Leite; Margarida Silva; Matheus Bissacot; Nuno Ferreira; Rúben Almeida; Rui Fonseca; Rui Nascimento; and Tiago Reis.
***Supervisor Teachers***: Sandra Eustáquio Ferraz, Marta Paz and Maria de Lurdes Alves

***Acknowledgments***

A word of appreciation to the Board of the Basic and Secondary School of Fontes Pereira de Melo (Porto, Portugal) and the coordination of the Support Unit for High School Performance (UAARE), for arranging space and facilities available in the unit's classroom; to Professor Filipe Pinto, for his help with choosing image editing software; to Professor Pedro Seco, for making students available during an academic period for recording the voice; to João Guedes, for his support in editing the audio-visual voice; and to Ricardo Guedes, for all his support in the final audio-visual edition.

***Side-Event Chairs***

Clara Vasconcelos (IAPG—Portugal, and FCUP), João Oliveira (APPBG and CITEUC), Ana Isabel Andrade (IAH-PC and CITEUC) and Manuel Abrunhosa (IAH-PC, IAPG and CITEUC).

# Foreword by Giuseppe Di Capua

Water is life. Water is everywhere on the planet, and humans search for it in the space trying to get indications about possible presence of extraterrestrial life. In future, we might fight for water.

The management of water resources, the access to drinking water and sanitation are issues that involve technical–scientific aspects, and also problems of social equity, intra- and inter-generational justice.

According to the 2019 World Water Development Report of United Nations, "*... the global water demand is expected to continue increasing at the current rate until 2050. Over 2 billion people live in countries experiencing high water stress, and three out of ten people do not have access to drinking water*". The same report also indicates that stress levels will continue to increase, as demand for water grows, and the effects of climate change intensify.

Water is an unalienable human right, a guarantee of the dignity of each individual. And even if each nation has the right to develop policies to safeguard its interests and priorities, nobody can contravene the fundamental right to access water that vital resource on which life on Earth depends.

Groundwater is considered a renewable resource of freshwater, sustaining human health and activities, and ecosystems. It needs to be managed carefully, and if we want it remains renewable, in a wise and responsible way, while respecting natural dynamics, cultural traditions, other living beings. Pollution, salinization and overexploitation are major threats to its usage, as well as climate change can bring groundwater to be depleted and lost as a resource.

Geoscientists and water-related practitioners have the (geo)ethical duty to support society in defining best ways to manage groundwater, but this implies competence in assuring high level of professionalism, accountability in applying scientific knowledge and providing sustainable solutions, a continuous dialogue with stakeholders and society, integrity in conducting their work.

From 2012, the International Association for Promoting Geoethics works to strengthen the awareness of geoscientists of the need for an ethical approach to georesources and in particular water resources. This means to carefully manage problems related to environmental impacts produced by human interventions on the natural processes that govern surface and groundwater resources, and also to develop strategies in order to harmonize expectations and requests of various stakeholders, including citizens, industry and policy-makers.

The Congress "Geoethics and Groundwater Management" has been a great step to make the hydrogeological community even more aware of its responsibilities and commitments towards society and the planet as a whole and a fundamental moment to bring together experiences, to analyse cases and to propose solutions. As Silvia Peppoloni, Secretary General of the IAPG-International Association for Promoting Geoethics, stated during her introductive speech at the Porto Congress (18–22 May 2020), "*We, as geoscientists and engineers, have ethical and social responsibilities, which arise from the fact of possessing specific knowledge and experience that are able to protect citizens and the environment, and to ensure the sustainable development of human communities. Whatever is our role, researchers, professionals, educators, in each circumstance we have to put our knowledge and experience at disposal of society to face and live with the environmental challenges of our times. And in doing this, we have great responsibilities. ... geoethics deals with those responsibilities. Acting geoetically presupposes the awareness of being responsible*".

Since geoethics "*...consists of research and reflection on the values which underpin appropriate behaviours and practices, wherever human activities interact with the Earth system (Bobrowsky et al. 2017, p. 5; Peppoloni and Di Capua 2015, pp. 4–5; Peppoloni et al. 2019, p. 30)... (and) deals with the ethical, social and cultural implications of geoscience knowledge, research, practice, education and communication, and with the social role and responsibility of geoscientists in conducting their activities (Di Capua et al. 2017; Peppoloni and Di Capua 2017)*", it is clear the importance to apply the geoethical thinking to groundwater management in order to achieve a more responsible use of water in general and groundwater specifically.

In recent years, many authors have already suggested and developed concepts related to the ethical, social and cultural aspects of hydrogeological knowledge and practice, such as "water ethics" (Groenfeldt 2019), "socio-hydrology" (Sivapalan et al. 2012) or "socio-hydrogeology" (Re 2015).

In addition to them and in line with the definition of geoethics, the concept of "hydrogeoethics" was proposed in 2017 by António Chambel, President of the IAH—International Association of Hydrogeologists, and Manuel Abrunhosa, President of the Portuguese Chapter of IAH and Chair of the Congress "Geoethics and Groundwater Management". It highlights the peculiarities of hydrogeology and groundwater studies and applications from a geoethical perspective (Abrunhosa et al. 2018). Hydrogeoethics can be considered the field of geoethics focused on ethical research and best practices related to responsible groundwater science and engineering, aimed at creating conditions for sustainable water resources management while respecting human needs and environmental dynamics. Its studying objects are related to transdisciplinary fields in geosciences, anthropological and social sciences, dealing with the relationship between humans and water cycle, cultural, aesthetic and historic traditions linked to water uses, legal frameworks, best practices and governance, groundwater management–society–policy interface.

Just like geoethics, hydrogeoethics is also founded on the principle of responsibility, the ethical criterion that should guide any human action on social–ecological systems. And just like geoethics, it can be defined through the same characteristics (Peppoloni et al. 2019): geoscience knowledge-based, contextualized in time and space, human agent-centric and shaped as virtue ethics.

This book represents a great asset and source of hydrogeological knowledge, professional experiences, case studies, practical solutions, social and cultural insights, inspired by a responsible approach: a valuable legacy authors are leaving to future generations.

Rome, Italy
July 2020

Giuseppe Di Capua
Istituto Nazionale di Geofisica e Vulcanologia (INVG)
International Association for Promoting Geoethics (IAPG)

## References

Abrunhosa MJ, Peppoloni S, Chambel A (2018) Introducing geoethics in groundwater science, technology, and management. In: Abstracts book, session T1.1 socioeconomic aspects of groundwater resources management, 45th IAH Congress, Daejeon, Korea, p 26

Bobrowsky P, Cronin VS, Di Capua G, Kieffer SW, Peppoloni S (2017) The emerging field of geoethics. In: Gundersen LC (ed) Scientific integrity and ethics with applications to the geosciences, Special Publication American Geophysical Union, John Wiley and Sons, 73:175–212

Di Capua G, Peppoloni S, Bobrowsky PT (2017) The Cape Town Statement on Geoethics. In: Geoethics at the Heart of All Geoscience. Annals of Geophysics, 60(7). https://doi.org/10.4401/ag-7553

Groenfeldt D (2019) Water ethics: a values approach to solving the water crisis, 2nd edn. Routledge, New York

Peppoloni S, Di Capua G, Bilham N (2019) Contemporary geoethics within the geosciences. In: Bohle M (ed) Exploring geoethics: ethical implications, societal contexts, and professional obligations of the geosciences. Palgrave Macmillan, Cham, pp 25–70

Peppoloni S, Di Capua G. (2015) The meaning of geoethics. In: Wyss M, Peppoloni S (eds), Geoethics: ethical challenges and case studies in earth sciences. Elsevier, Amsterdam, p 3–14

Peppoloni S, Di Capua G (2017) Geoethics: ethical, social and cultural implications in geosciences. In: Geoethics at the heart of all geoscience. Annals of Geophysics, 60(7). https://doi.org/10.4401/ag-7473

Re V (2015) Incorporating the social dimension into hydrogeochemical investigations for rural development: the Bir Al-Nas approach for socio-hydrogeology. Hydrogeol J 23:1293–1304

Sivapalan M, Savenije H, Bloeschl G (2012) Socio-hydrology: a new science of people and water. Hydrol Proc 26:1270–1276

# Foreword by Nabil Khélifi

With the rapid growth in global population, food demand, urbanization and industrialization, the need for water resources has surpassed all expectations. In particular, groundwater resources have become vulnerable to degradation and depletion even though more than 1.2 billion people are excluded from using these resources and remain without access to safe drinking water. Increased groundwater use and associated water pollution levels have crossed sustainable thresholds in many parts of the world. Today, 70% of the world's groundwater withdrawals are used for irrigation purposes as over 40% of global food is produced through irrigation. Moreover, because over 50% of the world's population now lives in urban areas, dependency on groundwater has dramatically increased. This situation has created an imbalance between the demand and the availability of this valuable resource, hence the need for effective, efficient and sustainable management and development of groundwater resources. However, decision-makers and planners face various ethical dilemmas since the importance of this fundamental resource means that there is a complicated relationship between water policies and ethical considerations. The lack of an ethical framework is indeed as big a hindrance as the other major factors causing the water crisis, i.e. increasing demand, zonal disparity in the distribution of water supply, major land-use changes, decline in long-term water level, and increased salinity and pollution. Accordingly, there is a growing consciousness of the need to address the key issues in the ethics of groundwater usage, especially given the paucity of responsible water administration due to limited knowledge of the groundwater situation.

Before we can address groundwater ethics, the main problem is that we still do not have a good understanding of the world's groundwater resources and how to sustainably manage aquifers despite the growing concern of a large number of scientists, practitioners and experts that global groundwater resources are threatened. The lack of information about groundwaters increases the risk of further negligence of the ethics surrounding groundwater extraction and supply. This will have significant repercussions on the sustainable management of water resources, thus will aggravate the tensions surrounding our dependence on groundwater and water and food security across the globe and will hamper the building of resilience to climate change. As concern in this field heightens, a large group of stakeholders are committed to addressing issues related to the ethics of groundwater use from various angles and at different temporal and geographic scales.

For example, UN-Water, at its 30th meeting in Rome, Italy, in January 2019, announced that the theme for World Water Day in 2022 will be *Groundwater—Making the Invisible Visible*. This is an important step in creating awareness of the essential role of global groundwater resources. This initiative has been further endorsed by UN-Water at its 32nd meeting in Rome, Italy, in January 2020 through the announcement of a proposal to hold a Groundwater Summit in 2022. This summit was proposed by the International Groundwater Resources Assessment Centre (IGRAC) along with several UN-Water members and partners, including partners from the Groundwater Solutions Initiative for Policy and Practice (GRIPP), such as the United Nations Educational, Scientific and Cultural Organization—Intergovernmental Hydrological Programme (UNESCO-IHP), International Association of Hydrogeologists (IAH) and the

International Water Management Institute (IWMI). The objective of the Groundwater Summit is to improve the science–policy–practice interface by highlighting the role of groundwater in the broader socio-economic and environmental context and by providing information on groundwater management and governance. The proposed Groundwater Summit and related initiatives and outcomes will be closely aligned with the United Nations 2030 Agenda for Sustainable Development and Sustainable Development Goals (SDGs), the Paris Agreement on Climate Change, Groundwater Governance—A Global Framework for Action, and the Sendai Framework for Disaster Risk Reduction 2015–2030.

In this context, the School of Engineering (ISEP), Polytechnic of Porto in Portugal, made the bold step of organizing the *1st International Congress Geoethics & Groundwater Management: Theory and Practice for a Sustainable Development (GEOETH&GWM'20)* in an online format from Porto in May 2020. It took an unprecedented global approach to the vast subjects of geoethics in groundwater management and to the recognized need for reflection on the correct and prudent actions by discussing theory and practice and by sharing values, knowledge, research, educational projects, best practices and strategies in order to institute responsible integrated management of groundwater resources for a resilient and sustainable future. In a world seeking answers, the aim of GEOETH&GWM'20 was to mobilize a courageous scientific and professional community capable of proposing synergetic scientific, cultural and practical answers to the complex problems affecting society in all its connections with groundwater and the hydrosphere in general. These issues were also raised by Dr. John Cherry, winner of the Stockholm Water Prize 2020, in his influential preamble. To address them, the leading institutions, International Association of Hydrogeologists (IAH) and International Association for Promoting Geoethics (IAPG), have embraced this cause by creating a unique synergy through the GEOETH&GWM'20. Lastly, the conference in Porto produced a landmark book on the new transdisciplinary concept of hydrogeoethics.

It seems clear that groundwater management and development is a topic that calls for the attention of worldwide water experts. In Springer, we share this concern as we have already launched an interdisciplinary publishing programme in the field of water resource management. We offer books, journals and book series on topics such as hydrology and water management; water industry and water technology; and water quality and water pollution. Our publications by international top authors highlight various aspects of the water sciences and advance the latest research results.

This edited volume will complement our Springer Water programme by highlighting case studies on the general topic of *Geoethics in Groundwater Use and Management*. It comprises over one hundred selected proceedings papers from the GEOETH&GWM'20. I would like to thank the editors for constructing a superb volume of work, as well as the reviewers and authors of the chapters for their efforts and confidence in Springer, the leading global publisher of academic books, by sharing their contributions to the new field of hydrogeoethics!

Heidelberg, Germany
July 2020

Nabil Khélifi
Senior Publishing Editor

# Preface

*Ethics are moral principles and values that govern the actions and decisions of an individual or group. Ethical behaviour comprises of honesty, trust, treating others fairly and loyally. Ethical perception may vary from person to person, among societies and countries. [...] Choices based on the best obtainable detailed scientific information, guided by ethical considerations, offer the best hope to protect groundwater from depletion and pollution (Datta 2005).*

Groundwater stored transiently in aquifers is, by far, the most abundant and widespread source of liquid freshwater on the planet (e.g. Shiklomanov 1998, Zektser and Everett 2004, Richts 2011). Its importance to societies is attested by the facts that worldwide about 50% of the public water supply, 40% of irrigation and 35% of industrial uses rely on groundwater. Rivers maintain a baseflow between sparse rainfall events in the basin because groundwater discharges invisibly and continuously to those water bodies that are often incorrectly considered as examples of surface waters originated by overland and run-off flows. Ecosystems dependent on groundwater constitute important repositories of biodiversity, areas of carbon sequestration and food production and have a significant role in local climate. Groundwater plays a determinant role in many engineering interactions with the subsurface, in seawater intrusion and in geothermal energy use, and a sink of energy needed for pumping water from underground. Groundwater is also an important part of climate change adaptation process and is often a solution for people without access to safe water. The quality of groundwater, natural of affected by pollution, has considerable impacts on human and crop health. Natural springs, besides being the most ancient direct access to groundwater, traditionally valued also by the quality of water, normal or mineral with attributes in health treatment, is also a symbol of purity that cultures and religions cherish and protect. However, when freshwater resources come forward, mostly rivers, lakes and artificial reservoirs are mentioned, forgetting groundwater that, by its nature, is a mostly hidden component of the water cycle. Gleetson et al. (2020) state an impressive thought: "holistically understanding, evaluating, and maintaining the water cycle's role for a resilient Earth System is extremely challenging and urgent in the Anthropocene, as the societal complexities interlock with the complex dynamics of the Earth System". In general, groundwater keeps being a disregarded subject by citizens, decision-makers and even scientists, other professionals and the citizens in some way related to water resources, ignoring its interlinkage and essential roles in the water cycle, the ecosystems and the functioning of society. Tortajada and Biswas (2017) highlighted a key issue focused on the quality of water as a human right and contributing to the balance of the ecosystems. Insufficient knowledge motivates a lack of proportional and responsible actions. This may be at the source of the threats to groundwater despite the importance of the economic, ecological, geological, health and cultural services it provides. As a consequence, and at their peculiar rhythms, the quantity and the quality of groundwater change due to intensive and inappropriate anthropogenic actions coupled to stresses coming from the natural dynamics of the Earth, climate change, population growth and patterning and health, economic development and also an insufficient investment in knowledge, public awareness, proper governance and management at all levels, from global to local. Meanwhile, there are aquifers that remain untapped in regions or periods of water scarcity. To raise global awareness about the roles played by the hidden groundwater, the UN-Water (2015) decided that "Groundwater: making the invisible visible" would be the theme for the World Water Day 2022. This is a promising step forward, but probably insufficient because nothing really new will be added. This situation configures a case of "the tragedy of the commons" because to "look for solutions in the area of science and technology only, the result will be to worsen the situation" (Hardin 1968), whereby an effective step forward can only be addressed if coupled with a shift in the paradigm of integrated (in substance and values) water resources management in sustainable development.

Hydrogeology is an established geoscience that studies the occurrence, movement and quality of groundwater as a basis for understanding this essential natural resource as a component of the water cycle and in the society, providing the scientific support for the management of its diverse environmental and anthropogenic uses (Freeze and Cherry 1979).

Geoethics is an emerging scientific field that deals with the ethical, social and cultural implications of geosciences knowledge, research, practice, education and communication, and with the relevant social role and responsibility of geo-professionals in conducting their activities while interacting with the Earth systems (e.g. Wyss and Peppoloni 2015, Bohle 2015, Bobrowsky et al. 2017, Peppoloni and Di Capua 2017, 2018, Bohle 2019 and references therein), where groundwater is one of its undisputed important components. Besides, the landmark publications related to the geoethics through the languages of the world and sharing ethical principles through cultural diversity (Peppoloni 2015, 2018) are an inspirational backbone aiming the scientific and technical integrity and culturally diverse approaches.

The ancestral relationship between early human evolution, settlements and water includes, among others, groundwater as a human evolution driver, pile dwellings on lakes and use of canals associated with rivers, rainwater-harvesting systems, wells, aqueducts, water mines, springs, and underground cisterns (e.g., Wittfogel 1956, Pétrequin 1984, Tempelhoff et al. 2009, Angelakis et al. 2012, Chaminé et al. 2014, Cuthbert and Ashley 2014, Lugo-Enrich and Mejías 2017, Ollivier et al. 2018). The wide diversity, scale, significance and increasing magnitude of the interactions of anthropogenic behaviour with aquifers and groundwater, sets the dilemma of ecocentric versus anthropocentric visions aggravated by lack of explicit consideration of the cultural and religious visions (Ribeiro 2017), involves some degree of conflict of budgets, and also of values or interests, decisions and demands from the all agents involved, calling for action for a water ethos grounded in eco-sociocultural responsibility, security concerns, technical-scientific integrity and societal approach to a sustainable groundwater use and management. Those needs of a responsible water ethics perspective are highlighted, among others, by Llamas (1975), Leopold (1990), Custodio (2000), Soromenho-Marques (2003), Llamas (2004), Datta (2005), Arrojo-Agudo (2010), Braga et al. (2014), Ribeiro (2017) and Abrunhosa et al. (2018). In a recent interview, Dr. John Cherry highlighted some impressive thoughts related to the key role of the water in society: "To make groundwater more visible, we need to get people to ask more questions about water and groundwater in particular", and also "We need more curiosity about water in the educational system" (SW 2020). In addition, the solutions must be sustainable and ethically designed with nature (e.g. McHarg 1992; Chaminé 2015; Chaminé and Gómez-Gesteira 2019). In fact, that transdisciplinary approach is an amazing opportunity to contribute decisively to a path to the sustainability of the hydrological cycle that could lead to a better future for all life on Earth (Attenborough 2020).

This Joint Congress emerges from an agreement for cooperation signed on 5 April 2017 about common grounds by the IAH—International Association of Hydrogeologists and IAPG—International Association for Promoting Geoethics. Following its terms, the International Congress "Geoethics and Groundwater Management: Theory and Practice for a Sustainable Development" (GEOETH&GWM'20) aims for the first global approach on the vast subjects of geoethics in groundwater management and its recognized need of reflection for correct and prudent actions. GEOETH&GWM'20 convenes specialists, scholars and professionals of distinct fields of science, engineering, humanities, law and culture as well as educators, students and early career colleagues in some way related to groundwater. They met and interacted online in May 2020 during the most frightening times of COVID-19 and in global lockdown for the first specialized world forum for discussing theory and practice, sharing values, knowledge, research, educational projects, best practices and strategies aiming at the responsible integrated management of groundwater resources for a resilient and sustainable future. In a world asking for answers, GEOETH&GWM'20 had the goal to stage-manage a courageous scientific and professional community that is capable of proposing synergetic scientific, cultural and practical answers to the complex problems affecting society in all its connections with groundwater.

This Joint IAH and IAPG Congress proposes to the scientific, the cultural community and the society stakeholders a moment of reflection and an opportunity for the foundation, in respect of their own deep roots, of a new logic resulting from the production of new trans-disciplinary scientific and cultural added value on geoethics of groundwater. It is considered that there is a real potential of development of a new transdisciplinary geoscience capable to produce its developments and to feedback positively into the root contributor sciences through its autonomous progress and contributions to a better world in peace, justice and sustainability. This growing concept has been named hydrogeoethics by António Chambel and Manuel Abrunhosa, since 2017. Its field is soundly grounded in hydrogeology and geoethical principles, including the engineering, socio-economic, legal, environmental, arts and cultural dimensions.

To the former motto of the congress "Leaving No One Behind" (United Nations World Water Day 2019), and given the dramatic times brought by COVID-19 pandemic risking to stall the ongoing efforts of implementing the conference as it was conceived, and mainly the need in groundwater progress in science and protection, we added "The Science Must Go On" (Fig. 1). This was the geoethical commitment for the groundwater community, related water fields and society.

This book comprises the selected proceedings during the 1st Congress "Geoethics and Groundwater Management" (GEOETH&GWM'20), Porto, Portugal, 18–20 May 2020. The groundwater community involved in science, exploration, abstraction, use and management of this evermore essential natural resource is becoming more and more aware that ethical issues pervade all our attitudes from concept to action and need to be addressed coherently. Diverse values and cultures, science and education, law and policies, human and natural environments, the public and the economic sectors foresee groundwater and its values and/or roles differently. We believe that in a globalization intertwined world a common ground must be discussed and agreed for peace, human development and sustainability. A multidisciplinary Scientific Committee from the science, engineering, law, social sciences, natural philosophy, geoethics, environment fields assured the quality of the event and the current publication by earlier proposing themes. That aims for discussion in the conference and assuming the peer review process that addressed scientific, philosophical and legal approaches, analysis of case studies from around the world, management models or proposals, educational views, innovative transdisciplinary knowledge, research or projects on responsible groundwater management, including decision-making under uncertainty and in neglecting groundwater functioning.

In this volume were considered 6 major topics to correspond to the main fields of theory and practice regarding the global combination between groundwater in all possible conceptual dimensions and the geoethical approach (Fig. 2):

1. Fundamentals of hydrogeoethics: cultures, principles and geoethical values on groundwater science and engineering
2. Lessons for a resilient and sustainable future with hydrogeoethics: case studies of geoethics in groundwater science engineering, profession and management
3. Scientific and humanistic components of hydrogeoethics in groundwater education and professional training
4. Socio-hydrogeology and ethical groundwater management

**Fig. 1** Some key moments on the GEOETH&GWM'20 Online Congress: (i) when live in inauguration day, May 18; (ii) address of Ken Howard, IAH Past President; (iii) a keynote lecture; (iv) an oral presentation; (v) a musical moment by the "Grupo de Fados" from ISEP; (vi) closing ceremony by the chairman Manuel Abrunhosa

**Fig. 2** Word cloud based on all abstracts of the special volume on "Advances in Geoethics and Groundwater Management: Theory and Practice for a Sustainable Development" (generated using http://www.wordle.net/)

5. Geoethics of decision-making under uncertainty and ethical issues in neglecting groundwater functioning
6. Groundwater: geological, legal, social and ethical challenges of a unique natural resource

The special volume has a core of 95 original proceedings grounded on the scientific sessions and 14 outstanding keynote lectures. The keynote speakers gave interesting insights from the philosophical principles in hydrogeoethics, to hydrological hazards focused on hydrogeomorphology and disasters, geotechnical hazards highlighting the role of groundwater, as well as landslide risks and flooding hazards and hydraulic design and the role of geoethics in groundwater modelling. The volume gathered over 227 authors of the academy, research centres, state laboratories or industry from 47 countries of all continents (Europe, Africa, America, Asia and Oceania).

The volume will be of interest to researchers and practitioners in the field of hydrogeology, hydrology, water resources management and groundwater engineering, as well as those engaged in earth sciences, environmental sciences, law, social sciences, natural philosophy, education and culture. Students, geoscientists, engineers, environmental lawyers, social scientists and water-related professionals beyond research in water, earth, environmental and social sciences will also find the book an inspirational and unique asset.

Porto, Portugal — Manuel Abrunhosa
Évora, Portugal — António Chambel
Rome, Italy — Silvia Peppoloni
Porto, Portugal — Helder I. Chaminé
July 2020

## References


Abrunhosa MJ, Peppoloni S, Chambel A (2018) Introducing geoethics in groundwater science, technology, and management. In: Abstracts book, session T1.1 socioeconomic aspects of groundwater resources management, 45th IAH Congress, Daejeon, Korea, p 26

Angelakis AN, Mays L, Koutsoyiannis D, Mamassis N (2012) Evolution of water supply through the millennia. IWA Publishing, London

Arrojo-Agudo P (2010) The ethical challenge of sustainability in water management. In: Environment at the crossroads: aiming for a sustainable future. Carcanet and Fundação Calouste Gulbenkian, Manchester, pp 32–60

Attenborough D (2020) A life on our planet: my wytness statement and a vision for the future. Witness books, Penguin Random House, London

Bobrowsky P, Cronin VS, Di Capua G, Kieffer SW, Peppoloni S (2017) The emerging field of geoethics. In: Gundersen LC (ed), Scientific integrity and ethics with applications to the geosciences, Special Publication American Geophysical Union, John Wiley and Sons, 73:175–212

Bohle M (2015) Simple geoethics: an essay on daily Earth science. In: Peppoloni S, Di Capua G (eds) Geoethics: the role and responsibility of geoscientists. Geological Society, London, Special Publications 419:5–12

Bohle M (ed) (2019) Exploring geoethics: ethical implications, societal contexts, and professional obligations of the geosciences. Palgrave Macmillan, Cham, Switzerland

Braga B, Chartres C, Cosgrove WJ, Veiga da Cunha L, Gleick PH, Kabat P, Kadi MA, Loucks DP, Lundqvist J, Narain S, Xia J (2014) Water and the future of humanity: revising water security. Gulbenkian think tank on water and the future of humanity. Calouste Gulbenkian Foundation. Springer, Berlin

Chaminé HI (2015) Water resources meet sustainability: new trends in environmental hydrogeology and groundwater engineering. Environ Eart Sci 73(6):2513–2520

Chaminé HI, Afonso MJ, Freitas L (2014) From historical hydrogeological inventory through GIS mapping to problem solving in urban groundwater systems. Eur Geol J 38:33–39

Chaminé HI, Gómez-Gesteira M (2019) Sustainable resource management: water practice issues. Sustain Water Resour Manag 5:3–9

Custodio E (2000) Some relevant ethical issues in relation to freshwater resources and groundwater. Bol Geol Min España 111(6):121–130

Cuthbert MO, Ashley GM (2014) A spring forward for hominin evolution in East Africa. PLoS ONE 9(9): e107358

Datta PS (2005) Groundwater ethics for its sustainability. Curr Sci 89(5):1–6
Freeze RA, Cherry JA (1979) Groundwater. Prentice-Hall Inc, Englewood Cliffs, New Jersey [https://gw-project.org/books/13/groundwater]
Gleeson T, Wang-Erlandsson L, Porkka M, Zipper S, Jaramillo F, Gerten D, Fetzer I, Cornell S, Piemontese L, Gordon L, Rockström J, Oki T, Sivapalan M, Wada Y, Brauman K, Flörke M, Bierkens M, Lehner B, Keys P, Famiglietti J (2020) Illuminating water cycle modifications and earth system resilience in the anthropocene. Wat Resour Res 56(4):10.1029/2019WR024957
Hardin G (1968) The tragedy of the commons. Science 162(3859):1243–1248
Leopold LB (1990) Ethos, equity and the water resources. Environ Sci Policy Sustain 32(2):16–41
Llamas MR (1975) Non economic motivations in ground water use: hydroschizophrenia. Ground Water 13 (3):296–300
Llamas MR (2004) Water and ethics: use of groundwater. UNESCO Series on Water and Ethics, Essay 7. UNESCO, New York
Lugo-Enrich LB, Mejías M (2017) The hydrogeological and paleoclimatic factors in the Bronze Age Motillas Culture of La Mancha (Spain): the first hydraulic culture in Europe. Hydrogeol J 25:1931–1950
McHarg IL (1992) Design with nature. 25th anniversary edition, Wiley series in sustainable design. Wiley, New York
Ollivier V, Fontugne M, Hamon C, Decaix A, Hatté C, Jalabadze M (2018) Neolithic water management and flooding in the Lesser Caucasus (Georgia). Quatern Sci Rev 197:267–287
Peppoloni S (ed) (2015) Sharing ethical principles through cultural diversity. Translations of the Montreal Statement on Research Integrity in Cross Boundary Research Collaborations. International Association for Promoting Geoethics [http://www.geoethics.org/translations-ms]
Peppoloni S (ed) (2018) Spreading geoethics through the languages of the world. Translations of the Cape Town Statement on Geoethics, International Association for Promoting Geoethics [https://f420cbad-ec08-4c39-902f-b0e5afecb44a.filesusr.com/ugd/5195a5_f80fe698039e4f6d9bc8c5e4a5f05898.pdf]
Peppoloni S, Di Capua G (2017) Geoethics: ethical, social and cultural implications in geosciences. In: Geoethics at the Heart of all Geoscience. Annals of Geophysics, 60(7). https://doi.org/10.4401/ag-7473
Peppoloni S, Di Capua G (2018) Ethics. In: Bobrowsky PT, Marker B (eds) Encyclopedia of Engineering Geology. Springer, Cham, pp 307–311
Pétrequin P (1984) Gens de l'eau, gens de la terre: ethno-archéologie des communautés lacustres. Hachette, Paris
Ribeiro L (2017) Implicações éticas das políticas hídricas. In: Neves MCP, Soromenho-Marques V (eds) Ética Aplicada: Ambiente, Edições 70, Lisboa, 4:247–265
Richts A, Struckmeier WF, Zaepke M (2011) WHYMAP and the groundwater resources of the world 1:25,000,000. In: Jones JAA (ed) Sustaining Groundwater Resources, International Year of Planet Earth, Springer, Dordrecht
Shiklomanov IA (1998) World water resources: a new appraisal and assessment for the 21st Century. UNESCO, Paris
Soromenho-Marques V (2003) A crise internacional do ambiente: balanço e perspectivas. In: Soromenho-Marques V (Coord), O desafio da água no século XXI: entre o conflito e a cooperação, IPRIS e Editorial de Notícias, Lisboa, pp 11–31
SW—Smart Water (2020) Interview Dr. John Cherry. Smart Wat Mag 1:20–27
Tempelhoff J, Hoag H, Ertsen M, Arnold E, Bender M, Berry K, Fort C, Pietz D, Musemwa M, Nakawo M, Ur J, van Dam P, Melosi M, Winiwarter V, Wilkinson T (2009) Where has the water come from? Wat Hist 1:1–8
Tortajada C, Biswas AK (2017) Water as a human right. Int J Water Res Develop 33:509–511
United Nations (2015) General Assembly Resolution A/RES/70/1. Transforming Our World, the 2030 Agenda for Sustainable Development. United Nations, New York [https://www.un.org/en/development/desa/population/migration/generalassembly/docs/globalcompact/A_RES_70_1_E.pdf]
Wittfogel KA (1956) The hydraulic civilizations. In: Thomas WL (ed) Man's role in changing the face of the earth. The University of Chicago Press, Chicago, pp 152–164
Wyss M, Peppoloni S (2015) Geoethics: ethical challenges and case studies in earth sciences. Elsevier, Amsterdam
Zektser IS, Everett LG (eds) (2004) Groundwater resources of the world and their use. IHP-VI, Series on Groundwater No. 6. UNESCO, Paris

# Acknowledgements

We are grateful to all authors, keynote speakers and presenters that were burdened by the demanding standardized pre-recorded high-quality contributions, by their exciting and insightful work covering all major themes proposed for discussion and, most of them, addressing them in transdisciplinarity that is at the core of methodologies proposed for hydrogeoethics. We would like to thank the international scientific committee for their in-depth reviews and great efforts in improving the overall quality of the contributions. They were essential in the quality assurance of all scientific contents.

A word of appreciation to all chairs of the thematic sessions and special sessions for their skilled support. Also, thanks are extended to Liliana Freitas, José Teixeira, Alexandra Paz dos Reis and João Paulo Meixedo for their extraordinary coordinated hard-working, dedication and competence, as well as to the designer Joel Vilas Boas (ISEP) that created a powerful logo that will remain a symbol of hydrogeoethics. Our thanks to Alexandra Aragão, Professor of Environmental Law, from the University of Coimbra, for the full support in the building common grounds related to the national and international colleagues from the field of law involved in the scientific committee. In addition, our thanks to the president of ISEP (Instituto Superior de Engenharia do Porto—School of Engineering of Porto), Maria João Viamonte, and the president of Polytechnic of Porto, João Rocha, for the enthusiastic full support to the event.

We are also honoured to count on the support from leading Portuguese and international professional and scientific organizations and research units with leading roles in fields related to the study and management of groundwater and water-related, and the High Patronage of the President of the Portuguese Republic. Many thanks also to John Cherry, Giuseppe di Capua and Nabil Khélifi for the insightful foreword. Finally, to all the participants for the resilience to participate in this unique COVID-19 context and all positive energy to build up an unforgettable online conference. In addition, our thanks to Ex Ordo: Intelligent Conference Management Software (Ireland), headed by Paul Killoran, that designed and developed, in a short time, an amazing interactive platform to run the conference and the continuous support of an enthusiastic skilled team (namely Eoghan Hughes and Roseann Dunne). Kellie Nicholson from IAH and Giuseppe Di Capua from IAPG were instrumental for their wise advice and kind vehicles of contact and support from both heading organizations.

Last but not least, the guest editors are grateful for the continuous and the passionate support of publishing staff of Springer headed by Nabil Khélifi, Senior Editor, Reyhaneh Majidi, Editorial Assistant and Springer production team, for their efforts in finalizing this volume. All the above-mentioned efforts were very considerable in making this landmark book a success.

The present book is dedicated to distinguished hydrogeologists Professor Partha Sarathi Datta (1950–2018) and Professor Luís Ribeiro (1953–2020), both amazing colleagues who promoted high standards and core ethical values in hydrogeology practice for teaching, science, engineering and society.

# Preamble by John Cherry

Groundwater makes up 99% of all liquid freshwater, 50% of river flow is groundwater seepage, and many ecological systems are groundwater dependent. Nearly 50% of the global population depends on groundwater for all or part of its drinking water, about 40% of our food comes from irrigation using groundwater, and much of the salinization loss of agricultural soil is due to groundwater. One-third of the major aquifers of the world are depleted beyond recovery, and about 25% of sea level rise is attributable to groundwater depletion. Of the nearly 8 billion people on the planet, two billion do not have access to safe drinking water largely because groundwater is not adequately accessed and this is a worsening problem because the population is ballooning by 2–3 billion in the poor countries by the end of the century. There is now a global water crisis largely due to the combination of groundwater depletion and groundwater pollution. Until recently, humanity solved its major water problems by building more dams for water storage and flood control, but this era is over and solving groundwater problems is now the challenge in most countries. But the growing awareness that groundwater is the essence of the global water crisis is only recent. The importance of groundwater is now recognized by the United Nations in its water theme for 2022: "Groundwater: making the invisible visible". Although there is now more recognition of groundwater's importance, there is a broad lack of understanding of the specific nature and magnitude of groundwater problems and how to effectively frame solutions. A root cause of this across the globe is that the knowledge published in the peer review literature is fragmented and specialized and not accessible to serve societal needs in water policy and management. The dialogs of the experts are siloed within many specialty subject domains, and this severely limits the collaborations needed across many expertise fields and the mingling of perspectives required for effective solutions.

The Geoethics and Groundwater Management Congress is an excellent and timely example of the types of multidisciplinary and transdisciplinary engagements needed for progress. This may be the first international conference focused on groundwater that has "geoethics" in the title. The use of this word is an appropriate reminder of how ethically dependent groundwater management and protection should be. A new approach was born as hydrogeoethics. The decisions made by one generation usually do not show up as beneficial or detrimental before many years or decades after they are taken because groundwater "happenings" take place so slowly, over many years or decades. This congress had the engagement of the spectrum of the expertise areas needed within the geoethics umbrella: hydrogeology, engineering, law, economics, sociology, sustainability, management and agriculture with many examples of types of issues and problems and the published proceedings are a service towards informing about many aspects of the spectrum.

Guelph, ON, Canada
July 2020

John Cherry
Adjunct Professor
G360 Institute for Groundwater Research
University of Guelph

Distinguished Professor Emeritus
University of Waterloo

# Contents

# About the Editors

**Manuel Abrunhosa** born in Porto in 1954 graduated in geology from the University of Porto, Portugal, in 1980. In 1986, he obtained a grade equivalent to a master's degree in the University of Porto after presenting a dissertation and public examination, and in 1988 a master's degree in groundwater hydrology from the Polytechnic University of Catalonia, Barcelona, Spain. He has several post-grade courses. He began a full-time professional career as Junior Geologist (hydrogeology and engineering geology) in 1977, before graduation, pursuing studies and continuing his first job as contracted helper to the practical classes of disciplines of his geology course in the University of Porto, from 1974 up to 1980. In 1981, he became a partner and director in a small company of Applied Geology and Hydrogeology Services. Since then, he directed and was Author and Co-author of innumerable projects aimed at the private and public sector until 2009. In 1981, after public examinations, he became Lecturer in geology at the Faculty of Sciences of the University of Porto, having taught theoretical and practical classes in hydrogeology, engineering geology, sedimentary petrology, geomorphology, geological cartographic methods, mineralogy, crystallography, structural geology and supervised pedagogic internships in secondary schools. He did research in hydrogeology of fractured media and participated in national and international research projects in this subject aiming water resources and groundwater management in the Minho region (NW Portugal), and in the optimization of groundwater exploitation in complex well field a thin coastal sandy aquifer. Other active interest in science includes natural heritage studies and geoarchaeology with collaborations beginning since as a student with research groups in archaeology, ethnology and history. He was involved in 1990 in the foundation of the first academic course on environmental health and hygiene that included curricula in geology and hydrogeology. He was also involved in a first master's degree in environmental marketing. He is often called as forensic geology expert witness and acts as pro bono scientific advisor to cultural and natural heritage and environmental advocacy associations.

Since 2006, he is an independent Consultant Geologist. He is Member of several national and international scientific and professional associations. At the Portuguese Association for the Study of the Quaternary—APEQ (publisher of the journal

Estudos do Quaternário/Quaternary Studies), he is Secretary of the Board and is President of the AIH-GP—Portuguese Chapter of the International Association of Hydrogeologists (IAH). As an active member of the International Association for Promoting Geoethics (IAPG), he was invited in 2017 to its Board of Experts (Corresponding Citizen Scientists) in Geoethics in Groundwater Management. In 2018, he was jointly designated by IAH and IAPG as chair of the International Congress "Geoethics and Groundwater Management: Theory and Practice for a Sustainable Development" (18–22 May 2020), to be held at ISEP, Porto, Portugal.

The ethics at the profession, in science production and in the applied earth sciences has always been a concern and a guideline for his activities, a challenge for a geologist who, being born from a school oriented to the exploration and exploitation of geological resources in what he calls predatory geology, has gradually changed his priorities to become a geoethics advocate, Environmental and Social Geologist for Sustainability.

**António Chambel** is a skilled Geologist and Professor of hydrogeology and water resources at the University of Évora, with over 35 years' experience in multidisciplinary groundwater research, consultancy and practice. He graduated in geology from University of Coimbra in 1984 and obtained an MSc in economic and applied geology from University of Lisbon in 1990 and Ph.D. in geology, specializing in hydrogeology, from University of Évora in 1999. His research interests are hydrogeological mapping, environmental hydrogeology, water resources management, groundwater modelling, urban groundwater, groundwater engineering, applied geology, among others. He has been a teacher of hydrogeology in the Department of Geosciences of University of Évora since 1985 and from 2003 to 2005 was President of the department. He has also been Invited Professor in the University of Algarve (Portugal), in the Universities Charles of Prague (Czech Republic), Huelva (Spain) and La Sapienza Rome (Italy), under the ERASMUS Programme, and in the Institute of Transport and Communication (master's degree in environmental impact studies) in Maputo, Mozambique. He supervised several master theses in Portugal, Mozambique and Angola and coordinated and worked in many scientific or applied projects in Portugal and Mozambique.

From 2002 to 2006, he was President of the South Chapter of the Portuguese Water Resources Association (APRH), and from 2004 to 2008 he was President of the Portuguese Chapter of the International Association of Hydrogeologists (AIH-GP). He has been Member of IAH since 1988, and from 2008 to 2012 he was IAH Vice-President for Finance and Membership, 2012–2016 IAH Vice-President for Programme and Science Coordination and President of IAH (2016–2020). In 2007, he was Chair of the 35th IAH Congress in Lisbon. Nowadays, he is past-President of IAH—International Association of

Hydrogeologists. His international experience has been with UNESCO-IHP, with the World Water Council (WWC), having participated in the last World Water Fora (WWF) in South Korea in 2015 and Brasilia in 2018 and, through the role of member of the Executive Committee of IAH in the last 12 years, he has organized and attended IAH congresses and other IAH regional or national meetings, as well as representing IAH in many other events around the world.

He has co-authored numerous publications in journals, conference proceedings/full papers, chapters, technical and professional papers, as well as co-edited several international special issues. In addition, he belongs to some editorial journal boards (e.g. Sustainable Water Resources Management, IAH +Springer, Journal of Groundwater Science and Engineering). He was on many scientific and organizing committees of national and international conferences.

**Silvia Peppoloni** is a skilled Ph.D. Geologist and Researcher at Italian Institute of Geophysics and Volcanology (INVG) in Rome, Italy, with over 20 years' experience in multidisciplinary applied geosciences research, consultancy and practice. Her professional activity covers the fields of engineering geology, geological hazards and risks, as well as geomorphology, geo-education and geoscience dissemination and communication. In addition, she is fully involved in the base research on geoethics, focusing on ethical, social and cultural issues related to geosciences and promoting sustainability, prevention and geo-education as key concepts of the relationship between geoscientists and society. Since 1999, she is engaged in some international projects on geological hazard and risks and recently in three H2020 European projects dealing with ethical issues: the Project ENVRI PLUS—Environmental Research Infrastructures Providing shared solutions for Science and Society, the Project GOAL: Geoethics Outcomes and Awareness Learning, and the Project EPOS SP: European Plate Observing System Sustainability Phase. She was Adjunct Professor in geology and applied geology at the Universities of Rome "La Sapienza" and Viterbo "Tuscia"; Member of the Experts Committee of the Ph.D. course: "Landscape and Environment", at the La Sapienza Rome University; Collaborator of some Italian universities and research institutes; and Teacher in courses on Disaster Risk Reduction (École Polytechnique Fédérale de Lausanne) and seismic damage on cultural heritage (European Project Interreg III C NOÈ).

She is Secretary General and Founder Member of the IAPG—International Association for Promoting Geoethics, Councillor of the Executive Committee of the IUGS—International Union of Geological Sciences, Founder and Director of the School on Geoethics and Natural Issues, and Editor in Chief of the series SpringerBriefs in Geoethics. Moreover, she is Member of the Executive Committee of the ICPHS—International Council for Philosophy and Human Sciences, Member of the Executive Council of the Italian Section of the IAEG—

International Association for Engineering Geology and the Environment, Member of the Executive Council of the Italian Geological Society and Coordinator of its Section on Geoethics and Geological Culture. She is Author and Co-author of several publications in indexed journals, conference proceedings/full papers and chapters, as well as Co-editor of several international special issues and books (e.g. Elsevier, Springer, Geological Society of London, Annals of Geophysics), and she is among the authors of the Geoethical Promise (the "Hippocratic-like Oath" for geoscientists) and the Cape Town Statement on Geoethics. In addition, she is Science Writer, Contributor of Italian newspapers/magazines and Author of books on earth sciences intended for the general public. In Italy, she has been awarded with prizes for science communication and natural literature.

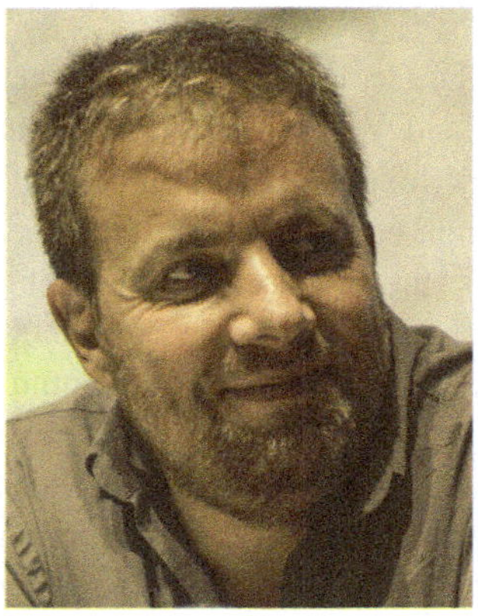

**Helder I. Chaminé** is a skilled Geologist and Professor of engineering geosciences at School of Engineering (ISEP) of the Polytechnic of Porto, with over 30 years' experience in multidisciplinary geosciences research, consultancy and practice. He studied geological engineering and geology (B.Sc., 1990) at the Universities of Aveiro and Porto (Portugal), respectively. He received his Ph.D. in geology at the University of Porto in 2000 and spent his postdoctoral research in applied geosciences at the University of Aveiro (2001–2003). In 2011, he received his Habilitation (D.Sc.) in geosciences from University of Aveiro. Before joining academy, he worked over a decade in international projects for mining, geotechnics and groundwater industry and/or academia related to geodynamics and regional geology, hard rock hydrogeology and water resources, engineering geosciences and applied geomorphology, rock engineering and georesources. His research interests span over fundamental to applied fields: GIS mapping techniques for applied geology, structural geology and regional geology, engineering geosciences and rock engineering, slope geotechnics, mining geology and hydrogeomechanics, hard rock hydrogeology, exploration hydrogeology, urban groundwater and hydromineral resources. He has interests on mining geoheritage, history of cartography, military geosciences and higher-education dissemination, skills and core values.

Presently, he is Head of the Laboratory of Cartography and Applied Geology (LABCARGA | ISEP), Senior Researcher at Centre GeoBioTec | U.Aveiro and Centre IDL | U.Lisbon, as well as belongs to the executive board of the M.Sc.+B.Sc. Geotechnical and Geoenvironmental Engineering Programmes (OE+EUR-ACE Label) and the Department of Geotechnical Engineering (ISEP). Currently, he belongs to the board of the Portuguese Chapter of the International Association of Hydrogeologists (AIH-GP), Portuguese Association of Geologists (APG), and Technical Commission of Environmental Geotechnics from SPG. He was Board Member of APGeom—Portuguese Association of Geomorphologists (2009–2013) and SPG—Portuguese Geotechnical Society (2016–2020). He was consultant and or responsible over 70 projects of applied geology, hydrogeomechanics, slope geotechnics, mining geology, exploration hydrogeology, hard rock hydrogeology, water resources, urban groundwater and applied mapping (Mozambique, Portugal and Spain).

He has co-authored over 200 publications in indexed journal articles, conference proceedings/full papers, chapters, technical and professional papers. He co-edited over 15 special volumes, as well as is presently involved in editing themed issues for some international journals (e.g. Environmental Earth Sciences—Springer, Springer Nature Applied Sciences, Water MDPI, Arabian Journal of Geosciences—Springer). He has a wide activity as a referee for several international journals. He served as Invited Expert Evaluator of Bologna Geoscience Programme for DGES (Portugal) and Scientific Projects Evaluation for NCST, 2017–2019 (Kazakhstan), and NRF | RISA, 2019 (South Africa), as well as Coordinator of "Geology on Summer/Ciência Viva" Programme at ISEP (2005–2019) for geoscience dissemination. He has been also active with teaching and supervising of many Ph.D., M.Sc. and undergraduate students.

He has been on the editorial board, among others, of Arabian Journal of Geosciences (SSG+Springer), Hydrogeology Journal (IAH+Springer), Euro-Mediterranean Journal for Environmental Integration (Springer), Springer Nature Applied Sciences (Springer), Mediterranean Geoscience Reviews (Springer), Discover Water (Springer), Geotechnical Research (ICE), Geosciences (MDPI), Revista Geotecnia (Portugal) and Geología Aplicada a la Ingeniería y al Ambiente (Argentina). He integrates as Moderator or Session Chair in several conferences, workshops and meetings. Currently, he is in organizing/scientific committees of the 3rd International Workshop on Natural Hazards—NATHAZ'22 (Terceira Island, Azores, May 2022), supported by Springer.

# Fundamentals of Hydrogeoethics: Cultures, Principles and Geoethical Values on Groundwater Science and Engineering

**Introductory Note**

Giuseppe di Capua (INGV & IAPG, Rome, Italy), José Teixeira (FLUP, Faculty of Arts of University of Porto, Porto, Portugal), Alexandra Aragão (FDUC, Faculty of Law of University of Coimbra, Coimbra, Portugal), Rute Saraiva (FDUL, Faculty of Law of University of Lisbon, Lisbon, Portugal)

The major theme *Fundamentals of Hydrogeoethics: Cultures, Principles and Geoethical Values on Groundwater Science and Engineering* provides insights on geoethics in water management and policies with inputs from environmental ethics and legal concepts, principles, and frameworks. Cultural implications and related values, societal concerns about water quality and water depletion were underlined by several authors. In fact, it is important to understand the sustainability needs, expectations, and demands by different stakeholders, including local and indigenous communities. In addition, numerous authors highlighted issues such as the water governance problems, public participation, best practices, and techniques in monitoring and database utilization.The goal of this part is to explore the complexity of problems associated with water and the importance of facing its management with multi-perspective approaches, assured by collaborative action and dialogue between all parties at stake. Only the awareness of this complexity and effective consideration of different points of view are capable to produce an effective advancement in developing a more responsible and sustainable use and protection of water in general, and particularly groundwater. The vital nature of this georesource requires a careful management, or else it will lead inevitably to the damage of natural environment and social structures of human communities.

**Highlights**

- Groundwater ethical issues;
- Legal frameworks;
- Societal perceptions and concerns;
- Water governance;
- Groundwater management–society–policy interface;
- Multi-perspective approaches in groundwater management.

# Relational Value as an Argument to Protect Geological and Hydrogeologic Goods

Alexandra Aragão

**Abstract**

What are the reasons to protect the geologic resources? Are there any ethical arguments to prevent extraction and support the "in situ" preservation of ordinary geological goods that are not covered by other legal protection regimes? And what about the geological goods which do not even have a market value because they are not economically interesting or because they are not in the market? Are not there any ethical arguments for "in situ" protection of these geological goods against development plans, programs, or projects? The legal rationale for the protection of geological goods—both with and without economic relevance—will be based on their relational value. The relational value is key to understand the fundamental reasons why people want to protect geological goods despite their non-biotic nature and regardless of their direct utility, economic value. The big question here is: how is it possible to know if and how much people care about certain geological objects or sites? The answer lays in cultural geo-ecosystem services. Recognizing the relational value of geological goods by using the ecosystem services language can help prevent conflicts and promote socially and environmentally sustainable development.

**Keywords**

Geological goods • Relational values • Ecosystem services • Water

A. Aragão (✉)
Faculty of Law, University of Coimbra, Coimbra, Portugal
e-mail: aaragao@ci.uc.pt

## 1 Introduction

Be it human, animal, or vegetal, life is a unique phenomenon in the Universe and a supreme legal value to be protected.[1]

This absolute prominence of life overshadows the ethical relevance of other environmental goods, whose legal protection is usually relegated to the background[2] and depends on purely instrumental reasons.[3]

Among the non-living environmental goods, water has a special statue. The preamble of the EU water framework directive clearly states: "water is not a commercial product like any other but, rather, a heritage which must be protected, defended and treated as such"[4] In the case of groundwater, the European Parliament and the Council recognize that it is "a valuable natural resource and as such should be protected from deterioration and chemical pollution".[5]

But all the other non-living environmental goods, usually called "geological resources" which do not have such an

[1]For human life, see article 3 of the 1948 Universal Declaration of Human Rights, article 2 of the 1953 European Convention for the Protection of Human Rights and Fundamental Freedoms, and article 2 of the 2000 Charter of fundamental rights of the European Union. For non-human life see the §2 of the Preamble of the United Nations Convention on biological diversity "Conscious also of the importance of biological diversity for evolution and for maintaining life sustaining systems of the biosphere" and article I 1 (a) of the Earth Charter "recognize that all beings are interdependent and every form of life has value regardless of its worth to human beings".

[2]The quantitative study carried out by van Ree et al. (2017).

[3]For instance, it is the case of the protection of wetlands as habitats for migratory birds, as in the 1971 Convention on Wetlands of International Importance especially as Waterfowl Habitat.

[4]First paragraph of the Directive 2000/60/EC of the European Parliament and of the Council of 23 October 2000 establishing a framework for Community action in the field of water policy (https://eur-lex.europa.eu/legal-content/EN/TXT/?uri=CELEX:32000L0060).

[5]Paragraph (1) of the Directive 2006/118 of 12 December 2006 on the protection of groundwater against pollution and deterioration (amended in 2014).



M. Abrunhosa et al. (eds.), *Advances in Geoethics and Groundwater Management: Theory and Practice for a Sustainable Development*, Advances in Science, Technology & Innovation,
https://doi.org/10.1007/978-3-030-59320-9_1

obvious and vital utility as water, are in the end of the scale of importance.[6] This does not mean that their value is not socially recognized or economically visible. Indeed, property and sovereignty over water and other geological resources have triggered serious conflicts at different scales throughout the globe. Besides, their non-renewable nature[7] explains why they are particularly sensitive natural elements. This includes both geological resources and non-renewable groundwater or "fossil aquifers"[8] as well.

What are the reasons to provide legal protection to geological resources[9]? The answer to this question shall contribute to the development of geoethics, the scientific field that studies the ethical value of the non-living environmental goods, known as geological resources.

## 2 The Value of Non-living Goods

Natural geological formations that are not human constructed or the result of human activity[10] may incorporate multiple values: aesthetic, cultural identity, religious, historic or all these values at the same time: Uluru Mountain (or Ayers Rock in Australia), the Grand Canyon (in The United States of America) or Sagres Point (the sacred promontory in Portugal, the southwestern most tip of Europe[11]).

Using the words of UNESCO, these are natural formations, natural areas or natural sites of outstanding universal value that constitute parts of the "world natural heritage".[12]

In fact, "natural heritage" is composed of "natural features consisting of physical and biological formations or groups of such formations, which are of outstanding universal value from the aesthetic or scientific point of view"; "geological and physiographical formations and precisely delineated areas which constitute the habitat of threatened species of animals and plants of outstanding universal value from the point of view of science or conservation"; "natural sites or precisely delineated natural areas of outstanding universal value from the point of view of science, conservation or natural beauty".[13]

Yet, beyond these examples of extraordinary geological goods enjoying a legal protection statute based on their widely recognized outstanding interest, there is a whole world of anonymous geological goods, protected only for the economic value of the extracted material placed on the market.

The legal challenge and the legal research question that is going to be to addressed is: are there any ethical arguments to prevent extraction and to support the in situ preservation of ordinary geological goods that are not covered by other legal protection regimes?

And what about the geological goods which do not even have a market value because they are not economically interesting or simply because they are not for sale in markets? Considering that they can, nevertheless, be affected by human development activities (namely construction of urban, transport or energy infrastructures[14]), aren't there any ethical arguments for in situ protection of these geological goods against destructive development plans, programs or projects?

The legal rationale for the protection of geological goods both with and without economic relevance will be based on their relational value.

[6]There is no specific European directive for the protection of geological objects and sites beyond the protection granted by the Habitats directive (Council Directive 92/43 of 21 May 1992). In 2006, there was a proposal for a Directive of the European Parliament and of the Council establishing a framework for the protection of soil (Brussels, 22.9.2006 com (2006) 232 final 2006/0086 (cod), but it was never adopted. Nevertheless, the European Union is showing a growing interest on the subject, namely on soil protection, and even created a European Soil Data Centre (ESDAC) a thematic centre for soil-related data in Europe (https://esdac.jrc.ec.europa.eu/).

[7]On the non-renewable character of the soil see the Communication from the Commission to the Council, the European Parliament, the European Economic and Social Committee and the Committee of the Regions on a thematic strategy for soil protection (Brussels, 22.9.2006 COM (2006)231 final).

[8]On the concept of fossil aquifers and their protection see Foster and Louck (2006) (https://eur-lex.europa.eu/legal-content/EN/TXT/?uri=CELEX:32000L0060).

[9]In the North American legal context, the discussion on whether natural objects should have legal standing and be allowed to go to court to defend their "subjective rights" became a classic in the history of environmental law. The trigger for this discussion was the seminal text by Stone (1972).

[10]This excludes geological structures having the highest historic heritage value such as Stonehenge and other megalithic monuments, but which are in fact cultural and not natural heritage.

[11]https://en.wikipedia.org/wiki/Sagres_Point.

[12]https://whc.unesco.org/archive/convention-en.pdf.

[13]Article 2 of the 1972 Convention concerning the protection of the world cultural and natural heritage.

[14]The Judgment of the European Court of Justice on 11 April 2013 (case C-258/11) on the criteria to be applied when assessing the likelihood that a plan or a project will adversely affect the integrity of a site is quite illustrative in this regard: The site Lough Corrib, a karstic limestone pavement, was being threatened by the authorization of the construction of a road. The Court declared that "the implementation of the N6 Galway City Outer Bypass road scheme would result in the permanent and irreparable loss of part of the Lough Corrib SCI's limestone pavement, which is a priority natural habitat type specially protected by the Habitats Directive".

## 3 Anthropometric and Relational Values

As explained by Sanna Stålhammar and Henrik Thorén, relational values are a third type of anthropometric and not merely instrumental values.[15]

The concept of anthropometric values is based on Luc Ferry theory of value distinguishing anthropocentric from anthropometric values.[16] According to Ferry's theory, values are identified, appreciated and balanced by humans. This always gives a human imprint to valuation even if the discourse is extensively favourable to the prioritization of ecocentric values. In fact, recognizing the anthropometric nature of the valuation process does not mean that the appraisal result is necessarily anthropocentric. Not every value estimation and prioritization performed by humans results in the advocacy of an anthropocentric scenario based on subjugation of non-human values to human ones. The result of an anthropometric valuation may be that humans are not in the centre of every value pyramid.

In the classic dichotomy, instrumental values are subjective considerations associated with the practical utility[17] or with the market value[18] of the goods, while on the contrary, intrinsic values emerge from the recognition of an absolute and objective worth, a value of the object in itself, regardless of the subject who is assessing.

Relational values go beyond the classic approach, overcome the antinomy between intrinsic and instrumental value and are key to understand the fundamental reasons why people want to protect geological goods regardless of their direct utility, their economic value and despite the fact that they are non-living things.

The big question here is how is it possible to know if and how much people care about certain geological objects or sites? How do we identify the value of a certain cave, mountain, cliff, shore, or even of a certain boulder, fossil, fossil water, sand dune or gravel deposit?

The answer lays in ecosystem services.

[15]Stålhammar and Thorén (2019).

[16]Ferry (1995).

[17]The 2006 Directive on groundwater stresses that protecting groundwater "is particularly important for groundwater-dependent ecosystems and for the use of groundwater in water supply for human consumption" (Paragraph (1) in fine of the Directive 2006/118 of 12 December 2006 on the protection of groundwater against pollution and deterioration (amended in 2014).

[18]Adam Smith's paradox of value is illustrated by two geological resources: diamants (having high market value and low utility) and water (having low market value, highest utility) (Smith 1776).

## 4 Geosystem Services

The movement aiming to give prominence and legal relevance to ecosystem services started at the turn of the millennium, after the call by the Secretary General Kofi Annan, during the first General Assembly of the United Nations,[19] when the so-called Millennium Ecosystem Assessment[20] was launched. A worldwide report (including global and sub global reports) on the status of the major ecosystems on the planet was prepared by a group of more than 1000 international experts. In Europe, since 2013, the initiative for Mapping and Assessment of Ecosystems and their Services (MAES)[21] has already produced four technical reports and a coherent analytical framework.[22]

In the European Union, ecosystem services are already an official legal concept. In environmental legislation, for instance, it is used in the context of the fight against invasive alien species,[23] for the promotion of renewable energy sources,[24] for maritime spatial planning,[25] in environmental liability[26] and in the Life program.[27]

The United Nations Statistical Division is currently working together with European Environment Agency to develop a Common International Classification of Ecosystem Services (CICES) for purposes of environmental accounting.

The latest version of the CICES[28] is now based on only three types of services: provisioning, regulation/maintenance and cultural.

The most obvious ecosystem services provided by the geologic resources—or geosystem services[29]—are the provisioning of materials for maintaining the economic productive/consumption processes, and the fundamental regulation (namely of the water and nutrient cycles) and maintenance (the major function of the soil) services. The cultural ecosystem services of abiotic resources are less

[19]https://www.un.org/en/events/pastevents/pdfs/We_The_Peoples.pdf.

[20]https://www.millenniumassessment.org.

[21]https://biodiversity.europa.eu/maes.

[22]https://ec.europa.eu/environment/nature/knowledge/ecosystem_assessment/index_en.htm.

[23]Regulation 1143/2014 of 22 October 2014 on the prevention and management of the introduction and spread of invasive alien species.

[24]Directive 2009/28/EC of 23 April 2009 on the promotion of the use of energy from renewable sources.

[25]Directive 2014/89/EU of 23 July 2014 establishing a framework for maritime spatial planning.

[26]Directive 2004/35/CE of 21 April 2004 on environmental liability with regard to the prevention and remedying of environmental damage.

[27]Regulation 1293/2013 of 11 December 2013 on the establishment of a Programme for the Environment and Climate Action (LIFE).

[28]The v.5.1., released on the 1st January 2018, available at https://cices.eu/.

[29]van Ree and van Beukering (2016).

obvious reasons to protect them but still can be decisive arguments for protection.

## 5 Relational Values and Cultural Geosystem Services

The Irish case decided by the European Court of Justice in 2013[30] illustrates the non-anthropocentric and non-instrumental protection of geological objects or sites. In a typical sustainable development decision, the Court had to choose between the construction of an important road and the protection of a karstic limestone pavement in Galway, Ireland. The fact that the geologic site was already protected by law (it was a Natura 2000 site) facilitated the Court's judgement but the argument that the loss of the pavement would be "permanent and irreparable" was decisive for the final verdict.

What would have happened if the site had not been legally protected as a "priority natural habitat"? What possible arguments could a national Court use to impose the protection of a geological element for which there is no market but whose very existence is threatened by progress? Once again, the answer should come neither from intrinsic value, nor from instrumental value, but rather from relational value of abiotic geosystems.

Considering the cultural geosystem services represents a more holistic view on the importance of natural elements,[31] contributing to better and more grounded decision-making processes.[32]

Relational values of abiotic resources are demonstrated by the existing connection between people and the geological goods or, in other words, by the cultural geosystem services.

There are many indicators[33] of cultural geosystem services (or relational value) of abiotic goods. For example, some indicators of existing personal (individual) and collective (social) connections between humans and the abiotic natural capital can be the number of:

- People living nearby;
- Visitors for tourism, leisure or recreational activities;
- Persons using it for education or science;
- Artistic representations (painting, sculpture, performative arts, literature, music, architecture, cinema)
- News in the newspapers;
- Times it is mentioned in political debates;
- Requests for access to geological information;
- References during public participation;
- Cases relating to geological objects or sites decided or pending in court;
- Social allusions and remembrances: writing about it (namely expressing positive emotional reactions) in social networks, wearing clothes, props and accessories displaying visual or graphic representations or statements.

If the decision-maker wants to take into account the relational value of geosystem services for future generations, it is important to consider the chronological evolution of these indicators to identify trends and to build scenarios based on these trends and possible influencing factors.

Even though the common classification of ecosystem services is not as developed for abiotic resources as it is for biotic resources, it is still the best "Rosetta stone" to decipher the value of nature for people.[34]

Considering the different types of cultural ecosystem services, it is possible to perform a *deciphering process* by making a correspondence between the cultural geosystem service of abiotic goods on one hand, and the human activity to be carried out both by groups or communities and by isolated individuals, on the other.

It is important to stress that this activity does not require either the direct contact of humans with the geological object or the physical presence of humans on the geological site. The activity can also be carried out in the presence of a material[35] or spiritual representation (a mental image) of the geological object or site.

Finally, it is possible to identify the human benefits generated by the geosystem service, which represents its relational value (Table 1).[36]

[30]Judgment of the European Court of Justice on 11 April 2013 (case C-258/11).

[31]Article 1 of the proposal for a Directive of the European Parliament and of the Council establishing a framework for the protection of soil (Brussels, 22.9.2006 com(2006) 232 final 2006/0086 (cod)) already described seven functions performed by the soil without precisely associating these functions to the ecosystem services: "this Directive establishes a framework for the protection of soil and the preservation of the capacity of soil to perform any of the following environmental, economic, social and cultural functions: (a) biomass production, including in agriculture and forestry; (b) storing, filtering and transforming nutrients, substances and water; (c) biodiversity pool, such as habitats, species and genes; (d) physical and cultural environment for humans and human activities; (e) source of raw materials; (f) acting as carbon pool; (g) archive of geological and archaeological heritage".

[32]This is also the opinion of van der Meulen et al. (2016).

[33]On the role of legal indicators for the development and effectivity of Law see Prieur (2018).

[34]Diaz et al. (2015).

[35]A picture, a movie, a painting, a sound recording, a small symbolic object, etc.

[36]For instance, for the identity value or "sense of place", see Ryfield et al. (2019).

**Table 1** Correspondence between cultural geosystem services, human benefits and relational values

| Cultural geosystem service | Collective action | Individual action | Human benefit and relational value |
|---|---|---|---|
| Identity | Community symbolism | Self-definition | Being |
| Heritage | Classified heritage | Inheritance | Having |
| Sacred | Cult | Pray | Venerating |
| Transcendent | Group meditation | Solo meditation | Balancing |
| Educational | Study (formal) | Learn (informal) | Knowing |
| Scientific | Group research | Individual research | Understanding |
| Performing | Actors | Attendance | Feeling |
| Aesthetic | Group contemplation | Solo contemplation | Admiring |
| Leisure | Group break | Solo break | Relaxing |
| Recreation | Group play | Solo play | Enjoying |

## 6 Concluding Remarks

Recognizing the relational value of geological goods by using the cultural geosystem services language allows decision-makers[37] to perceive the real importance of geological goods, thus ensuring an effective protection of the geologic capital and helping to prevent conflicts and promote socially and environmentally sustainable development.[38]

## References

Aragão A (2019) When feelings become scientific facts: valuing cultural ecosystem services and taking them into account in public decision making. In: Managing facts and feelings in environmental governance. Edward Elgar, pp 53–80

Diaz S, Demissew S, Joly CA, Lonsdale WM, Larigauderie A (2015) A rosetta stone for nature's benefits to people. PLoS Biol 13(1): e1002040

Ferry L (1995) The new ecological order. The University of Chicago Press.

Foster S, Daniel PL (eds) (2006) Non-renewable groundwater resources: a guidebook on socially sustainable management for water-policy makers. UNESCO, Paris

Prieur M (2018) Les indicateurs juridiques outils d'évaluation de l'effectivité du droit de l'environnement. Institut de la Francophonie pour le Développement Durable, Québec

Ryfield F, Cabana D, Branniganb J, Crowe T (2019) Conceptualizing 'sense of place' in cultural ecosystem services: a framework for interdisciplinary research. Ecosyst Serv 36:100907

Smith A (1776) Book I, Chapter V of the real and nominal price of commodities, or of their price in labour, and their price in money. In: An inquiry into the nature and causes of the wealth of nations, London

Stålhammar S, Thorén H (2019) Three perspectives on relational values of nature. Sustain Sci 14:1201–1212

Stone CD (1972) Should trees have standing. Toward legal rights for natural objects. South California Law Rev 45:450–501

van der Meulen ES, Braat LC, Brils JM (2016) Abiotic flows should be inherent part of ecosystem services classification. Ecosyst Serv 19:1–5

van Ree CCDF, van Beukering PJH (2016) Geosystem services: a concept in support of sustainable development of the subsurface. Ecosyst Serv 20:30–36

van Ree CCDF, van Beukering PJH, Boekestijn J (2017) Geosystem services: a hidden link in ecosystem management ecosystem services. Ecosyst Serv 26:58–69

[37]Aragão (2019).

[38]On ensuring sustainable use of soil see Communication from the Commission to the Council, the European Parliament, the European Economic and Social Committee and the Committee of the Regions on a thematic strategy for soil protection (Brussels, 22.9.2006 COM (2006)231 final).

# Ethical and Moral Issues Relative to Groundwater

Emilio Custodio

**Abstract**

Groundwater is the major freshwater continental resource and reserve. Its quantity and quality are characterized by the delayed and damped response to external action and the large storage with slow renovation rate. This is not known to everybody and is difficult to be experienced personally, as groundwater cannot be directly observed. Therefore, specific knowledge is needed to address correctly the ethical and moral issues related to aquifer use and management. Poor consideration cause deviated policies, social disturbance, inefficiencies and negative impact on the environment and the services it provides. The subject of ethics is humans, but not Nature and the environment as such. This is something that some powerful organizations and minorities try to reverse, by subverting the principles of ethics and trying to erase the God-related fundamentals of moral. Groundwater ethics deal with present circumstances, as well as with the future, represented by scenarios that, to be ethically acceptable, should be non-biased, scientifically feasible and free of pre-set orientations aimed at other objectives. Ethics play an important role in water policy making, especially for groundwater. Science and technology cannot produce unique solutions to existing human and environmental problems if conditions are not previously set by a well-informed society and administrative, legal, social and economic agreements bound the objectives. This involves deep ethical and moral implications. Science and technology contribute the means to convert objectives into assessments that help in decision making at a higher level.

E. Custodio (✉)
Royal Academy of Sciences of Spain, Madrid, Spain
e-mail: emilio.custodio@upc.edu

E. Custodio
Groundwater Hydrology Group, Department of Civil and Environmental Engineering, Technical University of Catalonia, Barcelona, Spain

**Keywords**

Environment • Ethics • Groundwater • Moral • Society

## 1 Introductory Background

Ethics and moral are close concepts, which are often considered synonymous, although they differ conceptually. Hereinafter, ethical behaviour refers to conformity with norms widely accepted by a given society to improve human well-being and to get a peaceful living and mutual respect. What is considered good and honest may vary from one group to another and change with time, according to local habits, fashion, and socio-economics. However, interest groups with other goals may condition them and set what is politically correct. Moral refers to conformity with that are partly non-dependent on humans but derived from religion, by recognizing that man is a transcendent being whose final destination is God. In the Judeo-Christian religion, as is considered hereinafter, norms are embedded in God's commandments, which are written down and imprinted in the right conscience of any human being. The fundamentals of moral are independent of what society may consider correct and of the objectives of pressure groups and doctrines in fashion.

Water ethical and moral principles have to be applied to make water available to humans and their activities to preserve water resources, to maintain a healthy relationship with Nature, the environment and the services it provides, and to consider other components that affect humans too, such as economy, health, energy, land use, employment, quality of life, and spiritual and religious values.

Ethics and moral apply to humans, who have rights and duties, but not to Nature in general and to the environment in particular, as they are not able to decide. The common home concept, associated to United Nations favoured modernism, is a biased concept, as Man is transcendent and



M. Abrunhosa et al. (eds.), *Advances in Geoethics and Groundwater Management: Theory and Practice for a Sustainable Development*, Advances in Science, Technology & Innovation,
https://doi.org/10.1007/978-3-030-59320-9_2

travels to its final destination in God. So, the "common home" is really a "common bus" to follow the path each human has to do.

What has been said before applies to water, as all humans have a right to it, as a vital asset and a provider of goods. Therefore, water quantity and quality have to be preserved, and make it available in space and time, considering that water use implies some degradation. Ethical behaviour means trying to compensate degradation, especially to those that suffer damage without participating in the benefits. Damage is not known beforehand, so precaution is needed, although an excess of precaution may led to inactivity, with possible worst consequences to society than errors.

Individuals, besides the right to receive the needed water quantity and quality, have the duty to get, contribute, obtain, preserve, and maintain the infrastructures, according to the capacity of each one. Both rights and duties are ethical issues. The right to water is not primary but a secondary one, as water is not everywhere, and in any moment, but has to be made available. Positive action is needed, which have to be compatible with other rights, such as those related to the environment and the services it provides, energy, land use, employment and health.

There is a close relationship between water ethics and social and economic issues. They depend on tradition, although a common behaviour code must be worked out when different cultures coincide in space and time. Ethics is different and more than good management, sound governance, and social benefits and accomplishments. Therefore, ethics is not a guide with rules and recommendations for action, but a set of supporting concepts and principles for human behaviour.

To some extent, ethics is an alternative to the neo-classical and neo-liberal economic vision. This refers not only to the utilitarian aspects, but also to the non-quantifiable ones, such as tradition, personal feelings, and religion (Custodio 2000; Llamas et al. 2005; Delli Priscolli and Llamas 2012).

Water ethics relies on the application of the precaution and subsidiarity principles, with transparent action and information, and considers the relationships with other ambits. Water is a common good, so the access to it and its use must be equitable and fair, adequately regulated, without clandestine abstraction and out of law activities, and has to be made available and secured by the authorities, although this does not exclude the participation of the private sector when it is able to contribute the efficiency and knowledge that the public administration may lack. The discussion about the public and private roles in water supply is sterile as they are complementary.

## 2 Groundwater Characteristics and Ethics

Groundwater is a part of the continental water cycle, the major freshwater volume aside from ice caps. It is more widespread and less variable than surface water runoff. Generally, groundwater renovation time is of many years or decades, the reserves greatly exceed the annual flow and recharge, and its quantity and quality often vary vertically. Surface runoff can be directly observed, although the large variations make natural availability very uncertain. Groundwater geographical distribution and yield are less uncertain, although aquifer properties are more difficult and costly to know and the effect of changes are often highly delayed and dampened. This is generally not known by citizens and even by many water managers and decision makers, and is out of the mind of politicians, who in many cases stay in duty shorter terms than the aquifer response delay. This is not an excuse to obviate ethical behaviour, which should emanate from civil society and based on non-biased data and information.

Small changes in groundwater storage may produce serious impacts in groundwater, which is often essential to humans and the environment. Besides a reserve, groundwater is a renewable water resource, depending on the relationships with the other components of the hydrological cycle. In arid and semiarid areas, reserves become important, not only to regulate yearly variations in recharge and exploitation, but as an asset which can be depleted (mined) to get temporal economic and social benefits. This introduces new ethical and moral considerations. Simplistic considerations that consider groundwater as a common mineral resource cannot be applied.

## 3 Are Groundwater Resources a Univocally Defined Concept?

A frequent question in water planning, especially in arid and semiarid areas with high water demand, is how much water resources could be safely taken from natural systems to supply demand. The European Water Framework Directive does not pose the question. This should be addressed in the Country Members water legislation and water planning norms. Water allocation by applying simple rules and looking for the fair access to water is a way to try to bypass ethical and moral responsibilities, assuming that ethics were already considered when the rules were set at a higher level. To ethically address the often slow response of groundwater to external action, specific knowledge at the different decision-making levels is needed.

Groundwater plays a role in any circumstance in the hydrological cycle. This has to be evaluated through complete water balances along time in a given area, basin or aquifer. Long-term average water balances only give the order of magnitude of the terms but not on their relevance, except for large aquifer systems. In a correctly done groundwater balance, besides recharge, abstraction and recycling, storage variation, use in natural processes, maintenance of ecological services and preservation of quality, local climate and landscape have to be included. This is different to the common practice of calculating beforehand how much water is needed for natural and human processes and assuming that what exceed the sum are the available additional water resources. When water is for crop irrigation, water needs depend on the year. This has to be considered in resources allocation.

Any human direct and indirect modification of one or several of the water balance terms modifies the others. For groundwater, these modifications are slow and balanced by storage changes. The right question is not how much water resources are available but how much the system will change as the result of externally introduced changes. There is not an available water resources quantity of a given quality but a decision from the top on how much can be disturbed the system to yield social net benefits, without unacceptable results. Thus, there is no unique solution, but a set. Deciding on what is advisable, desirable, or undesirable is the responsibility of water authorities, water users, the civil society and politicians, but not of hydrologists and hydrogeologists, whose role is to provide the analysis and synthetized information for decision makers. This analysis has to be supported by adequate knowledge, good data and consideration of unavoidable uncertainty. The commonly used concepts of hydrological disequilibrium and overexploitation are colloquial, but meaningless.

## 4 Aquifers as Natural Infrastructures

Aquifers store and distribute groundwater. Therefore, they can be used as water management structures, in the same way that they function naturally. This use modifies their natural behaviour and affects the terms of the water balance. However, this helps in adapting the water cycle to human needs without recurring to costly engineered infrastructures and often with less environmental changes and less energy consumption. Something similar applies to soil and ground use to modify and treat water quality, and of river stretches beds as channels to transport water. To decide on the acceptance of these possibilities, the complete water balance has to be analysed, not only a few terms of it. Direct effects as well as externalities have to be considered, but further to quantifiable tangible values, there are the non-measurable, intangible values, which may be significant to locals and visitors. Thus, the use of aquifers and the ground as natural infrastructures, which is often socially and environmentally favourable, is not a mere engineering activity and involves ethical considerations to take into account negative externalities.

## 5 Which Are the Subjects, the Humans or the Environment and Nature?

Utilitarianism, consumerism, and the search of increased benefits may lead to serious environmental damage. It includes an excessive use of non-renewable raw material resources, among which there are groundwater reserves and freshwater in coastal aquifers. To control this consumption and the associated energy demand, speculation and health problems, increasingly complex and sophisticated techniques and tools are needed to make the world more manageable and habitable. As this may go against what is often considered as sustainability, there are ethical and moral considerations involved. Human sensitivity has evolved at a slower rate than tools, after decades in which economic optimization of growing and productivity was the goals. The quality of human living is the result of a combination of knowledge, faith, will, and love.

Scientific and technological progress has reshaped the world's economy to be globally connected and increasingly urban. This is a paradigm that humans have still to assimilate. Nature and the environment are essential to human beings, although this is often out of focus and transformed into a new religion in which "Mother Earth" substitutes God and opens the door to of old myths and superstition. This man's centred vision has profound consequences in ethics and makes useless moral considerations.

The environment is not static. It evolves due to natural causes, although slowly and difficult to be observed during a human life. Going against these changes is difficult and costly, if not useless. This may require efforts that cut down the capacity to deal with other important human problems. Humans need Nature, but Nature does not need humans. Nature is not a god to be served in exchange of goods, but the provider of means to accomplish Man's role in the Earth. From the Judeo-Christian point of view, there are non-tradable principles. So, water and water resources cannot be ethically used to get political benefits, for coercion and as a mean to introduce ideologies.

The environment is a serious affair, but it is non-static and in the service humans. Adaptation seems a correct position in the long term, even if this implies migration and a different future world, provided, and there is time and will for a soft transition.

## 6 Other Groundwater Related Issues

Groundwater net recharge depends on climate, although there are other factors to be considered as well, such as river infiltration, changes in land use and soil, kind of vegetation and state of exploitation. Climate is variable due to natural processes, and it will continue to vary due to phenomena that are out of the human control. Therefore, ethics does not apply to them. However, there are factors affected by human activity, involving ethical and moral considerations. There are complex relationships with negative externalities derived from decisions, which are often disregarded or unknown beforehand. Water transfer to an area with scarce local water resources, to solve supply problems, may create mid- and long-term changes in local groundwater recharge, which result in salinization, high water tables and wetland deterioration.

Except in the rare situations in which groundwater is obtained by means of drainage galleries (tunnels) or natural outflows, it has to be pumped from the ground from wells and boreholes. Therefore, groundwater abstraction consumes energy. In areas with high water demand and deep groundwater levels, energy consumption may be a significant fraction of total energy, as in some of the intensely irrigated agricultural areas in arid environments. Therefore, there is a water-energy nexus, which has social implications and needs setting policies based on ethical principles, especially when energy supply is limited, other users cannot access it, and cost is subsidized.

Many people and scientists admit that climate is changing by human influence, related to an increased greenhouse effect due to fossil fuel combustion and related waste gases and particles. However, other generally well-informed scientists and aware of the Earth climatic history evaluate that this influence is small and less than the uncertainty of current climate knowledge and capacity of long-term prediction. This involves deep ethical and moral implications.

How climate change will affect groundwater can be evaluated with existing models, which are quite well advanced and good, but not all relevant processes are correctly incorporated; there is no long-term calibration, and downscaling is still a difficult and uncertain process. The main problem is how to obtain reliable data series for the coming with climate models. This is often done by setting scenarios, but to obtain and evaluate changes and their probability, these scenarios should be non-biased. Translating results into reports understandable by citizens, managers, mass media and politicians has to follow ethical and moral principles to avoid bias and manipulation.

The commonly large groundwater storage in aquifers relative to recharge allows the regulation of fluctuations, both natural and as the result of human action. This is a main asset in favour of groundwater and its integration in water resources systems that include the environment and its ecological services. At least, a part of storage should be preserved for this function, both under current and probable future circumstances, taking into account uncertainty.

When considering groundwater storage depletion (mining) as a solution to water supply problems, results have to be fair with current and future generations. The increasing cost of water, water quality impairment, possible land subsidence, impact on wetlands, discharge of water and solutes into the sea in the case of coastal aquifers, and capacity of regulating the water cycle fluctuations are significant ethical issues, which need data and transparency.

Virtual water may be important in many areas. Import of produced and manufactured goods implies that water and energy were used abroad and not in the area. Export to other areas means water and energy spent in the area, besides keeping the waste associated to this production and that derived from making available the water and energy consumed. This has profound ethical and moral implications. The future generations will inherit land changes, groundwater storage depletion, wetland disappearance, and a possible large passive as wastes that may affect groundwater in the mid and long term. More general scope ethical and moral considerations refer to the fair payment of virtual water-rich imported products, to guarantee that these goods are not obtained from impoverished areas, with destruction of non-recoverable land, environment, structures, and especially groundwater resources and reserves. It is surprising that virtual water is not considered in many agendas dealing with water and groundwater resources.

**Acknowledgements** Thanks to Dr. Andrés Sahuquillo (Royal Academy of Sciences and Academy of Sciences of Valencia) and to Ms. María Feliciana Fernández (School of Architecture, Technical University of Madrid) for the comments and suggestions.

## Reference

Custodio E (2000) Some relevant ethical issues in relation to freshwater resources and groundwater. Bol Geol Min, Madrid 111(6):121–130

Delli Priscoli J, Llamas MR (eds) (2012) The role of ethics in the globalization of water and food. Water Policy 14(1–2). https://iwaponline.com/wp/issue/14/S1

Llamas MR, Martínez-Cortina L, Mukherji A (eds) (2009) Water ethics, 3rd Marcelino Botín Foundation Water Workshop. CRC Press, Boca Raton

# Some Basic Considerations on the Applied Ethics to Water Resources Management

María Feliciana Fernández-García, Manuel Ramón Llamas, Emilio Custodio, and Francisco Javier Neila

**Abstract**

This study is a contribution to the analysis of ethical factors in water governance. Obviously, the issue is very broad, as water is a "polyhedral" resource, that is, it has many different facets according to the multiple roles it plays in the environment, in economic activities, in public health, in the liturgies of the different cultures and religions, and even in the current issue of gender difference. It is enough to remember that in poor countries women and girls are who normally take care of providing water to their family at the cost of losing education. These different facets have a diverse nature. Some are quantifiable like the amount of water needed in each climate and soil and with a specific technology to obtain a specific crop; on the other hand, there are others that are hardly measurable or even impossible to measure. Even in the measurable aspects, there is a great variety of circumstances. For example, the water supply and sanitation system of an area is very important for public health, whose care normally falls in almost every state into a health or similar title department.

**Keywords**

Water ethics • Human right to water • Integrated water resources management • Transparency

## 1 Introduction

Conflicts in relation to water are frequent. In some cases, they may lead to a situation of violence. However, the reality seems to indicate that initial situations of confrontation can often become situations of cooperation (Wolf 1998; Llamas 1999). One Stockholm Water Prize has argued that water is not an inductor of war but a catalyst for peace (Asmal 2000).

The situation seems to recommend, establish and spread a new culture or ethics of water, both socially and personally. Perhaps the fundamental need of water managers is to be trained in ethics, so that this allows sound environmental water management. This management must be far from technological triumphalism, in which almost everything is entrusted to the construction of infrastructures.

Another problem that may raise ethical issues is related to the public or private nature of property and/or the management of surface and groundwater. Today, there is an almost universal trend towards the privatization of water management and the creation of water markets, with the aim of achieving a more efficient allocation of this resource. However, there are relatively frequent voices that ask that this market does not exist, since they consider it incompatible with the concept of public domain of the waters, the non-commercial nature of a vital resource or religious or cultural principles. Other voices strongly recommend that this market be adequately regulated to avoid monopoly or other situations that make difficult, if not impossible, the operation of a true market economy in water. Still other voices demand what is called a "re-municipalization" of urban water services.

Water, in addition to being essential for life and having an economic or utilitarian value, also has an intangible value that is sometimes described as symbolic, cultural or religious. These characteristics of water make it practically impossible for this resource to be treated as a usual consumer good.

M. F. Fernández-García (✉) · F. J. Neila
Department of Construction and Technology in Architecture, High Technical School of Architecture of Madrid (ETSAM), Technical University of Madrid, Madrid, Spain
e-mail: marifeli.fdez@gmail.com

M. R. Llamas
Royal Academy of Sciences of Spain, Madrid, Spain

E. Custodio
Royal Academy of Sciences of Spain and Group of Groundwater Hydrology, Department of Civil and Environmental Engineering, Technical University of Catalonia, Barcelona, Spain


M. Abrunhosa et al. (eds.), *Advances in Geoethics and Groundwater Management: Theory and Practice for a Sustainable Development*, Advances in Science, Technology & Innovation,
https://doi.org/10.1007/978-3-030-59320-9_3

## 2 Methods: Some Basic Considerations on Ethics in Hydrogeology

It seems evident that individual humans are making value judgements about the actions of those around them. The act of anyone is almost automatically judged by the general people as good or bad. Now, by what criteria do we decide in our conscience whether something is good or bad? Is there a universal rule? Is the criterion the same for all men? Is it constant over time? Is it the same for all cultures?

Freedom is the basic ground for ethical or moral acts it is sometimes conditioned by external influences: what is "politically correct" (Part 2: Chapter LVIII, "Don Quixote"; Cervantes 1986).

Ethics and moral are disciplines that study these issues. With relative frequency, the ethical term is used for the analyses carried out from philosophy and the moral term in those carried out from theology. In practice, we considered that ethics and moral are often equivalent, although they may differ on the absolute and the relative values of supporting principles, these are philosophical in ethic and theological in moral.

The ethical movement of concern for nature is relatively recent, less than a century. Two fundamental milestones were the publication of a novel by the marine biologist Rachel Carson entitled "Silent Spring" in which she talks about the birds no longer singing, because insecticides have eliminated the insects that these birds fed on (Carson 1962). The second is the article by the English historian Lynn White ("The historical roots of our ecologic crisis"). White (1967) assured that the ecological disasters that appeared in those years were fundamentally due to the interpretation of the Judeo-Christian culture of the book of Exodus: "Multiply and Dominate the Earth‵.

This work was the catalyst for a very large series of articles on environmental ethics, including Aldaya et al. (2017), Ziegler and Groenfeldt (2017) and Fernández and Llamas (2017).

## 3 Discussion

"Social ethics" deals with the relations between human beings. It is an old concern, for example (a) Hammurabi Code (1800 BC); (b) Aristotle Ethics to Nikomacus (IV century BC).

On the other hand, "environmental ethics" is new. Less than a century before the humans were unable to significantly disturb the planet, ethical factors have great weight when making decisions in the processes of water resources allocation. In what follows, a classic classification is used, which is the same used in Llamas et al. (2001).

### 3.1 Water for People. Urban Uses: Supply and Sanitation

Urban water uses represent a relatively small share in total uses. Nevertheless, their political importance is very great and "in fact" in all countries they usually have a priority treatment in relation to uses in agriculture, industry, tourism and even the environment.

In principle, this approach seems correct to the authors', among other reasons, because the United Nations has integrated the right to water supply and minimal sanitation as a fundamental right of the human being (UN 2010). However, there is still a long way to implement truly this right (Gleick 1998).

### 3.2 Water for Agriculture: Agricultural Uses

Food production is by far the main user of blue (mobilizable) water on the planet and an important user of green (soil) water in irrigated and rain-fed crops. It is clear that many countries cannot produce the necessary food with their precipitation water. As the importation of blue water is generally prohibitive because of its high cost, the solution in these cases of water scarcity could be in the importation of virtual water through international food trade. However, the importance of international virtual water trade has been clearly presented by late professor Hoekstra from the Netherlands. A good application of Hoekstra´s principals has been done for the analysis of the Spanish situation (Aldaya et al. 2009) widely recognized nor the associated problems and externalities.

### 3.3 Industrial Uses

Industrial uses of water maybe usually different. They can be non-consumptive and simply limited to returning the water used with a higher temperature, or with a significantly modified flow rate.

In many cases, the main problem of industrial uses is the chemical or thermal pollution with which water returns to its natural cycle. In general, the universally proposed principle: "the polluter pays" only applies very partially. The fundamental cause of this situation is the low rates of urban supplies that are not enough to adequately treat urban water waste. However, in general this situation seems to be solved if urban water tariffs were enough for wastewater to receive tertiary treatment. Intensive agriculture is also a very important and sometimes dominant source of pollution. The solution to avoid this pollution is still far to be solved. Much work is still necessary to solution this problem. There are some questions when different groups or individual violate water quality standards, for example, what penalties should be exalted? And who should perform that function (Groenfeldt 2019). The answers are relative, and that case can be different each other, for that it is necessary to study specific situations. That is just because water ethics introduce a new discipline to consider in each of the different situations, according to social justice, economic efficiency and ecosystem functions.

### 3.4 Land Planning Uses

We should balance competing land use across different sectors in a rational, optimal and fair manner. Land use planning must include their impact on water resources. If it is not considered adequately, serious problems will probably arise, like the current ones in Mar Menor in Spain (Fernández 2019). The consequences are more visible during floods, but the influence of land use is directly related to water management. Deciding to make a change in the territory for a given economic use implies a change in the natural water cycle, for the benefit of some and by neglecting others. However, this is not always the case, since the economic and social aspects of a change in land use may influence lots of individuals. For that, land and urbanization study should be done case by case.

### 3.5 Water for Natural and Ecological Uses

The European Water Directive has made a very important change in water legislation in its Member States. The Directive requires an evaluation of all surface and underground bodies of water in each river basin or hydrographic demarcation. A detailed plan of measures must be included in the next hydrological plan. This detailed plan should also include the system of measures to try to correct the situation in those water bodies that do not have a good ecological state.

## 4 Concluding Remarks

The data nowadays provided by science and technology tell us that usually there is enough water if we work with nature and not against it, and if we cooperate with each other. One of the main elements to achieve this cooperation is not forgetting what negotiation experts usually call intangible values. These values go beyond the usual economic or utilitarian values that the conflicting parties can, in general, identify with relative ease. Ultimately, it is to recognize that the human intelligence and will are very important, but also are feelings, poetry, aesthetics, etc. If this is forgotten, water conflicts, like so many other problems, have a difficult solution.

## References

Aldaya MM, Martínez-Santos P, Llamas MR (2017) The relevance of ethical factors in the pursuit of integrated water resources management. In: Global water ethics. Routledge, pp 131–147

Aldaya MM, Garrido A, Llamas MR, Varela-Ortega C, Novo P, Casado RR (2009) Water footprint and virtual water trade in Spain. In: Water policy in Spain. CRC Press, pp 69–80

Asmal K (2000) Water: from casus belli to catalyst for peace. In: Address in the opening session in the Stockholm Water Symposium, vol 14

Carson R (1962) Silent spring. Houghton Mifflin Company

Cervantes SM (1986) The adventures of Don Quixote de la Mancha. New York, Farrar, Straus, Giroux

Fernández MF, Llamas MR (2017) Approach to theological, ethical and pastoral care of the Laudato Sion the right to the water from the scientific and technical point of view. https://hrw.encuentromundi.org; accessed in January 2020

Fernández MF (2019) El Mar Menor requiere proteger las aguas subterráneas y su planificación con las superficiales. iAgua Magazine. https://www.iagua.es/noticias/redaccion-iagua/mar-menor-no-se-podra-proteger-sino-se-protegen-aguas-subterraneas; accessed in January 2020

Gleick PH (1998) The human right to water. Wat Policy 1(5):487–503

Groenfeldt D (2019) Water ethics: a values approach to solving the water crisis. Routledge

Llamas M R (1999) El agua como elemento de cohesión social. In: Homenaje a D. Angel Ramos Fernández, Real Academia de Ciencias Madrid 1:197–215

Llamas MR, Fornés JM, Hernández-Mora N, Martínez-Cortina L (2001) Aguas subterráneas: retos y oportunidades. Fundación Marcelino Botín and Santander y Mundi Prensa, Madrid

United Nations (2010) The human right to safe drinking water and sanitation. General Assembly. Resolution A/RES/64/292

White L (1967) The historical roots of our ecologic crisis. Science 155 (3767):1203–1207

Wolf AT (1998) Conflict and cooperation along international waterways. Water Policy 1:257–265

Ziegler R, Groenfeldt D (eds) (2017) Global water ethics: towards a global ethics charter. Taylor & Francis, NY

# Groundwater Contamination Science and the Precautionary Principle

Ian G. Stewart, John Cherry, and Moira Harding

**Abstract**

Several kinds of uncertainty affect both the science of groundwater and, as a result, its management and governance. This work surveys groundwater contamination from the point of view of such uncertainties, focusing on aspects of ecosystem complexity, time, and space. We consider two cases of groundwater contamination: Per- and polyfluoroalkyl substances (PFAS) and hydraulic fracturing. We argue that to move towards sustainability of groundwater resources, there is urgent need for explicit inclusion of the precautionary principle (PP) at the level of both management and governance because the PP offers possibilities for dealing with various kinds of uncertainty more transparently. We conclude by noting that application of the PP is not just a matter of competent use of science for governance, but also a matter of central importance to the geoethical considerations advanced at this Porto conference on hydrogeoethics.

**Keywords**

Groundwater • Contamination • Uncertainty • Risk • Precautionary principle • Fracking • PFAS

## 1 Introduction

Societies seem long to have enjoyed a kind of intuitive confidence in groundwater 'purity' (Freeze and Cherry 1979). However, with increasing groundwater scientific capacity in recent decades, it is now known that pristine groundwater pumped from wells has become a rarity. Over the decades, each new contaminant type found in groundwater has appeared to come as a surprise, but things did not need to be this way. The surprise is testimony to societies' lack of dedication to proactively address this problem (Pankow and Cherry 1996; van der Gun 2017).

This does not mean that groundwater that contains anthropogenic chemicals is necessarily hazardous to human health or the environment; the science of remediation of groundwater pollution has been evolving for decades (Boulding 1995). But it does mean that groundwater has the potential to be hazardous—now, or in the future—in ways that are difficult to determine. It is arguably more challenging than ever to reliably determine the degree of risk associated with drinking water sourced from groundwater because the anthropogenic impacts on groundwater contamination appear to be increasing, perhaps even exponentially (van der Gun 2019). As we understand more about the complexities of how groundwater is a part of complex freshwater ecosystems, both our certainties and our uncertainties come into sharper focus. A key question is whether our capacity to manage and govern with respect to groundwater can keep pace with our evolving science and vice versa.

## 2 Methods

Our methodology is interdisciplinary in that we combine the perspectives of over 50 years of experience in the field of hydrogeology (J. Cherry) with the field of history and philosophy of science (I. Stewart) and the field of

I. G. Stewart (✉)
University of King's College, Halifax, Halifax, Canada
e-mail: igstewar@dal.ca

J. Cherry
G360 Institute for Groundwater Research, University of Guelph, Guelph, Canada

M. Harding
Schulich School of Law, Dalhousie University, Halifax, Canada



M. Abrunhosa et al. (eds.), *Advances in Geoethics and Groundwater Management: Theory and Practice for a Sustainable Development*, Advances in Science, Technology & Innovation,
https://doi.org/10.1007/978-3-030-59320-9_4

environmental law (M. Harding). In addition to expert judgement offering a general retrospective on the field of contaminant hydrogeology (J. Cherry) and applied to the two cases addressed in this work, we conducted a partial literature review (academic and grey literature) in the fields of groundwater contamination science, science governance, and science policy, with a focus on links between groundwater contamination and the precautionary principle.

## 3 Results

### 3.1 The Uncertainties Systems Complexity, Time, and Space in Groundwater Science

The relative hiddenness of groundwater compared with surface water has long been understood as a problem for the hydrological science of groundwater volumetric determinations and the related field of withdrawal management and governance (FAO 2016; Famiglietti 2014). The complex ways in which the hiddenness of groundwater as it impacts uncertainties concerning groundwater *contamination*, a second aspect of groundwater science and governance, is less mature. Its importance has long been recognized in the field of international law (Eckstein 2017). The connections between the two aspects are beginning to be considered in tandem in the context of the UN Sustainable Development Goals (UN-Water 2019).

The invisibility of groundwater is becoming more visible, as it were, particularly in the context of climate change research (Gupta and Conti 2017). The multiple feedback effects linking groundwater to ecosystems are identifiable, but fraught with the uncertainties attending all complex systems (Doak et al. 2008). These include multiple feedback effects depending on magnitude, intensity, seasonality, precipitation patterns, hydrogeological considerations, and varying anthropogenic influences (Jiménez Cisneros and Oki 2014).

Added to this baseline problem of complexity, two fundamental challenges face the science of groundwater contamination. The first kind of uncertainty problem concerns *time*. In contrast to surface water, the duration of groundwater contaminant residence for many deleterious recalcitrant chemicals can be decades, centuries, or even longer, well beyond the normal timeframe in which most scientific monitoring protocols or governance frameworks are maintained. Thus, for example, due to the long lag between cause and effects, what happens on private property to degrade groundwater can eventually harm public groundwater. As a result, detections through scientific monitoring systems usually do not happen or happen only after it is too late.

The second problem is *space*. Groundwater covers space that does not respect societal surface boundaries, introducing what the groundwater governance literature describes as a scalar problem. This scalar problem has been the focus of hydrologists interested in measuring groundwater volumes for the purpose of governing withdrawal rates (Cohen and Davidson 2011). Less is known about the spatial or scalar problem as it relates to science (and governance) of groundwater contamination, but the contours of the challenges are emerging (McCaffrey 2007). For example, even in the context of sophisticated science-based monitoring protocols, contamination occurrences are usually perceived as a separate problem because each is localized in a particular part of an aquifer. This can mean that each known contamination occurrence can be perceived by government as a separate problem because each is local in a particular part of a particular aquifer and hence each produces impacts relating only to a small subset of the population in each province, state or country (Villholth and Conti 2018).

Systems complexity, time, and space all introduce uncertainties that can influence both the capacity for contaminant hydrogeology to be carried out, and its capacity for influencing management and governance of the real or potential anthropogenic impacts on groundwater. The following case studies, drawn from recent history, provide two examples.

### 3.2 The Case of PFAS

The assumption that groundwater possesses strong assimilative capacity formed the basic justification for the septic systems that have been ubiquitous since the 1960s. The first inkling that some contaminants in groundwater are not readily assimilated came in the late 1970s when chlorinated solvents were found in thousands of public water supply wells across the United States and soon thereafter in Europe (Pankow and Cherry 1996). This discovery resulted from the 1974 USA safe drinking water regulations. In 1980, more regulations in the United States, and later elsewhere, mandated investigations of contaminated industrial sites. These showed persistent presence of chlorinated solvents (e.g. PCE, TCE, TCA, and daughter products) at many thousands of sites and demonstrated that some types of chemical compounds are not readily degraded and travel rapidly in many aquifers. Since the 1990s, more types of mobile and persistent industrial chemicals were discovered, the latest of which is a category of modern synthesized organic chemical known as per- and polyfluoroalkyl substances (PFAS). PFAS are a group of a several thousand individual chemical compounds used as oil and water repellents and coatings for common products including cookware, carpets, and textiles. Today, in many countries, PFAS chemicals are increasingly found in drinking water supplies near facilities where the chemicals have been used or disposed of. PFAS may be the worst of all the groundwater contaminants because they bioaccumulate and are least prone to degradation. Moreover, toxicological and analytical difficulties are large, and the safe drinking water levels are likely extremely low, less than one microgram per litre (Simon 2019).

### 3.3 The Case of Hydraulic Fracturing of Natural Gas

The rapid development of sophisticated engineering techniques for hydrocarbon mining known as hydraulic fracturing, or 'fracking', has changed the geo-political and economic landscapes on an international scale, impacting the USA in particular (Enerdata 2019). Fracking is also common in the western provinces of Canada and is growing rapidly in Russia and China, but it has been banned or delayed by moratoria in nearly all other jurisdictions around the world where it has been proposed. This was based in part on growing awareness of the uncertainties concerning impacts on groundwater due to fracking (Winter et al. 2016).

Fracking involves drilling holes up to kilometres deep, thus puncturing groundwater zones, and into which water mixed with chemicals ('fracking fluids') under intense pressure is pumped to crack open oil and gas contained within rock formations to extract what oil or natural gas they contain. Most importantly, the sealing of fracked wells, which relies on cement, is less reliable than for conventional petroleum wells that also have cement seals (Dusseault et al. 2000). The knowledge that fracked wells (during and after operational lifespans) leak methane, and that this might be a concern for groundwater, has been known for decades (Ingraffea et al. 2014). However, due to the structural features of ecosystem complexity, time, and space reviewed above, fundamental uncertainties emerge in the science of the impacts of natural gas leaking laterally into groundwater aquifers. Notably, although the fundamentals of groundwater geochemistry are well enough understood to raise concern about methane impacts on groundwater, disputes continue with respect even to basic methodologies of data collection in terms of impacts (Parker et al. 2012; Jackson and Heagle 2016). The science is thus relatively immature (CCA 2014). Important for this work, even if the appropriate groundwater monitoring was to take place, the uncertainties due to temporal and spatial scales suggest structural uncertainties will persist for some decades concerning our capacity to determine impacts on groundwater due to fracking. Recent field experiments, using methods only nascently applied to hydrogeological studies of fracking, suggest lateral mobility and persistence of methane in aquifers is far more extensive than anticipated (Cahill et al. 2017). Recent reviews in the field of public health research on populations near oil and gas operations have indicated similar states of immaturity of both research methods and data (Bamber et al. 2019).

## 4 Discussion

The key question addressed in this work is whether the evolving nature of the science and governance systems concerning anthropogenic impacts on groundwater can keep pace with each other in a context of increasing environmental degradation generally and contamination of groundwater specifically (van der Gun 2019).

Evidence from the past is not encouraging. In the case of PFAS, it was known as early as the 1970s that they would likely be toxic to humans if consumed in food or water. But as far as we are aware, substantial questions about consequences of PFAS entering groundwater were not addressed anywhere by the organizations charged with protecting human health and groundwater.

The case of hydraulic fracturing is similarly discouraging. Governance and regulatory bodies, in the context of the expanding dependence on hydraulic fracturing for socio-economic reasons, may have displayed questionable confidence in both engineering and scientific capacity to make robust risk assessments, given the sorts of uncertainties discussed above. In some jurisdictions, review of governance policy has concluded a noticeable absence of recognition on the part of regulators regarding the uncertainties in fields such as hydrology that have been reviewed above, despite mounting evidence in the academic and grey literature (Winter et al. 2016). In other jurisdictions, assessment of whether or not to frack in regions where it is not yet happening, even when the eventual judgement was favourable, has emphasized significantly new, rigorous, and not yet implemented governance and management regimes before development could take place (Government of Australia 2018).

These two cases are not anomalies. A recent review of such disconnects between science and governance in a broad range of sectors including environmental and human health, ecosystem sustainability and emergent technologies, has detected a pattern of a growing number of instances of misplaced certainty about the absence of harm, which has delayed preventive actions to reduce risks to human health, despite evidence to the contrary (Grosjeans 2013). It is a pattern well known to scholars in the field of environmental sociology and philosophy. Its flip side is the misplaced scientific scruple whereby impossibly high degrees of certainty were demanded before restricting or regulating industrial activities, but apparently justified under the scientific principle of wanting to *avoid* committing Type II errors, namely wrongly predicting harm where none in fact exists (Freudenburg et al. 2008). The evidence of many decades reinforces the need for robust implementation of the precautionary principle (PP).

Forged in the environmental policy circles of Germany in the 1970s, the PP received international attention in 1992, as Principle 15 of the UN Rio Declaration on Environment and Development. There it was formulated precisely as a way of dealing with both of these kinds of misuse of scientific certainty: 'In order to protect the environment, the precautionary approach shall be widely applied by States according to their capabilities. Where there are threats of serious or irreversible damage, lack of full scientific certainty shall not be used as a reason for postponing cost-effective measures to prevent environmental degradation.' (Stirling 2007). More recently, the principle has been included either explicitly or implicitly in the majority of international law concerning the physical and biophysical environment, environmental protection, and environmental management. It has been implemented domestically in federal, provincial, and even municipal laws of UN member states. And it is referenced in numerous policies and guidelines both domestic and international—albeit, in a variety of different forms (McCaffrey 2007; Eckstein 2017; Jaeckel 2017).

In its most developed forms to date, the PP provides a robust framework for transparently identifying the science and governance criteria needed for the sorts of science-based risk assessment and management that sound governance requires (Bourguignon 2015). It may not tell us what to do, but it at least forces us to ask the right questions (Stirling 2007). And when such criteria are not met, when uncertainties abound in relation either to probability or consequence determinations needed for risk assessment, it encourages frank disclosure and broader societal engagement in determining how to proceed (Stirling 2007; Jaeckel 2017). The PP encourages these more complex adjudications that are needed when uncertainties such as ambiguities within societal values for natural resource use abound, or when scientific ignorance is high enough to require difficult decisions about how to balance conflicting priorities in both scientific research and innovation funding prior to industrial development. Such messy adjudications are unavoidably also ethical ones, increasingly urgent as we face ecosystem degradations on local and global scales. The PP has encouraged recognition that interdisciplinary and multidisciplinary approaches to risk determinations and governance directions are needed to embrace such complex decisions. This is also why there are increased calls for evolving collaborations between the social and human sciences and the natural/biophysical sciences (Renn 2014). This is being recognized by hydrology scholars exploring, for example, the relatively new, hybrid, interdisciplinary field of 'socio-hydrology' (Sivapalan 2015). This conference's attention to groundwater as a space for ethical considerations attests to the need for such ongoing hybridities, consistent with the work of Abrunhosa et al. (2018).

But the PP has received much criticism, largely basing itself on the denial that the issues of uncertainty are such as to warrant the PP (Stirling 2007; Bourguignon 2015). However, for many fields of environmental science applied to anthropogenic ecosystem effects, appropriately robust risk assessment is often either not possible or not enough (Grosjeans 2013). We contend this applies to the field of groundwater contamination, given the complexity of the natural systems being studied, the relatively immature status of the necessary science, or the lack of requisite managerial and governance capacity to adequately deal with such complex adjudications in the face of temporal and spatial uncertainties. For groundwater contamination, as for other fields dealing with complex systems, this is a view that at times is resisted by scientists (Doak et al 2008) and even more so by politicians and regulators (Winter et al. 2016).

From a geoethics standpoint, however, the precautionary principle is of central importance. Its explicit formulations grew out of internationally recognized ethical norms of inter-generational justice, namely that future generations must not be made to shoulder the burden of today's poor decisions (UN 1992). Further (and as was also explored at this conference), governing under uncertainty is not just a technical question of risk assessment methodology, but also an inescapably ethical judgement, requiring collective deliberation about shared societal values (Oppenheimer et al. 2019). 'When do we know enough?' is not just a question for hydrogeologists in such cases. This is all the more true given temporal and spatial complexities of groundwater discussed above. Ethical deliberations are integral to even the most technical considerations impacting groundwater governance.

## 5 Concluding Remarks

The record surveyed in this work indicates that the profound uncertainties in science have proven very hard to delineate and translate into improved groundwater science and governance practice. This is especially troubling when the certainties required for confident predictions about our impacts on groundwater are not yet justified. However, frank discussion of these uncertainties, the concomitant embrace of requisite humility, and the use of flexible approaches to governance that incorporate uncertainties are urgently needed. It is such transparency that can lead, for example, to rigorous monitoring and adaptive management protocols, which history shows tend *not* to materialize when the PP is not observed. In the field of groundwater contaminant science and governance, the PP has barely made itself felt. As a matter both of scientific and geoethical integrity, it is high time for the PP to take a more central place in both word and deed.

**Acknowledgements** The authors gratefully acknowledge the helpful comments and corrections of four anonymous referees. Aspects of this work were also discussed with Douglas Mulhall and Katja Hansen, Research Fellows, Technical University, Munich.

## References

Abrunhosa MJ, Peppoloni S, Chambel A (2018) Introducing geoethics in groundwater science, technology, and management. In: Abstracts Book, Session T1.1 socioeconomic aspects of groundwater resources management, 45th IAH Congress, Daejeon, Korea, pp 26

Bamber A, Hasanali SH, Nair AS, Watkins SM, Vigil DL, Van Dyke M, McMullin TS, Richardson K (2019) A systematic review of the epidemiologic literature assessing health outcomes in populations living near oil and natural gas operations: study quality and future recommendations. Int J Environ Res Pub Healt 16:2123

Boulding JR (1995) Practical handbook of soil, vadose zone, and ground-water contamination: assessment, prevention, and remediation. Lewis Publishers, Boca Raton

Bourguignon D (2015) The precautionary principle: definitions, applications and governance. European Parliament Research Services, Paris

Cahill A, Colby MS, Olenka F, Olukayode KS, Emil R, Bernhard M, Mayer KU, Strous M, Ryan C, Cherry J, Parker BL (2017) Mobility and persistence of methane in groundwater in a controlled-release field experiment. Nat Geosc 10(4):1–9

Cohen A, Davidson S (2011) The watershed approach: Challenges, antecedents, and the transition from technical tool to governance unit. Wat Alternat 4:1–14

Council of Canadian Academies (2014) Environmental impacts of shale gas extraction in Canada. Ottawa (ON): the expert panel on harnessing science and technology to understand the environmental impacts of shale gas extraction. Council of Canadian Academies, Ottawa

Doak F, Estes J, Halpern B, Jacob U, Lindberg D, Lovvorn J, Monson D, Tinger MT, Terrie Williams J, Wootton T, Carroll T, Emmerson M, Michell F, Novak M (2008) Understanding and predicting ecological dynamics: are major surprises inevitable? Ecology 89(4):952–961

Dusseault MB, Gray MN, Nawrocki PA (2000) Why oilwells leak: cement behavior and long-term consequences. Society of Petroleum Engineers. In: International Oil and gas conference and exhibition in China, 7–10 Nov 2000, Beijing, China. https://doi.org/10.2523/64733-MS

Eckstein G (2017) The international law of transboundary groundwater resources. Routledge, New York

Enerdata (2019) Global energy statistical yearbook. https://yearbook.enerdata.net/natural-gas/world-natural-gas-production-statistics.html. Accessed May 2020

Famiglietti J (2014) The global groundwater crisis. Nat Clim Change 4:945–948

FAO (2016) Global framework for action to achieve the vision on groundwater governance. FAO, Rome

Freeze RA, Cherry JA (1979) Groundwater. Englewood Cliffs, Prentice-Hall

Freudenburg W, Grammling R, Davidson DJ (2008) Scientific certainty argumentation methods (scams): science and the politics of doubt. Sociol Inquir 78(1):2–38

Government of Australia (2018) Scientific inquiry into hydraulic fracturing in the northern territory. Canberra, Australia

Grosjeans (2013) Late lessons from early warnings: science, precaution, innovation. Publications Office European Environment Agency, Luxembourg

Gupta J, Conti K (2017) Global climate change and global groundwater law: their independent and pluralistic evolution and potential challenges. Water Int 42(6):741–756

Ingraffea RA, Wells MT, Santoro RL, Shonkoff SB (2014) Assessment and risk analysis of casing and cement impairment in oil and gas wells in Pennsylvania, 2000–2012. Proc Nat Acad Sci 111:10955–10960

Jackson RE, Heagle DJ (2016) Sampling domestic/farm wells for baseline groundwater quality and fugitive gas. Hydrogeol J 24:269–272

Jaeckel AL (2017) The international seabed authority and the precautionary principle: balancing deep seabed mineral mining and marine environmental protection. Leiden, Brill

Jiménez Cisneros B, Oki T (2014) Freshwater resources. In: Climate change 2014: impacts, adaptation, and vulnerability. part a: global and sectoral aspects. Contribution of working group II to the fifth assessment report of the intergovernmental panel on climate change, vol I, pp 229–269. Cambridge University Press, Cambridge

McCaffrey S (2007) The law of international watercourses. Oxford University Press, New York

Oppenheimer M, Oreskes N, Jamieson D, Brysse K, O'Reilly J, Shindell M, Wazeck M (2019) Discerning experts. The practices of scientific assessment for environmental policy. Chicago, University of Chicago Press

Pankow J, Cherry J (1996) Dense chlorinated solvents and other DANPLs in groundwater. Waterloo Press, Portland

Parker BL, Cherry JA, Chapman SW (2012) Discrete fracture network approach for studying contamination in fractured rock. AQUA Mundi 3:101–116

Renn O (2014) Toward a socio-ecological foundation for environmental risk research. In: Lockie S, Sonnenfeld D, Fisher D (eds) Routledge international handbook of social and environmental change. Routledge, New York

Simon JA (2019) PFAS experts symposium: statements on regulatory policy, chemistry and analytics, toxicology, transport/fate, and remediation for per-and polyfluoroalkyl substances (PFAS) contamination issues. Remediation 29(4):31–48

Stirling A (2007) Risk, precaution and science: towards a more constructive policy debate. EMBO Reports 8(4):309–315

Sivapalan M (2015) Debate—Perspectives on socio-hydrology: Changing water systems and the "'tyranny of small problems"'—Socio-hydrology. Wat Resour Res 51:4795–4805

UN-Water (2019) Climate change and water. UN Water Policy Brief. UN, Hague

van der Gun J (2017) Data, information, knowledge and diagnostics on groundwater. In: Villholth KG, López-Gunn E, Conti KI, Garrido A, van der Gun J (eds) Advances in Groundwater Governance. CRC Press, Leiden, pp 193–213

van der Gun J (2019) The global groundwater revolution. Oxford Research Encyclopedia of Environmental Science

Villholth K, Conti K (2018) Groundwater governance: rationale, definition, current state and heuristic framework. CRC Press/Balkema, Leiden

Winter J, Dobson S, Lorefice S (2016) Hydraulic fracturing and public policy. Final report: knowledge synthesis grant. Social Sciences and Humanities Research Council, Ottawa

# Geoethics for Operating in the Human Niche

Martin Bohle

**Abstract**

Geosciences co-shape the human niche; hence, geoethical thinking is pertinent for geoscientists. Within the human niche, geo-endowments, like water, are shared resources that are commons. As a societal context, the human niche is a planetary network of natural and cultural environments. Geoethical thinking explores cultural substrates that nurture the skills of human agents and the operational circumstances that they encounter in the human niche. Initially, geoscientists conceived geoethics for their professional circumstances. Subsequently, geoethics evolved into an epistemic, moral hybrid for citizens that are interacting with the Earth system. Furthering geoethics—that is, combining it with Kohlberg's 'hierarchy of moral adequacy' and Jonas's 'imperative of responsibility'—leads to formulating in a 'geoethical rationale', namely, to act *'actor centric, virtue-ethics focused, responsibility focused, knowledge based, all-actor inclusive, and universal rights based'*. Uniting geoethical thinking with thinking about moral adequacy and responsibility for future generations strengthens the applicability of geoethics. The geoethical rationale is formulated at a normative meta-level to apply in any societal or scientific context that is relevant for geosciences. Furthermore, the geoethical rationale supports any human agent (geoscientists or citizens) in navigating the human niche, for example, by framing how to handle a diversity of cultural, social and scientific circumstances.

M. Bohle (✉)
International Association for Promoting Geoethics (IAPG), Rome, Italy
e-mail: martin.bohle@ronininstitute.org

M. Bohle
Ronin Institute, Montclair, NJ, USA

**Keywords**

Geoethics • Moral adequacy • Imperative of responsibility • Sense-making • Complex adaptive • Social–ecological systems

## 1 Introduction

Geoethics is about how geo-professionals and citizens situate themselves as part of the Earth system (Bobrowsky et al. 2017; Bohle et al. 2019), that is, how to navigate the 'human niche'. The notion of human niche is a metaphor for the scientific concept of 'anthropogenic biome' (Ellis et al. 2016; Fuentes 2017). Essential parts of the human niche are shared resources like soil and water ('geo-endowments'), as well as the manners how societies do handle them (Leach et al. 2018).

This contribution takes a system dynamics perspective. The '*sense-making-action feedback loop*' of social–ecological systems relates the concepts of human niche and geoethics. Subsequently, a 'geoethical rationale' is sketched relating geoethics with Kohlberg's hierarchy of moral adequacy and Jonas' imperative of responsibility (Kohlberg 1981; Jonas 1984).

## 2 Geoethics: A Tool for Sense-Making

### 2.1 Complex-Adaptive Social–Ecological Systems

Anywhere at Earth, natural processes and human practices dovetail into local, regional and planetary social–ecological systems. Often these systems exhibit a dynamic that is marked by nonlinearity, threshold-dependent shifts of system stages and positive feedback loops. Subsequently, complex-adaptive system behaviour is emergent. The



M. Abrunhosa et al. (eds.), *Advances in Geoethics and Groundwater Management: Theory and Practice for a Sustainable Development*, Advances in Science, Technology & Innovation,
https://doi.org/10.1007/978-3-030-59320-9_5

practices or governance arrangements of people (human agents), corporations and institutions co-shape the given social-ecological systems (Fuerth and Faber 2012; Biermann 2014; Chaffin et al. 2016; Kowarsch et al. 2016; Bohle 2017). The rational and affective justifications of design choices for production systems and consumption patterns, as well as people's reactions to system behaviour, are parts of these systems (Hämäläinen 2015; Galaz et al. 2011; Head and Xiang 2016a, b; Preiser and Woermann 2018; Merwe et al. 2018). These (soft) parts or 'people-features' are system features in the same merit as any (hard) engineered artefact or natural process. In a social–ecological system, these 'people-features' make loops of positive feedbacks that, subsequently, may cause amplification of system responses and emergent complex-adaptive dynamics.

The ensemble of dynamical features, such as nonlinearity, threshold dependence and positive feedback loops, render any social–ecological system more adaptive and more complex. Subsequently, the system behaviour may be counter intuitive to people's expectations. Such counter-intuitive system behaviour includes (adapted from Preiser and Woermann 2018; Preiser et al. 2018): (1) multiple, parallel cause-and-effect pathways that couple local and system-wide behavioural patterns, which are resulting from networked causes; (2) outputs and inputs may not relate proportionally so that minor changes in the controlling driver can cause rapid, system-wide behaviour or significant changes in the controlling driver may cause a slow and limited system-wide response; (3) structural parts are multi-functional, so that the same function may be performed by different structural parts or the same structure can perform different functions; (4) dynamic interactions that amplify minor inputs to drive cascades of significant effects that cause surprise and uncertainty, and any local intervention may modulate the system-wide organisation.

## 2.2 Sense-Making in the Human Niche

When described from a systems perspective, the human niche is a network of tightly knotted process loops that exhibits non-separable societal and environmental dynamics (Colding and Barthel 2019; Crona et al. 2016; Schlüter et al. 2019; Donges et al. 2017). The human sense-making is an essential process within the system. To sketch it: people's (that is, individual, collective, corporate or institutional actors) perceptions lead to choices, for example, to deploy given technological schemes. Whatever the choice, it implies to undertake (tangible, physical) actions because of given (conceptual) aims. Subsequently, these actions alter the environment. Perceiving the altered environment leads the actor to undertake subsequent actions. The feedback loop '*sense-making* $\gg$ *action* $\gg$ *system behaviour* $\gg$ *sense-making*' is a feature of any social–ecological systems. Within that feedback loop, the design features of the human sense-making processes itself are essential.

Complex-adaptive system dynamics challenge human capabilities to make sense of them; see, for example (Termeer et al. 2019). People use less their rational sense-making capabilities when they face challenging circumstances, such as counter-intuitive system behaviour of complex-adaptive social–ecological systems. Instead, and as an alternative, people use their affective sense-making capabilities (Salvatore et al. 2018). Additional complexity arises through the processes of how individual agents coordinate their sense-making and action. Furthermore, the resulting governance system is a network of knotted process loops with feedbacks. The question arises how, in these circumstances, geoethical thinking can co-shape the rational and affective human sense-making.

## 2.3 Geoethical Thinking, Geo-endowments and the Human Niche

Nowadays, global supply chains amalgamate a planetary human niche (Folke et al. 2016; Rosol et al. 2017). Geosciences knowledge combined with engineering sciences enables this process (Bohle et al. 2019b). Sustainability means to secure that current and future generations can benefit from the geo-endowments of the human niche, such as air, water, soils, fuels, minerals as well as biodiversity or ecosystem services. It is ethically imperative that the individual, collective, corporate or institutional 'niche-builder' uses common resources in a manner that considers the needs of future generations. That is the essence of Jonas' 'imperative of responsibility' applied to geosciences and their societal applications.

The Cape Town Statement on geoethics outlines an actor centric virtue ethic. It promotes to act responsibly, knowledge based, all-actor inclusive (Capua et al. 2017). Geoethics can be strengthened by applying findings of Kohlberg and Jonas (Kohlberg 1981; Jonas 1984), that is, acting geoethically on a universal rights basis,[1] and considering the needs of needs of future generations. Combining these approaches, a 'geoethical rationale' emerges that calls for agents in the human niche to be '*actor centric, virtue-ethics focused, responsibility focused, knowledge based, all-actor inclusive, and universal rights based*' (Table 1).

[1]The highest level of moral adequacy, Kohlberg's 'upper post-conventional level', is described by a morality that is based on individual human rights and justice, by acts that are based on universal ethical principles, and by principled self-conscience and mutual respect.

**Table 1** Table concise meaning of categories of the geoethical rationale

| Category | Meaning of the category |
|---|---|
| Actor centric | To apply a normative framework that invests (empowerment) an individual/group to act to their best understanding in the face of given circumstances, opportunities and purposes |
| Virtue-ethics focused | A corpus of personal traits (honesty, integrity, transparency, reliability, or spirit of sharing, cooperation, reciprocity) of an individual/group that furthers operational (handling of things) and social (handling of people) capabilities of the individual/group |
| Responsibility focused | The outcome of a normative call (internal, external) upon an individual/group that frames decisions/acts in terms of accountability, as well for the intended effects as for unintended consequences and implications for future generations |
| Knowledge based | In the first and foremost instance, (geosciences/Earth system) knowledge acquired by scientific methods; experience-based ('indigenous/traditional) knowledge is a secondary instance; reproducibility of knowledge by third parties supports any claim of trustworthiness instead of allusion to faith or 'authorities' |
| All-actor inclusive | Achieve a practice of a 'shared social licence to operate' between various individuals/groups by mitigating differentials of power, voice etc. using participatory processes and capacity building |
| Universal rights based | Guide affective and rational sense-making of individuals/groups by universal human rights (life, liberty, justice) to strengthen secondary normative constructs such as utilitarian, sustainability or precautionary principles |

## 3 Concluding Remarks

Combining geoethical thinking with thoughts about moral adequacy and responsibility for future generations strengthens its operational guidance. The six categories of the *geoethical rationale* allow relating geoethics to various ethical norms. For example, the categories 'actor centric, virtue-ethics focused' give relevance to any human actor; the categories 'responsibility focused, knowledge based' qualify the action; the categories 'all-actor inclusive, universal rights based' call for participatory governance and may refer to ethics of equity or justice. The geoethical rationale is formulated at a normative meta level to apply in any societal or scientific context that is relevant for geosciences. Furthermore, the geoethical rationale supports any human agent (geoscientists or citizens) in navigating the human niche, for example, by framing how to handle a diversity of cultural, social and scientific circumstances.

**Acknowledgements** I like to thank Eduardo Marone (University of Paraná, Brazil) for having me introduced to the works of Kohlberg and Jonas.

## References

Biermann F (2014) The anthropocene: a governance perspective. Anthr Rev 1(1):57–61

Bobrowsky P, Cronin V, Di Capua G, Kieffer S, Peppoloni S (2017) The emerging field of geoethics. In Gundersen LC (ed) Scientific integrity and ethics: with applications to the geosciences. Wiley, New York, pp 175–212

Bohle M (2017) Ideal-type narratives for engineering a human niche. Geosciences 7(1):18

Bohle M, Marone E (2019) Humanistic geosciences and the planetary human niche. In: Bohle M (ed) Exploring geoethics: ethical implications, societal contexts, and professional obligations of the geosciences. Springer International Publishing, Cham, pp 137–164

Bohle M, Preiser R, Di Capua G, Peppoloni S, Marone S (2019) Exploring geoethics: ethical implications, societal contexts, and professional obligations of the geosciences. Springer International Publishing, Cham

Chaffin BC, Garmestani AS, Gunderson LH, Benson MH, Angeler DG, Arnold CA, Cosens B, Craig RK, Ruhl JB, Allen CR (2016) Transformative environmental governance. Annu Rev Environ Res 41(1):399–423

Colding J, Barthel S (2019) Exploring the social-ecological systems discourse 20 years later. Ecol Soc 24(1):2

Crona BI, Daw TM, Swartz W, Norström AV, Nyström M, Thyresson M, Folke C, Hentati-Sundberg J, Österblom H, Deutsch L, Troell M (2016) Masked, diluted and drowned out: how global seafood trade weakens signals from marine ecosystems. Fish Fisher 17(4):1175–1182

Di Capua G, Peppoloni S, Bobrowsky P (2017) The Cape Town statement on geoethics. Ann Geophys 60:1–6

Donges JF, Lucht W, Müller-Hansen F, Steffen W (2017) The technosphere in Earth system analysis: a coevolutionary perspective. Anthr Rev 4(1):23–33

Ellis EC, Richerson PJ, Mesoudi A, Svenning J-C, Odling-Smee J, Burnside WR (2016) Evolving the human niche. Proc Nat Acad Sci 113(31):E4436–E4436

Folke C, Biggs R, Norström AV, Reyers B, Rockström J (2016) Social-ecological resilience and biosphere-based sustainability science. Ecol Soc 21(3):41

Fuentes A (2017) Human niche, human behaviour, human nature. Interface Focus 7(5):2016–20136

Fuerth LS, Faber EMH (2012) Anticipatory governance practical upgrades: equipping the executive branch to cope with increasing speed and complexity of major challenges. Project on Forward Engagement, National Defense University Press, USA. https://ndupress.ndu.edu/Publications/Article/1216573/anticipatory-governance-practical-upgrades-equipping-the-executive-branch-to-co/

Galaz V, Moberg F, Olsson E-K, Paglia E, Parker C (2011) Institutional and political leadership dimensions of cascading ecological crises. Public Adm 89(2):361–380

Hämäläinen TJ (2015) Governance solutions for wicked problems: metropolitan innovation ecosystems as frontrunners to sustainable well-being. Technol Innov Manag Rev 5(10):31–41

Head BW, Xiang W-N (2016a) Why is an APT approach to wicked problems important? Landsc Urban Plan 154:4–7

Head BW, Xiang W-N (2016b) Working with wicked problems in socio-ecological systems: more awareness, greater acceptance, and better adaptation. Landsc Urban Plan 154:1–3

Jonas H (1984) The imperative of responsibility: in search of an ethics for the technological age. University of Chicago Press, Chicago

Kohlberg L (1981) The philosophy of moral development: moral stages and the idea of justice. Essays on moral development, vol 1. Harber & Row, San Francisco

Kowarsch M, Garard J, Riousset P, Lenzi D, Dorsch MJ, Knopf B, Harrs J-A, Edenhofer O (2016) Scientific assessments to facilitate deliberative policy learning. Palgrave Commun 2(1):16092

Leach M, Reyers B, Bai X, Brondizio ES, Cook C, Díaz S, Espindola G, Scobie M, Mark Stafford-Smith M, Subramanian MS (2018) Equity and sustainability in the Anthropocene: a social-ecological systems perspective on their intertwined futures. Glob Sustain 1:e13

Preiser R, Woermann M (2018) Conceptual and practical implications for understanding and engaging with complex adaptive systems. In: Galaz V (ed) Handbook on global challenges, governance, and complexity. Edward Elgar Publishing, Cheltenham

Preiser R, Biggs R, De Vos A, Folke C (2018) Social-ecological systems as complex adaptive systems: organizing principles for advancing research methods and approaches. Ecol Soc 23(4):46

Rosol C, Nelson S, Renn J (2017) Introduction: in the machine room of the Anthropocene. Anthr Rev 4(1):2–8

Salvatore S, Mannarini T, Avdi E, Battaglia F, Cremaschi M, Fini V, Veltri GA (2018) Globalization, demand of sense and enemization of the other: a psychocultural analysis of European societies' sociopolitical crisis. Cult Psychol 25(3):345–374

Schlüter M, Haider LJ, Steven L, Lindkvist E, Martin R, Orach K, Wijermans N, Folke C (2019) Capturing emergent phenomena in social-ecological systems: an analytical framework. Ecol Soc 24 (3):11

Termeer CJA, Dewulf A, Biesbroek R (2019) A critical assessment of the wicked problem concept: relevance and usefulness for policy science and practice. Policy Soc 38(2):167–179

van der Merwe SE, Biggs R, Preiser R (2018) A framework for conceptualizing and assessing the resilience of essential services. Ecol Soc 23(2):12

# Cross-Cutting Role of Groundwater in Achieving the SDGs and an Ethical Approach

Emilia Bocanegra

**Abstract**

Water is the basis of sustainable development. The SDGs recognize through SDG 6 the importance of water issues as human right and as a requirement to realize many of the other goals. This work presents an analysis of the cross-cutting role of water (with emphasis on groundwater) in the SDGs, shows some comparisons between the goals in Latin American countries and proposes an ethical approach of groundwater in the context of the SDGs. Water resources and the services that provide contribute from poverty reduction, food and human health to economic growth, reduction of inequalities and environmental sustainability: 33 social, environmental and economic targets including indicators directly or indirectly related to groundwater have been identified. Some comparisons between these indicators in Latin American countries show possible relationships or trends. To address the ethical aspects of groundwater, it is proposed to formulate a conceptual framework through the analysis and application of the ethical principles proposed by UN to SDG 6, the interaction of each SDG with groundwater and the creation of an interdisciplinary working group that integrates the social sciences and earth and water sciences.

**Keywords**

Cross-cutting role • Groundwater • SDGs • Ethical approach • Latin America

## 1 Introduction

In 2015, the United Nations General Assembly approved the 2030 Agenda whose Sustainable Development Goals (SDGs) reflect the global aspiration to transform the world before 2030, and it includes 17 main goals and 169 targets. The 2030 Agenda is universal, is aimed at all countries and all people and under the principle of "leave no one behind" insistently points out the need to reach the laggards first, placing inequality and social injustice in the heart of the Agenda (WWAP 2019).

The SDGs recognize the importance of water issues as human right through SDG 6 and as a requirement to realize many of the other goals. According to UN-Water (2014), the goals and targets should promote, among others, the following development outcomes related to water: healthy people, increased prosperity, equitable societies, protected ecosystems and resilient communities through the universal access to safe drinking water and sanitation, the sustainable use and development of water resources, robust and effective water governance, improved water quality and wastewater management and reduced risk of water-related disasters to protect vulnerable groups and minimize economic losses.

This work presents an analysis of the cross-cutting role of water (with emphasis on groundwater) in the SDGs, shows some relationships between goals in Latin American countries and proposes an approach to address the ethical aspects of groundwater in the context of SDGs.

## 2 Methods

The work methodology consists in carrying out an analysis of the SDGs targets in order to identify their relationship with the proposed targets for SDG 6, taking into account that the indicators corresponding to each target were directly or indirectly related to groundwater resources.

E. Bocanegra (✉)
Hydrogeology Group, National University of Mar del Plata, Mar del Plata, Argentina
e-mail: emilia.bocanegra@gmail.com



M. Abrunhosa et al. (eds.), *Advances in Geoethics and Groundwater Management: Theory and Practice for a Sustainable Development*, Advances in Science, Technology & Innovation,
https://doi.org/10.1007/978-3-030-59320-9_6

In order to know the status of Latin America, a compilation of indicators of different water-related targets for some countries of the region was made through the UNSTATS platform of the Global SDG Indicators Database (unstats.un.org/sdgs/indicators/database/) that provides access to data compiled through the UN System and other international organizations, and some comparisons between targets were made.

## 3 Results

### 3.1 Cross-Cutting Role of Groundwater in Achieving the SDGs

Groundwater resources in many parts of the world are the main or only source of water, providing urban supply services, development of productive activities and maintenance of ecosystems. There are some SDGs that are more related to water and in particular to groundwater. The main one is the SDG 6 dedicated exclusively to water. Then, there are some deeply involved in groundwater, such as SDGs 1, 2, 3, 4, 12, 13, 15, 16 and 17, and finally others that are less involved. The following targets related to groundwater have been identified:

SDG 1. End of poverty: Reduce poverty (1.2) and ensure access to basic services for all people, particularly the poor and vulnerable (1.4).

SDG 2. Zero hunger: Ensure that all people have access to healthy, nutritious and sufficient food (2.1), improve the agricultural productivity of small-scale food producers, through access to the necessary resources (2.3) and ensure sustainable food production systems and implement resilient agricultural practices (2.4).

SDG 3. Health and well-being: Reduce considerably the number of deaths and diseases caused by water pollution (3.9).

SDG 4. Quality education: Promote education at all levels for sustainable development (4.7).

SDG 5. Gender equality: Ensure women full and effective participation and equal leadership opportunities at all levels of decision-making (5.5).

SDG 6. Clean water and sanitation: Achieve universal and equitable access to drinking water for all (6.1), achieve access to sanitation and hygiene services (6.2), improve water quality by reducing pollution (6.3), significantly increase the efficient use of water resources in all sectors and ensure the sustainability of the extraction and supply of fresh water (6.4), implement integrated water resources management at all levels (6.5), protect and restore water-related ecosystems (6.6), expand international cooperation (6.a) and support and strengthen the participation of local communities in improving water management (6.b).

SDG 7. Affordable and clean energy: Ensure universal access to affordable, reliable and modern energy service (7.1).

SDG 8. Decent work and economic growth: Improve global resource efficiency in consumption and production and endeavour to decouple economic growth from environmental degradation (8.4).

SDG 9. Industry, innovation and infrastructure: Increase scientific research and improve technological capacity (9.5).

SDG 10. Reduction of inequalities: Achieve the economic growth of the poorest population (10.1) and empower and promote the social, economic and political inclusion of all (10.2).

SDG 11. Sustainable cities and communities: Ensure access for all to adequate, safe and affordable housing and basic services and upgrade slums (11.1), reduce the number of deaths caused by disasters, including those related to water (11.5) and reduce the adverse environmental impact of cities, paying special attention to waste management (11.6).

SDG 12. Responsible production and consumption: Implement a sustainable consumption and production patterns (12.1) and achieve the sustainable management and efficient use of natural resources (12.2).

SDG 13. Climate action: Strengthen resilience and adaptive capacity to risks related to climate and natural disasters (13.1) and incorporate climate change measures into national policies, strategies and plans (13.2).

SDG 14. Life below water: Prevent and significantly reduce marine pollution, particularly that caused by activities carried out on land (14.1).

SDG 15. Life on land: Ensure the conservation, restoration and sustainable use of terrestrial and inland freshwater ecosystems and their services (15.1).

SDG 16. Peace, justice and strong institutions: Develop effective, accountable and transparent institutions at all levels (16.6) and ensure responsive, inclusive, participatory and representative decision-making at all levels (16.7).

SDG 17. Partnerships for the Goals: Enhance North–South, South–South and triangular regional and international cooperation on and access to science, technology and innovation (17.6).

The targets contain some indicators, including several directly or indirectly related to groundwater, which are not yet in the implementation stage, so the UN database does not record information on these indicators.

### 3.2 Some SDG Indicators Related to Water in Latin America

Based on the analysis of the relations between the SDG of water and the other SDGs, a matrix of data obtained from the UNSTATS platform was prepared for different countries in

Latin America. The matrix has information on ten social, environmental and economic indicators in 18 countries, with some gaps.

As an example, some comparisons are presented between indicators that are intended to show some possible relationships or trends.

The relationship between SDG 2, Zero Hunger with the Water SDG was analysed through indicators 2.1.1 Prevalence of undernourishment (percentage of population) and 6.1.1 People using safely managed drinking water services (see Fig. 1). There are no data for 6.1.1 indicator for several countries in the region. It varies in a wide range and presents an inverse trend, so that a larger proportion of people using safely managed drinking water services corresponds to a decrease in the prevalence of undernourishment. Mexico is an exception, which could be due to the application of social policies to reduce the undernourishment or to cultural feeding practices.

The relationship with SDG 3 health and well-being was analysed using indicator 3.9.2 Mortality attributed to unsafe water, unsafe sanitation and lack of hygiene (see Fig. 2). A growing trend is observed, with the prevalence of undernourishment.

The comparison between SDG 6 and SDG 15 Life on land was made through indicators 6.4.2 Level of water stress: freshwater withdrawal as a proportion of available freshwater resources and 15.1.2 Terrestrial protected area relative to total land area (see Fig. 3). Most countries in the region have a water stress level less than 10% and cover a wide range of terrestrial protected area. Countries with water stress greater than 10% show a growing trend in this relationship.

It is emphasized that these comparisons between indicators are only intended to show some possible relationships. An adequate interpretation should be made by integrating the economic, social and environmental dimensions for each country, which exceeds the goal of this work.

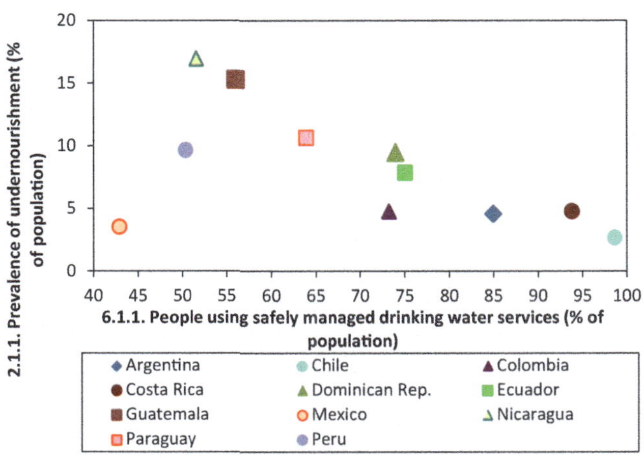


**Fig. 1** Comparison of indicators of SDG6 and SDG2 in Latin American countries (own elaboration)

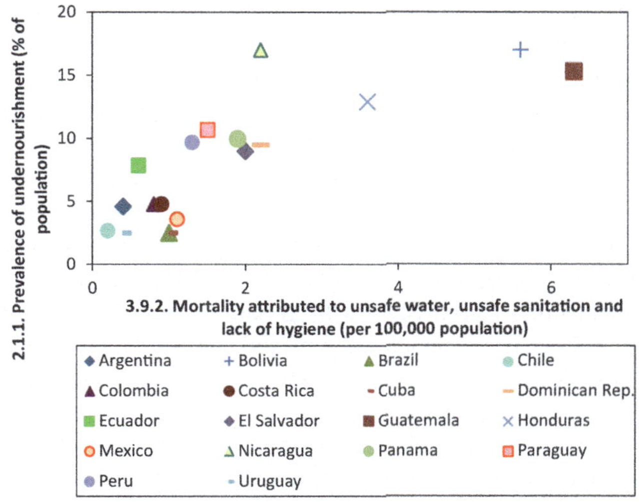


**Fig. 2** Comparison of indicators of SDG3 and SDG2 in Latin American countries (own elaboration)

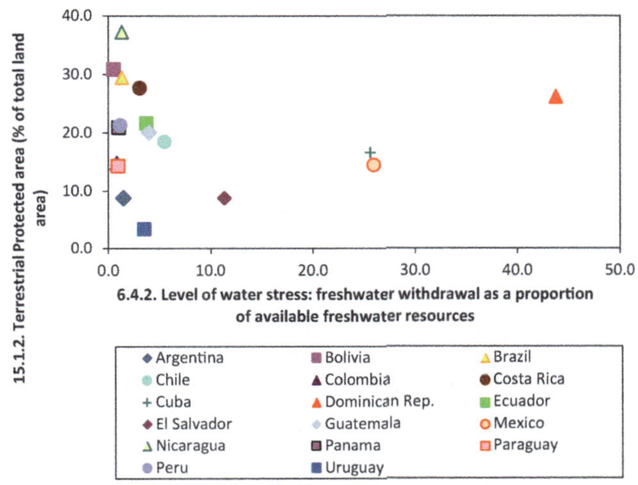


**Fig. 3** Comparison of indicators of SDG6 and SDG15 in Latin American countries (own elaboration)

## 4 Towards a Groundwater Ethics in the Context of the SDGs

One of the keys to achieving sustainable development is the use, management and protection of groundwater, which makes necessarily a new sense of water ethics at the personal and social level, that allows to revalue preconceived models and propose new paradigms.

A number of key UN bodies have identified ethical principles to integrate human concerns with those of various ecosystems affected by the global water cycle: Human

Dignity and Human Rights, Solidarity, Common Good, Frugality, Sustainability, Justice, Justice and International Transboundary Waters, Gender Equity, Research Integrity and Sharing Knowledge and Technology (Capacity Building) (COMEST 2018).

To address the ethical aspects of groundwater in the context of the SDGs, it is proposed to formulate a conceptual framework through the analysis and application of the ethical principles proposed by the UN to SDG 6 and to the interaction of each SDG with groundwater. In this way, a global perspective could be achieved that would allow, with an ethical sense, to establish priorities for decision-making now and in the future.

This proposal could be carried out through the creation of an interdisciplinary working group that integrates the social sciences and earth and water sciences. As operating criteria, it would be very important to use the available information to avoid bureaucracy and develop activities in a truly cooperative environment. The generated conceptual framework would initially be applied to pilot countries that have the necessary information, and, in this way, an evaluation and adjustment of the proposed methodology could be carried out.

## References

COMEST, (UNESCO World Commission on the Ethics of Scientific Knowledge and Technology (2018) Report of COMEST on water ethics: ocean, freshwater. Coastal Areas, Paris, UNESCO

UNSTAT Homepage, https://unstats.un.org/sdgs/indicators/database/. Last accessed 2020/01/20

UN-Water (2014) A post-2015 global goal for water: synthesis of key findings and recommendations from UN-Water. Final Report. UN-Water Publications

WWAP, (UNESCO World Water Resources Assessment Program) (2019) United Nations World Report on the Development of Water Resources 2019: leave no one behind. UNESCO, Paris

# Inclusion of Indigenous Communities in Water Resources Management in the Middle West of Brazil: A Proposal

Sandra Garcia Gabas, Giancarlo Lastoria, and Denise Aguema Uechi

**Abstract**

The relationship between indigenous people and water is beyond the use and quality criteria usually present at techno-scientific standards. Indigenous communities commonly assign to water, some culture and spiritual values, which are normally neglected in Brazilian watershed management planning. Historically, these people have close relationship with surface water, such as rivers and lakes, an abundant resource in the country. However, due to the reduction and delimitation of their land, in conjunction with some non-traditional land use, indigenous communities are mainly supplied by groundwater. In this paper, it is presented the case of Ivinhema River watershed, in Middle West of Brazil. This watershed is one of the Parana Basins. It has the major indigenous population in the region, with frequent land-use conflicts between indigenous people and farmers. The main economic activities are agriculture and livestock. The state's water resources and Ivinhema River watershed planning are from 2010 and 2015, respectively. In both cases, indigenous population and their water knowledge were not considered. The paper examines watershed diagnosis and management proposals according to geoethics concepts and United Nations Guidelines for water resources development in order to base a future revision of the management actions.

**Keywords**

Watershed planning • Vulnerable population • Indigenous communities' engagement • Inclusive development

## 1 Introduction

Although Brazilian legal instruments that regulate water resources management in the country have more than twenty years (Brasil 1997), few states have an active structure and effective actions to allow good water management, both in quantitative and qualitative terms. Thus, discussions about water resources management system (WRMS) have been restricted to the most developed states of the country, focusing on their local specificities, usually based on scientific and technical data (hydrology and hydrogeology). The debate normally excludes social and cultural aspects. In this context, the precepts of geoethics (Peppoloni and Di Capua 2017) can assist in assessing WRMS, broadening the debate and assisting in the construction of water resources plans (WRP) with environmental, social, and economic responsibility.

Besides Brazilian constitution establishes that all Brazilians have the right to access to water in quantity and quality, the water law does not address actions to vulnerable communities, for environmental, social, and ethnic reasons.

The purpose of this paper is to discuss the current water resources management situation of a watershed in the Midwest of Brazil, which has some indigenous communities and the inclusion of this population in the WRMS and in formulating policies that affect their areas.

S. G. Gabas (✉) · G. Lastoria
Engineering, Architecture and Geography Faculty, University of Mato Grosso do Sul, Campo Grande, Brazil
e-mail: sandra.gabas@ufms.br

D. A. Uechi
Environmental Technology Graduation Program, University of Mato Grosso do Sul, Campo Grande, Brazil


M. Abrunhosa et al. (eds.), *Advances in Geoethics and Groundwater Management: Theory and Practice for a Sustainable Development*, Advances in Science, Technology & Innovation,
https://doi.org/10.1007/978-3-030-59320-9_7

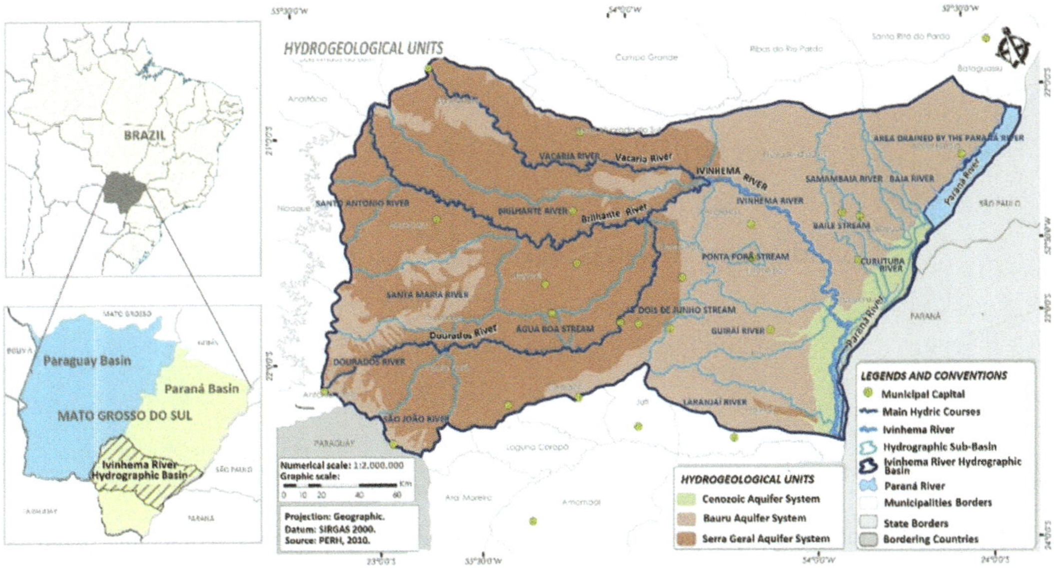


**Fig. 1** Location of Ivinhema River watershed (IRW) and hydrogeology units

## 2 Settings

The studied basin, Ivinhema River watershed (IRW), is in Mato Grosso do Sul State, Middle West of Brazil, a sub-basin at the right margin of Parana River (Fig. 1). It has an area of 46.688.75 km$^2$ and comprises 25 municipalities (Semac 2010; Demeter 2015). Within this area, Ivinhema River has 16 main drainage as its affluent, and some of them are directly drained to Parana River.

The number of rural properties in IRW is 30.731, with 75% as private owners, 24.7% as rural settlements, and 0.3% as indigenous villages. The main economic activities are agriculture and livestock. The area distribution of land use and occupation is characterized by 56% of agriculture, 27% of livestock, 10% of forest, 4.7% of wet areas, 2,1% of water bodies and 0.2% of urban areas (Demeter 2015).

Despite having 15 conservation area in the basin, only three of them are classified as integral conservation unity. Most of them are occupied by agriculture or livestock activities (69%), and the remaining 31% of the area correspond to forests, water bodies, and wet areas (Demeter 2015).

The total population is 675 million people, most of them (82%) living in urban areas. IRW has a water resources committee (IRWC), which is composed by representatives from different society sectors, including government and society organizations.

## 3 Results

### 3.1 Surface Water

There are approximately 3.500 drainages in the basin varying from 0.1 to 595 km in extension (Fortes et al. 2004).

The terrain has a low slope relief, less than 8%, with some large and extensive river terraces. Some of them have archaeological sites from older population such as hunter-gatherers and primitive farming people (Kashimoto and Martins 2009; Kashimoto and Martins 2018).

Most of the withdrawn water (97%) is assigned to irrigation, as the main economic activity is agriculture. The withdrawal flow rate is 80.716.8 m$^3$/h (Demeter 2015), mainly from superficial water.

### 3.2 Hydrogeological Units

Geological units outcropping in the area are basalts from Jurassic volcanic spill (Serra Geral Formation) and Cretaceous sandstones (Bauru Group) deposited above the basalt unity. At fluvial terraces, there are gravel deposits due to recent alluvial deposition.

According to the geological setting, three aquifers are present in the watershed. Two of them are primary porosity aquifer type, Cenozoic Aquifer System (CAS) and Bauru

Aquifer System (BAS), and a fractured one, the Serra Geral Aquifer System (SGAS) (Fig. 1). In terms of groundwater reserves, BAS has the major water reserve, followed by SGAS and CAS, respectively, 3.015.14, 2.731.37, and 657.85 million $m^3$/year (Demeter 2015).

The main water source in the basin is groundwater, both for rural and urban areas. 23 from 25 municipalities within the watershed have their water supply system 100% provided by groundwater. In the major city, Dourados, formerly supplied only by surface water, is nowadays 47% water supplied by SGAS. At countryside, except some farms that exploit springs, the majority depends on groundwater.

### 3.3 Indigenous Communities

Human presence in Mato Grosso do Sul State (MS) has been described longer before Europeans arrival. The oldest traces of human beings are to the end of the latest glacial period, about 12.000 b.p (Kashimoto and Martins 2018).

MS has the second greatest indigenous population in Brazil with 67.433 people (Urquiza 2019) from eleven indigenous groups that live currently in its territory: Terena, Kinikinau, Kaiowá, Guarani, Kadiwéu, Ofaié, Guató, Chamacoco, Ayoreo, Atikum, and Camba (Chamorro and Combès 2018).

The relationship between indigenous people and water is beyond the use and quality criteria usually present at techno-scientific standards. Indigenous communities commonly assign to water some culture and spiritual values, which are neglected in Brazilian water resources management system. Historically, these people have close relationship with surface water in daily life, such as rivers and lakes, an abundant resource in the country. However, for some ethnic groups, such as Mbyá and Kaiowá, groundwater source is related to their origin, which is called *Jasuka*, a beautiful spring from underground also called "mother" (Chamorro 2008).

Although traditional indigenous life is dependent on rivers and lakes, all the communities are mainly supplied actually by groundwater due to the reduction and delimitation of their land, in conjunction with some non-traditional land use, such as land leasing for intensive agriculture activities, which brings about some quality problems in superficial water. The change in indigenous territoriality may be the reason for the several cases of suicide among young people in this population (Staliano and Modardo 2019).

There are twenty indigenous lands in IRW (FUNAI 2019), most of them situated nearby Dourados municipality, the biggest city of the basin (Table 1). The watershed concentrates the remaining of original indigenous people distributed in a large area of Middle West of Brazil and the North of Paraguay, before the Portuguese and Spanish arrival. The total area of indigenous land is 1261.9 $km^2$, but there is not a safe estimation of indigenous population in the basin.

## 4 Discussion

The Brazilian water resources management system (Brazil, 1997) defines the state assignment for water resources planning, its implementation, and the monitoring of water bodies. It also prescribes the superficial and groundwater integrated management. Although it has been more than twenty years, most of the Brazilian states do not have their water resources plan, which ones normally focus mainly superficial water, and groundwater is usually neglected.

Water resources planning elaboration process considers different stakeholders, and citizen participation is encouraged. Representatives from different sectors of the society

**Table 1** Indigenous communities in the Ivinhema River watershed and their respective ethnic group (FUNAI 2019)

| Indigenous land | Ethnic group | Area ($km^2$) | Situation |
|---|---|---|---|
| Buriti | Terena | 72.0 | Declared |
| Buriti | Terena | 20.9 | Regularized |
| Buritizinho | Guarani Kaiowá | 0.1 | Regularized |
| Caarapó | Guarani Kaiowá | 35.9 | Regularized |
| Dourados | Guarani Nhandeva, Terena | 34.7 | Regularized |
| Dourados-Amambaipeguá I | Guaraní | 556.0 | Delimitated |
| Guaimbé | Guarani Kaiowá | 7.2 | Regularized |
| Guyraroká | Guarani Kaiowá | 114.4 | Declared |
| Jarara | Guarani Kaiowá | 4.8 | Approved |
| Jatayvari | Guarani Kaiowá | 88.0 | Declared |
| Ñande Ru Marangatu | Guarani Kaiowá | 93.2 | Approved |
| Panambi—Lagoa Rica | Guarani Kaiowá | 122.0 | Delimitated |
| Panambizinho | Guarani Kaiowá | 12.7 | Regularized |

are normally included in the planning discussion group, such as industry, agriculture, and livestock farmers with different scale production, entrepreneurs, public agents, university representatives, and civil society organizations.

The MS water resources plan was formulated in 2010, coordinated by the Brazilian Environment Ministry, and done in partnership to OEA (Semac 2010). There is any mention to the indigenous land and population in the plan. Although IRW plan was done five years later, in 2015, and the IRWC has a representative from Brazilian indigenous agency, it also did not consider this people. Guideline 9 from IRW plan state that there must have grant requirements for groundwater exploitation in special areas, which are described in the document as water bodies with local economic relevance (Demeter 2015). Some questions could help in interpreting the guideline: Are indigenous land considered as special areas? Is groundwater included in mentioned water bodies?

As groundwater conditions vary, sometimes, greatly from place to place, according to geological heterogeneities, and there are some other factors such as environment diversity, socioeconomic, and political settings, we should aim our actions related to groundwater/water resources development, management, and governance to be highly context-specific (FAO 2016). Water balance is usually helpful to understanding interactions involving different water uses and demand across a river basin (FAO 2015), provided that groundwater is effectively considered.

Furthermore, we should consider ethics and social aspects of water resources issues, even though they are not determined in the water resources system. These aspects are beyond the human rights to water and sanitation (Unesco 2019) as usually expressed in most of the specialized publication. In order to do so, we could follow the reasoning of some professionals concerned about ethics, social, and cultural implications in geosciences (Peppoloni and Di Capua 2017; Marone and Peppoloni 2017) as our activities could have strong impacts not only on the natural environment but also on the population. We believe that universities can contribute to a change of approach not only by assisting in the debate but also mainly by educating professionals better prepared with a broader vision of sustainable and inclusive development. Thus, we can avoid professional bias or vested interests with a lack of hydrogeological know-how (Chaminé 2015).

For IRW plan future revision, we suggest some actions that might help to include indigenous people in the MS water resources management system: (a) to stimulate and encourage the indigenous people representation and participation in the IRW Committee; (b) to include water resources and their situation in the basin and in their lands as a subject to be lectured in indigenous schools, and (c) to revise the IRW plan considering the indigenous people view and proposals.

## 5 Concluding Remarks

MS is the second-most indigenous populated state in Brazil, and IRW is the MS's sub-basin with indigenous land major concentration. This population was not considered in both the state's water resources and watershed plans.

Groundwater has great relevance to the IRW, considering public and private supply and the diverse economic activities. There are three aquifer systems in the basin; two of them are the most important in terms of exploited volume. All the indigenous villages are supplied by groundwater. Besides its importance, few guidelines and actions are devoted to groundwater and any one to indigenous communities.

Indigenous representatives should be stimulated in IRW Committee in order to include not only their water demand in terms of quantity and quality but also their cultural aspects. And as soon as possible, the water resources plan should be revised considering their need and view.

## References

Brazil (1997) Lei n° 9433. https://www.planalto.gov.br/ccivil_03/LEIS/L9433.htm

Chaminé HI (2015) Water resources meet sustainability: new trends in environmental hydrogeology and groundwater engineering. Environ Earth Sci 73:2513–2520

Chamorro G (2008) Terra Madura. Yvy Araguyje: 1st edn. UFGD, Dourados

Chamorro G, Combès I (2018) Povos indígenas de Mato Grosso do Sul: história, cultura e transformações sociais, 1st edn. UFGD, Dourados

Demeter Engenharia (2015) Plano de recursos hídricos da bacia do rio Ivinhema. 383p. https://www.imasul.ms.gov.br/plano-de-recursos-hidricos-rio-ivinhema-2/

Food and Agriculture Organization of the United Nations-FAO (2016) Global diagnostic on groundwater governance, 189p. FAO, Rome

Food and Agriculture Organization of the United Nations-FAO (2015) Towards a water and food security—critical perspectives for policy-makers, 76p. FAO, Rome and Marseille

Fortes E, Volkmer S, Steveuax JC (2004) Controles hidrológicos e hidrográficos dos processos erosivos do baixo curso do rio Ivinhema. In: Anais V Simpósio Nacional de Geomorfologia, pp 1–12. Santa Maria

Fundação Nacional do Índio-FUNAI (2019) homepage. https://www.funai.gov.br/. Last accessed 2019/10/12

Kashimoto E, Martins G.R (2009) Arqueologia e paleoambiente do rio Paraná em Mato Grosso do Sul. 1. ed Campo Grande - MS: Life Editora, v 1. 335 p

Kashimoto E, Martins GR (2018) Arqueologia da bacia do Alto Paraná em Mato Grosso do Sul. Reference 4:93–100

Marone E, Peppoloni S (2017) Ethical dilemmas in geosciences. we can ask, but, can we answer? Anna Geophys 60, FAST TRACK 7. https://doi.org/10.4401/AG-7445

Peppoloni S, Di Capua G (2017) Geoethics: ethical, social and cultural implications in geosciences. Ann Geophys 60, Fast Track 7/ag-7473

Semac (2010) Plano de Recursos hídricos do Estado de Mato Grosso do Sul. Campo Grande/MS. UEMS, Editora

Staliano P, Mondardo ML (2019) Onde e como se suicidam os Guarani e Kaiowá em Mato Grosso do Sul: confinamento, Jejuvy e Tekoha. Psicologia – Ciência e Profissão, 39:9–21

Unesco (2019) The United Nations World Water Development Report, Paris

Urquiza AHA (2019) Povos indígenas de Mato Grosso do Sul – alguns aspectos antropológicos. https://www.mcdb.org.br/materias.php. Last accessed 2019/10/18

# Evaluating Public Opinion on Groundwater Extraction from Public Comment Submissions and Google Trends

Simon Gautrey

**Abstract**

When hydrogeologists evaluate new groundwater takings, they frequently must consider whether such takings are in the public interest. This often means that hydrogeologists will consider public opinion in their evaluation. But how can hydrogeologists understand public opinion, when groundwater is rarely a subject of properly designed public opinion surveys? In lieu of survey data, hydrogeologists might turn to comments submitted as part of a formal environmental assessment process. However, hydrogeologists might suspect that these comments were submitted by only the most motivated individuals and may not reflect the views of the general public. This paper includes a study of thousands of public comments regarding bottled water takings in Ontario, which is arguably the largest recent groundwater conflict in Canada. The paper compares these results to data from Google Trends and other sources to evaluate how those comments might reflect opinions of the wider population. The results highlight the roles geographic proximity and droughts might play in forming public opinion.

**Keywords**

Public opinion • Google trends • Nestle • Nestlé • Groundwater • Bottled water

## 1 Introduction

Groundwater is rarely on the mind of the general public, but in 2016, an application by Nestlé Waters (Nestlé) to undertake a pumping test of a well in Ontario, Canada, for bottled water was followed by a frenzy of media interest. The government responded by proposing a moratorium of new takings for bottled water to which over 20,000 public comments were submitted. Public comments on subsequent government proposals to extend this moratorium in 2018 and 2019 each received more than 7,000 comments. To put this into perspective, the Inquiry into the Walkerton Tragedy, a water contamination event in the year 2000 in Ontario that resulted in the tragic death of seven persons, and serious illness befalling thousands of others, lists less than 200 public submissions (Walkerton Inquiry 2001). The purpose of this research is to seek insight into the value of these public comments concerning bottled water by comparison with publicly accessible data from Google Trends and other sources. This paper will not distinguish between the pros and cons of bottled water.

In 2019, there were 16 active permits for groundwater takings for bottled water in Ontario. Of these, Nestlé has the largest permitted allowance representing 45% of the total permitted allowance for bottled water in Ontario (MECP PTTW database).

Nestlé has two well supplies for bottled water in Ontario near Guelph, one at Aberfoyle and the other at Erin. Both Guelph and Erin are communities dependent on groundwater for drinking water. According to the annual reports prepared by their consultants in 2019, Nestlé took an average of 822,581 $m^3$/year in the previous 5 years, although this decreased to 746,352 $m^3$/year in 2018 (Golder 2019a, b).

## 2 Data Sources and Methods

The data used in this study comes from public databases and websites from the following: Ontario Ministry of Environment, Conservation and Parks (MECP); the MECP Environmental Registry of Ontario (ERO, formerly Environmental Bill of Rights (EBR)); Nestlé; and Google Trends.

S. Gautrey (✉)
Wood Environment and Infrastructure Solutions, Hamilton, ON, Canada
e-mail: simon.gautrey@woodplc.com



M. Abrunhosa et al. (eds.), *Advances in Geoethics and Groundwater Management: Theory and Practice for a Sustainable Development*, Advances in Science, Technology & Innovation,
https://doi.org/10.1007/978-3-030-59320-9_9

The MECP is the government body responsible for regulating groundwater taking in Ontario. The ERO website describes the proposed moratorium and provides access to public comments submitted online about proposals. A first proposal for a two-year moratorium was posted for public comment in 2016. Further proposals to extend the moratorium were posted in 2018 and 2019. Table 1 summarizes the number of comments received for these proposals. The last moratorium discussed in this paper expired on October 1, 2020.

Only comments visible online were used in this study. Comments from the 2016 proposal were previously analyzed by Gautrey (2017). Gautrey (2017) randomly selected 377 comments as a representative sample and analyzed the sample for common themes. The sample size was selected using a confidence level of 95% and a confidence interval 5, using the entire 21,276 comments received as the population size, but only analyzing the available online comments.

Comments from the moratorium extensions in 2018 and 2019 were analyzed using a similar methodology to Gautrey (2017). However, the population size was reduced to online comments only, and the confidence interval increased to 10. The sample size of randomly selected comments for both the 2018 and 2019 proposals was 81.

In addition to the analysis of randomly selected comments for common themes, the database of all accessible online comments for the 2018 and 2019 proposals was searched for popular words.

Google Trends is an online tool by Google which can generate data on the frequency and location of Google searches by search term. Gautrey (2017) identified several search terms for the 2016 moratorium, and these same terms were used here. Google Trends provides relative interest data, with the most common search term assigned a value of 100 at the time of the most searches for that term, and other terms rated in proportion to that search.

**Table 1** Summary of comments received on government ERO webpage for moratorium

| ERO (or EBR) number (and reference) | Comment closure date and comment period length | Number of comments received: | | |
|---|---|---|---|---|
| | | Online | In writing[a] | In total |
| 012-8783 (EBR 012-8783) Initial 2-year moratorium | December 1, 2016, after 45 days | 8156 | 13120 | 21276 |
| 013-3974 (ERO 013-3974) First (1-year) extension | November 28, 2018, after 30 days | 537[b] | 6412 | 6949 |
| 019–0913 (ERO 019–0913) Second (9-month) extension | December 18, 2019, after 30 days | 598[c] | 8105 | 8703 |

[a]Either by mail or email for 012-08783, and email only for 13-3974 and 19-0913
[b]Only 527 could be viewed online, of which 518 are comments by individuals, not organizations
[c]Only 505 could be viewed online, of which 496 are comments by individuals, not organizations

**Table 2** Analysis of most common themes expressed in comments on moratorium proposals

| Theme | 2016 Moratorium (%) | 2018 extension (%) | 2019 extension (%) |
|---|---|---|---|
| Supported moratorium to water takings for bottled water for any reason | 98.1 | 98.8 | 100.0 |
| Concerned about the plastic waste | 15.6 | 39.0 | 33.3 |
| Opposed to the sale of water for profit | 40.4 | 36.6 | 21.0 |
| Concern about water availability for future generations | 9.4 | 23.2 | 9.9 |
| Supported increasing the cost of water taken by bottled water companies | 11.6 | 20.7 | 17.3 |
| Prioritize water takings for local communities (including agricultural) | 14.8 | 18.3 | 12.3 |
| Opposed to private ownership of water on basis of access to water as a human right | 17.3 | 17.1 | 17.3 |
| Concerned about climate change | 5.1 | 9.8 | 12.3 |
| Concerned about drought or the threat to stressed aquifers from the perceived large magnitude of the taking | 13.2 | 8.5 | 6.5 |
| Prioritize water for the environment | 4.0 | 3.7 | 1.2 |

## 3 Results

### 3.1 Analysis of Comments Submitted to the Government Regulator (MECP)

The results of the analysis of the comments grouped into common themes is summarized in Table 2. Very few comments opposed the moratorium. The most common reasons for supporting the moratorium were opposition to the sale of water for profit and the generation of plastic waste. From 2016 to 2019, there is an apparent increase of interest in: plastic waste; availability of water for future generations; a desire to see companies pay more for the water they take; and climate change. There also appeared to be a decrease in concern about drought impacts or stressed aquifers.

The recognition of a theme in a comment is partly qualitative, and to validate the assignment of comments to theme from the samples, the entire set of 2018 and 2019 online comments was ranked by common words related to the themes (Table 3). These results are generally similar to the rankings of themes, with plastic, profit, future and climate as commonly identified words that might indicate an interest in the theme topics. Other identified common words were corporation and environment.

### 3.2 Analysis of Google Trends Data

A Google Trends dataset from 2015 to 2020 was queried for frequency of search term results that might be associated with the bottled water moratorium, based on previous work (Gautrey 2017), plus the term groundwater (Fig. 1). Only Google searches originating from Ontario were queried. Results indicate that the greatest interest was in the term Nestlé, occurring the late summer of 2016, during a period when the terms drought and water ban were also more frequently searched than at other times. The results for search term Nestlé were also queried for the origin location of the searches (Table 4). Guelph, a large population center located immediately north of Nestlé's main water bottling plant in Aberfoyle was the most common origin of searches for the word Nestlé.

## 4 Discussion

The datasets collected could be analyzed various ways. The Google Trends data indicates a strong interest in Nestlé in 2016, which was immediately prior to the MECP's first request for comment on the moratorium, when media interest in Nestlé's bottled water operation was high following Nestlé's announcement to buy a well that the local municipality had expressed interest. This was also a drought summer.

Most searches for Nestlé originate from Guelph, which is a community dependent on groundwater and close to Aberfoyle where Nestlé obtains much of its water. Other nearby communities also score highly when the search results are normalized on a per capita basis. These results are interpreted to suggest that the conflict over bottled water is a local issue and in 2016, exacerbated by drought conditions that summer.

Since 2016, the interest of the commenting public appears to shift to a greater interest in plastic waste, concern about water for future generations, making bottled water companies pay more for water, prioritizing water for local communities, and climate change.

**Table 3** Frequency of words as percentage of all available downloaded comments

| Word | 2018 extension (%) | 2019 extension (%) | Average (%) |
|---|---|---|---|
| Plastic | 29.3 | 32.3 | 30.8 |
| Nestlé (or Nestle) | 25.2 | 36.0 | 30.4 |
| Profit | 28.2 | 26.7 | 27.5 |
| Corporate or corporation | 22.7 | 22.7 | 22.7 |
| Environment | 22.3 | 22.9 | 22.6 |
| Groundwater | 17.8 | 23.7 | 20.6 |
| Future (as in future generations) | 18.0 | 15.0 | 16.5 |
| Climate | 8.5 | 12.8 | 10.6 |
| Aquifer | 8.3 | 7.3 | 7.8 |
| Finite (as infinite resource) | 6.6 | 7.5 | 7.0 |
| Policy | 3.6 | 8.9 | 6.2 |
| Drought | 6.4 | 4.0 | 5.3 |
| Human right | 3.2 | 6.7 | 4.9 |
| Regulation | 4.2 | 5.3 | 4.7 |
| Research | 4.2 | 2.0 | 3.1 |

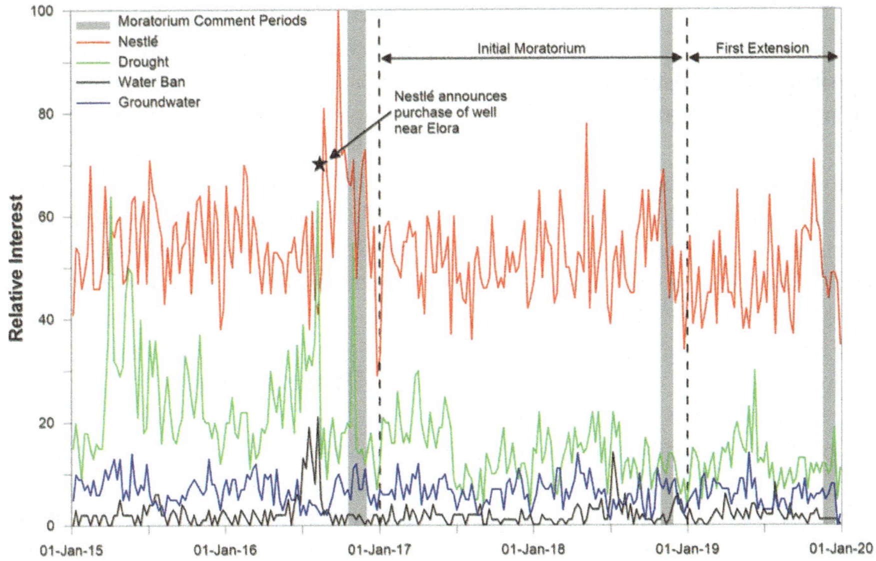


**Fig. 1** Google Trends relative interest results for searches from Ontario (2015–2019)

**Table 4** Origin location of Google searches for "Nestlé" (2015 to 2020 results)

| City of origin for search as recorded by Google Trends | Dominant municipal water source | Google relative interest score (0 to 100) | Google maps driving distance from Aberfoyle (km) | Google maps approximate mid-day driving time from Aberfoyle (minutes) | Population 2016 Canada census data[1] | Google score normalized to Guelph on a per capita basis |
|---|---|---|---|---|---|---|
| Guelph | Groundwater | 100 | 13.8 | 16–30 | 132,000 | 100.0 |
| Brampton | Great Lakes | 96 | 56.2 | 35–45 | 594,000 | 21.3 |
| Mississauga | Great Lakes | 87 | 55.7 | 35–50 | 722,000 | 15.9 |
| London | Great Lakes | 85 | 123 | 70–100 | 384,000 | 29.2 |
| Vaughan | Great Lakes | 80 | 77 | 45–105 | 306,000 | 34.5 |
| Milton | Great Lakes | 77 | 27 | 18–30 | 110,000 | 92.4 |
| Markham | Great Lakes | 76 | 88 | 50–65 | 329,000 | 30.5 |
| Richmond Hill | Great Lakes | 75 | 103 | 50–70 | 195,000 | 50.8 |
| Cambridge | Groundwater | 72 | 25 | 26–35 | 130,000 | 73.1 |
| Toronto | Great Lakes | 71 | 78 | 50–75 | 2,730,000 | 3.4 |

[1]Statistics Canada website. Census boundaries may not align with Google Trend areas

## 5 Concluding Remarks

The results of this work indicate that public interest in Nestlé's bottled water operations in Ontario varies with both events (increasing after droughts or unpopular actions by the proponent) and location (local is important, with public interest appearing to decrease with distance from the Nestlé sites). Furthermore, analysis of public comments over time shows that the commenting public are more likely to mention general environmental issues such as plastic waste or non-groundwater issues such as profit than to mention groundwater, with the relative interest in individual concerns changing slightly over time.

The drawing of conclusions about public interest in groundwater is typically stymied by a lack of data. The results of this work indicate the potential usefulness of the tools available with Google Trends for analysis of public opinion on groundwater issues, providing context not only on the timing of public interest, but also general location of the commentators. These tools open a potential new avenue for researchers interested in groundwater conflict. However, the approach might only be useful where public interest is both large enough and sufficiently focused to be recognizable above the "noise" of everyday web search activity, and in locations where the public has ready access to the Internet.

## References

EBR 012-8783. https://www.ebr.gov.on.ca/ERS-WEB-External/displaynoticecontent.do?noticeId=MTMwMjU1&statusId=MTk4OTEw. Last accessed 2020.01.11

ERO 013-3974 https://ero.ontario.ca/index.php/notice/013-3974#comments-received. Last accessed 2020.01.11

ERO 019-0913 https://ero.ontario.ca/notice/019-0913. Last accessed 2020.01.11

Gautrey (2017) Contrasting public participation in two recent groundwater governance initiatives: insights into public opinions about groundwater. In: 70th Canadian Geotechnical Conference, Ottawa, pp 1–8

(https://www.researchgate.net/publication/325386914_Contrasting_public_participation_in_two_recent_groundwater_governance_initiatives_insights_into_public_opinions_about_groundwater)

Golder (2019a) Nestlé Waters Canada Aberfoyle Site. 2018 Annual Monitoring Report

Golder (2019b) Nestlé Waters Canada Erin Springs Site. 2018 Annual Monitoring Report

Google (2019) www.google.com/trends. Last accessed 2020.01.11

Nestlé website. https://www.nestle-waters.ca/en/sustainability. Last accessed 2020.01.11

MECP PTTW database. Last accessed 2020.01.11

Statistics Canada. https://www12.statcan.gc.ca. Last accessed 2020.01.12

Walkerton Inquiry (2001) https://www.archives.gov.on.ca . Last accessed 2020.01.11

# Public Perceptions and Attitudes Towards Groundwater and Climate Change: The Case of the Barbate River Basin

Mercedes Vélez-Nicolás, Santiago García-López, María Jesús Pacheco-Orellana, Verónica Ruiz-Ortiz, and Alex Fernández-Poulussen

**Abstract**

This work presents some of the results obtained from a survey research that is currently being conducted among water users in the Barbate River basin area. The purpose of the survey questionnaires is to evaluate the use, opinions and extent of knowledge of the local population about water resources, especially groundwater, in the region. The questionnaires are also aimed at assessing citizens' perception of potential impacts on surface and groundwater resources, their evolution in the last decades regarding aspects such as quantity and quality, or the effects of climate change among others through the lens of their own experience. This information is of major importance in a space such as the Barbate River basin, where the marked seasonal pattern of precipitation, the prevalence of the primary sector (agriculture and livestock farming) as the main economic engine of the region and the increasing pressures and demands on hydrogeological systems are decisive factors in the development of management strategies. First-hand knowledge of citizens' perception is crucial to promote the cooperation between water users, manage conflicts and create synergies in a context of global change.

**Keywords**

Groundwater • Perception • Geoethics • Surveys • Adaptation

M. Vélez-Nicolás (✉) · S. García-López · M. J. Pacheco-Orellana
Department of Earth Sciences, Faculty of Marine and Environmental Sciences, University of Cádiz, Cádiz, Spain
e-mail: mercedes.velez@uca.es

V. Ruiz-Ortiz
Department of Industrial and Civil Engineering, Polytechnic School of Algeciras, University of Cádiz, Cádiz, Spain

A. Fernández-Poulussen
Good Stuff International B.V, Den Bosch, The Netherlands

## 1 Introduction

Surveys are a useful tool to evaluate aspects such as the public perception and level of knowledge about different scientific and social issues. Recent works on water, drought and climate change using survey questionnaires can be cited: March et al. (2013), La Jeunesse et al. (2016) and Gholson et al. (2019). The present study is one of the research lines included in a project on water resources and climate change in the Barbate River basin conducted by the University of Cádiz (UCA 2019).

The Barbate River basin, with an area of 1329 km$^2$, is located in the south of the Iberian Peninsula, in the province of Cádiz, near the Strait of Gibraltar. It comprises most of the region known as "La Janda", whose demographic dynamics has been marked in recent decades by migrations, rural depopulation, low growth rates and population ageing (Sánchez 2001). The productive model of this area has been traditionally based on agriculture and livestock farming. In particular, irrigated agriculture is of major importance and covers a surface of 15,000 ha that are supplied with both surface water (84%) and groundwater (16%).

Surface water in the basin comes from the Barbate, Celemín and Almodóvar reservoirs (joint capacity of 277 hm$^3$), which are exploited to supply irrigation mainly. A Water Users Association (WUA) is responsible for the management of surface resources, ensuring a planned distribution of water and controlling the required operations. The groundwater resources come from two groundwater bodies (GWB 062.13 and GWB 062.14) with a joint extension of 145 km$^2$, which are intensively exploited for human supply and irrigation. Unlike surface water, groundwater resources lack of entities aimed at its management and water is extracted by individuals in an unplanned manner, under poor administrative control. This situation can be partly explained by the fact that some of the farms are covered by the Water Law of 1879 and are included in a "Private Water Catalogue". These GWB are



M. Abrunhosa et al. (eds.), *Advances in Geoethics and Groundwater Management: Theory and Practice for a Sustainable Development*, Advances in Science, Technology & Innovation,
https://doi.org/10.1007/978-3-030-59320-9_10

declared in poor quantitative and chemical status by the Hydrological Planning documents (Junta de Andalucía 2015), which demand to adopt measures aimed at achieving a sustainable use of the resource in the medium term. The poor quantitative status was declared because the rate of water extraction exceeded the limits set by the current legislation. The poor chemical status responds to the evidence of nitrate contamination linked to agricultural activities. In recent years, urban supplies of groundwater are being replaced or complemented with surface water transfers from the adjacent river basin, which is managed by the entity known as "Consorcio de Aguas de la Zona Gaditana".

The aim of the current work is to assess citizens' perception regarding the local hydric resources and the implications of their exploitation and climate change. This is of major importance in the management of groundwater bodies and the improvement of their qualitative and quantitative status.

## 2 Materials and Methods

This study is based on a survey conducted among the local population in 2019. The number of surveys carried out was 98. The study area was the Barbate River basin, with special regard to the municipalities settled on the GWB 062.014 (Benalup) and northern area of GWB 062.013 trying to encompass the main productive sectors in the area. Surveys were classified according to different criteria, such as municipality, age intervals, gender and occupation.

The questionnaire used in this work (Table 1) consists of nine questions selected from a more extensive questionnaire that was aimed at exploring citizens' perception regarding water resources, pressures and other adverse impacts related to climate change. The surveys were anonymous and carried out in situ, orally, in public spaces or workplaces. The language used was colloquial and suitable for the educational level of the interlocutor. The responses were noted down by the interviewer. The questionnaire consisted of closed-ended questions with several possible answers. The respondents could choose several answers for the same question and to suggest alternative ones.

## 3 Results and Discussion

Question 1 shows an evident dichotomy between citizens who were supplied with water from the local reservoirs, mostly farmers belonging to the Water Users Association (37.2%) and those whose farms and households were supplied with groundwater (47.8%) from the GWB 062.14 (Benalup) in most cases. Some participants also claimed to combine the use of surface and groundwater (depending on the activity). This has been computed in the previous categories. The option "other alternative sources" was chosen by a minority of respondents (8.9%), although this choice could be attributed to a lack of knowledge about the origin of water from the supply network of their own dwellings. Most respondents who were supplied by the public network attributed its origin to surface water, when in fact it is groundwater.

The scores given to evaluate the quantity and quality of water (question 2) were similar, with average values of 6.5 and 6.3, respectively. This could be interpreted as the general perception that water availability in the region is acceptable to good, although a high standard deviation is observed regarding this question.

Question 3 explores the possible existence of conflicts arising from water exploitation in the region. Almost three-fourths of respondents pointed out the absence of conflicts. On the other hand, those who confirmed the existence of conflicts mentioned the high demand for water of rice crops and a minority of respondents, the illegal pumping wells or the need for authorizations for their exploitation. This question evidences that most citizens are not aware of the declaration of the GWB 062.014 as in poor quantitative and qualitative status and the administrative course of action that this may entail.

The origin of groundwater (question 4) elicited different opinions. Groundwater was mistakenly conceived as "underground rivers/lakes" by a 36.2% of the respondents, while a 40.7% rightly considered that its origin was rainfall infiltration. The third most frequently chosen option (14.1%) was "I don't know." This highlights that a considerable fraction of citizens lacks knowledge regarding the functioning and dynamics of hydrogeological systems, and the potential adverse implications that this may have concerning individual behaviour and resource conservation.

On the other hand, a large part of respondents considered that groundwater is susceptible to pollution processes (question 5). The 53.6% of them pointed out the diffuse spills of agrochemicals on the soil surface as potential sources of pollution, followed by those who considered point sources of pollution like septic tanks of irregular settlements (22.8%), which are common in the study area. It is significant, however, that some respondents considered that aquifers are not susceptible to pollution (10.6%), that groundwater only becomes contaminated at well installations (8.1%) and that some of them didn't know it (4.9%).

Question 6 addresses citizens' perception of local changes regarding the presence of water and other related aspects. A 38.7% of the respondents, generally older adults, agreed that there is now less water in the region, and recalled springs and small streams that had disappeared. This is a verifiable situation and a result of the adaptation of the aquifer's natural discharge to groundwater pumping for

**Table 1** Question wording and response set

| Code | Question | Response set |
|---|---|---|
| 1 | Do you know where the water used in your activity/house comes from? | (a) Surface water<br>(b) Groundwater<br>(c) Other sources (rain, reused water…)<br>(d) I don't know |
| 2 | Rate from 0 to 10 the availability and quality of the water resources in the region | (a) Quality<br>(b) Quantity |
| 3 | Do you know if there are any conflicts derived from water use in the region? | (a) Yes<br>(b) No<br>(c) I don't know |
| 4 | Do you know which is the origin of groundwater? | (a) Underground river/lake<br>(b) Local mountain ranges<br>(c) It comes from the sea, due to salt filtration<br>(d) It comes from the rocks, it has always been there<br>(e) Rainfall infiltration/percolation<br>(f) I don't know<br>(g) Other |
| 5 | Do you think that groundwater can undergo contamination processes? | (a) Groundwater is not susceptible to contamination<br>(b) Yes, when pollutants are poured into a well<br>(c) Yes, when pollutants leach from the surface<br>(d) Yes, due to the presence of septic tanks<br>(e) I don't know |
| 6 | Have you noticed any changes in the quantity or quality of the local water resources or in other related aspects during the last 20–30 years? | (a) There is more water nowadays<br>(b) There is less water nowadays<br>(c) The quality has improved in recent years<br>(d) The quality has worsened in recent years<br>(e) There is more pollution nowadays<br>(f) Changes in the local flora and fauna<br>(g) There has been no changes<br>(h) Other |
| 7 | Do you think there will be problems with water in the next 20–30 years? | (a) I don't think so<br>(b) Climate change will result in scarcer rain<br>(c) The quality will become worse due to pollution<br>(d) There will be more droughts<br>(e) Wells will run out of water<br>(f) Other |
| 8 | Do you think it is necessary to take measures in order to ensure the future availability of water and its uses? | (a) I don't think it is necessary<br>(b) Improve Administration's control over water use<br>(c) Improve the cooperation and transparency among water users<br>(d) Increase the investments in water infrastructure<br>(e) Control and set limits to groundwater pumping<br>(f) Allow more groundwater pumping<br>(g) Reinforce the protection of the natural environment<br>(h) Replace the current crops with other types that consume less water<br>(i) Other |
| 9 | Have you noticed changes in the local rainfall pattern in the recent years? | (a) There has not been any changes<br>(b) Rainfall is becoming more unpredictable<br>(c) Rainfall is becoming more torrential<br>(d) I don't know<br>(e) Other |

more than 40 years. The second option in importance (16.9%) was the absence of changes, which can be interpreted as the result of the stabilization of groundwater exploitation levels during the last years. The third most chosen option pointed to an impoverishment of water quality, what is difficult to legitimize as the respondents did not put forward solid arguments to back this assertion; however, it may reflect a general concern about human activities contribution to the pollution of the environment.

Regarding the medium term, in the horizon of the next 20–30 years (question 7), the answers given by the respondents were also uneven: a 37.5% of them claimed that climate change would lead to a decrease in the amount of rainfall. A 16.4% believed that droughts would increase, a

10.2% that water quality would worsen and a 8.6% that wells would run dry. On the contrary, a 19.5% of the respondents thought that there would be no changes beyond the cyclic character of climate. These different outlooks may be the result of the influence of mass media on the population regarding this issue.

Question 8 suggests potential course of action aimed at the conservation of water resources. A stricter control of water use by the Administration (20.3%), limitations on groundwater withdrawals (17.7%) and further investment in water infrastructure (17.7%) were the most frequent answers, along with the option "other" (19.0%), which basically included considerations about the need for awareness and citizenship education in aspects related to water. Only a 9.2% believed that it is unnecessary to take any measures.

With regard to the changes perceived in the local climate (question 9), many respondents claimed that the occurrence of rain has become more unpredictable in recent years (52.1%), torrential (21.8%) and only the 6.7% considered that there have been no significant changes.

## 4 Concluding Remarks

This study constitutes an approximation to a complex reality that comprises several interrelated aspects such as the perception and knowledge of water resources (with special regard to groundwater), aquifer dynamics, status and potential impacts linked to climate change among the population of the Barbate River basin. Although the surveys reveal a significant disparity of opinions and outlooks among the respondents, the information gathered enables to reach the following conclusions:

The local population assumes that water resources in the region are sufficient in both quantity and quality, and that there are no significant supply problems despite the variability of climate. Besides, most of the respondents know the source of the water they consume (surface water or groundwater).

The extent of knowledge about the nature, dynamics and vulnerability of hydrogeological systems is uneven and sometimes remarkably poor. An important segment of the population has not assumed that groundwater is a renewable resource, susceptible to be degraded in terms of quantity and quality. In particular, there is a widespread lack of knowledge regarding the declaration of the GWB 062.14 as in poor condition by the competent institutions, and the consequences that this may involve. Any action aimed at recovering the good condition of the aquifers will require education and awareness campaigns in order to engage the local population and encourage their role as effective actors in water conservation. At this regard, it is also necessary that stakeholders assume the economic and social restrictions and costs that this could entail in order to achieve a sustainable use of water resources.

The sampled population is familiar to some extent with the concept of climate change and associate it in a more or less direct manner with water availability and occasionally, with the adverse effects of these phenomena on the hydrogeological systems of the region.

The results evidence that most population understands that it is necessary a stricter control on water resources by the Administration (with special regard to groundwater), along with a reinforcement of environmental protection and further efforts in citizen education and awareness campaigns. However, there is still a fraction of respondents who consider that it is unnecessary to undertake any course of action, or that the measures to be adopted are related to investment in water infrastructure.

**Acknowledgements** This work has been developed within the framework of the REMABAR project, supported by the Biodiversity Foundation of the Ministry for the Ecological Transition. The authors also want to express their gratitude to the citizens who have kindly participated in the surveys.

## References

Gholson DM, Boellstorff DE, Cummings SR, Wagner KL, Dozier MC (2019) A survey of public perceptions and attitudes about water availability following exceptional drought in Texas. Universities Council on Water Resources. J Contemp Water Res Educ 166:1–11

Junta de Andalucía (2015) Consejería de Agricultura, Pesca y Medio Ambiente. Plan Hidrológico del Guadalete Barbate 2015–2021. Tomo I (Memoria) y Tomos II, II, IV, V y VI (Anejos)

La Jeunesse L, Cirelli C, Aubin D, Larrue C, Sellami H, Afifi S, Bellin A, Benabdallah S, Bird DN, Deidda R, Dettori M, Engin G, Herrmann F, Ludwig R, Mabrouk B, Majone B, Paniconi C, Soddu A (2016) Is climate change a threat for water uses in the Mediterranean region? Results from a survey at local scale. Sci Total Environ 543:981–996

March H, Doménech L, Sauri D (2013) Water conservation campaigns and citizen perceptions: the drought of 2007–2008 in the Metropolitan Area of Barcelona. Nat Hazards 65(3):1951–1966

Sánchez DG (2001) Aproximación a los fenómenos demográficos relevantes de la comarca de La Janda en el siglo XX. Revista De Estudios Vejeriegos 4:130–164

UCA-Universidad de Cádiz (2019) Informe sobre acciones de sensibilización y divulgación. Proyecto REMABAR: Análisis de estrategias para reducir las pérdidas por evaporación y mejorar el estado de los acuíferos bajo contexto de cambio climático en la cuenca del Río Barbate. https://proyectoremabar.blogspot.com/

# Ethical Issues of Intensive Use of Groundwater in Stressed Spanish Aquifers

Emilio Custodio

**Abstract**

Intensive groundwater abstraction in arid and semiarid areas of the world often approaches and even overcomes available resources. The consequences are important groundwater level drawdown, loss of water quality, salinization and other negative externalities, such as land subsidence, increasing cost of water and social changes. There are positive and negative consequences, both direct and as externalities. Each situation is different and requires a detailed study. However, there is a background, whose knowledge helps in understanding other cases. Changes involve ethical issues as they affect humans and their activities, as well as the current and the future generations, given the slow reaction of groundwater to changes. These ethical issues go beyond hydrogeological considerations and include economic, administration, management and policy aspects, which should back sound water governance. Some of the driest areas of the European Union are in Spain, where there is long a tradition of groundwater use, mostly for intensive crop irrigation. Here, part of the ethical and moral point contents of this experience is considered, based on two recent reports, the experience in water planning, and the dominant use of groundwater resources in irrigated agriculture.

**Keywords**

Current knowledge • Ethical issues • Intensive development • Groundwater • Spain

E. Custodio (✉)
Royal Academy of Sciences of Spain, Madrid, Spain
e-mail: emilio.custodio@upc.edu

Groundwater Hydrology Group, Department of Civil and Environmental Engineering, Technical University of Catalonia, Barcelona, Spain

## 1 Introduction

In arid and semiarid areas having good aquifers and scarce allochthonous rivers, groundwater provides most of available natural water resources, including brackish water and additional (engineered) water resources from wastewater reclamation, seawater desalination, and enhanced and managed artificial recharge. This is especially true when there are good soils and climate, as intensive agricultural development may take place, demanding large annual water volumes for irrigation. Irrigated agriculture puts pressure on water resources, both in quantity and in quality, which is generally greater than that due to the supply to other uses, including the growing tourism in coastal areas.

Results of intensive development vary conspicuously from one area to another, but there are commonalities. Worldwide, well-documented cases to obtain general and specific knowledge are scarce, but they exist, as in the Central and Western USA, Mexico and Australia. In Europe, good studies derive from the application of the Water Framework Directive after its transposition into the water norms of each of the European Union Member States. Intensive groundwater use happens mostly in the South, and especially in Spain, in the Mediterranean area of the Iberian Peninsula and in the Balearic and Canary Islands.

## 2 Methods

The considerations that follow refer to the dry part of Spain and benefit from knowledge in other areas of the world. They are based on reports whose contents do not derive from specific research but from the collection, analysis and synthesis of published documents and papers, complemented with a large number of author's studies carried out since the 1960s and the comments of more than 50 experts. Most of the available significant material has been used. The contents include the general scientific hydrogeological principles and



M. Abrunhosa et al. (eds.), *Advances in Geoethics and Groundwater Management: Theory and Practice for a Sustainable Development*, Advances in Science, Technology & Innovation,
https://doi.org/10.1007/978-3-030-59320-9_11

afterwards their application to Spain. Economic, administrative, social and ethical aspects are included. Groundwater reserve depletion (groundwater mining) is considered in MASE (2015), coastal aquifers in SASMIE (2017) and complementary knowledge on aquifer recharge to support conclusions derive from RAEMIA (2019). These are the only references here included as they contain the references that support what is presented hereinafter.

## 3 Results of Intensive Groundwater Development in Arid and Semiarid Areas of Spain

Both in the South-east of the Iberian Peninsula and in the Canary Islands, there is intensive development of groundwater that in many areas of Spain has produced valuable social benefits by allowing storage depletion and salinization of coastal aquifers. However, there are serious negative aspects. MASE (2015) deals with intensive development of groundwater resources and storage depletion in the South-eastern Iberian Peninsula and the Gran Canaria and Tenerife, in the Canary Islands, and SASMIE (2017) refers to coastal aquifer's exploitation in the Spanish Mediterranean area and the Balearic and Canary Islands.

Intensive groundwater development allows the water supply to a fast-growing population and tourist sector. At the same time, there is an important development of intensive greenhouse, orchard and fruit-tree irrigated agriculture, significant worldwide, which has transformed a poor socio-economy into one with income greater than the national average. This happens in a complex system with areas supplied almost exclusively with surface water, others dependent only on groundwater and still others using a mix of surface and groundwater, as well as reclaimed wastewater, seawater desalination and brackish groundwater after salinity reduction. The public sector played and plays an important role in surface water management, while the situation is more complex for groundwater. Groundwater is currently a public domain, according to the Spanish Water Act of 1985, but many of former groundwater right owners retain a conditioned property. This explains partly the current situation of depletion of groundwater reserves and of coastal aquifer salinization and the often-serious ethical issues around.

In Gran Canaria and Tenerife, in the Canary Islands, groundwater-intensive exploitation for cash crop irrigation started in the 1920s, developed especially in the 1960s, and peaked in the 1980s. Private water markets regulated exploitation. Currently, the system is evolving due to increasing urban and tourist demand, decrease of groundwater availability due to hydrodynamic conditions and the introduction of seawater desalination and to a lesser extent by wastewater reclamation, mostly as an initiative of the public sector. The water infrastructure has been economically supported by local inhabitants by subscribing bonds. The large private investors did and do not show interest in water supply, except to supply the urban sector with industrial water. In South-eastern Spain, intensive development started in the 1960s and was well established in the 1970s, for irrigated intensive agriculture.

In the considered areas, the groundwater level is deep enough to significantly affect the cost of pumping the water, due to energy consumption and maintenance and reposition of the wells. In the case of Tenerife, the part of groundwater obtained by water galleries (tunnels) is not affected by energy cost, but continuous investment is needed to maintain the flow, as dynamic reserves are progressively depleted. The high-water cost favours the increase of the water use efficiency, but also the cost of the pressurized water for the modern irrigation methods. In some areas, other problems refer to increasing soil water salinity and decreased recharge from return flows to aquifers placed down flow. An important collateral result is the disposition of water users to join efforts for collective action. However, the actual result is often an increased pressure on politicians to import water and get subsidies for the agricultural sector. This has important ethical consequences, as the cost is paid by society.

Despite the seriousness of groundwater reserves depletion, in neither the South-east of Spain nor the Canary Islands, physical depletion has neither been produced yet, nor it is expected in the near future. Some aquifers still have significant reserves, while in others exploitation is decreasing due to increasing cost and quality deterioration, thus making actual abstraction close to actual recharge. In this last case, a sustainable situation can be attained, but with high cost to exploiters and the population. This has ethical implications. In Tenerife and to a lesser extent in Gran Canaria, the groundwater depletion produced by high elevation water galleries is actually a physically and economically irreversible reduction of usable underground space for regulation. This has ethical implications as well, as it affects future generations and their heritage.

When water is scarce, water quantity has been a main concern. However, quality issues will become more important and acute in the near future. This implies another kind of ethical concerns due to health implications, increase of treatment cost and use limitations of currently available water resources. Intensive groundwater use may be associated with growing quality problems due to an excess of concerning solutes. Changes appear slowly and may grow inadvertently when monitoring does not detect them. In some cases, there is a recovery after some years, but in others damage becomes permanent and correction measures, if feasible at all, may disappear after the period of direct benefit generation. These externalities involve serious ethical concerns, as is the case of fluoride excess in some water galleries in Tenerife or the increased salt dissolution from geological formations in some areas of South-eastern Spain.

The intensive use of groundwater has produced serious environmental impacts in Spain, such as wetland desiccation, dry out of springs and some stretches of rivers, and conspicuous subsidence, as in Lorca (Murcia). Some of them happened many years ago, so they are not in the mind of local inhabitants. The cost and damage have been largely supported by society and by individuals that did not received the benefits of used water. The Mar Menor-Campo de Cartagena (Murcia Region) is currently a paradigmatic case of groundwater reserves depletion and non-foreseen externalities of a large water transfer. In other cases, intensive groundwater exploitation in the Llobregat and Besós deltas, Barcelona, produced a valuable underground space for urban development and transportation, but the subsequent salinization and urban occupation produced a water table recovery and a series of costly inundation problems.

## 4 Ethical Issues and Concluding Remarks

The main ethical and moral aspects of intensive use of groundwater, coastal aquifers and reserve depletion refer to damage to current and future third persons, as well as the environment and the services it provides to humans. An important human issue associated with intensive depletion of groundwater resources and coastal aquifer salinization is the loss of living resources of locals, which may force human displacement from self-sufficient rural areas to areas of dependent employment. This may cause family and social disorganization and the associated marginalization. This is independent of poverty.

Conventional mining ends when the physical mineral reserves are exhausted or decrease their economic value. Groundwater mining is essentially different as water is needed for life, is at least partly removable, lasts longer and may come from different sources. Therefore, groundwater exploitation costs or administrative and legal limitations do not necessarily lead to an end of the activity. As groundwater mining is strictly unsustainable, it must be followed by a change in the water use paradigm, forced by circumstances or as the result of management. However, society, and especially the rural environment and the public water administration, tends to resist the change. Well-designed government incentives are needed to overcome this, but the private enterprise has to be involved, as it was in the past, under flexible and well-conceived regulations. Ethical behaviour is here a key element to guarantee non-skewed information, data acquisition and diffusion, sound planning, and social participation and involvement, including the development of civil society institutions.

In water scarce areas, temporal groundwater reserve depletion is not necessarily bad. It helps in the process of changing the water use paradigm, including the compensation of negative externalities. The European Water Framework Directive rejects groundwater reserve depletion of freshwater. Its strict interpretation by many water authorities and environmentalists may produce negative social consequences and insufficient positive results when considering the well-being of Europeans, which is the main objective of the European Union regulations. To avoid misinterpretations, good monitoring and adequate studies are needed, designed case by case. This technical issue is really an ethical and moral issue as it implies devoting human and economic resources. However, this is a weak point when other pressures exist, such as a short-sighted interpretation of economic crises.

Slow renovation aquifers, with depleted reserves or salinized, degraded or penetrated by poorly constructed and designed boreholes, will not be available in the future. When allowing and carrying out action, sustainability must be considered. This is an ethical issue.

Water saved today and stored for the future has the potential to produce economic benefits to future generations. However, the evaluation is highly controversial and depends on the discount rate. While a low rate makes valuable the future benefits and a high rate favours depletion, as the future value is low. The value of the discount rate is a social decision that involves ethical and moral considerations, but there is not agreement on the discount rate values to be applied, which may change with time, the current economic situation, dominant policies and political orientation.

The collapse or fast degradation of human activities related to coastal aquifers is an inheritance loss. Government members, politicians, water users and the civil society must act ethically. This is the framework of action in the Metropolitan Area of Barcelona to maintain in operation the Baix Llobregat aquifer system and its integration in the water resources system. This is also the framework to understand the degradation that happened some decades ago in the Besós River delta and the coastal areas of the Baix Camp of Tarragona. Current concerning situations are those of the Campo de Cartagena and Campo de Dalias, which are highly important areas from the social and economic point of view and risk a breakdown. Solutions need to set first a framework with a higher rank than water norms and water planning, with rules that must be followed by the water administration and government officials. As humans are involved, ethical and moral principles must be applied from the beginning, regarding not only water, but energy, land use, employment and social equity as well.

In the groundwater stressed areas, there is an increasing use of imported goods, mainly fodder for the actually very important animal rising activities. This can be translated into virtual water import. It should be added the export of products obtained, which includes virtual water from the country's groundwater resources and the current and especially the future loss of groundwater quality due to the wastes associated with the activity. However, this is not considered in water planning and actual values are not known. This involves serious ethical considerations on the actual water benefits, on food security and the distortion that may appear as a consequence of malfunctioning and manipulation of international markets. The ethical considerations have to be considered independently on been in favour or against virtual water trade.

**Acknowledgements** The results and comments presented can be found in the free-access eBooks listed in the references, produced by the Groundwater Group of the Department of Civil and Environmental Engineering, Technical University of Catalonia, Barcelona, with the support of Aqualogy/Suez-Spain and Cetaqua.

## References

MASE (2015) Aspectos hidrológicos, ambientales, económicos, sociales y éticos del consumo de reservas de agua subterránea en España: minería del agua subterránea en España. [Hydrogeological, environmental, economic, social and ethical aspects of groundwater reserves consumption in Spain]. Prepared by E. Custodio for UPC, Aqualogy and Cetaqua: 1–488. Technical University of Catalonia, Barcelona. https://hdl.handle.net/2117/111272

RAEMIA (2019) Recarga natural a los acuíferos, metodología y soporte de la isotopía del agua. Aplicación a la planificación hidrológica y conocimiento de las aguas subterráneas en España. [Aquifer natural recharge, methodology and water isotopes support. Application to water planning and groundwater knowledge in Spain]. Prepared by E. Custodio for UPC, SUEZ-Spain and Cetaqua, revised by J. Jódar and with contributions from J.V. Giráldez and A. Sahuquillo: 1–1206. Technical University of Catalonia. https://hdl.handle.net/2117/182282

SASMIE (2017) Salinización de las aguas subterráneas en los acuíferos costeros mediterráneos e insulares españoles. [Groundwater salinization in Spanish Mediterranean and island coastal aquifers]. Prepared by E. Custodio for UPC, SUEZ-Spain and Cetaqua: 1-852. Technical University of Catalonia, Barcelona. https://hdl.handle.net/2117/111515

# Public Geospatial Data for Groundwater Governance: The Brazilian Case

César de Oliveira Ferreira Silva and Rodrigo Lilla Manzione

**Abstract**

The present work presents two databases available in Brazil related to groundwater governance applications, exposing their applicability, gaps and potential geoethical discussions. The first database is the Rural Environmental Registry (CAR), which is a national electronic public record, mandatory for all rural properties. The second database is the Groundwater Information System (SIAGAS) developed by the Geological Survey of Brazil (CPRM), which consists of a continuously updated database of wells. CAR has a record of 1,892,067 water springs located within rural properties and SIAGAS has 322,922 registered wells. Publicity of CAR data is ensured by law. Both are public databases, so their use does not generate legal conflicts. CAR data is self-reported (by landowners) and requires validation by the government. For this reason, its use currently requires care to avoid distortions caused by errors in the georeferencing of springs. SIAGAS also has a small number of wells registered, and it is estimated that 88% of wells in Brazil are clandestine. SIAGAS and CAR are examples of databases of great applicability to groundwater governance, both by government initiatives, which require validation and improvements in order to be expanded and applied with greater reliability.

**Keywords**

Springs • Wells • Computation • CAR • SIAGAS

César deO. F. Silva (✉)
School of Agriculture (FCA), São Paulo State University (UNESP), Botucatu, SP, Brazil
e-mail: cesaroliveira.f.silva@gmail.com

R. L. Manzione
Biosystems Engineering Department, School of Sciences and Engineering (DEB/FCE), São Paulo State University (UNESP), Tupã, SP, Brazil

## 1 Introduction

Water resources need a management view that covers their multiple uses, stakeholders involved and their connections. An effective management should not be limited to governmental decision-making institutions, as this would nullify their public and political character, just as it should not be limited to water resources allocation and quantitative distribution. It is also necessary to identify and evaluate the interaction between society and government in order to build an effective management of water resources. Thus, a concept called “water governance” was created, which aimed to analyse and assess the political, social, economic and legislative means to develop and manage water resources and water services’ delivery at different levels of societies.

According to Bellaubi and Bustamante (2018), there are basic principles for water management like autonomy, equity and responsibility in the water allocation, considering the great physical and social, geographical and historical variability of water needs and values. Llamas (2003) emphasizes that the application of these principles may generate legitimate but different practical solutions because of the complexity of water uses (urban, irrigation, energy, etc.). According to Hirata et al. (2019), the exploitation of groundwater in Brazil is still very small compared to the potential of groundwater resources, estimating that the 172 million Brazilian citizens who have access to the public water network, 30.4 million (17.7%) are served by groundwater; the remaining 141.6 million (82.3%) are served by surface water sources. Thus, there is an opportunity to build the governance of groundwater resources as the basis for the sustainable use of this resource of great potential.

The planning and rational use of groundwater requires extensive knowledge about the physical characteristics of the aquifers, their spatial distribution, the relationship with the surrounding environments, as well as adequate information about water quality and quantity. The data type that support this information is not always available in an organized way,



M. Abrunhosa et al. (eds.), *Advances in Geoethics and Groundwater Management: Theory and Practice for a Sustainable Development*, Advances in Science, Technology & Innovation,
https://doi.org/10.1007/978-3-030-59320-9_12

as it requires specific knowledge to obtain and validate it. Thus, efforts to create such databases are extremely valuable for the construction of water governance strategies and their effective application within society. Mogk and Bruckner (2020) state that explicit training is required to allow the community to recognize, prevent and mitigate ethical issues and dilemmas. The application of water allocation rules can be impaired by a lack of acceptance and implementation problems (Walker et al. 2015).

In Brazil, the central institution responsible for the integrated water resources management is the Brazilian National Water Agency (ANA), which follows the Sustainable Development Goal n. 6 (SDG 6), as defined in the United Nations (UN) 2030 Agenda. ANA (2019) contributes to the monitoring process of the eight SDG 6 targets. The goal deals with sanitation and water resources in an integrated perspective. In any case, it is not clear how ethical aspects involving data gathering, sharing and public policies elaboration are considered in the scope of groundwater management, protection and exploitation.

The present work presents two databases available in Brazil related to applications in groundwater governance and discusses their applicability, gaps and potential geoethical aspects.

## 2 Relevant Databases for Groundwater Governance in Brazil

### 2.1 Requirements for Data in Groundwater Governance

Water resources management depends on regulatory practices for controlled use and protection measures according to the rules and compliance with the relevant legislation. In practice, groundwater is more complex to manage than surface water bodies, because it is not visibly in the environment: so, this is more difficult to understand its dynamics and detect harmful processes. For this reason, the main difficulty in establishing an effective groundwater governance is the lack of reliable data. The following two databases are examples of relevant efforts to catalogue geospatial environmental data in Brazil.

### 2.2 Rural Environmental Registry (CAR)

The Brazilian Law of Protection of Native Vegetation (Federal Law nº 12.651/2012), commonly called in Brazil as the "new Forest Code", became the main regulatory framework of Brazilian forestry policy, creating guidelines for land use and occupation in rural properties and establishing rules for the environmental regularization of rural producers. In this legislation, some land use classes are defined as places for environmental preservation, such as the Permanent Preservation Areas (APP), which are strips of riparian forests of any river or stream (even when the river is intermittent), lakes, hill tops, places higher than 1800 m and water springs. In 2018, the Brazilian Supreme Court decided that intermittent water springs also had to be included in the APP (CPI and Agroicone 2018).

A device for controlling and monitoring the compliance of rural producers to this legislation is the Rural Environmental Registry (CAR). CAR is a national electronic public record, mandatory for all rural properties, in order to integrate the information about rural properties and possessions referring to APPs, restricted use, legal reserve, remnants of forests and other forms of native vegetation, and consolidated agricultural areas. It is also a database for control, monitoring, environmental and economic planning and fighting deforestation.

According to the last CAR Bulletin (SFB 2020), the number of rural properties registered in the CAR, until January 2020, exceeded 6.4 million across the country, spread over a registered area of more than 540 million ha. An amount of 1,892,067 water springs throughout Brazil was registered in CAR's database.

### 2.3 Groundwater Information System (SIAGAS)

The SIAGAS is the "Groundwater Information System" developed by the Brazilian Geological Survey (CPRM), which consists of a continuously updated database of water wells. It contains modules capable of perform queries, research, extraction and generation of reports. For each well, information such as constructive aspects, geological and hydrogeological data, water level monitoring, pumping test and chemical analysis is recorded and updated frequently. Among the constructive data record in this database, there are the initial and final depth of drilling, diameter, coating, filter, annular space and mouth of the well tube, as well as the drilling method. Among the geological data, the description of the geomorphological features, depth of the geological formation and the ranges of each lithology can be registered. Among the hydrogeological data, aquifer characteristics can be included, like porosity, water level (m) and flow rates ($m^3\ h^{-1}$). Among the pumping test data can be registered static level (m), dynamic level (m), specific flow ($m^3\ h^{-1}\ m^{-1}$), storage coefficient, free flow ($m^3\ h^{-1}$), permeability ($m\ s^{-1}$), transmissivity ($m^2\ s^{-1}$) and flow after stabilization ($m^3\ h^{-1}$). Among the data of chemical analysis, the researcher can retrieve the date of sample collection, electrical conductivity ($\mu s\ cm^{-1}$), water quality ($pt\ co^{-1}$), water flavour, water quality (odour), temperature (°C), turbidity (NTU), suspended solids ($mg\ L^{-1}$), solids sediments (mg $L^{-1}$) and pH. The last update of SIAGAS was made on 9 April 2020, accounting for a total of 322,922 registered wells.

In SIAGAS (2020), it is possible to perform simple and complex searches of the database, produce reports in different ways, download existing wells in the database, up to a certain limit or even without limit, depending on the level of access allowed to the user. In addition, it is possible to access the specific window of the Integrated Groundwater Monitoring Network (RIMAS), in order to obtain data and information, and view the wells registered with thematic maps and satellite images. Figure 1 presents the water springs record in CAR and the water wells record at SIAGAS.

## 3 Applicability, Gaps and Potential Geoethical Discussions

Geospatial data from CAR and SIAGAS can be associated with information on land use change, proximity to potentially polluting sources, assessment of groundwater availability in terms of springs and wells density. A geoethical analysis of groundwater governance requires considering physical, sociocultural, economic and political factors in a network of complex relationships. Most limitations of groundwater governance reflect a combination of technical, social, behavioural and organizational limitations that are compounded by social "conflicts" over access to the resource and the way it is used. The main problem arising from the lack of groundwater monitoring in agriculture is the overuse. Advances in drilling technology allowed the spreading of intense groundwater abstraction in agriculture around the world since the 1970s, but it was not accompanied simultaneously by the evolution of institutional arrangements and investments in management agencies, resulting in a laissez-faire mode of groundwater use (Kemper 2007). That is the case of Brazilian irrigation, where no attention is given to the sustainability of the resource. Schlager (2007) lists other potential problems, such as well owners being able to place their wells too close together, interfering with each other's pumping or a farmer being able to install a deep pipe very close to another farmer's shallow well, drying it out. These types of decisions go beyond the limits of rural properties and create problems for the community. Adoption of strict controls on aquifer pumping, to conserve water for the future, is a very difficult issue because the current irrigation water user can suffer immediate economic loss, raising an ethical question about safe aquifer yield (Peck 2007).

The main gap in the CAR database is the need for data validation, since they are initially declared by landowners, but in a second stage, they require to be analysed by the government. This process is being done gradually. Thus, before using them for regional studies, it is necessary to verify whether the region's data has been already validated, otherwise it is necessary to use satellite images and digital terrain models to verify that springs have been located with sufficient geographical accuracy. The main gap of SIAGAS is that only part of wells has all data fields registered, where most of them have only the lithological, pumping and hydrogeological data completed, and incomplete data regarding construction, geological and chemical analysis. Hirata et al. (2019) consider that the number of registered wells is small and estimate that in Brazil, there are around 2.5 million tubular wells, 88% of which are outside the

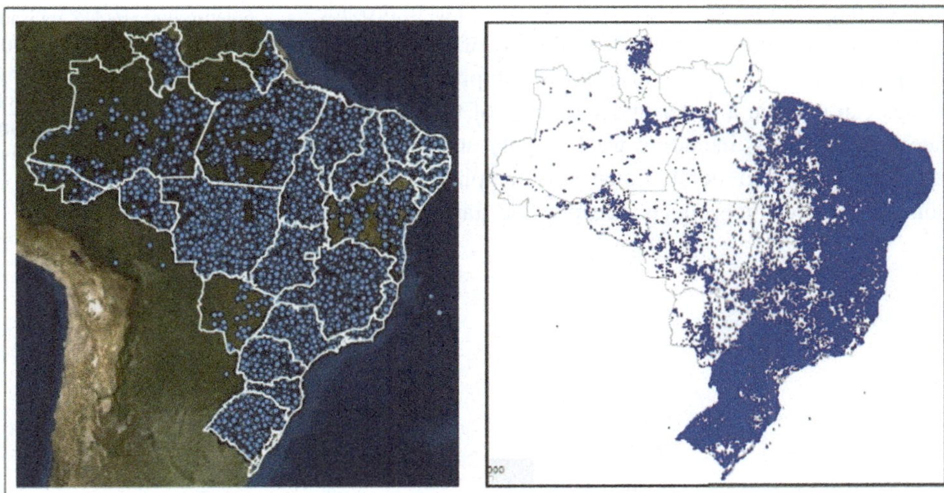

Water springs from Rural Environmental Registry (CAR in Portuguese acronym) of the Brazilian Forest Law

Wells from Groundwater Information System (SIAGAS in Portuguese acronym) of the Geological Survey of Brazil

**Fig. 1** Overview of geospatial data of interest for groundwater governance. CAR has a record of 1,892,067 water springs located within rural properties (left) and SIAGAS has 322,922 registered wells (right)

official records of the government. The total amount of water extracted from those wells would be 17.580 $mm^3$ $year^{-1}$ (557 $m^3$ $s^{-1}$), enough to supply the entire Brazilian population for one year. The validation of CAR data using satellite images is easier by comparing with SIAGAS data, when considering the possibility to hide or not declare a spring or a small dam for irrigation purposes and a tubular well. The punitive sentiment felt by local people about the declaration of water still dominates rural areas in Brazil. Therefore, it was necessary to impose sanctions, such as blocking access to rural credit, to landowners who did not register with the CAR, in order to push farmers to provide correct and reliable information. The development of socio-hydrological strategies, as proposed by Re (2015) and Limaye (2017), is the path to be followed to achieve such goal: it promotes the structured interaction between members of society, especially rural communities and water technicians. Law ensures publicity of CAR data. Both are public databases, so their use does not generate legal conflicts. CAR data is self-reported (by landowners) and requires subsequent verification and validation by the government, for this reason currently its use requires particular care to avoid distortions caused by errors in the georeferencing of springs.

## 4 Concluding Remarks

CAR and SIAGAS are examples of government databases with great applicability to groundwater governance in Brazil. Both require validation under a geoethical scope in order to be expanded and applied with greater reliability. In addition, both provide qualitative and quantitative information about the use and availability of groundwater in agricultural and urban environments. This data is necessary to diagnose conflicts between users of multiple water sectors, as well as to provide data to find solutions for conflicts in the use of groundwater. Geospatial and monitoring data from pumping wells and springs can provide a better understanding of the water balance in watersheds, including water supply and demand in both spatial and temporal dimensions, helping mitigation future climate change effects. Despite the gaps, CAR and SIAGAS are tools of great value for groundwater governance and deserve attention to be improved for future applications in order to provide accurate data for policymaking.

## References

ANA (National Water Agency) (2019) SDG 6 in Brazil: ANA's vision of the indicators. Brasília, National Water Agency

Bellaubi F, Bustamante R (2018) Towards a new paradigm in water management: Cochabamba's water agenda from an ethical approach. Geosciences 8:177

CPI and Agroicone (2018) The Brazilian Forest Code finally can be implemented. What's next? https://www.inputbrasil.org/wp-content/uploads/2018/03/CPI_Agroicone_Insights_Forest_Code_-_Whats_Next.pdf. Last accessed 2020/04/11

Hirata R, Suhogusoff AV, Marcellini SS, Villar PC, Marcellini L (2019) A revolução silenciosa das águas subterrâneas no Brasil: uma análise da importância do recurso e os riscos pela falta de saneamento. São Paulo, Trata Brasil

Kemper KE (2007) Instruments and institutions for groundwater management. In Giordano M, Villholth KG (2007) The agricultural groundwater revolution: opportunities and threats to development. Cambridge, CABI

Limaye S (2017) Socio-hydrogeology and low-income countries: taking science to-rural society. Hydrogeol J 25(7):1927–1930

Llamas MR (2003) Ethical Considerations in water management systems. Water Nepal 10(1):15–33

Mogk DW, Bruckner MZ (2020) Geoethics training in the Earth and environmental sciences. Nat Rev Earth Environ 1:81–83

Peck J (2007) Groundwater management in the High Plains Aquifer in the USA: Legal problems and innovations. In Giordano M, Villholth KG (eds) The agricultural groundwater revolution: opportunities and threats to development. Cambridge, CABI

Re V (2015) Incorporating the social dimension into hydrogeochemical investigations for rural development: the Bir Al-Nas approach for socio-hydrogeology. Hydrogeol J 23(7):1293–1304

Schlager E (2007) Community management of groundwater. In Giordano M, Villholth KG (2007) The agricultural groundwater revolution: opportunities and threats to development. Cambridge, CABI

SFB. CAR Bulletin. https://www.florestal.gov.br/documentos/car/boletim-do-car/4418-revisao-boletim-car-encaminhar-07abril2020-1/file. Last accessed 2020/04/11

SIAGAS. Sistema de Informações de Águas Subterrâneas, https://siagasweb.cprm.gov.br/layout/index.php. Last accessed 2020/04/11

Walker WE, Loucks DP, Carr G (2015) Social responses to water management decisions. Environ Process 2:485–509

# An Initiative for Protection of the Transnational Guarani Aquifer System Based on Geoethics

Celso Dal Ré Carneiro and Luciana Cordeiro de Souza-Fernandes

**Abstract**

A contemporary question interesting to Brazilian society refers to the access and use of groundwater, especially for the strategic Guarani aquifer system (GAS). The name identifies a transnational groundwater reservoir occupying an area of over one million km$^2$, comprising portions of central-south-southeastern Brazil, northern Argentina, eastern Paraguay and northwestern Uruguay. The "GAS Science Diffusion Program" intends to make people aware that water is exhaustible, although being abundant in Brazil. The program is strongly based on geoethics, which represents a recognition of the human responsibility toward the Earth (or an ethics before the planet). To raise people awareness on the subject, an open debate about related issues is required. It is also urgent to promote a movement for expanding awareness among governments and public agents about the singular geological characteristics of the critical reserves. Ignoring the geological dynamics can lead to irreparable human tragedies, as documented in recent hazards scattered on many Brazilian municipalities.

**Keywords**

Population • Underground water • Geology • Geodiversity • Brazil

## 1 Introduction

The Brazilian population growth over 200 years [1822–2022] is expected to be "more than 6 times" higher than the world population growth rate in the same interval (Alves 2017). The relationship between population size and drinking water demand is one of the critical issues facing humanity in the first half of the twenty-first century. The most relevant variables relate to the way water resources are used (water policy), patterns of population dynamics, population flows and population densities (Hunter 2000a, b), and the effects of climate change, partly caused by human interference.

The finitude of the planet's resources, spaces and ecosystems sets limits to population growth. For water scarcity, however, the existing links between population and water "are more complicated than an apparently simple 'more population = more water resource problems'" (Carrol 2010). For the case of Brazil, the Amazon region is a typical case of plenty water abundance, if one compares the low density of population living there with the huge amounts of precipitation and river discharge along the year. By the other hand, people living in southeastern Brazil exerts a strong pressure on water resources, due to high population concentration, living mainly in big cities and state capitals.

The Guarani aquifer system (GAS) designation identifies a transnational groundwater reservoir occupying an area of over one million km$^2$, comprising most of the south-central and southeastern regions of Brazil (almost ¾ of the total SAG area). The name honors, as a direct reference to, the broad spatial domain of the ancient nation of the Guarani American Indians, who occupied the entire southern central portion of South America (Fig. 1) before colonization. The GAS has a wide distribution (still undivided today) in the northeastern territory of Argentina and a smaller portion in eastern Paraguay and northwestern Uruguay. Hosted in porous and permeable rocks of diverse Mesozoic sandstones deposited under dry conditions, evolving from fluvial up to

C. D. R. Carneiro (✉)
Geosciences Institute, University of Campinas,
Campinas, SP, Brazil
e-mail: cedrec@unicamp.br

São Paulo State Council of Geologic Monuments (CoMGeo),
São Paulo, SP, Brazil

L. C. de Souza-Fernandes
Faculty of Applied Sciences, University of Campinas,
Limeira, SP, Brazil



M. Abrunhosa et al. (eds.), *Advances in Geoethics and Groundwater Management: Theory and Practice for a Sustainable Development*, Advances in Science, Technology & Innovation,
https://doi.org/10.1007/978-3-030-59320-9_13

desertic conditions, the waters are unprotected, although they can supply countless cities in these four South American countries (Rocha 1997), particularly outcrop areas of the lithostratigraphic units: Pirambóia Formations and Botucatu Formations (Fig. 1).

The authors actively contribute to the "GAS Science Diffusion Program" (Gonçales et al. 2016, 2018), an education and teaching initiative dedicated to protect the reserves of the Guarani aquifer system. The core concepts of the program strongly rely on geoethics.

## 2 Changing Attitudes Toward Water Use and Conservation

Contamination of surface water resources in south-southeastern Brazil and the seasonal variations of rain intensity are critical factors controlling the availability of drinking water resources. Thus, the region presents a clear example of undesirable interaction of three variables: poor water quality, seasonal availability restrictions and bad sanitation conditions. These factors are among the environmental causes of a significant "burden of death, disease and disability—particularly in developing countries" (World Health Organization 2020). The main question is: "what would be better to learn? Facing water scarcity or avoiding it?" (Rosa Filho 2005).

A solution considered by local and national government authorities for public water supply refers to the appropriation of the pristine underground reserves of GAS. Gonçales et al. (2018) show that this water, as a public good, has been seriously threatened, especially given the hypothesis of intensification of the use of GAS reserves in Brazil.

The environmental challenge represented by scarcity of water resources requires "an urgent need for changes in attitudes towards water use and conservation through sustainable use of this resource" (Cardoso et al. 2015). Initiatives addressed to promote attitudinal changes must affect the perception and sensitization not only of the population but also of the public managers.

## 3 Geoethics

The "GAS Science Diffusion Program" plans to make society aware that water is not inexhaustible, although being abundant in Brazil. Some environmental education programs are included in this initiative. The inspiration on geoethics is clear. Geoethics can be understood as a (social and individual) responsibility toward the Earth; in other words, it is an ethics before the planet (Peppoloni and Di Capua 2017). In spite of the fact that an "ethical behavior is mainly about making the right choices" (Marone and Peppoloni 2017),

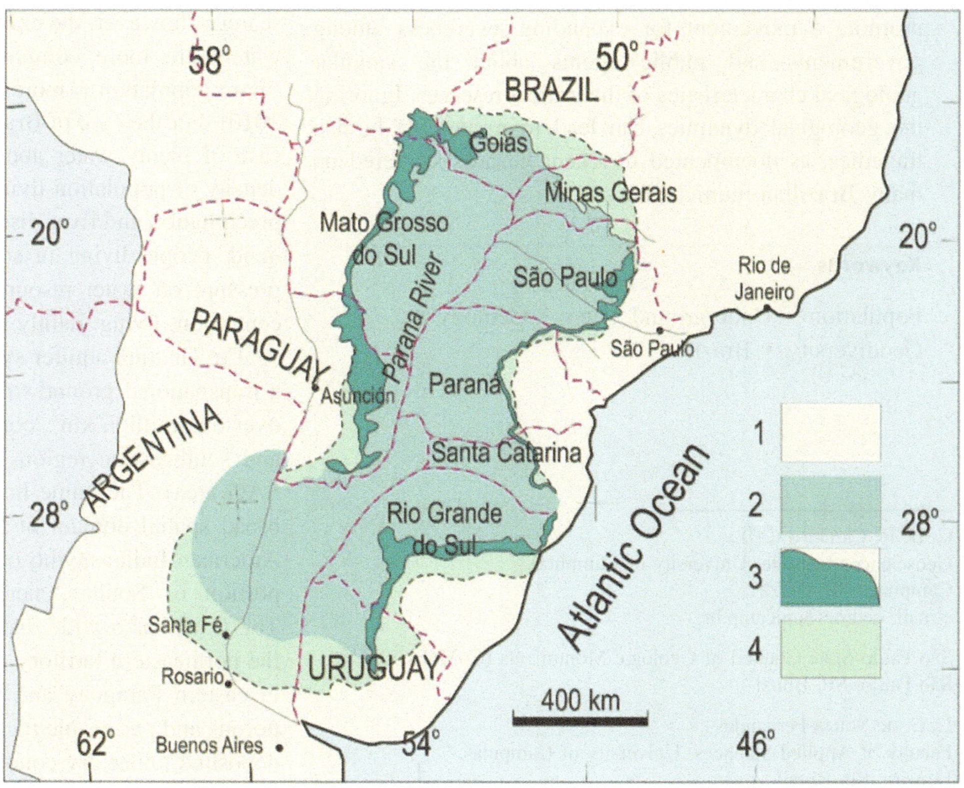


**Fig. 1** Distribution of the Guarani aquifer system in the South America. Conventions: (1) Basement: igneous and metamorphic rocks of pre-Silurian age; (2) Serra Geral Formation: flood basalts confining sedimentary beds, which host the Guarani aquifer system waters; (3) Outcrops of the Pirambóia and Botucatu Formations (fluvial to desertic sandstones), and their equivalents outside the Brazilian border; (4) Limits of the Paleozoic–Mesozoic-age Parana Basin. Modified from Assine et al. (2004), Carneiro and Campos (2012)

reaching a solution for the specific question of supplying water to a large population is not simple, nor immediate.

Two factors of great importance can explain the heterogeneous distribution of surface and underground water resources in Brazil: the highly diversified geological framework, added to the particular geographical position of the territory (Carneiro and Campos 2012, p. 797). Both factors result from the Brazilian geological evolution, which determines a special geological heritage, and, therefore, the singular geodiversity of the country.

Moreover, the GAS reserves can be considered part of a continental-wide geological heritage. In the twenty-first century, the preservation of geo-biodiversity and geological heritage has gained relevance, giving rise to new professional challenges for geologists and geological engineers on fields such as geotourism, geoethics, geoconservation, territorial planning, teaching, geosciences dissemination and education (Brilha 2009).

Many environmental decisions put on evidence an urge to balance consequences of each alternative considered. Choosing between distinct options is not always simple: "We cannot make the mistake of considering Geoscience knowledge as a universal law, thinking that we could solve any geoethical dilemma based on it and only in it" (Marone and Peppoloni 2017). As long as the economic appropriation of accumulated water in underground spaces is an issue of general interest to society, it is important to open a debate as well as to raise awareness among governments and public agents about geological and environmental issues. Ignoring the geological dynamics can lead to irreparable human tragedies, as documented in many Brazilian municipalities.

This is a continental-size country; so, the geological conditions (which constitute its rich heritage) are distinct from one region to another. Since the majority of the population is concentrated in a range of 200 km from the coastline, education and teaching on geology and geosciences contribute to the debate and decision making on such subjects.

## 4 Discussion

This paper focuses on critical issues related to the teaching and learning of underground water reserves, a specific geology subject. The analysis focuses on the importance of geoethics, geological heritage and geodiversity, fields which attract the attention of legal experts. Due to the undeniable usefulness of this learning for the exercise of Environmental Law, the authors believe that these concepts are of broad interest.

## 5 Concluding Remarks

The concept of geoethics recognizes the human responsibility toward the Earth, a consequence of the increasing human interference with ecosystems and life forms. The rich regional geodiversity of Brazil reflects geological conditions, which result from a heritage built over millions of years.

Supporting research on the underground water dynamics, especially due to the expected population increase in the twenty-first century, must not be left aside the immense education challenge, especially about geology. Reintroducing geology and Earth science into the Brazilian basic school curricula can develop in children, youth and adults a better understanding of how the planet works and prepare them effectively for a full exercise of citizenship.

**Acknowledgements** The first author, CDR Carneiro, thanks the National Council for Scientific and Technological Development (CNPq) for a Research Productivity grant. The authors also acknowledge the Scientific Initiation Scholarship Program of CNPq for Scientific Initiation grants, which enabled the research and the academic maturation of several undergraduate students in geology.

## References

Alves JED (2017) População e economia nos 200 anos da Independência do Brasil: 1822-2022. Juiz de Fora: Univ. Fed. Juiz de Fora, Lab. Demografia e Estudos Populacionais. https://www.ufjf.br/ladem/2017/05/05/populacao-e-economia-nos-200-anos-da-independencia-do-brasil-1822-2022-artigo-de-joseeustaquio-diniz-alves/

Assine ML, Piranha JM, Carneiro CDR (2004) Os paleodesertos Pirambóia e Botucatu. In: Mantesso Neto V, Bartorelli A, Carneiro CDR, Brito-Neves BBde (orgs) (2004) Geologia do Continente Sul-Americano: Evolução da obra de Fernando Flávio Marques de Almeida. Ed. Beca, São Paulo, pp 77–93. (Cap. 5)

Brilha JBR (2009) A Importância dos Geoparques no Ensino e Divulgação das Geociências. Geologia USP, Publ Espec 5:27–33. https://www.journals.usp.br/gusppe/article/download/45388/49000. Access 04 Feb. 2019

Carneiro CDR, Campos HCNS (2012) Recursos hídricos subterrâneos. In: Hasui Y, Carneiro CDR, de Almeida FFM, Bartorelli A (eds.) Geologia do Brasil. Ed. Beca, São Paulo, pp 797–813. (Cap. 31)

Cardoso FA, Vieira SR, de Morais JL (2015) Recursos hídricos como temática para Educação Ambiental: reflexões a partir de um curso empregando Tecnologias da Informação e Comunicação. In: III Congresso Internacional de Educação Ambiental dos Países e Comunidades de Língua Portuguesa, Torreia, Murtosa, Portugal, 8 a 12 de julho, 2015. Livro de Resumos. Comunicações Orais. pp 590–591

Carrol L (2010) Population and water: where does it matter? People Place 18(3):14–26. https://theaustralianpopulationresearchinstitute.files.wordpress.com/2014/12/v18n3_2carroll.pdf

Gonçales VG, Carneiro CDR, Balsalobre B, Pereira SY, Fernandes LCS, Gastmans D, Bartorelli A, Vilela RCCL, Mantesso Neto V, Anelli LE, Piranha JM (2018) Programa Aquífero Guarani:

unindo água subterrânea e história da Terra à consciência ambiental. Campinas: Programa de Pós-Graduação em Ensino e História de Ciências da Terra (PEHCT) do Instituto de Geociências, Unicamp. Ed. CRV, 108p. (Livro, Série Ensino e História de Ciências da Terra, 2). (ISBN 978-85-444-2556-5)

Gonçales VG, Carneiro CDR, Balsalobre B, Pereira SY, Fernandes LCS, Vilela RCCL, Mantesso Neto V, Bartorelli A, Simonato MD, Anelli LE (2016) Programa Aquífero Guarani: difusão de hidrogeologia nas praças públicas e escolas. In: Congresso Brasileiro de Águas Subterraneas, 19, Campinas, Campinas, ABAS. http://docplayer.com.br/29736375-Xix-congresso-brasileiro-de-aguas-subterraneas-programa-aquifero-guarani-difusao-de-hidrogeologia-nas-pracaspublicas-e-escolas.html

Hunter LM (2000a) Population and environment: A complex relationship. Santa Monica, CA: RAND Corporation. https://www.rand.org/pubs/research_briefs/RB5045.html

Hunter LM (2000b) The environmental implications of population dynamics. Santa Monica, CA: RAND Corporation. https://www.rand.org/pubs/monograph_reports/MR1191.html

Marone E, Peppoloni S (2017) Ethical dilemmas in geosciences. We can ask, but, can we answer. Ann Geophys 60, Fast Track 7. https://doi.org/10.4401/ag-7445

Peppoloni S, Di Capua G (2017) Geoethics: ethical, social and cultural implications in geosciences. Ann Geophys 60, Fast Track 7. https://doi.org/10.4401/ag-7473

Rocha GA (1997) O grande manancial do Cone Sul. Estudos Avançados USP 11(30):191–212. http://dx.doi.org/10.1590/S0103-40141997000200013

Rosa Filho EF (2005) Águas do Guarani II. Ernani Francisco da Rosa Filho. Ecoa—Ecologia e Ação. https://riosvivos.org.br/a/Noticia/Aguas+do+Guarani+II+++Ernani+Francisco+da+Rosa+Filho/6730. Access 20.01.2018

World Health Organization (WHO) (2020) Environment and health in developing countries. WHO. https://www.who.int/heli/risks/ehindevcoun/en/. (Health and Environment Linkages Policy Series)

# Geoecological Mapping to Identify Groundwater Ecosystem Services Conflicts in a Brazilian Municipality

João Vitor Roque Guerrero, Alberto Gomes, José Augusto di Lollo, Reinaldo Lorandi, and Luiz Eduardo Moschini

**Abstract**

This study applies geoecological-based mapping techniques to identify the natural potential of the landscape to promote groundwater ecosystem services. In addition, this diagnosis analyzed its relationship with land use to identify groundwater geoethical conflicts in Brotas municipality, Brazil, which is completely included in the Guarani aquifer system, one of the largest world natural groundwater reservoirs. In the analysis, we used spatial data of geology, soils, land use, terrain forms, DEM and lineament density. The results of land use analysis indicate that the main economical drive force for the study area is the sugarcane production. On the other hand, the analysis shows that local geoenvironmental conditions of this region are favorable to the groundwater production. Finally, the geoethical conflicts chart, produced from the interaction between land use and the potential to provide groundwater ES's chart, showed that 59% of the study area has geoethics groundwater conflicts, i.e., places where anthropic activities endanger the quality and availability of groundwater ecosystem services. The results obtained can support decision making in Guarani aquifer areas for several countries as Brazil, Argentina, Uruguay and Paraguay.

**Keywords**

Geoethics • Ecosystem services • Groundwater management • Guarani aquifer system

J. V. Roque Guerrero (✉) · R. Lorandi · L. E. Moschini
Department of Environmental Sciences, Federal University of São Carlos, São Carlos, Brazil
e-mail: jvguerrero2@gmail.com

A. Gomes
Department of Geography—CEGOT, University of Porto, Porto, Portugal

J. A. di Lollo
Department of Civil Engineering, São Paulo State University, Ilha Solteira, Brazil

## 1 Introduction

In Brazil, since the 1970s, land use changes, mainly promoted by the expansion of agriculture and urbanization (Lollo et al. 2019), have been the main driving force that induces the degradation of ecosystem services balance, considering the benefits that natural areas and resources provide to humans (MEA 2005; Lollo et al. 2019).

Among the main services provided by ecosystems in Brazil, there is the provision of drinking water by the Guarani aquifer system (GAS), one of the largest groundwater reservoirs on the planet (Hirata et al. 2007).

Despite the large availability of water provided by the GAS, the ES's delivered by this natural resource are increasing the vulnerability due to intensive use of the physical environment without any planning and the slackening of environmental laws in favor to increased profits from agricultural activities (Covre et al. 2017; Guerrero et al. 2019) and situation that reveal the existence of potential geoethical conflicts.

Geoethical conflicts can be defined as reflections about the negative results of the interaction of anthropic activities with the geosphere (Peppoloni and Capua 2015).

Cartography based on landscape geoecology like Rodriguez et al. (2007) is defined as a set of research methods, techniques and procedures that consist of obtaining knowledge about the natural environment and establishing operational diagnoses which emerges as a tool to provide support for geoecological planning of landscapes and ecosystem services to mitigate geoethical conflicts, including those related to groundwater exploration and vulnerability.

Thus, the objective of this work was to use geoecological techniques using GIS to identify the potential of the landscape to produce groundwater ecosystem services and identify geoethical conflicts that occur in Brotas, Brazil. The results obtained are expected to aid decision making in Guarani aquifer areas of Brazil.



M. Abrunhosa et al. (eds.), *Advances in Geoethics and Groundwater Management: Theory and Practice for a Sustainable Development*, Advances in Science, Technology & Innovation,
https://doi.org/10.1007/978-3-030-59320-9_14

## 2 Materials and Methods

### 2.1 Study Area

The municipality of Brotas is located in the central region of São Paulo state, southeastern Brazil (Fig. 1), with an area of 1103 km$^2$ and an estimated population of 24,160 inhabitants. The municipality is located over the sedimentary basin of Paraná, and the local morphostructure refers to the western plateau of São Paulo. The geologic–geomorphological framework reflects the lithological units and aquifer units described at Fig. 1.

### 2.2 Methods

In order to map the landscape potential to promote ecosystem services of groundwater production, geoecological-based mapping techniques were used based on the adaptation of the methodology proposed by Andualem and Demeke (2019).

The production of the Groundwater Ecosystem Services Potential Chart (GESPC) first required the weights assignment representing the ability of each attribute of the parameters (e.g., geology, soils) to assist in groundwater production. In addition, values of relative importance were also attributed between the parameters themselves, indicating which ones have the greatest influence on the production of groundwater ES's. All the values assigned are shown in Fig. 2a.

Parameters' overlap was performed using the analytic hierarchy process (AHP), a mathematical theory that allows to organize and evaluate the relative importance between criteria and to measure the consistency of judgments (Saaty 1990; Moreira et al. 2001).

In order to understand how human appropriation of local territory occurs, we used land use data from the MAPBIOMAS project (https://mapbiomas.org/).

Finally, we produced a geoethical conflict chart following the lines proposed by Brown and Reymond (2014), which illustrates the results from the paired comparison between the landscape potential of providing groundwater supply

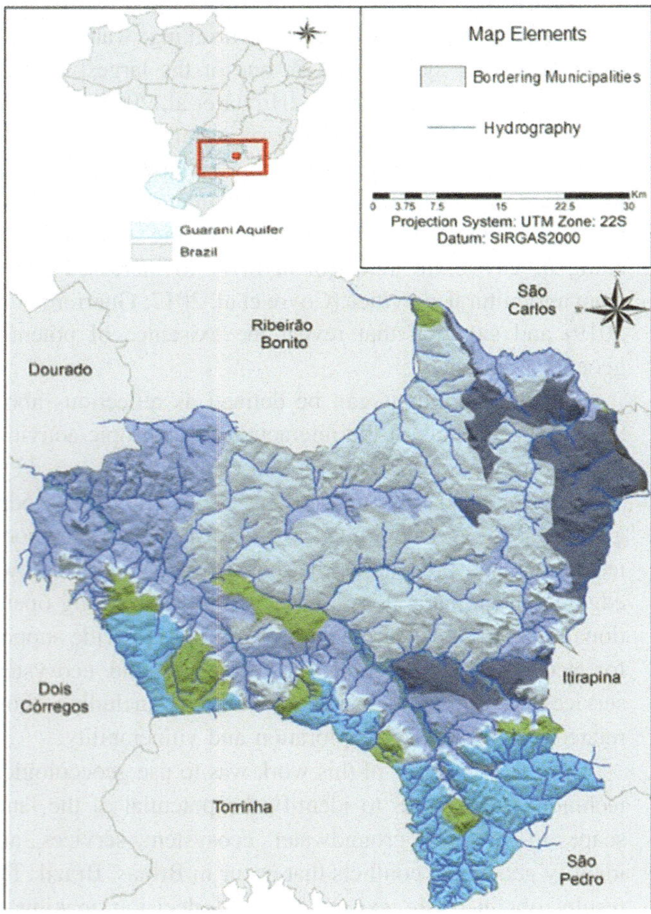


Granular Aquifer Units

| | | |
|---|---|---|
| Aluvionar | Sedimentary, Porous, Free | 26.27 km² - 2% |
| Eluvionar | Sedimentary, Porous, Free | 124.9 km² - 11% |
| Itaqueri | Sedimentary, Porous, Free | 149 km² - 14 % |
| Botucatu | Sedimentary, Porous, Free | 460.09 km² - 42 % |
| Pirambóia | Sedimentary, Porous, Free | 274.14 km² - 25 % |

Fractured Aquifer Units

| | | |
|---|---|---|
| Serra Geral | Fractured, Partially Free | 66.025 km² - 6% |

| Geological Units | Description |
|---|---|
| **Botucatu** | Sandstones formed in desert environment, composed by fine to medium clasts of quartz well rounded. Thickness varies between 20 and 280 meters. Formation with high porosity and high capacity to store and supply water. |
| **Pirambóia** | Deposits of wind and fluvium-wind genesis formed in desert environment, linked with a field of wind dunes. Composed by thick reddish, yellowish and whitish sandstones, with fine to medium clasts and high porosity and high groundwater stores capacity |
| **Itaqueri** | Silicified and stratified sandstones and conglomerates. Composed by rudaceous deposits of alluvial fans comprising the southern region of the municipality of Brotas. its porosity and groundwater storage capacity are lower than the Guarani aquifer formations. |
| **Serra Geral** | Basaltic spills, interspersed sandstones of the Botucatu Formation, forming the so-called Serra Geral Basin; such formation also incorporates intrusive rocks associated with the volcanism process occurred in the Paraná sedimentary basin, consisting of dykes and diabase sills. Presents low porosity and storage capacity, where infiltration is restricted to basalt fractures |
| **Alluvial Deposits** | Alluvial deposits formed primarily by transported and deposited sandy sediments located in plains allocated in the valley bottoms and surrounding areas. Its material is so porous and disintegrated that it is not able to store or transmit water |
| **Eluvial Deposits** | Eluvial deposits composed by chaotic and poorly selected sediments ranging from blocks and fragments to clays. Deposits resulting from mechanical breakdown of rocks and slopes. Its hydrogeological characteristics depend on its original material. However, they have high porosity and low storage capacity |

**Fig. 1** Study area corresponding to Brotas municipality, Brazil

a

| *Atributes* | *Groundwater ES's Potential* | | |
|---|---|---|---|
| | Low | Medium | High |
| Rock Units | Elluvial and Alluvial Deposits | Serra Geral, Itaqueri | Botucatu , Piramboia |
| Soil Types | Lithic Entisols; Quartzipsamments Entisols; Histosols | Red Oxisoils Sandy Oxisoils; Inceptisols (Glei) | Ultisols Red-Yellow Oxisoils Médium Oxisoils |
| Slope | >20 % | 5 – 20 % | 0 – 5 % |
| Drainage Density | 0.63 – 0.75 | 0.57 – 0.63 | 0 – 0.57 |
| Terrain Forms | Convergent-Concave; Rectilinear-Convergent; Convergent-Convex | Rectilinear planar; Convex-Planar; Concave-Planar | Divergent-Concave; Divergent-rectilinear; Divergent-Convex |
| Linement Density (km²) | 0 – 0.1 | 0.1 – 0.2 | 0.2 – 0.25 |

b

| Land Use Type | Value |
|---|---|
| Forests | 1 |
| Savanna | 2 |
| Water | 3 |
| Grassland | 4 |
| Pasture | 5 |
| Forestry | 6 |
| Non Vegetaded | 7 |
| Urban | 8 |
| Agriculture | 9 |

+

| ES Potential | Value |
|---|---|
| Low | 10 |
| Medium | 100 |
| High | 1000 |

=

| Result Value | Conflict | Result Value | Conflict | Result Value | Conflict |
|---|---|---|---|---|---|
| 11 | * | 101 | * | 1001 | * |
| 12 | * | 102 | * | 1002 | * |
| 13 | * | 103 | * | 1003 | * |
| 14 | * | 104 | ** | 1004 | ** |
| 15 | * | 105 | ** | 1005 | *** |
| 16 | * | 106 | ** | 1006 | **** |
| 17 | * | 107 | *** | 1007 | **** |
| 18 | * | 108 | *** | 1008 | **** |
| 19 | * | 109 | *** | 1009 | **** |

Where :

| | | | |
|---|---|---|---|
| * | No Conflict | *** | High Conflict |
| *** | Low Conflict | **** | Highest Conflict |

**Fig. 2** Groundwater potential table (**a**) and geoethical conflict workflow (**b**)

ES's and the current pattern of land use. For this, we assign dummy values to the attributes of each map and analyze the result of the sum between them from the perspective of what kind of conflict it represents (Fig. 2b).

The geoethical conflict chart produced analyzes spatially the resultant relationships between the interaction of anthropic activities (land use) with the geosphere (groundwater ES's potential), considering negative results as potential conflicts.

## 3 Results and Discussion

From the analysis of the study area physical environment dynamics, we conclude that 50% of the territory as high potential to provide significant groundwater ecosystem services (Fig. 3b), meaning that in these areas, the combination of landscape elements results in maximum groundwater potential. This result is compatible with those published by Guerrero et al. (2019) and Hirata et al. (2007), which can be explained by the predominant presence of the Botucatu and Pirambóia formations (units that makes up the Guarani aquifer system), terrain slopes and shapes that facilitate infiltration rather than runoff (Costa et al. 2019; Wendland et al. 2015).

The areas mapped as medium potential indicate that landscape can produce groundwater; however, the elements analyzed show that the capacity is lower than those previously presented. These sites represent 44% of the total and occur mainly for two conditions: (1) presence of the geological formations Itaqueri and Serra Geral (forming the Serra Geral aquifer), which, due to their porosity regulated by discontinuities as the main hydrodynamic feature, the recharge is slower than in the Botucatu and Pirambóia sandstones (Giampá and Souza 1982); (2) where despite favorable slopes and terrain, extremely sandy rock units such as alluvial and eluvial deposits occur, which have very low water retention capacity (IAC 2015).

Finally, we have areas where there is little or no groundwater storage capacity which is classified as low potential, making up 6% of the territory and occurring by the combination of low water retention capacity of soils and rock units, high slopes and terrain forms that increase runoff.

From the land use chart (Fig. 3a), we identified main driving force of the municipality which is farming activity, considering that 78% of the territory is occupied by agriculture (mainly sugar cane), pasture and forestry (Eucalyptus). Natural formations such as forests, savannah and grasslands cover only 22% of the area, indicating the high degree of anthropization of the municipality toward unsustainable activities.

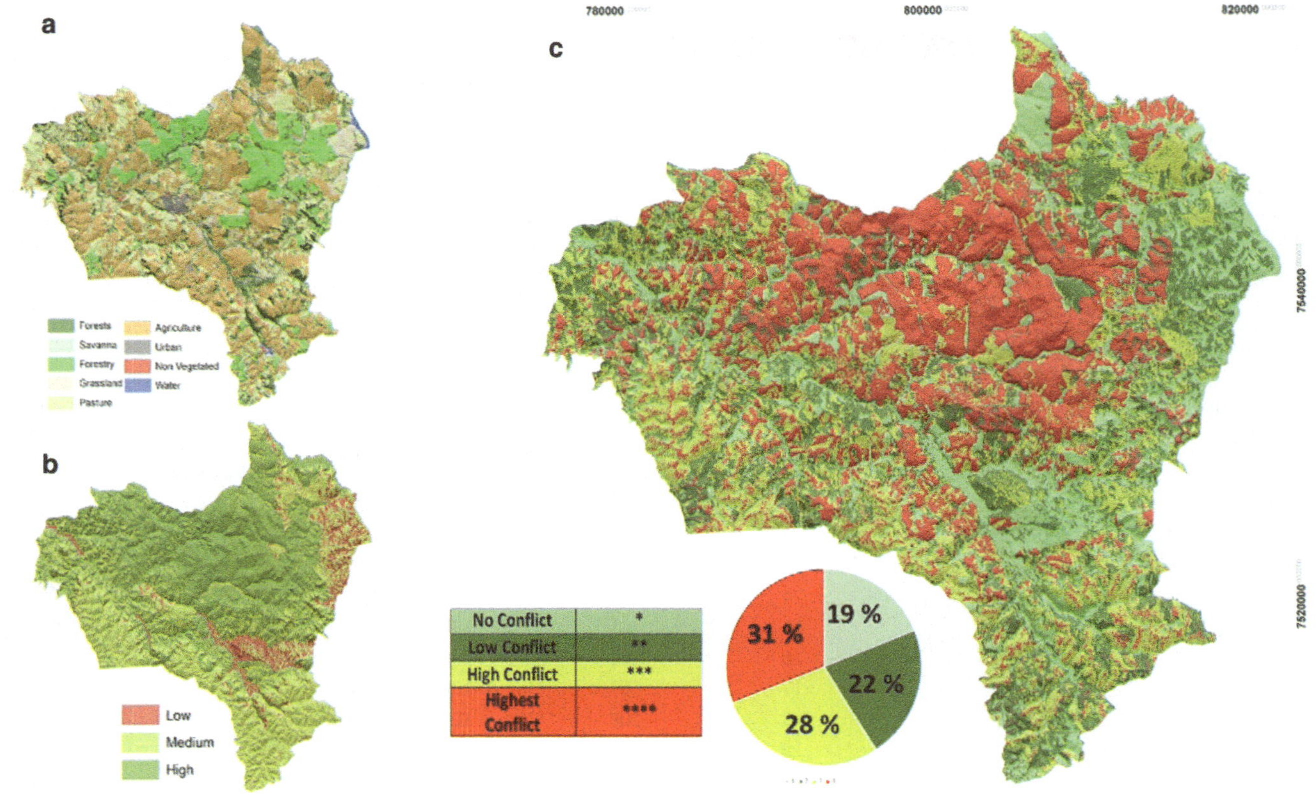


**Fig. 3** Results: charts of land use (**a**) and potential to provide groundwater ES's (**b**) and geoethical conflicts (**c**)

Our analysis shows that 59% of the territory consists of areas with high and higher geoethical conflict. These geoethical conflicts essentially occur in areas where the potential for landscape to promote ecosystem groundwater services is maximum and land use is comprised of sugarcane, urban, exposed soil and forestry.

Sugarcane (which covers 42% of the Brotas municipality) is one of the largest commodities produced in the state of São Paulo (Carvalho et al. 2013), not only responsible for sugar production, but also for the production of ethanol, one of Brazil's main fuels (Kohlhepp 2010).

Despite the strategic importance of sugarcane to the local economy, its production becomes the main geoethical conflict in the area from the unrestricted use of pesticides (Acayaba 2017; Fialho et al. 2018), overexploitation of groundwater for irrigation and the flexibility of environmental legislation in favor of higher productivity (Covre et al. 2017), which puts great pressure on surface and underground water resources (Acayaba 2017).

Areas with low (22%) or no conflict (19%) generally occur in locations where the potential for ecosystem service provision is medium or low and/or where land use is composed of natural formations, like forests, savanna and grasslands.

## 4 Concluding Remarks

The municipality of Brotas has natural conditions that favor the availability of ecosystem services related to groundwater supply, especially related to the presence of rock units of the Botucatu and Pirambóia formations, part of the GAS. However, we identified agricultural activity as the major driving force that governs the local economic dynamics, especially large areas of sugarcane production.

Our study identified that sugarcane cultivation without adequate territorial planning, combined with the lack of a conservationist policy in the municipality of Brotas, generated geoethical conflicts related to groundwater, jeopardizing the availability of this essential resource, including for agriculture itself.

Spatial identification of geoethical conflicts is an important process in the management of groundwater ecosystem services, making it possible to direct territorial planning actions to areas where conflicts endanger the quality of this essential resource for human well-being.

The results obtained here are expected to assist decision making in municipalities with conflicts between land use, groundwater dynamics and ES's provided by groundwater, especially for the multinational Guarani aquifer system.

**Acknowledgements** This work was supported by grant #2016/19020-0, São Paulo Research Foundation (FAPESP).

## References


Acayaba FD (2017) Ocorrência de agrotóxicos usados na cana-de-açúcar em corpos d'água do Estado de São Paulo. Campinas State University, Campinas (MSc Thesis)

Andualem TF, Demeke GG (2019) Groundwater potential assessment using GIS and remote sensing: a case study of Gunatana landscape, upper Blue Nile Basin, Ethiopia. J Hydrol Reg Stud 24:100610 (2019)

Brown G, Raymond CM (2014) Methods for identifying land use conflict potential using participatory mapping. Landsc Urban Plann 122(2014):196–208

Carvalho LC, de Freitas Bueno RCO, Carvalho MM, Favoreto AL, Godoy AF (2013) Cana-de-açúcar e álcool combustível: histórico, sustentabilidade e segurança energética. Enciclopédia Biosfera 9 (16):530

Costa CW, Lorandi R, Lollo JA, Santos WS (2019) Potential for aquifer contamination of anthropogenic activity in the recharge area of the Guarani Aquifer System, southeast of Brazil. Groundwater Sust Dev 8:10–23

Covre J, Clemente F, Lirio VS (2017), New Brazilian forest code: changes and prospects, Revista de Desenvolvimento Econômico—RDE—Ano XIX 3(38):267–278

Fialho ML, Adorno PA, Lima JIC, Reis KP, Oliveira RB (2018) O uso de agrotóxicos na cultura de cana-de-açúcar e os principais riscos à saúde do trabalhador rural. Revista Científica Integrada v2 n4

Giampá CEQ, Souza JC (1982) Potencial Aquífero dos Basaltos da Formação Serra Geral. Revista DAE v1n1xxxx

Guerrero JVR, Lorandi R, Moschini LE, Gomes AAT, Lollo JÁ (2019) Carta de unidades aquíferas do município de Brotas, SP, Brasil: subsídios ao planeamento territorial. In Carta De Unidades Aquíferas Do Município De Brotas, Sp, Brasil: Subsídios Ao Planeamento Territorial, 2° Seminário Águas Subterrâneas p1-4, Coimbra University, Coimbra

Hirata R, Suhogusoff A, Fernandes A (2007) Groundwater resources in the State of São Paulo (Brazil): the application of indicators. Anais Da Academia Brasileira De Ciências 79(1):141–152

IAC (2015) Solos do estado de São Paulo. https://www.iac.sp.gov.br/solossp/. Last accessed 2019/09/27

Kohlhepp G (2010) Análise da situação da produção de etanol e biodiesel no Brasil, estudos avançados 24 (68)

Lollo JA, Guerrero JVR, Abe ACP, Lorandi R (2019) Land change, soil degradation processes, and landscape management at the Clarinho River watershed, Brazil. In: IAEG/AEG Annual Meeting Proceedings, San Francisco, California, 2018, vol 2

MEA- Millenium Ecosystem Assessment (2005) https://www.millenniumassessment.org/en/Synt. Last accessed 2019/09/27

Moreira FR, Barbosa C, Camara G, Almeida Filho R (2001) Inferência Geográfica e Suporte à decisão, In: Introdução à ciência da Geoinformação, 1 edn, São José dos Campos

Peppoloni S, Di Capua G (2015) The meaning of geoethics. In: Geoethics: ethical challenges and case studies in Earth Sciences pp 3–14

Rodriguez JMM, Silva EV, Cavalcanti APB (2007) Geoecologia das Paisagens, 2nd edn. UFC edições, Fortaleza

Saaty TL (1990) How to make a decision: The analytic hierarchy process. Eur J Oper Res 48(1):9–26

Wendland E, Gomes LH, Troeger U (2015) Recharge contribution to the Guarani Aquifer System estimated from the water balance method in a representative watershed. Anais Da Academia Brasileira De Ciências 87(2):595–609

# Current Status and Future of Groundwater Management in Japan

Makoto Nishigaki, Shusuke Oji, and Hironori Hara

**Abstract**

In Japan, groundwater has been used as a water resource since the 1900s. As a result, land subsidence occurred due to consolidation of urban clay layers. To deal with this issue, each city set conditions for groundwater pumping control to suppress land subsidence. This has raised the groundwater level. The Basic Law on the Water Cycle was established in 2014 for groundwater management. Under this law, groundwater in groundwater basins that are managed and monitored is treated as "public water". However, at present in Japan, groundwater is still "private water." In 2019, the government granted permission for the use of groundwater to heat and cool buildings in the Osaka area. Such use is expected to be developed throughout Japan. Under such circumstances, how to conserve and manage groundwater for each groundwater basin is a major issue in Japan. This paper describes the followings: (1) Japan's past groundwater use situation and countermeasures; (2) research on measuring method of seepage characteristics and monitoring method of groundwater for multiple aquifers; (3) establish effective groundwater use plan and management.

**Keywords**

Japan • Groundwater • Management • Land subsidence

M. Nishigaki (✉)
Graduate School of Environmental and Life Science, University of Okayama, Okayama, Japan
e-mail: n_makoto@cc.okayama-u.ac.jp

S. Oji · H. Hara
Chuo Kaihatsu Corporation, Tokyo, Japan

## 1 Introduction

Since the early 1900s, people have dug deep wells and used groundwater in alluvial plains throughout Japan (Miyake 2019). As a result, land subsidence occurred in various parts of Japan. In order to prevent land subsidence due to excessive pumping of groundwater, the following laws were set and implemented:

1. Industrial water law (1956)
2. Revised industrial water law (1962)
3. Law on the regulation of building groundwater extraction (1962).

With such laws, land subsidence due to groundwater use in large cities has been reduced (Ota 2019). However, the government has not made any efforts such as long-term artificial recharge for effective use of groundwater. The administration has not yet considered the relationship with the river, which is the source of groundwater.

This work discusses the current state of such groundwater management in Japan. Then, the future direction is also outlined.

## 2 Examination of Guidelines for Groundwater in Japan

### 2.1 Background of the Study

The Basic Act on the Water Cycle was promulgated by the Government of Japan on April 2, 2014, and came into effect on July 10, 2014. At the same time, the government established the Cabinet Secretariat Water Cycle Policy Headquarters in order to promote the water cycle in a concentrated and comprehensive manner.

After that, the Cabinet decided on July 1, 2015, by the Japanese government, the Basic Plan for the Water Cycle, a



M. Abrunhosa et al. (eds.), *Advances in Geoethics and Groundwater Management: Theory and Practice for a Sustainable Development*, Advances in Science, Technology & Innovation,
https://doi.org/10.1007/978-3-030-59320-9_15

basic plan for comprehensive and systematic promotion of water cycle measures.

Japan's "Groundwater Conservation" Guidelines are designed to support groundwater conservation measures by local governments and other organizations with a focus on groundwater conservation, which is a component of the water cycle, in line with these movements. It was compiled by the Ministry of the Environment in 2016 to summarize the issues and the basic concept of groundwater conservation.

The Basic Plan for Water Cycle will be reviewed in 2020 for priority measures and new perspectives. Among them, necessary changes are made by the government, assuming the issues related to the water cycle in the region, the risk of drought and floods due to the effects of climate change, and environmental changes surrounding the water cycle.

For this reason, it is necessary to review the groundwater conservation guidelines assuming future groundwater conservation considering increasing awareness of the water cycle, climate change, technological development, and water cycle measures.

In examining the items to be revised in these guidelines, the five basic principles of the Basic Law on Water Cycles [(i) importance of water cycle, (ii) publicity of water, (iii) consideration on sound water cycle, (iv) comprehensive basin management, (v) international cooperation on water cycles) regarding recent Ministry of the Environment measures (basic plans, etc.)]. As a direct work, it will be coordinated with the detailed examination of the measures of the Water Cycle Basic Plan and the Fifth Environmental Basic Plan of 2018, but also the review of the Water Cycle Basic Plan by the Water Cycle Policy Headquarters, groundwater management cases and national land. It is necessary to review guidelines with reference to the Ministry of Transportation's data on water circulation.

#### Reference Materials

In reviewing the guidelines for groundwater conservation, the following are the main documents that refer to measures related to water circulation, the current status of initiatives, the current state of groundwater conservation, and ideas.

1. ***Ministry of the Environment***
   - Groundwater Conservation Case Studies March 2015
   - Fifth Basic Environmental Plan April 2018
   - Toward the formation of a regional circulation zone October 2018.
2. ***Water Cycle Policy Headquarters***
   - Basic Plan for Water Cycle July 2015
   - A collection of case studies on the water cycle April 2016
   - Recommend introduction of groundwater management April 2017
   - 2018 Water Cycle Measures (White Paper on Water Cycle) June 2018
   - Recent water cycle measures initiatives July 2018
   - Basic Management Guide, Case Studies July 2018
   - How to promote consensus on groundwater management July 2018
   - Reviewing the Basic Plan for Water Cycle September 2018
   - The 1st meeting of experts on promoting water cycle measures October 2018.
3. ***Ministry of Land, Infrastructure, Transport, and Tourism***
   - Results of a survey on groundwater-related ordinances October 2018.

## 3 Outline of Basic Water Cycle Plan: a Conceptual Law Approach

The Basic Plan on Water Cycle in Japan shows the basic policy of the water cycle and the measures that the government should take comprehensively and systematically in relation to the water cycle (Tanaka 2014). That approach aims a holistic approach toward realizing a balanced Integrated Water Resources Management (IWRM). The main contents related to the groundwater field are summarized below.

The Basic Plan for Water Cycle consists of three parts. Part 1 corresponds to the five basic principles of the Basic Act on Water Cycles, and Part 2 corresponds to the basic measures of the Basic Act on Water Cycles.

In particular, efforts in the field of groundwater, which is the purpose of this project, will focus on measures related to "promotion of sustainable conservation and use of groundwater."

The Basic Plan for Water Cycle is a comprehensive plan of water basin water circulation for both surface water and groundwater. For groundwater, in Part 1, "Basic Policies of Water Cycle Measures" are addressed. "Ensuring the proper use of water and the enjoyment of the benefits of water" describes the following as the promotion of sustainable conservation and use of groundwater.

In the first part, groundwater is used for (1) diversity of uses, (2) importance of groundwater conservation and promotion of groundwater use (it is difficult to recover groundwater damage once it has occurred), (3) analysis/visualization, conservation (quality/quantity), recharge. It describes the behaviour of groundwater such as sampling, the actual situation of groundwater used, the establishment of a system for implementing measures, the formation of consensus, and the importance of groundwater management.

Similarly, in Part 2, "Measures to be taken comprehensively and systematically by the government in relation to the water cycle", the government and local governments should take measures in "Promotion of appropriate and effective use of water" (land subsidence, groundwater pollution, prevention of salinization, preservation of ecosystem, utilization as water resources).

In the promotion of sustainable groundwater conservation and use, the government and local authorities shall work together to implement groundwater management in accordance with local conditions and to conduct monitoring, survey, and data analysis.

In addition, based on the guidelines such as prevention of land subsidence, the observation data will be shared with related municipalities and utilization of these wide-area groundwater management mechanisms will be considered.

Based on such basic measures of the Basic Plan for Water Cycle, each local government is formulating a water basin plan for watershed and is promoting measures according to the actual situation in each region.

## 4 Survey on Measures and Activities Related to the Watershed Plan in the Basin

Under the current Water Cycle Basic Plan, local governments, national branch offices, businesses, organizations, and residents will work together to set up a watershed council and formulate a comprehensive "watershed water circulation plan."

As of December 2018, a total of 35 "basin water cycle plans" have been announced. Under the Basic Policy, the two parties are cooperating with each other on measures related to the water cycle, and the measures are being implemented. The measures and activities related to the watershed water circulation plan are summarized below: (1) arrangement of watershed water circulation plan measures; (2) implementation status of measures based on the watershed plan.

## 5 Survey on Measures and Activities Related to Groundwater Management

From the viewpoint of promoting sustainable conservation and utilization, it is necessary to understand the actual conditions such as the amount and behavior of groundwater and to coordinate the formation of consensus among the stakeholders. The water cycle is a concept that integrates surface water, such as river water and lake water, with groundwater. However, groundwater is specifically clarified as "groundwater management" and is controlled by local authorities such as local governments. It is necessary to take necessary surveys and examinations and make consensus efforts while paying attention to groundwater.

As of November 2018, 23, "groundwater use countermeasures councils" were established as members of the groundwater use optimization survey conducted by the Ministry of Economy, Trade, and Industry. In addition, there are groundwater councils for groundwater use and groundwater conservation, such as the Tottori Prefecture Sustainable Groundwater Utilization Council. By investigating and organizing the purpose of establishment and efforts of these councils on groundwater management, information that could be used for studying measures was compiled.

Many of these groundwater management initiatives were established between 1965 and 1975 when land subsidence had become a social problem and were implemented before the enforcement of the Basic Water Cycle Plan, particularly:

1. Organizing the groundwater council including groundwater sampling companies and groundwater users.
2. Land subsidence area and establishment of groundwater use countermeasures council.
3. Arrangement of survey items by the Groundwater Council.
   (a) Land subsidence.
   (b) Information on initiatives and measures of the Groundwater Council.
4. nnSorting out measures and initiatives of representative groundwater councils.
   Substitute groundwater council measures and initiatives include:
   (a) Yamagata Prefecture (Yamagata Basin, Yonezawa Basin).
   (b) Gifu and Aichi Prefectures (Nobi Plain).
   (c) Niigata Prefecture (Minami Uonuma area).

## 6 Concluding Remarks

In this study, we discussed what laws have recently been enacted for groundwater in Japan and what direction the groundwater is going to take in the whole basin. These main results are:

1. Comprehensive management of groundwater basins has not yet been implemented nationwide.
2. Groundwater environmental risk management is controlled by ground subsidence by using groundwater, but the other is not performed much.

3. There is a water source tax on the management of recharge sources, but forest management is not actively pursued.
4. Underground heat utilization as renewable energy tends to increase. However, since there is very little information on the three-dimensional geological model of the ground and the characteristics of seepage and mechanical properties, many issues remain in this field.

**Acknowledgements** Special thanks are due to the valuable guidance from the Ministry of the Environment.

## References

Miyake N (2019) Confined groundwater recovery in mainly Tokyo alluvial lowland of metropolitan area. J Hydrol Syst 114:19–24

Ota K (2019) The current state of the land subsidence, problem and the measure. J Hydrol Syst 114:9–18

Tanaka T (2014) Groundwater governance in Asia: present state and barriers to implementation of good governance. In: Castellarin A, Ceola S, Toth E, Montanari A (eds) Evolving Water Resources Systems: Understanding, Predicting and Managing Water–Society Interactions, Proceedings of ICWRS2014, Bologna, Italy, June 2014, IAHS Publ 364:470–474

# How im(Moral) is the "Nimby" stand? Elements to the Ethics of "environmental" Conflicts

José Rodrigues dos Santos

**Abstract**

The extensive use of the "Not In My Back Yard" (NIMBY) acronym by social scientists and society at large (i) is supposed to describe an existing phenomenon, even if it is admittedly ill-defined and elusive and (ii) became the indicator of a set of social, ideological and moral choices, and is used as a tool to depreciatively qualifying the resistance and protest of communities against the implementation of projects affecting them (namely their water resources). The argument relies on two oppositions: (i) at the social level, between "private" (namely individual) interest and "general interest" or "common good" and (ii) on a moral perspective, the split between an egocentric or selfish attitude and an altruist one. From a methodological perspective, NIMBYism tends to be analysed at the individual level: attitudes, motivations, self-interest calculation, rationality. We will consider that complex phenomenon at a social scale, considering convergence of interests and solidarity among concerned individuals, propagation of protests beyond the local level, and the process of aggregation of actions to constitute genuine social movements. A reappraisal of the Ethics of so-called "NIMBYism".

**Keywords**

NIMBY • Protest • Common good • Rationality • Selfishness

J. R. dos Santos (✉)
Centro Interdisciplinar de História, Culturas e Sociedades (CIDEHUS), University of Évora, Évora, Portugal
e-mail: jose.rds3@gmail.com

## 1 Introduction

The "NIMBY" acronym has been extensively "used by social scientists since the end of the 1970s to describe the resistance of communities to the siting of controversial facilities and locally unwanted land uses" (Borell and Westermark 2018). The lexical use these authors refers to has been acknowledged by number of other scientists, together with its diffusion in society at large: institutional actors, project promoters, enterprises and of course, the media. We must qualify theses uses not as "descriptive of a phenomenon", but rather as a label imposed on certain local situations, mainly those which imply, as Borell and Westermark (2018) aptly indicate, "resistance movements of communities" against controversial projects affecting them. Two attributes characterise such label: it is imposed on the situations from outside the community, and it is strongly depreciative. The negative load carried by the notion in virtually all its occurrences is best summarized by (Burningham et al. 2014): "In everyday use, the term is a pejorative shorthand to denote irrational, selfish, and obstructive individuals". They confirm a fact described long before: whenever "[t]he issue of public responses to technological risks has attracted attention from social and behavioural scientists, and also from policy spokespersons, three main viewpoints can be discerned: the public as ignorant/irrational, selfish, and prudent" (Freudenburg and Pastor 1992) write. An implicit assumption is that, by contrast, policymakers and project promoters are expert, informed, rational, and altruist. This dichotomy was adopted without a serious critical examination of its moral implications by a first wave of writings. A new trend emerged in studies on local opposition to "locally unwanted land uses", whose LULU acronym, due to (Freudenburg and Pastor 1992), has confronted the implicit or explicit assumptions of nimbyism phrasing, both with the theoretical flaws of the concept and the lack of empirical evidences that should sustain it. Burningham et al. (2014) writes "[I] conceptualize this



M. Abrunhosa et al. (eds.), *Advances in Geoethics and Groundwater Management: Theory and Practice for a Sustainable Development*, Advances in Science, Technology & Innovation,
https://doi.org/10.1007/978-3-030-59320-9_16

discourse as embodying an array of deficit models of the public and public knowledge". We proceed to briefly examining the "array of deficits" "the nimby discourses" the first trend of studies ascribed to opponents to technological and industrial transformations of a given territory.

## 2 An Inquiry into a Flawed Concept

First of all, are those opponents to local or regional projects "ignorant"? The opposite happens. The high degree of information (including hard technological data) the actors have, as empirical research has demonstrated in a number of cases, either is massively ignored by the tenants of the nimbyism phrasing or has been taken as a huge surprise (Wolsink 2000). People, groups, and communities opposing such projects as urban waste incinerators, high-speed railroads, wind farms, mines, hydraulic projects, etc., appear to be well informed. They are sometimes able to produce reliable information on certain aspects of the projects and their impacts that their promoters either are not aware of or fail to publicly disclose. As time passes after the unwanted project has been designed and decided (typically by outside powerful actors), local communities are able to acquire knowledge and enrol experts who may support their claims at a high technical and scientific level. Ignorance has been demonstrated to become the problem not of the concerned public, but rather that of some project promoters and may simply end stopping the projects on the ground of scientific flaws, along with social considerations, as was the case with the Alaska Pebble Mine project (Kozacek 2014).

As for the second type of "deficits", what leads the NIMBY discourse assumptions to qualify their opponents' action as "irrational"? Two examples may illustrate their reasoning. The first may be found in renewable energy projects (REP). REP is supposed to unambiguously contribute to a common good-replacing fossil energy sources by "clean" ones. Several opinion polls demonstrate a large support to this kind of solutions. When asked if they will accept a wind farm in a given spot which may be near the place they live in, people who had shown support to REP projects in general, will strongly refuse their construction in certain places. This, according to the nimbyist phrasing would be a clear disruption of a rational thinking. But, as Fischel suggests, "NIMBYism is a rational response to the uninsured risks of homeownership", when a person or a community evaluates the material consequences of a given project on her property (Fischel 2001).

The second set of examples for the use of "Nimby" when labelling resistance movements concerns opposition to mining projects and is certainly instructive as well. Mining has been one of the most impacting industries on environment (US EPA). Amongst several potential hazard factors, like toxic dusts, noise, waste rock and tailings disposal, those connected to water resources are assuming a highest relevance. Classically, a great number of studies have dealt with local groundwater pollution, stemming out of Acid Rock Drainage, of toxic mud dams failure, etc. Those risks have motivated a huge (and growing) number of protests and conflicts between mining companies and communities at a local or national levels (Del Campo 2015). Another set of conflicts has emerged, not about possible accidents, but about water resources consumption during "normal" mining operation. Mining industries are using enormous amounts of fresh water for ores treatment. Gold mining is at the top of water consumption per ton of metal produced (Stoltenborg and Boelens 2016). But such a fact is by far not exclusive of gold mining, as copper and lithium mines, to name just a few, also demand huge water resources for their operation. This is where conflicts emerge between companies and the communities: competition for alternative (incompatible) uses of a limited resource (Bridge 2004; Martsynkevych 2012; Tsatsaros et al. 2018). Present examples of these conflicts are those opposing lithium mining companies and local communities in the Chilean Atacama Desert, where the situation has been described as a "water war" (Sherwood 2018). As communities oppose mining projects perceived as profoundly harming their interests and even their survival by dramatically depleting water resources in their territory, are they acting as irrational "NIMBYS"? This was clearly not the case with Rio Tinto's Pebble Mine project, abandoned after a strong opposition by the Native communities, and the final regulation by US EPA (Kozacek 2014). Such is the nature of mining: the mine must be implemented in the very place where the mineral resource is located or will have to be dismissed. From the point of view of local communities, those places are not an abstract section space defined by GPS coordinates: they are their unique territory. To destroy it, is to destroy the community, as surely as if a deportation would force them out of it. Space may be fungible; community territories are not.

The third "deficit", the moral one, remains unsolved so far. It would stem out of the fact that a person approves a certain type of projects whenever they are conducted far away from her place but refuses it in her backyard. These choices are blatantly "selfish": are they? First, we should clarify what is meant by "backyard". In a restrictive reading, this would mean the immediate proximity to one's home. In this acceptation, the person is only defending her personal, almost intimate space, something that you are compelled to consider legitimate, but may be (although not necessarily) seen as ignoring her neighbour's interests and a fortiori, "general interest". If the meaning of "backyard" is larger, and describes a neighbourhood, valley, river basin, or region to which the person claims having some right and/or duty of conservation or some other interest in maintaining a certain state of affairs, to label her stand as "selfish" is much more problematic. The larger scope entails necessarily the interests of some other people (neighbours, inhabitants, users) and may be seen as a question, not of property, by sovereignty of a community on her territory. Such is the case when local communities resist mining projects which endanger water resources, compromising their very survival. But let us examine the "worst" case: the person is just refusing the construction of the project in her "backyard"; how could she accept its construction in the backyard next to her own? Or say, three or four backyards away? What is the distance at which she would think that her specific interest would cease being damaged? Of course, if we accept this as her individual reasoning, we must admit that her neighbours may follow the same. The "narrow selfishness" of a given person may thus propagate across space and society. It is then possible that instead of a "nimby" attitude, the resistance to a given project may be better described as a "NIABY" (Not In Anybody's Backyard) stand (Soutra 2016). A stand which, apart from the case of mining, may mean: far away from any densely inhabited place (e.g., off-shore REP). In every single instance (wind farm, mine, hydroelectric dam, waste dump…), it is probable that the propagation of such a kind of shared or "empathetic selfishness" will reach a frontier beyond which the feeling, the stand and willingness for action ceases to spread. Interestingly, as we verified in our fieldwork, propagation from a dense nucleus of action readiness fed by self-interest and empathy with neighbours' interest, while reaching a spatial limit, may literally project itself to some social groups and spaces which are not contiguous to the starting point (Santos 2016). This happens every time when external groups, congregating people who, while not being directly affected at that time, feel that their situation might become similar and join the action and/or negotiation. This led (Schaffer Boudet 2011) to devote "special attention to the key mechanisms of frame bridging, relational diffusion, brokerage and certification". Further, groups of people whose personal, material, interests are not, and most likely will not be, threatened, may join the criticism of a given project for the sake of a given interpretation of what is, and what is not, the "common good", or "general interest" (Conde 2015; Soutra 2016). Analysis of the kind of horizontal process described above is interesting because it endangers interpretations that reduce people's resistance to a moral deficit and calls attention to the shared "good reasons" they may have to resist a given project implying "Locally Unwanted Land Use" (LULU), as (Freudenburg and Pastor 1992) phrased it. A vertical analysis must complete the former, in order to account for the asymmetric relationships between the local, impacted interests, and the tenants of the project against which resistance arises. These are always powerful stakeholders, either private or public, who are invariably able to in a position to take the decisions about what is to be done, where and how a project must be implemented, including, as it may be, against local will.

## 3 "Common Good", "General Interest" and Private Profits

This is where the rhetoric of "common good" or "general interest" plays a major role in justifying the decisions to implement the controversial projects, and dismiss resistance, by simply taking for granted without further examination the qualification of the project by its promoters as corresponding to "common good", as explicitly does, among others (O'Hare 2010). The research thus has to deal with two problems (i) the competition between local and external actors around the definition what is, and what is not the "common good" or "general interest", and conversely, (ii) what is or is not a private, particular interest and whose interest it is. In several cases, locally affected people will tend to bring to the light the particular egoistic interests moving the project's promoters, which may entail the defense of private profits of an enterprise, political interests of a party or politicians, etc. In many cases, like in urban waste disposal projects, politicians and enterprises will have decided a trade-off between alternative technical solutions and among several possible locations, and have populations being confronted with their decisions (Hou et al. 2019). As (Wolsink 2006) puts it, "The conflicts are actually about fairness (Sjöberg and Drottz-Sjöberg 2001 in Wolsink 2006); my emphasis). Use of the "NIMBY" label is thus unfair and has been challenged. "Since the early 1990s, there has been an increasing trend for studies to require full clarification of the concept, to avoid it, or completely discard it as an analytic tool" (…), one of a kind of "ill-defined, lazy concepts that are easily used to legitimize simplistic politics" (Wolsink 2006). The use of the NIMBY label "is likely to exacerbate conflict and result in those so labelled feeling excluded and aggrieved" (Burningham et al. 2006). We may hypothesize that the strong dual opposition between "common good" and "particular interest" (as problematic as it may be) is a basic cognitive schema for social categorization of actors, actions and motivations and is consequently a "good" tool or "weapon" for actors involved in social conflicts. A particularly clear counterexample was documented by (Jakobsen 2008), in her study on the causes of success to the construction of wind-farms in the Danish island of Samsø. She points out mainly "knowledge sharing", "involvement of local people in the decision-making process and the organisation of ownership". This is also what (van Veelen and Haggett 2017) found in their Scottish Highlands' research on locally managed industrial energy projects: a common ground may take form only after a long, agonistic, process and eventually at the cost of radical changes in the initial project design.

It's only by deconstructing the definition of social utility and general interest that resistance movements may avoid the "nimby" qualification. This is done by unveiling the nature of involved private interests, analysing and confronting the planned sharing of costs and benefits between promoters and impacted stakeholders and eventually constructing an alternative definition of a common good. Such a definition is ever highly problematic because it includes extremely complex components and can by no means avoid political, questionable choices, namely those between short term and long term and between economic growth ("development") and ecological preservation.

## 4 Trust, Power and Legitimacy

Accusing a group or community to engage in "Nimbyism" because they resist the realization of a project decided by an external power, private or public amounts to ignore the deep variables at play in these situations. It has been shown that a major factor is at stake in these conflicts: trust. Resistance arises when people don't trust the enterprise or public power's discourse, promises and accountability (Hou et al. 2019). Such distrust may be, and often is, motivated by prior experiences of institutional failure to meet their engagements towards affected people. In her extensive review of the literature on mining resistance movements (Conde 2017) concludes: "The literature reveals that local communities react not only to perceived environmental impacts but also to their lack of representation and participation in decisions concerning their development path, lack of monetary compensation and distrust with the mining company and the state". We retrieve all three elements of the situational schema we have been analysing so far: risk, powerlessness, distrust. In most cases, the resistance to the localisation entails that there may be alternative locals where to build it. This is obviously not the case with mining projects because mineral resources are located where they are, the remaining alternative being between extracting them and renouncing to do it. This makes conflict setting much more difficult. In spite of promoters' efforts to meet some of the locals' concerns (mainly by negotiating modalities of implementation, promises of economic compensation of material losses, etc.), communities may simply refuse to accept any mine which would imply the destruction of their human and natural habitat (water resources, villages, fields, valleys, etc.). When it comes to mining activities, local resistance to a project may mean, as mentioned above, the overall rejection of any mining at all in that given place. The problem raised by this resistance may be phrased as a typical "nimby" one: those communities reject a given metal mine, but they still want to use products that could not be produced without that metal: so where would its extraction be legitimate? Private mining companies may thus draw on the "common good" rhetoric to promote their own interest. Public institutions, in turn, may, and in most cases do, recognise the mining project as coinciding with the general interest: state income (taxes, royalties), employment, economic growth. The states may decide to allow a company to prospect and eventually exploit a mineral deposit, giving to the local communities some information about the decision, and promising to the communities that some care will be taken to mitigate the inevitable damages to their territory: biodiversity, water and land resources, homes, health, and well-being.

Especially since three decades, there was a considerable rise in the number of mines, in their dimension, and further localisation in more populated areas, aggravating their impact on environment and on communities' lives and territories. This fostered the rise in the number and in intensity of resistance movements (Conde 2017). A major factor has been identified to explain the origins of conflict and the new forms it takes: lack of credibility of government information and general mistrust of public institutions. "Additional factors are a decline of confidence in the ability of government and industry to make informed, prudent and equitable decisions about risky technologies and statutory creation of new opportunities for public participation in administrative and judicial processes (Mazmanian and Morell 1990). There is no reason to believe these conditions will be altered appreciably anytime soon" (Kraft and Clary 1991). Almost thirty years later, drawing on an important corpus of works, Banning concludes that "[a]ll of the above-mentioned results point towards a deep mistrust of local citizens towards their governments to responsibly consider the siting of risk-generating facilities" (Banning 2016).

## 5 Concluding Remarks: Positive Collective Outcomes of Apparently "nimbyist" Resistance

The spreading of protest movements had at least two consequences. On the one hand, they exerted a very efficient pressure towards the reinforcement of legal frameworks, imposing new and stricter technical constraints on extractive activities. Water resource's consumption and management, which was seen decades ago as a secondary or lateral issue, became integral to any project involving land transformation, and its incidence on watersheds, streams and their biological health. On the other hand, the process of decision has tended—in the best cases—to integrate some participation of the locally affected communities, eventually since the very beginning of any project. They draw on a better inventory of effective damages, better compensation, and mining companies' insurance against industrial risks. It is only apparently paradoxical that each local resistance movement against mines, that some would disqualify as "nimby" because they defend a particular place and community, has very positive outcomes to the situation of many other communities that are eventually confronted with mining projects, and to society at large. Better laws, better decision processes, better compensation of damages and insurance against hazards, which would probably not have happened without resistance movements, are undoubtedly outcomes that correspond to a consensual acceptation of a "common good".

## References


Banning F (2016) Is NIMBY behaviour morally justifiable? an introductory paper to rights-based NIMBY ethics. Climate, energy, ethics p. 26. Institute of Philosophy, Bochum

Borell K, Westermark A (2018) Siting of human services facilities and the not in my back yard phenomenon: a critical research review. Commun Dev J 53(2):246–262

Bridge G (2004) Contested terrain: mining and the environment. Annu Rev Environ Resour 29(1):205–259

Burningham K, Barnett J, Thrush D (2006) The limitations of the NIMBY concept for understanding public engagement with renewable energy technologies: a literature review. School of Environment and Development, University of Manchester, Manchester

Burningham K, Barnett J, Walker G (2014) An array of deficits: unpacking NIMBY discourses in wind energy developers' conceptualizations of their local opponents. Soc Nat Resourc 28(3):246–260

Conde M (2015) Resistance to mining enabling factors and control of knowledge in uranium mining conflicts in Africa. Institut de Ciència i Tecnologia Ambientals, Universitat Autònoma de Barcelona, Barcelona, p. 187 (Ph.D. thesis)

Conde M (2017) Resistance to mining: a review. Ecol Econ 132:80–90

Del Campo C (2015) Research on company–community conflict. Social and economic development. International Council on Mining and Metals (ICMM), London, p 8

Fischel WA (2001) Why are there NIMBY's. Land Econ 77(1): 144–152

Freudenburg WR, Pastor SK (1992) NIMBYs and LULUs: stalking the syndromes. J Soc Issues 48(4):39–61

Hou G, Chen T, Ma K, Liao Z, Xia H, Yao T (2019) Improving social acceptance of waste-to-energy incinerators in china: role of place attachment, trust, and fairness. Sustainability 11(6):1727

Jakobsen I (2008) The road to renewables: a case study of wind energy, local ownership and social acceptance at Samsø. Innovation systems, social and ecological change, ESST MA. University of Oslo

Kozacek C (2014) Following EPA action, Rio Tinto Exits Alaska pebble mine project. North America, Water News, Water Policy & Politics

Kraft M, Clary B (1991) Citizen participation and the NIMBY syndrome: public response to radioactive waste disposal. Western Polit Quart 44(2):299–328

Martsynkevych V (2012) The conflict of mining and water resources in Kyrgyzstan. CEE Bankwatch Network, Brussels, p 3

O'Hare M (2010) Environmental and other co-benefits of developing a high-speed rail system in California: a prospective vision 2010–2050. In: U.C.f.E.P. Policy, Ed. University of Berkeley, Berkeley

Santos JR (2016) Evitar o impensável: a destruição irremediável do quadro de vida. Uma análise a partir do projecto de Mina de Ouro da Boa Fé. In: Guimarães PE, Cebada JDP (eds) Conflitos Ambientais na Indústria Mineira e metalúrgica. O Passado e o Presente. CICP (Univ. Évora) e CETEM (Univ. Huelva), pp 185–213

Schaffer Boudet H (2011) From NIMBY to NIABY: regional mobilization against liquefied natural gas in the United States. Environ Polit 20(6):786–806

Sherwood D (2018) In Chilean desert, global thirst for lithium is fueling a 'water war' . World News, Reuters Thompson

Soutra H (2016) Conflits d'aménagements : du «NIMBY» au «NIABY». La Gazette des Communes

Stoltenborg D, Boelens R (2016) Disputes over land and water rights in gold mining: the case of Cerro de San Pedro. Mexico. Water Int 41 (3):447–467

Tsatsaros JH, Wellman JL, Bohnet IC, Brodie JE, Valentine P (2018) Indigenous water governance in australia: comparisons with the United States and Canada. Water 10:2–18

van Veelen B, Haggett C (2017) Uncommon ground: the role of different place attachments in explaining community renewable energy projects. Sociol Ruralis 57(S1):533–554

Wolsink M (2000) Wind power and the NIMBY-myth: institutional capacity and the limited significance of public support. Renew Energy 21:49–64

Wolsink M (2006) Invalid theory impedes our understanding: a critique on the persistence of the language of NIMBY. Trans Instit Br Geograph 31(1):85–91

# Recharging Groundwater Security by Ensuring Polycentric Governance and Social Learning Strategies

Maria Paula Mendes and Nuno Barreiras

**Abstract**

The natural water cycle has been altered by human actions. Groundwater is an essential part of this cycle and an important source for human consumption. Climate change effects carry also ethical implications, where most-favoured nations are able to cope better with water scarcity issues than the other countries. Here phenomena such as "urban sprawl", the "tyranny of small decisions", the "Jevon paradox", and "intergenerational injustice" are presented, showing that it is impossible to have groundwater security without taking into consideration the human nature itself. Several examples are presented, showing that many individual minor actions can induce big transformations in the water cycle, and, therefore, in groundwater security. Effective groundwater management processes must incorporate polycentric governance and social learning strategies. Moreover, citizens should be involved emotionally. Public participation in the groundwater governance is a trend that needs to be cherished and fostered, since geoethics is dependent of the knowing. Due to its nature, groundwater ethical exploration depends greatly on the regular water users' will, introducing challenges in its governance. However, politicians and decision-makers point out that water management has been poorly conscious.

**Keywords**

Groundwater • Urban sprawl • Jevons paradox • Water security • Governance

M. P. Mendes (✉) · N. Barreiras
CERIS, Civil Engineering Research and Innovation for Sustainability, Instituto Superior Técnico (IST), University of Lisbon, Lisbon, Portugal
e-mail: mpaulamendes@tecnico.ulisboa.pt

## 1 Introduction

In the second half of the twentieth century, human activity exploded, causing a strong impact on the earth's system. This post-Second World War period marks the beginning of the so-called great acceleration (Steffen et al. 2015). One of the main consequences was the decline of ecosystem services, primarily affecting water and food supply. During the 2001–2010 decade, although a levelling of the trajectory of water use in OECD countries, the middle class's growth of BRICS nations leads to a global increase in its use (Steffen et al. 2015). For instance, the groundwater overexploitation in Beijing has declined its quantitative and qualitative status (UNESCO 2016). In Europe, urban sprawl and soil sealing threaten biodiversity and increase the risk of both flooding and water scarcity (EC 2011). Raising public awareness for (ground)water resources is an essential means of empowering citizens, enhancing the societal resilience to extreme weather events (GWP 2017; John and Kagan 2014).

Urban water resilience can be tackled through more efficient use and landscape management. At the scale of the individual plot, small changes occurred *en masse* can impact urban groundwater resources. For countries with moderate to high per-capita income, domestic investments can be adequate to find solutions to water shortage if sufficient political will can be found (Mcdonald et al. 2011). Some groundwater-related issues are presented and debated, illustrating public perspective should be integrated into the urban groundwater resource planning, and citizens have also responsibilities to create a more sustainable water management.

## 2 Urban Sprawl and the "Tyranny of Small Decisions"

Urban sprawl is a specific form of land take, resulting from the spread of low-density settlements. Urban sprawl contributes to the loss of agricultural land and, to the



M. Abrunhosa et al. (eds.), *Advances in Geoethics and Groundwater Management: Theory and Practice for a Sustainable Development*, Advances in Science, Technology & Innovation,
https://doi.org/10.1007/978-3-030-59320-9_17

degradation of ecosystems (EC 2011), causing an increase of soil sealing, which in turn, leads to disturbances in the hydrological cycle (Bricker et al. 2017; Mendes and Ribeiro 2014, 2017).

During the period 1990–2000, urbanization grew along the coastal regions of Europe faster than inland areas, with the highest rates of increase (20–35%) in Portugal, Ireland and Spain (AEA 2006). Overexploitation of coastal aquifers can increase groundwater salinity. In Graciosa and Pico islands (Portugal), groundwater salinization is due to seawater intrusion. Although there are natural factors associated with groundwater salinization, high pumping rates and well completion problems were also driven forces (Cruz and Andrade 2017).

Individual gardens can play an attenuating role in urban floods and can promote groundwater recharge, since, as a whole, they form a green complex inside urban space (Verbeeck et al. 2011). In Portugal, in Alfama district (Lisbon), we have been witnessing the paving inside these small green spaces (Ribeiro et al. 2008). In Flanders's region (Belgium), in five residential areas, top-down measures have been established for preventing the pavement or construction inside private domestic gardens; nevertheless, the imperviousness area increased from 38% (date of original building plans) to 56% in 2008 (Verbeeck et al. 2011). Goddard et al. (2010) used the term "tyranny of small decisions" for this cumulative outcome of many garden-scale management decisions. Here is extended this term for all those situations where the small decisions of many can have a strong impact on the quality and quantity of groundwater.

## 3 Groundwater Use and Jevon Paradox

In southern Europe, water competition can be enhanced, especially during the summer months. Moreover, Mediterranean climate regions face periodic droughts. Improving the efficiency of groundwater use does not necessarily mean a reduction in its consumption (Jevons paradox).

Urban agriculture and green spaces need water for their maintenance and compete with other urban uses of water. Perth (Australia) was a growing city that was facing an increased water supply deficit. Private citizens started to use groundwater as the source to irrigate their gardens, reducing the consumption of treated water for human consumption. However, citizens who used this water source consumed more water than those who used other sources. In the end, Perth Water Authorities imposed water-use restrictions, regardless of its source (Saayman and Adams 2002). Improving agricultural water efficiency can sometimes translate into more land under irrigation and/or cultivation of water-intensive crops (Perry and Steduto 2017). In Morocco, drip irrigation is related to higher crop density, a shift to more water-intensive crops, and the reuse of "saved water" to expand cultivated areas (Molle and Tanouti 2017).

## 4 Climate Change and Intergenerational Justice

The effects of human-made climate change also carry ethical implications, with a particular focus on the issue of global and intergenerational injustice. Many works have been done to address this problem, but the legal implementation of intergenerational justice remains a challenge. While the normative obligations to future generations are greater than we commonly assume, the empirical probability that we will leave behind a world with better or at least equal opportunities for future generations has dropped over the past decades (Tremmel 2009). The path towards a common vision for sustainable development took a major step with the publication of the UN Sustainable Development Goals (SDGs 2015). Nevertheless, even if the acknowledgement of the relevance of fair distribution principles in the implementation of sustainable development strategies is unanimous, distant future generations' requirements are not minimally ensured by the current formulation of the SDGs (Vasconcellos Oliveira 2018). The destabilization of the hydrological cycle caused by climate change tends to exacerbate these problems.

Regarding Portugal, the perception of the political class is that water resources are being depleted due to climate change (namely in the southern regions) and to pollutant activities and other serious socio-economic pressures. Various opinions of Portuguese politicians have been heard on this topic, in the review report made by Moury (2018), and the general opinion is quite worrying: "(…) In fact, we are not thinking about future generations, we are only thinking about current ones. And even in these, I also have many doubts". Several decision-makers point out that water management has been poorly conscious, with negative effects on future generations, and no reference to groundwater resources has been made. Another opinion states that "In Portugal there is a major problem of lack of supervision, and we know that there are repeated permanent polluting behaviours without any punishment for these entities, and we know that this also jeopardizes people's quality of life". However, there is also the perception that there is no area as water-related regulations where so many advances in so short period of time had been made in the last 30 years. Moury (2018) points out the difficulty in solving these problems. The inability to deal with them is mainly related to the lack of political incentives to do so, followed by unfit political agendas and unconscious socio-economic driving forces.

## 5 Collective Actions to an Improved Groundwater Governance

Water governance is critical to water security, and to the long-term sustainability of the freshwater systems. Water security is defined as "the capacity of a population to safeguard sustainable access to adequate quantities of acceptable quality water for sustaining livelihoods, human well-being and socio-economic development, for ensuring protection against water-borne pollution and water-related disasters and for preserving ecosystems in a climate of peace and political stability"- (UN-Water 2013).

Bakker and Morinville (2013) debated the governance dimensions of water security that can be applied to groundwater governance: adaptive governance, multi-level governance and the political and institutional dimensions. Briefly, water security enhances the uncertainty inherent to the management of complex socio-ecological systems, combined with economic driving forces. Therefore, the adaptive management paradigm is raised, wherein polycentric governance fosters social learning. The multi-level governance is enhanced for the necessity of multi-scalar linkages within and beyond the groundwater bodies. Important as well is to "Mind the Gaps, Bridge the Gaps", as it is documented by OECD (2013), that identified seven categories of governance deficits related to (ground)water management, relevant to all countries regardless of their institutional organizations, water availability and demand. The third aforementioned dimension is related to the centrality of social power, in a variety of modalities (both legal and illegal) and scales (from the local to supranational). Social power emphasizes negotiating conflicts generated by tensions at the water–energy–food nexus. In this context, environmental management processes that incorporate polycentric governance and social learning strategies are considered to be desirable.

When looking to phenomena as urban sprawl and the "tyranny of small decisions", Jevon paradox and intergenerational justice, it is clear that social (and ethical) learning has to be enabled and improved through the involvement of a greater diversity of actors in on-the-ground management and decision-making processes. Part of this diversity is the citizens, as individuals, which in fact for some phenomena (i.e. water use) are the problem and the solution as well. For example, calls for public participation in the environment (and water) governance are a trend that is increasing due to multiple factors, including awareness of the expertise available outside of government agencies, new approaches to citizen participation, and the socio-economic restructuring. Nevertheless, there are very few examples of effective public participation in groundwater governance. So the question remains, how to deal with the ethic and socio-economical phenomena directly related to groundwater in order to achieve long-term sustainability? Cutting the story short, a resilience collective action is required for an integrated long-term socio-ecological policy towards groundwater governance.

## 6 Final Remarks

Some ideas are presented by Re and Misstear (2017) that contribute to the development of resilient collection actions, ensuring more ethical and sustainable use of groundwater resources. Education and capacity development are crucial, and two different actors and educational needs are considered: (1) Academic and groundwater professionals, (2) policymakers and the general public, including well owners (key stakeholders).

For the success of the education, it must be considered that different communication strategies are required for different stakeholders, and there are opportunities for groundwater scientists to make more use of social media and visual art in their outreach activities. Awareness raising of the groundwater issues is also only possible by including the public. The informing about success stories around groundwater can help to bring these issues to public eyes. But in order to engage massive public participation, and towards optimal results, in collective action, citizens should be involved emotionally. This is quite visible nowadays regarding climate change, where especially the youngsters which are indeed the heirs of the environmental actions we take today, are mobilized in unforeseen manifestations of will. Though, from manifestation to real action on the day-by-day living, the outcomes are not quite visible.

The same goes for the regular water users, which keep using water above the replenishment levels based on the rooted traditions, practices and live standards. If real collective action is to be achieved, stronger actions need to be implemented in the society, technology and citizens' lives. Educational and establishment of actions and practices made by the governing environmental agencies would be very recommendable towards a sustainable and intergenerational groundwater use and conservation. Therefore, from our view, not only an urban policy needs to be understood and to operate in a multi-scalar context but also new governance modes based on citizens' empowerment, the participation of all relevant stakeholders and innovative use of social capital are strongly needed.

## References


AEA (2006) Urban sprawl in Europe: the ignored challenge. Copenhagen. Retrieved from https://www.eea.europa.eu/publications/eea_report_2006_10/eea_report_10_2006.pdf

Bakker K, Morinville C (2013) The governance dimensions of water security: a review. Phil Trans R Soc A 371: 20130116

Bricker SH, Banks VJ, Galik G, Tapete D, Jones R (2017) Land use policy accounting for groundwater in future city visions. Land Use Policy 69:618–630

Cruz JV, Andrade C (2017) Groundwater salinization in Graciosa and Pico islands (Azores archipelago, Portugal): processes and impacts. J Hydrol Reg Stud 12:69–87

European Union (2011) Cities of tomorrow: challenges, visions, ways forward. European Commission, Directorate General for Regional Policy

Goddard MA, Dougill AJ, Benton TG (2010) Scaling up from gardens: biodiversity conservation in urban environments. Trend Ecol Evol 25(2):90–98

GWP (2017) IWRM toolbox: raising public awareness (C8.02). Retrieved from https://www.gwp.org/en/learn/iwrm-toolbox/Management-Instruments/Promoting_Social_Change/Raising_public_awareness/

John B, Kagan S (2014) Extreme climate events as opportunities for radical open citizenship. Open Citizenship 5(1):60–75

Mcdonald RI, Green P, Balk D, Fekete BM, Revenga C, Todd M (2011) Urban growth, climate change, and freshwater availability. Proc Natl Acad Sci 108(15):6312–6317

Mendes MP, Ribeiro L (2014) The importance of groundwater for the delimitation of the Portuguese national ecological reserve. Environ Earth Sci 72(4)

Mendes MP, Ribeiro L (2017) A água subterrânea e a agricultura em áreas urbanas. In: Oliveira R, Amâncio S, Fadigas L (eds) Alfaces na avenida. Estratégias para (bem) alimentar a cidade. Universidade de Lisboa, Lisboa, Colégio Food, Farming and Forestry, pp 55–60

Molle F, Tanouti O (2017) Squaring the circle: agricultural intensification versus water conservation in Morocco. Agric Water Manag 192:170–179

Moury C (2018) Perceções da classe política portuguesa sobre Justiça Intergeracional. https://content.gulbenkian.pt/wp-content/uploads/sites/46/2019/02/13164401/Perceções-da-classe-política-portuguesa-sobre-a-Justiça-Intergeracional.pdf

OECD (2013) Water governance in OECD countries: a multi-level approach. IWA Publishing. https://doi.org/10.2166/9781780406046

Perry C, Steduto P (2017) Does improved irrigation technology save water? A review of the evidence. https://www.fao.org/3/I7090EN/i7090en.pdf

Re V, Misstear B (2017) Education and capacity development for groundwater resources management. In: Villholth KE, Lopez-Gunn K, Conti A, Garrido J, Van Der Gun (eds) Advances in groundwater governance, p 620

Ribeiro L, Quintino M, Alfaiate T, Mendes MP (2008) Al-Hama—methodology for integrating groundwater in sustainable planning and management of urban landscapes. Lisboa, The Case-Study of Alfama, Lisboa

Saayman IC, Adams S (2002) The use of garden boreholes in Cape Town, South Africa: lessons learnt from Perth, Western Australia. Phys Chem Earth Parts A/B/C 27(11):961–967

SDGs (2015) Sustainable development goals. https://sustainabledevelopment.un.org/sdgs

Steffen W, Broadgate W, Deutsch L, Gaffney O, Ludwig C (2015) The trajectory of the anthropocene: the great acceleration. Anthropocene Rev 2(1):81–98

Tremmel J (2009) A theory of intergenerational justice. Earthscan, London

UN-Water (2013) Water security and the global water agenda. Policy and analytical briefs. https://www.unwater.org/app/uploads/2017/05/analytical_brief_oct2013_web.pdf

UNESCO (2016) Water, megacities and global change: portraits of 15 emblematic cities of the world. In: S. and C. O. United Nations Educational (ed) International conference on water, megacities and global change. UNESCO/ARCEAU IdF, Paris, p 95

Vasconcellos Oliveira R (2018) Back to the future: the potential of intergenerational justice for the achievement of the sustainable development goals. Sustainability 10(2)

Verbeeck K, Van OJ, Hermy M (2011) Landscape and urban planning measuring extent, location and change of imperviousness in urban domestic gardens in collective housing projects. Landsc Urban Plann 100(1–2):57–66

# Sharing Knowledge and Data About Groundwater in EU: The EIGR Metadata Inventory of the KINDRA Project

Marco Petitta, Balazs Bodo, Adrienn Cseko, Isabel Fernandez, Clint García Alibrandi, Mercedes Garcia Padilla, Eva Hartai, Klaus Hinsby, Viktoria Mikita, Peter Szucs, and Peter van der Keur

**Abstract**

Groundwater knowledge and research in the European Union is often scattered and non-standardised, because of different subjects involved and different approaches from member states. The Horizon 2020 project KINDRA has conducted an EU-wide assessment of existing groundwater-related practical and scientific knowledge based on a new Hydrogeological Research Classification System, identifying more than 280 keywords related to three main categories (namely operational actions, research topics and societal challenges) to be intersected in a 3D-diagram. The classification is supported by a Web service, the European Inventory of Groundwater Research, which acts not only as a knowledge repository but also as a tool to help identify relevant research topics, existing research trends and critical research challenges. The metadata included in the inventory at the end of the project are about 2300, and the analysis of the results is considered useful for producing synergies, implementing policies and optimising water management in Europe. Using the three-axes classification, occurrence and relationship of different topics in groundwater research have been highlighted. The EIGR inventory is a powerful and useful took for sharing information and data about groundwater issues following the FAIR principle, in agreement with an ethical approach.

**Keywords**

Groundwater framework directive • Inventory • European Union • Metadata • Classification

M. Petitta (✉)
Earth Sciences Department, Sapienza University, Rome, Italy
e-mail: marco.petitta@uniroma1.it

B. Bodo · A. Cseko
La Palma Research Center for Future Studies, Santa Cruz de La Palma, Spain

I. Fernandez · E. Hartai
European Federation of Geologists, Bruxelles, Belgium

C. G. Alibrandi · M. G. Padilla
Agencia de Medio Ambiente y Agua de Andalucia, Sevilla, Spain

E. Hartai · V. Mikita · P. Szucs
Faculty of Earth Science and Engineering, University of Miskolc, Miskolc-Egyetemvaros, Hungary

K. Hinsby · P. van der Keur
Geological Survey of Denmark and Greenland, Kobenhavn, Denmark

## 1 Introduction

Water is a key-topic in modern society, requiring several interconnections in research and in practical applications, as balancing human needs and environmental requirements, governing the water–food–energy–climate nexus, impacting the concepts of circular economy and smart cities, applying nature-based solutions, among others. Groundwater is the largest and most widely used resource for drinking water supply, globally, and is an important part of the hydrological cycle sustaining life on earth. The general understanding of the importance of groundwater is often low, even within related natural science disciplines partly because it is hidden below ground and the interactions with surface water are difficult to quantify. Nevertheless, groundwater role is frequently underestimated, and practical and scientific knowledge are scattered amongst various actors in Europe.

In this context, the mission of KINDRA (EC framework program H2020, Grant Agreement No. 642047) was to make groundwater visible by demonstrating its transdisciplinarity and importance to all the grand societal challenges of Horizon 2020 and EU water policies and by providing an overall view of the scientific knowledge that exists across Europe. This approach has also the aim to raise the awareness of citizens of science affecting their daily lives and, at the same time, will allow the correct management and policy development of groundwater at EU scale, as recommended



M. Abrunhosa et al. (eds.), *Advances in Geoethics and Groundwater Management: Theory and Practice for a Sustainable Development*, Advances in Science, Technology & Innovation,
https://doi.org/10.1007/978-3-030-59320-9_18

also by the Blueprint Document (European Commission 2012). In addition, the project is based on the geoethical principle of sharing information and data to improve groundwater management and protection, providing a geo-database (named EIGR) useful for geoscientists, stakeholders and users.

A dedicated groundwater research and knowledge classification system was developed to enable population of the new European Inventory of Groundwater Research (EIGR) with **F**indable, **A**ccessible, **I**nter-operable and **R**eusable metadata on groundwater research according to the "FAIR" principles (Wilkinson et al. 2016). The EIGR repository was filled by metadata by the useful support of the National Associations of Geologists throughout Europe, under the umbrella of the European Federation of Geologists. Collected metadata include both scientific resources, i.e. peer-reviewed literature as well as reports and other resources with little, uncertain or no peer review ("grey literature"). The latter category of resources includes a large number of data reports, maps and other relevant work published by, e.g. authorities, consultants and geological surveys, which contribute substantially to the knowledge pool, but which is often difficult to find (e.g. Lawrence et al. 2015).

In order to develop a database system for population with heterogeneous data resources and determine the degree of (peer) review and other types of quality assurance within the KINDRA project, keywords and categories have been identified to allow for an effective and useful classification according to the FAIR principles, i.e. with easy access to and reuse of the resources. This effort has enabled networking, mutual recognition, trust and visibility across the hydrogeology communities. At the same time, collected metadata, including scientific papers, research projects, articles and data, represent a concrete manifestation of Open Science (Fernandez et al. 2017). In this sense, KINDRA can be considered as a reference project in implementing Open Science which could be copied by other scientific areas.

The aim and the results obtained by the KINDRA project show a good fitting with the ethical and social requirements that geoethics indicates as indispensable for scientific research to transform in a real benefit for society (Peppoloni and Di Capua 2016; Bobrowsky et al. 2017; Peppoloni et al. 2019). Moreover, the methodological approach followed in this project is adherent with several geoethical values such as cooperation, respect for colleagues, openness to multidisciplinarity and sharing. Furthermore, it is evident the long-term usefulness of such a tool in the protection of groundwater that are considered as a resource and common good, not only at local but also global level (in this case relating to the European community).

## 2 Methods and Materials

The KINDRA project proposes a comprehensive approach with a new classification system (Van der Keur et al. 2018) specifically developed for groundwater field, tested and approved by the research community, the professional community in geology and the wider public at large. It is a multistep venture from a well-defined thematic categorisation to the complete roll-out as an open searchable service.

The classification starts from the definition of a keyword list, able to comprehend the different topics of groundwater research and at the same time to allow searching options among the database. To establish a common terminology and approach to carry out the analysis presented here, various academic, industrial and research classification schemes have been reviewed to create a hierarchical structure and a selected list of keywords from relevant EU directive documents. Notably, these documents include the Water Framework Directive (WFD, European Commission 2000), its daughter directive the Groundwater Directive (GWD, European Commission 2006) and the Blueprint to Protect Europe's Water Resources (BWR, European Commission 2012). Supplementary identification of relevant keywords and topics from the most important scientific journals publishing groundwater research has been performed. The combined final keyword list counts about 284 keywords to be considered in the classification system.

To compare and intersect research products in groundwater, the KINDRA project group categorised all groundwater research according to a 3D-approach, identifying three main categories (Petitta et al. 2015): (1) Horizon 2020 societal challenges (which represents the impact of groundwater in the society) (2) Operational actions (corresponding with the main activities of hydrogeologists and other groundwater scientists and practitioners) and (3) Research topics (resuming the interdisciplinary role of groundwater).

From the original seven societal challenges (SC) identified by H2020 European programme, five societal challenges relevant for groundwater have been selected: (1) Health; (2) Food; (3) Energy; (4) Climate, environment and resources; (5) Policy, innovation and society. The five main operational actions (OA) characterising the activity of geologists have been identified by the help of an end-user survey performed during the project, as follow: (1) Mapping, (2) Monitoring, (3) Modelling, (4) Water Supply, (5) Assessment and Management. The Research topics (RT) correspond to the higher number of keywords grouped into the main natural science disciplines: (1) Biology, (2) Chemistry, (3) Geography, (4) Geology, (5) Physics and Mathematics.

## 3 Results and Discussion

After definition of the classification system KINDRA European Inventory on Groundwater Research (EIGR), a tool for classifying information sources regarding Research and Knowledge product in Europe has been realised (Garcia Alibrandi 2018). The EIGR is a relevant new tool for mapping and analysis specifically of groundwater research and knowledge that range somewhere between the peer-reviewed research tools (Web of Science and Scopus) and other broader databases and tools such as Google Scholar, Mendeley and ResearchGate. The added value of EIGR is to provide direct access to important metadata about available groundwater research and knowledge publications including "grey literature", which has been classified according to a policy relevant classification system based on the grand societal challenges of the EU Horizon 2020 programme. The EIGR is a permanent resource, publicly still available at https://kindraproject.eu/eigr/ as open access searchable repository also after the end of the KINDRA project.

As stated above, the EIGR allows for the insertion of different information products (papers, technical reports, database, project, maps, guidance, books and proceedings, etc.). A fundamental novelty in KINDRA, respect with other existing scientific databases, is the inclusion of documents different from peer-reviewed papers. In fact, the inventory aims to consider not only the research products, but also the knowledge products, which includes national reports and documents and the grey literature in general.

For this reason, in the process of inserting information in the EIGR, users are guided to classify the uploaded information and distinguish between "research" and "knowledge" according to four different classes identified by the level of the performed quality assurance the uploaded work has received. The "research" classes include not only the papers in peer-reviewed journals (class 1) but also conference proceedings and books included in international scientific databases as Web of Science and Scopus (class 2). In the "knowledge" group, there are Class 3, including reports and information having a review, which act as a quality assurance (QA), as reports from national official organisations, and Class 4 which refers to reports, journals and newsletter with no certain quality assurance.

As KINDRA is most concerned to explore gaps within societal challenges, the approach of exploring resources at the intersection between research topics and operational actions for societal challenges is adopted. Clearly, in this project developed EIGR has a population which is much smaller compared to large scientific databases as Scopus (or Web of Science) and also contains mainly Class 3 and 4 information. However, inspecting and analysing EIGR data give insight in research topics and operational action, associated to groundwater management and operation, which is not available in the Scientific Scopus database, i.e. Class 3 + 4 resources.

Based on the data provided by the National Associations of EFG, the EIGR has been populated in the form of a Web service, with a total of more than 2300 records, most of them completely public available (2200 records). Remaining unpublished records inserted in the EIGR are about 130. These records cannot be published because they contain incomplete information and consequently are not useful for the gaps and trends analysis. Actually, it is mandatory when inserting records to specify the category and the overarching group of the classification system HRC-SYS; by this way, all records contain at least this information.

An in-depth analysis of the EIGR inserted information, crossing the three main categories is described in detail in Petitta et al. (2018).

## 4 Concluding Remarks

The performed analysis of the records inserted during the KINDRA project in the EIGR has demonstrated the advantages and the potential of the inventory and has led to recommendations for improvement and further development of the groundwater research and knowledge repository covering all relevant research disciplines. The EIGR database includes additional data and has significant advantages, benefits and added value compared to existing research databases, as follow:

- Easy access to metadata on research projects, reports, databases and maps without formalised peer review, produced at international and national scales, not easily available, which have been classified according to European Societal Challenges as defined in Horizon 2020;
- EIGR focuses exclusively on groundwater research and knowledge, increasing the precision and relevance of information retrieval conducted in the database;
- Access to a dedicated platform, integrating information on groundwater research and knowledge of relevance to the whole community of researchers and practitioners, extending across groundwater relevant disciplines and EU member states organisations and sectors to the benefit of all;
- Improvement of the overview of the vast amount of groundwater research, knowledge and data, especially within the non-peer-reviewed segment, conducted in Europe, making access to related studies easier and duplication of work less likely;
- Dynamical features, like adding new relevant keywords by users as they emerge in the future.

In terms of geoethical approach, both the aims and the obtained results of this project are promoting and applying geoethical values such as cooperation, respect for colleagues, openness to multidisciplinarity and sharing information and data. At the same time, the subject of groundwater protection and management has a dramatic relevance also from the societal point of view, requiring not only technical solutions, but common visions and widespread knowledge of scientific principles and practical solutions, as previewed by the KINDRA project.

**Acknowledgements** The KINDRA project has received funding from the European Union's Horizon 2020 research and innovation programme under grant agreement No 642047 (call WATER-4a-2014—Coordination and Support Action).

## References

Bobrowsky P, Cronin V, Di Capua G, Kieffer S, Peppoloni S (2017) The emerging field of geoethics. In: Gundersen L (ed), Scientific integrity and ethics in the geosciences. Special publications 73. American Geophysical Union, Washington, Wiley, Hoboken

European Commission (2000) Directive 2000/60/EC (Water framework directive) of the European parliament and of the council of 23 October 2000 establishing a framework for Community action in the field of water policy. OJ L 327, 22.12.2000, pp 1–51

European Commission (2006) Directive 2006/118/EC (Groundwater Directive) of the European parliament and of the council, Official Journal of the European Union L 372/19

European Commission (2012) Communication from the commission to the European parliament, the council, the European economic and social committee and the committee of the regions—a blueprint to safeguard Europe's water resources, COM (2012) 673. SWD 2012:381–382

Fernandez I, Petitta M, Hinsby K, Cseko A, Szucs P, Garcia Padilla M, Hartai E, Biševac V, Stein A, Bodo B, van der Keur P, Mikita V, Van Leijen G, García Alibrandi C (2017) The KINDRA project—towards open science in hydrogeology for higher impact. Eur Geol J 44:38–43

Garcia Alibrandi C (2018) KINDRA deliverable 1.6—European inventory of groundwater research (https://kindraproject.eu/wp-content/uploads/D1-6-WEB-revised-1.pdf, 66 pages) (last access 07/01/2020)

Lawrence M, Thomas J, Houghton J, Weldon P (2015) Collecting the evidence: improving access to grey literature and data for public policy and practice. Aust Acad Res Libr 46(4):229–249

Peppoloni S, Bilham N, Di Capua G (2019) Contemporary geoethics within the geosciences. In: Martin B (ed) Exploring geoethics: ethical implications, societal contexts, and professional obligations of the geosciences, vol 14, Palgrave Pivot, Cham

Peppoloni S, Di Capua G (2016) Geoethics: ethical, social, and cultural values in geosciences research, practice, and education. In: Wessel G, Greenberg J (eds) Geoscience for the public good and global development: toward a sustainable future. Geological Society of America, Special Paper 520, pp 17–21

Petitta M, Bodo B, Caschetto M, Colombani N, Correia V, Csekő A, Di Cairano M, Fernández I, Garcia Alibrandi C, Hartai E, Hinsby K, Madarász T, Mikita V, García Padilla M, Szűcs P, van der Keur P (2015) The KINDRA project: a tool for sharing Europe's groundwater research and knowledge. Eur Geol J 40:5–8

Petitta M, Bodo B, Cseko A, Del Bon A, Fernandez I, García Alibrandi C, Garcia Padilla M, Hartai E, Hinsby K, Muller P, Mikita V, Szucs P, van der Keur P (2018) The KINDRA project: sharing and evaluating groundwater research and knowledge in Europe. Acque Sotterranee Ital J Ground 7(AS24-324):07–16

Van der Keur P, Caschetto M, Hinsby K, Di Cairano M, Mikita V, Petitta M, Szűcs P (2018) The new EU-harmonised hydrogeological research classification system, KINDRA. Deliverable 1.2.—HRC-SYS (https://kindraproject.eu/wp-content/uploads/D1-2_versionrevisedFINAL-WEB.pdf, 47 pages) (last access 07/01/2020)

Wilkinson M, Dumontier M, Aalbersberg Ij, et al. (2016) The FAIR guiding principles for scientific data management and stewardship. Scientific data 3: sdata201618. www.nature.com/sdata. doi: https://doi.org/10.1038/sdata.2016.18)

# Groundwater Conjunctive Use

Andrés Sahuquillo

**Abstract**

The volume of water stored in aquifers varies from tens of times the annual recharge in small aquifers, to thousands of times in big ones, and additionally, the volume of aquifer storage provided by a relatively small fluctuation in the piezometric head in most unconfined aquifers exceeds considerably the available or economically feasible storage in dams. In many cases, aquifers used in conjunction with surficial components meet water needs, avoiding more expensive structures, less appropriate for the environment. Two conjunctive use methods exist: artificial recharge and alternate conjunctive use (ACU). Artificial recharge stores volumes of water that cannot be used directly. With ACU, more surface water is used during wet periods and more groundwater in dry ones. This allows the increase of water supply with lower dam storage and important socioeconomic and environmental advantages. Modelling of groundwater and surface water components jointly is needed in the most contentious or complex cases, particularly in systems of various dams, aquifers and conduits, with multiple alternatives for periods of several tens of years. Variability of the surface contributions aggravated by the uncertainties of climate change requires simulating more alternatives. The eigenvalue method simplifies modelling greatly. Aquifer heads can be explicitly calculated only at given points, the aquifer–river flow interaction in various river stretches, volumes stored in the aquifer or different areas of the aquifer and flow exchanges between several couples of two contiguous areas. But even in many developed countries, groundwater is not duly considered by decision makers and government officials, so the problems caused by intensive exploitation (overexploitation) or misuse of aquifers are exaggerated, thus down placing the important role of aquifers in favour of water works that are more expensive and less environmentally friendly. Consequently, this involves ethical considerations.

A. Sahuquillo (✉)
Royal Academy of Sciences of Spain and Technical, University of Valencia, Valencia, Spain
e-mail: asahq@hma.upv.es

**Keywords**

Conjunctive use • Modelling • Eigenvalues • Ethical use

## 1 Introduction

Groundwater use has increased dramatically worldwide in recent decades, especially for irrigation and urban uses in arid and semiarid regions. A decisive reason for the increased exploitation of groundwater, in addition to its ubiquity, is that it is cheap, easy to exploit and with low variability in quality. Intensive pumping has produced in some areas significant decrease in aquifer levels, in the water volume stored in them, and in the base flow of the rivers, seawater intrusion, land subsidence and has affected wetlands intensely. The increase in water demand for urban, industrial and agricultural uses and the possible influence of climate change on the decrease in recharge of aquifers and river flows in some regions can aggravate socioeconomic, environmental and legal problems (Llamas and Custodio 2003; Sahuquillo et al. 2005). This is a concern that extends to the future availability of water resources. The main ethical and moral aspects of intensive use of groundwater, coastal aquifers use and reserve depletion refer to damage to current and future third persons, as well as the environment and the services it provides to humans. Aquifer overexploitation, which often is perceived as something ethically bad, is not necessarily detrimental if it is not permanent. Custodio (2002) provides a more adequate view of the problem and proposes replacing the denomination of overexploitation by that of intensive use. In this work, it is considered the integrated management of surface and underground water



M. Abrunhosa et al. (eds.), *Advances in Geoethics and Groundwater Management: Theory and Practice for a Sustainable Development*, Advances in Science, Technology & Innovation,
https://doi.org/10.1007/978-3-030-59320-9_19

resources that affects the well-being of people and the environment in a decisive way.

In addition to their annual recharge, small aquifers store, a volume of water tens of times greater than their annual mean recharge above the level of the rivers, being thousands of times their recharge in large ones, increasing in little penetrating rivers and poorly connected with the aquifers. It is the so-called one-time reserve, in which volume stored below channels are not counted. Additionally, storage provided by relatively small fluctuations in piezometric heads in unconfined aquifers exceeds considerably the available or economically feasible storage in dams. Key aspects in conjunctive use systems are (a) the flow variability of river flows aggravated by climate change uncertainties and (b) the slowness of changes in the aquifer due to the big water volumes of water involved, which can only be known after years of observations and careful analysis of their behaviour provided by a well-planned monitoring network and detailed knowledge of the aquifer's exploitation. Monitoring that must be permanently maintained at an adequate level similar with that happens with the flows and quality of rivers and volumes in reservoirs.

It is necessary to keep in mind the relationship of aquifers with rivers, in which sections the rivers are winners, in which they are losers or the sections that being winners become losers due to the exploitation of the aquifers.

## 2 Conjunctive Use of Groundwater and Surface Water

### 2.1 Alternate Conjunctive Use and Artificial Recharge

The more direct solution to solve some of the problems raised by aquifer intensive development and to increase the availability of water is in many cases the conjunctive use of surface and groundwater (Foster and Steenbergen 2011; Sahuquillo 2002, 2010). The more intuitive action is through artificial recharge of the aquifer. It has been used in many parts of the world to store local or imported water in aquifers. The second is alternate conjunctive use (ACU). With ACU, groundwater pumping is reduced in wet periods and the surface water use increases. On the contrary, during periods of reduced surface water availability, groundwater pumping increases. Groundwater storage increases in wet periods, decreasing in dry ones. In many cases, aquifers have being pumped well above their recharge in Israel in a planned way, to defer building of the National Water Carrier. In California, south-eastern Spain and in many other places, the large water volume in aquifers has allowed, through unplanned overexploitation, the development of primary economic activities, which have been the origin of a further economic growth (Sahuquillo 1985, 2002, 2010; Custodio et al. 2017). Aquifer water content and storage allows both the use of water during dry seasons or droughts and the use of the subsurface space (van der Gun and Custodio 2018) for storing imported water. In fact, groundwater use has traditionally been augmented worldwide to back up supply in times of shortage or drought. In most of the Mediterranean river basins in Spain, vulnerability to droughts is reduced by the increased exploitation of aquifers, as it is also done in many other countries. This involves a certain type of conjunctive use. To meet the water demands in droughts, the water agencies of the Júcar and Segura basins, in Spain, built batteries of large capacity wells. A relatively small sporadic pumping of groundwater from aquifers, instead of the construction of a new reservoir, can be used to increase the resources of many low guarantee irrigation systems. The same can be done, with obvious economic and environmental advantages, to alleviate possible declines in river flows due to climate change in areas prone to become dry. In California, ACU was suggested in the San Joaquin River Basin as early as in the 1930s. Later, in the California Water Plan in the late 1950s, it was projected to apply ACU massively by storing local and imported surface water, adding new reservoirs to the existing ones with a total storage to 24 $km^3$, and a foreseen subsurface storage of 37 $km^3$. The idea was later replaced by artificial recharge and was forgotten, but somehow re-emerged later in what has been called "in lieu recharge", using sporadic surface water runoff instead of pumping from aquifers. Artificial recharge and ACU use are fully compatible (Jaquette 1981). A local example of ACU is the Mijares River Basin, in eastern Spain. Two of the three existing dams were built in karstified limestone and have important leaks that recharge the Plana de Castellon aquifer. The river also recharges the aquifer. About one-third of the local irrigated surface area is supplied from surface water, another third and the urban water with the aquifer, and the other third alternately depending on the river flow and the water stored in the reservoirs. Storage achieved by applying the ACU concept would be the groundwater volume stored at the end of a wet cycle less the volume at the end of a dry one, over 700 $hm^3$ of the Plana de Castellón aquifer. This is more than four times the existing surface storage. This makes it possible to use a very high percentage of the annual average river flow. It is important to note that the operating rules of the system were proposed by the farmers' association and legally approved in 1973. Throughout other Spanish Mediterranean basins, it is usual to increase groundwater pumping in dry years (Sahuquillo 2002, 2003, 2017). Adequate operation rules of an ACU scheme with storage in dams and aquifers is important to improve results but is much more complex than in the case of systems with only surface reservoirs. The decrease in the river flow due to groundwater pumping is slow, especially in large aquifers, and is not perceived immediately by users. In some sections of the river, years or decades after intense pumping, there are changes from a gaining river to become a losing one and can even contribute to recharge the aquifers. The use of the UCA has the potential to solve current or future scarcity problems and those that may pose climate change, in many cases without the need to build new reservoirs. But those are solutions that need an important knowledge of groundwater. Something that is not usual on the agenda of decision makers because aquifers cannot be opened, at less at the usual time frame politicians usually remain in the government.

### 2.2 Relief of Drainage and Salinization Problems in Irrigation Areas in Arid and Semiarid Zones

Aquifer recharge has increased because of water losses from unlined distribution canals, in addition to the infiltration of return irrigation water. Increments in aquifer recharge augment the potential for groundwater development, and in some zones, they also create drainage and salinity problems by rising groundwater levels. This is a customary problem of large irrigation projects in arid and semiarid lands. Increased and more efficient use of groundwater can help to prevent these problems.

A classic example is the irrigation in the Indus River Basin (Sahuquillo 1985, 2002, Custodio 2010), which started to be intensively implemented in the late nineteenth century under British colonial rule. The irrigation canals are fed by several big dams. The largest ones are the Mangla Dam and Tarbela Dam, with 5.5 and 10.6 $km^3$ of storage, respectively. Most canals are unlined, having significant losses that feed the huge aquifer below. During the last 80–100 years, water levels in this aquifer rose between 20 and 30 m, and up to 60 m in some places. In the middle of the last century, 25,000 ha were abandoned every year. In 1960, 2 million hectares of a total 14 were abandoned. A Salinity Control and Reclamation Project was initiated by the Pakistani government to drill high capacity wells to lower the water levels and pump the saline groundwater water out of the aquifer, to be disposed downstream. Irrigators began a very intense exploitation of the aquifer, producing an important water table drawdown, which prompted the government to paralyse the project. The problems of drainage decreased in upstream areas, but drainage and salinization problems persisted in the area and downstream.

Similar drainage and salinity problems exist in most arid countries, jointly with subsidence problems. Losses in canals and distribution systems can be alleviated by lining the conductions. However, if infiltrated water feeds usable aquifers and conjunctive use is practiced, it could be more convenient to leave canals unlined, unless drainage problems persist, and water losses produce excessively high aquifer levels. Conjunctive use can contribute to improve water use efficiency and regional environment of irrigated areas.

Foster et al. (2010) comment some impediments to promote a more rational conjunctive use in areas commanded by irrigation canals. These impediments are the socio-political dominance of farmers in areas excessively endowed with surface water. They refuse to reduce canal intakes. Water resource and irrigation engineers have an inadequate understanding of conjunctive use and the potential role of groundwater. The responsibility for surface water and groundwater development and/or management is split, with organizations and agencies that tend to 'mirror' historical irrigation water supply realities and perpetuate the status quo.

## 3 Analysis of Conjunctive Use Systems

### 3.1 General Issues

Water storage provided by conjunctive use allows: increasing water availability without augmenting surface reservoirs, inclusion of aquifer as an additional water source, advantages in aquifer storage and distribution, increasing water supply guarantee, drought mitigation, decrease of aquifer exploitation and in some arid areas the reduction of drainage problems caused by irrigation return flows. Alternatives of conjunctive use for the development of water resources or the expansion of existing ones should be considered instead, or in addition, to the construction of new dams.

The possibility of storage in aquifers is significantly higher than that in existing or possible dams. Lund et al. (2014) give an economically viable capacity in California of about 180 $km^3$ for storage in aquifers, compared to about 50 $km^3$ in surface reservoirs. The ground and surface storage used in the relatively small Mijares Basin amounts respectively to 0.7 and 0.15 $km^3$. The ability to utilize additional water storage in California varies greatly with its location, the availability of water conveyance capacity and the operation of the system to integrate groundwater, surface water and conveyance facilities,= and brings out ACU, emphasising that many conjunctive use efforts rely on in lieu recharge. The lack of conveyance capacity to bring stored water to or from reservoirs or aquifers, water rights, contract constraints and regulatory limitations limits the effective use of available storage space.

## 3.2 Models as Water Planning and Decision-Making Tools

Models have become essential tools for analysing complex systems to adequately evaluate their performance. The analysis of conjunctive use must include the physical components, the flow interchanges between groundwater and surface water, the supply facilities for different in-stream and off-stream uses and the operating rules that are crucial for the different alternatives. The same set of structural facilities can produce very different yields, depending on the operation of the system. In the case that the variability of river flows increase in the coming decades, in addition to flow decrease in some basins, could cause more intense floods and droughts. The latter has two important consequences: (1) it will be necessary to have a greater water storage capacity, where conjunctive use would be useful, and (2) it will be crucial to perform much more simulations of the behaviour of each system, with different operating rules, for dry and wet time series and large periods, as well as for more intense droughts.

The California Department of Water Resources (CDWR) has been developing models for the management of the California Water Plan. The lack of a groundwater component was noted as a major deficiency in the first CALSIM (California simulation model) Peer Review, and in the 2009 upgrading of the California Water Plan (CWP), it was considered that models were not sufficient to model the impact of pumping on surface flows. Only in 2013 was finished the complex surface and groundwater flow model of the Central Valley of California, with a surface area of 51,000 km$^2$, 35,000 nodes and 35 stream reach where exchange of aquifer–river flow were determined.

## 3.3 The Eigenvalue Solution

In classical groundwater models, the algebraic linear system of equations obtained using finite differences (FD) or finite elements (FE) are solved in sequential time steps, providing water heads in all cells of the aquifer. The simultaneous simulation of the surface and underground systems is cumbersome if many alternatives of a few decades have to be analysed, considering adequately the aquifer–river relationships. To calculate the flow exchanges between the river and the aquifer in a period of time, it is necessary to calculate them from the aquifer heads in the period that are being calculated. The eigenvalue method solves explicitly the linear partial difference (PD) flow equation, considers constant in time transmissivity, storativity and boundary conditions and computes directly and very easily the heads, flows and volumes required. It uses the same parameters of a calibrated model in FD or FE. The eigenvalues method discretizes the space in N nodes or cells, like in usual FD or FE models, but not discretizes time. It solves the flow equation explicitly and continuously in time, obtaining $N$ ***modal orthogonal components***. Each ***mode*** has an eigenvalue $\alpha_i$, always positive, and their corresponding eigenvector $a_{i,1}$, $a_{i,2}$, $a_{i,3}$, … $a_{i,N-2,}$, $a_{i,N-1}$, $\alpha_{i,N}$. The eigenvalue computation has to be performed only once, and for any recharge or pumping, the solution is explicitly given by a vector $\bar{\mathrm{L}}_{(t)}$ of the intensity of the ***modes***, crucial in case of many alternatives and long simulation periods. It is an exact solution of the problem but given on a different orthogonal basis than that of the heads in the aquifer cells. If the storage vector of the aquifer model is $\mathrm{SF}_1$, $\mathrm{SF}_2$, $\mathrm{SF}_3$, … $\mathrm{SF}_N$) orthogonality is expressed by

$$\sum_{k=1}^{k=N} a_{i,k}a_{j,k}\mathrm{SF}_k = \delta_{i,j};\, 1 \text{ if } j = k, \quad \text{and } 0 \text{ if } i \neq j \qquad (1)$$

The orthogonal base in the classic solutions of FD or FE is (1,0,0 …0.0,0,0), (0,1,0,0, …0,0),…. (0,0,0, … 0,0,1). As in other physic problems, the solution could be given with enough accuracy considering a limited number of eigenvectors. What is an additional important help to simplify the solution. The calculations to determine in each time interval, the J levels, flows and volumes, would be those of the product of a $J \times E$ matrix, $E$ being the number of modes considered. that in most cases is an important improvement.

There are a small number of dominant modes, and the others are much less important. With a much-reduced number of components, the results are almost identical to the FD or FE solution, especially when calculating the aquifer–river flow. From the previous situation $\bar{\mathrm{L}}_{(t)}$ and pumping or recharge in the period, a new $\bar{\mathrm{L}}_{(t+\Delta t)}$ vector is obtained explicitly by a very simple state equation. Moreover, not all $N$ heads values at each time period are needed. What is needed are the heads in a few points, the flow between aquifer and river at given stretches, the ground water volume stored in the aquifer or in different areas of it and the flows between some adjacent zones. For each mode, the components of those heads, flows and volumes are linear combinations of the piezometric heads and can be calculated directly (Sahuquillo 1983, 2013; Andreu and Sahuquillo 1987; Sahuquillo et al. 2010; Alvarez-Villa 2014).

In the management problems, it is convenient to define the external actions on the aquifer with a limited number of elementary unit actions $J$, (top left of Fig. 1). The component li of the vector $\overline{L}$ of each elementary action of intensity $P_j$ is distributed among all modes by the pre-calculated coefficients $b_{i,j}$, whose sum is unity (bottom left of the Fig. 1). The component of the recharges of each node would be:

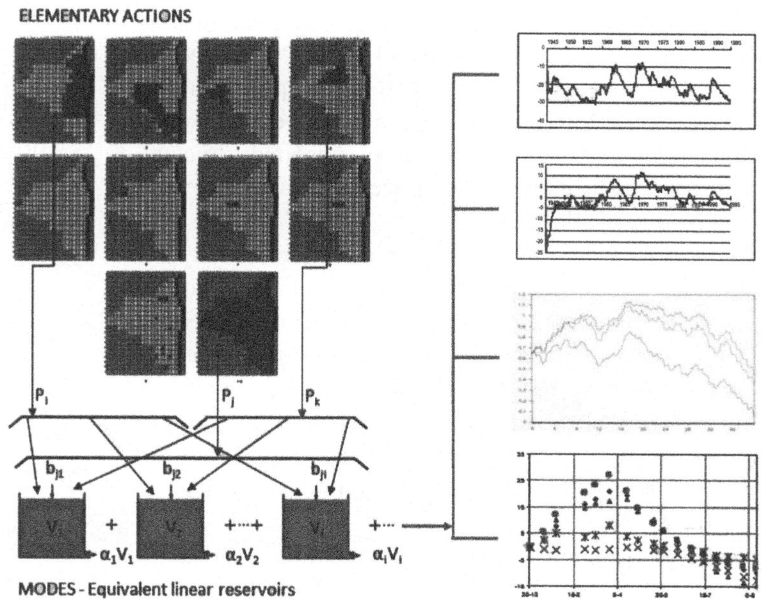


**Fig. 1** Simulation of elementary actions, modes, heads, flows and volumes

$$R_i(t) = \sum\nolimits_j b_{i,j} P_j \tag{2}$$

It would behave as a single-cell deposit with a volume $V_i(t)$ at time $t + 1$ would become

$$V_i(t+1) = V_i(t)\mathrm{e}^{-\alpha_i \Delta t} + R_i(t)\Delta t \tag{3}$$

Vector $\overline{V(t)}$ or their corresponding $\overline{L(t)}$ easily provides the heads, flows and volumes for each mode, as commented above and shows the right column of the Fig. 1. In each time interval, the calculations to determine, the $J$ levels, flows and volumes of the aquifer would be those of the product of a $J \times E$ matrix, $E$ being the number of modes considered. This is an important improvement in most cases.

For simple geometries, there are analytical solutions, which can be used when the knowledge of the aquifer is limited (Sahuquillo and Andreu 1986; Sahuquillo et al. 2010). Until now, the eigenvalues method has been applied only in some aquifers, due to the limited knowledge of the relationship of many aquifers with surface water in most of our aquifers. It seems obvious that its greatest advantage is in applying it to complex systems of conjunctive use with multiple components.

## 4 Related Ethical Issues and Final Remarks

The conjunctive use of surface and groundwater is mostly an improved aquifer management engineering method which is preferably oriented towards water quantity, although groundwater salinity and contamination problems can and should be addressed as well. This implies the technical, legal and administrative capacity to implement actions that do not disturb other rights and preserves the environmental values and services that are socially advisable under the point of view dominating at a given moment.

Groundwater is a neglected and generally misused resource. It can solve and is solving innumerable water demands for irrigation and industrial problems, an essential component in the environment, an integral part of the hydrological cycle and a component of water resource systems. Groundwater pumping produces a reduction in the flow of rivers or springs and other negative externalities, such as subsidence of the land or deterioration of wetlands and aquatic habitats. In addition to renewable water resources on an annual scale, it has reserves of tens or thousands of times its annual resources in large aquifers. To the resources and reserves available in the aquifers, it has to be added their capacity to store local or imported water through artificial recharge or ACU. It is a moral and ethical obligation to know, use and protect them, estimating in each case the results of their use and limiting the degradation of groundwater quality as much as possible by avoiding contamination. When using aquifers, the relationship between surface and groundwater should be considered. A quantitative analysis must be carried out, which can be done using simple tools when the recharge of the aquifer used exceeds demand widely. But in cases where the use of aquifers is important, advanced management models must be used. In any case, quantitative assessments should be accompanied by estimating the possible uncertainty. The possibility of increasing available resources or increasing the guarantee of supply with small sporadic pumping was discussed above. In cases of important use, temporary information on aquifer pumping and water quality is required and an adequate piezometric network should be established, as well gauging stations to determine river–aquifer flow exchanges. It is advisable to have annual records of surface water diversions and aquifer pumping. Data must be of permanent open access, and periodically published and updated. An independent, stable and economically supported structure is required, which allows maintenance, data collection, process and regular publication.

Unfortunately, in most countries, conjunctive use schemes that could provide obvious economic and social advantages have not been planned and water agencies lack the human and material resources to management and knowledge of the systems. In addition, the lack of employees with hydrogeological training in water agencies is notoriously insufficient and the institutional concern about the protection of groundwater against pollution is small.

**Acknowledgements** The author acknowledges the comments, revisions and important contributions to the manuscript of Professor E. Custodio.

## References

Alvarez-Villa OD (2014) Efficient simulation of the river-aquifer relations in conjunctive use systems applying reduction techniques on linear time invariant models. Technical University of Valencia, Spain, PhD Thesis (in Spanish)

Andreu J, Sahuquillo A (1987) Efficient aquifer simulation in complex systems. J Water Resourc Plann Manag Div 113(1):110–129

California Water Plan Update (2009) Department of Water resources. State of California

Custodio E (2010) Intensive groundwater development: a water cycle transformation, a social revolution, a management challenge. In: Martínez–Cortina L, Garrido A, López–Gunn E (eds) Rethinking water and food security. Fundación Botín and CRC Press, pp 259–298

Custodio E, Albiac J, Cermerón M, Hernández M, Llamas MR, Sahuquillo A (2017) Groundwater mining: benefits, problems and consequences in Spain. Sustain Water Resour Manag 3:213–226

Foster S, van Steenbergen F, Zuleta J, Garduño H (2010) Conjunctive use of groundwater and surface water from spontaneous coping strategy to adaptive resource management. World Bank, Washington

Foster S, van Steenbergen F (2011) Conjunctive groundwater use: a 'lost opportunity' for water management in the developing world? Hydrogeol J 19:959–962

Jaquette DJ (1981) Efficient water use in California: conjunctive management of ground and surface reservoirs. In: Beard LR (ed) Water for survival. J Hydrol 51:187–204

Lund J, Munévar A, Taghavi A, Hall M, Saracino A (2014) Integrating storage in California's changing water system. U.C. Davis California. S.D. Bechtel, Jr. Foundation https://watershed.ucdavis.edu/library

Llamas MR, Custodio E (2003) Intensive use of groundwater. Challenges and opportunities. Balkema

Sahuquillo A (1983) An eigenvalue numerical technique for solving unsteady groundwater models continuously in time. Water Resour Res 12:87–93

Sahuquillo A (1985) Groundwater in water resources planning: conjunctive use. Water Int 10(2):57–64

Sahuquillo A, Andreu J (1986) Quantitative analysis of the interaction between ground and surface water by the eigenvalues method. In: Water C (ed) Gorelick SM. IAHS Press, Use, pp 51–60

Sahuquillo A (2002) Conjunctive use of surface water and groundwater. In: Silveira L (eds) UNESCO Encyclopaedia of Life-Support Systems. EOLSS, Cambridge

Sahuquillo A, Capilla J, Martinez Cortina L, Sanchez-Vila X (eds) (2005) Groundwater intensive use. International Association of Hydrogeologist Selected Papers, Balkema

Sahuquillo A, Cassiraga E, Solera A, Murillo JM (2010) Modelos de uso conjunto de aguas superficiales y subterráneas. Instituto Geológico y Minero de España y Universidad Politécnica de Valencia, pp 1–328

Sahuquillo A (2013) Potential and merits of the eigenvalue method to simulate the conjunctive use of groundwater and surface water. Br J Environ Clim Change 3(3):480–498

Sahuquillo A (2017) Water storage and conjunctive water use. In: Ziolkowska JR, Peterson JM (eds) Experiences and management approaches in the USA and Europe Elsevier, pp 382–402

van der Gun J, Custodio E (2018) Governing extractable subsurface resources and subsurface space. In: Villholth K-G, López-Gunn E, Conti K, Garrido A, van der Gunn J (eds) Advances in groundwater governance. CRC Press, pp 389–409

# Enforcement of Groundwater Drilling and Abstraction Sites: Beyond the Backyard

Jeroen November, Rita Van Ham, and Sigrid Raedschelders

**Abstract**

Enforcement plays an important role in groundwater governance and management. Not only does it provide a crucial end-of-pipe view on the success or failure of imposed regulation, but also it acts as an advocate for good groundwater governance and management. As groundwater drilling or abstraction is often not the core business of inspected operators, a lack of knowledge on good installation and maintenance of the installations is often reported. A 'not in my backyard' or NIMBY vision is also often encountered with stakeholders living near to inspected installations. At the same time, however an 'only my backyard' or OMBY vision comes into play with regard to groundwater supply and use. A broadening of these visions is necessary, not only in the field, but also at the governance level to facilitate best practices. Since enforcement translates this governance and best practices to the field, it is also important to broaden the view to an international level regarding is successes and failures in doing so. Cooperation and sharing information about lessons learned during enforcement of groundwater drilling and abstraction sites aid in getting the enforcement business out of its own backyard. Resulting adaptation of legislation and enforcement tools or methods are a benefit for good groundwater governance and management, as this valuable resource does not adhere to these administrative boundaries.

**Keywords**

Environmental compliance • Groundwater • Enforcement

J. November (✉) · R. Van Ham · S. Raedschelders
Enforcement Division, Department of Environment and Spatial Development, Government of Flanders, Brussels, Belgium
e-mail: jeroen.november@vlaanderen.be

## 1 Introduction

Enforcement plays an important role in groundwater governance and management. Commonly, it is seen as the translator for groundwater legislation into the field, checking compliance and enforcing illegal drilling or abstraction when necessary. But that is only one aspect of enforcement. Not only does it provide a crucial end-of-pipe view on the success or failure of imposed regulation, but also it acts as an advocate for good groundwater governance and management.

Within Flanders, enforcement of groundwater abstractions at the regional level is carried out by the Enforcement Division (ED). These inspections are mainly focused on the larger (or class 1) establishments. At the local level, or for smaller installations, enforcement responsibility sits within the municipalities. This differentiation is not always beneficial regarding groundwater drilling and abstraction sites, due a lack in personnel and/or knowhow. Also, the environmental risks associated with illegal exploitation of these sites are not always directly linked to the size or capacity of the installations themselves. For example, erroneous installation of abstractions wells poses an equal risk to groundwater short circuiting or contamination, no matter the volume abstracted on a yearly basis.

The ED therefore also provides support to the municipalities when able, aiding them in augmenting groundwater compliance promotion within their jurisdiction, as groundwater flow does not adhere to the administrative boundaries set in place by governments.

On the other hand, the ED is also active at an international level within the European Union Network for the Implementation and Enforcement of Environmental Law (IMPEL), IMPEL Homepage (2020). Within this network, the ED raises attention to environmental compliance at groundwater drilling and abstraction sites within a 2 year project starting in 2021.



M. Abrunhosa et al. (eds.), *Advances in Geoethics and Groundwater Management: Theory and Practice for a Sustainable Development*, Advances in Science, Technology & Innovation,
https://doi.org/10.1007/978-3-030-59320-9_20

Both movements, at the local and international level, were inspired by inspection results acquired over the years by the ED and the subsequent feedback to policy makers.

## 2 Inspection Results and Adaptation of Legislation

Groundwater abstractions have been regularly inspected by the ED over the years. This included site visits as well as office reviews on administrative groundwater requirements by operators (e.g., yearly reporting of abstracted volumes, groundwater levels and qualities).

Results of these inspections showed a frequent on-site lack of good groundwater management (November et al. 2013). The installation of production wells and their maintenance and monitoring were frequently below standards. Abstraction of groundwater more often than not is just a side event of business operations. With some exceptions, e.g., drinking water companies, groundwater wells are seldom opened for inspection unless there is an obvious defect like a pump malfunction. As long as the waterflow is maintained, there is often no sense of urgency when waterflow meters stop performing correctly, or some trend in water quantity or quality could potentially be observed.

During inspections, the ED therefore often acts as an advocate for good groundwater management, next to hard enforcement when necessary. It is remarkable to notice how often operators think that imposed groundwater measures (e.g., monitoring of groundwater levels and quality, well maintenance, adequate sampling during drilling, etc.) are solely for the benefit of the government who in turn will have to report this to some higher authority. Simply explaining how these same data, when looked at properly, can improve their own operation and at the same time reduce costs and help the environment can do wonders. Changes in management over time within these same companies do however mean that a single quick visit and explanation will not suffice. Raising groundwater awareness is a constant job.

As a lot of the reported that deficiencies were related to bad construction and maintenance, feedback was given by the ED to the policy makers.

An accreditation scheme for drilling companies was suggested and subsequently installed since 2017. This was an improvement as some major issues could now be tackled at the source, namely the effective drilling and installation of production wells. As a consequence, permitting requirements for operators were reduced or even removed and transferred to the drilling companies.

An accreditation scheme however needs its own enforcement to be effective in the field. In this specific case of drilling companies, a lot of competition is present in the field, depending on the specific area, e.g., dewatering sites, or geothermal closed-loop systems. Having best practices and good legislation in place therefore does not guarantee adherence to these in a tight competitive business. In recent years, the ED, in cooperation with the Flemish Environment Agency, started enforcement on the accreditation itself. At first, focus was put on compliance promotion by providing extra courses on best practices and legislation, on-site clarification of requirements during inspections, etc. (Anonymous 2017). After 2 years, enforcement became stricter when drilling companies proved to be non-compliant after several exhortations. This may eventually lead to fining, suspension or revoking of the accreditation. Especially, the latter two instruments proved to be an eye-opener for the companies involved, as this would basically mean suspension or termination of their business within Flanders.

## 3 The Backyard Principles

Some other noticeable specific aspects during groundwater inspections are the backyard principles. The 'not in my backyard' or NIMBY principle is widely known. Nobody wants any major installation in their backyard which may pose a threat to their well-being or business.

On the other hand, however, the 'only my backyard,' or OMBY principle as it could be called, can also be observed within the groundwater topic. This became more apparent in recent years as Flanders was subject to prolonged periods of drought during the last three summers. Therefore, limitations on water use were set for individuals and/or specific sectors (e.g., irrigation for agricultural use or gardening). As people realized that rainwater supply may not be enough in Flanders in the future and tap water use could be restricted, demand for own groundwater supply boomed. Vision became narrowed to the own backyard or production site.

Neither NIMBY or OMBY is beneficial in the long run. On the one hand, groundwater should be able to be abstracted or drilled through at delineated water bodies for specific requirements (e.g., drinking water, industrial or geothermal use). Groundwater governance can guide this through permitting systems, levies and subsoil planning. On the other hand, synergies should be found and implemented so that water co-use, reuse and looking beyond the backyard become the standard mode of operation instead of the individual or 'first come first served' approach. The former gets more attention in recent years because the droughts and opportunities in that direction can also be highlighted during on-site inspections by enforcers.

## 4 A Double Way Forward

Considering the results and principles mentioned above, the ED will proceed with a double way forward starting 2020 in the groundwater topic.

As groundwater abstractions and drilling sites are not structural part of inspection tasks at the municipal level, the ED will provide specific support regarding these topics. This may include extra information or clarification on specific technical legislation via different channels as well as distribution of usable instruments (e.g., checklists, online documentation, etc.). Joint inspections can be carried out with the local enforcers at specific sites or sectors to invoke a level playing field over the territory of Flanders, cross-cutting municipal or provincial boundaries.

Within the IMPEL network, the ED will start a project 'Tackling Illegal Groundwater Drilling and Abstractions' (TIGDA). The aim of TIGDA is sharing knowledge and good practices on how to manage groundwater drilling and abstractions on EU-level. This would include specific permitting conditions, accreditations as well as enforcement tools and methods in place in different member states to reduce illegal drilling and groundwater abstraction, thus helping in attaining a good quantitative and qualitative state of groundwater bodies. Important to note that by 'illegal' in this sense non-permitted installations as well as permitted but non-compliant installations are meant (e.g., over-abstraction, faulty installation (depth, grouting)). Subsequent drafting guidance documents on groundwater drilling and abstraction legislation and enforcement will be carried out in the next phases. These guidance documents should be practical and usable in the field by policy makers and enforcers. This poses a good opportunity to get groundwater scientists, enforcers, policy makers and other stakeholders around the table on the subject of groundwater drilling and abstractions. The resulting guidance could act as an incentive for better cross-boundary groundwater governance.

Thus, with this double way forward, the ED aims to exit its own backyard and look beyond the borders for knowledge sharing, compliance promotion and optimizations.

## 5 Concluding Remarks

Having an efficient and performant groundwater enforcement scheme in place on different levels is beneficial for good groundwater governance and management. This enforcement, like the operators they focus their attention on, should be aware of the backyard principles. Looking further than those limits is beneficial in the long run to protect groundwater recourses for all stakeholders involved.

Since enforcing is not solely about the act of enforcing itself, it remains important to have this instrument in the field as an actor to aid compliance promotion by highlighting risks of bad management to the environment. Providing support and information based on know-how can be a groundwater awareness incentive for inspected companies.

By including a double way forward in groundwater enforcement, the ED aims to put this incentive in place at the local as well as the international level, thus protecting groundwater recourses. Tackling illegal groundwater drilling and abstractions will become more important in the future, working in an environment under a changing climate.

## Reference

Anonymous (2017) Milieuhandhavingsrapport. Department Omgeving, Brussels

November J, Baert R, Blondeel M (2013) Groundwater extractions in Flanders: an enforcement review. In: Cobbing J, Adams S, Dennis I, Riemann K (eds) Assessing and managing groundwater in different environments. IAH selected papers on hydrogeology, vol 19, pp 135–145

IMPEL Homepage, https://www.impel.eu. Last accessed 2020/02/14

# On the Relevance of Environmental Law Evolution as a Fundamental Pillar of the European Union

António J. Dinis Ferreira

**Abstract**

Environmental law is a main pillar of European Union construction. It has changed throughout the decades to encompass the new visions and strategies for Europe, towards a more sustainable future. We analyse a number of EU legal documents to assess the development of several concepts throughout the decades, pinpointing the key challenges throughout the evolution. If at the beginning of the period in appreciation key principles of openness, equity and accountability are important in the legal documents, they seldom appear in more recent documents, where the recently coined challenges of circular economy, nature-based solution and new green business models are more dominant. Climate change and risk assessment and management are pervasive topics, witnessing the European Union concern with these matters for several decades. The vertical and horizontal integration with international institutions and with the member countries and European regions is also a dominant worry of European environmental legislation, providing the freedom for adaptation to the local and national cultural and societal characteristics. In fact, in contrast to the commonly believed by the citizens, due to the subsidiary principle, the enforcement of European legislation is performed, in an adapted manner at country level.

**Keywords**

Environmental law • European Union • Enforcement • Challenges • Evolution

A. J. Dinis Ferreira (✉)
CERNAS, Instituto Politécnico de Coimbra, Escola Superior Agrária, Bencanta, Portugal
e-mail: aferreira@esac.pt

## 1 Introduction

Environmental law is considered a fundamental pillar to the European Union construction. Partly because the resolution of environmental problems back in the 1960s and 1970s' coincided with the European Union affirmation and widening the scope of it's' action, partly due to the lack of a coherent environmental legislation at national level, prior to that, and partly because the environmental legislation soon became a fundamental tool towards a vision of sustainability.

The environmental sector has gained an increasing role at EU level, given its' transversal characteristic and potential capacity to bind together all the sectors and stakeholders, in increasingly tighter, more consolidated and integrated connections, around common visions, strategies, plans and actions.

In response to the mounting environmental problems posed by an intensification of industrial linear activity in a post II world war, the 1970s witnessed a reactive response with reliance on abatement, based on end-of-pipe solutions and little consideration of resources consumption. This was encompassed by a regulation-driven pollution control. In the 1980s started the concerns with pollution prevention aiming at reducing the quantity of pollutants produced and the reduction of the amount of materials used by implementing the first systems of recycling and reuse. In the 1990s, a new generation of environmental tools popped up, such as the first steps of life cycle analysis and eco-design, environmental certification and the concept of extended product responsibility. This witnesses a more proactive attitude and acting beyond compliance.

The beginning of the XXI century witnessed the enforcement of individual and corporate responsibility in a wide range of areas with environmental impact, namely in terms of energy, economy, social and on the use of resources, and the implementation of concepts such as eco-industrial parks, industrial metabolism, industrial



M. Abrunhosa et al. (eds.), *Advances in Geoethics and Groundwater Management: Theory and Practice for a Sustainable Development*, Advances in Science, Technology & Innovation,
https://doi.org/10.1007/978-3-030-59320-9_21

ecology. These matured to the current state at the dawn of the third decade of the XXI century, where a new concept, Circular Economy, was coined by the European Union as a vision of a more sustainable future, to avoid a civilization collapse while maintaining acceptable levels of well-being. Concomitantly, the new vision gets hold to a set of new concepts, such as nature-based solutions or short supply chains to strength the path to a more sustainable future.

This work witnesses this evolution, from the analysis of a set of circa 150 European legislation documents, on biodiversity, environmental management and regional and urban planning. The work establishes a direct relation between the European Union sustainable development paradigm and the evolution of the criteria set to evaluate the European legislation for the last half century.

## 2 Methods

The survey was performed on documents extracted from the European Legislation Platform EUR-Lex database (https://eur-lex.europa.eu/). The search spreads from the 1970s onwards and includes several types of legal documents, namely directives, regulations, decisions, resolutions, opinions, action programs, Commission legislative proposals, Council common positions, European Parliament legislative and budgetary resolutions and initiatives, European Economic and Social Committee opinions, Committee of the Regions opinions and other preparatory documents.

To address the evolution of the European Environment legislation body, a set of 30 criteria were selected to check the evolution and coherence of the legislation, providing a sound basis to gauge the evolution of concepts and visions throughout the last 50 years of European Construction.

The criteria derived from EC (2013a; b) and Campbell (2016) include 4a subgroup on (1) "Relations to other relevant EU policies and funding", that included five criteria, including among other the relation with other programs, funds and policies, such as the CAP and territorial planning. (2) "Relations to other environment issues", with seven criteria, such as circular economy, climate change and natural disaster risks, public participation and citizen science. (3) "Relations to key EU legislation requirements" with 15 criteria, including openness, equity or accountability and the vertical and horizontal integration of policies. (4) "Relations to factors fostering impact" with seven criteria, including the potential to foster business, conflict prevention and resolution and the capacity to enable power networks.

## 3 Results and Discussion

In brief, current legislation is more consequent in what concerns the links between different policies and working areas, including the definition of programs funding; the existence of a multiple scale and integrated approach, including vertical and horizontal integration, namely with customary and formal governance or the connectivity with national and international developments, the creation of a sense of community, the management of risk and the provision to accommodate climatic change, the fostering of natural capital and effective green solutions, nature conservation and the fostering of circular economy and the bio-economy.

In what concerns those issues, some of them are dominant, such as the integration and funding aspects; the integration issues depend on the type and scope of the document, while some of the newest topics, such as the bio-economy and circular economy, recently coined definitions, only came apparent in recent years. Climate change and risk assessment and management are a pervasive topic.

More ethic issues such as openness, equity and accountability, together with conflict prevention and resolution are seldom referred to, although they are important issues central to the European legislation construction and largely embodies the action of public institutions within the European space. As if those values were already part of the agreed underlying principles and do not need to be stated.

Innovation and knowledge sharing and leading, the involvement of NGOs, the fact that they can enable negotiations, power networks and the construction of hegemonic visions towards the future is a worry that appears in more recent years.

Finally, the need to make the environment and biodiversity economically viable starts to appear in more recent documents, namely in what concerns the fostering business and employment, witnessing a recently framed vision of economy diversification and increasing the value of the environment and nature (including ecosystem functioning and biodiversity) that is embodied in another touchstone of European development for the future, the nature-based solutions.

## 4 Concluding Remarks

Environmental legislation is one of the main pillars of European construction. Synchronic with the beginning of the European Economic Community and afterwards with the European Union, it started by consolidating the sparse environmental legislation at national level, by providing a coherent structure, that soon started to embody a vision for a sustainable future.

The documents analysed witness the evolution of environment concepts and visions throughout the decades, and the pivotal role of the environment in pursuing a more sustainable development in a context of limited resources. In this context, the work will address in particular the questions of groundwater management that is of utmost importance, namely for the built environment, where the impacts on human and ecosystem health are more stringent.

**Acknowledgements** This work was performed under project URBAN-GAIA Managing Urban Biodiversity and green infrastructure to increase city resilience. JPI BiodivERsA3-2015-143/FCT-BIODIVERSA/0008/2015.

## References

Campbell LK (2016) Getting farming on the agenda: planning, policymaking, and governance practices of urban agriculture in New York City. Urban Fores Urban Green 19:295–305

EC (2013a) Communication from the commission to the European Parliament, the Council, the European economic and social committee and the committee of the regions on green infrastructure (GI)—enhancing Europe's natural capital COM (2013) 249 final, 11p

EC (2013b) Commission staff working document. Technical information on green infrastructure (GI) SWD (2013) 155 final, 24p

# Geoethics in the Design and Implementation of Temporary Groundwater Control Systems

James Watson and Stephen Thomas

**Abstract**

Geoethics is required at almost every stage of the design and implementation of temporary groundwater control systems. Although these systems are designed for temporary purposes, they can have long-lasting impacts on the environment if they are not designed and installed with geoethics in mind. These impacts can include (i) transfer of contaminants, (ii) saline intrusion and (iii) damage to local watercourses. Engaging effectively with geoethics during the implementation of temporary groundwater control operations involves more than simply having a caring attitude. Effective engagement requires training, extensive knowledge, careful assessment of risks and meticulous planning and management. Experience shows that in cases where temporary groundwater control operations are implemented without active engagement in geoethics, there are not only environmental impacts, but also later consequential costs and project delays. Therefore, careful consideration of geoethics in temporary groundwater control can be viewed as positive both morally and also commercially. This paper documents some of the key stages in the design and implementation of temporary groundwater control systems at which issues of geoethics are encountered. The paper also discusses the importance of a design risk assessment (DRA) and an inspection testing and monitoring (ITM) plan in addressing identified issues of geoethics.

**Keywords**

Geoethics • Groundwater engineering • Groundwater control • Design risk assessment • Inspection testing and monitoring plan

J. Watson · S. Thomas (✉)
OGI Groundwater Specialists, Durham, UK
e-mail: Stephen.Thomas@ogi.co.uk

## 1 Introduction

Due in part to the world's expanding population and resultant increasing demand for geological resources, geoethics is becoming an increasingly important global discipline. This discipline has influence and implications in a number of sectors including hydrogeology, water resources and groundwater engineering and control.

Despite the fact that temporary groundwater control systems are by definition non-permanent in nature, issues of geoethics are encountered at almost every stage of their design and implementation. If these issues of geoethics are not carefully considered, then there can be significant (and in some cases permanent) impacts on the environment including (i) transfer of contaminants, (ii) saline intrusion, (iii) damage to watercourses and (iv) potential settlement of local structures.

The potential detrimental impacts of temporary groundwater control systems are usually experienced as a result of the system being designed by inexperienced personnel who do not fully appreciate the relevant issues of geoethics and consequent environmental risks. To mitigate these risks, the designer must be more than just competent and caring. If the behaviour of geoscientists is not guided by geoethical principles, even scientifically competent individuals may not find and choose solutions that are respectful of the natural dynamics of our planet (Bobrowsky et al. 2017).

Effective engagement with geoethics during the design and implementation of temporary groundwater control systems involves training, extensive knowledge, careful assessment of risks and meticulous planning and management. The development of a design risk assessment (DRA) and inspection, testing and monitoring (ITM) plan are key steps which ensure that issues of geoethics are identified and considered.

Engaging with geoethics does not only have environmental benefits. Careful planning and management can reduce later consequential costs and project delays.



M. Abrunhosa et al. (eds.), *Advances in Geoethics and Groundwater Management: Theory and Practice for a Sustainable Development*, Advances in Science, Technology & Innovation,
https://doi.org/10.1007/978-3-030-59320-9_22

Therefore, there is often an economic benefit to engaging with geoethics during the design and implementation of temporary groundwater control systems.

This paper documents some key stages in temporary groundwater control at which issues of geoethics are encountered, including (i) abstraction well design, (ii) pumping system design, (iii) discharge and recharge and (iv) decommissioning. The importance of a DRA and an ITM plan in identifying issues of geoethics early on is also discussed.

## 2 Issues of Geoethics in Temporary Groundwater Control

Careful and meticulous planning during the preliminary design stages of a temporary groundwater control system is essential if geoethics is to be engaged with effectively. The available site information must be reviewed with geoethics in mind.

It is critical that adequate site information is available so that issues of geoethics and resultant environmental risks can be identified and planned for at an early stage (e.g. contamination risks, nearby sensitive watercourses, vulnerable aquifers, etc.).

The following subsections document some, but not all, of the key stages in temporary groundwater control at which geoethical issues are encountered.

### 2.1 Abstraction Well Design and Installation

Without appropriate design and installation, abstraction wells utilised in temporary groundwater control operations can have detrimental environmental impacts. Poorly designed abstraction wells will pump silty water which (dependent on the discharge location) can cause pollution, damage to habitats and clogging of culverts or sewers.

Effective engagement with geoethics in the design and installation of abstraction wells involves taking steps to mitigate the potential environmental risks they pose.

Abstraction wells are to be designed and installed to include (i) a suitably sized filter annulus, (ii) a suitable filter material which is selected based on the ground conditions, (iii) a suitable base cap and (iv) a suitable casing slot size. If the above items are taken into account during appropriate well design, then the risk of pumping silty water is minimised, resulting in the abstraction of clear, filtered groundwater (Fig. 1).

### 2.2 Pumping System Design

If the pumping system of a temporary groundwater control system is over-designed or over-conservative, then this can result in excessive drawdown and over-abstraction. If this occurs, then there are a number of environmental risks including (i) reduction in baseflow to surrounding surface water features,

**Fig. 1** Example of clear, filtered abstracted groundwater

(ii) reduction in water quality from reversing flow gradients from a salt water source and (iii) settlement of local structures.

Mitigation of the environmental risks of excessive drawdown and over-abstraction is achieved by ensuring that the pumping system design is not over-conservative. Furthermore, careful management on site is essential so that pumping can be reduced when appropriate (e.g. by throttling back or switching off certain pumps).

Where feasible, pressure relief wells can be considered as a means of temporary groundwater control. Pressure relief wells do not need to be pumped directly and can be designed and installed such that the abstracted groundwater is intercepted before it mixes with surface water and soil so that the filtered water remains clear and silt free.

## 2.3 Discharge and Recharge of Abstracted Groundwater

Discharge of silty or contaminated groundwater can have a severe impact on the local environment, including wildlife and water resources. Discharge operations therefore need to be planned and managed in a geoethical way.

To prevent silt and contaminant pollution, groundwater is first filtered using carefully designed wells, then, if required, the groundwater is chemically treated prior to discharge. Clear abstracted groundwater is also to be kept separate from collected perched or surface water, as mixing reduces the quality of the abstracted water. This can be achieved using separate tanks, treatment facilities and discharge methods.

In England and Wales, legislation means that temporary groundwater control operations are now licensable activities (Water Abstraction and Impounding (Exemptions) Regulations 2017). This means that permits are now required for both abstraction and discharge of groundwater in many parts of the UK. Licence applications should be made at an early stage in the design of a temporary groundwater control system and must be strictly adhered to for both ethical and environmental reasons.

Where feasible, temporary groundwater control recharge systems can be considered as an alternative to conventional means of groundwater discharge (i.e. discharge to sewer or watercourse). Artificial groundwater recharge during temporary groundwater control involves the recirculation of abstracted groundwater back into the same geological strata by reinjection (via wells) or infiltration (via trenches or shallow wells) (Preene and Fisher 2015). This approach is geoethical and can mitigate environmental risks by (i) reducing drawdown at distance and (ii) reducing net abstraction from the aquifer to negligible levels. The concept of temporary groundwater control recharge systems is presented in Fig. 2. This process is described in '*Groundwater Recharge Systems to Mitigate the Impact from Construction Dewatering*' (Thomas et al. 2020).

## 2.4 Decommissioning

The decommissioning of temporary groundwater control systems is also an important issue of geoethics. If the groundwater control system is not decommissioned appropriately, then there can be permanent damage to the environment. Wells can provide pathways for groundwater to travel between different aquifers and/or geological strata. This can result in the transfer of groundwater and potential contamination risks.

Groundwater control wells must be decommissioned so that pathways for groundwater transfer are sealed to prevent permanent seepage (i) between different aquifers, (ii) to the ground surface or (iii) into buildings. Furthermore, wells must be decommissioned with inert materials which do not pose environmental risks themselves.

In the UK, Environment Agency Guidance on the decommissioning of groundwater control wells should be

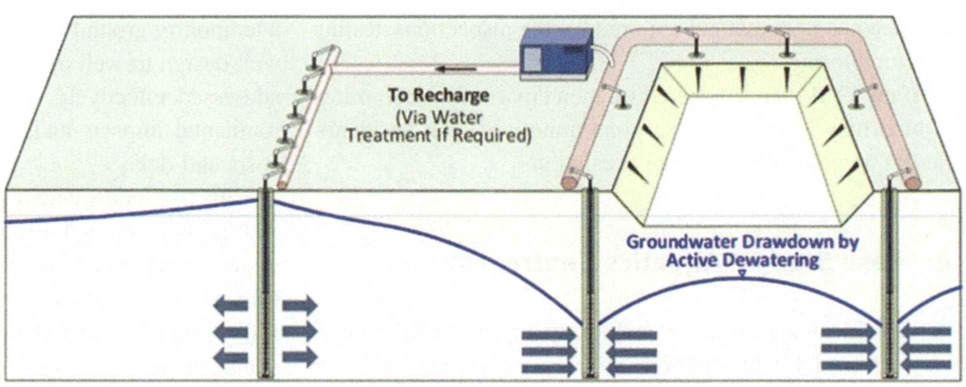


**Fig. 2** Concept of dewatering and artificial recharge system

followed (Environment Agency—Good Practice for decommissioning redundant boreholes and well 2012).

## 3 Design Risk Assessment and Inspection, Testing and Monitoring

### 3.1 Design Risk Assessment (DRA)

The development of a design risk assessment (DRA) is a critical step in designing a temporary groundwater control system if engagement with geoethics is to be effective. The purpose of the DRA is to identify potential hazards, quantify the risks they pose and to document mitigation and control measures.

While a DRA is not necessarily limited to environmental hazards and risks, it is strongly recommended that this category of hazards is included. Environmental risks are then documented in the DRA together with geoethical mitigation measures.

A number of environmental risks are common to almost all temporary groundwater control operations (e.g. migration of contaminated water), but there are other project-specific hazards and risks which can be included in the DRA.

### 3.2 Inspection, Testing and Monitoring (ITM) Plan

The development of an inspection, testing and monitoring (ITM) plan is considered good practice for the successful implementation of a temporary groundwater control system if geoethics is to be engaged with effectively. The purpose of the ITM plan is to document inspections, tests and monitoring activities required during the temporary groundwater control works, from the drilling stage through to decommissioning.

The ITM plan needs to be developed prior to the drilling, installation, cleaning and testing of the groundwater control system and needs to document the relevant frequencies and competence of personnel required for the inspections, testing and monitoring.

If the ITM plan is adhered to, then environmental hazards (which may have otherwise gone unnoticed) can be identified at an early stage, before escalation.

## 4 Case Study—Aquatics Centre, UK

A geoethical approach to temporary groundwater control was adopted by the authors on a major recent UK construction project. This project involved the construction of an aquatics centre which required excavation to 7 m below ground level. A high water table, together with artesian groundwater pressures meant that groundwater control was essential, as the risk from groundwater was considered a health and safety critical issue.

Early stakeholder engagement meant that a number of environmental concerns were identified during the preliminary stages of the project including (i) the regional water resource status, (ii) the protection of trees surrounding the site and (iii) potential contamination in the made-ground deposits.

A design risk assessment (DRA) was developed during the concept design stage which provided confidence to stakeholders that environmental risks were first identified and then considered. The DRA documented the environmental hazards, potential risks and proposed geoethical mitigation measures. Risks to water resources were quantified as part of the DRA through groundwater modelling and test pumping.

The risks to the water resources were mitigated through the design and construction of a temporary groundwater control recharge system. This system enabled recharge of the abstracted groundwater back into the same geological strata.

Wells were installed at distance from marked out '*root protection zones*' so as to mitigate damage to the trees surrounding the site. Furthermore, soakaway trenches were excavated close to some trees to ensure that any drawdown did not starve them of water.

Finally, potential contamination risks were mitigated through the design and installation of bentonite seals for both the abstraction and recharge wells. These seals ensured that pathways for groundwater flow were not created between the made-ground deposits and underlying geological strata.

The early development of a DRA and implementation of an ITM plan meant that the project was managed geoethically with environmental risks identified and mitigated.

## 5 Concluding Remarks

Issues of geoethics are encountered at almost every stage of a temporary groundwater control operation, from abstraction well design to well decommissioning. If these issues are not addressed effectively, then there can be detrimental environmental impacts and, in many cases, later consequential costs and delays.

Effective engagement with geoethics during the design and implementation of a temporary groundwater control system involves more than simply having a caring attitude. Effective engagement requires appropriate training, technical skill, extensive experience and meticulous planning and management.

The development of a DRA and an ITM plan can help to ensure that geoethical issues and consequent environmental risks are identified and planned for at an early stage in the design and installation process. This results in the mitigation of environmental risks and increases the likelihood of project success, both geoethically and economically.

## References

Bobrowsky P, Cronin VS, Di Capua G, Kieffer SW, Peppoloni S (2017) The emerging field of geoethics. In: Gundersen LC (ed) Scientific integrity and ethics with applications to the geosciences. Special Publication American Geophysical Union. Wiley

Environment Agency (2012) Good practice for decommissioning redundant boreholes and wells. At: https://webarchive.nationalarchives.gov.uk/20140328154120/http://cdn.environment-agency.gov.uk/LIT_6478_8cbe6f.pdf. Accessed: 2020/04/28

Preene M, Fisher S (2015) Impacts from groundwater control in urban areas. In: Proceedings of the XVI European conference on soil mechanics and geotechnical engineering. ICE Publishing, London, pp 2847–2852

Thomas S, Goodfellow T, Watson J (2020) Groundwater recharge systems to mitigate the impact from construction dewatering. INCREaSE 2019. Springer, Cham

Water Abstraction and Impounding (Exemptions) Regulations (2017) Available at: https://www.legislation.gov.uk/uksi/2017/1044/made. Last accessed: 2019/05/07

# Remote Sensing for Irrigation Water Use Control: The Case of the Benalup Aquifer (Spain)

Alex Fernández-Poulussen, Mercedes Vélez-Nicolás, Verónica Ruiz-Ortiz, María Jesús Pacheco-Orellana, and Santiago García-López

**Abstract**

In recent years, the Copernicus programme developed by the European Space Agency (ESA) has enabled to improve our knowledge on the functioning of hydrological systems and water management. The added value of Copernicus lies in its open access and adequate time–space resolution that in combination with other datasets on land use, climate or hydrology makes it a powerful tool to monitor irrigated areas, assess irrigation performance or identify illegal abstractions. The present work is focused on the exploitation of the Benalup aquifer, which is located in the Barbate River Basin (Southern Spain) and has been declared as in poor quantitative and qualitative status by the Public Administration. The use of remote sensing techniques has allowed to identify the time–space distribution of irrigation in agricultural plots supplied with groundwater, the only available hydric resource in the area. The results obtained reveal a marked seasonality in the existing crops and a mismatch between the irrigated plots that have been identified by remote sensing techniques and those with irrigation license. This study evidences the potential of remote sensing for promoting a sustainable water use through the identification of illegal groundwater pumping that is of major importance in a context of climate change.

**Keywords**

Agriculture • Water • Satellites • Index • Stress • Abstraction

## 1 Introduction

Satisfying humanity's water demands while ensuring the qualitative, quantitative and ecological integrity of freshwater systems is one of the major challenges for the twenty-first century. Agriculture plays a decisive role in this context, as it is responsible for roughly 70% of water abstractions globally (FAO 2017). Thus, in the current context of climate change, with increasing pressures on hydric resources and aquatic ecosystems and a growing food demand, the ethical dimension of water management has increased in importance among farmers, supply chain partners and consumers. This situation calls for a water use in accordance with the regulations legitimately enacted by states that establish administrative procedures to allocate water rights and guarantee environmental sustainability. The ultimate objective of these procedures is to ensure an equitable and fair distribution of the resource among water users in order to promote social welfare and ecosystem conservation. Nevertheless, in recent decades, many countries (especially in Mediterranean and/or semi-arid climates) have undergone what some authors have termed "silent revolution" (Llamas 2005). This revolution, usually carried out by modest farmers, has consisted in the exploitation of groundwater by drilling thousands of boreholes through which large volumes of water are extracted for irrigation with little or no technical or financial assistance from administration, usually infringing the law.

Although this uncontrolled exploitation has resulted in substantial benefits to society and has enabled many less-developed areas to thrive, it has also led to a general deterioration of aquifers due to the depletion of groundwater reserves and the degradation of their quality. To reverse this

A. Fernández-Poulussen (✉)
Good Stuff International B.V, Den Bosch, The Netherlands
e-mail: alex@goodstuffinternational.com

M. Vélez-Nicolás · M. J. Pacheco-Orellana · S. García-López
Department of Earth Sciences, Faculty of Marine and Environmental Sciences, University of Cádiz, Cádiz, Spain

V. Ruiz-Ortiz
Department of Industrial and Civil Engineering, Polytechnic School of Algeciras, University of Cádiz, Cádiz, Spain





M. Abrunhosa et al. (eds.), *Advances in Geoethics and Groundwater Management: Theory and Practice for a Sustainable Development*, Advances in Science, Technology & Innovation,
https://doi.org/10.1007/978-3-030-59320-9_23

situation, the public administrations need to implement appropriate control mechanisms in order to adjust the extraction rates to what is established in the water planning documents. Any initiatives in this regard should be gradual and adapted to the local socio-economic reality, encouraging the stakeholders to become involved in water conservation. On the other hand, the administration needs knowledge-based tools to identify and evaluate groundwater extractions, and remote sensing offers a wide range of possibilities in this regard (Psomas et al. 2016; Alexandridis et al. 2018; Chance et al. 2017; Gao et al. 2018).

The present work is focused on the Benalup aquifer (Ground Water Body 062.014), located in the Barbate River Basin (province of Cádiz), which is currently declared as in poor quantitative and qualitative status. To a large extent, this situation is the result of the agricultural activity developed on the hydrogeological system; in the area, there are numerous boreholes from which large volumes of water are extracted annually, exceeding the limits set by current regulations. These regulations specify that the exploitation rate (ratio between the volume extracted and the available resource once the ecological water demands have been met) should be lower than 80%.

This study aims to highlight the potential of remote sensing for water management and the usefulness of the information generated by the Copernicus programme. Its observations, combined with other datasets and geographic information systems, provide key information to improve the use of irrigation water, to understand the catchment dynamics and to support the development of control mechanisms in the fields of water, agriculture and land use both at public and private levels. This work identifies and analyzes the agricultural use of groundwater, its time–space distribution and supports initiatives on efficient water management and control.

## 2 Methods

The study period comprised the hydrologic years 2016/17 and 2017/18. In the first stage, georeferenced information of the aquifer boundaries and land uses was gathered. Datasets from official sources were incorporated in a geographic information system to generate a detailed map of land uses and afterward, to carry out a quantitative analysis that allowed to discriminate between agricultural and non-agricultural plots and between plots with irrigation license and those unauthorized (Fig. 1).

Subsequently, satellite images from the ESA Copernicus programme were acquired. These images were obtained from the MultiSpectral Instrument (MSI) aboard the Sentinel-2 satellites. The Sentinel-2 mission comprises a constellation of two "twin" satellites (2A and 2B, launched in 2015 and 2017, respectively) placed in the same solar synchronous orbit. The aim of this mission is to monitor changes in Earth's surface, (vegetation, soil and water coverage) with a large strip width (290 km) and a high revision time (10 days in the equator with 1 satellite in cloudless conditions, which shortens to 2–3 days at mid latitudes when 2 satellites are operating). The MSI sensor produces 12 bands with a spatial resolution between 10 and 60 m and a radiometric resolution of 12 bits. The free access EO browser of the Sentinel Hub platform was used to download the images. The processing level chosen for this study was the Level-2A (orthorectified Bottom-of-Atmosphere (BOA) reflectance).

Daily data on precipitation and reference evapotranspiration (ETo) during the study period were obtained from the climate station of Vejer de la Frontera, which is located approximately 2 km away from the aquifer.

The Normalized Soil Moisture Index (NDMI) was used as an indicator of the plots that were irrigated during periods of absence of rain. The NDMI provides information about the level of water stress in crops; when the index reveals the existence of healthy crops and the absence of stress during dry periods, the situation is only attributable to irrigation. In the area of the Benalup aquifer, the only resource available for irrigation is groundwater; therefore, it is possible to identify the agricultural plots where the aquifer is pumped for irrigation. The product that indicates the moisture level of the crops was obtained from the combination of the NIR—near infrared band and the SWIR—short wavelength infrared band: NDMI = (NIR − SWIR)/(NIR + SWIR).

In principle, high values of the NDMI index are indicative of soil moisture. For this study, the following NDMI ranges have been considered (Table 1).

The analysis was focused on dry periods with the absence of rain. The reason is that during periods with rainfall, soil moisture and therefore the NDMI index are usually high, and consequently, the absence of water stress in crops makes irrigation unnecessary. Accordingly, for this study, a total of 20 images were selected from two temporal windows coinciding with dry periods (Fig. 2).

Once the pixel values of the crops were obtained and isolated from the pixels corresponding to non-agricultural land, the surface of monthly irrigated plots was estimated using geographic information systems and geospatial processes. Subsequently, the surface of irrigated plots was compared with the surface of land with authorized agricultural uses.

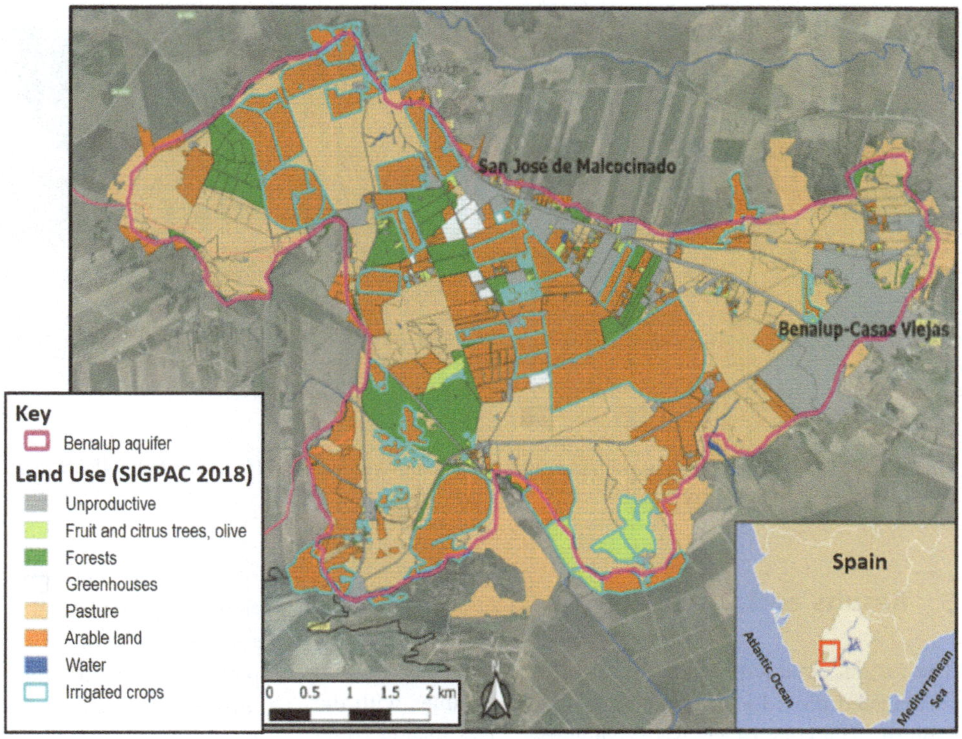


**Fig. 1** Land uses on the Benalup aquifer and plots with irrigation license. *Source* SIGPAC 2018 (Junta de Andalucía 2018)

**Table 1** Soil moisture levels interpretation according to NDMI from Antognelli (2018)

| NDMI | Interpretation |
|---|---|
| <0 | Low canopy cover, dry areas, bare soil |
| 0–0.2 | Average canopy cover: low water stress/mid-low canopy cover: low water stress |
| 0.2–0.4 | Mid-high canopy cover: low water stress/average canopy cover: low water stress |
| >0.4 | High canopy cover, no water stress/waterlogging |

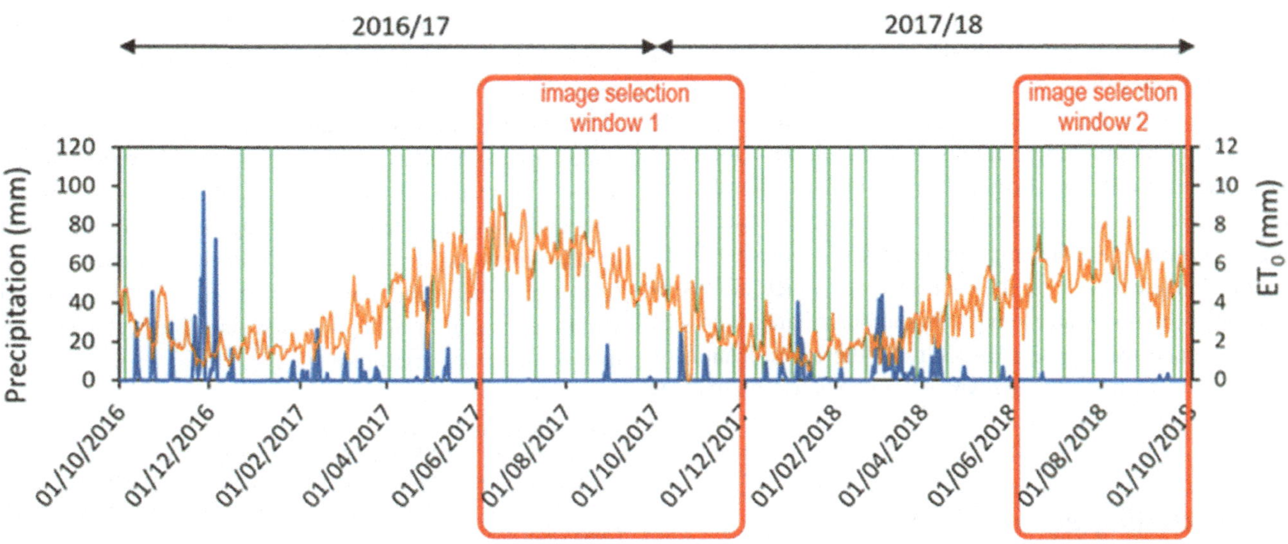


**Fig. 2** Daily rainfall during the study period (blue lines), reference evapotranspiration (orange lines) and dates of images obtained (green bars). Driest periods marked in red

**Fig. 3** Image showing irrigated plots in the central area of the aquifer (June 11, 2017). Left: NDMI image; Right: True color image, in both cases with overlapping land use layer

## 3 Results and Discussion

The results obtained include maps of the spatial distribution of irrigated plots for each date analyzed and tables with the monthly statistics of areas by categories. The results show similar values of irrigated surface both in 2017 and 2018 evidencing that on average irrigation affects a surface of 296 ha, of which 213 ha (72%) hold irrigation licenses, whereas 83.5 ha (28%) affects land with no declared license which could be classified as rainfed land or other uses (forest, pastures, urban land, etc.). On the other hand, only a 27% of the total irrigated area that is authorized was being irrigated. Figure 3 (true color images) shows an example of the distribution of irrigated plots, some of them lacking the corresponding license. Since in the previous 30 days to the image capturing there was no rainfall, the presence of soil moisture clearly indicates that irrigation was occurring in the area.

Although the mentioned results are referred to average values, this methodology allows to monitor on a monthly basis the evolution of irrigated crops and to identify the number of harvests gathered in a same plot. It also enables to assess groundwater consumption by crossing these data with the ones taken in situ on the type of crop, which allows to estimate irrigation water needs. On the other hand, it should be mentioned the intrinsic difficulty in the identification of irrigation in fruit tree crops whose canopies are permanent (citrus, grapefruit, etc.).

## 4 Concluding Remarks

The methodology presented in this work, when applied to an aquifer that has been declared as in poor quantitative and qualitative status, allows to acquire detailed and updated knowledge on land use, with special regard to irrigated agriculture. Besides, it enables to get information about pressures derived from agricultural practices and the exploitation to which the aquifer is subjected. In this case, it is striking that only a 27% of the area authorized for irrigation was actually being irrigated and that, of the total area that was irrigated, the 28% were plots without authorization for this activity. This situation evidences, therefore, a mismatch between reality and what is regularised by administration.

The existence of a real-time monitoring tool promotes transparency and self-control among water users and, in case of illicit exploitation of groundwater resources, enables the adoption of coercive measures by the authorities, Water Users Associations or the corresponding administrations. This is especially relevant in semi-arid regions where groundwater overexploitation is a major problem or in areas severely affected by climate change.

Additionally, this technique offers the possibility of being scaled up and applied to different contexts and geographical areas. The results of this study highlight the versatility of remote sensing tools and methodologies developed by public organisations such as the European Spatial Agency (ESA) and their potential contribution to public or private initiatives aimed at achieving a sustainable water use, land use planning or agriculture.

**Acknowledgements** This work has been developed within the framework of the REMABAR project, supported by the Biodiversity Foundation of the Ministry for the Ecological Transition. The authors also want to express their gratitude to Derk Kuiper and Erika Zárate of Good Stuff International for their support and advice.

## References

Alexandridis TK, Zalidis GC, Silleos NG (2008) Mapping irrigated area in mediterranean basins using low cost satellite earth observation. Comput Electron Agric 64:93–103

Antognelli S (2018) Agricolus. NVI and NDMI vegetation indices: instructions for use. https://www.agricolus.com/en/indici-vegetazione-ndvi-ndmi-istruzioni-luso/. Accessed 1 June 2019

Chance EW, Cobourn KM, Thomas VA, Dawson BC, Flores AN (2017) Identifying irrigated areas in the snake river plain, Idaho: evaluating performance across composting algorithms, spectral indices, and sensors. Remote Sens 9:546

ESA-European Spatial Agency. Sentinel Hub. https://apps.sentinel-hub.com/eo-browser/. Accessed: 15 Nov 2019

ESA-European Spatial Agency. Sentinel Hub Custom Scripts. https://custom-scripts.sentinel-hub.com/custom-scripts/sentinel-2/ndmi_special/. Accessed 10 June 2019

FAO-Food and Agriculture Organization of the United Nations (2017) Water for sustainable food and agriculture

Gao O, Zribi M, Escorihuela MJ, Baghdadi N, Quintana SP (2018) Irrigation mapping using statistics of Sentinel-1 time series at field scale. Remote Sens 10:1495

Junta de Andalucía (2018) Visor SIGPAC Andalucía. https://www.juntadeandalucia.es/organismos/agriculturaganaderiapescaydesarrollosostenible/servicios/sigpac/visor.html. Accessed: 1 June 2019

Llamas MR (2005) Una causa radical de los conflictos hídricos en España. Tecnol Del Agua 259:72–76

Psomas A, Dagalaki V, Panagopoulos Y, Konsta D, Mimikou M (2016) Sustainable agricultural water management in Pinios river basin using remote sensing and hydrologic modeling. Procedia Eng 162:277–283

# Considerations About Wastewater Reuse in Areas Subjected to Strong Pressures in the South of Spain

José Manuel Nieto-López, Matías Mudarra-Martínez, and Bartolomé Andreo

**Abstract**

Treated wastewater reuse is essential for water supply in areas where the pressure over water resources is high. This is the case of the Costa del Sol and the Guadalhorce Valley, where a lot of water is required for urban, agricultural and recreational uses. Most of this water is coming from natural sources instead of being reused, which percentage varies over Spain. In these areas, located in the Málaga province (Andalucía region), less than 6% of treated wastewater is assigned for any kind of use (garden irrigation and golf courses is the major one). Trying to solve this lack of reclaimed water, a wetlands restoration project was carried out in the Guadalhorce River Mouth, near the city of Málaga. Several lagoons were created using treated wastewater. Results were satisfactory with a clear increase in biodiversity, but employing a reduced amount of replenished water, due to the very strict requirements of the regional government. So, a great effort must be done to improve water reusing, especially in parts where freshwater is scarce (like S Spain and Mediterranean area), but also to hire well-trained technicians.

**Keywords**

Reuse • Spain • Restoration • Wastewater • Wetland

## 1 Introduction

The Costa del Sol (Málaga, S Spain) is one of the main Spanish and European tourist destinations, which concentrates most of its visitors in summer. Besides, there are agricultural areas with high water needs, as the Guadalhorce Valley. Consequently, water demand is higher in this period and presents strong seasonal fluctuations over the year (Armagasilla 2017).

In this region, water demand can be divided mainly in urban supply, garden/golf courses irrigation and agriculture, which is usually pumped from aquifers, although there exist a lot of surface water sources like reservoirs and water diversion from rivers. However, these water requirements can also generate a negative impact on groundwater-dependent ecosystems, so it is necessary to look for different ways to satisfy the water needs of population for different uses. Thus, the objective of this work is to make a short review about water reuse in Spain and the performance of a study case where a wetland was created using treated wastewater.

## 2 Reclaimed Water in Spain

Alternatives for satisfying a higher water consumption are necessary due to population and tourism increasing, as well as the impact of climate change in this area (Gutiérrez et al. 2006; Ruiz Sinoga et al. 2010; among others). These studies predict that rainfall will suffer a reduction of 10–40 mm each season, but also an increment of 6 °C in the average temperature to the end of the twenty-first century. Moreover, in Spain only a 10.4% of treated water is reused (INE 2016a), decreasing this percentage to 5.9% in the region of Andalucía and even 0% in Extremadura, Navarra, La Rioja, Ceuta and Melilla (Fig. 1). In the opposite, more than 45% of treated wastewater is reused in the Murcia and the Comunidad Valenciana regions. These values are possible because of the investment and the involvement of the regional governments, although they are not the same everywhere in Spain.

Reclaiming of treated wastewater is regulated by the law *Real Decreto 1620/2007, por el que se establece el Régimen*

J. M. Nieto-López (✉) · M. Mudarra-Martínez · B. Andreo
Department of Geology, Centre of Hydrogeology of the University of Málaga (CEHIUMA), Málaga, Spain
e-mail: nietolopezjm@uma.es





M. Abrunhosa et al. (eds.), *Advances in Geoethics and Groundwater Management: Theory and Practice for a Sustainable Development*, Advances in Science, Technology & Innovation,
https://doi.org/10.1007/978-3-030-59320-9_24

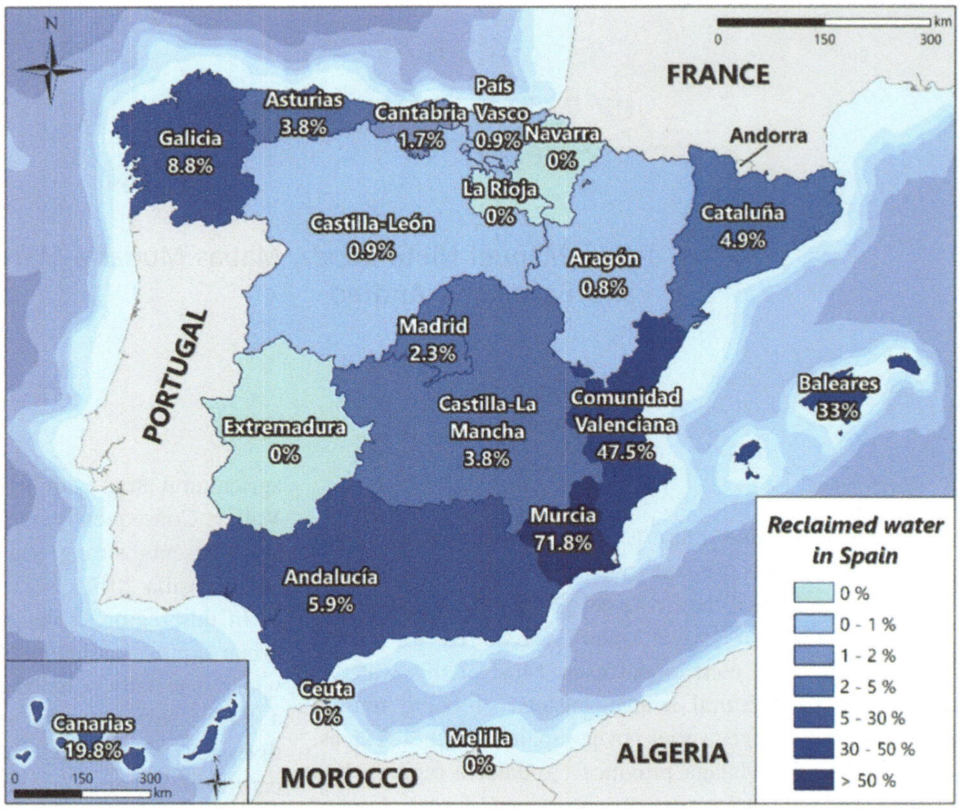


**Fig. 1** Percentage of reused treated wastewater in Spain (INE 2016a)

*Jurídico de la Reutilización de las Aguas Depuradas* (Order 1620/2007), which stablishes the legal rules for treated wastewater reusing. This law sets the quality status that a treated wastewater has to have for its use, as well as the allowed and forbidden uses. It also develops some rules for determining responsibilities related to the maintenance of this quality (MMAMRM 2010). The considered uses by the legislation are urban, agricultural, industrial, recreational and environmental, and their distribution varies significantly between regions (Fig. 2) (INE 2016b). In Spain, reclaimed water is employed for gardening and recreational particularly in Andalucía, Galicia and Madrid, whereas in regions as Murcia, Comunidad Valenciana or Canarias, major treated wastewater reusing is for agriculture.

## 3 A Pilot Project to Reuse Water in the South of Spain

The Guadalhorce River Mouth is located in the vicinity of the city of Málaga (Fig. 3), where a wetland restoration project was carried out for 3 years by reusing treated wastewater coming from a near wastewater treatment plant (WWTP). Restoration works were supported by the Coca-Cola Foundation (Atlanta, USA). The aims of the project were to create wetlands over some existing depressions, flooding them with treated wastewater, which would have facilitated the settlement of new animal and vegetal species, as well as serving as a hydraulic barrier against marine intrusion due to the dome of recharge created through the subjacent alluvial aquifer.

However, very strict water quality requirements, imposed by the regional government, made very difficult to achieve a nice performance of the project. So, total phosphorus and *Escherichia coli* were the selected parameters to regulate the replenish, with maximum allowed values of 2 mg/l and $10^4$ CFU/100 ml, respectively. Treated wastewater had a mean concentration of 3.2 mg/l of total phosphorus and $10^6$ CFU/100 ml of *E. coli*. These values were set by the Administration after considering the project as "Environmental use 5.4" into the current treated wastewater legislation, which includes wetlands maintenance, minimum outflows and similar ones. Here, the law establishes that the minimum quality required will be studied case by case, so it will be in charge of the public technicians' criteria.

It was very complicated to accomplish the regional government's requirements due to the low required values and to the budget limitations. The solution was the installation of a UV disinfection system in the WWTP to reduce the amount of *E. coli*. In addition, the limitation for total phosphorus was removed under a strict monitoring protocol for this parameter. These difficulties were responsible to not reaching the initial estimates of 650,000 $m^3$ of replenished water to the wetlands; only 73,775 $m^3$ were spilt ($\sim$10%).

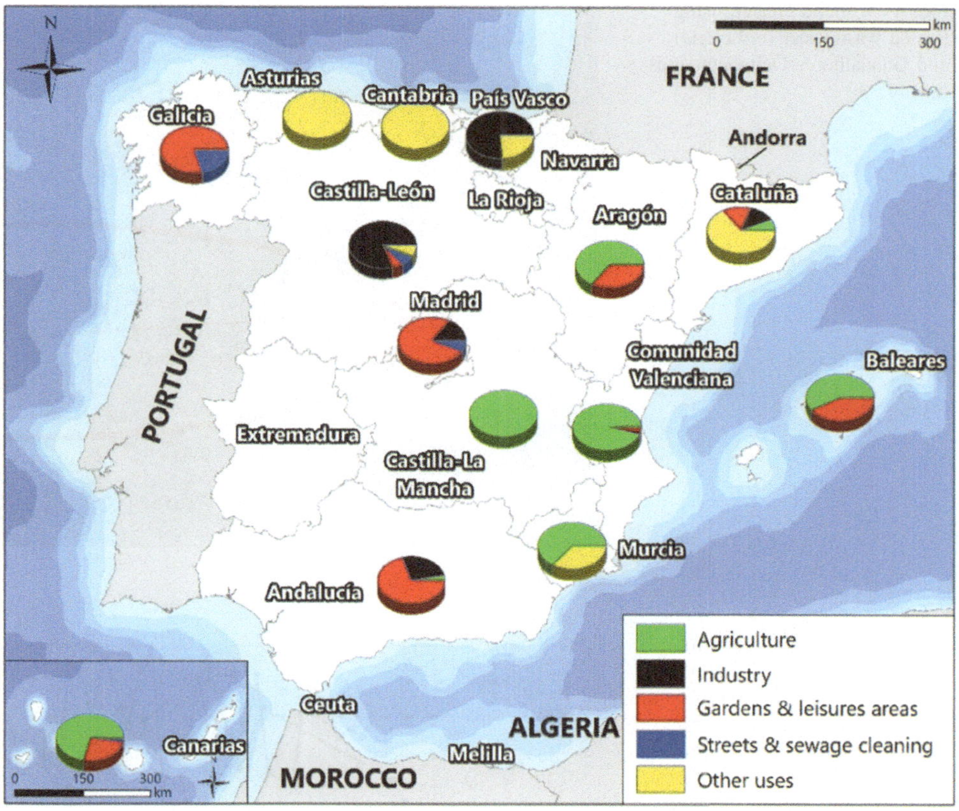


**Fig. 2** Uses distribution of reclaimed water in Spain (INE 2016b)

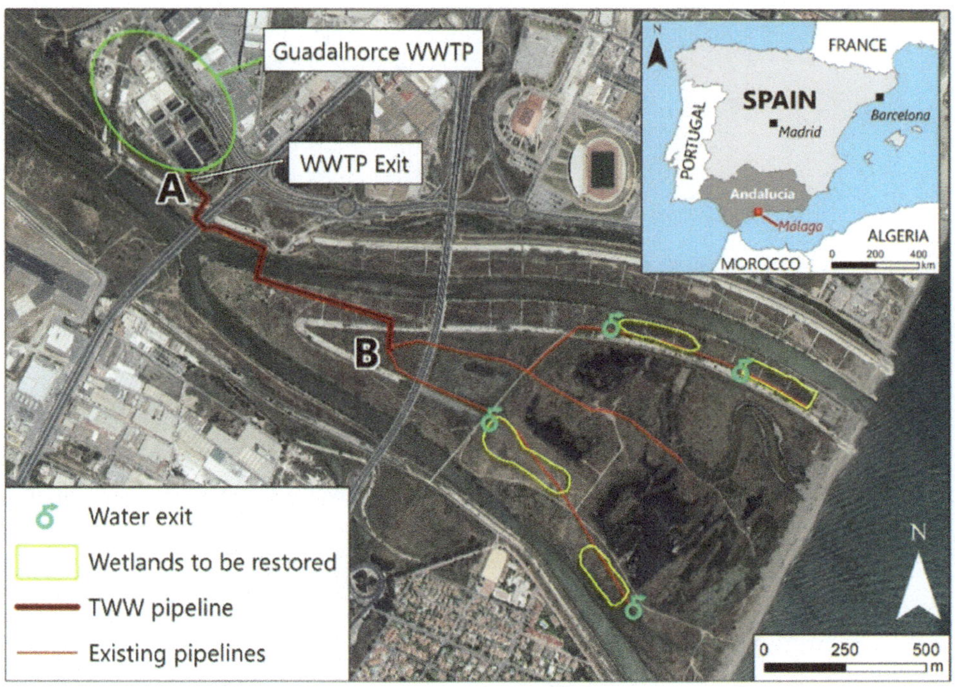


**Fig. 3** Location and schematic explanation of the wetland's restoration project in Málaga, reusing treated wastewater. TWW: Treated wastewater; A: WWTP exit; B: connection point with existing pipelines

Despite of these issues, a two-pool wetland was created (Fig. 4) and new animal and vegetal species started to use it as refuge and feeding spot. In general, an improvement in the water quality of the wetland was also detected. Total nitrogen and phosphorus values were below the WWTP concentration in a lot of samples (Fig. 5), excepting after the spring of 2018, which can be related to a higher vegetal productivity in the wetland. *E. coli* concentration was always

**Fig. 4** Wetland created using treated wastewater (red circle) and Guadalhorce Delta Wetlands (up)

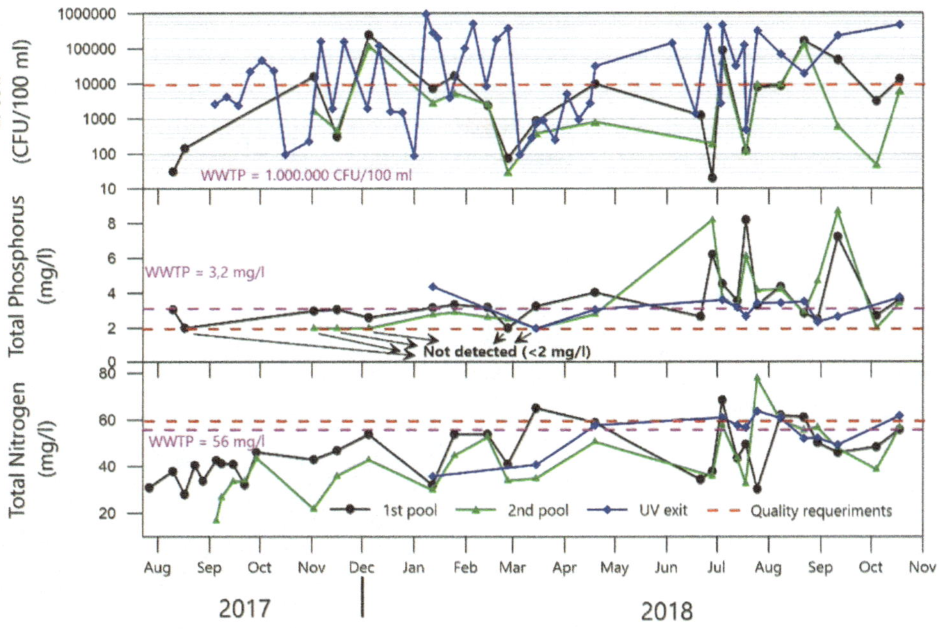


**Fig. 5** Temporal variation of total nitrogen, total phosphorus and *E. coli* concentration in the restored wetland for 15 months. Pink line represents the average concentration of the parameters in the wastewater treatment plant exit

below 1,000,000 CFU/100 ml, showing a combined cleaning effect of vegetation, infiltration and sun radiation over the bacteria amount.

## 4 Concluding Remarks

This work puts the focus on the Public Administration that must adapt to the new water culture, as well as to the increasing water scarcity, and to make a big effort in promoting the reuse of treated wastewater, but also political issues will be needed. These efforts must be even bigger in the Andalucía region, where groundwater is a commonly used resource and the reclaiming percentages are very low. Furthermore, the environmental uses do not exist, practically. This is a result of the ambiguity and the relativity of the maximum allowed values criteria assignment for the considered parameters. In turn, this is in consonance with the lack of well-trained technical supervisors in the government, who should have an interdisciplinary point of view for assessing the required parameters in the water destined for a specific environmental use.

**Acknowledgements** This work is a contribution of the Research Group RNM-308 of the Andalusian Government to the project "Hydrological and environmental restoration of wetlands in the delta of

Guadalhorce River (Málaga, Spain) reusing treated wastewater," funded by the Coca-Cola Foundation (Atlanta, USA).

## References

Argamasilla M (2017) Desarrollo metodológico para la planificación y gestión del agua en áreas turísticas costeras sometidas a clima mediterráneo y a grandes variaciones estacionales de la demanda. Caso de la Costa del Sol Occidental. Thesis Dissertation, University of Málaga, Spain, 444 p

Gutiérrez JM, Pons MR (2006) Modelización numérica del cambio climático: bases científicas, incertidumbres y proyecciones para la Península Ibérica. Rev De Cuaternario Y Geomorfol 20:15–28

INE (2016a) Destino de las aguas residuales tratadas por comunidades y ciudades autónomas, lugar de destino y periodo. Instituto Nacional de Estadística, Madrid

INE (2016b) Usos del agua reutilizada por comunidades y ciudades autónomas, tipo de uso y periodo. Instituto Nacional de Estadística, Madrid

MMAMRM (2010) Guía para la Aplicación del R.D. 1620/2007 por el que se establece el Régimen Jurídico de la Reutilización de las Aguas Depuradas. Ministerio de Medio Ambiente y Medio Rural y Marino. Centro de Publicaciones, Gobierno de España, Madrid, 145 p

Ruiz Sinoga JD, García Marín R, Martínez Murillo JF, Gabarrón Galeote MA (2010) Precipitation dynamics in sourthern Spain: trends and cycles. Int J Climatol 31(15):2281–2289

# Stakeholder View of Efficient Risk Communication in Contaminated Sites

Uilians Vieira de Oliveira, Juliana Gardenalli de Freitas, and Rosangela Calado Costa

**Abstract**

Environmental risk communication comprises an essential step in the management of contaminated sites. However, in Brazil there are no legal guidelines specifying how this communication should be performed. This research aims to identify the relevant aspects for risk communication in contaminated sites, from the perspective of stakeholders, such as responsible for contaminated sites that performed risk communication, environmental agencies, consultancies, and people affected by risks. The Q-technique was used, a methodology capable of identifying people's point of view and their subjectivities. The Q-set consisted of 67 statements that were judged and organized by 24 individuals in a value matrix, according to the opinion of each respondent, representing a group of actors involved. Five factors were identified representing the view of the research subjects: Factor 1 demonstrates concern about the health of those affected by the risk; factor 2 demonstrates the importance of safe communication, being responsible for the liability responsible for the process; factor 4 is strongly related to the legal issues that permeate the process; factor 5 exposes the concern to communicate aspects directly related to risk. All factors show concern with social factors and the rejection of forms of communication through social networks.

**Keywords**

Contaminated site • Risk • Q-technique • Environmental remediation • Environmental liability

U. V. de Oliveira (✉) · J. G. de Freitas · R. C. Costa
Institute of Environmental, Chemical and Pharmaceutical Sciences, Universidade Federal de São Paulo, Diadema, SP, Brazil
e-mail: uilians.wo@hotmail.com

## 1 Introduction

Land contamination is a problem in both developed and developing countries. In Brazil, the identification and characterization of contaminated sites are concentrated in the southeast region, in the states of Minas Gerais, Rio de Janeiro and São Paulo, pioneer state in this regard (Araujo 2014). It is noteworthy that there is one nationwide specific legislation for guidance or management of contaminated sites at the national level, the resolution National Council of the Environment No. 420/2009.

Contaminated sites management aims to reduce contamination to levels determined by law as acceptable. One of the activities involved in contaminated land management is risk communication, defined by Di Giulio (2010) as a dynamic process in which all stakeholders are informed about the risks and involved in the decision making. The requirement of risk communication highlights the importance of this activity for the management of contaminated areas. However, there are notable failures in conducting and developing risk communication (Di Giulio et al. 2012). In addition, although risk contamination is required, the national and state regulation do not specify how they should be conducted.

This study investigates what would be successful a risk communication in the perspective of different stakeholders: responsible for contaminated sites; affected or potentially affected by the contamination; consultants working with contaminated sites and regulators.

## 2 Methodology

The present study consists of qualitative and quantitative research using the literature review and the Q-technique, a specialized methodology for subjectivity analysis. In Q-technique, survey participants are invited to organize statements in a value matrix, according to their point of



M. Abrunhosa et al. (eds.), *Advances in Geoethics and Groundwater Management: Theory and Practice for a Sustainable Development*, Advances in Science, Technology & Innovation,
https://doi.org/10.1007/978-3-030-59320-9_25

view. The generated data are analyzed through factor analysis. It is noteworthy that this factor analysis methodology seeks patterns in the study situations and not in people, that is, there is an inversion of conventional factor analysis (Couto et al. 2011; Webler et al. 2009).

Initially, a set of statements was defined, known as concourse, created from interviews with people related to the management of contaminated areas, as well as bibliographic research. The concourse subsidized the selection of statements to be analyzed by the interviewees, the Q-set. A Q-set should be balanced, appropriate for the study, simple and easy to understand, but comprehensive enough to reflect the full range of concourse views so that it is representative (Couto et al. 2011). Q-set statements aimed to answer the following question: "Given that risk communication is one of the basic principles of contaminated area management, what statements below represent an efficient risk communication from your point of view?".

The Q-set underwent validation interviews with experts in contaminated sites and/or risk communication. A total of 67 statements were selected and categorized into five analysis groups: (1) communication strategies; (2) transparency of information; (3) economic aspects; (4) legal aspects; (5) environmental, social and human health aspects.

The Q-set was printed on 5 cm × 3 cm cards and then plasticized, ensuring greater durability. A matrix of values was made, a diagram with 67 statements divided into 11 columns, each column receiving a value between + 5 (indicating greater agreement) and −5 (indicating lower agreement).

The interviews were scheduled by email and performed in person from August 2018 to July 2019. The data generated in the interviews were analysed using the PQMetod® application, following the methodology proposed by Brown (1982).

## 3 Results

### 3.1 Participant Profile

Twenty-eight individuals were interviewed between July 2018 and July 2019, three during the validation of the Q-set and 24 for data collection. An interview was excluded from the analysis for not following the proposed method. In the final analysis, data collected during the validation stage were excluded. Thus, the P-set analyzed was composed of 24 individuals, 12 males and 12 females.

The choice of the number of respondents followed the recommendations of the Q methodology, in which few participants are needed. What needs to be broad is the number of statements within the universe of possibilities for the subject addressed. The proportion of three statements for each individual (3:1) is considered enough for the Q technique. This study achieved a proportion of 2.7 statements for each individual, a value within the expected for this technique (Webler et al. 2009).

Individuals comprised four analysis groups: affected or potentially affected by contamination (*G*1), contaminated site consultants (*G*2), responsible for contaminated site (*G*3) and regulators (*G*4). The number of respondents is not equal among the four analysis groups. *G*2 was the group with the largest number of respondents—eight in total—followed by *G*4 (six individuals), while the other groups had five individuals each. Despite presenting equal numbers of individuals for both sexes, it is noted that there is no uniform distribution by analysis group. The number of respondents per analysis group is shown in Table 1. The three participants in the instrument's validation were excluded from the final analysis.

### 3.2 Factors Analysis

Statistical analysis revealed eight factors; however, following significance criteria, five factors were selected for further analysis. Each factor represents a common or very close point of view of a group of individuals, distinct from the others. Thus, from the interpretation of the factors, it is possible to establish the vision of the different groups involved.

Seven individuals did not fit into any of the factors or fell into two or more factors. Interestingly, four individuals in the environmental agency (*G*4) did not fit into any of the factors.

The five selected factors were named as follows: factor 1—understand contamination; factor 2—communicate safely; factor 3—focus on compliance with legislation; factor 4—transparent communication and; factor 5—risk and health. Factors and their respondent information are listed in Table 2.

## 4 Discussion

Factor 1 is common to four individuals. Only two stakeholder groups identified with this factor: *G*2 with one individual and *G*3 with three. The individuals who compose this factor considered the amount of resources spent on risk communication as irrelevant, as well as the beginning of communication parallel with the site investigation. They also rejected the availability of printed material as a strategy of communication. However, maintaining a trust relationship between stakeholders, the interest in solving the problems presented, and communicating and minimizing the exposure to contaminants were considered important.

**Table 1** Number of participants in this research per analysis group

| Analysis group | Number of individuals | Female | Male |
|---|---|---|---|
| Affected or potentially affected by risk (*G*1) | 5 | 4 | 1 |
| Contaminated area consultants (*G*2) | 8 | 2 | 6 |
| Responsible for contaminated area (*G*3) | 5 | 4 | 1 |
| Environmental agency (*G*4) | 6 | 2 | 4 |
| Total | 24 | 12 | 12 |

**Table 2** Individuals' distribution per factor

| Factor | Total of individuals | Total of individuals per group | | | |
|---|---|---|---|---|---|
| | | *G*1 | *G*2 | *G*3 | *G*4 |
| 1 | 4 | 0 | 1 | 3 | 0 |
| 2 | 3 | 1 | 1 | 1 | 0 |
| 3 | 3 | 0 | 2 | 0 | 1 |
| 4 | 2 | 1 | 1 | 0 | 0 |
| 5 | 5 | 2 | 2 | 0 | 1 |

*Note G*1—risk-affected group; *G*2—consultants in contaminated areas; *G*3—group of managers and responsible for environmental liabilities; and *G*4—representatives of the environmental agency

Factor 2 represents a common view among three respondents from three distinct analysis groups: *G*1, *G*2 and *G*3. Communication with security was considered very important, so mechanisms should be sought to avoid causing panic in those affected. These individuals do not discern the best way to communicate risk, but give preference to face-to-face communication, supported by representatives of stakeholders. The aspects related to the health of those involved were considered unimportant or irrelevant. There is also a relevant concern with the image of the institution that owns the contaminated area.

Factor 3 is composed of three individuals; two individuals distributed in *G*2 and one in *G*4. Participants in this factor are very concerned about complying with the current legislation. For this to happen, expert remediation advice, where possible supported by communication specialist, is important. They also consider that the institution holding the liability is responsible for the communication.

There is appreciation of presential communication so that a relationship of trust is established between the parties involved. Although they value the relationship of trust between stakeholders, they are not open to dialogue.

Factor 4 is composed of only two individuals from *G*4. Noteworthy in this regard is the concern with transparency of information and the rejection of the adoption of specific rules that guide how communication should occur. Individual communication and the opportunity to express their views on the issue were considered irrelevant. Individuals are interested in analysing risk perception of those affected and pathways of exposure to the contaminant and ways to minimize health risks.

Factor 5 is composed of five respondents from three different analysis groups: *G*1 and *G*2, with two individuals, and *G*4, with one individual. The set of individuals that makes up this factor considers it important to communicate aspects directly linked to risk, understanding the current state of risk and which contaminants are present in the area. Also considered important is knowing the time required to reduce the risk to levels considered acceptable, as well as the aspects related to the health of those affected.

The individuals presented different points of view. However, they also shared similarities in some respects. The research subjects did not consider it important to make information available through e-mail and through the institution's website, as well as to provide information to different mass media, such as radio, TV, print and electronic newspapers, and reject communication by social media. However, the adequacy of language to different audiences is of considerable importance, fact that enables the understanding information to individuals who do not specialized in contaminated area.

## 5 Concluding Remarks

The four stakeholder groups involved in the risk communication process, as already described, have different perspectives about efficient risk communication. This also occurs within the same stakeholder group: The *G*1 shares the view on three different factors (factors 2, 3 and 5). However, there is a predominance of factor 5, which shows this group prioritizes health-related aspects related to exposition to risks from the contaminated area. The *G*2 individuals are distributed in all factors, so there is no predominant view for this group of participants. The *G*3 comprises two factors (one and two), but there is a predominance with a view represented by factor 1. *G*4 is distributed in two factors. However, it is noteworthy that most individuals in *G*4 did not fit in a single factor.

Individuals who are more concerned with understanding the risk, that is, how the exposition to risk occur and how to minimize exposure to these risks have greater affinity with factor 1. Factor 2 is composed of individuals who value the safety of risk communication and are concerned about the image of the institution responsible for the contaminated site. The concern with the legal aspects, appreciation presential communication and the maintenance of trust are characteristics of the individuals that compose factor three. The individuals that compose factor 4 show greater concern with communication transparency and with analysis of the risk perception of those affected. The concern with the health of those involved and aspects directly linked to risk, such as information about contaminants in the site, is a view that represents factor 5. In summary, the views are very different and not directly related to which group the individual belongs to.

## References


Araujo M (2014) O Brasil no contexto do gerenciamento de áreas contaminadas: Um Olhar Crítico para as Regiões Desiguais do País. Universidade Federal de Alagoas, Maceió (MSc thesis)

Brown S (1982) Political subjectivity, applications of Q methodology in political science. Yale University Press, Yale

Couto M, Farate C, Ramos S, Fleming M (2011) A metodologia Q nas ciências sociais e humanas: o resgate da subjectividade na investigação empírica. Psicologia. Edições Colibri Lisboa 25 (2):7–21

Di Giulio GM (2010) Comunicação e governança do risco: exemplos de comunidades expostas à contaminação por chumbo no Brasil e Uruguai. Universidade Estadual de Campinas, Campinas (PhD thesis)

Di Giulio GM, Figueiredo BR, Da Costa Ferreira L, Dos Anjos JASA (2012) Communication and risk governance: the Brazilian experience in areas contaminated by lead 13(2):283–297

Webler T, Danielson S, Tuler S (2009) Using Q method to reveal social perspectives in environmental research. Social and Environmental Research Institute, Greenfield, vol 54, pp 1–45

# Urban Planning and Aquifer Management Using Recession Curves Method

Giselly Peterlini and André Celligoi

**Abstract**

The expansion of cities and population growth can intensify the use of water resources. The management of these resources presents several interfaces with urban planning, since water is a fundamental element in urban and regional management, due to its potential to induce or hinder social and economic development. An assessment of the regulatory reserves in the Cafezal stream was carried out using the recession curve analysis method, as well as the current groundwater consumption. Hydrographs were made to observe the recession period based on the selected date from 2008, considered a dry year. Data were obtained from 88 wells drilled in the area with a total discharge of 1005.39 $m^3/h$, with an average of 11.42 $m^3/h$ and annual consumption of 4.04 × 106 $m^3/year$. Regulatory reserves total are 39.48 × 106 $m^3/year$, with 10.23% of their potential being used through deep wells. Thus, the aquifer has a current exploitation under its potential, and the discharge is considered safe, causing no damage to the aquifer system.

**Keywords**

Recession curves • Water resources management • Environmental planning • Groundwater • Regulatory reserves

## 1 Introduction

The expansion of cities and population growth, as highlighted by Braga and Carvalho (2003), Tundisi (2003, 2008), as well as Almeida et al. (2006) may intensify the use of water, which implies an increasing use of water resources for the development of the most diverse activities of society, whether agricultural, industrial or domestic. To meet this demand, there is a growing need for surface water sources and increased use of groundwater. Braga and Carvalho (2003) clarified that water resources management presents several interfaces with urban planning, since water is a fundamental element to be considered in urban and regional management, due to its potential to induce or hinder social and economic development.

The authors also argue that water management is essential to ensure its quantity and quality at appropriate levels. In this sense, the present work aims to conduct a study of groundwater water potentiality in the Cafezal stream watershed using the recession curve analysis method to assess its aquifer reserves and the current groundwater consumption situation. The study also intends to demonstrate the importance of proper management of water resources in urban environmental planning for the municipalities located in the study area.

## 2 Study Area Characterization

The Cafezal stream watershed is located in the municipalities of Londrina, Cambé and Rolandia, in the northern state of Paraná, between the coordinates 23°16′17″ and 23°24′36″ south latitude and 51°23′27″ and 51°07′2″ west longitude, as can be seen in Fig. 1. This basin was chosen for the study because it has a regional location, as it covers the areas of three neighboring municipalities, being a predominantly urban basin.

According to IAPAR (2018), the climate in the area is of the mesothermal humid subtropical type (Cfa). The historical average temperature from 1976 to 2017 was 21.1 °C with average maximum temperature of 27.3 °C and minimum average of 16.1 °C, as well as average rainfall of 1641 mm. The geological base belongs to the São Bento Group,

G. Peterlini · A. Celligoi (✉)
Department of Geosciences, State University of Londrina, Londrina, Brazil
e-mail: celligoi@uel.br



M. Abrunhosa et al. (eds.), *Advances in Geoethics and Groundwater Management: Theory and Practice for a Sustainable Development*, Advances in Science, Technology & Innovation,
https://doi.org/10.1007/978-3-030-59320-9_26

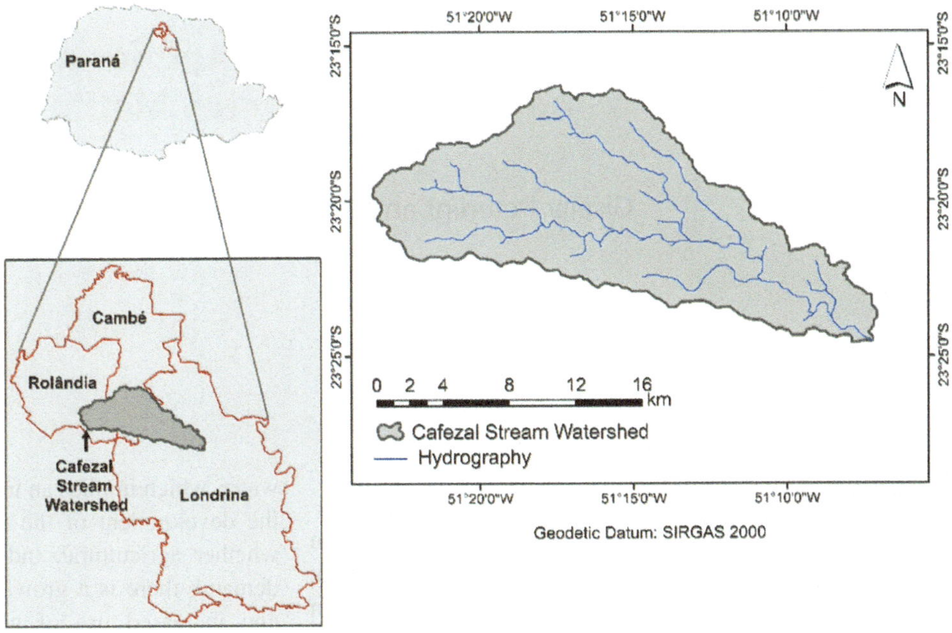


**Fig. 1** Location map of the Cafezal stream watershed

represented by the Serra Geral Formation which, according to Celligoi and Duarte (1990), consists of a series of overlapping volcanic sequences. Hydrogeologically, the basin is located over the Serra Geral Aquifer System (SASG), characterized as fractured, in which according to Rosa Filho et al. (2006), water circulation and storage are conditioned to the physical discontinuities of rocks (joints, faults and fractures).

Regarding the recharge aspects of the SASG, Viana and Celligoi (2002) pointed out that in the city of Londrina and its region, due to the thick soil layer and the humid climate, the recharge and discharge area of the aquifer is related to the topography, where the recharge occurs at topographically higher locations. Water is supplied through surface catchments, as well as groundwater pumping through deep tubular wells, since the Cafezal stream watershed has 88 tubular wells that exploit water in the SASG.

## 3 Materials and Methods

Flow data and daily pumping period were obtained from the Paraná Water Institute from 88 wells, located in the contributing basin of Cafezal upstream of the fluviometric station point. For the analysis of precipitation, data were obtained from historical averages of rainfall collected in the São Luiz station, located in Londrina. The period used for the calculations was from 2006 to 2016, which presented an average of 1680.4 mm, with a minimum of 1258.5 mm in 2008 and a maximum of 2332.1 mm in 2015.

One of the main objectives when studying groundwater resources is to determine maximum rates that can be withdrawn according to the hydrogeological environment. Freeze and Cherry (1979) point out that yields must be considered in terms of balancing the benefits of groundwater extraction and the undesirable changes that will be induced by this pumping, since overexploitation of the maximum rates may result in overexploitation of reserves causing damage to the aquifer. Feitosa and Filho (2000) define as regulatory reserves the amount of water that is stored in the aquifer during natural recharge and, because of this, are subject to the seasonal or interannual effects of precipitation.

For the calculation of the regulatory reserves, data were obtained from historical averages of the river flow of the Cafezal stream from the Paraná Water Institute, taken by the fluviometric station upstream of the Londrina Water Treatment Plant. The data used correspond to the period from 2006 to 2016. The year 2008 was selected because it presented less precipitation. The methodology, based on Santos and Celligoi (2002), generates a log-scale curve that represents the groundwater recession curve, based on the Maillet Eq. (1):

$$Q = Q_0 \cdot e^{-kt} \quad (1)$$

where $Q$ = Discharge ($m^3/s$) after a period $t$ (days); $Q_0$ = Discharge at the beginning of the recession ($m^3/s$); $k$ = Recession constant, which is calculated using Eq. (2):

$$k = (\ln Q_0 - \ln Q)/t \quad (2)$$

Subsequently, the volume stored in the basin ($V_0$) was calculated using Eq. (3):

$$V_0 = (Q_0.86,400)/k \quad (3)$$

Equation (4) was used to calculate the restitution ($h$), where $A$ is the basin area.

$$h = V_0/A \quad (4)$$

The value of the regulatory reserves is obtained through Eq. (5), which uses the values of the area of aquifer occurrence in the basin ($A_a$) and the restitution of precipitation ($h$).

$$R_R = A_a.h \quad (5)$$

## 4 Results

The data obtained regarding the location of the fluviometric station and the tubular wells allowed them to be located in the watershed and to determine the contributing basin, as can be observed in Fig. 2.

The hydrograph was elaborated by obtaining the average data of the river flows of the Cafezal stream. From the recession curve observed for 2008 (Fig. 3), the discharge value at the beginning of the recession ($Q_0$) was 2.54 $m^3$/s and after a period of 139 days the flow value ($Q_t$) was of 1.12 $m^3$/s.

The value of the recession constant $k$ was calculated to give a value of 0.0059. Subsequently, the volume of the basin ($V_0$) was determined, which was 37.19 × $10^6$ $m^3$/year. Since the basin area is 153.02 $km^2$, the value of the average water restitution for the year 2008 was calculated, with $h_m$ = 0.243 m/year or 243 mm/year. In this sense, knowing that the average rainfall in the Cafezal stream watershed in 2008 was 1258.5 mm, it is observed that the average restitution was 19.31%. With the average restitution value, as well as the value of the aquifer occurrence area in the basin, it was possible to obtain the volume of the regulatory reserves (RR), which was 37.18 × $10^6$ $m^3$/year.

From the flow data and pumping time of the tubular wells, it was possible to determine the current scenario of exploration of SASG groundwater reserves in the basin under study. The total flow of the 88 wells in the area is 1005.39 $m^3$/h, with an average of 11.42 $m^3$/h. By averaging the pumping time of the wells, which was equal to 11 h/day and considering the period of one year, the wells' production estimate is about 4.04 × $10^6$ $m^3$/year. Therefore, it is estimated that the wells use about 10.86% of the potential of the regulatory reserves.

## 5 Discussion

The wells located in the Cafezal stream watershed exploit an annual volume of groundwater lower the potential of the aquifer for regulatory reserves.

In this sense, this exploited volume may not cause damage to the aquifer, such as a decrease in the water level or a reduction in the flow of surface water, since they represent a small portion of the regulatory reserve, calculated from the

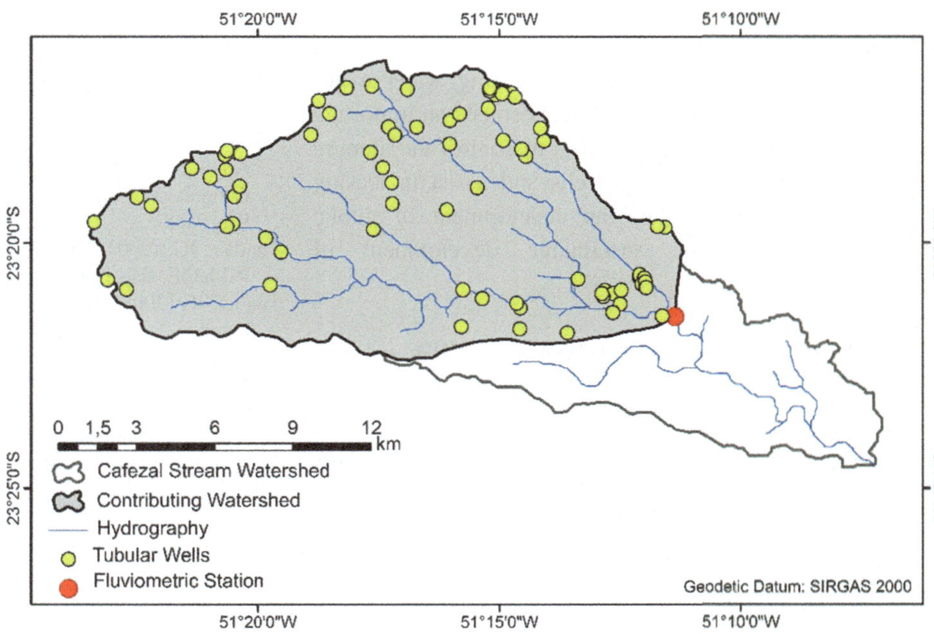


**Fig. 2** Location of the contributing basin, fluviometric station and tubular wells in the Cafezal stream watershed

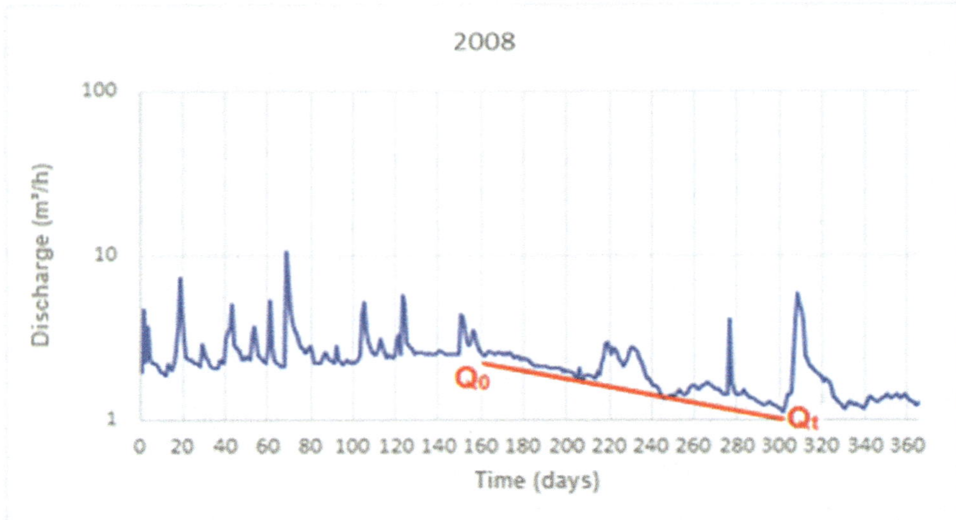


**Fig. 3** Hydrograph and recession curve of the Cafezal stream for the dry year

base flow observed in the Cafezal stream for a dry year, that is, considering an unfavorable precipitation scenario in which the river remains perennial thanks to the flow from the aquifer. However, as explained by Viana and Celligoi (2002), some areas of the basin may use a larger volume of regulatory reserves as there is a higher concentration of wells and, therefore, a larger volume of exploited water.

## 6 Concluding Remarks

Understanding the current situation of groundwater exploitation is fundamental for decision making and management of water supply and demand. The realization of this study and the application of the recession curve method allows the comparison between the existing water availability and what is being used, which becomes an important management tool, which can be used to control the amount of water exploited by existing wells and assist in leasing future wells, contributing to the development of public policies aimed at the sustainable development of municipalities.

**Acknowledgements** Special acknowledgements to CNPq and Fundação Araucária do Estado do Paraná, Brasil, for financial support through the projects of the Agreement CNPq/Fundação Araucária: 61.0088/06-8 and Convênio Fundação Araucária/UEL: 063/08; to CAPES for granting a master's degree scholarship in the Graduate Program in Geography of the State University of Londrina.

## References

Almeida L, Resende L, Rodrigues AP, Campos JEG (2006) Hidrogeologia do Estado de Goiás. Secretaria de Insdústria e Comércio. Superintendência de Geologia e Mineração, Goiânia

Braga R, Carvalho PF (2003) Recursos hídricos e planejamento urbano e regional. Laboratório de Planejamento Municipal—Deplan—Unesp—IGCE, Rio Claro

Celligoi A, Duarte U (1990) Hidrogeologia da cidade de Londrina-PR. Águas Subterrâneas 6:64–71

Feitosa FAC, Filho JM (2000) Hidrogeologia: Conceitos e Aplicações, 2nd edn. Serviço CPRM/REFO, LABHID-UFPE, Fortaleza

Freeze R, Cherry JA (1979) Groundwater. Prentice Hall, New Jersey

Instituto Agronômico do Paraná – IAPAR (2018) Médias históricas, https://www.iapar.br/arquivos/Image/monitoramento/Medias_Historicas/Londrina.htm. Last access 2018/07/08

Rosa Filho EF, Hindi EC, Mantovani LE, Bittencourt AVL (2006) A importância do Sistema Aquífero Serra Geral para a cultura da soja no estado do Paraná. Rev Águas Subterrâneas ABAS 20(2):49–56

Santos MM, Celligoi A (2002) Utilização da metodologia de análise da curva de recessão para o cálculo das reservas reguladoras do Aquífero Caiuá na sub-bacia do Rio dos Índios-PR. In XII Congresso Brasileiro de Águas Subterrâneas, Águas Subterrâneas, São Paulo

Tundisi JG (2003) Água no século 21: enfrentando a escassez. RiMa/iiE, São Paulo

Tundisi JG (2008) Recursos hídricos no futuro: problemas e soluções. Estudos Avançados 22(63):7–16

Viana TR, Celligoi A (2002) Análise das reservas de águas subterrâneas do Aquífero Serra Geral em Londrina: Recarga e Consumo. In XII Congresso Brasileiro de Águas Subterrâneas, Águas Subterrâneas, São Paulo

# Lessons for a Resilient and Sustainable Future with Hydrogeoethics: Case Studies of Geoethics in Groundwater Science-Engineering, Profession, and Management

**Introductory Note**

Rodrigo Lilla Manzione (UNESP, São Paulo, Brazil), José Manuel Marques (IST, University of Lisbon, Lisbon, Portugal), Manuela Simões (FCT, Nova University Lisbon, Lisbon, Portugal) and Tiago Abreu (ISEP, Polytechnic of Porto, Porto, Portugal)

This major thematic part counted over 30 contributions addressing many aspects of geoethics in groundwater science and practice, mostly from key studies around the world (Africa, Asia, Europe, and South America). The case studies bring an overview of topics where geoethical thinking and behaviour are essential, like groundwater quality assessment, wastewater treatment, agriculture practices vs. groundwater pollution, the role of groundwater in the sustainable development of urban areas, hydrodynamics and hydro-ecological effects, groundwater environments (e.g. coastal areas), conservation of natural resources and learning.

The contributions deal with the ethical, integrity, eco-responsibility, cultural and social implications of geosciences knowledge, education, research, practice, and communication, enhancing the concern of geoscientists, engineers, and water experts related in conducting their activities. Risk- and hazard-based assessment is key to deliver feasible solutions for groundwater quantity and quality maintenance for cities, rural communities, and ecosystems. High-level analysis should be encompassed with geoethical principles in order to ensure it. Considering different groundwater settings (confined/unconfined aquifers, karst/porous/hard rock media, and surface/underground storage), it is also a critical requirement for successful actions and police-making.

In a changing world, threatened by global warming and competing claims for water resources, the role of geoethics and groundwater management emerges as a compulsory posture for the actual and next generation of groundwater experts if a resilient and sustainable future is desired.

**Highlights**

- Challenges in groundwater use, quality, and management in coastal regions;
- Wastewater treatment;
- Agriculture vs. groundwater pollution;
- Managed artificial recharge and groundwater conservation;
- Groundwater and sustainable development of urban areas;
- Scientific and technical integrity, eco-responsibility, and ethics in groundwater science and practice.

# Ethical Aspects of Water Use in the Campo de Cartagena and the Associated Impacts on the Mar Menor

María Feliciana Fernández-García, Emilio Custodio, and Manuel Ramón Llamas

**Abstract**

In recent years, the Mar Menor, a large and emblematic hypersaline lagoon on the Spanish Mediterranean coast, has seriously deteriorated. This lagoon and the entire surrounding area is one of the main sources of wealth in the Spanish south-east and is an ecological asset. Consequently, it has generated great controversy, confronting researchers, managers and citizens in general. At a general level, the lagoon undergoes a eutrophication process (excess of nutrients) caused by different factors, which go beyond merely scientific ones. In contrast to the intense fall in groundwater levels detected in most aquifers in the area, the water table level in the Campo de Cartagena aquifer has a rising trend and water quality loss, mainly due to irrigation returns. However, the lack of clarity in knowledge, enough data and good cooperation are major handicaps for sound management. This is compounded with poor ethical behaviour in what refers to increasingly limited water resources, insufficient attention to scientific knowledge and data availability, maintaining corporate and institutional privileges, and looking for illusory fast politically worth results instead of medium- to long-term solutions to problems that have been cooked for decades.

M. F. Fernández-García (✉)
Department of Construction and Technology in Architecture (DCTA), High Technical School of Architecture of Madrid (ETSAM), Technical University of Madrid (UPM), Madrid, Spain
e-mail: marifeli.fdez@gmail.com

E. Custodio
Department of Civil and Environmental Engineering, Royal Academy of Sciences and Groundwater Hydrology Group, Technical University of Catalonia, Barcelona, Spain

M. R. Llamas
Royal Academy of Sciences of Spain, Madrid, Spain

**Keywords**

Integrated water resources management • Hydrogeoethics • Intensive groundwater exploitation • Socio-economy • South-eastern Spain

## 1 Introduction

The Mar Menor is the main wetland in the Region of Murcia, in SE Spain. It is a large lagoon closed by a continuous sand bar, the largest in the Mediterranean Sea, with an area of 135 km$^2$. It stores 591 hm$^3$ of water, with an average depth of 4 m (SASMIE 2017). The Mar Menor (Small Sea) is included within the Ramsar Convention List of wetlands and holds other protection legal figures under the European Habitat and Birds Directives (Navarro Caballero 2019).

In recent years, the Mar Menor has suffered a remarkable ecological deterioration due to the cumulative massive inflow of nutrients, coming mainly from diffuse pollution from agriculture and from the growing human settlement in the inland basin, which has led to eutrophication problems. The eutrophication state reached its first maximum in 2016, leading to an ecological collapse. A second breakdown happened in September 2019, after an extraordinary rainfall and flooding event that triggered the mobilization of the accumulated nutrients into the water column and a peak in phytoplankton communities that caused the death by anoxia of thousands of fish, which were carried and deposited by the currents onto the lagoon beaches. These beaches have historically attracted an intense and important touristic development, which is currently one of the main sources for the economy of the Region, as well as an icon of natural and patrimonial wealth for its citizens.

The region has a semiarid-to-arid climate, with sporadic torrential rainfall events that can cause severe flooding, carrying with it high amounts of pollutants and sediments



M. Abrunhosa et al. (eds.), *Advances in Geoethics and Groundwater Management: Theory and Practice for a Sustainable Development*, Advances in Science, Technology & Innovation,
https://doi.org/10.1007/978-3-030-59320-9_27

dragged on their way through greatly modified areas by urban settlements and agricultural activity, to finally end at the lagoon.

This article aims to give an overview of the situation from an ethical standpoint, by analysing how the deterioration of the Mar Menor has been reached by obviating externalities in water management decision-making during the last 50 years, ignoring groundwater, as well as the ethical consequences involved in these technical decisions.

## 2 Connection Between the Lack of Groundwater Management in the Campo de Cartagena Aquifer and the Associated Effect on the Deterioration of the Mar Menor

The deterioration of the Mar Menor began approximately half a century ago. This is due to several interrelated causes that have affected the complex multilayer aquifer system of Campo de Cartagena (SASMIE 2017; Velasco et al. 2018). The Mar Menor is just at the down-flow boundary of the upper Quaternary aquifer, which near the coast is poorly connected with the heavily exploited deep aquifers. This upper aquifer has experienced the water table elevation due the well-enhanced recharge by return irrigation flows, as the consequence of water import to the area through a main transfer canal from the Atlantic Tajo basin in central Spain. This water import was to stop the intense aquifer exploitation and water mining in the area for irrigating an economically very important area, although further development of irrigated area happened and intense groundwater exploitation continued (MASE 2015; Custodio et al. 2016; Jiménez-Martínez et al. 2016).

### 2.1 Agricultural Activities

The aquifer hosts over 30,000 ha of intensively irrigated agriculture, applying 190 $hm^3$/year, of which the share of groundwater goes from 30% in a wet year to up to 75% in a dry year. The remaining portion is derived from the Spanish Government-controlled Tajo-Segura Transfer (TTS), since 1979, with an average contribution of 59 $hm^3$/year. Most of the resulting agricultural activity, which is actually close to an industry, is concentrated in the Campo de Cartagena. However, this contribution of water and its use in agriculture has led to negative externalities that were not taken into account, such as enhanced aquifer recharge and high nitrate pollution, up to 300 mg/L $NO_3$ in the return flows and even higher in the rejection from the numerous small desalination plants used to freshen the brackish water in the upper aquifer (SASMIE 2017; Aldaya et al. 2019).

### 2.2 Touristic Activities

Hydrological changes in the hypersaline lagoon have been conditioned by the development of mass-tourism centres, with uncontrolled urban growth in the last 50 years. Between 1986 and 2016, the urban surface of the Mar Menor area increased by some 6000 ha. Such a development has led to a total transformation of the landscape and to the destruction of much of the natural and ecological heritage, with the associated spills, land filling, opening and dredging of channels in "La Manga" (sand bar), construction of marinas and artificial beaches. The lack of control in the coastal urban developments of the Mar Menor has led to a decrease in the lagoon salinity, from 70 to 47–52 g/L, because the "golas" (channels of connection with the sea) have been enlarged, allowing a faster renovation with the less saline seawater, besides the increased inflow of surface and ground continental water. Surface water is a variable source of nitrate and phosphorous, untreated until recently, and groundwater contributes nitrate at a more continuous rate. Part of the phosphorous is trapped in the lagoon as organic matter sediments if water is oxygenated. Otherwise, as happened recently, it can be explosively released when the lagoon becomes anoxic. The situation can be suddenly worsened by large rainfall events. So, the floods of 2019 dragged between 500 and 1000 tons of nitrates, 35 tons of ammonia and more than 100 tons of phosphate, in dissolved form, plus more than 100,000 tons of suspended organic carbon and phosphorus (Ruiz-Fernández et al. 2019).

## 3 Ethical Aspects of the Use of Water Resources in the Campo de Cartagena and the Mar Menor

The ethical aspects of water here considered refer to the environmental, economic, social and cultural circumstances of the Mar Menor, taking into account that they are different from the technical, engineering and managerial criteria, although, as transparency is involved, there is a link to ethical behaviour.

In the multilayer aquifer of the Campo de Cartagena, before the arrival of the TTS, there was already intensive groundwater exploitation, with large piezometric level drawdown. In the area, local groundwater, imported surface water, seawater desalination, wastewater reclamation and salinity reduction of brackish shallow groundwater are used, in a complex time and space framework. This allows getting notable economic results and employment and quality of life. But the vision is sectorial; many different, poorly related institutions are acting; there is incomplete understanding of the system, poor data and selfish behaviour. Analyses disregard negative externalities in the short- and especially in the mid-term. All these aspects involve ethical failures.

Agricultural production and mass tourism have contributed a lot to fast development at regional and national levels, so continuing this trend is highly favoured, and external resources are demanded to foster what is considered a benefit to all, even at European level. However, this consumerist vision is short-term and often misleading, since the economy is not distributed equitably among the population, whereas environmental problems do affect everybody equally, and besides limited natural resources are exhausted and taken from the heritage of coming human generations. The example of the Mar Menor shows the fragility of the economy in the medium and long terms when there are environmental imbalances, which are already affecting today and will affect future generations.

Ethical issues rise in relation to sectorial approaches, insufficient knowledge, poor data, poorly cooperative institutions that often collide looking for self-interest, privileges of some professionals, excessive politically conditioned human behaviour that overcomes social interest, and the use of administrative rights to get additional benefits in stressed situations, such as groundwater rights to sell water to those without this rights in droughts and when the water transfer is not allowed (MASE 2015).

## 4 Concluding Remarks

The existing problem in the Mar Menor would have appeared anyway with groundwater being ignored, but these problems would have been mitigated and controlled by considering externalities in an ethical context.

The deterioration of the lagoon affects natural capital and thus also the citizens who live in its environment and the embedded intrinsic value for them. There are not enough studies detailing what happens, and for those existing, the data is not available for everyone. This is a problem of corporatism, with unsuccessful struggles between different organizations, institutions, universities and research groups, environmental groups, and companies. Each one works actively looking only for their particular interests, which is unethical, as well as not helping a joint and consensual solution. This is made difficult by stoking political struggles between different governments (central and regional). Using water resources for political gain is not ethically acceptable. Furthermore, the situation is not being dealt properly.

It is a clear example of hydrogeoethics, where if they do not consider groundwater in planning, there would be no solution for Mar Menor. Considering that any solution requires a long time and sustained corrective action, it has the ethical aspect of returning part of the gross profit.

## References

Aldaya MM, Custodio E, Llamas R, Fernández MF, García J, Ródenas MA (2019) An academic analysis with recommendations for water management and planning at the basin scale: a review of water planning in the Segura River Basin. Sci Tot Environ 662:755–768

Custodio E, Andreu-Rodes JM, Aragón R, Estrela T, Ferrer J, García-Aróstegui JL, del Villar A (2016) Groundwater intensive use and mining in south-eastern peninsular Spain: hydrogeological, economic and social aspects. Sci Tot Environ 559:302–316

Jiménez-Martínez J, García-Aróstegui JL, Hunink JE, Contreras S, Baudron P, Candela L (2016) The role of groundwater in highly human-modified hydrosystems: a review of impacts and mitigation options in the Campo de Cartagena-Mar Menor coastal plain (SE Spain). Environm Rev 24(4):377–392

MASE (2015) Aspectos hidrológicos, ambientales, económicos, sociales y éticos del consumo de reservas de agua subterránea en España: Minería del Agua Subterránea en España. Aqualogy y Cetaqua, UPC, Barcelona https://hdl.handle.net/2117/111272

Navarro Caballero MT (ed) (2019) Una mirada global sobre el Mar Menor. Editorial Diego Marin, Universitat d'Alacant, Alicante, Spain

Ruiz-Fernandez JM, León VM, Marín-Guirao L, Giménez-Casalduero F, Alvárez-Rogel J, Esteve-Selma MA, Martínez Fernández J (2019) Informe de síntesis sobre el estado actual del Mar Menor y sus causas en relación a los contenidos de nutrientes. In: Projects of sustainability and conservation of Mar Menor Lagoon and Its Basin; Universidad de Alicante, Alicante, Spain

SASMIE (2017) Salinización de las aguas subterráneas en los acuíferos costeros mediterráneos e insulares españoles. Suez Solutions y Cetaqua, UPC, Barcelona. https://hdl.handle.net/2117/111515

Velasco AM, Pérez-Ruzafa A, Martínez-Paz JM, Marcos C (2018) Ecosystem services and main environmental risks in a coastal lagoon (Mar Menor, Murcia, SE Spain): the public perception. J Nat Conserv 43:180–189

# Water Resources Management Under Climate Change Pressure in Limpopo National Park Buffer Zone

Francesca Andrei, Maurizio Barbieri, Paulino Vincente Muteto, Lorenzo Ricolfi, Giuseppe Sappa, and Stefania Vitale

**Abstract**

This study aims to give an insight and reflection about water resources as key elements of human civility and nature. Despite Scientific Community has developed excellent concepts such as equity and sustainability, water it is not only a natural and environmental question, but it is also an ethic question. The 2030 Agenda for Sustainable Development issues an ambitious challenge to change our world and leave no one behind: Sustainable Development Goals (SDGs) include targets for access to safe drinking water, sanitation, and better water management. Three out of ten people do not have access to safe drinking water. Almost half of these people live in sub-Saharan Africa. The starting point of this study is the outcoming of hydrogeological and hydrogeochemical studies carried on the framework of the SECOSUD Phase II ("Conservation and equitable use of biological diversity in the Southern African Development Community—SADC") project, supported by the Italian Ministry of Foreign Affairs in South Africa Development Countries. More people in sub-Saharan Africa has lack access to clean water than anywhere in the world, and they continue to depend mainly on rivers, lakes, and ponds, but some of these reservoirs are naturally polluted or vulnerable to anthropogenic pollution.

**Keywords**

Human rights and ethic question • Climate change • Groundwater quality and availability • Hydrological cycle • Wet and dry season

## 1 Introduction

The need of water and the search of availability has always led man: water has a complex nature and the relationship between water and man has always been configured as a complex, problematic and sometimes contradictory relationship.

Water is not only a resource, having multiple values, and, for the environment, it is an important tool to ensure natural systems survive and thrive for the benefit of all. Water, for human being, is the source of life, but it can be source of death, when phenomena such floods occur. Southern Africa faces water scarcity challenges due to drought recurrence. Globally, the inequalities between those having access to safe water, in urban and rural areas, have decreased but large gaps remain, especially in less developed countries. In fact, eight out of ten people without access to safe drinking water live in rural areas and nearly half of them live in sub-Saharan Africa (WASH Programme 2019).

The focus area is Limpopo National Park and its buffer zone. The UNESCO defines buffer zone as "an area that should ensure an additional level of protection to areas recognized as a World Heritage sites" (Vitale et al. 2017). Mozambique is a SADC country affected by tropical cyclones and storms almost every year, due to its extensive coastline; in fact, in coastal areas, two out of three people are vulnerable to disasters such as cyclones, storms, and flash floods. In 2019 Spring, the impact of two cyclones hitting Mozambique in one season was devastating, leaving many people in need of basic supplies like food and water.

F. Andrei · G. Sappa
Department of Civil Building and Environmental Engineering, Sapienza University, Rome, Italy

M. Barbieri (✉) · L. Ricolfi · S. Vitale
Department of Earth Science, Sapienza University, Rome, Italy
e-mail: maurizio.barbieri@uniroma1.it

P. V. Muteto
Faculdade de Ciências, Universidade Eduardo Mondlane, Maputo, Mozambique



M. Abrunhosa et al. (eds.), *Advances in Geoethics and Groundwater Management: Theory and Practice for a Sustainable Development*, Advances in Science, Technology & Innovation,
https://doi.org/10.1007/978-3-030-59320-9_28

Unfortunately, there is an inadequate water resources information. Women and children spend their time walking to collect water. In rural areas, water point sources include shallow handy dug wells with hand pumps, natural springs without fencing protection, shallow alluvial aquifers with highly mineralized water are putting Mozambican people at risk for a variety of negative health outcomes (Fig. 1), but groundwater supplies are increasingly threatened by contamination due to various sources. The 48% of all rural Mozambicans use other unimproved sources, and 15% use surface water from lakes, dams, rivers, and other sources as their drinking water, decreased by half since 1990 (UNICEF 2015). Moreover, climate change is a growing-up factor of pressure on groundwater resources availability and water quality protection.

In the framework of this project were investigated groundwater and shallow resources in the buffer zone of Limpopo National Park. In the aim of contributing to the baseline knowledge of aquifers recharge and to identify which phenomena (climate change, drought cycles) can be connected to it, the hydrogeological inverse budget was applied (Sappa et al. 2015).

This profile of research is to understand how water resources are already stressed by climatic conditions and how several populations of Southern Africa use groundwaters for their own survival and irrigation uses.

## 2 Materials and Methods

Mozambique can be divided into three major hydrogeological regions: a basement complex, volcanic terrain and the Mozambique sedimentary basins. The latter, made of textured arkosic sandstone and clay layers, covers about 21% of the country inland, and its morphology is characterized by extensive erosion plains gently dipping coastward (Barbieri et al. 2018). The geology of the study area is characterized by unconsolidated quaternary sedimentary rocks and sedimentary rocks of the Cretaceous–Tertiary age. The most common soil classifications in the study area are arenosols and cambisols (Africa Groundwater Atlas 2019).

**Fig. 1** People living Limpopo National Park buffer zone

It has been set up a digital geological map, using open-source software QGis and considering the lithostratigraphic units. It was obtained by the Soil and Terrain SOTER Database for Southern Africa to give a simplified interpretation aimed to emphasize hydrogeological feature (Van Engelen et al. 1992).

They considered that data were coming from the extensive meteorological station's network of neighbouring Kruger National Park, due to the lack of a meteorological station network inside of Limpopo National Park (Vitale et al. 2017). A total of 25 water samples were collected during two surveys carried out in October 2016 and March 2017 along the Limpopo River and Elephants River:

- In October, at the end of the dry season, 16 water samples: 7 surface water samples (lake and rivers) and 9 groundwater samples (depth <30 m).
- In March, at the end of the wet season, 9 water samples: 3 surface waters and 6 groundwater samples.

Water samples were collected and moved to Geochemistry Laboratory of Sapienza University in Rome. The trace elements concentrations were measured using an ICP-MS (X Series 2; Thermo Fisher Scientific) following filtration (0.45 μm) and acidification in the field. The results of our previous research (Barbieri et al. 2018) identified the presence of groundwater with long residence times, under past climatic and geological conditions in southern Limpopo National Park.

The presence of groundwater with long residence times, associated with the effects of climate change, denounces serious problems related to water supply in Mozambique.

## 3 Results

### 3.1 Water Quantity

The Hydrogeological Inverse Budget technique is a spatial spread data method for the evaluation of aquifer recharge, taking in consideration the mean precipitation and the outcropping rocks properties. In the aim of estimate climate change effects on precipitations and, consequently, on groundwater recharge, they were considered different range of years, included in the historical series, going from 1965 to 2018 (Table 1).

Figure 2 shows a decreasing trend of infiltration, rainfall, and evapotranspiration especially from 1995 to 2018, but with different values.

**Table 1** Values of infiltration, rainfall, and evapotranspiration in the considered range of years

| Range of years | Infiltration (Million m$^3$) | Rainfall (Million m$^3$) | Evapotranspiration (Million m$^3$) |
|---|---|---|---|
| 1965–1974 | 785 | 4834 | 4049 |
| 1975–1984 | 836 | 5120 | 4284 |
| 1985–1994 | 629 | 3855 | 3227 |
| 1995–2004 | 901 | 5549 | 4648 |
| 2005–2014 | 724 | 4454 | 3730 |
| 2015–2018 | 543 | 3329 | 2785 |

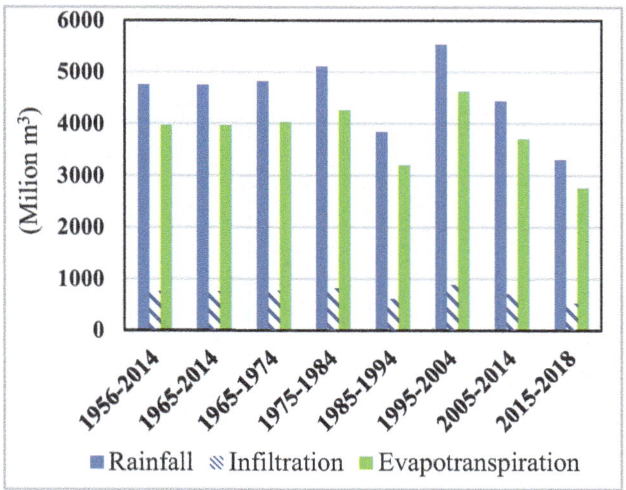


**Fig. 2** Trend of rainfall, infiltration, and evapotranspiration in the considered range of years

The Hydrogeological Inverse Budget (Fig. 2) highlights very low infiltration values, with high evapotranspiration. It is possible to verify that the evaporation rate is very high and exceeds the rainfall rate, probably due to temperature increase.

## 3.2 Water Quality

The groundwater vulnerability to contamination is closely related to its age. Several populations of southern Africa use, for drinking and irrigation purposes, groundwater with long residence times for their own survival and irrigation use. Apparently, these waters could be identified as immune to anthropogenic contamination. Really, they and the aquifers from which they derive are more vulnerable to pollution from contaminants than previously thought (Jasechko et al. 2017).

Results show higher concentrations of mercury, uranium, and boron in groundwater (Fig. 3a) than in surface waters (Fig. 3b), with values higher in wet season respect in dry season. In fact, there is a great separation during the year between wet season (or WS) and the dry season (or DS).

These results support the presence of groundwater with long residence times, due to heterogeneity aquifer with mix between younger groundwater, through high permeability, and older groundwater, in less permeable parts of aquifer system (Jasechko et al. 2017).

# 4 Discussion

Quantitative elaborations highlight two essential aspects. The first one is the tiny part of precipitations, becoming infiltration in the study area (Table 1). This is due to the very low hydraulic conductivity of outcropping rock masses in the area under study. On the second, as it has been noticed for rainfall evolution within the considered range of years, there is no regular trend of decreasing in infiltration, but, in this case, it is very evident that in the last considered years, from 2005 to 2018, it has been output a sensitive decreasing of infiltration, as in comparison with the previous decade, from 1995 to 2004, as with the whole considered period from 1965 to 2018.

On the other hand, trace elements' result highlights the exceeding of the threshold of mercury (Hg) imposed by the World Health Organization in several groundwater samples, in every season. The health risk is significantly higher in the wet season than the dry season. Total dissolved solid (TDS) values are higher during the dry season than the wet season, contrary to the trace elements concentration trend which are higher in the wet season than in the dry season.

# 5 Concluding Remarks

The evaluation of the qualitative characteristics of waters, linked with the anomalous variations of precipitation and temperatures, evaluated from 1960 to 2012 in wet and dry seasons, compared to the output coming from groundwater recharge evaluation, which is very low, gives necessary to set up a system able to capture meteoric water to be used for drinking purposes and thus to reduce the consumption of groundwater standing on meteoric waters collected in little dams, or by rain harvesting works.

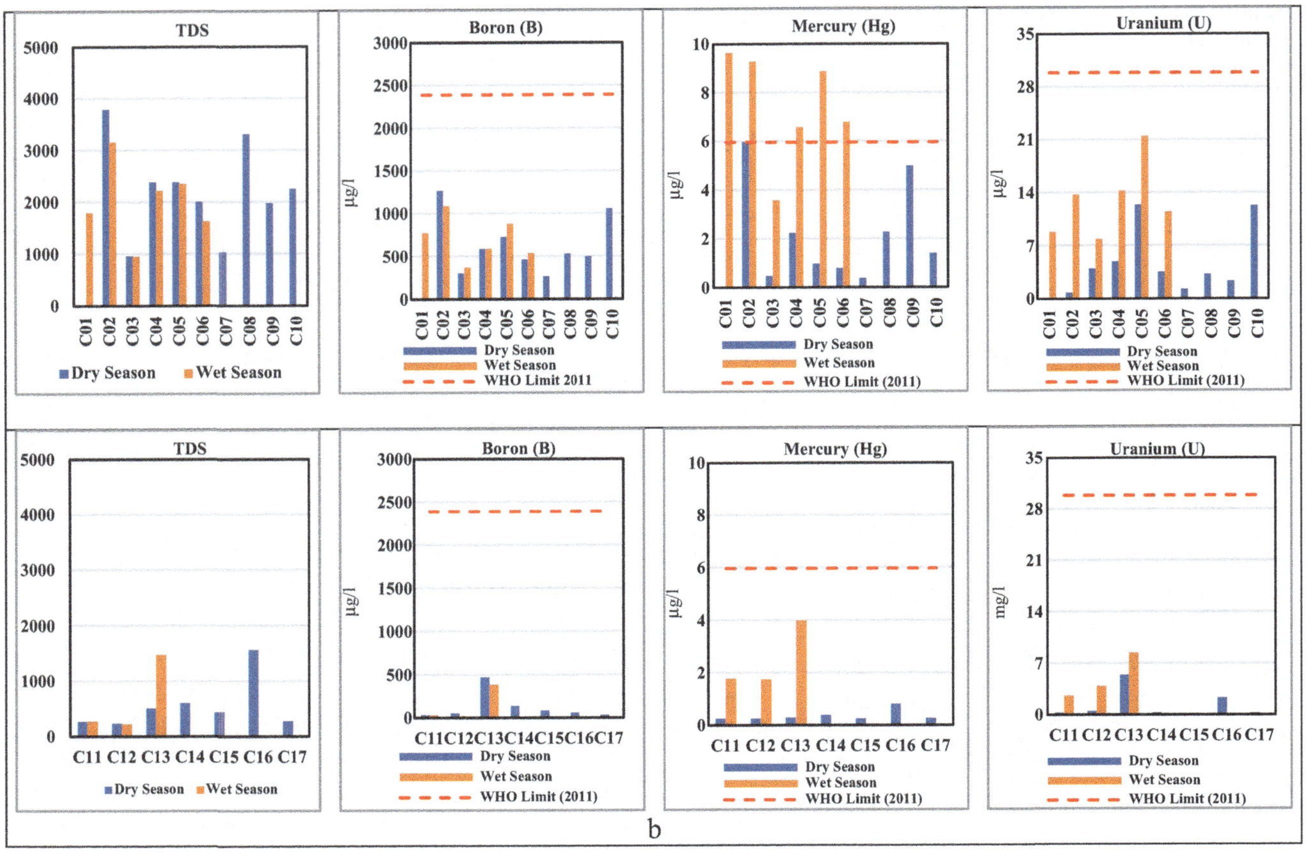


**Fig. 3** Values of TDS and trace elements (B, Hg and U) concentrations in dry season (DS) and in wet season (WS) in groundwater (**a**) and in superficial water (**b**)

**Acknowledgements** The authors thanks Italian Agency for Development Cooperation (AICS https://www.aics.gov.it).

## References

Africa Groundwater Atlas (2019) Climate. British Geological Survey. Accessed 19/12/2019. https://earthwise.bgs.ac.uk/index.php/Climate

Barbieri M, Ricolfi L, Vitale S, Muteto PV, Nigro A, Sappa G (2018) Assessment of groundwater quality in the buffer zone of Limpopo National Park, Gaza Province, Southern Mozambique. Environ Sci Pollut Res 26(1):62–77

Jasechko S, Perrone D, Befus KM, Cardenas MB, Ferguson G, Gleeson T, Luijendijk E, McDonnell JJ, Taylor RG, Wada Y, Kirchner JW (2017) Global aquifers dominated by fossil groundwaters but wells vulnerable to modern contamination. Nat Geosci 10(6):425–429

Sappa G, Trotta A, Vitale S (2015) Climate change impacts on groundwater active recharge in coastal plain of Dar es Salaam (Tanzania). Engineering geology for society and territory—Volume 1: climate change and engineering geology, pp 177–180

UNICEF (2015) WHO: Progress on sanitation and drinking water—2015 update and MDG assessment

Van Engelen, VWP, Wen TT (1992) The SOTER manual: procedures for small scale digital map and database compilation of soil and Terrain conditions, (draft). ISRIC

Vitale S, Barbieri M, Sappa G (2017) Integrated groundwater characterization for biodiversity protection in Limpopo National Park (Mozambique). Int Multidisc Sci GeoConf Survey Geol Min Ecol Manag SGEM 17(31):411–420

Water, Sanitation and Hygiene (WASH) Programme in Mozambique 2019 https://www.unicef.org/mozambique/en/water-sanitation-and-hygiene-wash. Last accessed 19/12/2019

# Spatiality of Ethical Challenges in Use and Management of Groundwater in Coastal Regions, Sri Lanka

Ashvin Wickramasooriya and M. M. G. S. Dilini

**Abstract**

Agglomeration of anthropogenic activities in coastal zone is remarkable in Sri Lanka as an island nation. Since water is essential for human survival, coastal communities of Sri Lanka utilize surface water as well as groundwater. In this concern, anthropogenic drivers on groundwater contamination induced by saltwater intrusion have become critical issue during the recent past. Hence, this study was conducted to identify extent of this issue and its spatial characteristics. Studies carried out during the recent two decades have utilized as database to construct this review paper. According to the study, it has been found out that the major reason associated with saltwater intrusion is overextraction of groundwater through pumping. The groundwater extraction is concentrated more in Jaffna peninsula as majority of the population is dependent on groundwater for their sustenance. The natural geographical location with micro-climatic setting of lagoons in the area makes the problem a major one. The wet zone coastal stretch has plenty of surface water for potable and irrigation needs, but extracting groundwater raises the improper utilization of water as a resource. Therefore, other than negotiate, this problem between authorities and general public is better and more effective to follow suggested mitigatory measures to overcome future challenges.

**Keywords**

Anthropogenic impact • Coastal management • Environmental ethics • Groundwater • Salinity intrusion

A. Wickramasooriya (✉)
Department of Geography, University of Peradeniya, Peradeniya, Sri Lanka
e-mail: awickramasooriya@gmail.com

M. M. G. S. Dilini
Postgraduate Institute of Agriculture, University of Peradeniya, Peradeniya, Sri Lanka

## 1 Introduction

A set of concepts and principles that guide in determining which type of activities help or harmful for sentient creatures has implicated as ethics by Richard et al. (2006). As per the Stanford Encyclopedia of Philosophy (2015), environmental ethics determines moral relationship of human beings to the environment and its non-human contents. When it focused on the current study, groundwater ethics is the fundamental concept which denotes moral relationship of human beings to the groundwater. Vast array of uses of groundwater is not negligible where water scarcity is an overwhelming issue in the world. There have been identified four key international groundwater issues by Fienen et al. (2016): depletion of groundwater, degradation of water quality, the water-energy nexus, and transboundary water conflicts.

Given that, the current study has been narrow down to the groundwater aquifers in coastal regions of Sri Lanka. With the study carried out by Custodio and Bruggeman (1987), it can be proved that there is a saltwater-freshwater interrelationship under natural conditions when aquifer locates near seafront. Moreover, saltwater-freshwater relationship under anthropogenic circumstance also has described in the same study. Groundwater abstraction with urbanization, industrialization, and extensive irrigation, surface water losses to the ground through extensive water distribution networks, improvement in runoff river infiltration, and establishment of artificial recharge works could be identified as major aspects among those. Therefore, ethical challenges over coastal groundwater use and management are an inevitable concern.

Within the South Asian Seas (SAS) region as well as its geographical setting as an island nation, it is imperative to look ethical challenges in coastal groundwater use and management in Sri Lanka. Since coastal community is dependent on groundwater for several needs, it is been necessitating to identify regions where this issue is more



M. Abrunhosa et al. (eds.), *Advances in Geoethics and Groundwater Management: Theory and Practice for a Sustainable Development*, Advances in Science, Technology & Innovation,
https://doi.org/10.1007/978-3-030-59320-9_29

demanding and what are the mitigatory measures have been taken to overcome challenges with special focus on groundwater quantity and quality reduction induced by saltwater intrusion due to anthropogenic drivers.

## 2 Methodology

Available literature sources found through the search engines have been utilized to conduct this study. Accordingly, coastal groundwater-related studies carried out in Sri Lanka during the recent two decades have been supported to construct this review article. Saltwater intrusion due to human interventions was another specific concern in article selection. Studies carried out in northern and southern regions of Sri Lanka have been considered separately in order to identify the spatial characteristics of the issue. Accordingly, this article critically looks over the issues and suggests possible mitigation measures to be applied to face ethical challenges in groundwater use and management.

## 3 Results and Discussion

### 3.1 Ethical Challenges in Coastal Groundwater Use and Management in Sri Lanka

Jaffna peninsula located in northern part of Sri Lanka has been concerned in several studies within the collected database due to these communities are more dependent on groundwater for their day-to-day needs. In fact, there is an uncertainty over surface water due to karstic conditions as karst bedrocks are much permeable and connected to the surface, less rainfall caused by geographical setting of the area, as well as due to the fact that there are no perennial rivers. Definitely, saltwater intrusion can become a serious problem for local communities in relation to drinking and irrigation water availability. According to Kumara et al. (2013), overextraction of water in Jaffna from limestone aquifers pulls saltwater upward and mix up with the fresh water. The study carried out by Uthayashangar et al. (2019) highlighted that dug wells near two major saltwater lagoons (Vadamarachchi and Uppuaru) in Jaffna peninsula have been affected by saltwater intrusion. More than 60% of the wells in Irupalai, Navathkuli, Chemmani, Neerveli, and Madduvil villages are highly affected: Both EC (>2.5 S/cm) and salinity (>900 ppm) are exceeded the maximum tolerable limits for human consumption. The study carried out by Janen et al. (2014) proved the saltwater intrusion susceptibility of land areas near Thondaimanaru, Upparu, and Valukiaru lagoons. Furthermore, it shows a positive relationship between the level of saltwater intrusion in Jaffna peninsula and the operation of the gates of Thondaimanaru, Ariyali and Arali barrages. Hence, it can be justified the improper management of existing barrages at the lagoon mouths. Gunaalan et al. (2018) revealed the deterioration of groundwater quality with the progression of the dry season. Consequently, communities in these areas facing limited access to safe water for drinking and agriculture uses are more at risk during dry period from February to September. As per the study, western zone of the Jaffna peninsula shows good groundwater quality and quantity compared to eastern, northern, and southern zones of the peninsula. The study carried out by Jayasekera et al. (2010) showed seawater intrusion due to unmanaged water pumping resulted by high water demand with expansion of agricultural and domestic uses in the Kalpitiya peninsula located in the northwestern area of Sri Lanka. The same study highlighted that salinity from seawater intrusion produces water quality deterioration with an increasing level of chloride content up to 471 mg/l. Karunarathne et al. (2011) also reported deterioration of groundwater quality in the Puttalam limestone aquifers in northwestern Sri Lanka due to unlimited extraction of freshwater from those aquifers.

According to the Sustainable Water Management Policy Project (2007), it could be identified human interaction with groundwater in Colombo district in Western Coastal Plain. Colombo district consists of high density of drainage paths especially with the Kelani river; the third largest watershed in the country as the main surface water source.

However, freshwater lens aquifers that float on saline water are often contaminated with salt water due to excessive abstraction, especially during prolonged droughts. Instead of that, laterite aquifers which are located away from the flood plain and used for medium-scale water supply schemes were being used which further result in lowering of the water table. Industrial zone in Ekala is a best example for that. Apart from that, semi-confined rock aquifers are used in high-volume for industrial uses and water supply schemes which also lead to rapid lowering of water levels. On the other hand, though there are several surface water supply schemes upstream of Colombo, importance of salinity control has emerged in such schemes also. As an instance, salinity problems at Ambatale estimate that a minimum flow of 30 $m^3$/s is required for salinity control. As Ekaratne (2011) revealed, there is an increasing attention on groundwater in the southeastern part of Sri Lanka after the civil unrest due to rapid development in the area. The study on Panama lagoon area shows results of exposed influx of salt water due to overextraction of groundwater. Areas in southern coastal zone are also suffering due to saltwater intrusion. As it has found out through study carried out by Perera et al. (2018), 25 km distance along the Bentota River has been affected by seawater intrusion exceeding tolerable limits indicated according to WHO standards. The study carried out by Ranasinghe et al. (2019) uncovered the condition of very high salinity in water bodies during dry months. Saltwater intrusion in Koggala lagoon in southern coast of Sri Lanka is also a matter of concern. Jayasiri et al. (2012) mentioned that, there are preventive measures to avoid saltwater intrusion and to regulate water level in the catchment area of lagoon. There is a groin built at the lagoon mouth and it has become a matter of concern for local resource users due to salinity increase in the lagoon. Furthermore, it has impact on drinking water wells of surrounding areas.

As per the referred studies which have been carried out in northern and southern coastal regions of Sri Lanka, it can be found out that there are different contexts of ethical challenges in coastal groundwater use and management in Sri Lanka. It is true that people in Jaffna peninsula have to rely on groundwater due to uncertainty with surface water which is further making the area suppressing. People in the northwestern, northeastern, southwestern, southeastern, and southern regions having plenty of surface water resources with vast array of river network. However, why they have to seek for groundwater for their uses is due to pollution of surface water sources and increasing demand for water as major urban hubs have concentrated near to the coastal zone. Consequently, overextraction of groundwater has arisen the issue of quantity and quality reduction of groundwater. Challenges have already emerged, and these several studies make sense in having more challenges in near future; especially for the communities in Jaffna peninsula as they don't have any other natural water sources to survive.

### 3.2 Mitigatory Measures to Overcome Challenges

It is very clear that, ethical challenges in coastal groundwater use and management are outstanding where there is no other option to access water under natural circumstances. That is why most of the studies have been directed on the issue of groundwater in Jaffna peninsula. Therefore, there is a dire need for researches where the research problems are critical in concern. As the problem in Jaffna peninsula has been narrowed down to the areas near lagoons, construction of saltwater barrages may be more effective. Additionally, studies carried out in this geographical region show that importance of groundwater recharge. Surface ponds could be proposed as a measure in this case. When it concerned about other regions where research requirements have arisen and conducted on different studies to clarify the issue of coastal groundwater quantity and quality, it is very clear that the major problem is improper resource utilization. Especially in the southern and western wet zone coastal stretches. This is because naturally they have plenty of surface water sources. However, people have tendency of using groundwater due to improper use and management of surface water. Even in the cases where relevant authorities involve in groundwater abstraction for water supply schemes, it is clear that they do not have concern for emerging problems and issues with overextraction. It is the real cause to drag salt water to the freshwater aquifers during dry seasons.

Furthermore, people have utilized groundwater as an alternative water source in the urbanized coastal zones in wet zone of Sri Lanka to fulfill their drinking water requirements due to poor pipeline coverage. It needs timely concern, since water received in wet zone with southwest monsoon is more than enough to fulfill water requirements of dwellers in southern and western coastal regions. In this situation, what can be the most magnificent mechanism is to design a water storage mechanism, maybe in the form of small-scale ponds in small tributary outlets before water is discharged to the main branch of Kelani, Kalu, Gin, and Nilwala rivers. It may not only provide relief for the associated problem with groundwater extraction but would also be better in flood control during rainy seasons. Another imperative aspect to consider is the continuous monitoring mechanisms for groundwater levels and maintenance of salt trapping structures in the lagoon mouths.

## 4 Concluding Remarks

The important reason outstanding for saltwater intrusion in coastal regions of Sri Lanka is mostly associated with overextraction of groundwater by the method of pumping. The situation is more visible in Jaffna peninsula due to bounding reasons with its geographical setting and micro-climatic conditions. At the same time, seawater intrusion put them into a more critical situation while they do not have natural water sources for their daily water necessity. Though it is possible to arrange water supply scheme, it is a doubtful whether these communities can bear tariff structures for both drinking and irrigation water uses. It is not easy to find potential water source from nearby area, and therefore, cost accrued with making this solution may not be easy for the natives to afford. Another outstanding situation is challenging with seawater intrusion for groundwater users in wet zone coastal stretch. However, they may not be susceptible under this concern if they can manage surface water sources for their needs. Moreover, in all cases, continuous assessments, maintenance, awareness, and capacity building should be done practically keeping into consideration the issue of intrusion of salt water with groundwater.

## References


Ekaratne K (2011) Investigating groundwater quality in panama. https://www.mangrovesforthefuture.org/grants/small-grant-facilities/sri-lanka/phase-2-sgf-projects-lk/cycle-1-sgf-projects-lk/investigating-groundwater-quality-in-panama/. Last accessed 05 Apr 2020

Fienen MN, Arshad M (2016) The international scale of the groundwater issue. In: Jakeman AJ, Barreteau O, Hunt RJ, Rinaudo JD, Ross A (eds) Integrated groundwater management. Springer, Cham, pp 21–48

Custodio E, Bruggeman GA (1987) Groundwater problems in coastal areas. Studies and Reports in Hydrology, vol 45, UNESCO, Paris [https://unesdoc.unesco.org/ark:/48223/pf0000074510. Last accessed 05 Apr 2020

Janen SS, Sivakumar SS (2014) Groundwater quality improvement of Jaffna Peninsula of Sri Lanka by regulating water flow in the lagoon mouths. Int J Scient Eng Res 5(4):973–978

Jayasekera DL, Kaluarachchi JJ, Villholth KG (2010) Groundwater stress and vulnerability in rural coastal aquifers under competing demands: a case study from Sri Lanka. Springer J Environ Monit Assess 176:13–30

Jayasiri HB, Dahanayaka DDGL (2012) Saltwater Intrusion and Its Impact on Koggala Lagoon and Associated Waters, Southern Coast of Sri Lanka. Asian J Water Environ Pollut 9(2):73–84

Gunaalan K, Ranagalage M, Gunarathna MHJP, Kumari MKN, Vithanage M, Srivaratharasan T, Saravanan S, Warnasuriya TWS (2018) Application of geospatial techniques for groundwater quality and availability assessment: a case study in Jaffna Peninsula. Sri Lanka. ISPRS Int J Geo-Inf 7(1):20

Karunarathne GRR, Samarakoon ASMNB, Wikramanayake DGL, Edirisingha EANV (2011) An investigation on the trends of water quality deterioration in Northwestern Limestone Aquifer system in the Puttalam Region based on Stable Isotope Composition. In: Proceeding of the workshop on challenges in groundwater management in Sri Lanka. Water Resources Board and Dam Safety & Water Resources Planning Project, Colombo, pp 53–59

Kumara IGCI, Ratnayaka SSK, Mayadunne MMCM, Rajapakse RRGR (2013) Assessment of groundwater quality in Jaffna Peninsula. J Geol Soc Sri Lanka 15:137–146

Perera MDND, Ranasinghe TKGP, Piyadasa RUK, Jayasinghe GY (2018) Risk of seawater intrusion on coastal community of Bentota river basin Sri Lanka. In: 7th International Conference on Building Resilience; using scientific knowledge to inform policy and practice in disaster risk reduction, ICBR2017, 27 – 29 November 2017 Bangkok, Thailand, p. 699–705. Elsevier Ltd

Ranasinghe TKGP, Piyadasa RUK (2019) Visualizing the spatial water quality of Bentota, Sri Lanka in the presence of seawater intrusion. J Groundwater Sci Eng 7(4):340–353

Richard P, Linda E (2006) The Miniature Guide to Understanding the Foundations of Ethical Reasoning. United States: Foundation for Critical Thinking Free Press. p. NP. ISBN 978–0–944583–17–3. https://www.criticalthinking.org/files/Concepts_Tools.pdf. Last accessed 05 Apr 2020

Stanford Encyclopedia of Philosophy (2015) Environmental Ethics. The Metaphysics Research Lab, Center for the Study of Language and Information (CSLI), Stanford University Library of Congress Catalog Data: ISSN 1095–5054. https://plato.stanford.edu/entries/ethics-environmental/. Last accessed 05 Apr 2020

Sustainable Water Management Policy Project (2007) https://www.iges.or.jp/en/publication_documents/pub/policyreport/en/654/08_chapter3-5srilanka.pdf. Last accessed 05 Apr 2020

Uthayashangar S, Nanthakumaran A, Devaisy S (2019) Status of the saltwater intrusion in Jaffna. Sri Lanka. Ceylon J Sci 48(3):235–241

# Rural Water Supplies in Galicia

Acacia Naves, Javier Samper, and Bruno Pisani

**Abstract**

A large part of the water catchments in Galicia do not comply with the sanitary standards of supply water due to bacteriological contamination and concentrations of nitrates and some metals which in some cases exceed the admissible limits. This situation is due to shortcomings in: (1) infrastructure; (2) system maintenance; and (3) territorial planning to take into account the protection of water catchments. This situation poses a public health problem that needs to be addressed. Who is responsible and how should this problem be tackled is here analyzed. The administration has the responsibility to provide all residents with a water supply suitable for human consumption and should find ways to improve the small autonomous rural supply systems groundwater quality. This includes to raise awareness among the population about the need to obtain and preserve water that meets the quality standards so that the owners of the systems are the ones who are concerned about and work together with the administration. Hydrological planning and groundwater protection measures should be established in a coordinated manner between the different administrations, the productive sectors involved and neighbors and with a multidisciplinary point of view that integrates environmental, hydrogeological, economic and social aspects.

**Keywords**

Rural water supplies • Water quality • Groundwater management • Galicia

A. Naves (✉) · J. Samper · B. Pisani
Centro de Investigaciones Científicas Avanzadas (CICA), University of A Coruña, A Coruña, Spain
e-mail: acacia.naves@udc.es

## 1 Introduction

Groundwater is an important source of potable drinking water, particularly for rural and dispersed population (Fornés et al. 2005). In the Galicia-Costa hydrologic district, which covers all the watersheds completely located inside Galicia (NW of Spain), more than one quarter of the total population rely on groundwater supply through autonomous traditional solutions (Romay and Gañete 2007). The technical and economic feasibility of centralized infrastructures in this region is severely limited by the distance to highly populated areas and the large required investments to undertake them. Local people developed spontaneously autonomous systems in response to the historic shortcomings of local administrations. They consist of private wells and small rural supplies which collect water from springs and supply one or several houses. Users are the only responsible for infrastructure maintenance and supplied water quality. Furthermore, some people refuse the connection to a centralized supply system when it is available or combines both systems, due to the lower price of water in the traditional systems.

There are also many spring catchments which feed public fountains and traditional wash houses. In the Galician rural environment, public fountains can be found even in unpopulated areas. The wash houses have lost their use with few exceptions. In some cases, they are well maintained as a heritage asset, but in most cases, they are abandoned. In contrast, many of the public fountains continue to be used. Neighbors and even people from nearby urban areas fill up their vessels of water in the public fountains and take them home in the belief that, despite not receiving any treatment or having any kind of sanitary guarantee, the water is of especially good quality.

The granitic and metamorphic rocks that dominate the Galicia-Costa hydrologic district have traditionally been considered almost impervious or to have very low permeability, and their groundwater resources have usually been



M. Abrunhosa et al. (eds.), *Advances in Geoethics and Groundwater Management: Theory and Practice for a Sustainable Development*, Advances in Science, Technology & Innovation,
https://doi.org/10.1007/978-3-030-59320-9_30

neglected in planning and water management (Raposo et al. 2012). However, the traditional use of shallow groundwater proves that the weathered and fractured rock stores significant water volume that should not be neglected in the water resources evaluation. The few studios of groundwater resources in this region estimate the recharge to be between 8 and 18% of the precipitation rate (900–2500 mm $yr^{-1}$) (Soriano and Samper 2000; Xunta de Galicia 2011; Raposo et al. 2012; Samper et al. 2020).

Many water abstractions contrast with the little knowledge and study of the groundwater in Galicia. An accurate characterization of the aquifer and a quantitative analysis of rural private uses are crucial for the efficient groundwater resource management. The LIFE AQUA-PLANN project (Ameijenda 2013) and LIFE Rural Supplies project (Ameijenda 2017), founded by UE, aimed to analyze the groundwater uses in the Abegondo municipality (A Coruña, Galicia) and increase the knowledge of the hydrogeology of the study area. Here, some of the conclusions of those projects are presented that indicate the need of improvements for the rural water supplies and pose several ethical dilemmas in which should be the role of the different stakeholders to guarantee an appropriate service.

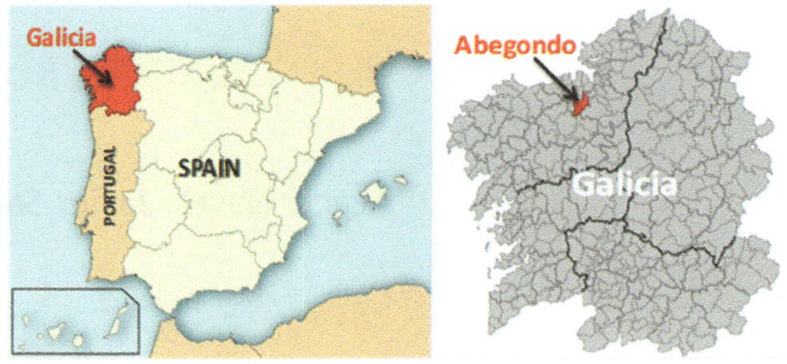

**Fig. 1** Location of Abegondo municipality in Galicia (Spain)

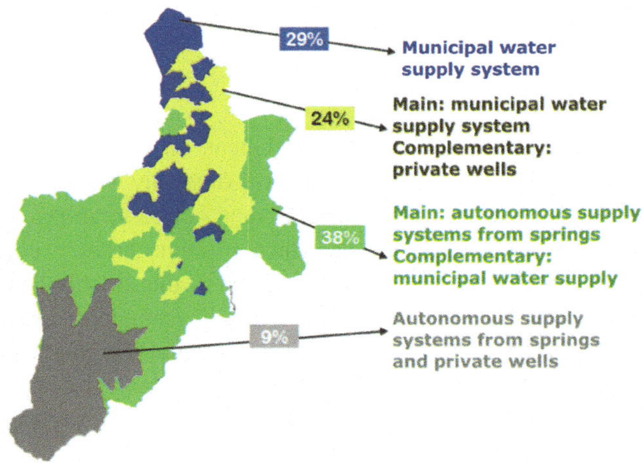

**Fig. 2** Map of Abegondo municipality in which different water supply systems are distinguished

## 2 Analysis of Water Supply Systems in Abegondo

Abegondo is a municipality of Galicia in northwest Spain (Fig. 1). It has an area of 84 $km^2$. It is a rural area with a low density and very dispersed population (~68 inhabitants per $km^2$). The territory is covered with areas of pasture, crops, forest and scrub. Its economy depends on agriculture, livestock and forest plantations, as well as small industries derived from them. Traditionally, the population was supplied with water through small autonomous water supplies from springs and private wells. In 2001, a municipal supply system started to operate which takes water from a reservoir and whose distribution network supplies 90% of homes. However, only 40% of the homes in the municipality have a connection to the municipal system and a part of these homes use water from the municipal system as a complement to the water from the wells and from the autonomous water supplies (Fig. 2). As a result, around 70% of the population is supplied with groundwater by means of autonomous infrastructures. On the other hand, spring catchments also feed many public fountains and traditional washing houses throughout the municipality.

The Abegondo municipality developed an inventory which compiles nearly a hundred public fountains and traditional wash houses within the framework of the LIFE AQUA-PLANN project (Ameijenda 2013). A large part of those public-use springs in Galicia do not meet the sanitary standards for water supply due to bacteriological and nitrate contamination. This situation is propitiated by an inappropriate maintenance of the infrastructures and the lack of adequate water and land use plans in the spring catchments. The recovery of water quality and the assurance of the potability of public fountains in the municipality of Abegondo were also undertaken in the AQUA-PLANN project (Naves et al. 2019). Demonstrative actions of groundwater quality restoration were implemented in five pilot springs to clean and disinfect the fountains, restore the spring catchments and define and implement spring protection zones.

The small autonomous rural water supplies in the south of Abegondo, where only this kind of supply systems are available (Fig. 2), were studied in depth within the framework of the LIFE Rural Supplies project (Ameijenda 2017) aiming to strength their sustainability in spite of the following threats: (1) low guarantee of water supply; (2) water quality deterioration; and (3) deficiencies in the governance of groundwater use and waste treatment and disposal. Improving the knowledge of the hydrogeology is essential for the groundwater protection and management. Available topographic, meteorological, geological, piezometric, streamflow and hydrochemical data have been complemented with:

(1) the update of the groundwater inventory; (2) the collection of piezometric and spring flow data; (3) the construction of three piezometers for online monitoring of the hydraulic head; (4) several sampling campaigns for chemical and bacteriological analyses; (5) hydraulic tests; and (6) electrical resistivity tomographies. All this information has been used to elaborate the conceptual hydrogeological model of the study area and assess the groundwater chemical and microbial status. Extensive evidences of bacteriological contamination have been found in the study area. This contamination is simply removed by water chlorination. The water quality analysis shows bacteriological and nitrate contamination in a large part of the study area. It has been established that nitrate pollution is a consequence of inadequate management of the manure in the fields and the occasional discharge of large volumes of slurry from local pig and mink farms. Concentrations of some metals exceed parametric values for drinking water at some points. However, they are related to suspended matter in many cases, and, therefore, the pollution can be prevented simply by filtration of groundwater.

The analysis of the groundwater resources carried out using a hydrometeorological balance model indicates that the recharge is of the order of 20% of the precipitation and that the sustainable resources are significantly greater than the demand required in the rural areas (Samper et al. 2020). The average recharge for the period 2006–2015 is 187 mm/year. The problems of guaranteed supply are generally more related to deficiencies in infrastructure and management than to the availability of resources. The calibrated hydrological model has been used to assess the effects of climate change at the end of the twenty-first century using the predictions of the regional climate model PROMES-HadAM3H for two IPCC emissions scenarios: one more pessimistic (A2) and one more optimistic (B2) (Nakicenovic et al. 2000). Estimated decrease in groundwater resources (6–12%) should not affect the availability of water resources to meet the demand. Other expected effects of climate change at the end of the century include a possible decline in the water table and a concentration of recharge in shorter periods (Samper et al. 2020). As a consequence, supply problems could arise in some areas which, however, could be overcome by deepening the water catchment or drilling wells.

## 3 Discussion and Concluding Remarks

The studies carried out in Abegondo have shown the large number of small groundwater catchments in the rural areas of Galicia and their importance for the population. A large part of those water catchments does not comply with the sanitary standards of supply water due to bacteriological contamination and concentrations of nitrates and some metals such as manganese which in some cases exceed the admissible limits. This situation is due to shortcomings in: (1) infrastructure design; (2) system maintenance; and (3) territorial planning to take into account the protection of water catchments. Similarly, conclusions have been derived from other research related to impact of civil engineering works in groundwater such as those by Li (2012) and Samper et al. (2019). This situation poses a public health problem that needs to be addressed, but who is responsible and how should this problem be tackled?

The administration has the responsibility to provide all residents with a water supply suitable for human consumption. However, it cannot assume the construction and management of centralized supply systems in the rural areas. Furthermore, even if a municipal supply is available, the population would refuse to be connected to it and continues to use autonomous solutions, which are the cheapest, or combines both services. Thus, the administration should find ways to improve the small autonomous rural supply systems, incorporating a filter and a chlorination system and establishing protection measures in the catchment areas, and to monitor water quality of the groundwater.

Nowadays, any water catchment legally requires a permit from the basin district administration which involves its inclusion in a register. However, most of the rural water catchments are not registered. It is estimated that more than 70% of existing wells and spring catchment of autonomous rural supplies have not been registered. In many municipalities, there is not even an inventory of public fountains and traditional washing places. If the water catchments do not exist from a legal point of view, the administration cannot establish their monitoring, directly implement or subsidize their improvement. Measures should be taken to encourage the registration of catchments.

The requirement from the administration to the owners of the registered rural water supplies for a control of the state of the installations and the quality of the water would be very complex. On the one hand, it would have as a consequence a greater rejection to the registration of the catchments. On the other hand, the administration is not in a position to demand a quality of service from the neighbors, who are trying to overcome a deficiency in their functions. It is necessary to raise awareness among the population about the need to obtain and preserve water that meets the quality standards for human consumption so that the owners of the systems are the ones who are concerned about and ask for help to implement improvements in the systems and water quality.

Finally, the previous experiences have shown the need for hydrological planning and groundwater protection measures to be established in a coordinated manner between the different administrations (local, regional, national and European), the productive sectors involved (farms, farmers, petrol stations, etc.) and neighbors and with a multidisciplinary point of view that integrates environmental, hydrogeological, economic and social aspects.

**Acknowledgements** This work has been performed within the framework of the projects: LIFE07 ENV/E/000826 "AQUA-PLANN PROJECT—Integrated management of water resources and their application to the local planning of the SCI Abegondo—Cecebre" (Partners: Augas de Galicia, the Rural Environment Department of Xunta de Galicia, EMALCSA and the Municipality of Abegondo) and LIFE12 ENV/ES/000557 "Sustainable solutions for small networks of supply" (Partners: Augas de Galicia, the Health Department of Xunta de Galicia and the Municipality of Abegondo) of the LIFE program of the European Commission. The work has been funded also by projects of the aid 2012/181 of the "Consolidation and structuring of competitive research units- Competitive Reference Groups" program from the Xunta de Galicia. We also acknowledge the support provided by Carlos Ameijenda from the Municipality of Abegondo and Raquel Piñeiro, Mónica Velo and Roberto Arias from Augas de Galicia (Xunta de Galicia), Elvira Íñiguez from Consellería de Sanidade (Xunta de Galicia) and SAI-UDC Services staff. This work could not have been performed without the priceless collaboration of the Abegondo neighborhood.

## References

Ameijenda C (2013) Aqua-plann project—Integrated management of water resources and their application to the local planning of the SCI Abegondo—Cecebre (LIFE07 ENV/E/000826). LIFE project's webpage: https://ec.europa.eu/environment/life/project/Projects/index.cfm?fuseaction=home.createPage&s_ref=LIFE07%20ENV/E/000826. Last accessed 27 Jan 2020

Ameijenda C (2017) Life rural supplies—sustainable solutions for very small water supplies (LIFE12 ENV/ES/000557). LIFE project's webpage: https://ec.europa.eu/environment/life/project/Projects/index.cfm?fuseaction=search.dspPage&n_proj_id=4685. Last accessed 27 Jan 2020

Fornés JM, de la Hera A, Llamas MR (2005) The silent revolution in groundwater intensive use and its influence in Spain. Water Policy 7:253–268

Li Y (2012) Hydrodynamic and geochemical models for environmental applications in civil and mining engineering. University of A Coruña, Spain (PhD thesis)

Nakicenovic N, Alcamo J, Davis G, de Vries B, Fenhann J, Gaffin S, Gregory K, Grübler A, Jung TY, Kram T, La Rovere EL, Michaelis L, Mori S, Morita T, Pepper W, Pitcher H, Price L, Riahi K, Roehrl A, Rogner H-H, Sankovski A, Schlesinger M, Shukla P, Smith S, Swart R, van Rooijen R, Victor N, Dadi Z (2000) IPCC special report on emission scenarios. Cambridge University Press, Cambridge, United Kingdom and New York, NY, USA

Naves A, Samper J, Mon A, Pisani B, Montenegro L, Carvalho JM (2019) Demonstrative actions of spring restoration and groundwater protection in rural areas of Abegondo (Galicia, Spain). Sustain Water Resour Manag 5:175–186

Romay D, Gañete M (2007) Las comunidades de usuarios de agua como ejemplo de autoabastecimiento en las cuencas de Galicia-Costa [The water users communities as an example of self-sufficiency in Galicia-Costa]. In: Molinero J, Ortuño F, Valverde M, Lambán LJ (eds) Las aguas subterráneas en España ante las directivas europeas: Retos y perspectivas. IGME, Madrid, pp 239–245

Raposo J, Molinero J, Dafornte J (2012) Parametrization and quantification of recharge in crystalline fractured bedrocks in Galicia-Costa (NW Spain). Hydrol Earth Syst Sci 16:1667–1683

Samper J, Naves A, Mon A, Fernández L, Calvo J (2019) Conceptual and numerical groundwater flow models to evaluate construction, geotechnical and environmental effects of the Brión railway tunnel for the Ferrol Outer harbor. In: 46th IAH Congress Málaga 2019, Groundwater management and governance, coping with scarcity, 22–27 Septiembre 2019, Málaga (España)

Samper J, Naves A, Pisani B, Dafonte J, Montenegro L, García-Tomillo A (2020) Sustainability of groundwater resources of weathered and fractured schists in the rural areas of Abegondo (Galicia, Spain). Hydrogeol J (Accepted for publication)

Soriano G, Samper J (2000) Hidrogeología de una pequeña cuenca piloto en medios graníticos: Cuenca del Valiñas (A Coruña) [Hydrogeology of a small pilot basin on granitic terrain: Valiñas basin (A Coruña)]. In: Molinero J, Ortuño F, Valverde M, Lambán LJ (eds) Las aguas subterráneas en España ante las directivas europeas: Retos y perspectivas. IGME, Madrid, pp 73–82

Xunta de Galicia (2011) Proposal for hydrological plan of Galicia-Costa [in Spanish]

# Assessment of Groundwater Balance and Importance of Geoethical Approach for Upper Kabul Sub-basin, Afghanistan

Asadullah Farahmand, Mohammad Salem Hussaini, and Sayed Waliullah Aqili

**Abstract**

To conduct an integrated groundwater management, understanding the situation of recharge, discharge, balance, and potential of groundwater resources is required. Therefore, purpose of this study was assessment of upper Kabul sub-basin and finds out balance and potential condition of groundwater resources in this sub-basin. During data collection, meteorological and hydrogeological information including water table measurements, storage coefficient, effective porosity, and transmissibility of aquifer has been collected from 2008 to 2018. Method of balance calculation has been developed according to the available information and physical–geographical condition of the study area, and then a modified balance equation has been applied. Results of this study indicate that recharge was 11.17 million cubic meters (MCM), discharge was 25.74 MCM, and groundwater balance is negative significantly (−14.75 MCM) in upper Kabul sub-basin in 2018. This negative balance value proves a significant difference between recharge and discharge in the study area. Moreover, urban utilization is the main discharge parameter. Therefore, to reduce water demand and usage by population in the study area, developing geoethical approach is a key factor. Geoethical approach can affect on reducing water usage rate and help to manage groundwater balance in upper Kabul sub-basin positively.

**Keywords**

Groundwater balance • Recharge • Discharge • Geoethical approach • Kabul aquifer

## 1 Introduction

In this research, groundwater balance and potential of upper Kabul aquifer (Darulaman and Paghman sectors) have been estimated and assessed. The results of the study can help to understand availability of groundwater resources, discharge, and recharge situation in the subject aquifer.

However, balance and discharge of groundwater resources can be reduced and controlled by developing geoethical approach effectively. Therefore, this research tries to state the importance of hydrogeoethics aspects on groundwater management in upper Kabul aquifer.

The study area is located in western part of Kabul city between longitudes 34° 27′ 30″ and 34° 32′ 00′ and latitudes 69° 4′ 30″ and 69° 9′ 30″ (Fig. 1). In addition, two rivers including Paghman River and Maidan River are developed in this area (Thomas 2010). Totally upper Kabul sub-basin is extended in an area of 158 km$^2$, and its aquifer area is around 89 km$^2$ (Broshears 2005).

## 2 Methods and Materials

To carrying out this study, meteorological and hydrological information such as monthly precipitation, runoff, monthly temperature, and also hydrogeological information including storage coefficient, hydraulic conductivity, and transmissibility of aquifer has been collected and used. In addition, information regarding water level depth has been gathered from 18 observation wells with minimum depth of 40 m and maximum of 80 m.

A. Farahmand (✉) · S. W. Aqili
Groundwater Resources Directorate, National Water Affairs Regulation Authority, Kabul, Afghanistan
e-mail: asadullahfarahmand18@gmail.com

M. S. Hussaini
Department of Engineering Geology and Hydrogeology, Kabul Polytechnic University, Kabul, Afghanistan



M. Abrunhosa et al. (eds.), *Advances in Geoethics and Groundwater Management: Theory and Practice for a Sustainable Development*, Advances in Science, Technology & Innovation,
https://doi.org/10.1007/978-3-030-59320-9_31

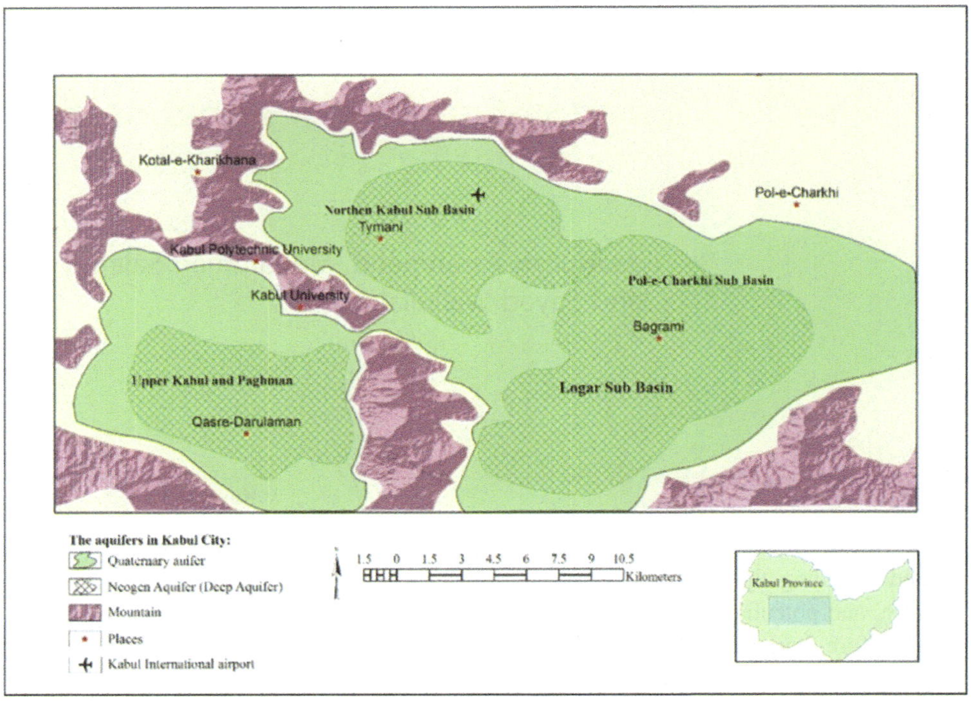


**Fig. 1** Location of upper Kabul sub-basin in Kabul basin (JICA 2011)

The previous studies on groundwater balance and potential of Kabul basin were conducted by BGR in 2005 and JICA in 2010. Some information such as hydraulic conductivity of aquifer, effective porosity, and transmissibility coefficient of materials has been taken from them (JICA 2009). Meteorological information for estimation of recharge and discharge has been collected from 2018 to find out up-to-date results.

The data used in this study has been extracted from many reports and documents from 2008 to 2018 (Table 1).

According to the literature review in this research, there is no specific equation for determination groundwater balance. All suggested equations in the previous studies have been proposed based on the site condition and availability of information for study areas (Kumar 2012). Therefore, according to the availability of information and physical—geographical condition of the study area, below balance equation has been developed. The basis of this equation is a model which was suggested by Maurya (2017) for Pindra Block Varanasi, India.

**Table 1** Summarized data set which are used in this study

| Data | Year | Source |
|---|---|---|
| Groundwater level | 2008-2018 | MEW, JICA |
| Characteristics of aquifer transmissibility and effective porosity, storage coefficient | 2005, 2010 | BGR, JICA |
| Hydraulic conductivity | Different years | DACAAR, MEW, MRRD, BGR & private companies |
| Population | 2018 | NSIA |
| River flow | 2008-2018 | MEW |
| Precipitation and temperature | 2018 | MEW, AMD |
| Shallow aquifer thickness | Different years | MEW, MoMP, JICA,USGS, BGR, AUWSSC, LM |

*JICA* Japan International Cooperation Agency
*DACAAR* Danish Committee for Aid to Afghan Refugees
*MEW* Ministry of Energy and Water
*MRRD* Ministry of Rural Rehabilitation and Development
*NSIA* National Statistics and Information Authority
*USGS* United State Geological Survey
*LM* Landell Mills
*BGR* Federal Institute for Geosciences and Natural Resources
*MoMP* Ministry of Mines and Petroleum
*AMD* Afghanistan Meteorological Department
*AUWSSC* Afghanistan Urban Water Supply and Sewerage Corporation

Generally, balance equation can be presented as below:

$$\sum Q_{\text{in}} - \sum Q_{\text{out}} = \text{change in storage} \quad (1)$$

where

$Q_{in}$ recharge parameters.
$Q_{out}$ discharge parameters.

For this study, below equation has been modified and applied for calculation of groundwater balance.

$$R_f + R_r + R_q + R_g = Q_{pop} + Q_g + Q_{out} + \Delta S \quad (2)$$

where

$R_f$ recharge from rainfall
$R_r$ recharge from river seepage
$R_q$ recharge from Qargha Lake
$R_g$ recharge from green land
$Q_{pop}$ discharge for urban purposes
$Q_g$ discharge for green land
$Q_{out}$ subsurface outflow
$\Delta S$ change in storage.

All the above parameters have been calculated separately based on the collected information and available equations in the literatures, and finally, groundwater balance has been calculated for the study area.

# 3 Results

## 3.1 Groundwater Balance and Potential

Recharge and discharge parameters of balance equation have been obtained based on the calculations, which summarized in Tables 2 and 3:

Amount of change in storage for upper Kabul aquifer can be calculated as following:

$$\Delta S = \sum Q_{in} - \sum Q_{out} = 11.17 - 25.74 = -14.57 MCM/year$$

According to the above calculation, groundwater balance is negative for the study area in 2018. It shows that discharge is more than recharge and difference is significant (more than twice).

## 3.2 Importance of Geoethical Approach

According to the previous section, urban utilization is the main parameter in groundwater discharge, and it is around 88 percent of total discharge in the study area. Mack et al. (2010) have found rate of 40 (liter/day/person) for water demand in Kabul basin area, and population in upper Kabul sub-basin area was announced 1,549,366 in 2018. A simple calculation shows that if water demand rate decreases from 40 to 35 L/d, urban discharge can be decreased to 19.79 MCM/year and urban discharge can be changed to 16.9 MCM/year with 30 L/d water demand rate.

To reduce water demand and usage by population in the study area, geoethical approach is a key factor. Geoethics is a theory about ethical relations of humans with inorganic nature, based on the perception of this nature as a member of moral community, moral partner (subject), based on the principles of equality and equivalence of inorganic matter and on limitation of the rights and needs of humans in relation to inorganic nature (Nikitina 2016). Basis of this definition, authorities, scientists, and managers taking roles on groundwater management in Kabul basin must consider hydrogeoethical aspects in their all duties.

In the current situation, most people are digging and drilling more and more wells, and it becomes typical that every house has deep water well. Extraction of groundwater is carrying out without any ethical considerations, irrespective of any attention of the effects that the new well might have on the yields of existing wells in the vicinity. While this unmoral approach needs to be changed into geoethical approach, so people should believe that groundwater resource is reducing, and they should have self-commitment

**Table 2** Recharge parameters in upper Kabul sub-basin for 2018

| Parameter | Symbol | Value | Unit |
|---|---|---|---|
| Rainfall | $R_f$ | 7.80 | MCM/year* |
| River seepage | $R_r$ | 3.264 | MCM/year |
| Qargha Lake | $R_q$ | 0.107 | MCM/year |
| Green land | $R_g$ | N.A | |
| Total recharge | | 11.17 | MCM/year |

**MCM* million cubic meters

**Table 3** Discharge parameters in upper Kabul sub-basin for 2018

| Parameter | Symbol | Value | Unit |
|---|---|---|---|
| Urban purposes | $Q_{pop}$ | 22.6 | MCM/year |
| Green land | $Q_g$ | 2.8 | MCM/year |
| Subsurface outflow | $Q_{out}$ | 0.34 | MCM/year |
| Total recharge | | 25.74 | MCM/year |

regarding extraction of groundwater. On the other hand, geoethical approach is important for both quantity and quality aspects of groundwater. Many people do not care about groundwater pollution; they are drilling absorption sewage wells in their ownerships, for instance, since absorption sewage wells are one of the main sources for increasing groundwater pollution.

## 4 Discussion

With comparison of the results obtained from this study and the previous studies, it can be concluded that discharge in 2018 is more than discharge amount in the previous years. Because population is growing up in the study area significantly, utilization and extraction of groundwater are raising as a consequent.

Amount of potential and inflow is calculated 11.17 MCM/year in the study area in 2018. But groundwater balance was negative with a significant difference between discharge and recharge. If this balance situation continues in the future, it will cause a very critical problem on management of groundwater resources in the study area.

On the other hand, extraction of water from production wells for urban utilization is the most effective parameter on discharge and outflow of groundwater from this aquifer.

People must be aware of this fact that water discharge is twice more than water recharge and balance is negative in upper Kabul sub-basin so that they will comprehend the current critical situation of groundwater resources. Government and authorities should develop geoethical approach including behaviors and values such as commitment, responsibility, and humanity regarding the ways of protecting groundwater against pollution and exceeding discharge through school, mosque, TV, social media, broacher, street banner, etc.

## 5 Concluding Remarks

Considering the results of this study, below concluding remarks can be found out:

- Upper Kabul aquifer is recharging by rainfall, Qargha Lake and two riverbeds. Total amount of recharge in this aquifer was 11.17 MCM in 2018.
- Groundwater of upper Kabul aquifer is outflowed by three factors: urban usage, green land, and outflow to other aquifers. Total amount of discharge from this aquifer was 25.74 MCM in 2018.
- Groundwater balance of upper Kabul aquifer is negative (−14.57 MCM) in 2018.
- The major reason for negative balance in the aquifer is exceeding discharge through huge urban utilization.
- Authorities, scientists, and managers must consider hydrogeoethical aspects in their all duties for management of groundwater in Kabul basin.
- Geoethical approach development in society can affect on reducing water usage rate positively, and it will help to manage groundwater balance in upper Kabul sub-basin.

## References

Broshears RE (2005) Inventory of groundwater resources in the Kabul Basin. Afghanistan. U.S, Geological Survey, Reston, Virginia

JICA—JI (2009) The study for the development of the master plan for the Kabul metropolitan area in the Islamic Republic of Afghanistan sector report 4. Kabul: Ministry of Urban Development (MoUD)

JICA—JI (2011) Study on groundwater resources potential in Kabul basin final report 1. Kabul: Ministry of Mine and Petroleum

Kumar CP (2012) Assessment of groundwater potential. Int J Eng 1 (1):64–79

Maurya D (2017) Assessment of Groundwater Potential Using Water Balance Approach in Pindra Block Varanasi. Int J Res Appl Sci Eng Technol (IJRASET) 5(9):1–7

Mack TJ, Akbari MA, Ashoor MH, Chornack MP, Coplen TB, Emerson DG, Hubbard BE, Litke DW, Michel RL, Plummer LN, Rezai MT, Senay GB, Verdin JP, Verstraeten IM (2010) Conceptual model of water resources in the Kabul Basin. Afghanistan. U.S, Geological Survey, Reston, Virginia

Nikitina NK (2016) Geoethics: theory, principles, problems. 2nd edition, Geoinformmark Lda, Moscow. https://www.icog.es/iageth/files/Nikitina_Geoethics.pdf

# Water-Level Dynamic and Hydro-Ecological Effects of Typical Karst Wetland in Southwest China

Danni Zhu, Shengzhang Zou, Yongsheng Lin, Li Lu, and Liankai Zhang

**Abstract**

Huixian wetland is the largest karst wetland in China, which has important ecological and environmental functions. Based on field fixed-point monitoring data, the seasonal water level change and the response to precipitation in Huixian wetland were studied by statistical methods and the influence of wetland water-level fluctuation on ecological environment was further discussed. The results showed that there were significant spatial differences between the surface and groundwater levels in Huixian wetland, both of which responded quickly to precipitation and have significant characteristics of multi-peaks and multi-valleys. The surface and groundwater in the area are frequently changed, and the hydraulic connection is close. Wetland groundwater supplies surface water with a stable intensity all the year round, which is an important recharge water source for Huixian wetland. Pearson correlation analysis showed that there was a significant negative correlation between wetland surface water level and water conductivity (R = - −.7808), and the concentration of nutrients in water increased during the decline of water level, which might induce eutrophication of water body. However, the correlation between wetland groundwater level and shallow soil water content was not significant, and the disturbance to wetland habitat was small.

**Keywords**

Karst wetland • Surface water level • Groundwater level • Ecological effect

## 1 Introduction

Huixian wetland is the largest karst wetland in China. Recently, due to the influence of climate change and human activities, the wetland area is gradually shrinking, and the ecological environment is deteriorating (Xu et al. 2019). At present, many studies have been carried out in Huixian wetland, including preliminary discussion of hydrological process (Li 2008), biodiversity (Wei 2010), chemical characteristics and nutrient content of soil (sediment) (Li et al. 2017; Zhang et al. 2016) and nitrate pollution in water (Peng et al. 2018). However, the fine characterization of water-level dynamic characteristics in Huixian wetland is still slightly insufficient. Therefore, the real-time monitoring data of nearly a hydrological year from April 2018 to February 2019 was selected, and descriptive statistical analysis and Pearson correlation analysis were used to study the dynamic change characteristics of surface and groundwater levels and the conversion relationship, and then the potential impact of wetland water-level changes on wetland ecology was discussed. The research results can provide basic theoretical support for the protection and restoration of karst wetlands.

## 2 Materials and Methods

### 2.1 Introduction of the Study Area

Huixian karst wetland is located in Guilin City, Guangxi Province, which is a typical karst wetland in China and the largest natural wetland in the middle and low latitudes of China with a total area of about 120km$^2$. Four aquifer

D. Zhu · S. Zou (✉) · Y. Lin · L. Lu · L. Zhang
Institute of Karst Geology, Chinese Academy of Geological Sciences, Guilin, China
e-mail: zshzh@karst.ac.cn

D. Zhu · S. Zou · Y. Lin · L. Lu · L. Zhang
Key Laboratory of Ecosystem and Rocky Desertification Treatment, Ministry of Natural Resources, Guilin, China

D. Zhu · S. Zou · Y. Lin · L. Lu · L. Zhang
Key Laboratory of Karst Dynamics, Ministry of Natural Resources & Guangxi, Guilin, China



M. Abrunhosa et al. (eds.), *Advances in Geoethics and Groundwater Management: Theory and Practice for a Sustainable Development*, Advances in Science, Technology & Innovation,
https://doi.org/10.1007/978-3-030-59320-9_32

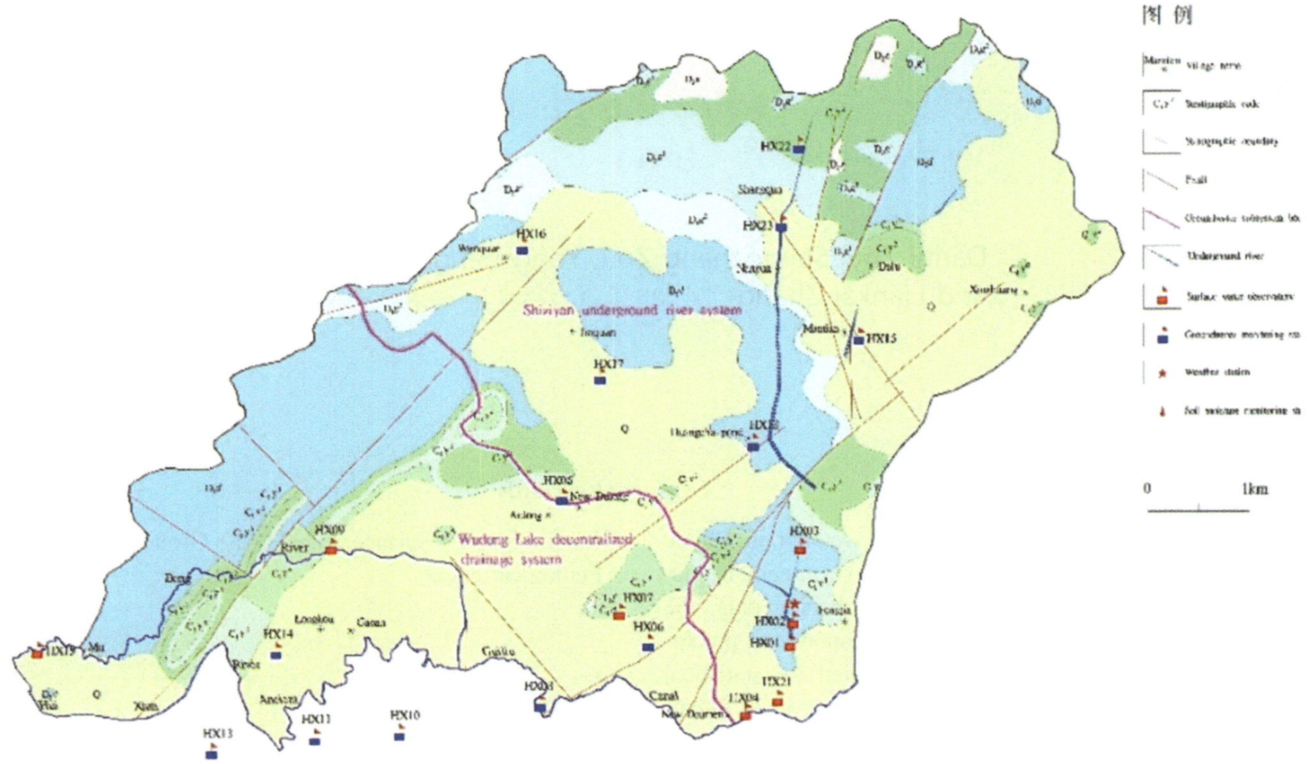


**Fig. 1** Hydrological geological sketch map of study area

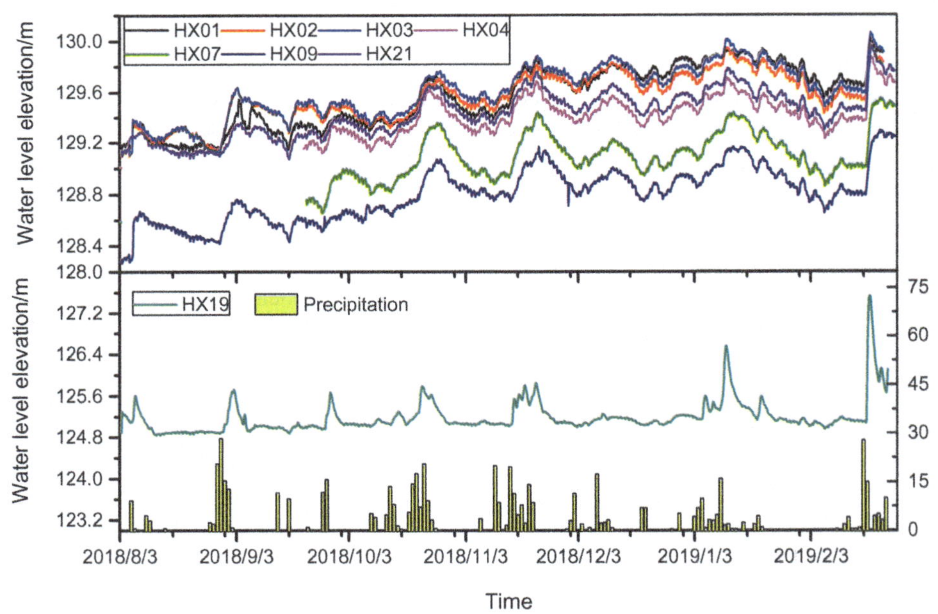


**Fig. 2** Time series variations of surface water-level response to precipitation

mediums are distributed in the area, namely loose pore, strong karst aquifer, medium karst aquifer and clastic karst aquifer (see Fig. 1). According to the division of groundwater system, it can be divided into Shiziyan underground river system and Wudong Lake decentralized drainage system (Fig. 2).

## 2.2 Station Layout and Data Acquisition

There are 22 observation stations in Huixian karst wetland, including 8 surface water observation stations, 13 groundwater observation stations and 1 weather station (see Fig. 1).

LTC Level-Logger three-parameter water-level gauge was used for dynamic monitoring of surface and groundwater water level, water temperature and electrical conductivity. Onset HOBO soil temperature and humidity wireless transmission system were used to automatically monitor wetland soil moisture. The Onset HOBO rain gauge was used to observe the dynamic rainfall.

## 2.3 Data Analysis

The daily time series data from April 2018 to February 2019 was selected for statistical analysis, and OriginLab 9.1was used for data analysis and map production.

# 3 Results

## 3.1 Dynamic Characteristics of Wetland Surface Water Level

The surface water level responds rapidly to precipitation. Within 48 h after a large concentrated precipitation, the surface water level can reach a peak value. After the precipitation stopped, the water level dropped rapidly, and the rise and fall of the water level were equivalent, reaching 0.24 (m•d$^{-1}$). The surface water level has obvious characteristics of multi-peaks and multi-valleys. After each large concentrated precipitation, the surface water level will have a peak value and then fall back rapidly. Judging from the whole observation period, the surface water level showed a slow upward trend, which reflects the function of wetland in water conservation. Water storage in wet years plays a role in regulating and buffering the water cycle of the ecosystem in the region (Chen and Lu 2003). The dynamic variation characteristics of surface water level in Huixian wetland are consistent with the research results of Li Yuanyuan (Li 2008), Cai Desuo (Cai et al. 2012), indicating that the hydrological process in Huixian wetland in the past ten years is relatively stable.

## 3.2 Dynamic Characteristics of Wetland Groundwater Level

As can be seen from Fig. 3, the wave crest of the groundwater-level dynamic curve is narrow and steep, and the wave trough is wide and slow. It indicates that the precipitation pulse input is faster, while the water release is slower, and the karst groundwater system has significant water storage and regulation functions. There are two obvious stages in the fluctuation curve of groundwater level: the first stage is from April to August, the rainy season precipitation and frequency in the study area are relatively large, and the groundwater level fluctuates greatly under the influence of precipitation frequency and intensity. The second stage is from September to February of the following year. With the decrease of precipitation frequency and intensity, the fluctuation of water level also decreases. In addition, groundwater level responds rapidly to precipitation, but the sensitivity of water level to precipitation response varies from station to station.

## 3.3 Characteristics of Wetland Surface Water and Groundwater-Level Transformation

The typical surface and underground observation points of Shiziyan underground river system were selected to analyze the transformation relationship between surface and underground water levels.

The groundwater level was obviously higher than the surface water level throughout the year, and there was a significant hydraulic gradient between the two. The

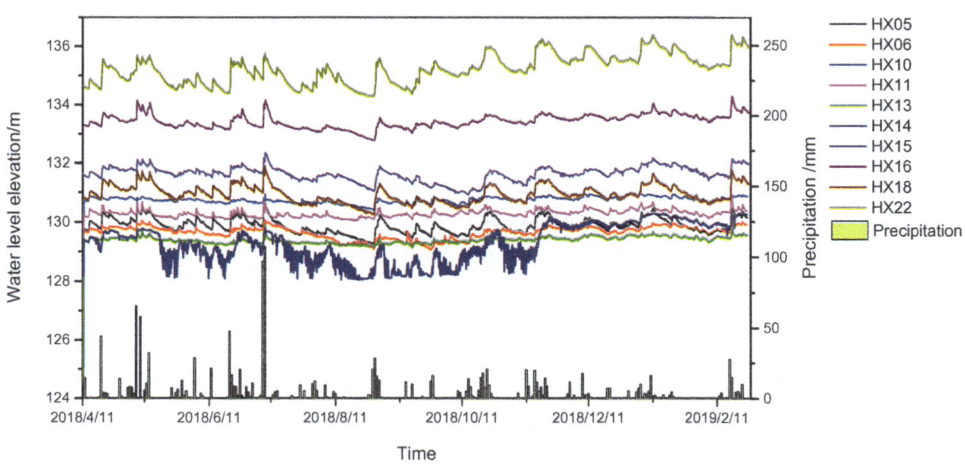


**Fig. 3** Time series variations of groundwater-level response to precipitation

water-level difference between the two would lead to groundwater recharging wetland surface water at different rates throughout the year. Taking HX01 as an example, the influence of groundwater levels at HX15, HX16 and HX18 on surface water levels was fitted. The groundwater levels at HX15, HX16 and HX18 were significantly positively correlated with the surface water levels at HX01. From the fitting equation, it can be found that the response rate of HX01 point to the change of groundwater level at HX15 point is 0.68, the response rate to HX18 point is 0.66, and the response rate to HX16 point is 0.94. The results show that the groundwater at HX15, HX18 and HX16 points belongs to two runoff systems, while the surface water at the outlet of Shiziyan underground river (HX01) is more closely related to the groundwater at HX16 point in runoff direction. The hydraulic gradient from HX15, HX18 to HX01 is $0.66 \times 10^{-3}$, which is lower than $0.86 \times 10^{-3}$ between HX16 and HX01. This shows that the permeability of aquifer in the former is higher than that in the latter. According to the hydrogeological characteristics of the study area (Fig. 1), the karst aquifer medium at HX15 and HX18 are mainly underground river pipelines, while there are mainly karst caves and fissures in HX16 direction. Therefore, the former has better aquifer permeability and smaller hydraulic gradient, which further reflects the multi-directional heterogeneity and complexity of karst aquifer.

## 4 Discussion

Wetland water level can disturb wetland ecological conditions by affecting wetland water depth, water area, water quality, soil water content and other factors. In this paper, Pearson correlation fitting analysis was carried out on the data of water conductivity and soil water content obtained from monitoring and wetland water level to further explore the impact of water-level fluctuation on the ecosystem in Huixian wetland.

The surface water level of Huixian wetland had a significant negative correlation with the conductivity of the water, and Pearson correlation coefficient was -0.7808, indicating that the conductivity decreases with the increase of the surface water level of the wetland. During the decline of wetland water level, the conductivity of the water increases and the concentration of nutrients in the water is sufficient, which is conducive to the growth and reproduction of aquatic plants and microorganisms. However, the temperature is suitable for the growth of algae in autumn, coupled with the drop of wetland water level and the increase of nutrient concentration in the water, which may induce large-scale reproduction of algae and lead to eutrophication of the water body, thus affecting the ecological health of the wetland.

The correlation between groundwater level and soil water content in karst wetland was poor. The special geological conditions in karst area make its soil and water pattern different from that in non-karst area. The surface-underground double-layer spatial structure has formed a water and soil distribution pattern of "soil on the upper floor and water on the lower floor". The soil layer is shallow and discontinuous and is mostly embedded with bedrock. However, the depth of karst groundwater is usually large, and the hydraulic connection between karst cave-fissure karst aquifer and vadose zone soil layer is weak. Soil water content in karst areas is mainly closely related to precipitation, topography and other factors (Guo et al. 2016), but less controlled by groundwater level.

Huixian wetland is the largest karst wetland in China, which has important ecological environmental functions. In recent years, due to climate warming, precipitation reduction and intensified human activities, the wetland has a significant trend of shrinking and degradation. The total population of Huixian wetland is 30 000, and the density is as high as 717 people per square kilometer. Groundwater is the major water source for agricultural irrigation, fishery and drinking in the region. The overexploitation of groundwater has led to the increasing depth of groundwater, resulting in the groundwater depth being more than 2 m. Therefore, the limited supply of groundwater to wetland surface water leads to the further reduction of wetland area in dry season.

## 5 Concluding Remarks

Based on detailed field monitoring data, descriptive statistical analysis and correlation analysis methods were used to study the dynamic characteristics of water-level and hydrological ecological environment effects in Huixian wetland. On the basis of data analysis, a more detailed wetland hydrological process is outlined.

## References

Cai DS, Ma ZL, Jiang ZC (2012) Study on Karst Wetland Ecosystem in Huixian. Geological Publishing House, Beijing

Chen YY, Lu XG (2003) The wetland function and research tendency of wetland science. Wetl Sci 1(1):7–11

Guo XJ, Gong XP, Tang QJ (2016) The response processes of moisture at soil profile to precipitation in typical karst hillslopes. Carsol Sin 35(6):629–638

Li YY (2008) A preliminary study on hydrological process of Huixian karst wetland in Guilin. Guilin: Guilin university of technology (MSc thesis)

Li ZY, Jin ZJ, Li Q (2017) Changes in land use and their effects on soil properties in Huixian karst wetland system. Pol J Environ Stud 26 (2):699–707

Peng C, Pan XD, Jiao YJ (2018) Identification of nitrate pollution sources through various isotopic methods: A case study of the Huixian wetland. Envir Sci 39(12):5410–5417

Wei F (2010) Study on biodiversity and protection of Karst wetland in Huixian, Guilin. Guilin: Guangxi normal university (MSc thesis)

Xu GP, Li YQ, Shen YY, Duan CY (2019) Soil organic carbon distribution and components in different plant communities along a water table gradient in the Huixian karst wetland in Guilin. Envir Sci 40(3):1491–1503

Zhang Y, Liu C, Song A (2016) Relationship between soil physico-chemical properties and soil enzyme activities in Huixian karst wetland system based on Canonical Correspondence Analysis. Carsol Sin 35(1):11–18

# The Role of Temporary Groundwater Control Recharge Systems in Water Conservation

Stephen Thomas, James Watson, and Thomas Goodfellow

**Abstract**

The conservation of water is a prominent issue of geoethics and is becoming increasingly important due to widespread groundwater quality and quantity deterioration in a number of regions worldwide. This deterioration in groundwater quality and quantity is often caused by over-abstraction from aquifers and has resulted in global groundwater resource sustainability concerns. Construction groundwater control operations are usually temporary in nature; however, the impacts of these operations on groundwater resources and water conservation can be significant. This is especially true when temporary groundwater control operations are required in areas where groundwater resources are already strained. In the light of the water conservation issues faced, groundwater control recharge systems are increasingly being designed and implemented in order to mitigate the impact of temporary groundwater abstraction on groundwater resources. Where used, these systems (i) reduce the net abstraction of the temporary groundwater control and (ii) provide an alternative approach to the conventional means of groundwater discharge (i.e. discharge to watercourse or sewer, etc.). This paper discusses the principle of utilizing temporary groundwater control recharge systems as a means of water conservation along with their advantages and disadvantages. A recent case study from the UK is also presented.

**Keywords**

Groundwater • Geoethics • Water conservation • Artificial recharge • Construction dewatering • Groundwater control

S. Thomas (✉) · J. Watson · T. Goodfellow
OGI Groundwater Specialists, Durham, UK
e-mail: Stephen.Thomas@ogi.co.uk

## 1 Introduction

Communities across the world face water supply challenges due to a number of factors including increasing demand, drought and contamination of groundwater (Miller 2006). The challenges faced mean that water conservation and the protection of groundwater resources are now major issues of geoethics.

Groundwater abstraction is relied upon heavily for the supply of water. In 2010, total global groundwater abstraction was estimated to be circa 1000 km$^3$/a of which 72 km$^3$/a was within Europe (Margat and van der Gun 2013). The world's dependence on groundwater abstraction means that temporary groundwater control for the construction industry needs to be planned and managed carefully so that water resources are protected.

One method of mitigating the impact of temporary groundwater control operations on water resources is the design and implementation of *groundwater control recharge systems*. Although this method is not regularly utilized in the UK, the implementation of groundwater control recharge systems during temporary construction groundwater control operations is not a new concept (Bock and Markussen 2007, Cashman and Preene 2013, Powrie and Roberts 2013, Preene et al. 2016).

Historically, the additional monetary costs of implementing artificial recharge during groundwater control operations have been a deterrent to construction companies, unless there is a site-specific requirement (Thomas et al. 2020). However, the increasing importance of water conservation and associated issues of geoethics mean that this approach during temporary dewatering operations is being considered more frequently.

Furthermore, recent legislation in England and Wales means that temporary groundwater control dewatering operations are now licensable and regulated activities (Water Abstraction and Impounding (Exemptions) Regulations 2017). In some areas of the UK, this legislation means that



M. Abrunhosa et al. (eds.), *Advances in Geoethics and Groundwater Management: Theory and Practice for a Sustainable Development*, Advances in Science, Technology & Innovation,
https://doi.org/10.1007/978-3-030-59320-9_33

temporary dewatering operations cannot be undertaken unless artificial recharge is implemented as a means of mitigating potential negative impacts on regional water resources. This has been experienced by the authors in some parts of the UK where license conditions have meant that one hundred percent of the abstracted groundwater must be recharged back to the same geological strata.

This paper aims to present the principle, advantages and disadvantages of utilizing temporary groundwater control recharge systems as a means of water conservation. One recent case study is also presented for a major UK tunnel construction project.

## 1.1 The Principle of Utilizing Groundwater Control Recharge Systems as a Means of Water Conservation

Temporary groundwater control for construction involves the abstraction of groundwater for the purpose of lowering the water table to a target level. This process mitigates groundwater-related risks so enabling construction in safe and stable ground conditions.

The groundwater abstracted during temporary groundwater control operations is typically discharged to local watercourses or sewers. These discharge mechanisms are generally straightforward and cost-effective; however, they can impact negatively upon the aquifer as they result in a reduction in the quantity of groundwater stored in the aquifer. This in turn can place excessive strain upon local water resources.

Artificial groundwater recharge during temporary construction dewatering involves the recirculation of abstracted groundwater back into the same geological strata by reinjection (via wells) or infiltration (via trenches or shallow wells) (Preene and Fisher 2015), generally without any intervening use. If the groundwater recharge system is designed and implemented appropriately, the groundwater quality remains constant, drawdown at distance is reduced and net abstraction from the aquifer is negligible.

The recharge wells need to be designed to enable groundwater to readily discharge back into the aquifer, whilst at the same time preventing groundwater entering or interfering with overlying or underlying geological strata. Careful design of the recharge well screens and appropriate design and installation of any required seals are therefore essential.

Artificial groundwater recharge generally works best in permeable ground conditions where the recharged water can readily soak away. It is also favourable if water levels are not flowing artesian or close to the ground surface (i.e. water level at depth).

The temporary groundwater control recharge concept is illustrated in Fig. 1.

## 1.2 Advantages of Utilizing Groundwater Control Recharge Systems as a Means of Water Conservation

Utilizing an artificial recharge system on a temporary groundwater control project reduces a number of environmental and water resource risks.

The net abstraction from the aquifer is reduced to a negligible quantity, so preventing any impact to the local and regional groundwater balance.

Drawdown at distance is reduced preventing the alteration of head gradients, which could lead to reduction in baseflow to surface water systems (e.g. streams, wetlands or other similar water bodies), deterioration of existing abstractors and contamination/saline interface migration.

## 1.3 Disadvantages of Utilizing Groundwater Control Recharge Systems as a Means of Water Conservation

Recharge systems are not always a feasible groundwater management strategy and will be dependent on the hydrogeological conditions encountered.

Where recharge is feasible, the advantages of a recharge system can only be realized if the groundwater control system is appropriately designed and monitored by competent and experienced personnel.

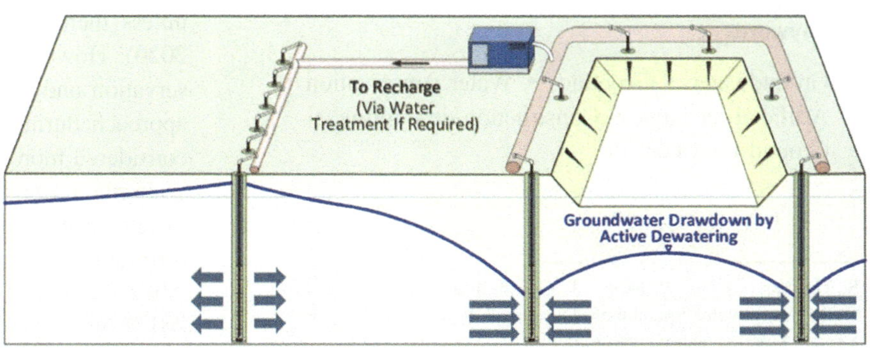


**Fig. 1** Concept dewatering and artificial recharge system

Without appropriate design and monitoring, the recharge system can become overwhelmed. This can result in localized flooding around the recharge wells which in turn can impact negatively on local water resources.

Furthermore, without appropriate design and installation, recharge wells can provide pathways for groundwater to travel between different aquifers and/or geological strata. This can result in the transfer of groundwater and potential contamination risks.

## 2 Case Study—Tunnel Boring Machine Drive Pit, England

The River Humber gas pipeline replacement by National Grid is a major infrastructure project to replace an existing gas pipeline which stretches circa 5 kms beneath the River Humber, a large tidal estuary on the eastern coast of Northern England.

To construct the pipeline, a tunnel boring machine was driven from a deep drive pit on the southern bank of the Humber. The construction of the drive pit required temporary groundwater control to lower the artesian head and water table to below excavation level.

During the Development Consent Order (DCO) stage (planning permission for developments categorized as Nationally Significant Infrastructure Projects), the sensitivity of the chalk aquifer on the southern bank of the Humber to any groundwater abstraction was identified. The primary concerns of the Environment Agency were (i) the risk of the saline interface mobilizing inland and (ii) the regional water resource status. The Catchment Abstraction Management Strategy (CAMS) status for the southern bank of the Humber states that "*no water is available for abstraction except at extremely high flows*". Groundwater resources in the Lincolnshire Chalk are fully committed to existing users and the environment (Environment Agency 2013).

The geology of the site comprised low-permeability alluvial and glacial deposits to circa 10 m below ground level. These deposits were underlain by the Flamborough Chalk, which was highly weathered in the top several metres and recovered as a chalk gravel. This in turn was underlain by intact, albeit fractured chalk. The weathered chalk was highly permeable with sub-artesian groundwater encountered during the site investigation. The underlying chalk was also relatively permeable, but significantly less permeable (i.e. greater than an order of magnitude) than the weathered chalk.

To mitigate the impact on water resources, a groundwater control strategy was developed utilizing a partial physical cut-off wall, together with groundwater abstraction from pumping wells inside the structure, with groundwater recharged back into the chalk at distance outside the wall. This resulted in abstraction being non-consumptive and prevented drawdown towards the river which could have resulted in saline intrusion.

As the upper weathered chalk aquifer was of very high permeability, a cut-off wall was required to penetrate through these deposits. Full cut-off of the weathered and underlying fractured chalk strata was originally considered; however, this was deemed impractical due to the fractured chalk extending circa 40–50 m below ground level.

An internal groundwater abstraction system was specified within the drive pit (borehole pumps within designed wells). These internal wells abstracted groundwater from the more competent and less weathered underlying fractured chalk.

The abstracted groundwater was then recharged back into the weathered chalk outside the pit, reducing the radius of influence of the groundwater control operation and reducing the net abstraction from the aquifer to effectively zero (see Figs. 2 and 3). The recharge system was fully enclosed, preventing the groundwater from coming into contact with the atmosphere so minimizing oxidation.

Water quality was monitored for the duration of the project through the utilization of monitoring wells installed towards the estuary. Data loggers were installed in these monitoring wells which recorded conductivity at regular time intervals. This monitoring, together with quarterly laboratory analysis of electrical conductivity and chloride concentration, enabled continuous assessment of any saline migration.

The project is described in more detail in the Proceedings of the Chalk 2018 Conference (Goodfellow and Thomas 2018; Holmes et al. 2018).

## 3 Concluding Remarks

The conservation of water and protection of groundwater resources are important global issues of geoethics.

Without careful planning and management, temporary groundwater control operations can have major detrimental impact on water resources.

The design and implementation of groundwater recharge systems can mitigate the risks of over-abstraction and have proved to be both successful and buildable in practice.

Groundwater recharge systems involve the recirculation of abstracted groundwater back into the same geological strata by reinjection or infiltration. This process has a number of advantages including: a reduction in the net abstraction from the aquifer, a reduction in drawdown at distance and minimization of environmental impact.

The recent case study presented in this paper demonstrates that when appropriately designed, installed and managed,

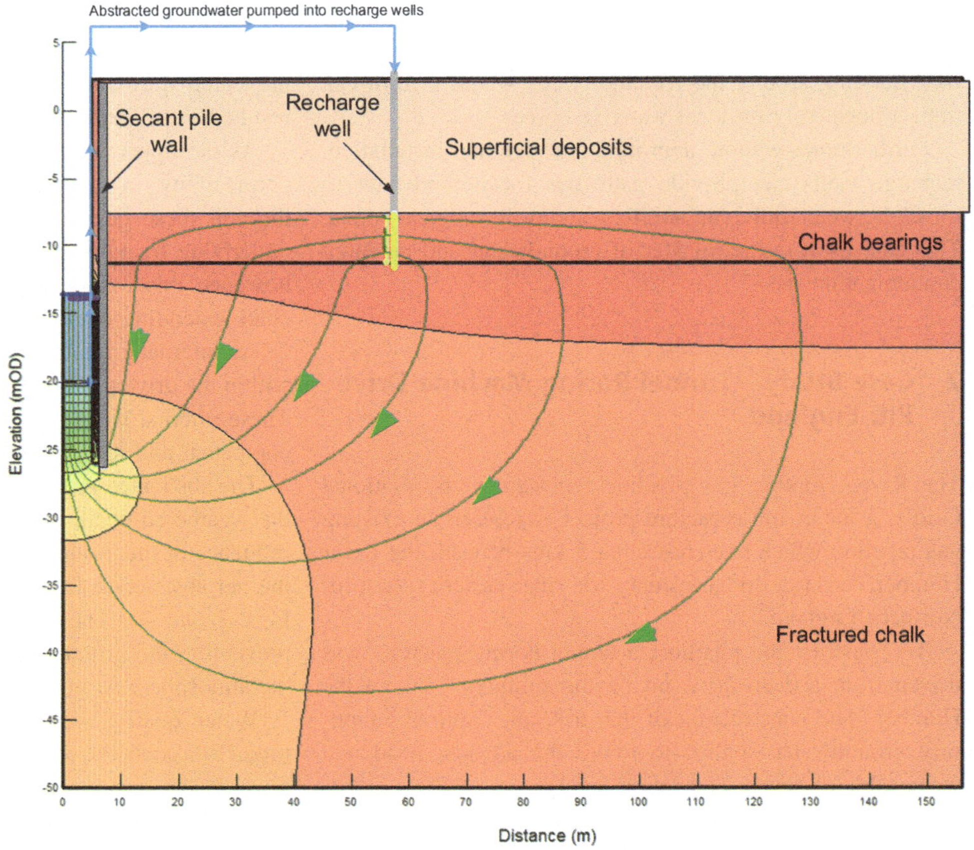


**Fig. 2** Finite element modelling (section) of abstraction and recharge system

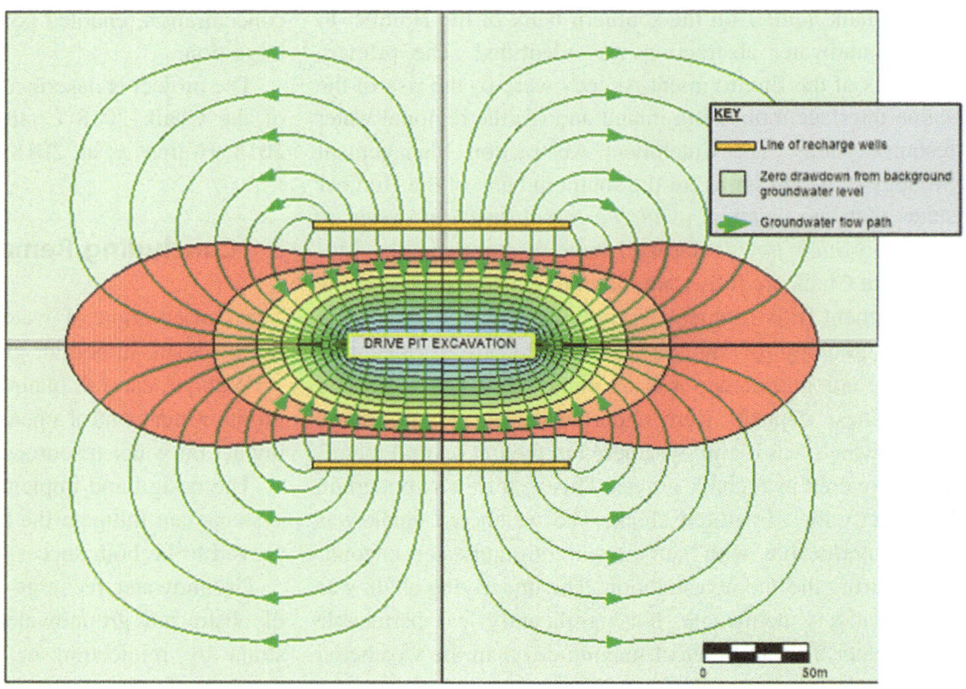


**Fig. 3** Finite element modelling (plan) of groundwater abstraction and recharge system

groundwater recharge systems can be implemented as a practical and successful means of protecting groundwater resources.

**Acknowledgements** The authors would like to acknowledge National Grid who continually challenges us to provide innovative engineering solutions and technical expertise.

## References


Bock M, Markussen LM (2007) Dewatering control in central Copenhagen. Geotechnical Engineering in Urban Environments. In: Proceedings 14th European conference on soil mechanics and geotechnical engineering 2, pp 715–720, IOS Press, Amsterdam

Cashman PM, Preene M (2013) Groundwater lowering in construction: a practical guide to dewatering, 2nd edn. CRC Press, Boca Raton

Environment Agency (2013) Grimsby, Ancholme and Louth catchment abstraction management strategy

Goodfellow TL, Thomas SD (2018) Groundwater resource protection from construction groundwater control in Lincolnshire chalk. In: Engineering in chalk: proceedings of the chalk 2018 conference, pp 669–674. ICE Publishing, London

Holmes GJ, Roberts TOL, Lee M (2018) A case study of construction dewatering in Northern Province Chalk. In: Engineering in chalk: proceedings of the chalk 2018 conference, pp 147–153. ICE Publishing, London

Margat J, van der Gun J (2013) Groundwater around the world. Balkema, Lisse, The Netherlands

Powrie W, Roberts TOL (1995) Case history of a dewatering and recharge system in chalk. Geotéchnique 45(4):599–609

Preene M, Fisher S (2015) Impacts from groundwater control in urban areas. In: Proceedings of the XVI european conference on soil mechanics and geotechnical engineering, pp. 2847–2852. ICE Publishing, London

Preene M, Roberts TOL, Powrie W (2016) CIRIA Report C750—groundwater control: design and practice, 2nd edn. Construction Industry Research and Information Association, London

Thomas S, Goodfellow T, Watson J (2020) Groundwater recharge systems to mitigate the impact from construction dewatering. INCREaSE 2019. Springer, Cham

Wade Miller G (2006) Integrated concepts in water reuse: managing global water needs. In: Desalination, vol 187, pp 65–75. Elsevier

Water Abstraction and Impounding (Exemptions) Regulations (2017) Available at: https://www.legislation.gov.uk/uksi/2017/1044/made. Last accessed 07 May 2019

# Groundwater Contribution to Alpine Ponds Recharge in Serra Da Estrela Natural Park, Portugal

Alexandre M. Almeida, Paula M. Carreira, José Manuel Marques, Jorge Espinha Marques, Marina Paiva, Alexandra Carvalho, and Catarina Mansilha

**Abstract**

Groundwater data from 15 springs at Serra da Estrela mountain, located at the Seia-Torre-Covilhã sector, were collected between 2010 and 2011 along six fieldwork campaigns. Isotopic ($^2H$; $^{18}O$, and $^3H$) and chemical determinations were performed. The data were compared with the isotopic and chemical composition of seven alpine ponds located within the same area. The local meteoric water line ($\delta^2H = 6.58(\pm0.29) + 1.31(\pm1.23)$) was calculated using groundwater isotopic data. The results show that the springs located at the highest altitudes are more depleted than those located downhill. A slightly different isotopic fractionation with altitude was found between the western and eastern slopes of Serra da Estrela Natural Park (0.12 ‰ versus 0.10 ‰ /100 m in $\delta^{18}O$, respectively). The groundwaters are hyposaline, mainly Cl-Na-type waters, with slightly acidic pH reflecting a shallow and short flow path and small residence time. The alpine ponds also show Cl-Na facies and are mainly recharged by direct precipitation. However, in two alpine ponds (L2 and L6), evidences for groundwater recharge were identified. The physical, chemical, and isotopic composition allowed the identification of: (i) contribution of de-icing into the groundwater and alpine ponds systems; (ii) contribution of the shallow groundwater flow toward the alpine ponds systems recharge.

**Keywords**

Environmental isotopes • Springs • Alpine ponds • Contamination • De-icing

## 1 Introduction

Serra da Estrela region, which integrates a natural park and a UNESCO Geopark, is located in the ENE-WSW oriented Iberian Central Mountain range, which extends for 500 km toward Spain and has a maximum width of 40 km. The geology of the study area is mainly composed by granitic rocks, schists, alluvial, and colluvial deposits, as well as by glacial and fluvioglacial deposits (Teixeira et al. 1974; Migón and Vieira, 2014). The granitic rocks were formed during the Variscan Cycle, and the metasedimentary rocks are included in the Beiras Group of the Ante-Ordovician Schist-Greywacke Complex (Teixeira et al. 1974; Ferreira and Vieira 1999). Geomorphological evidences of a glacial period are widespread in the region, particularly glacial landforms, namely the U-shaped Zêzere glacial valley. Serra da Estrela is located between two regional tectonic accidents, namely: the Bragança-Vilariça-Manteigas fault zone and the Seia-Lousã fault zone.

The annual precipitation reaches 2800 mm in the highest zones, while the minimum values are found to the NE (between 700 and 800 mm). The monthly average temperature varies between 2.5 and 17.4 °C, although the months of December, January, and February have negative

A. M. Almeida · J. M. Marques
Instituto Superior Técnico (CERENA/IST), Lisbon University, Lisbon, Portugal

P. M. Carreira
Centro de Ciências e Tecnologias Nucleares, Departamento de Engenharia E Ciências Nucleares, Instituto Superior Técnico, Lisbon University, Lisbon, Portugal

J. Espinha Marques (✉) · A. Carvalho
Instituto de Ciências da Terra, Porto University, Porto, Portugal
e-mail: jespinha@fc.up.pt

J. Espinha Marques
Departamento de Geociências, Ambiente e Ordenamento Do Território, Faculdade de Ciências, Porto University, Porto, Portugal

M. Paiva
TARH—Terra, Ambiente E Recursos Hídricos, Lda., Lisbon, Portugal

C. Mansilha
Instituto Nacional de Saúde Dr. Ricardo Jorge (INSA), Porto, Portugal



M. Abrunhosa et al. (eds.), *Advances in Geoethics and Groundwater Management: Theory and Practice for a Sustainable Development*, Advances in Science, Technology & Innovation,
https://doi.org/10.1007/978-3-030-59320-9_34

temperatures and snow precipitation which usually occurs above 1500 m a.s.l.

Six fieldwork campaigns were performed between 2010 and 2011 by Carvalho (2013), and 15 selected springs were sampled and analyzed for physico-chemical and isotopic parameters. In this work, we intent to discuss the results obtained and make a reinterpretation of the data acquired by Carvalho (2013), comparing these results with the chemical data obtained by Paiva (2016) and with the isotopic signatures (unpublished data) of seven alpine ponds, within the same research area (Fig. 1). The spring selection was established considering a WNW-ESE cross section of the area, and the altitude distribution, which varies between 565 and 1990 m a.s.l. at Marrão and Mina da Torre springs, respectively.

## 2 Results and Discussion

### 2.1 Groundwater Systems

All groundwater samples show low mineralization. Mina da Torre spring (at the highest altitude) presents the higher mineralization, (TDS around 300 mg/L), while the lower TDS content (around 3.5 mg/L) is found at Perús and Saibreira springs. The relative high calcium content (average = 4.47 mg/L) found in Espinhaço de Cão, Lagoa Comprida, Mina da Torre, and Nave de Santo António Sul springs suggests an anthropogenic origin for calcium, considering the geological background of the region. The most commonly used salt in the de-icing processes is Na-Cl; however, the presence of salt impurities cannot be excluded, thus justifying the presence of Cl-Na-Ca-Mg facies in Mina da Torre and the Espinhaço do Cão springs.

The spring samples collected during winter period normally exhibit Cl-Na facies with the exception of Covão do Boi with a $HCO_3$-Na facies. The weak mineralization recorded in all water samples indicate low water–rock interaction. Anomalous $Na^+$ and $Cl^-$ concentrations were observed in several springs (Lagoa Comprida, Mina da Torre, Espinhaço do Cão and Nave Santo António Sul), pointing to pollution originated by the salts added in the roads, at high altitude sites, to induce the snow melting (Espinha Marques et al. 2019). Furthermore, the $HCO_3$-Na facies observed at Covão do Boi and the $HCO_3$-Cl-Na facies

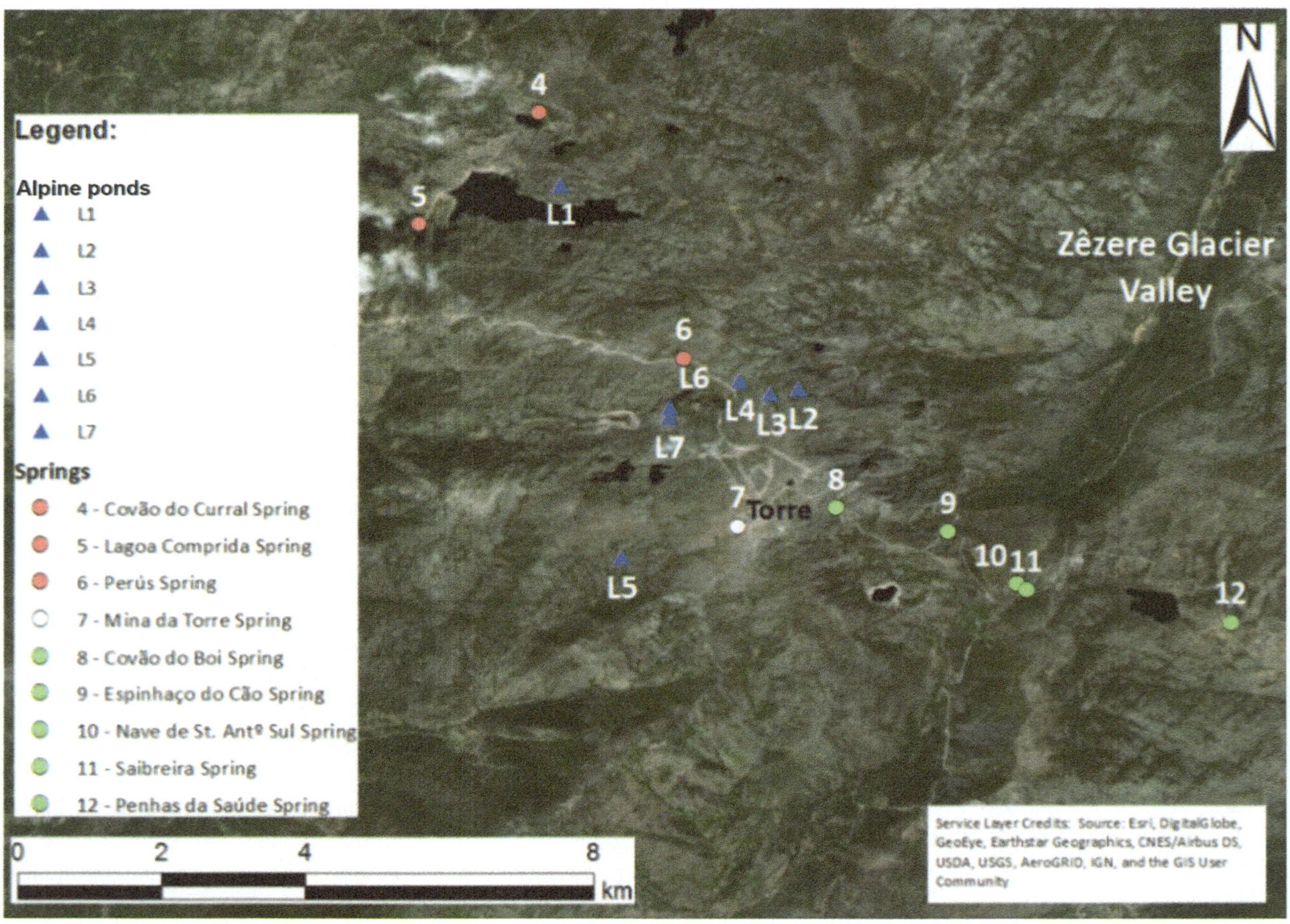


**Fig. 1** Geographic localization of the higher altitude shallow groundwater sampling sites (springs) and alpine ponds (Almeida 2019)

at Fonte Santa can probably indicate a relatively longer flow path favoring Na-feldspars hydrolysis.

The stable isotopic composition of all groundwater samples was plotted in an orthogonal diagram (Fig. 2). Two groups can be recognized (A and B). The isotopic composition of the groundwater samples is predominantly controlled by the preferential recharge altitudes. Group A is composed by the springs located at lower altitudes, which have a more enriched isotopic composition, while Group B is related to the spring waters found at higher altitudes, presenting more depleted $\delta^{18}O$ and $\delta^{2}H$ values. Nevertheless, these two clusters are also related to the movement of water vapor air masses in the region, preferentially moving from W to E, with a clear isotopic depletion in the eastern slope. A slightly different isotopic fractionation with altitude was found between the western and eastern slopes of Serra da Estrela, 0.12 ‰ / 100 m versus 0.10 ‰ / 100 m in $\delta^{18}O$, respectively.

In Fig. 3, an evolution in the isotopic composition of the groundwater samples towards more depleted values can be explained by the contribution of snowmelt to the groundwater systems. In some samples, the most depleted content was found in May 2010 and April 2011 campaigns, and not during the winter samples (December 2010). A similar pattern ascribed to the increase of the mineralization was also noticed, due to the addition of NaCl in the roads for promoting de-icing.

A relation between $^{3}H$ content and altitude of sampling sites was observed; in the springs located at the highest altitudes, the $^{3}H$ content range from 2.16 TU to 3.52 TU, indicating a short underground flow path. The tritium content in the water samples also showed seasonality, with higher concentrations during spring campaigns (May 2010 and April 2011).

## 2.2 Alpine Ponds

Paiva (2016) performed, in the same sector of Serra da Estrela, four fieldwork campaigns (January, March, June, and October 2016) in order to collect surface waters from seven alpine ponds (see Fig. 1). The L1 alpine pond is located at the lowest altitude (1645 m a.s.l.) and L5 at the highest sampled point (1854 m a.s.l.). According to Paiva (2016), the water temperature of the alpine ponds is close to the mean monthly air temperature.

All water samples from the studied alpine ponds present low mineralization and Cl–Na facies, suggesting that these waters are the result of direct recharge by precipitation. However, considering the low mineralization of these waters, an "anomalous salt concentration" was easily identified (case of L3 and L6 alpine ponds), suggesting a contamination by an anthropogenic source. This hypothesis is corroborated by the higher electrical conductivity of these waters. This type of contamination is also well-matched with the dispersion of de-icing salts (NaCl) on the roads.

The superficial systems are strongly affected by evaporation mechanisms inducing a shift toward enriched δ values (Fig. 4c). Such evaporation processes are responsible for the deviation of these water samples from the global/local meteoric water lines. The evaporation phenomenon is more

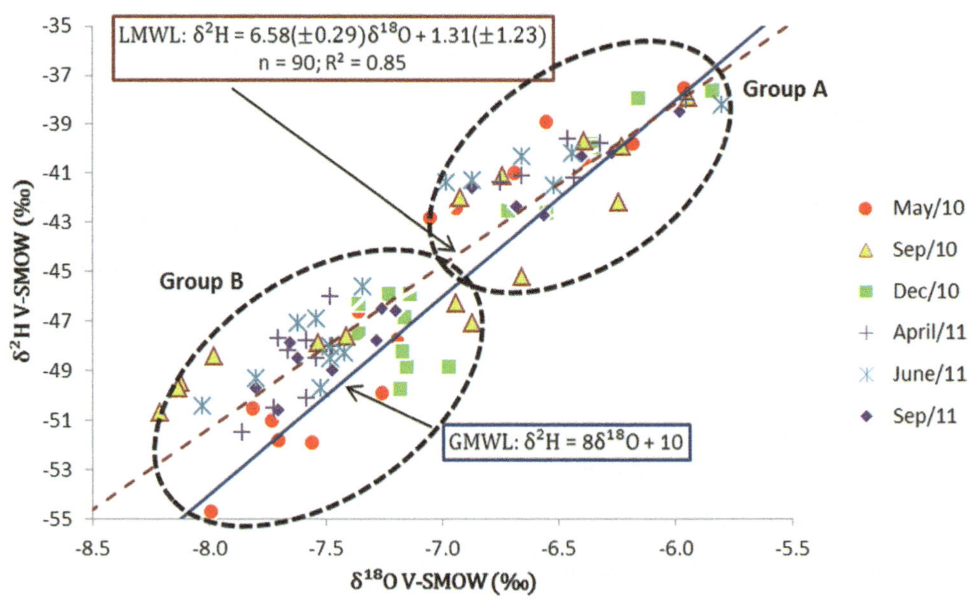


**Fig. 2** Local meteoric water line equation of Seia-Torre-Covilhã sector (Almeida 2019). Data from Carvalho (2013)

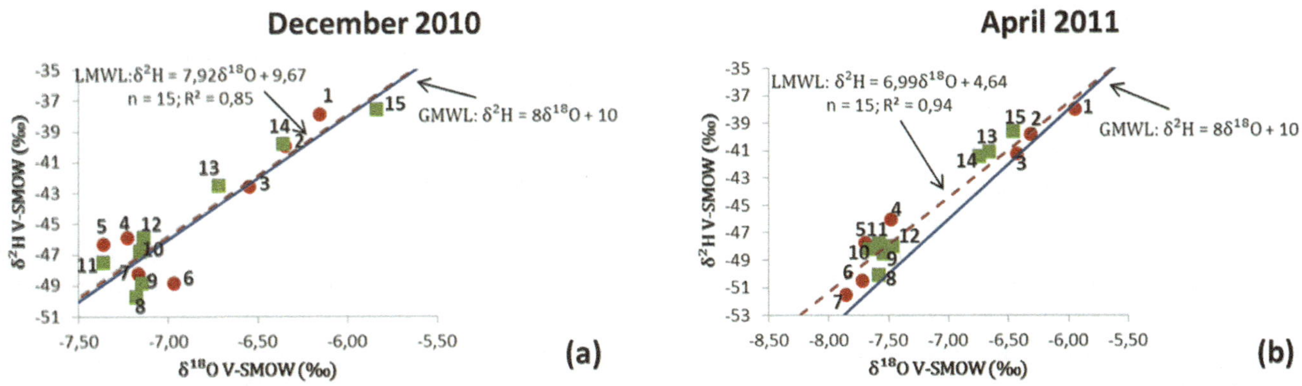


**Fig. 3** Projection of $\delta^{18}O$ and $\delta^{2}H$ data of the groundwater samples collected during December 2010 and April 2011 campaigns. Adapted from Almeida (2019). Data from Carvalho (2013)

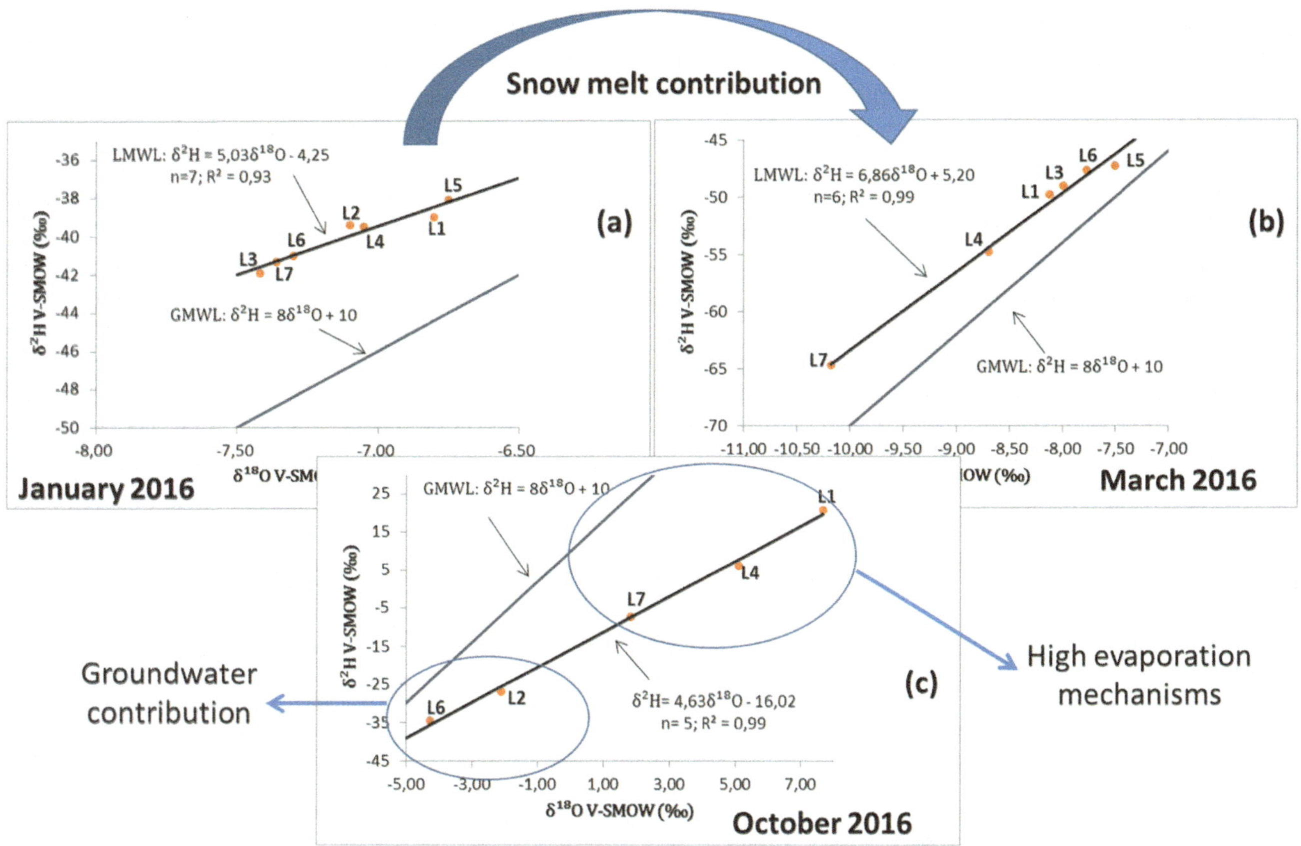


**Fig. 4** Isotopic composition of alpine ponds, enhancing an isotopic depletion due to the addition of melted snow and contribution of groundwater recharge. Adapted from Almeida (2019)

evident in October when the most enriched isotopic values reach positive values, associated to the scarce rainfall. However, in some cases, the water table is higher than the bottom of the alpine ponds, originating a groundwater contribution (L6 and L2) confirmed by the evolution of δ values (Fig. 4c). March 2016 campaign recorded the most depleted isotopic composition in the alpine ponds (Fig. 4b). One possible explanation to these depleted compositions is the addition of snowmelt, leading this mixing process to an isotopic depletion of the ponds.

## 3 Concluding Remarks

The isotopic study carried out in 15 springs from SENP allowed the determination of the local meteoric water line ($\delta^2H = 6.58\ (\pm\ 0.29)\ \delta^{18}O + (1.31 \pm 1.23)$). In addition, the $^3H$ content measured in the spring samples increase with the altitude of the sampling site, indicating a short underground flow path. The mean tritium content in these groundwater systems varies between 3.52 TU at Mina da Torre and 2.60 to 2.16 TU at the lowest altitude sampling sites, supporting the hypothesis of shallow and short residence time of the studied groundwaters. Based on the total dissolved solids concentration, the water–rock interaction processes are very low. Anthropogenic contamination due to road de-icing operations using NaCl was identified by an increase in water mineralization and by the depletion in the isotopic composition ($\delta^2H$ and $\delta^{18}O$).

The alpine ponds are predominantly recharged by direct precipitation. The contribution of de-icing water in the isotopic composition of the ponds is either the result of surface runoff or shallow groundwater flow. Indeed, L2 and L6 ponds show anthropogenic contamination identified by the shift toward more depleted $\delta^2H$ and $\delta^{18}O$ values, possibly ascribed to groundwater recharge affected by road de-icing salts.

Given the high socioeconomic and environmental importance of Serra da Estrela hydrological systems and water resources, a more sustainable alternative for road de-icing is required in order to prevent the contamination of aquifers, streams, and alpine ponds. A possible choice, with a more ethical balance between human activities and environmental protection would be to promote road de-icing exclusively with mechanical processes and heated water.

**Acknowledgements** CERENA authors acknowledges the FCT support through the UID/ECI/04028/2019 Project. $C^2TN$ author acknowledges the FCT support through the UID/Multi/04349/2019. FCUP and ICT authors acknowledges the support through the contracts UIDB/GEO/04683/2020 with FCT and COMPETE POCI-01-0145-FEDER-007690.

## References

Almeida AAM (2019) Contribuição das assinaturas isotópicas ($^2H$, $^{18}O$ e $^3H$) na caracterização da dinâmica de águas subterrâneas e sua relação com as lagoas: Parque Natural da Serra da Estrela—Setor Seia-Torre-Covilhã. Instituto Superior Técnico, Universidade de Lisboa, Lisboa (MSc Dissertation)

Carvalho A (2013) Hidrogeologia ambiental da região do Parque Natural da Serra da Estrela (Setor de Seia-Torre-Covilhã): uma abordagem multidisciplinar. Faculdade de Ciências, Universidade do Porto, Porto (Ph. D. thesis)

Espinha Marques J, Marques JM, Carvalho A, Carreira PM, Moura R, Mansilha C (2019) Groundwater resources in a Mediterranean mountainous region: environmental impact of road de-icing. Sustain Wat Res Manag 5:305–317

Ferreira N, Vieira G (1999) Guia geológico e geomorfológico do Parque Natural da Serra da Estrela. Locais de interesse geológico e geomorfológico. Parque Natural da Serra da Estrela. Ed. Instituto da Conservação da Natureza/Instituto Geológico e Mineiro. Lisboa

Migón P, Vieira G (2014) Granite geomorphology and its geological controls, Serra da Estrela, Portugal. Geomorphology 226:1–14

Paiva M (2016) Estudo morfométrico e hidrogeoquímico de Lagoas da Serra da Estrela. Faculdade de Ciências da Universidade do Porto. (MSc Dissertation)

Teixeira C, Paula Santos J, Carvalho HF, Peinador Fernandes A, Margarida Vairinho M, (1974) Carta geológica de Portugal na escala de 1/50 000. Notícia Explicativa da Folha 20-B—Covilhã. Serviços Geológicos de Portugal, Lisboa

# An Analysis of Spatio-Temporal Variability of Precipitation and Contours of Water Management in the Upper Teesta River Basin of Sikkim Himalaya, India

Pawan Kumar

**Abstract**

In mountainous regions, it is problematic to calculate the rainfall distribution because of the rugged, hilly terrain as well as poor obtainability of rainfall points. Rainfall estimation is vital for land use patterns and zoning and resource analysis. In this context, geostatistical techniques have been developed for mapping rainfall in the Upper Teesta River Basin with the help of data extraction using DEM and long-term mean monthly rainfall (MMR) data of 10 rain gauge stations. Interpolation technique was applied to standardized rainfall depths associated with elevation, as the primary varietal with the help of ArcGIS. The study has been successful in bringing out the rainfall pattern in the Teesta region which is helpful for the locales because Teesta is considered as "The Lifeline of Sikkim". Assessing the variability in rainfall will consent the people to choose the precise technique for management of the river water also serving them with ideas to regulate flash floods. This paper will provide an insight to the people of Sikkim who are reliant on the river water to step a foot towards development.

**Keywords**

Rainfall variability • Climate change • Geostatistical techniques • Watershed management

## 1 Introduction

No parts of the world are untouched from the variation in precipitation, and hence, Teesta River Basin is not an exception of it. Comprehensive, accurate, quantitative knowledge and documentation regarding the spatio-temporal variation of rainfall is a central requirement for designing and establishing cost-effective engineers' structures, mitigating and management of drought and flood-like situation (Bhalme and Mooley 1980; Narain et al. 2006). Such knowledge and information also help in designing and establishing rain gauge centre, weather and hydrological forecasting centre, initialization and validation of numerical weather prediction models and modelling of watershed regions (Sinha and Srivastava 1997). The direction of the wind, humidity, air temperature and pressure and topographic features such as elevation, relief, degree of slope, distance from the sea, continentality, etc., other geomorphic and geological barriers are the important factors that affect the distribution of rainfall (Dore 2005; Batisani and Yarnal 2010).

The facts on climatic erraticism over the past eras on various scales are necessary to understand the different climatic pattern and their impact on the environment and society (Oguntunde et al. 2006). To understand soil infiltrability and successive propagation of surface runoff as well as to predict flash flood, it is necessary to have precise and sound knowledge of spatio-temporal distribution of precipitation. Also, the long-term spatial variability of rainfall is an important attribute for agroecological zoning and also demarcating groundwater availability regions (Milly et al. 2005; Malhi et al. 2015). The rainfall pattern decides the amount of water draining into the river and also seeping underground. The perceived increase in rainfall has been a topic of debate for humankind as it proves to be an alarming cause for climatic variability (Oriola 1994).

Since, Teesta is an important river of Sikkim and most of the state owes its livelihood to this river, therefore, the broad and comprehensive study will lead to insight for the people residing around the Upper Teesta River Basin for management and proper utilization of the river water. They can use the excess rainwater appropriately by harvesting it. This will lead to the sustainable development of water resources of the

P. Kumar (✉)
Department of Geography, Delhi School of Economics, University of Delhi, Delhi, India
e-mail: pawank.dse@gmail.com


M. Abrunhosa et al. (eds.), *Advances in Geoethics and Groundwater Management: Theory and Practice for a Sustainable Development*, Advances in Science, Technology & Innovation,
https://doi.org/10.1007/978-3-030-59320-9_35

Teesta River Basin and helps in bringing more area under irrigation. The protection of water resources is a central critical issue in recent days (Parthasarathy 1984; Jain et al. 2012). The upsurge in flood risk is now posing an overwhelming threat to man especially those who are residing near the mainstream. Moreover, recent studies have proposed that in 2014, already this part of Sikkim had been struck with a torrential flood. This had led to an enormous loss in life and property. Since there is limited information on general trends of rainfall within the basin and no study has investigated detail variability and trends of rainfall in the Upper Teesta River Basin. Therefore, projecting the variability and trends in rainfall will provide an outreaching method and scope to regulate water availability.

## 2 Materials and Methods

The Mann–Kendall test is used for trend analysis in variables like rainfall, temperature and streamflow. These parameters are extremely important for watershed modelling, studying catchment characteristics which in turn can help in determining water resources planning strategies for short term, medium and long term for any region. The benefit of this test is that data need not confirm any particular distribution. In this test, each data value in the time series is compared with all subsequent values. Initially, the Mann–Kendall statistics (S) is assumed to be zero, and if a data value in subsequent time periods is higher than a data value in the previous time period, S is incremented by 1, and vice-versa. The net result of all such increments and decrements gives the final value of S. This is tested against the alternative hypothesis H1, which assumes that there is trend. The Mann–Kendall statistics is computed as follows:

$$S = \sum_{k=1}^{n-1} \sum_{j=k+1}^{n} sgn(xj - xk) \quad (1)$$

The trend test is applied to a time series X $_k$, which is ranked from $k$ = 1,2,3…, n − 1, which is ranked from $j$ = $i$ + 1,$i$ + 2,$i$ + 3…..$n$. Each of the data points x $_j$ is taken as a reference point,

$$sgn(xj - xk) = 1 \; if \; xj - xk > 0 \quad (2)$$

$$= 0 \; if \; xj - xk = 0 \quad (3)$$

$$= -1 \; if \; xj - xk < 0 \quad (4)$$

This particular test has been calculated using XLSTAT 2017 software. A very high positive value of S is an indicator of an increasing trend and a very low negative value indicates a decreasing trend. The presence of a statistically significant trend is evaluated using Z value.

It is used for the detection of a statistically significant trends. After that rainfall variability is interpolated based on Kriging and mapped in ArcGIS 10.5. Quality controlled monthly rainfall data has been calculated of 10 rain gauge stations present in Upper Teesta River Basin. The periodic rainfall variability graph has been plotted using monthly and yearly rainfall data in SPSS software.

## 3 Results

The rainfall modelling denotes the amount of rainfall concerning for a decadal period. The rainfall analysis has been carried based on 30 years of data. The pattern of rainfall has been shown with the help of the interpolation method in ArcGIS 10.5. The analysis shows that June and July receive the highest rainfall as high as 1175 mm. The lowest is seen in the months December as 7–8 mm only. The temporal range is decadal at the interval year of 1985, 1995, 2005 and 2015.

From the data results of the different station, it can be said that there is no significant trend observed across the study period. From the results, the rainfall datasets of the different station located in Sikkim, the rainfall is evenly distributed across the state in the last 35 years, moreover a positive trend. Analysing the monthly trend of rainfall, it is evident that majority of the year the rainfall shows a positive trend across different months, except in the months of March (−0.666, ss-0.019- significant), November (-4.22, ss-0.029-significant) and December (-13.22, ss- -0.081), which shows a negative trend. It indicates that the rainfall is showing a positive trend in the rest of the months and the negative trend in these three months. The values obtained from this test indicate that there is no significant trend ($p < 0.05$). About more than 75% of the stations showed a positive trend, i.e. increasing trend in rainfall was observed.

## 4 Discussion

These results further suggest an increasing trend in the higher altitude and fluctuating trend in the lower altitude or leeward side of the region. Further to understand the seasonal trend, the entire year was divided into four seasons, i.e. March–May (summer), June–August (monsoon), September–November (autumn) and December (winter)–February. It was found that in all the season except December-February saw a negative trend in the rainfall pattern in the last 35 years. The rest 75% of the season saw a positive trend. Sen slope analysis depicts a negative trend (−0.074), though not significant. Similarly, looking at the annual trend of the last 35 years (95%), a positive trend indicates an increasing trend in the overall rainfall taking

**Table 1** MK Sen test for all stations, (Monthly), 1980 and 2015

| | January | February | March | April | May | June |
|---|---|---|---|---|---|---|
| 1985 | 5.659 | 55.430 | 36.185 | 23.718 | 4.163 | 84.419 |
| 2015 | 7.098 | 86.662 | 8.456 | 36.360 | 129.990 | 474.921 |
| | July | August | September | October | November | December |
| 1985 | 301.445 | 280.291 | 139.152 | 68.379 | 13.929 | 32.169 |
| 2015 | 738.471 | 644.150 | 243.823 | 263.141 | 0.235 | 2.537 |

place across these different stations located in the state. The result so obtained shows that the probability of climatic factors and geophysical elements driving the rainfall variability across the different months and seasons has not been changed to a large extent. The normal trend of monsoon tends to occupy the entire state, except during the onset and at the time of withdrawal (Table 1).

Observing the gradual trend in the rainfall across the entire 35 years, a normal liner trend analysis is done for the years 1985 and 2015, as 1985 is considered as the base year and 2015 is taken as the last year of observation. If we see the linear trend for 1985 and 2015, it is clear that there is an increase in rainfall across the year. Perusal of the monthly rainfall for 1995 clearly shows that the rainfall for July is 301.445 mm in comparison to 738.471 mm for July in 2015 (Fig. 1). Similarly, rainfall for August has increased from 280.91 mm in 1985 to 644.150 mm in 2015, which clearly indicates the increasing trend in rainfall across the different station located in the state. This increasing trend in the rainfall is clearly against the general notion that over the year's rainfall is decreasing and temperature is increasing.

As discussed earlier, changing rainfall pattern and trend is highly driving the paddy cultivation and area under them, which are also continuously declining. There is also other thing to note that, though there has been increase in the overall rainfall during the rainy season, the timing and the duration of rainfall highly influences the agriculture pattern and crop type in the study region. If there is adequate rainfall

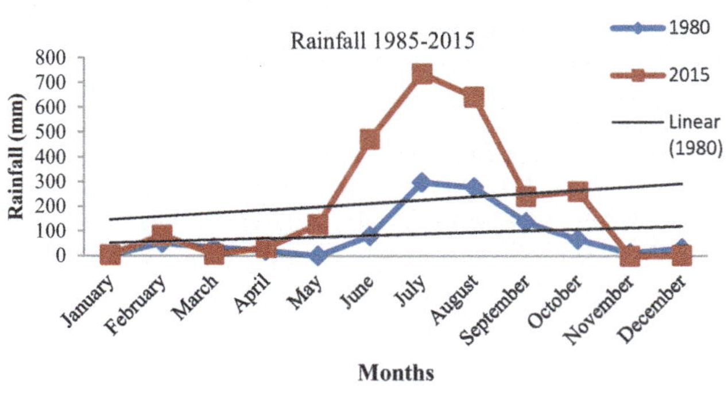


**Fig. 1** MK Sen test for all stations, from the month of January to December, for the year 1985–2015

over all but the rainfall takes place after the germinating time or season, it would have negative impact on the agrarian economy of the state. Hence, in the era of climate change, it becomes essential for the policy maker to make these rural folks highly dependent on agriculture to adapt to these changing patterns. The present trend is making them more and more vulnerable to climate crisis, which needs to be addressed in the near future (Fig. 2).

In order to further visualize the detected trends, a trend line was drawn for the year 1985 and 2015. The trend line for 1985 shows that somewhat a flat/horizontal line indicating normal distribution in rainfall across the year, but the trend line for the year 2015 shows an upward trend in increasing rainfall during the monsoon season. It can be concluded from the graph that the annual average rainfall at different stations exhibited an increasing trend. It also seems to register an increase in the amounts of annual rainfall in the last 35 years, with fluctuations in between in some of the years under observation. All the stations under observation show a positive trend across the years and across all months except the winter months.

## 5 Concluding Remarks

This study enlightens the spatial and temporal variability in the rainfall pattern of the Upper Teesta River Basin. Rainfall plays an important role in sustainable water management. Water is subject to life, to the economy, to religious thought and to cultural attitudes. It proposes that a virtue ethics approach that explicitly focuses on the cultivation of an attitude of respect for water founded on three key principles (eco-responsibility participation, hydro-philanthropy and proactive engagement) is the best solution for Sikkim. The results reveal that there is a significant increase in rainfall during June and July. The same is seen to decrease during January, and December. Also, rainfall is seen to change on a yearly basis. The anomaly of rainfall over the 10 rain gauge stations shows that dry and wet years intermingled with each other. The variation in rainfall might be due to change in a squall line, thunderstorms which have a significant effect on 70% of rainfall over a region annually. Analysis of data from

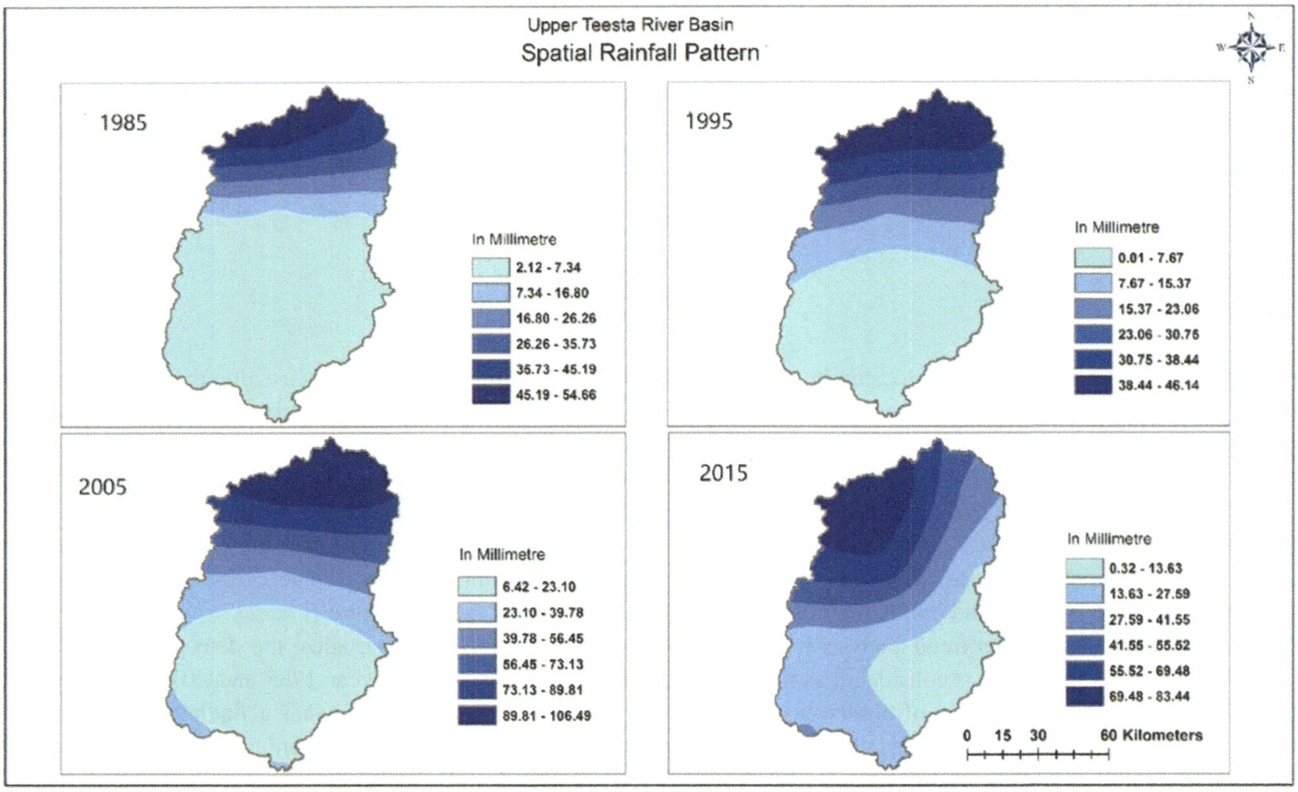


**Fig. 2** Spatial rainfall pattern in Upper Teesta River Basin

1985–2015 reveals that the amount of rainfall is more in the month of July followed by August and September. There is variation in the amount of rainfall during the inception and retreat of monsoon season, i.e. June and September. In all the four months of monsoon/rainy season, rainfall is more consistent in the month of July and August. Except in winter season (December–February), the rainfall pattern is towards increasing in all the rest three season, i.e. summer, monsoon and autumn, particularly in the upper reaches of Sikkim Himalayas in the last 30 years. A slight increase or decrease in the existing rainfall pattern may alter the agricultural eco-system of the state, as it is highly dependent on agriculture as their mainstay. Excess or deficient soil moisture may lead to crops failures. Apart from agriculture, longer duration of rainfall may hamper its tourism potential and activities, which in turn may have negative impact in the overall socio-economic growth of the state. Hence, it would be advised to the researchers and academic fraternities to help the locals in cope up with changing patter and devise a means for sustainable growth in the future.

**Acknowledgements** Author is extremely thankful to the Head of the Department of Geography, Delhi School of Economics, University of Delhi, for providing facilities. In addition, he is also grateful to the Government of Sikkim and India Metrological Department, Government of India and thankful to University Grants Commission for providing financial help.

## References

Batisani N, Yarnal B (2010) Rainfall variability and trends in semi-arid Botswana: implications for climate change adaptation policy. Appl Geogr 30(4):483–489

Bhalme HN, Mooley DA (1980) Large-scale droughts/floods and monsoon circulation. Month Weat Rev 108(8):1197–1211

Dore MH (2005) Climate change and changes in global precipitation patterns: what do we know. Environ Int 31(8):1167–1181

Jain SK, Kumar V, Saharia M (2012) Analysis of rainfall and temperature trends in northeast India. Int J Climat 33(4):968–978

Malhi RKM, Kiran S (2015) Analysis of trends in area, production and yield of important crops of India. Int J Agr Agricult Res 7(1):86–92

Milly PC, Dunne KA, Vecchia AV (2005) Global pattern of trends in streamflow and water availability in a changing climate. Nature 438 (7066):347

Narain P, Rathore LS, Singh RS, Rao AS (eds) (2006) Drought assessment and management in arid Rajasthan. Central Arid Zone Research Institute, Jodhpur, India. https://www.cazri.res.in/publications/KrishiKosh/109-(DROUGHT.pdf

Oguntunde PG, Friesen J, van de Giesen N, Savenije HH (2006) Hydroclimatology of the Volta River Basin in West Africa: trends and variability from 1901 to 2002. Phys Chem Earth, Parts a/B/C 31 (18):1180–1188

Oriola EO (1994) Strategies for combating urban flooding in a developing nation: A case study from Ondo Nigeria. Environmentalist 14(1):57–62

Parthasarathy B (1984) Interannual and long-term variability of Indian summer monsoon rainfall. Proc Indian Ac Sci-Earth Planet Sci 93 (4):371–385

Sinha Ray KC, Srivastava AK (1997) Is there any change in extreme events like droughts and heavy rainfall. INTROPMET-97, IIT New Delhi, pp 2–5

# Analysis of Phreatic Levels in Riparian Forest and Pasture in an Agricultural Watershed, Santa Catarina, Brazil

Mateus Melo, Adilson Pinheiro, Edson Torres, Gustavo Piazza, and Vander Kaufmann

**Abstract**

This study analysed the short-term behaviour of groundwater levels under two different conditions, riparian forests and pasture, in Concórdia watershed, Lontras, SC, Southern Brazil. Data referring to 3 piezometers were used. PZMC is installed in a riparian forest and PZ2127 and PZ3 in pasture areas. PZ2127 is located near the watershed's outlet. Water levels of the river outlet installed near PZ2127 were also used. Storm events were selected considering the historical record, from 23 April 2012 to 2 September 2015. Water levels that exceeded 60% of the maximum amplitude recorded in all piezometers were the criteria for the storm event selection. Ascension and recession coefficients were calculated for the entire duration of each event, for phreatic levels and water levels, at 2 h intervals. The difference between water and phreatic levels peaks was calculated. Decay of the mean coefficients of ascension and recession can be observed as consequence of the increase of distance and depth. Ascension movements are faster than recession movements for piezometers and water level. This result is important to develop the knowledge about local hydrological processes that involves phreatic aquifers, in view of watershed management.

**Keywords**

Agricultural watershed • Phreatic levels • Short-term analysis

## 1 Introduction

Family-farmed watersheds in Southern Brazil have cultivated and pasture areas, changing water quantity due to changes in land use. Monitoring and analysis of the groundwater behaviour of phreatic aquifers can generate knowledge for conservation, environmental protection projects and sustainable development (OAS/GEF 2001). The behaviour of these systems and the way human activities influence the geosphere can be identified, principles of geoethics (Peppoloni and Di Capua 2012).

This study analysed the short-term behaviour of groundwater levels under two different land uses, riparian forests and pasture, in Concórdia watershed, Lontras, SC, Southern Brazil. The study aims to generate knowledge of the local groundwater dynamics and, consequently, the hydrological cycle, to support the planning and decision-making involving water resources.

## 2 Materials and Methods

The study was developed in the Concórdia watershed, in the municipality of Lontras, Santa Catarina, Southern Brazil. The watershed has a surface area of 30.93 $km^2$ and is located at latitude 27° 11′ 17.0″ S and longitude 49° 29′ 40.1″ W. The Concórdia river flows from east to west.

The area of the watershed is occupied by agricultural activities approximately 50% of the total area (Piazza et al. 2014). No-tillage, minimum tillage and conventional tillage practices are adopted as soil management by almost all farmers (Pinheiro et al. 2009).

The climate, according to Strahler, is subtropical, with hot and humid summers, and cold and dry winters. The mean annual temperature and precipitation are 20 °C and 2000 mm, respectively. The climate can be classified by Köeppen as Cfa, humid subtropical climate, with mean temperature in the coldest month below 18 °C and mean

M. Melo (✉) · A. Pinheiro · E. Torres · G. Piazza · V. Kaufmann
Environmental Engineering Postgraduate Program, Regional University of Blumenau, Blumenau, Brazil
e-mail: mateusnvm@gmail.com



M. Abrunhosa et al. (eds.), *Advances in Geoethics and Groundwater Management: Theory and Practice for a Sustainable Development*, Advances in Science, Technology & Innovation,
https://doi.org/10.1007/978-3-030-59320-9_36

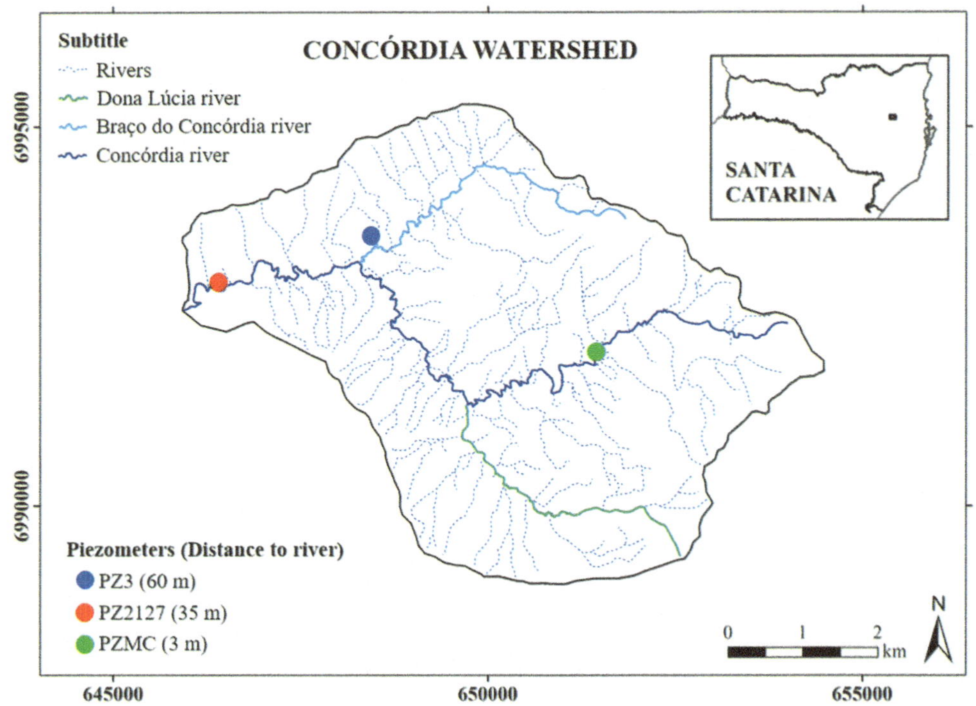


**Fig. 1** Location of PZMC, PZ2127 and PZ3 piezometers in the Concórdia watershed, Lontras, Santa Catarina, Southern Brazil

temperature in the warmest month above 22 °C (Alvares et al. 2013).

For analysis of the groundwater levels, data referring to three piezometers installed in the watershed were used. PZMC is in a riparian forest and PZ2127 and PZ3 in pasture areas. PZ2127 is located near the watershed's outlet (Fig. 1), the depths are 2.4, 3.8 and 5.1 m, and they are 3, 35 and 60 m away from the river, respectively. Their elevation is 433, 367 and 372 m, respectively.

The piezometers are equipped with hydrostatic level sensors connected to dataloggers. Levels were recorded, and the estimated mean of 2-h data was used. Also, it was used the water level obtained in the fluviometric section installed in the direction of PZ2127. This section features an automatic level sensor with a float-operated axis encoder. The estimated mean of 1-h intervals was used.

Storm events were selected from the historical series, from 23 April 2012 to 2 September 2015. Water levels that exceeded 60% of the maximum amplitude recorded in all piezometers were the criteria for selection, covering events with different durations.

Ascension and recession coefficients were calculated for the entire duration of each event, for phreatic levels and river quotas, every 2 h, considering an exponential curve. Their numerical determination was made using the Maillet equation (Eq. 1) (Dewandel et al. 2003):

$$\alpha = (\log \mathrm{NA} - \log \mathrm{NA}_0)/(0.4343\ t) \qquad (1)$$

where α is the coefficient of ascension or recession ($d^{-1}$), NA is the water level, $NA_0$ is the water level at the initial instant and t is time (days). The results obtained were submitted to ANOVA test and the means compared by Tukey test, with a significance level of 5%.

The difference between water levels and phreatic levels peaks periods was calculated according to Peláez (1978). Negative values indicate that the phreatic level peak occurred earlier than the river quota peak, and positive values indicate the opposite. Data were organized in Excel software, and statistical tests were performed in PAST (version 3.25) software.

## 3 Results

Table 1 shows the 18 selected storm events, with their start and end dates. Events ranged from 3 to 30 days of duration, thus encompassing different intensity and duration characteristics related to rainfall. Table 2 shows the statistics of river and groundwater levels of the events.

Figure 3 shows ascension and recession mean coefficients calculated for each compartment, based on the mean of each event. The mean ascension coefficients detected were 0.77, 1.49, 1.01 and 0.48 and recession of −0.12, −0.50, −0.34 and −0.10 for the river, PZMC, PZ2127 and PZ3 piezometers, respectively.

Box plot of the difference between water and phreatic levels peaks can be seen in Fig. 4. PZMC differences ranged from −4 to 16 h. PZ2127 differences ranged from −4 to 10 h. PZ3 differences ranged from −4 to 12 h. Differences in peak times oscillated for the three piezometers (Fig. 4).

**Table 1** Start and end dates of storm events in the Concórdia watershed, Lontras, Santa Catarina, Brazil

| Event | Start date | End date | Event | Start date | End date |
|---|---|---|---|---|---|
| 1 | 28/04/2012 | 22/05/2012 | 10 | 03/04/2013 | 03/05/2013 |
| 2 | 23/05/2012 | 02/06/2012 | 11 | 04/05/2013 | 19/05/2013 |
| 3 | 04/06/2012 | 09/06/2012 | 12 | 30/05/2013 | 16/06/2013 |
| 4 | 10/06/2012 | 16/06/2012 | 13 | 05/12/2013 | 08/12/2013 |
| 5 | 17/06/2012 | 05/07/2012 | 14 | 16/12/2013 | 30/12/2013 |
| 6 | 17/07/2012 | 24/07/2012 | 15 | 29/03/2015 | 19/04/2015 |
| 7 | 10/09/2012 | 19/09/2012 | 16 | 20/07/2015 | 17/08/2015 |
| 8 | 10/10/2012 | 18/10/2012 | 17 | 21/10/2015 | 30/10/2015 |
| 9 | 20/01/2013 | 31/01/2013 | 18 | 07/11/2015 | 17/11/2015 |

**Table 2** Statistics of river and groundwater levels (m) of storm events in the Concórdia watershed, Lontras, Santa Catarina, Brazil

| Event | River | PZMC | PZ2127 | PZ3 |
|---|---|---|---|---|
| Mean | 2.08 | −1.87 | −3.13 | −2.74 |
| Minimum | 1.80 | −2.11 | −3.31 | −3.38 |
| First quartile | 1.87 | −1.99 | −3.22 | −3.00 |
| Median | 1.95 | −1.88 | −3.17 | −2.76 |
| Third quartile | 2.17 | −1.76 | −3.06 | −2.43 |
| Maximum | 3.57 | −1.51 | −2.60 | −2.08 |

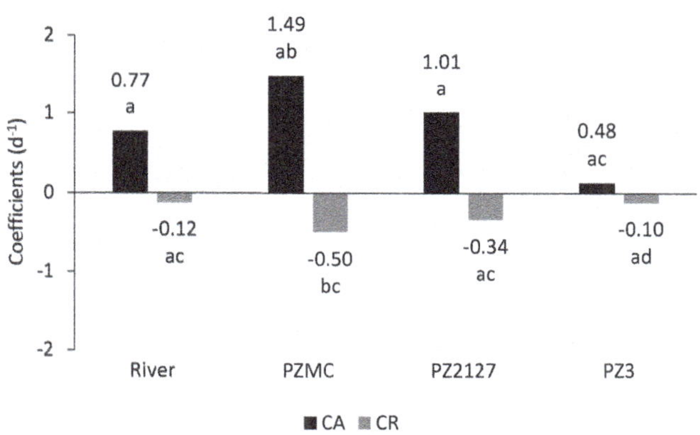


**Fig. 3** Ascension and recession mean coefficients calculated for each compartment in the Concórdia watershed, Lontras, Santa Catarina, Brazil. *Note* Values followed by the same letter in the line, between different compartments, do not differ significantly from each other, according to ANOVA, post-hoc Tukey (significance level = 5%). CA and CR mean ascension and recession coefficients, respectively

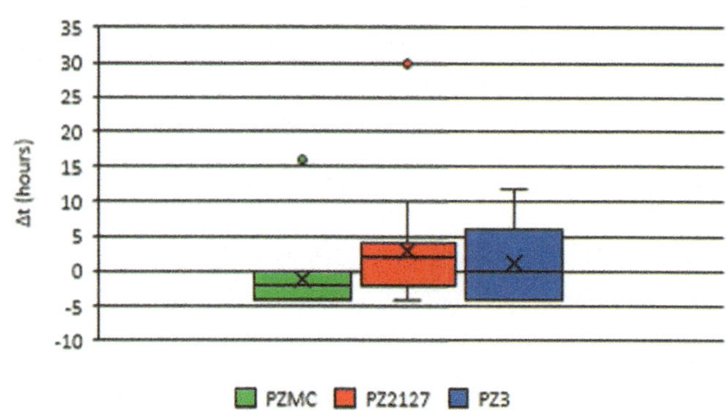


**Fig. 4** Difference between water levels and phreatic levels peaks periods. Note: Negative values indicate that the phreatic level peak occurred earlier than the river quota peak, and positive values indicate the opposite

For PZMC, except for one outlier, the peak of the phreatic level was before than the peak of the water level. The peak of phreatic level of PZ2127 and PZ3 was, in general, after the peak of river quota.

## 4 Discussion

Decay of the mean coefficients of ascension and recession can be observed as consequence of the increase of this distance and depth. A longer lowering of the phreatic level, evidenced by the lower coefficients, can be explained by the weaker river–aquifer interaction, given the greater distance between the piezometer and the watercourse.

PZMC and PZ2127 presented a recession curve more like the river, possibly because of their shorter distance from it.

Another noteworthy point is the differences between the ascension and recession movements, which are larger for the phreatic level ascension. It suggests a contribution of aquifer water to river during a drought (Manzione 2015).

The alternations between differences of water and phreatic levels peaks can be explained by the different characteristics of intensity, duration and location of the precipitations that originated the events.

## 5 Concluding Remarks

The dynamics of phreatic levels in piezometers can change according to distance from the river, depth and location relative to the watershed. The calculation of the ascension and recession coefficients is a good tool to compare the characteristics of the velocities for each monitoring point of phreatic levels.

Studies of soil characteristics may also reveal important information for understanding the phenomena of ascension and recession phreatic levels, as well as studies that relate hydroclimatic variables.

This kind of analysis is important to develop the knowledge about local hydrological processes that involves phreatic aquifers. Increasing demand for groundwater makes these kinds of studies even more important in view of watershed management. In the context of geoethics, this study provides knowledge and affirms the commitment of science to citizens and institutions.

**Acknowledgements** This work was carried out with the support of the Coordination of Improvement of Higher Education Personnel (CAPES) —Brazil (Financing Code 001 and Process 88881.143991/2017-01). The authors thank CAPES for granting the master and doctoral scholarship to the first and third authors, respectively, and to National Council for Scientific and Technological Development (CNPq) (Process 309980/2017-8) for the second author's research productivity scholarship. The fourth author thanks Foundation of Support for Research of Santa Catarina (FAPESC) for granting a research scholarship.

## References

Alvares CA, Stape JL, Sentelhas PC, Gonçalves JLM, Sparovez G (2013) Koppen's climate classification map for Brazil. Meteorol Z 22(6):711–728

Dewandel B, Lachassagne P, Bakalowicz M, Weng PH, Al-Malki A (2003) Evaluation of aquifer thickness by analyzing recession hydrographs. Application to the Oman ophiolite hard-rock aquifer. J Hydrol 274:248–269

Manzione RL (2015) Águas subterrâneas: Conceitos e aplicações sob uma visão multidisciplinar. Jundiaí

Organization of American States/Global Environment Facility (2001) Proyecto para la Protección Ambiental y Desarrollo Sostenible Integrado del SAG. Montevideo

Piazza GA, Depine H, Pinheiro A, Kaufmann V, Gonçalves Júnior AC, Oliveira MHC (2014) Spatial distribution of soil attributes in the Concórdia river watershed in Southern Brazil. Environ Qual Manage 24:1–12

Pinheiro A, Teixeira LP, Kaufmann V (2009) Capacidade de infiltração de água em solos sob diferentes usos e práticas de manejo agrícola. Revista Ambiente Água 4(2):188–199

Pelaéz JRT (1978) Cálculo hidrometeorológico de caudales máximos en pequeñas cuencas naturales. Madrid

Peppoloni S, Di Capua G (2012) Geoethics and geological culture: awareness, responsibility and challenges. Ann Geophys 55(3):335–341

# Evaluation of Protective Capacity of Unconfined Aquifers Using Geoelectric Techniques: A Case Study from North India for Supporting Sound Geoethics

Dinesh Chandra Singhal

**Abstract**

In this work, a technique is proposed for evaluating the protective capacity of unconfined aquifers by using a geoelectrical parameter viz. total longitudinal conductance of the unsaturated vadose zone overlying the aquifer. The approach has been tested in Saharanpur town of Uttar Pradesh, India, from available data of electrical resistivity soundings. Interpretation of the resistivity soundings yielded data of total longitudinal conductance of the unsaturated sedimentary overburden (vadose zone) which ranged between 0.03 and 0.74 mho. The perusal of the total longitudinal conductance map of the area and its comparison with a potentially hazardous activities map of the area was found to match quite well indicating the effectiveness of the technique in finding degree of protection of the unconfined aquifer vis a vis the hazardous chemicals percolating downwards with the infiltrating runoff in the area. This approach when used in combination with groundwater vulnerability estimates can prove to be of considerable help in planning groundwater protection and governance issues in alluvial and coastal areas of the country.

**Keywords**

Groundwater protection • Unconfined aquifers • Total longitudinal conductance • Overburden • Saharanpur town

## 1 Introduction

Groundwater use has undergone dramatic expansion throughout the world over the past 50 years, with global abstraction rates increasing from an estimated 100–150 cubic km in 1950 to 950–1000 cubic km in the year 2000 (Shah et al. 2007). In India, through the construction of millions of private wells, there has been a phenomenal growth in the exploitation of groundwater during the last five decades (Garduno et al. 2011). The government has no direct control over the groundwater use of millions of private well owners, both in rural and urban areas. This often results in unregulated growth of open wells and boreholes especially in the industrial areas of the country, with the Saharanpur town of the western part of Uttar Pradesh province being a classic example. The surface drainage in the town often directly receives the municipal and civic wastes in the absence of a proper sewer system. Such wastewaters thus get direct access to the open drainage system of the region, infiltrating directly into the unconfined aquifer of the area. Exploitation of groundwater through shallow wells and handpumps constructed in these areas for meeting drinking water requirements of the large populace is found to be contaminated from the hazardous, often toxic chemicals and faecal coliforms drawn from the wastes/effluents percolating with the infiltrating runoff directly into the unconfined aquifer(s) of the region. In this study, a new technique is highlighted for evaluating the thickness of the unsaturated overburden (vadose zone), computed from the data of electrical resistivity soundings and to evaluate the degree of groundwater protection of the unconfined aquifer.

It was found by Singhal et al. (2016) that the magnitude of 'total longitudinal conductance' of the unsaturated zone overlying an unconfined aquifer can be used for estimating degree of its protection from the pollutants infiltrating and percolating into the aquifer. The ability of the overburden to retard and filter percolating waste effluents is a measure of its protective capacity (Belmonte et al. 2005). The longitudinal conductance (S) gives a measure of the permeability of the confining clay/shale layers. Such layers have low hydraulic conductivity and low resistivity. Braga et al. (2006) suggested that protective capacity (Pc) of the overburden layers is proportional to its longitudinal conductance(S). So,

D. C. Singhal (✉)
Former Professor, Department of Hydrology,
Indian Institute of Technology, Roorkee, India
e-mail: dcsinghal7@gmail.com



M. Abrunhosa et al. (eds.), *Advances in Geoethics and Groundwater Management: Theory and Practice for a Sustainable Development*, Advances in Science, Technology & Innovation,
https://doi.org/10.1007/978-3-030-59320-9_37

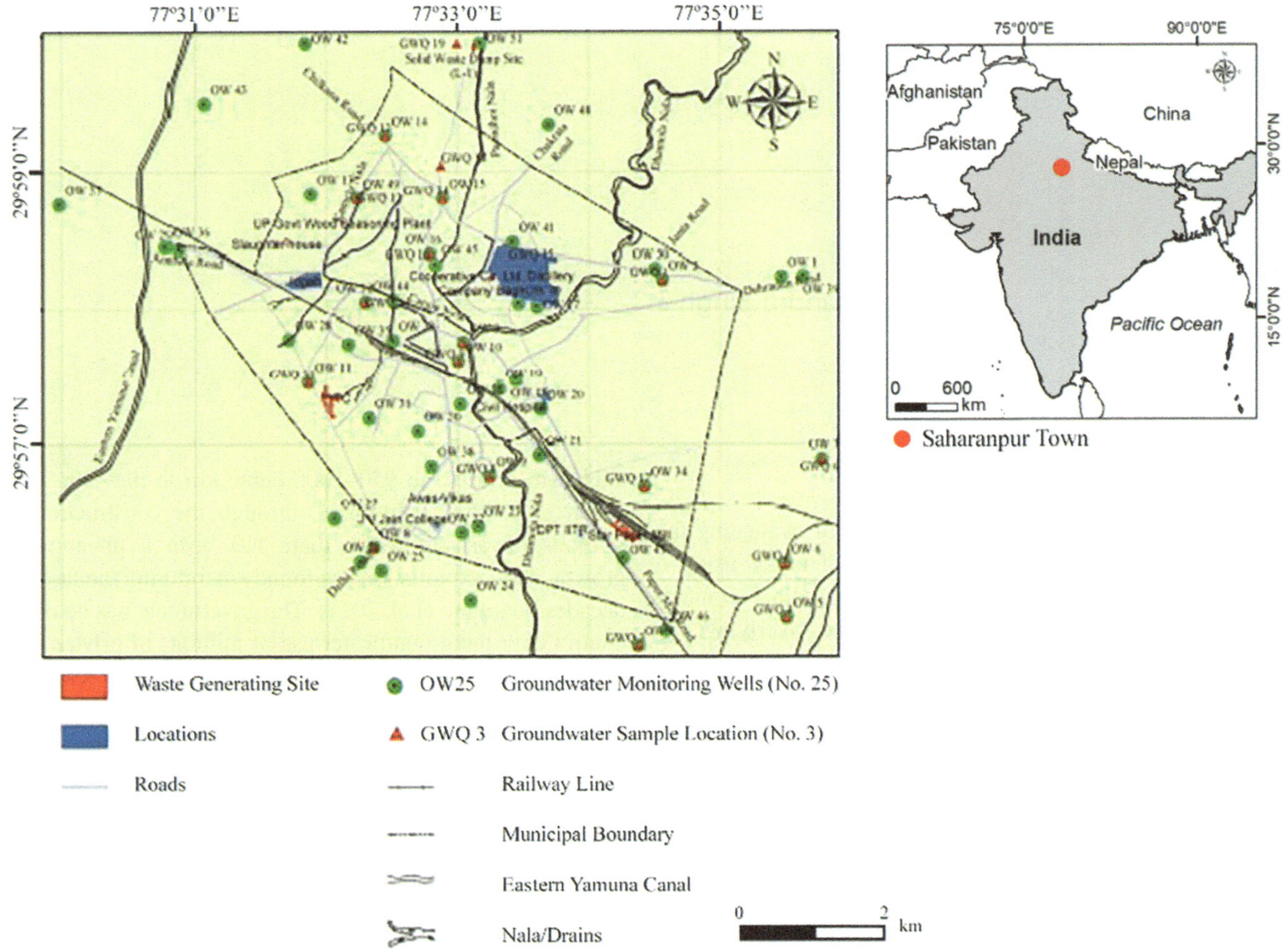


Fig. 1 Location map of Saharanpur town

$$\text{Pc} = \text{S} = \sum_{i=1}^{n} {h_i}/{\rho_i} \qquad (1)$$

The use of 'Total Longitudinal Conductance' in ascertaining groundwater protection capability of the unconfined aquifer was tested in the Saharanpur town of Uttar Pradesh (Singhal et al. 2016). The Saharanpur town falls in the Ganga river plain between latitudes 29° 55′ and 30° 0′ North and longitudes 77° 30′ and 77° 36′ East (Fig. 1). There are many notable industrial units in the area such as a large paper mill, a tobacco company, distilleries, besides several cardboard manufacturing units, electroplating, meat products and chemical units. These industrial units mostly discharge untreated waste effluents directly into the nearby drains, which finally meet the river Hindon. Further, the municipal effluents often carry sewage from the town and meet the river system adding to the waste load of the river.

## 2 Risk of Groundwater Pollution and Vulnerability

Based on the hydrogeological setting of the study area, drastic method of aquifer vulnerability assessment was applied to find out its risk of pollution in different parts of the Saharanpur town using the approach suggested by Aller et al. (1987). The calculation of the drastic index (DI) values indicated that some central and southern localities of Saharanpur city are in medium-risk (D.I.: 160–179) and high-risk zones (D.I. > 180).

Based on the chemical quality of groundwater vis-à-vis possible sources of surface pollution, it has been inferred that the main sources of pollution are indicated to be point sources (outfalls of sewage, paper mill effluent, etc.) and line sources. A potentially hazardous activities map (PHAM)

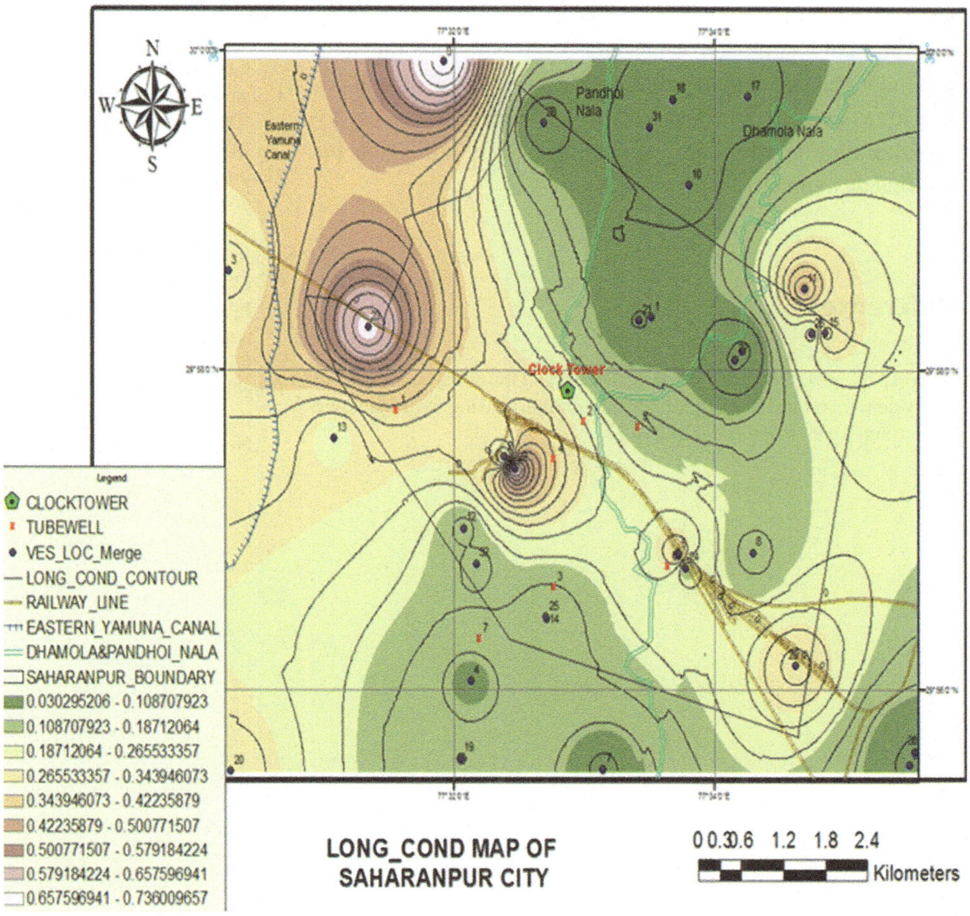


**Fig. 2** Total longitudinal conductance map

was generated using ArcGIS software for the town for comparison with the groundwater vulnerability map. It is suspected that the localities inferred to be hydrogeologically more vulnerable for groundwater pollution are more prone to contamination by hazardous pollutants in groundwater and thus stand greater risk of pollution than the areas with lower vulnerability.

## 3 Evaluation of Protective Capacity of the Vadose Zone

Data of 32 vertical electrical resistivity/induced polarization soundings recorded in the study area was interpreted by curve matching techniques as well as by available software for generating geoelectrical sections for evaluating the 'total longitudinal conductance' of the overburden over the unconfined aquifer. It was observed from the data interpretation that the total longitudinal conductance of the unsaturated overburden using Eq. (1) varied between 0.03 and over 0.7 mho (Singhal et al. 2016).

Figure 2 showing the contours of total longitudinal conductance of the unsaturated overburden for the Saharanpur town, indicates that area towards central, western and NW parts possesses highest longitudinal conductance. This implies that these parts of the town offer relatively higher protection to the underlying unconfined aquifer from the infiltrating pollutants. On the other hand, the localities towards south, southeast and the north-eastern parts offer relatively lesser degree of protection to the unconfined aquifers from infiltrating pollutants. Such differentiation of the area based on the total longitudinal conductance can prove to be of immense value in the groundwater management, especially for planning protection of groundwater resources.

## 4 Concluding Remarks

The above study has demonstrated that the natural protection capacity of unconfined aquifers in alluvial or coastal areas can be evaluated by adopting a geoelectrical approach involving analysis of data of total longitudinal conductance of the unsaturated vadose zone overlying the aquifer. The estimation of this parameter gives a fair idea of the protective capacity of the unconfined aquifer, thus indicating the need of necessary corrective measures for ensuring its protection. This approach when used in combination with groundwater vulnerability data can prove to be of considerable assistance in groundwater governance issues by incorporating appropriate hygiene prior to its utilization for various designated uses.

## References

Aller L, Lehr J, Petty RJ, Bennet T (1987) Drastic: a standardized system for evaluation of groundwater pollution using hydrogeologic setting. J Geol Soc India 29:23–27

Belmonte JS, Enrigueze JC, Zamora M (2005) Vulnerability to contamination of the Zaachila aquifer, Oaxaca Mexico. Geof Internat 4(3):283–300

Braga ACO, Walter MF, Filho WM, Dourado JC (2006) Resistivity (DC) method applied to Aquifer Protection Studies. Rev Bras Geof 24(4):573–581

Garduno H, Romani S, Sen Gupta B, Tuinhof A, Davis R (2011) India groundwater governance case study (water papers). Water Partnership Program, World Bank, NY

Shah T, Burke J, Villholth K, Angelica M, Custodio E, Daibes FH (2007) Groundwater: a global assessment of scale and significance. In: Molden D (ed) Water for food, water for life: a comprehensive assessment of water management in agriculture, Earthscan. London, UK, pp 395–419

Singhal DC, Sharma VK, Tungariya M (2016) Role of groundwater protection planning. In: groundwater governance: a case study from North India. J Geol Soc India 88:449–463

# Geoethics from Geophysical Exploration to a Glass of Mineral Water, Including Iberian Thermal Medicine

Carla Sofia Rocha

**Abstract**

Often, when we speak about mining exploration, we resort rapidly to the idea of the exploration relating to metallic and no-metallic minerals, and we omit, though in an unconscious way, the hydrogeological exploration, as if this one was not forming a part of the geological resources. We forget the water as a raw material to explore. The thermal spas are an example of the hydrological exploration. In the whole of the territory of the Iberian Peninsula, there is a heritage and a thermal waters tradition that often is not known and needs major valuation. The thermal and natural mineral springs are the result of a topographic, a geological, and tectonic setting that regarding its therapeutic properties and benefits for the health can be exploited. In the Iberian Peninsula, there are numerous testimonies of these thermal practices, from the Roman times, more 2000 years ago. We are favored and responsible of the conservation and diffusion of this Iberian heritage. It is extremely relevant for a better management, knowledge, and sustainability of the aquifers, the application of non-destructive geophysical methods, with detailed reach and deeper scales, as well as chemical control associated with the characteristic microbiology of each water in particular, in addition to scientists recognizing the pertinence of geoethics and geoliteracy of the populations within this geological wealth.

C. S. Rocha (✉)
Center for Natural Resources and Environment (CERENA), University of Lisbon—Instituto Superior Técnico, Lisbon, Portugal
e-mail: carla.s.rocha@ist.utl.pt

C. S. Rocha
Center for Nuclear Sciences and Technologies (C2TN), University of Lisbon—Instituto Superior Técnico, Lisbon, Portugal

**Keywords**

Geoethics • Iberian Peninsula heritage • Geophysical methods • Natural mineral water • Legislation

## 1 Introduction

The characterization of the groundwater resources aims at defining good practices for mineral water exploration. The development of non-destructive geophysical methods, such as electrical resistivity and electromagnetic exploration in the frequency domain, contributes to a better knowledge of the geology of the subsoil, which will provide an optimization and better management tool for exploitation of the groundwater resources. The development of legislation appropriate to the different water uses and consumption and specify contemplate non-destructive geophysics methods, as well as hydrochemistry and microbiology, is fundamental for the preservation and enhancement of our Iberian thermal heritage, with awareness and geoethics.

## 2 Background

The tradition of using medicinal baths and thermal treatments has survived the historical, political, and social evolution of the Iberian Peninsula, which highlights its great importance. In Spain, the most notable work is the "Espejo cristalino de las aguas de España: hermoseado y guarnecido con el marco de variedad de fuentes y baños cuyas virtudes, excelencias y propiedades se examinan, disputan y acomodan à la salud, provecho, y conveniencias de la vida humana," pioneer of general character, from the Spanish Hydrology, printed in 1697, as the authorship of the Professor Alfonso Limon Montero from the at University of Alcala de Henares, considered the "father of Hydrology in Spain" (Maraver et al. 2003). In Portugal, the first



M. Abrunhosa et al. (eds.), *Advances in Geoethics and Groundwater Management: Theory and Practice for a Sustainable Development*, Advances in Science, Technology & Innovation,
https://doi.org/10.1007/978-3-030-59320-9_38

manuscript record with the inventory of the occurrences of mineral natural springs and their health benefits is reported in the eighteenth century, "Aquilegio Medicinal, em que se dá noticia das agoas de Caldàs, de fontes, rios, poços, lagoas e cisternas do Reyno de Portugal e dos Algarves que, ou pelas virtudes medicinaes que tem, ou por outra alguma singularidade, são dignas de particular memoria" (1726) by the medical doctor Francisco da Fonseca Henriques (1665–1731) also known as Dr. Mirandela. Associated with the occurrence of mineral natural water springs, villages, towns, and cities created in the university laboratories, expertise's dedicated to thermal waters studies developed in the fields of medicine and chemistry. That Hydrological Institutes were founded in Lisbon, Porto, and Coimbra Universities. Scientific cooperation was promoted between Geologists like Ernest Fleury (1878–1958) and Paul Choffat (1849–1919), Chemists like Charles Lepierre (1867–1945) (see Fig. 1), and António Ferreira da Silva (1853–1923) and Medical Doctors like Ricardo Jorge (1858–1939).

The first Thermal Hospital in the world is located in Portugal, in Caldas da Rainha, Leiria, founded in the fifteenth century in 1485 by Queen Dona Leonor, wife of King Don João II who, seeing the poor and the sick, plunged into those waters, cured themselves sufferings (https://termascentro.pt/pt/termas/distrito-de-leiria/caldas-da-rainha). It is thanks to the different reports by patients who see their illnesses cured, for example skin, digestive system, reproductive system, respiratory system, among others, that they communicate to their medical doctors, and make it possible to survey the beneficial properties corresponding to the different chemical signatures of the mineral natural waters and allow to correlate the geological provenance, including water–rock interaction and health effects.

In December 2019, the Hydrogenome Project (https://hidrogenoma.javali.pt/) resulting from the cooperation of General Directorate of Energy Geology (DGEG) and National Laboratory of Energy and Geology (LNEG) makes possible the scientific recognition of the therapeutic properties of mineral water, relating microbiology, chemistry, and geology, resulting in the best efficiency of thermal medicine, using natural mineral water and promoting the different spas and different therapeutics.

## 3 Geographic and Geological Localization

The mineral natural water springs of the Iberian Peninsula (see Fig. 2) may be concentrated at high altitudes associated with mountainous regions, which result from deep geological faults and other tectonic structures that provide the infiltration of rain water at depth and the discharge of mineral natural water in these areas.

The structural discontinuities, favoring the infiltration of groundwater and water–rock interaction, strongly contribute to the physical and chemical compositional characteristics of mineral natural water and the development of specific microbiology. The physical and chemical composition, as well as the specific microbiology of natural mineral waters, results in the medicinal properties of the mineral natural water and with that the cure of pathologies.

**Fig. 1** Ernest Fleury the left (n.°1) y Charles Le Pierre on the right (n.°2). Collection of the Instituto Superior Técnico of the University of Lisbon

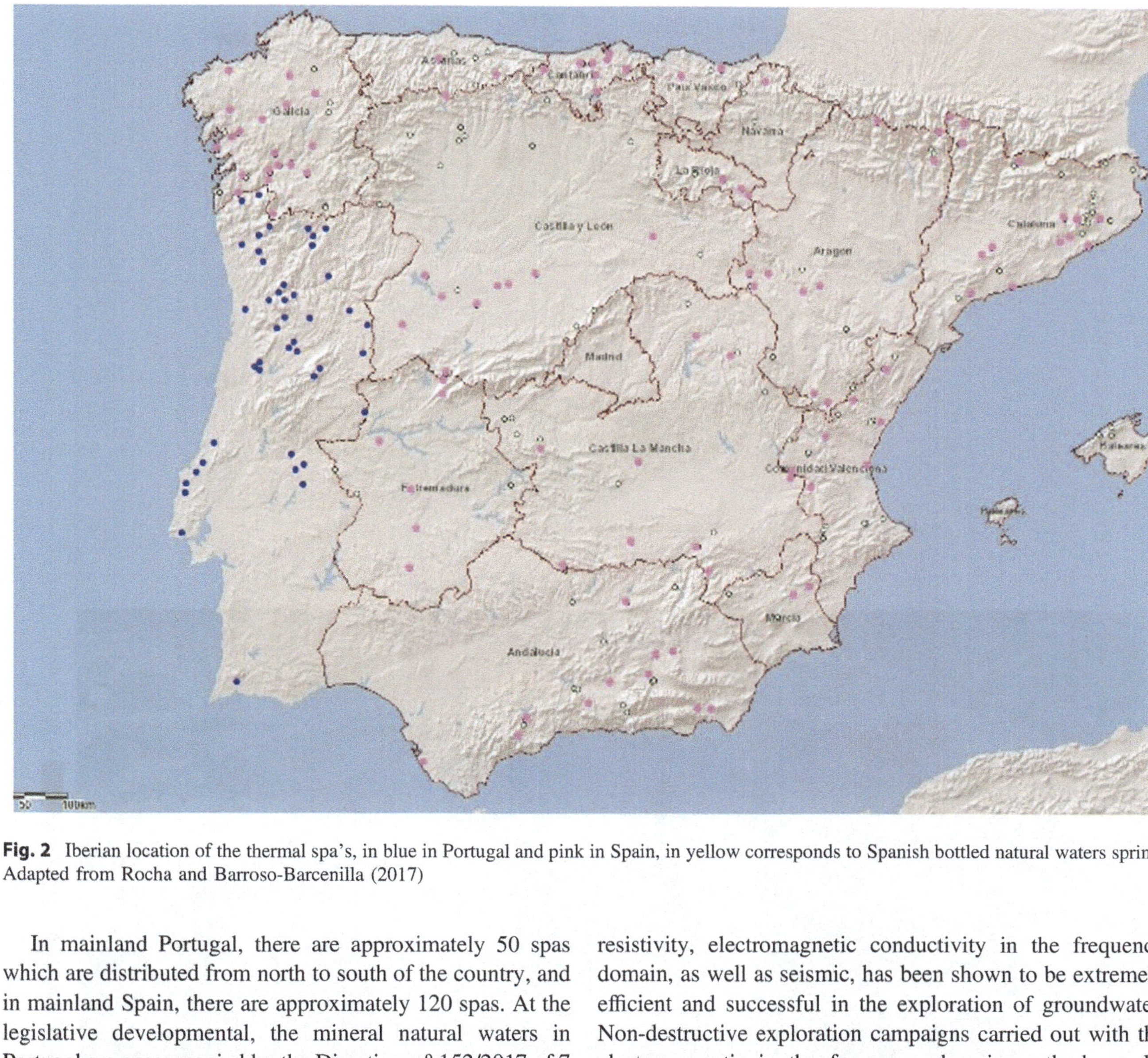


**Fig. 2** Iberian location of the thermal spa's, in blue in Portugal and pink in Spain, in yellow corresponds to Spanish bottled natural waters spring. Adapted from Rocha and Barroso-Barcenilla (2017)

In mainland Portugal, there are approximately 50 spas which are distributed from north to south of the country, and in mainland Spain, there are approximately 120 spas. At the legislative developmental, the mineral natural waters in Portugal are accompanied by the Directive n.° 152/2017 of 7 of December and in Spain, legislation is adapted to the different autonomous regions of the country. The compliance with legislation promotes in an ethical and scientific way the preservation of natural mineral water and surrounding natural places.

## 4 Geoethics and the Importance of Geophysical Exploration for Water Resources

### 4.1 Non-destructive Geophysical Exploration

The structural characterization of aquifers reservoirs is very important for their studies and preservation. The application of non-destructive geophysical methods, such as electrical resistivity, electromagnetic conductivity in the frequency domain, as well as seismic, has been shown to be extremely efficient and successful in the exploration of groundwater. Non-destructive exploration campaigns carried out with the electromagnetic in the frequency domain method can be complemented with the electrical resistivity method, with the dipole–dipole and Wenner–Schumberger measurements devices, which allow the identification of vertical and horizontal, structural discontinuities (see Fig. 3), where groundwater circulates Rocha et al. (2019).

These methods, as well as the refractive seismic that reaches an approximate depth of 100 m of investigation, are non-destructive methods suitable for the detection of structural discontinuities, filled with clay minerals and saturated oxides due to the percolation of groundwater. Non-destructive geophysical exploration translates into a conscious, responsible, and geoethical approach, for the geostructural characterization in the field of geological resources and, in particular, water resources.

In our view, it is necessary that Portuguese and European legislation include the mandatory deployment of

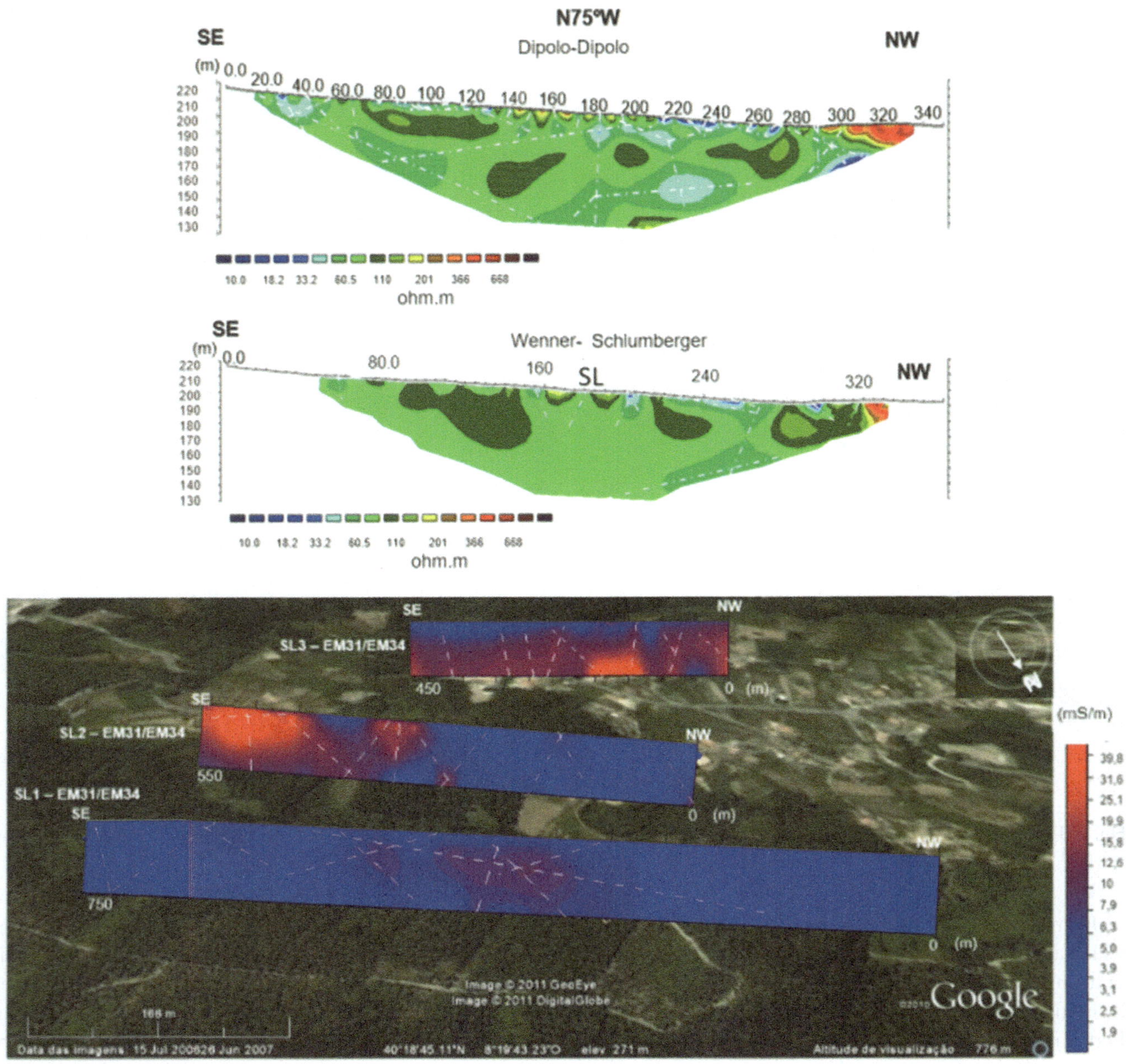


**Fig. 3** Example of the interpretation of the electrical resistivity section for the dipole–dipole and Wenner–Schlumberger measuring device with topographic correction, obtained with the Res2Dinv 3.55 software; and conductivity image processed with Surfer 7.0 software. The white dashed lines show the possible discontinuities through which the groundwater can be percolated. Adapted Rocha et al. (2019)

non-destructive geophysical research campaigns for the characterization of aquifers for the purpose of exploring natural mineral water, or natural water spring, for bottling, or for therapeutic treatments. For the recognition of aquifers and groundwater, there is no law with mandatory application of non-destructive geophysical surveys, neither in Portugal, nor in Spain, nor in the European Union, which is fundamental for a responsible and geoethical good practice.

For free access information, there is the website European Geological Data Infrastructure (EGDI) (https://www.europe-geology.eu/map-viewer/) which contains an interesting information from the European community, where there are available data from geophysical, hydrochemistry, and hydrological maps. By a geoethical approach, this information should be filled in by all the countries of the European Union, through the responsible entities of each country.

### 4.2 Geoethics in Exploration and Surveys of Geological Resources

Geoethics must also get involved in the exploration and surveys through the drilling wells and boreholes. For the best performance of the drill works, these must be accompanied by geological sample collection, drill cores, and the drilling log records, so that they can be consulted for future works and research. These drilling log and geological samples, drill cores, should be incorporated by official government institutions and be freely accessible for scientists. At the end of exploration works by drilling, the holes must be filled in their entirety by inert materials and adequately covered and sealed, with the purpose of preventing accidents or being potential sources of contamination. Unfortunately, the mandatory collection of the geological sampled material, as well as the complete filling of the hole by inert materials, is not contemplated in Portuguese or European legislation and must be outlined in accordance with geoethics principles.

## 5 The Importance of the Information Contained in the Labeling

The information included on the bottles waters labels complies with legal standards defined by the European Union by Directive 2009/54/EC of the European Parliament and of the Council of June 18, 2009. On the label of bottled is mandatory the indication of natural mineral water or natural water spring. In addition to the commercial name, the place of exploitation, the name of the source or borehole extraction, indicates the typical chemical composition of the water contained in the bottle, the pH value, the total mineralization (total dissolved salts), and the quantities of the elements that are present in it in greater quantity are indicated.

However, the labels lack information on the chemical facies that characterize water, as well as the advice and advise against of ingestion by consumers who may have health restrictions. This information makes it possible to distinguish a particular water from other similar waters. Geoethics must also be present in the need to revise the normative regulation that legislates labeling, for better information and consumer choice.

## 6 Concluding Remarks

Water is the most valuable, noble, and rich natural resource that exists, without which the existence of life is not possible. In our vision, to protect and better sustainable management of groundwater's and aquifers, the legislation needs some amendments that contemplate the mandatory development of non-destructive geophysical campaigns with specific methodologies, such as electrical resistivity, electromagnetic conductivity, and seismic. In drilling work, the legislation should also mandatorily include the collection and records of the cores drilled materials, as well as at the end of the work, the total filling of the holes by inert materials, and subsequent sealing. The labeling of natural waters spring and mineral natural water and also the legislation should contemplate the chemical specificity and health care in its consumption. Historical and cultural dissemination and scientific literacy are essential to perpetuate the extraordinary Iberian heritage of thermal culture. Currently, it is essential that geologists, chemists, biologists, biochemists, pharmacists and medical hydrologists, work and cooperate as a team, for the best management and sustainability for a mineral natural water.

**Acknowledgements** PhD Research Scholarship, financed by MCTES Portuguese funds, from the Foundation for Science and Technology (FCT SFRH/BD / 105622/2015). Publication supported by FCT- project FCT UID / ECI / 04028/2019 - CERENA/IST, and FCT UID/Multi/04349/2013 - $C^2$TN.

## References

Direção-Geral de Energia e Geologia https://www.dgeg.gov.pt/. Last accessed 01 Mar 2020

EGDI—European Geological Data Infrastructure https://www.europe-geology.eu/map-viewer/. Last accessed 01 Mar 2020

European Union Law https://data.europa.eu/eli/dir/2009/54/oj. Last accessed 08 Mar 2020

Hidrogenoma https://hidrogenoma.javali.pt/. Last accessed 01 Mar 2020

Maraver F, Aguilera-Lopez L, Armijo F, Martín-Megías A, Meijide-Faílde R, Soto J (2003) Vademecum de aguas mineromedicinales Españolas, Instituto de Salud Carlos III, Madrid

Rocha C, Barroso-Barcenilla F (2017) El patrimonio geológico y minero, identidad y motor de desarrollo, Almadén (Ciudad Real). In: XVII International Congress on Geological Heritage and Mining XXI Scientific Session of La SEDPGYM Sociedad Española para la Defensa del Patrimonio Geológico y Minero

Rocha C, Figueiredo F, Lopes A, Catarino L, Azevedo JM (2019) Aplicação da prospecção geofísica para a caracterização da massa de água mineral das "corgas do Buçaco", Penacova. In: Book of Abstracts of 12th Seminar on Groundwater Coimbra, APRH, pp 110–114

Termas de Portugal https://termascentro.pt/pt/termas/distrito-de-leiria/caldas-da-rainha. Last accessed 10 Feb 2020

# Variation of Salinity Levels in Water Bodies in and Around Weligama Bay Due to Effect of Hydrological Processes

Ashvin Wickramasooriya and Viran Samarawera

**Abstract**

Waligama Bay in Sri Lanka is an important coastal region where there are many social and economic activities taking place. However, people living in closer to Weligama Bay are facing a serious problem related to water quality. Surface water bodies and groundwater in this area highly contaminate with saline water. This study has focused on understanding the influence of natural hydraulic processes and geological conditions on the variation of salinity levels in water bodies in the study area. Fifty sample points which are radially distributed away from the bay and are located at different distances away from the coastline are selected (distances to theses sample points including dug wells, tube wells and canals from the coastline are measured using GPS coordinates). Water samples were collected from each sample point, and their pH, EC, salinity levels, depth to water table in wells were measured twice a month for four months from July to October 2017 (after southwest monsoon period). Digital thematic maps were produced to understand the distribution pattern of these water quality parameters in the study area for different periods. Also variation of groundwater table with respect to precipitation data and tidal heights for these periods was recorded. According to the analysis of data, it has been observed that salinity levels decrease with the increase in precipitation while increases with high tidal periods. Also, it is significant that the salinity levels decrease away from the coastline towards land. However, at some locations within low salinity levels record high, salinity especially where there is high porous and high permeable subsurface material. Therefore, there is a clear relationship between salinity levels of the area with natural hydraulic processes such as precipitation, infiltration and tidal activities.

**Keywords**

Salinity • Tides • Groundwater • Precipitation

## 1 Introduction

Weligama is located in Matara district of the southern province of Sri Lanka (Fig. 1). Weligama experiences a distinct bimodal rainfall pattern and has a tropical monsoonal climate. Maximum rainfall for the area receives from southwest monsoon during May to September. The main river in the Weligama is Polwatta River which discharges its water into the Weligama Bay at Polwatumodera. Even though tide dominates estuaries mixing saline water and causes more or less vertical homogeneity, Polwatta River records weak tidal influence as it has a narrow and shallow mouth (e.g. Boyd et al. 1992; Dalrymple and Choi 2007).

The Weligama Bay which has the average elevation is about 2–3 m consist of sandy beach is about 10–30 m width, and beyond the bay, towards the land, it was found that 4–5 m thick weathered overburden. At few locations, highly fractured granitic gneiss and garnet biotite gneiss outcrops can be found. It also observed that the river has lower elevation than the bay area. Therefore, it has recorded that within last couple of years nearly 5000 acres of paddy lands have been destroyed in Polwatta, Jamburugodayaya, Pelena areas in Weligama due to contamination of surface water and groundwater due to saline water. Sea water intrusion and landward migration of a saltwater wedge due to density differences in sea water and inland water bodies contribute to increase the salinity levels in water bodies in the area (Alwis and Dassanayake 1993). Saltwater intrusion, sea-level change, reclamation of low-lying areas and tidal effect involve migrating of saline water into freshwater aquifers closer to coastal areas (Freeze and Cherry 1979).

A. Wickramasooriya (✉) · V. Samarawera
Department of Geography, University of Peradeniya, Peradeniya, Sri Lanka
e-mail: awickramasooriya@gmail.com





M. Abrunhosa et al. (eds.), *Advances in Geoethics and Groundwater Management: Theory and Practice for a Sustainable Development*, Advances in Science, Technology & Innovation,
https://doi.org/10.1007/978-3-030-59320-9_39

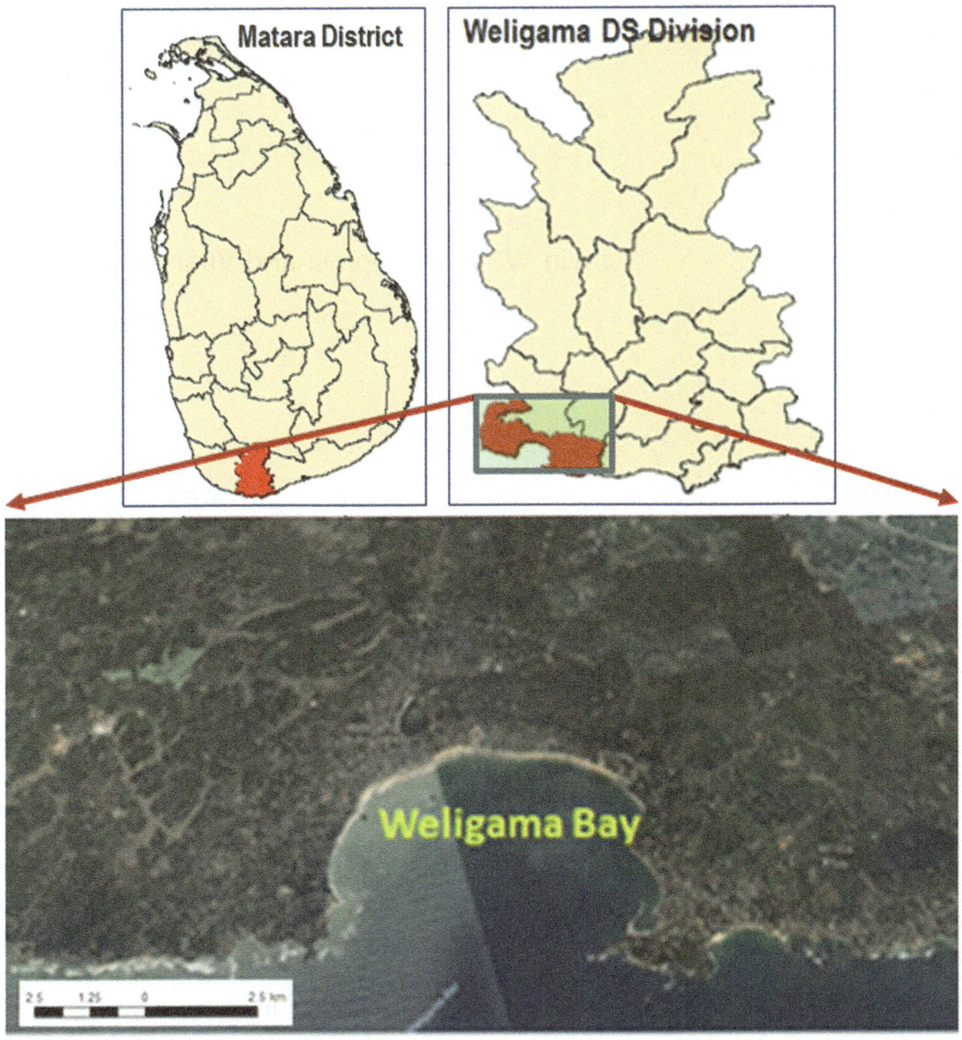


**Fig. 1** Weligama Bay, Sri Lanka

## 2 Materials and Methods

Using systematic sampling method, radially distributed fifty sample points including dug wells, tube wells and streams which are located at different distances away from the coastline towards the land at Weligama Bay are selected as sampling points. The study area extends up to two kilometres landwards from the coastline, and topographically, the height difference between bay area and the high elevated location in the study area is about 8 m above M.S.L. The sample points were selected based on radially distributed and located approximately 50 m, 100 m, 500 m, 1 km and 2 km distances away from the coastline (Fig. 2). Water samples were collected in every two weeks for four months from July to October 2017 and analysed for salinity, EC and pH. Also, corresponding precipitation, temperature and tidal heights were recorded for the study period. In addition to these data, depths to water table (on the date of water samples are collected), elevation of sample locations, geological conditions and soil characteristics surrounding the sampling locations are recorded.

## 3 Results

### 3.1 Variation of the Salinity Levels in the Study Area During the Study Period

It has been clearly noted that the average salinity levels decrease towards the land from the coastline throughout the study period (Table 1 and Fig. 3). However, there are few locations which are located closer to the streams and deep tube wells do not show this general trend. Distribution of the salinity levels calculated based on EC for four months July, August, September and October in 2017 is shown in Fig. 4.

Highest salinity concentrations were recorded during September where the area affected by high tidal heights. During October and July, there were the area which had recorded the lowest tidal heights and highest precipitation,

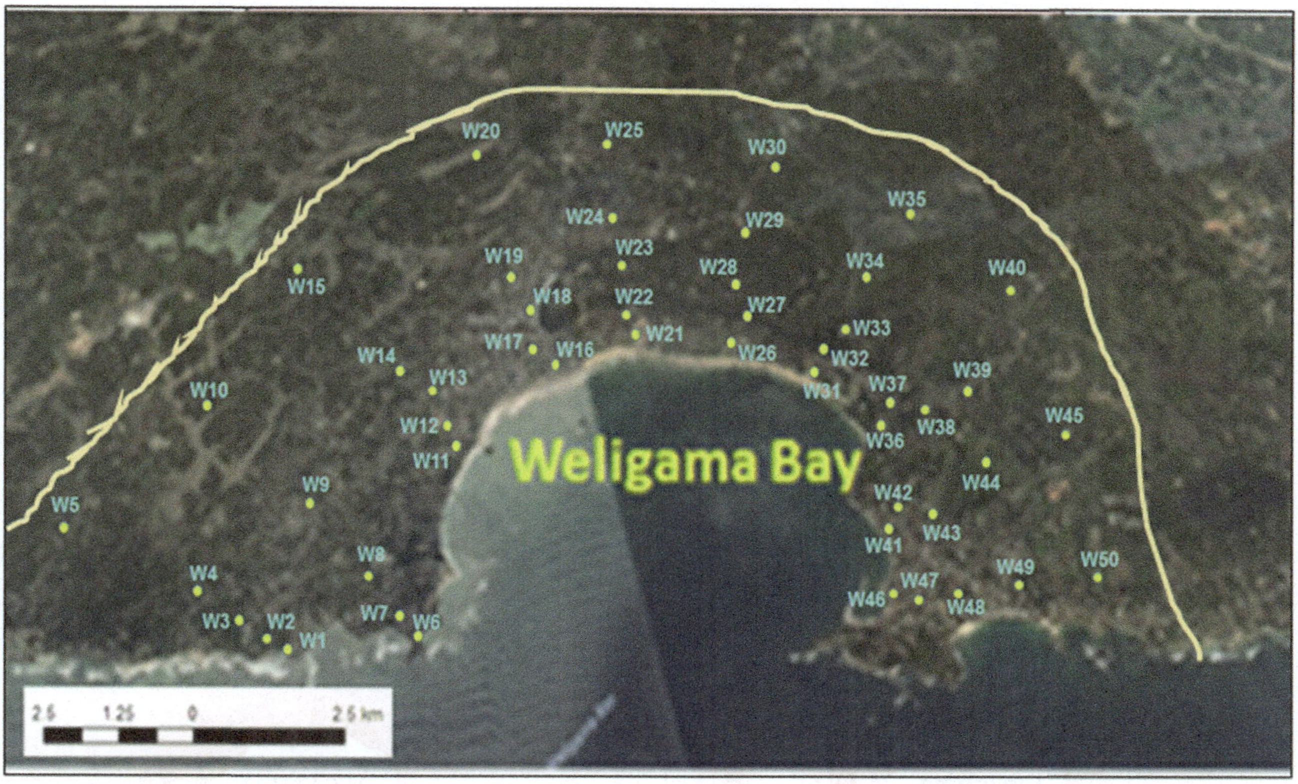


**Fig. 2** Sample points in the area

**Table 1** Variation of average salinity levels (ppt) from July to October 2017 in the study area

| | Distance away from the coastline towards land and average | | | | |
|---|---|---|---|---|---|
| Month | 50 m | 100 m | 500 m | 1000 m | 2000 m |
| July | 22.69 | 18.07 | 14.20 | 9.35 | 7.60 |
| August | 30.76 | 20.00 | 10.81 | 7.36 | 5.82 |
| September | 55.30 | 37.92 | 21.00 | 17.50 | 14.54 |
| October | 18.25 | 13.46 | 8.23 | 7.63 | 5.69 |

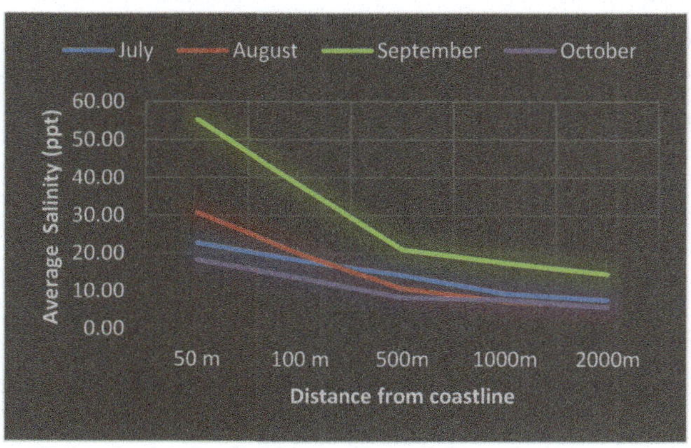


**Fig. 3** Variation of the salinity levels with the distance away from the coastline towards land

respectively (Table 2). Further, it was observed that the salinity levels in sample locations had been increased during the periods with high temperature. However, in deep tube wells, the variation of salinity levels due to increase in temperature was negligible. Also, it was noted salinity levels reduced at sample sites in and around the stream during the area had received high precipitation.

## 3.2 Variation of Depth to Groundwater Level during the Study Period

Depth to groundwater table in the study area varies between 1 and 13 m during the study period. September recorded higher depths to groundwater table, while October recorded

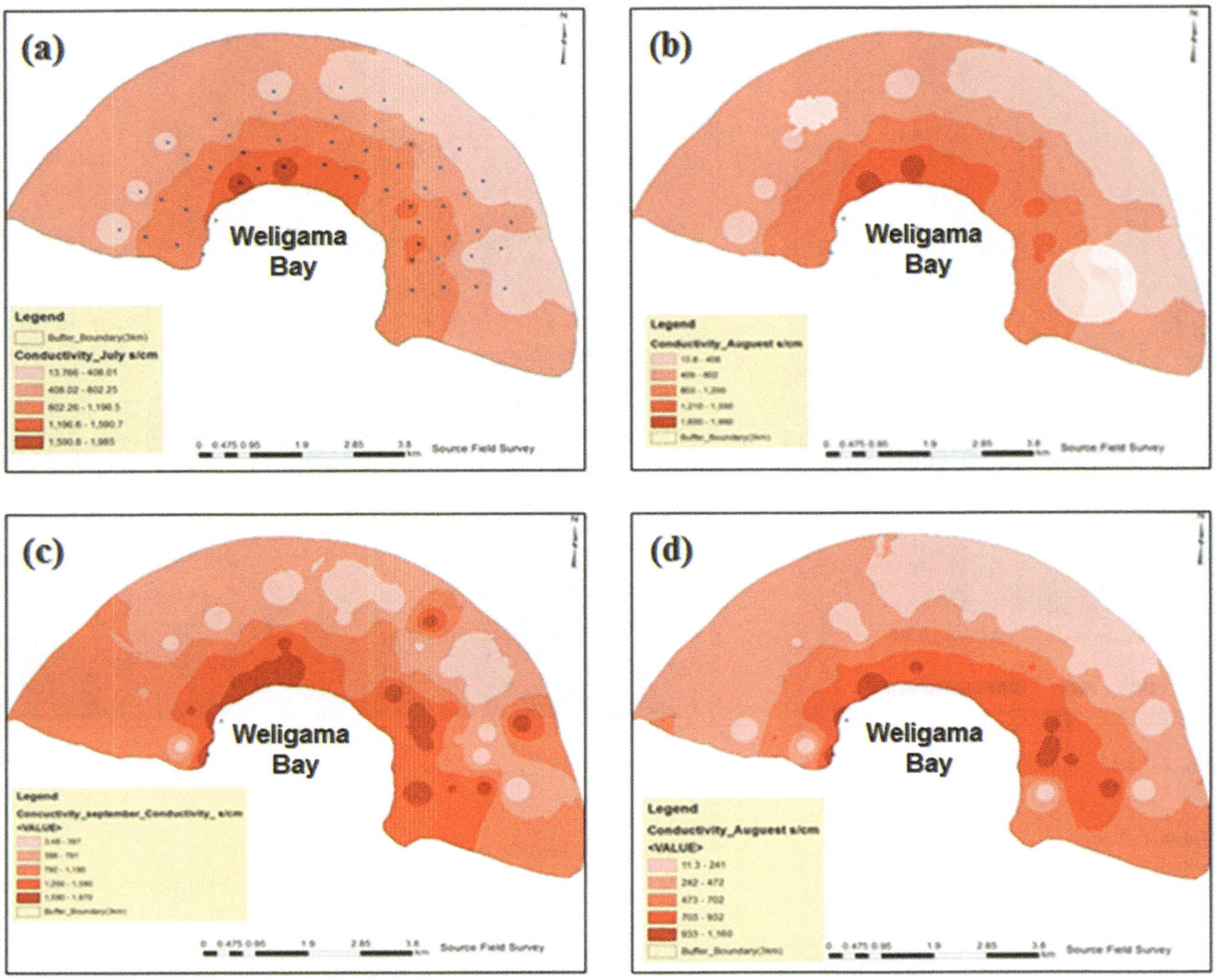


**Fig. 4** Distribution of the salinity levels in the study area **a** July—2017, **b** August—2017 **c** September—2017 and **d** October—2017

**Table 2** Variation of average rainfall and tidal heights in the study area

| Month | Average rainfall (in.) | Maximum tidal height (ft) |
|---|---|---|
| July | 36.48 | 6.15 |
| August | 15.36 | 5.12 |
| September | 18.24 | 5.49 |
| October | 73.68 | 3.16 |

the shallow depths (Fig. 5). Based on the groundwater depths, it was found that the groundwater flow is towards the North and Northeast of the study area.

## 4 Discussion

Weligama Bay has become a significant issue for people live in and around that area. During October and July, this area had recorded the lowest tidal heights and highest precipitation. It was observed that the levels of salinity decreases away from the coastline towards land. Also, compared to periods of low tidal heights, salinity levels in water increased during high tidal height periods. Therefore, sea water intrusion can be identified as the main source of concentration of salinity in water bodies in the area. However, few tube-well which are located less than 100 m away from the coastline are recorded low salinity levels (W7, W41 and W42). When considering the subsurface geological conditions around those wells, it was found that there is an

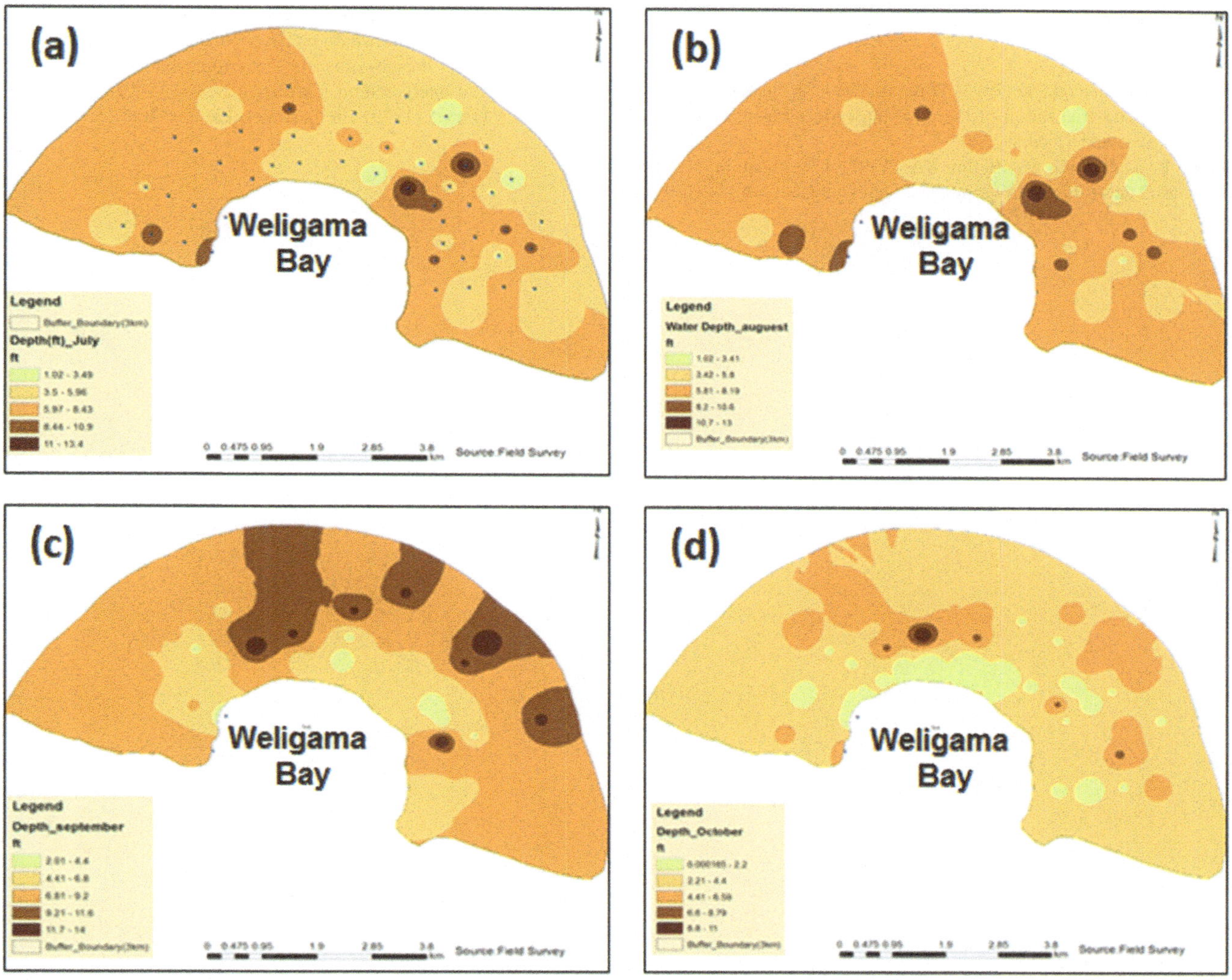


**Fig. 5** Depth to groundwater table **a** July—2017, **b** August—2017 **c** September—2017 and **d** October—2017

impervious 2.5–3.5 m clay layer which exist about 1 m depth from the surface. However, some wells which are located greater than 500 m away from the coastline and even when their water levels go very deep recorded very high salinity levels (W30 and W45). It was found that these two tube wells are constructed within highly fractured rocks, and therefore, sea water can be directly seepage through these fractures and is the main reason to increase salinity in those wells.

## 5 Concluding Remarks

It was obvious that during high precipitated periods recorded higher concentration of salinity than periods got low precipitation indicate that surface water bodies with high salinity mix with rain water flow towards the river and then distributed to other low elevated areas. When there is high tides, water can be moved towards land along the Polwatta River which causes to increase the salinity levels in groundwater and surface water bodies closer to the river. Therefore, by establishment of a filter dam across the river, this situation can be minimized and the effect of agricultural lands from saline water can be reduced. Also, organic fertilizers like compost can be utilized for agricultural purposes to reduce the influence of agricultural lands from high salinity. Further, some locations in the study area recorded high salinity levels because of subsurface conditions such as available soil type, fractures in rocks. Therefore, seepage and groundwater flow also contribute to change the salinity levels in water bodies surrounding the Weligama Bay. It can be observed that the salinity of both surface water bodies and groundwater decreases away from the coastline and also with the depth of the groundwater table. According to this study, it can be concluded that there is a direct relationship between concentration of salinity in surface and groundwater water bodies in the area and hydrological processes such as precipitation, tides, stream flow, seepage, groundwater flow.

## References

Alwis P, Dassanayake H (1993) Water quality of Koggala lagoon prior to the industrialization of the area. https://localhost/handle/1/1373. Last accessed 12 Mar 2020

Boyd R, Dalrymple R, Zaitlin BA (1992) Classification of clastic coastal depositional environments. Sediment Geol 80(3–4):139–150

Dalrymple RW, Choi K (2007) Morphologic and facies trends through the fluvial–marine transition n tide-dominated depositional systems: a schematic framework for environmental and sequence-stratigraphic interpretation. Eart Sci Rev 81:135–174

Freeze A, Cherry J (1979) Groundwater. Prentice-Hall Inc., London

# Variation of Victoria Reservoir Water Level and Its Effect on Fluctuation of Groundwater Level Closer to the Reservoir

Ashvin Wickramasooriya and Nirmala Rajapaksha

**Abstract**

Water level of dug wells can be fluctuated due to changes in water level of a reservoir, especially which are located lower elevation and closer to the reservoir. To identify the relationship between water level fluctuation of the Victoria reservoir and how it can be effected to water levels of dug wells in Wewegama village which is situated closer to the Victoria reservoir, ten dug wells having different depths to water level and located different distances to reservoir are selected. Water levels of dug wells were measured weekly for five months during SW monsoon and inter-monsoonal period from May to October in 2016, and water levels of the same date of the reservoir are obtained from the Mahaweli Authority of Sri Lanka. By plotting reservoir water levels, corresponding well water levels and by correlation analysis, it was found that there is a close relationship between fluctuation of reservoir water levels and water levels of some wells throughout the study period. However, a correlation analysis which was conducted between closest distance between reservoir and each well and the change in water levels of reservoir in each week, it was found that there is no such strong relationship. Therefore, according to the results of this study, it can be concluded that with the variation of reservoir water level, well water levels which are located closer to the Victoria reservoir also vary accordingly.

**Keywords**

Water level • Fluctuation • Distance • Correlation

A. Wickramasooriya (✉) · N. Rajapaksha
Department of Geography, University of Peradeniya, Peradeniya, Sri Lanka
e-mail: awickramasooriya@gmail.com

## 1 Introduction

Natural phenomena such variation of long-term and short-term climatic changes can be influenced to fluctuate water level of a reservoir (Gribovszkia et al. 2010). Also, some anthropogenic activities such as destroy the forest cover closer to reservoirs can be influenced to reduce the water level of a reservoir. It is expected that the water level of a reservoir increases during heavy rainy period while the water level goes down during dry period. Fluctuation of water level of a reservoir can be affected its surrounding environment as well as surface and groundwater resources closer to the reservoir. Although Taylor and Alley (2001) showed that the variation of groundwater level monitoring should be based on long-term water level data, this study has mainly focused to analyse whether there is a relationship between fluctuation of reservoir water level and the variation of groundwater level in a short period. To achieve this task, five months during Southwest and inter-monsoonal period from May to October 2016 has selected as the study period. Victoria reservoir which is one of the major reservoirs in Sri Lanka and Wewegama village which is situated closer and at lower elevation than Victoria reservoir has been selected as the study area (Fig. 1). The study area belongs to the central highlands of Sri Lanka, and highly fractured crystalline limestone and garnet biotite gneiss rocks can be found in the area. The overburden thickness of the area varies from about 1–7 m, depths to groundwater table in the area vary with the climatic conditions, and generally, it varies from 1 m above the ground level of the wall during wet period while goes down up to about 10 m during dry periods.

Tharani et al. (2015) and McMillan and Srinivasan (2014) stated that the interaction between groundwater and surface water depends on the aquifer characteristics such as presence of pore spaces, their size and storage capacity, permeability, and transmissivity. However as the extent of the study area is small, many of these characteristics are almost similar in the study area other than the overburden thickness.



M. Abrunhosa et al. (eds.), *Advances in Geoethics and Groundwater Management: Theory and Practice for a Sustainable Development*, Advances in Science, Technology & Innovation,
https://doi.org/10.1007/978-3-030-59320-9_40

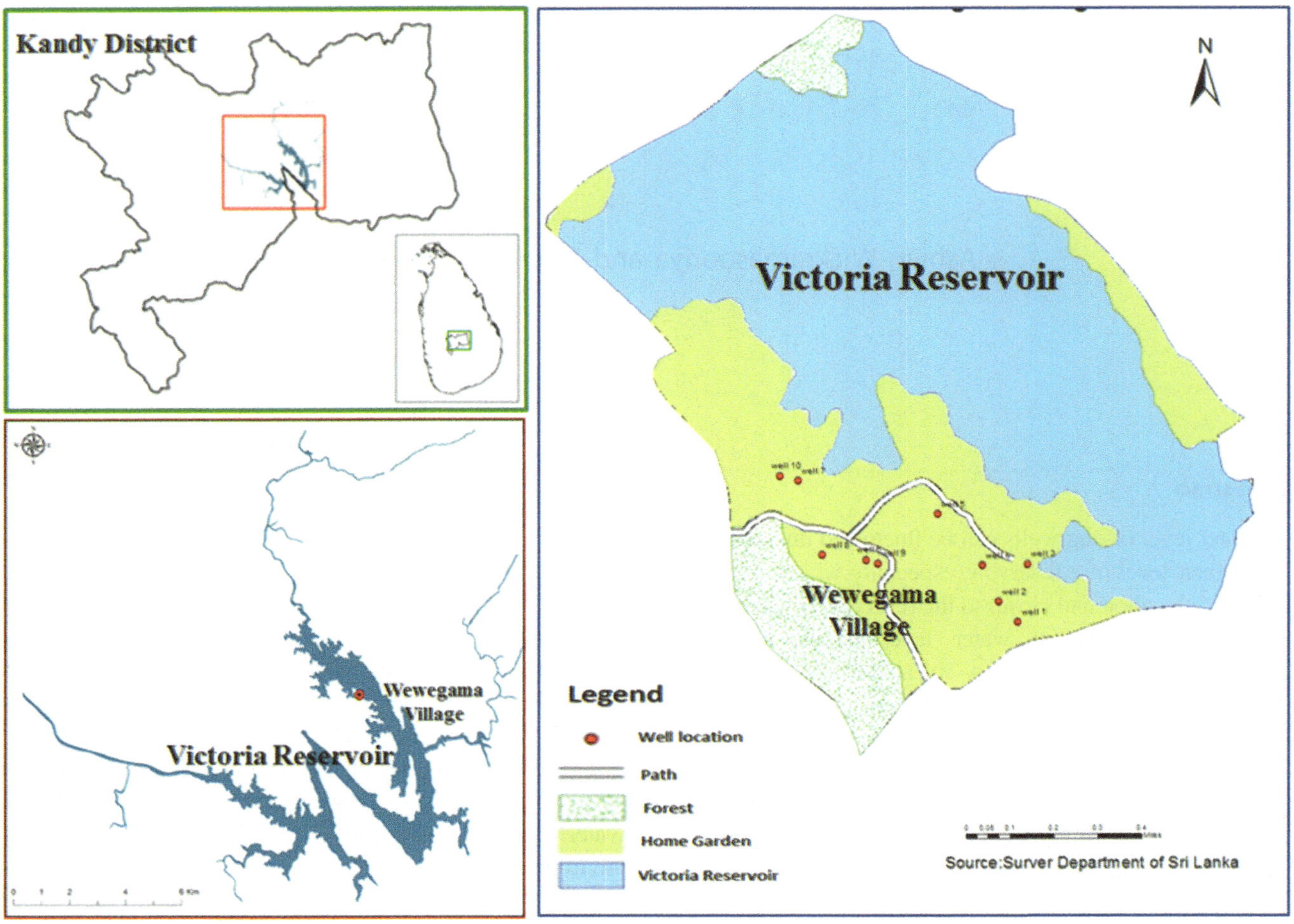


**Fig. 1** Well locations in Wewegama village situated closer to Victoria reservoir

## 2 Materials and Methods

Shortest distance to the reservoir from each well has been calculated using GPS coordinated obtained at each well location and relevant locations at the reservoir using the following equation.

$$\text{Shortest distance to the reservoir} = \sqrt{(x_1 - x_2)^2 + (y_1 - y_2)^2}$$

($x_1$, $y_1$)—GPS coordinates at the well, ($x_2$, $y_2$)—GPS coordinates at reservoir.

Elevation difference between reservoir and each well location has measured using survey equipment. Both primary data such as depths to water levels of each well from the ground level as well as secondary data such as water level, storage and discharge of the Victoria reservoir are used in this research. Depths to water levels at each well from the wall level of the well were measured using a probe, and a measuring tape is used to measure the height of the wall of the well Fig. 2.

The reservoir water levels are obtained from Mahaweli Authority of Sri Lanka weekly from the second week of May to mid-October 2016, while on the same day, water levels of wells are measured. To identify the relationship between fluctuation of Victoria reservoir water level and depths of wells, a Pearson correlation analysis was conducted (Eq. 2) for each well separately.

$$r = \frac{\sum (x - \bar{x})(y - \bar{y})}{\sqrt{\sum (x - \bar{x})^2 (y - \bar{y})^2}} \quad (2)$$

$r$—Correlation coefficient.

$x$—Water level of the Victoria reservoir.

$y$—Depth to well water level.

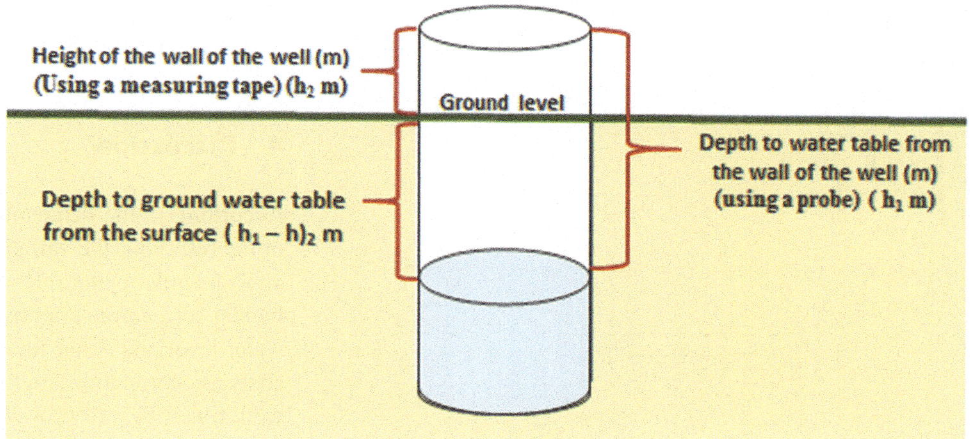


**Fig. 2** Calculation of the depth to water level at wells from the ground surface

## 3 Results

### 3.1 Measurements of Closest Distance to Each Well From the Reservoir and the Elevation Difference Between the Reservoir and Each Well Location

The distance to each well from the reservoir and the elevation difference between reservoir and each well location has calculated using the methods mentioned in the above. It was noted that the minimum to maximum distance from the reservoir to wells is 15–194 m, respectively, while the minimum and maximum elevation difference between reservoir and the well location is 8–33 m respectively (Table 1).

### 3.2 Fluctuation of the Water Level at the Victoria Reservoir

Based on the Victoria reservoir water levels, the fluctuation of the water level of each week of the reservoir for five months has plotted as shown in the Fig. 3.

### 3.3 Measuring of Depths to Well Water Level from the Ground Surface and Fluctuation of Well Water Levels

Well water levels from the ground surface was monitored for the study period, i.e. from second week of May to mid-October 2016 (22 weeks), as shown in Table 2. Variation of well water levels below the ground surface is plotted

**Table 1** Distance to reservoir and elevation difference between well and the reservoir

| Well no | East coordinates | North coordinates | Distance from the reservoir (m) | Elevation difference between reservoir and the well (m) |
|---|---|---|---|---|
| W1 | 199,386 | 231,643 | 75 | 26 |
| W2 | 199,316 | 231,718 | 74 | 18 |
| W3 | 199,420 | 231,859 | 15 | 33 |
| W4 | 199,256 | 231,854 | 86 | 17 |
| W5 | 199,089 | 232,045 | 102 | 19 |
| W6 | 198,830 | 231,876 | 175 | 17 |
| W7 | 198,577 | 232,171 | 40 | 27 |
| W8 | 198,666 | 231,894 | 190 | 16 |
| W9 | 198,871 | 231,862 | 194 | 10 |
| W10 | 198,511 | 232,190 | 64 | 8 |

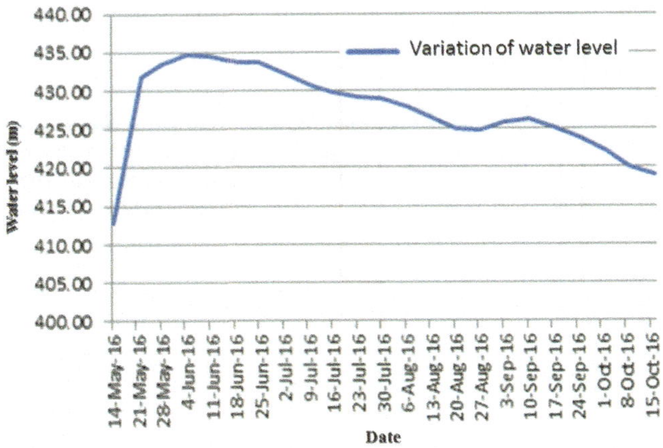


**Fig. 3** Fluctuation of the water level at the Victoria reservoir

in each week as shown in Fig. 4. to compare the behaviour of the water level fluctuation curve of the reservoir.

## 4 Discussion

According to the behaviour of water level fluctuation curve of the reservoir, the variation of water depths of wells shows almost similar pattern. However, to confirm this behaviour, a Person correlation analysis was carried out between reservoir water level and water levels of each well. According to this analysis, correlation coefficients of well 1, well 2, well 3, well 4 well 5, well 6, well 7, well 8, well 9 and well 10 are calculated as 0.87, 0.77, 0.82, 0.88, 0.82, 0.73, 0.85, 0.86,

**Table 2** Depths to water level below the surface (in metres) of ten wells for five months

| | Well 1 depth | Well 2 depth | Well 3 depth | Well 4 depth | Well 5 depth | Well 6 depth | Well 7 depth | Well 8 depth | Well 9 depth | Well 10 depth |
|---|---|---|---|---|---|---|---|---|---|---|
| Week 1 | −4.3 | −4.0 | 0.2 | −4.6 | −3.5 | −2.6 | −0.3 | −4.4 | −3.1 | −3.6 |
| Week 2 | −4.6 | −4.6 | 0.8 | −3.5 | −3.2 | −2.3 | 0.8 | −5.5 | −2.7 | −4.7 |
| Week 3 | 0.8 | 0.4 | 0.8 | −3.2 | 0.8 | −1.7 | 0.8 | −3.2 | −2.0 | 0.8 |
| Week 4 | 0.8 | −4.9 | 0.8 | −2.6 | 0.8 | −1.9 | 0.8 | −3.6 | −1.3 | 0.7 |
| Week 5 | 0.8 | 0.6 | 0.8 | −2.1 | 0.8 | −2.0 | 0.8 | −4.0 | −2.4 | 0.3 |
| Week 6 | 0.8 | 0.7 | 0.8 | −2.2 | 0.8 | −2.2 | 0.8 | −4.2 | −2.6 | 0.9 |
| Week 7 | 0.8 | 0.6 | 0.8 | −3.2 | 0.8 | −2.3 | 0.8 | −4.5 | −2.8 | 0.6 |
| Week 8 | 0.8 | 0.5 | 0.8 | −4.4 | −0.6 | −3.5 | 0.8 | −2.7 | −3.0 | 1.2 |
| Week 9 | −1.2 | −2.1 | 0.8 | −2.7 | −0.8 | −2.3 | 0.8 | −4.9 | −3.2 | −2.8 |
| Week 10 | −5.4 | −6.1 | 0.1 | −2.0 | −1.1 | −0.9 | 0.8 | −5.0 | −3.4 | −5.0 |
| Week 11 | −6.3 | −7.0 | 0.8 | −3.2 | −1.5 | −3.2 | −0.1 | −5.3 | −3.7 | −5.5 |
| Week 12 | −7.0 | −7.2 | 0.8 | −3.5 | −1.7 | −3.3 | −0.1 | −5.3 | −2.8 | −5.5 |
| Week 13 | −5.3 | −5.4 | 0.8 | −3.7 | −2.0 | −3.5 | −0.2 | −5.5 | −4.1 | −5.8 |
| Week 14 | −3.6 | −3.8 | 0.3 | −4.1 | −2.3 | −3.7 | −0.3 | −5.5 | −3.2 | −4.1 |
| Week 15 | −4.9 | −5.0 | 0.2 | −4.2 | −2.6 | −3.5 | 0.6 | −5.8 | −4.4 | −3.3 |
| Week 16 | −3.2 | −8.3 | 0.1 | −4.5 | −2.9 | −4.0 | −0.7 | −5.9 | −4.6 | −10.8 |
| Week 17 | −3.5 | −8.6 | 0.0 | −4.8 | −3.2 | −4.1 | −0.9 | −5.9 | −4.8 | −10.8 |
| Week 18 | −3.9 | −8.8 | 0.0 | −5.0 | −3.4 | −4.3 | −1.0 | −5.8 | −5.0 | −10.1 |
| Week 19 | −4.1 | −9.1 | −0.1 | −5.2 | −3.6 | −4.5 | −1.2 | −5.8 | −5.1 | −10.1 |
| Week 20 | −4.3 | −9.4 | −0.2 | −5.6 | −3.9 | −4.7 | −1.4 | −5.8 | −5.4 | −10.1 |
| Week 21 | −4.5 | −9.6 | −0.3 | −5.6 | −4.1 | −4.9 | −1.5 | −5.8 | −5.5 | −10.1 |
| Week 22 | −4.7 | −9.7 | −0.4 | −5.8 | −4.3 | −5.2 | −1.6 | −5.8 | −5.8 | −10.1 |

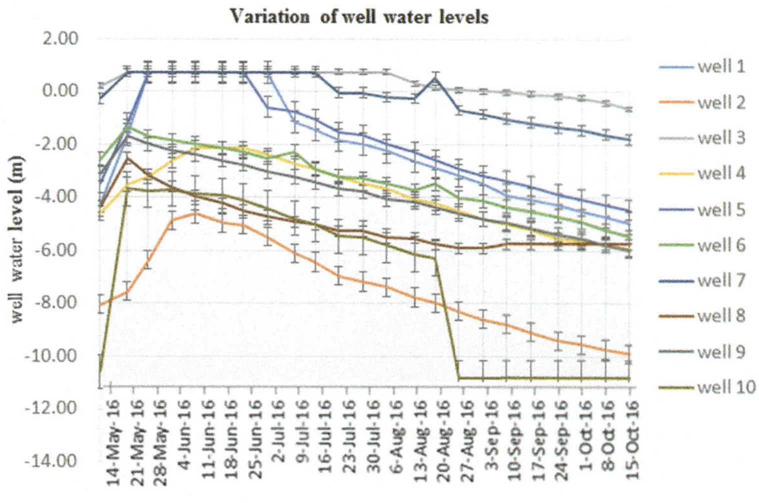


**Fig. 4** Variation of well water levels during the study period

0.86 and 0.87, respectively. The average correlation coefficient is 0.83. Therefore, there is a close relationship between water level of the reservoir and the depth to the water level of well. Further, a Person correlation analysis was carried out between distance to the well from the reservoir and the average change in water depth of each well. This analysis showed that the correlation coefficient is 0.48, and therefore, there is no clear relationship between distance away from the reservoir and the change in water levels of wells in different periods.

## 5 Concluding Remarks

According to the results and discussion mentioned in the above, it is very clear that fluctuation of Victoria reservoir water level was influenced for the variation of water levels of wells which are located closer to the reservoir in Wewegama village. However, distance away from the reservoir is not a significant factor that affects to change in water levels of wells. As Wewegama village is a small area, *total extent less than 3km$^2$, some external factors such as underground geological conditions, soil types can be negligible for this study compared to the main factor, i.e. fluctuation of water level of the Victoria reservoir. Therefore, it can be concluded that there is a direct relationship between the fluctuation of the water level in the Victoria reservoir and depths to water level of wells in study area which are very closer to the reservoir.

**Acknowledgements** The Mahaweli Authority of Sri Lanka and the Irrigation Department of Sri Lanka are acknowledged for providing secondary data used in this research.

## References

Gribovszkia Z, Szilágyi J, Kalicz P (2010) Diurnal fluctuations in shallow groundwater levels and streamflow rates and their interpretation: a review. J Hydrol 385:371–383

McMillan HK, Srinivasan MS (2014) Characteristics and controls of variability in soil moisture and groundwater in a headwater catchment. Hydrol Earth Syst Sci 19(4):1767–1786

Taylor CJ, Alley WM (2001) Ground-water-level monitoring and the importance of long-term water-level data. Circular 1217, U.S Geological Survey, USA. https://pubs.usgs.gov/circ/circ1217/pdf/circular1217.pdf

Tharani G, Nanthakumaran A, Karunainathan T (2015) Hydrological interaction between surface water and groundwater in Vavuniya District of Sri Lanka. Int Res J Environ Sci 4(9):39–44

# Modeling Short-Term Groundwater-Level Fluctuations Using Multivariate Adaptive Regression Spline

Ozgur Kisi and Hadi Sanikhani

**Abstract**

The study investigates accuracy of two machine learning methods, neuro-fuzzy system with grid partition (ANFIS-GP) and multivariate adaptive regression spline (MARS) in prediction of 1-day- to 6-day-ahead groundwater levels (GWLs) using data from two wells, USA. The outcomes indicate that the ANFIS-GP provides inferior results compared to regression-based simple MARS method. The MARS method which is much simpler than the ANFIS-GP is recommended for short-term GWL prediction.

**Keywords**

Groundwater prediction • Neuro-fuzzy with grid partition • Multivariate adaptive regression spline

## 1 Introduction

Groundwater as a main component of water has a vital importance for domestic, agricultural, and industrial purposes especially in arid and semi-arid regions (Jolly et al. 2008). Generally, fluctuation of groundwater level can be estimated using physically based numerical approaches (Yoon et al. 2011). In recent decades, data-driven techniques have been successfully applied in various fields of water science and engineering (Valizadeh et al. 2017). Emamgholizadeh et al. (2014) reported successful application of artificial neural networks (ANNs) and adaptive neuro-fuzzy inference system (ANFIS) for predicting groundwater levels of Bastam plain in Iran. Rezaie-balf et al. (2017) used wavelet-coupled multivariate adaptive regression splines (MARS) and M5 model trees to simulate the groundwater fluctuations of three wells.

The main aim of this study is to investigate the accuracy of MARS method for daily groundwater level forecasting and to compare it with the ANFIS with grid partitioning (ANFIS-GP). The MARS models were developed and compared with ANFIS-GP models for 1-, 2-, 3-, and 6-day-ahead forecasting of groundwater fluctuations using previous groundwater levels as inputs in two wells located in Montgomery and Tuscaloosa counties, Alabama state, USA, that operated by the USGS.

## 2 Materials and Methods

### 2.1 Description of ANFIS-GP

Jang (1993) introduced the learning procedure for the fuzzy inference system (FIS) that uses the neural networks learning algorithm for constructing a set of fuzzy if–then rules with appropriate membership functions (MFs) from specified input–output pairs. ANFIS-GP is obtained using combination of ANFIS and grid partition. Grid partition divides the input space into rectangular subspaces using a number of local fuzzy regions by axis-paralleled partition based on a predefined number of MFs and their types in each dimension. For calculating fuzzy sets and parameters, the least square method according to the partition and MF type is used. During the construction of the fuzzy rules, consequent parameters in the linear output MF are set to zero. Therefore, by using ANFIS, parameters are identified and refined. GP and its combination with ANFIS are described with details in Jang (1993). By increasing the number of input variables, the number of fuzzy rules is exponentially increased. For instance, if there are n input variables and m MFs for each

O. Kisi (✉)
Civil Engineering Department, Ilia State University, Tbilisi, Georgia
e-mail: ozgur.kisi@iliauni.edu.ge

H. Sanikhani
Water Sciences and Engineering Department, Faculty of Agriculture, University of Kurdistan, Sanandaj, Iran


M. Abrunhosa et al. (eds.), *Advances in Geoethics and Groundwater Management: Theory and Practice for a Sustainable Development*, Advances in Science, Technology & Innovation,
https://doi.org/10.1007/978-3-030-59320-9_41

input variable for the problem, the total number of fuzzy rules equals $m^n$ (Wei et al. 2007).

## 2.2 Description of MARS

Friedman (1991) presented the MARS model as a new method for nonlinear regression modeling of high-dimensional data. Piecewise basis functions are used to define relationships between predictors and a response variable. An inflection point along the range of predictors is defined by using a value of a variable or knot that defines basis functions. For a given variable $X_i$ and a given value c, two linear splines function can be defined as $(X_i\text{-}c)+$ and $(c\text{-}X_i)+$, where ' + ' refers the positive part. To satisfy the basis function's continuity, two adjacent splines will touch each other at a knot. Then by using a subset of all possible linear spline's functions, MARS model is constructed. Least squares regression is used to calculate the coefficients of the newly generated model. Cost of the building the model is calculated by the number of the least squares regressions that are performed because computing of least squares coefficients is the most computationally intensive part. The MARS algorithm includes two steps, namely forward and backward stepwise steps. In the first step, a set of suitable input variables is chosen. Pair of basis functions is defined by placing knots at any positions in the range of each predictor variable. At each step, knots and its corresponding pair of basis functions are chosen by the model to obtain the maximum decrease in the residual sum of squares. In the second step, backward stepwise was used to increase the precision of primary MARS model. During this step, inessential basis functions using the generalized cross-validation (GCV) measure are removed. Finally, by evaluation of GCV, the MARS model with the best-fitted basis function and efficient parameters was selected.

## 3 Results

The capability of two machine learning methods, ANFIS-GP and MARS, was evaluated in prediction of 1-day- to 6-day-ahead groundwater level (GWL) fluctuations based on root mean square error (RMSE), mean absolute error (MAE), and determination coefficient ($R^2$). The equations of the evaluation statistics are as follows:

$$RMSE = \sqrt{\frac{\sum_{i=1}^{N} (GWL_{i,o} - GWL_{inm})^2}{N}} \quad (1)$$

$$MAE = \frac{\sum_{i=1}^{N} \left| GWL_{i,o} - GWL_{i,m} \right|}{N} \quad (2)$$

where N is data quantity, $GWL_{i,o}$ is observed mean groundwater level, $GWL_{i,m}$ is modeled mean groundwater level. The ANFIS-GP method was implemented using fuzzy toolbox of MATLAB software. For the MARS method, free toolbox ARESLab was utilized (Jekabson 2011).

GWL data were obtained from two wells in USA (one is USGS 322,047,086,214,301 Well K 107 MTG-3 Montgomery County Al, latitude 32°20′47", longitude 86°21′43", Montgomery County, Alabama, hydrologic unit 03,150,201, well depth: 270 feet, hole depth: 271 feet, land surface altitude: 167.20 feet above NGVD29, and second is USGS 333,204,087,324,601 Well (TW5 DRY BRANCH WELL), latitude 33°32′04", longitude 87°32′46", Tuscaloosa County, Alabama, hydrologic unit 03,160,112, well depth: 60 feet, hole depth: 60.00 feet, land surface altitude: 519.00 feet above NGVD29). Data were first divided into three stages, training (%50 of the whole data, 3936 values from 5/14/1998 to 2/19/2009), validation (%25 of the whole data, 1968 values from 2/20/2009 to 7/11/2014), and test (%25 of the whole data, 1968 values from 7/12/2014 to 12/6/2019). Table 1 gives the brief statistics of the used groundwater data for each well. It is obvious from the table that the second well data have more skewed distribution than the first well. In applying ANFIS-GP method, various types of membership functions (MFs) were employed and the best one which provided the lowest RMSE in validation stage was selected. In Table 2, the MF types are seen in the 2nd column. In this column, first numbers indicate the number of MFs used for each input and the second stands for the MFs type (gaussmf: Gaussian MF, psigmf: p sigmoid MF and dsigmf: d sigmoid MF). Inputs of the models consist of previous values of GWL. The optimal inputs were identified with respect to correlation analysis, and 5 lags (from GWLt-1 to GWLt-5) were found to be optimal.

Table 2 compares the ANFIS-GP and MARS models in prediction of GWL of well 1 for various time lags. It is apparent from the table that the both models have almost same accuracy for the 1-day- to 3-day-ahead prediction cases, while the MARS model (RMSE: 0.644 ft, MAE: 1.931 ft and $R^2$: 0.946 for validation and RMSE: 0.379 ft, MAE: 1.252 ft and $R^2$: 0.957 for testing) performs superior to the ANFIS-GP model (RMSE: 0.556 ft, MAE: 1.759 ft and $R^2$: 0.959 for validation and RMSE: 0.369 ft, MAE: 1.218 ft and $R^2$: 0.960 for testing) in prediction 6-day-ahead GWL in both validation and test stages. The neuro-fuzzy and MARS models are compared in Table 3 for prediction of GWL of well 2. In this well, also the difference between the ANFIS-GP and MARS is very little for the 1-day- to 3-day-ahead GLW prediction. Here, also the MARS (RMSE: 0.840 ft, MAE: 1.460 ft and $R^2$: 0.772 for validation and RMSE: 1.042 ft, MAE: 1.644 ft and $R^2$: 0.773 for testing) has superior accuracy compared to neuro-fuzzy

**Table 1** Statistical parameters of the applied data

| | Well 1 (feet) | | | Well 2 (feet) | | |
|---|---|---|---|---|---|---|
| | Train | Valid | Test | Train | Valid | Test |
| Mean | 25.8 | 25.8 | 23.7 | 41.2 | 42.0 | 41.6 |
| Minimum | 17.7 | 15.5 | 7.12 | 33.5 | 35.6 | 32.8 |
| Maximum | 31.9 | 30.9 | 19.9 | 43.6 | 44.3 | 43.9 |
| Skewness | 0.03 | −1.30 | −0.20 | −1.23 | −1.37 | −1.50 |
| Std. dev | 2.56 | 2.69 | 1.82 | 2.04 | 1.72 | 2.19 |

**Table 2** Results of neuro-fuzzy methods in prediction groundwater fluctuation—Well 1

| Method | Structure | Validation | | | Test | | |
|---|---|---|---|---|---|---|---|
| | | RMSE | MARE | $R^2$ | RMSE | MARE | $R^2$ |
| *1 day ahead* | | | | | | | |
| ANFIS-GP | 2, gaussmf | 0.113 | 0.351 | 0.998 | 0.091 | 0.304 | 0.997 |
| MARS | | 0.113 | 0.350 | 0.998 | 0.091 | 0.306 | 0.997 |
| *2 day ahead* | | | | | | | |
| ANFIS-GP | 2, psigmf | 0.242 | 0.756 | 0.992 | 0.184 | 0.617 | 0.990 |
| MARS | | 0.239 | 0.749 | 0.992 | 0.184 | 0.618 | 0.990 |
| *3 day ahead* | | | | | | | |
| ANFIS-GP | 2, dsigmf | 0.348 | 1.090 | 0.983 | 0.254 | 0.854 | 0.981 |
| MARS | | 0.343 | 1.071 | 0.984 | 0.254 | 0.852 | 0.981 |
| *6 day ahead* | | | | | | | |
| ANFIS-GP | 2, gaussmf | 0.644 | 1.931 | 0.946 | 0.379 | 1.252 | 0.957 |
| MARS | | 0.556 | 1.759 | 0.959 | 0.369 | 1.218 | 0.960 |

model (RMSE: 0.832 ft, MAE: 1.293 ft and $R^2$: 0.775 for validation and RMSE: 1.026 ft, MAE: 1.636 ft and $R^2$: 0.780 for testing) in both stages.

Figures 1 and 2 illustrate the time variation graphs and scatter diagrams of the ANFIS-GP and MARS models in prediction of 6-day-ahead GWL of both wells. It is obviously seen that the both neuro-fuzzy and MARS models can catch the general trends of observed GWL values, while the first model cannot well predict in some cases (see the red line in the time variation graphs). Scatter diagrams reveal that the MARS model has higher $R^2$ which means having less scattered predictions.

**Table 3** Results of neuro-fuzzy methods in prediction groundwater fluctuation—Well 2

| Method | Structure | Validation | | | Test | | |
|---|---|---|---|---|---|---|---|
| | | RMSE | MARE | $R^2$ | RMSE | MARE | $R^2$ |
| *1 day ahead* | | | | | | | |
| ANFIS-GP | 2, gaussmf | 0.149 | 0.211 | 0.993 | 0.166 | 0.237 | 0.994 |
| MARS | | 0.147 | 0.214 | 0.993 | 0.169 | 0.252 | 0.994 |
| *2 day ahead* | | | | | | | |
| ANFIS-GP | 2, gaussmf | 0.329 | 0.477 | 0.964 | 0.383 | 0.557 | 0.970 |
| MARS | | 0.326 | 0.468 | 0.964 | 0.381 | 0.566 | 0.970 |
| *3 day ahead* | | | | | | | |
| ANFIS-GP | 2, gaussmf | 0.495 | 0.732 | 0.921 | 0.591 | 0.871 | 0.928 |
| MARS | | 0.481 | 0.722 | 0.923 | 0.589 | 0.875 | 0.928 |
| *6 day ahead* | | | | | | | |
| ANFIS-GP | 2, trimf | 0.840 | 1.430 | 0.772 | 1.042 | 1.644 | 0.773 |
| MARS | | 0.832 | 1.293 | 0.775 | 1.026 | 1.636 | 0.780 |

**Fig. 1** Time variation graph and scatterplots of the ANFIS-GP and MARS models in predicting 6-day-ahead groundwater fluctuation–Well 1

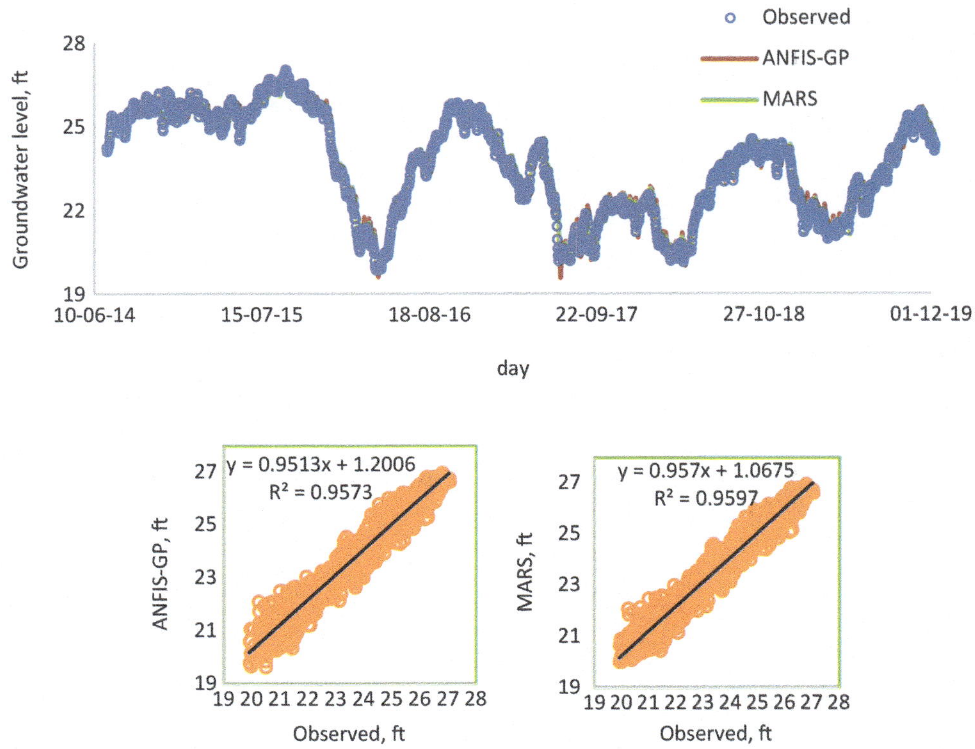


**Fig. 2** Time variation graph and scatterplots of the ANFIS-GP and MARS models in predicting 6-day-ahead groundwater fluctuation–Well 2

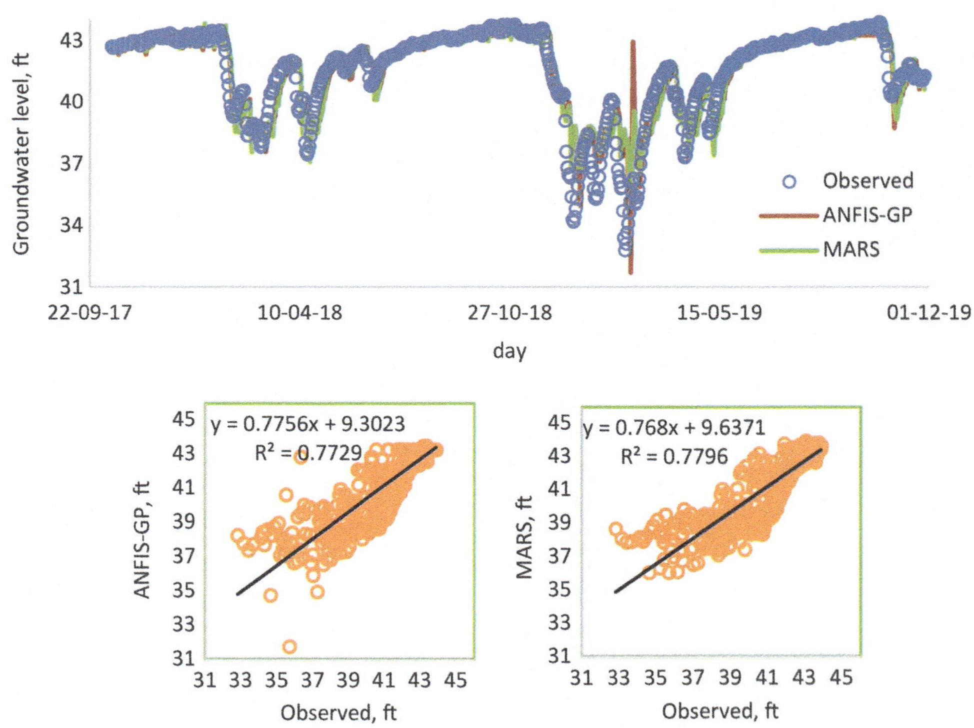

## 4 Discussion

It is found from the tables and figures that the MARS models performed better than the ANFIS-GP model in prediction of 6-day-ahead GWL. It should be noted that the MARS model is much simpler structure and faster than the neuro-fuzzy model. The other advantage of the MARS is that it provides explicit equations and can be easily used in practical applications. Comparison of results indicated that the both models have less accuracy in prediction of second well compared to first one. The main reason of this might be the fact that the second well has higher skewed distribution compared to other well data of two wells (see skewness values of data in Table 1).

## 5 Concluding Remarks

The study compared the ability of two different machine learning methods in prediction of short-term GWL fluctuation. Data from two wells, USA, were utilized for model inputs. The outcomes of the models showed that there is a slight difference between the ANFIS-GP and MARS models in prediction 1-, 2- and 3- day ahead GWL, while the latter model performs superior to the first model in prediction 6-day-ahead GWL. The study recommends the MARS method in modeling daily GWL prediction in the study area. This study may be useful for the water managers and decision-makers to optimally plan available groundwater resources. New methods for better prediction are required to manage available resources and to meet the resources needs of future generations.

## References

Emamgholizadeh S, Moslemi K, Karami G (2014) Prediction the groundwater level of Bastam plain (Iran) by artificial neural network (ANN) and adaptive neuro-fuzzy inference system (ANFIS). Water Resour Manage 28(15):5433–5446

Friedman JH (1991) Multivariate adaptive regression splines. the Annals of Statistics 19(1):1–67

Jang JSR (1993) ANFIS: adaptive-network-based fuzzy inference system. IEEE Transactions on Systems, Man, and Cybernetics 23(3):665–685

Jekabson G (2011) ARESLab: Adaptive regression splines toolbox for Matlab/Octave; 2011. https://www.cs.rtu.lv/jekabsons

Jolly ID, McEwan KL, Holland KL (2008) A review of groundwater–surface water interac-tions in arid/semi-arid wetlands and the consequences of salinity for wetland ecology. Ecohydrol-Ogy 1(1): 43–58

Rezaie-balf M, Naganna SR, Ghaemi A, Deka PC (2017) Wavelet coupled MARS and M5 Model Tree approaches for groundwater level forecasting. J Hydrol 553:356–373

Valizadeh N, Mirzaei M, Allawi MF, Afan HA, Mohd NS, Hussain A, El-Shafie A (2017) Artificial intelligence and geo-statistical models for stream-flow forecasting in ungauged stations: state of the art. Nat Hazards 86(3):1377–1392

Wei M, Bai B, Sung AH, Liu Q, Wang J, Cather ME (2007) Predicting injection profiles using ANFIS. Inf Sci 177:4445–4461

Yoon H, Jun SC, Hyun Y, Bae GO, Lee KK (2011) A comparative study of artificial neural networks and support vector machines for predicting groundwater levels in a coastal aqui-fer. J Hydrol 396(1–2):128–138

# For a Better Understanding of Recharge and Salinization Mechanism of a Cenomanian–Turonian Aquifer

Otman El Mountassir, Mohammed Bahir, Driss Ouazar, and Paula M. Carreira

**Abstract**

The Cenomanian–Turonian aquifer of the Ouazzi basin is the source of drinking water for the local population, as well as irrigation water for agricultural needs. The aim of the present work was to understand the origin, geochemical evolution of deep groundwater by using a group of years of piezometric, hydrochemistry and isotopic data. The deterioration in groundwater quality of the study area is explained by the decrease in the precipitation due to climate change and overexploitation. The hydrochemical study shows that the chemical composition of groundwater in the study area consists of Cl-Ca-Mg, $HCO_3$-CaMg, Cl-Na, and $SO_4$Ca chemical facies for the years 1995, 2007, 2012, and 2019. Results show that electrical conductivity increased from 1995 to 2019, and this could be explained by a decrease in annual precipitation, in relation to climate change and the characteristics of water–rock interaction. Geochemical and environmental isotopic data show that the water–rock interaction and cation-exchange process are the major geochemical mechanisms controlling hydrochemical evolution of groundwater in the Cenomanian–Turonian aquifer. The diagram of $\delta^2H = f\ (\delta^{18}O)$ shows that the isotopic contents are close or above to the Global Meteoric Water Line, which suggests that the aquifer is recharged by precipitation of Atlantic origin.

**Keywords**

Groundwater quality • Essaouira • Environmental isotopes • Climate change • Hydrochemistry

## 1 Introduction

The shortage of water became a crucial problem lived by all societies, and particularly those of the developing countries. Indeed, the growth of the populations and the development of the agglomerations, the industrial units, and the cultivated earths had for corollary a deterioration of the quality of the groundwater and a very meaningful decrease of the reserves that sometimes represent the only resources of water for the food of the populations (Carreira et al. 2018; Bahir et al. 2019).

The Meskala- Ouazzi Basin is used as an example in this study. It is in the east and north of the Essaouira city, with an area of 1196 km$^2$ (Fig. 1). It is limited in the north by Hadid anticline, by Meskala region in the south, by Mramer wadi in the east and Atlantic Ocean in the west. This basin is controlled by a semi-arid climate, with an average annual rainfall of 300 mm and temperatures of 20 °C (El Mountassir et al. 2020). The aim of this study is to: (A) better understanding of groundwater recharge mechanism using environmental isotopes ($^{18}O$ and $^2H$) in conjunction with geochemical data (1995, 2007, 2012, and 2019); (B) assessing salinization sources of deep groundwater (Cenomanian–Turonian aquifer).

## 2 Methodology

Four water sampling campaigns (151 boreholes, wells, and springs) were performed in 1995 (40 samples), 2007 (25 samples), 2012 (28 samples), and 2019 (58 samples), it was taken in Meskala-Ouazzi region of Essaouira basin (Fig. 1).

O. El Mountassir (✉) · M. Bahir
High Energy and Astrophysics Laboratory, Faculty of Sciences Semlalia, Cadi Ayyad University, Marrkesh, Morocco
e-mail: otman.elmountassir@ced.uca.ma

M. Bahir · D. Ouazar
International Water Research Institute, Mohammed VI Polytechnic University, Ben Guerir, Morocco

P. M. Carreira
Centro de Ciências e Tecnologias Nucleares, CTN/IST, University of Lisbon, Lisbon, Portugal



M. Abrunhosa et al. (eds.), *Advances in Geoethics and Groundwater Management: Theory and Practice for a Sustainable Development*, Advances in Science, Technology & Innovation,
https://doi.org/10.1007/978-3-030-59320-9_42

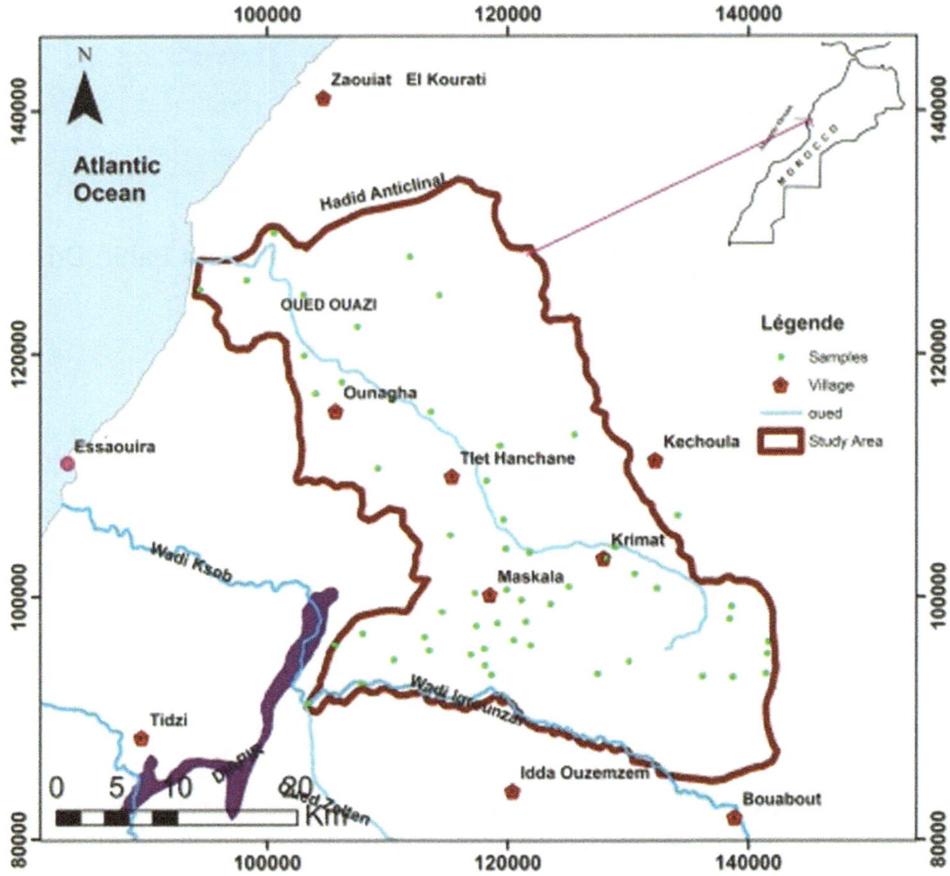


Fig. 1 Study area

Physico-chemical parameters (temperature, electrical conductivity, pH) were measured in situ. The alkalinity and analyses of concerned major chemical elements were carried out at the Laboratory of Geosciences and Environment-ENS at the Ecole Normale Supérieure of Marrakech (Morocco). The analyses of stable isotopes ($^{2}H$ and $^{18}O$) were carried out at Centro de Ciências e Tecnologias Nucleares (CTN/IST, Universidade de Lisboa) laboratory using a mass spectrometer. The methodology adopted thus makes it possible to explore the potentials of a multidisciplinary approach, centered on the use of stable isotopes, to determine the recharge areas of the main sources studied. The interpolation technique via a Geographical Information System (GIS) was used to compile spatiotemporal distribution maps of electrical conductivity.

## 3 Results and Discussion

### 3.1 Piezometry

Four piezometers (A1–A4) were used to monitor the piezometric level of the Meskala-Ouazzi aquifer (Fig. 2). The peaks of intense precipitation are accompanied by a rise in the piezometric level, reflecting the dependence of groundwater levels on precipitation. Despite the slight rise in the piezometric level with intense rainfall, the four piezometers studied show a general downward trend (Bahir et al. 2020). Like other basins in coastal areas, the Essaouira basin has not been spared the effect of climate change, which is added to the effect of the overexploitation of groundwater. This effect is manifested by (a) an increase in population of 9% leading to an increase in irrigated land (1300 ha); (b) upward trend in temperatures with 1.5° C warming and a general downward trend in precipitation of 12%; (c) a deterioration in the quality of groundwater with an increase in salinity. This degradation is due to marine intrusion and the decrease in the recharge rate of aquifers caused by the decrease in precipitation as a result of climate change (Ouhamdouch et al. 2019).

### 3.2 Hydrochemistry

The results of chemical analysis are plotted on the Piper diagram (Fig. 3a). This diagram shows that the groundwater of the Meskala-Ouazzi has four chemical facies: Cl–Ca–Mg, $HCO_3$–Ca–Mg, Cl–Na, and $SO_4$–Ca with the dominance of the Cl–Ca–Mg. This transition from one facies to another emphasizes the complexity of the hydrogeochemical

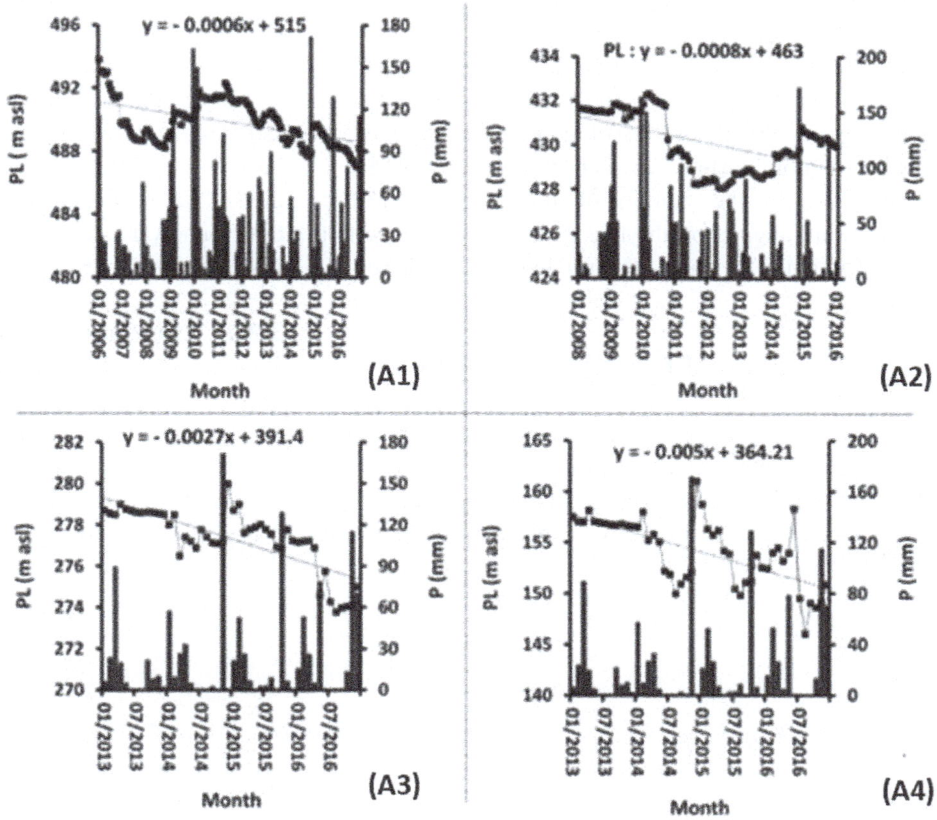


**Fig. 2** Piezometric levels for the period 2006–2016 in the study area

processes that govern the salinity of the waters of this aquifer. During the four years, 1995 to 2019, groundwater of the Turon aquifer keeps the same facies.

The $Ca^{2+} + Mg^{2+}$ versus $SO_4^{2-} + HCO_3^-$ diagram (Fig. 3b) shows that the cation exchange is not the only process controlling the composition of groundwater. For most samples in the Cenomanian–Turonian aquifer, samples located above the 1:1 line indicate the $Ca^{2+} + Mg^{2+}$ are in excess with respect to $SO_4^{2-} + HCO_3^-$ indicating the dissolution of carbonate (calcite, dolomite and aragonite) and evaporate minerals (gypsum and anhydrite). High contents can be related to the dissolution of gypsum and anhydrite as indicated by saturation indexes (SI) varying between -6.93 and -2.11 for all samples. Similarly, the dissolution of carbonate minerals is believed to be responsible for high levels of calcium and bicarbonates in the groundwater.

Figure 4a shows that the majority of the points are aligned on the straight line of slope 1. This reflects the contribution of the dolomite dissolution in the groundwater mineralization, especially for waters sampled in 1995, 2007, and 2012. The $Na^+$ vs $Cl^-$ diagram (Fig. 4b) shows that some samples are aligned on the line 1:1, reflecting the

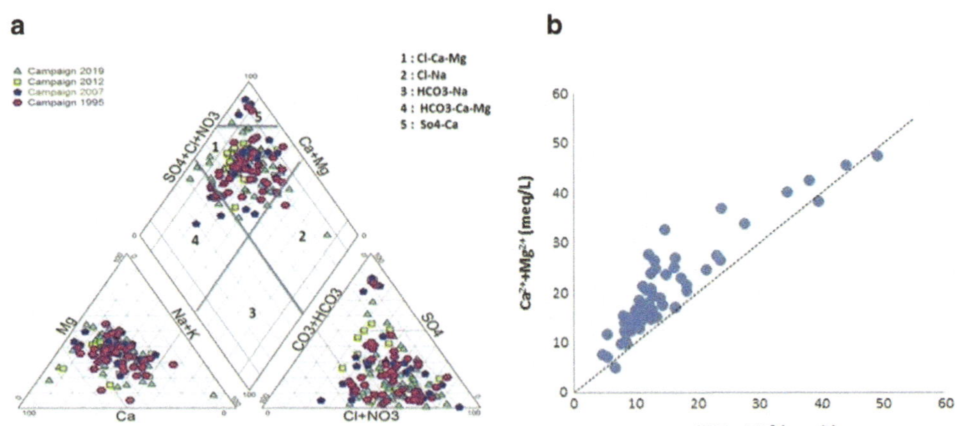


**Fig. 3** Hydrochemistry in the study area. **a** Piper diagram; **b** Correlation between $Ca^{2+} + Mg^{2+}$ (meq/L) and $HCO_3^- + SO_4^{2-}$(meq/L) for 2019

contribution of halite dissolution to the groundwater salinization of the study area.

The correlation between $Ca^{2+}$ and $HCO_3^-$ (Fig. 4c) is very weak for the four campaigns. This reflects that the participation of the calcite dissolution in the groundwater mineralization of the study area is negligible. $Ca^{2+}$ vs $SO_4^{2-}$ diagram (Fig. 4d), the majority of the points for the four campaigns are aligned on the line with slope 1, suggesting the contribution of gypsum and/or anhydrite dissolution.

## 3.3 Spatiotemporal Evolution of Electrical Conductivity

The electrical conductivity (EC) values vary between 616 and 5729 μS/cm with an average of 2453 μS/cm for the samples collected in 2019, between 703 and 5550 μS/cm with an average of 2129 μS/cm for the waters of the 2012 campaign, between 966 and 3855 μS/cm with an average of 2046 μS/cm for the 2007 campaign waters, and between 350 and 3061 μS/cm with an average of 1125 μS/cm for the samples of 1995. However, an increase in EC values has been observed from 1995 to 2019. The spatial distribution of the EC values (Fig. 5a–d) for the four campaigns shows that the groundwater salinity has increased over time in the area where groundwater discharges to the Atlantic Ocean.

### 3.3.1 Environmental Isotopes ($^2H$ and $^{18}O$)

Isotopic contents of sampled boreholes were ranging from −6.01 to −3.28‰ versus V-SMOW for $\delta^{18}O$ and from −34.4 to −20.2‰ versus V-SMOW for $\delta^2H$. The mean values are, respectively, −4.89 and −28.01‰ versus V-SMOW for $\delta^{18}O$ and $\delta^2H$. Figure 6a shows two types of water: a water not evaporated, located above or on the

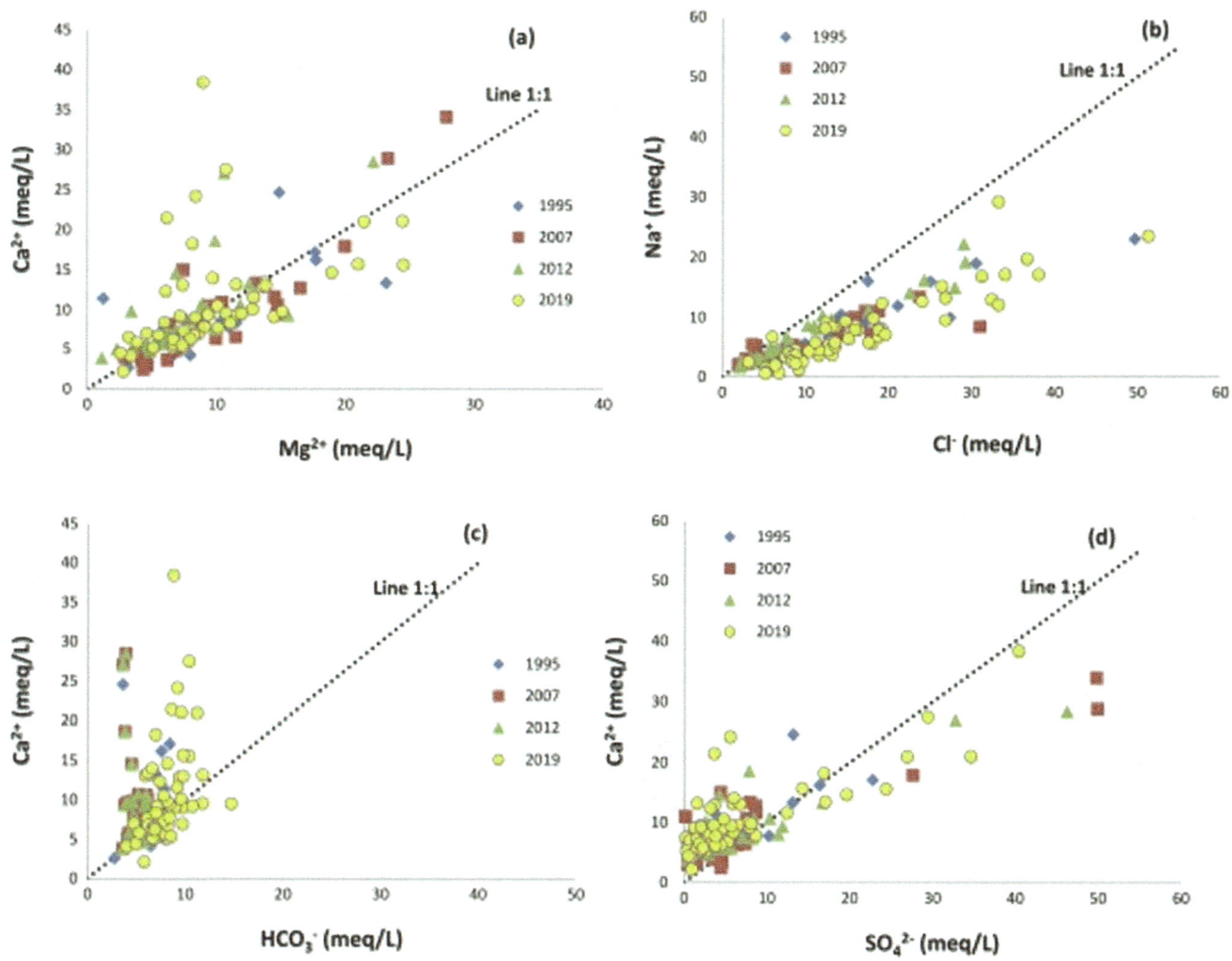


**Fig. 4** Correlation diagram: **a** $Ca^{2+}$ vs $Mg^{2+}$, **b** $Na^+$ vs $Cl^-$, **c** $Ca^{2+}$ vs $HCO_3^-$, **d** $Ca^{2+}$ vs $SO_4^{2-}$

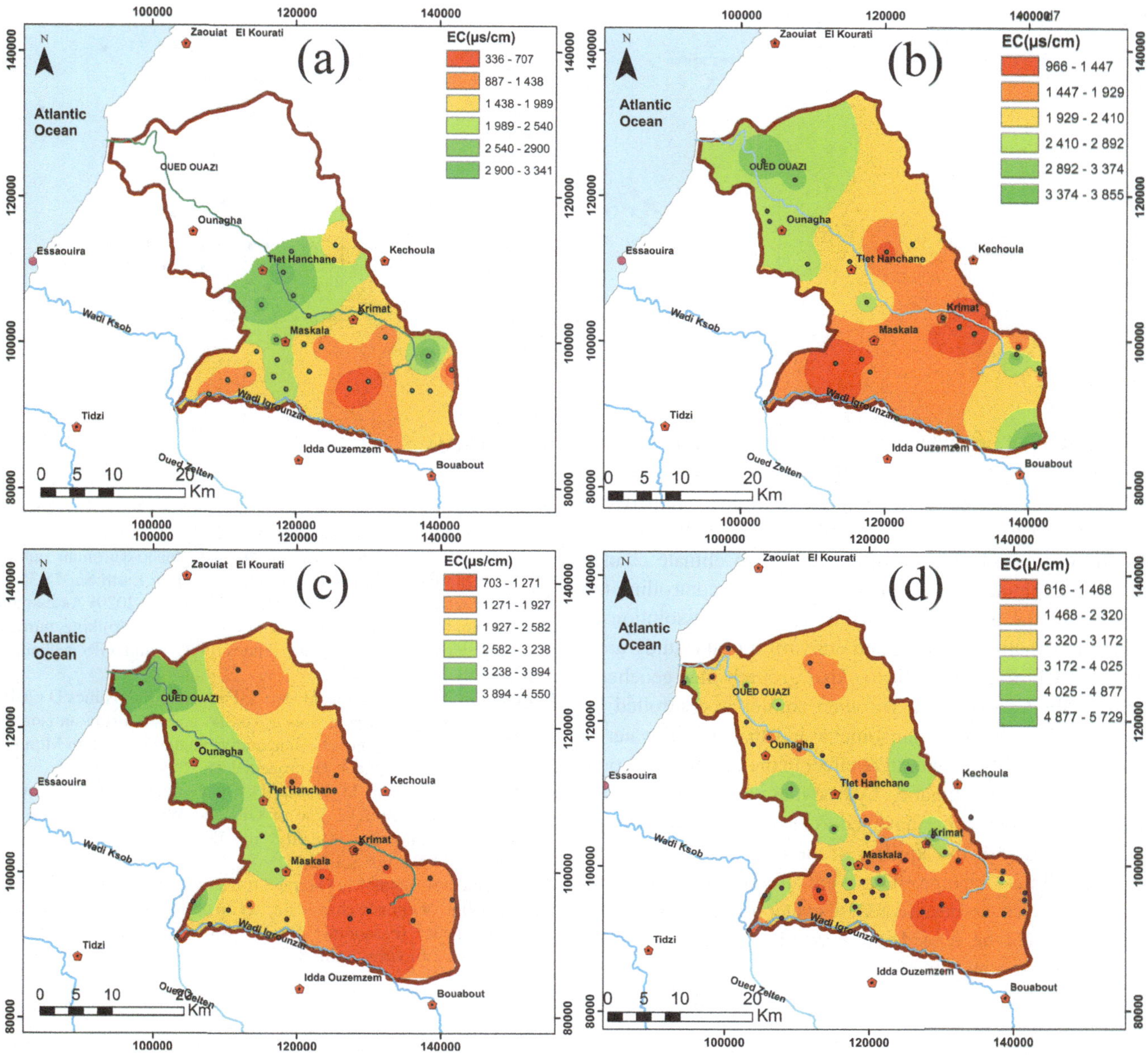


**Fig. 5** Spatial distribution of EC in 1995 **a**, 2007 **b**, 2012 **c**, and 2019 **d**

Global Meteoric Water Line (GMWL). These waters come from rapid meteoric water infiltration without any modification of their content isotopic. B shows slightly evaporated water located below of the Global Meteoric Water Line (GMWL). Evaporation can be produced either before recharging when precipitation passes thorough the atmosphere with low water content; or after recharging by leaching heavy isotopes accumulated during the dry season in the unsaturated zone. The $^{18}O/Cl^{-}$ diagram indicates that groundwater geochemistry in the Cenomanian–Turonian aquifer (deep aquifer) of the Essaouira basin is controlled by water–rock interaction phenomena and that groundwater is not affected by the evaporative effect in the unsaturated zone (Fig. 6b).

## 4 Concluding Remarks

In the region of Meskala-Ouazzi, the Cenomanian–Turonian aquifer is the source of drinking water for the local population. The combination of piezometry, hydrochemical and isotopic tools applied in this study allowed a better understanding of the functioning of the Cenomanian–Turonian aquifer. Concerning hydrochemistry, the chemical

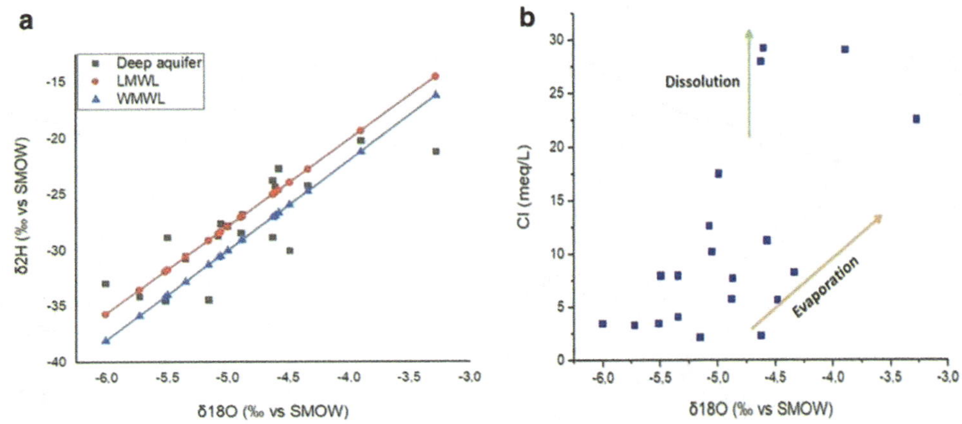


**Fig. 6** Isotopic data in the study area. **a** $^{18}O$ versus $^{2}H$ for 2016 campaign; **b** $^{18}O$ versus $Cl^{-}$ diagram

composition of groundwater at any given sampling location did not change with the observed salinity increases. Groundwaters are mineralized, since EC values are between 1125 μS/cm for 1995 and 2453 μS/cm for 2019, explained by the decrease in the precipitation due to climate change and overexploitation. The main phenomenon controlling the salinity of the groundwater seems to be the dissolution of evaporites (halite, gypsum, and anhydrite) in the study area. The $^{18}O/Cl^{-}$ diagram indicates that groundwater geochemistry in the Cenomanian–Turonian aquifer is controlled by water–rock interaction phenomena and that groundwater is not affected by the evaporative effect in the unsaturated zone. In general, hydroclimatic, hydrodynamic, hydrochemical, and isotopic approaches have led to the diagnosis of the state of vulnerability of the aquifers of the Essaouira basin in the face of climate change. Thus, the use of unconventional resources, such as desalinated seawater for drinking water supply, should be considered a priority in order to avoid triggering a serious situation of water scarcity.

## References

Bahir M, Ouazar D, Goumih A, Ouhamdouch S (2019) Evolution of the chemical and isotopic composition of groundwater under a semi-arid climate; the case of the Cenomano-Turonian aquifer within the Essaouira basin (Morocco). Environ Earth Sci 78:353

Bahir M, Ouhamdouch S, Ouazar D, Chehbouni A (2020) Assessment of groundwater quality from semi-arid area for drinking purpose using statistical, water quality index (WQI) and GIS technique. Carbonat Evapor 35:27

Carreira PM, Bahir M, Ouhamdouch S, Fernandes PG, Nunes D (2018) Tracing salinization processes in coastal aquifers using an isotopic and geochemical approach: comparative studies in western Morocco and southwest Portugal. Hydrogeol J 26:2595–2615

El Mountassir O, Bahir M, Ouazar D, Ouhamdouch S, Chehbouni A, Ouarani M (2020) The Use of GIS and Water Quality Index to Assess Groundwater Quality of Krimat aquifer (Essaouira; Morocco). SN Appl Sci 2:871

Ouhamdouch S, Bahir M, Ouazar D, Carreira PM, Zouari K (2019) Evaluation of climate change impact on groundwater from semi-arid environment (Essaouira Basin, Morocco) using integrated approaches. Environ Earth Sci 78:449

# Salinization as Groundwater Contamination in Estarreja Shallow Aquifer, Aveiro (Portugal)

Ana Carolina Marques, Rosário Carvalho, and Eduardo Ferreira da Silva

**Abstract**

Agriculture, chemical industry and livestock activity are the main activities in Estarreja since 30s of the twentieth century. In this century, particularly in the 50s, the installation of the Estarreja Chemical Complex (ECC) took place with a huge investment in the chemical industry mainly in the production of ammonia for the manufacture of nitro-ammoniacal fertilizers used in agriculture. Despite the relevance of agricultural and industrial sectors in this region, these activities have been generating an increase of Estarreja shallow aquifer vulnerability, particularly surface water and groundwater, due to contamination by raw materials and/or final products, as well as wastewater, nitrogen and natural fertilizers such as animal manure. Besides these kinds of contamination, groundwater can be also liable to salinization as Estarreja is a coastal zone, near to Ria de Aveiro. The main goal of the present study is the identification of salinization phenomenon in Estarreja shallow aquifer through mapping of variables as well as the application of stable isotopes such as hydrogen and oxygen isotopes. Results show that two different types of salinization could happen, one related to natural phenomenon, the saline intrusion, and the other associated with industrial activities.

**Keywords**

Groundwater • Salinization • Mapping • Stable isotopes • Estarreja

Ana C. Marques (✉) · E. F. da Silva
GeoBioTec Research Centre, Department of Geosciences, University of Aveiro, Aveiro, Portugal
e-mail: a.carolina.marques@hotmail.com

R. Carvalho
Department of Geology, Faculty of Sciences and IDL, University of Lisbon, Lisbon, Portugal

## 1 Introduction

Salinization is a type of contamination which can have a natural or anthropogenic origin. It can occur by saline intrusion, a typical phenomenon in coastline zones where aquifers can come into contact with seawater. As freshwater flows into the sea, saltwater, denser water, tends to penetrate the aquifer and to form a wedge under freshwater. When this phenomenon occurs near to the shoreline, it may be accentuated and accelerated with serious consequences if a large volume of fresh groundwater is extracted by pumping. It can cause an advancement of saltwater inside the aquifer and thus the consequent salinization of water from wells or boreholes that catch in it. On the other side, salinization can be triggered by industrial activities through liquid effluent discharges with high electric conductivity values and/or presence of brines, or by agriculture activities through the application of fertilizers enriched, namely in sodium or chlorides.

The mapping of variables allows to distinguish the two different types of contamination components such as salinization component and industrial component. The $\delta^2$H and $\delta^{18}$O isotopes in the water are useful to understand the evaporation/condensation processes as a first approach and to identify the possible salinization origin. The present work aimed to study the salinization as groundwater contamination in Estarreja shallow aquifer through mapping of variables and application of $\delta^2$H and $\delta^{18}$O isotopes, as also to integrate the results in terms of geoethics, protection and management of water resources.

## 2 Materials and Methods

The study area is characterized by Holocene and Pleistocene detrital sedimentary deposits according to Teixeira (1962) (Fig. 1). This area has two important aquifer systems (Marques da Silva 1990): (a) the Aveiro Cretaceous



M. Abrunhosa et al. (eds.), *Advances in Geoethics and Groundwater Management: Theory and Practice for a Sustainable Development*, Advances in Science, Technology & Innovation,
https://doi.org/10.1007/978-3-030-59320-9_43

Multi-Aquifer System (ACMAS) and (b) the Aveiro Quaternary Aquifer System (AQAS) which is essentially made up of three main aquifer units with different hydrogeological and hydraulic characteristics (Ordens 2007). The upper aquifer unit is set in modern deposits (Holocene) comprising dune formations or dune sands and alluvial deposits (Almeida et al. 2000). The second aquifer unit, a semi-confined aquifer, is set in Pleistocene deposits, at Quaternary base aquifer (Ordens 2007). Cretaceous sandstones and clays, as well as schists from Precambrian, characterize the third aquifer unit.

A field campaign was carried out in Estarreja for water sampling on the wet season (4th and 5th May 2018). A total of 35 wells or small diameter boreholes were sampled with a maximum depth of 7 m on Estarreja shallow aquifer (Fig. 1). Field investigations of water included in situ physical and chemical parameters measurements: pH, Temperature (*T* in °C), Electric Conductivity (EC in µS $cm^{-1}$), Oxidation Reduction Potential (ORP in mV) and Dissolved Oxygen (DO in mg $L^{-1}$), with a HI769828 multisensor probe coupled to a HANNA Instruments HI9828 multiparameter metre®. The total alkalinity was analysed by $H_2SO_4$ (0.16 N) titration. Ionic Chromatography (IC) was used to determine anions content (Cl and $SO_4$), by a DIONEX 2000 SPI equipment. Inductively Coupled Plasma Mass Spectrometry (ICP-MS) was used in order to determine cations, with an Agilent Technologies 7700 Series ICP-MS equipment. The $\delta^2H$ and $\delta^{18}O$ measurements (vs. V-SMOW—Vienna-Standard Mean Ocean Water) were performed by Mass Spectrometry (MS), with an Isoprime (Micromass, UK) SIRMS equipment, at Stable Isotopes and Instrumental Analysis Facility (SIIAF) in the Faculty of Sciences, University of Lisbon (Portugal).

The mapping of variables was done by ArcMap 10.6.1 ArcGis® and natural neighbour was chosen as interpolation method. Minimum, P5, P10, P25, P50, P75, P90, P95 and maximum values of each variable established the maps' classes. Warm colours (red) were associated with high values and cold colours (green) to low values defining colour's scale.

The groundwater quality was analysed under the Annex V of Decree-Law (DL) no. 77/2006 of 30 March and DL no. 208/2008 of 28 October, which transposes for the Portuguese Law the European Directive no. 2006/118/EC of 12 December, and the quality standards proposed in the Annex VII of APA (2016).

## 3 Results and Discussion

### 3.1 Components of Salinization

The salinization of the sampled water is reflected by the EC that varies between 62 and 1690 µS/cm. The higher values were observed in the East part of the study area, near to Estarreja Chemical Complex (ECC), and in the West and

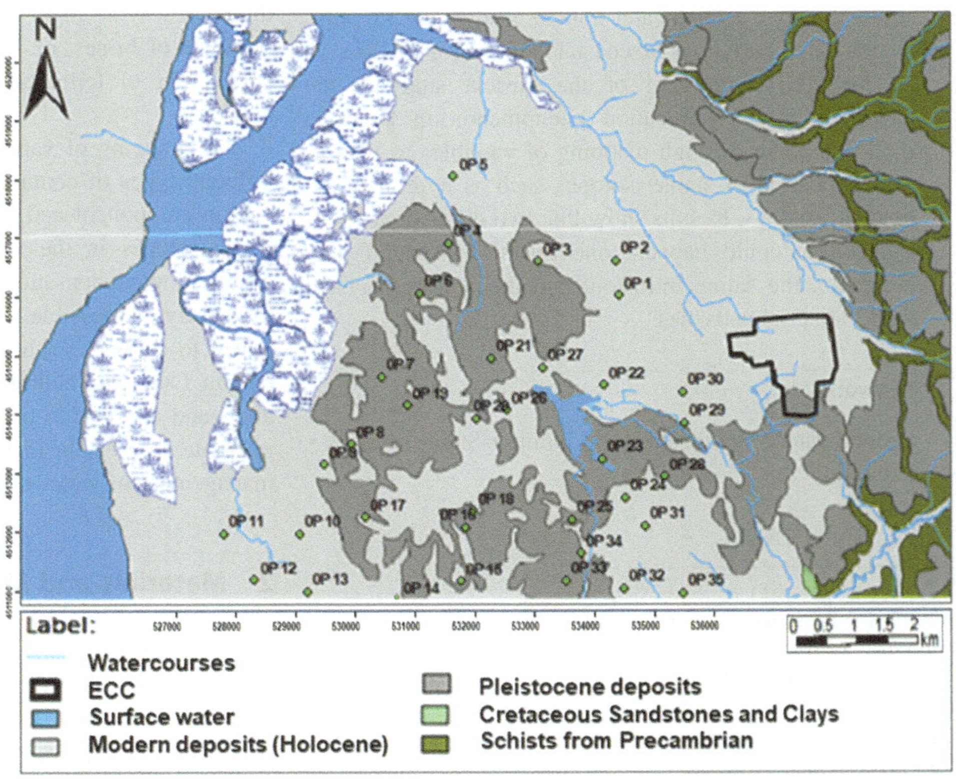


**Fig. 1** Location of the sampled groundwater on a simplified geological map. Adapted from Ordens (2007)

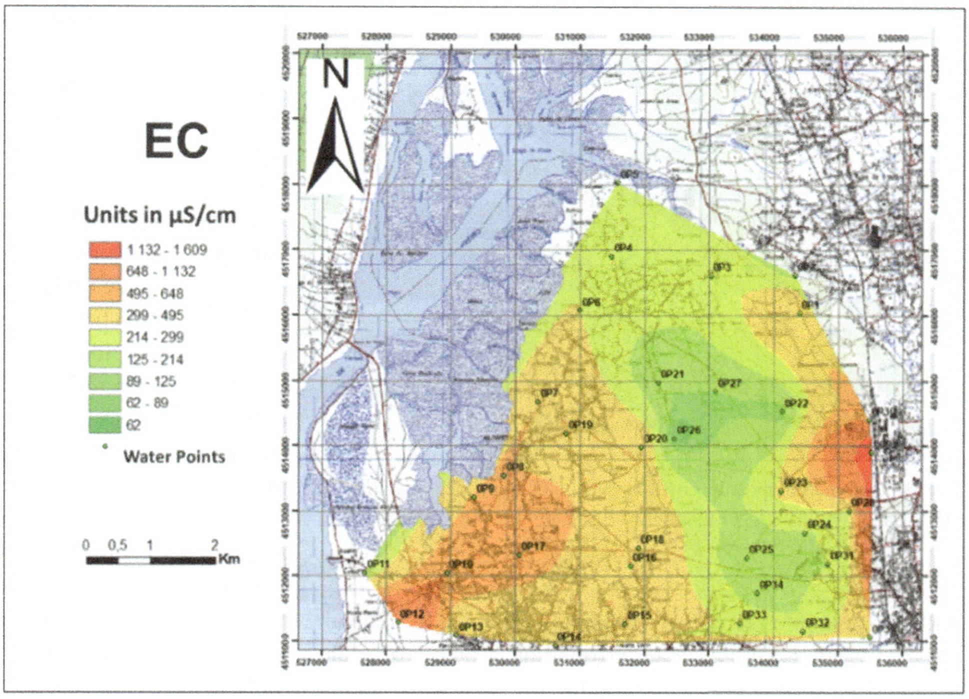


**Fig. 2** Spatial distribution for electric conductivity (EC—μS/cm) in groundwater

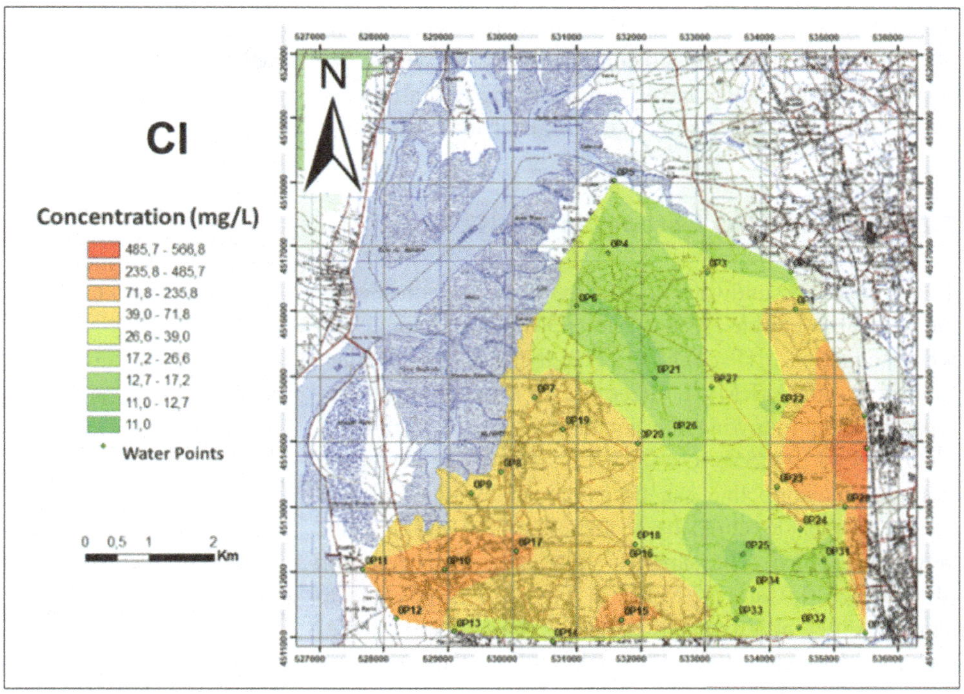


**Fig. 3** Spatial distribution for Cl concentration (mg/L) in groundwater

Southwest areas, close to Ria de Aveiro (Fig. 2). The salt components, Cl and Na ions have the same spatial distribution, showing high relation between both ions and the EC (Fig. 3). The lower Cl concentration is close to rainwater value (11 mg/L) and the higher concentration, measured in the East limit of the study area, is 566.8 mg/L. $SO_4$ is also a relevant industrial indicator. The higher concentrations (140.2–311.4 mg/L) are in Eastern zone, near to ECC, as some trace elements such as Al, As, Co, Cr, Cu, Li, Ni, V and Zn. In the study area, pH values range from 4.41 to 7.63

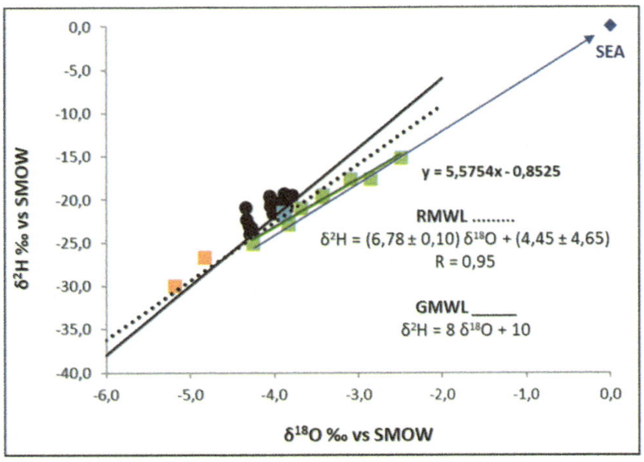


**Fig. 4** Isotopic composition of $\delta^2H$ and $\delta^{18}O$ from Estarreja shallow aquifer and its location in relation to the GMWL (Craig 1961) and to the RMWL (Carreira et al. 2006)

and most samples have pH values less than 7. In general, plus-acid-samples are in ECC surroundings (4.41–5.65), and less-acid-samples are near to coastline (5.86–7.36) pointing out small contribution of the seawater (pH $\approx$ 8).

Considering the quality standards proposed in the Annex VII of APA (2016), the EC values are below the limit of 2500 µS/cm, but the Cl concentrations are above the 250 mg/L limit in some areas, namely in the Eastern part of the research area, but also at Western, near to coastline.

### 3.2 $\delta^2H$ and $\delta^{18}O$ Isotopes in the H2O

The sampled groundwater has meteoric origin considering the Global Meteoric Water Line (GMWL) (Craig 1961) whose equation $\delta^2H = 8\ \delta^{18}O + 10$ or the Meteoric Water Line in Portugal (RMWL) (Carreira et al. 2006) with isotopes relation following the equation $\delta^2H = (6.78 \pm 0.10)\ \delta^{18}O + (4.45 \pm 4.65)$, R = 0.95 (Fig. 4). All the deviations from the meteoric line can be explained by physical processes as the slope of the evaporation line has a value between 3 and 6. The samples show excess of deuterium (d) between 4.9 and 13.8 ‰. In Fig. 4 it is possible to identify different groups of water: $\delta^2H$ enriched samples probably associated with a great humidity at the recharge area (black dots); samples enriched in heavy isotopes showing a shift to the right of the water lines that could be affected by evaporation or mixed with seawater (blue and green dots); samples depleted in heavy isotopes that can be related to far way recharge, representing condensation phenomena (orange dots).

## 4 Concluding Remarks

The mapping of variables shows that salinization can occur in W and SW areas, close to Ria de Aveiro, through saline intrusion, a natural process. However, salinization phenomenon can also have an industrial origin at E area, near to Estarreja Chemical Complex (ECC). In the study area, industrial salinization process could be triggered by discharges and/or leaching effluents from raw materials and/or final products. These discharges could be associated with a contamination plume spread, observed in the mapping of variables, which follows the radial direction of groundwater flow network was the main source is ECC (Ordens 2007). In terms of $\delta^2H$ and $\delta^{18}O$ isotopic composition, it is verified that blue and green samples suggest salinization phenomenon as they can be mixed with seawater, assuming $\delta^2H$ and $\delta^{18}O$ isotopic values for seawater are close to zero. EC values are below quality standards unlike Cl concentrations values that are above in some areas, namely in East area. In this way, it must be considered some environmental remediation measures in order to a better protection of water resources in the research area.

## References

Almeida C, Mendonça JJL, Jesus MR, Gomes AJ (2000) Sistemas aquíferos de Portugal continental. INAG, Lisboa

APA (2016) Plano de Gestão de Região Hidrográfica Vouga, Mondego e Lis (RH4) 2016–2021. Parte 2–Caracterização e Diagnóstico. Agência Portuguesa do Ambiente. Lisboa, pp 204

Carreira PM, Valério P, Nunes D, Araújo MF (2006) Temporal and seasonal variations of stable isotopes ($\delta^2H$ and $\delta^{18}O$) and tritium in precipitation over Portugal. In: IAEA conference and symposium papers 26/P. Isotopes in environmental studies—aquatic forum, pp 370–373

Craig H (1961) Isotopic variations in meteoric waters. Science 133 (3465):1702–1703

Decree-Law no. 77/2006 of 30 Mar

Decree-Law no. 208/2008 of 28 Oct

European Directive no. 2006/118/EC of 12 Dec

Marques da Silva MA (1990) Hidrogeología del sistema multiacuífero Cretácico del Bajo Vouga—Aveiro (Portugal). University of Barcelona, Barcelona (Ph.D. thesis)

Ordens CMM (2007) Estudo da contaminação do aquífero superior na região de Estarreja. Faculty of Sciences and Technology, University of Coimbra, Coimbra (MSc Dissertation)

Teixeira C (1962) Carta Geológica de Portugal 1/50000. Notícia Explicativa da Folha 13-C (Ovar). Serviços Geológicos de Portugal, Lisboa

# How to Control Groundwater Quality Degradation in Coastal Zones Using Mar Optimized by GALDIT Vulnerability Assessment to Saltwater Intrusion and GABA-IFI Models

João Paulo Lobo-Ferreira

**Abstract**

To counteract harmful, eventually with catastrophic consequences, today and future groundwater quality degradation due to saltwater intrusion into coastal aquifers, managed artificial recharge (MAR) is considered the best solution, a sound, safe and sustainable solution. MAR, in coastal areas, depends on the availability of water including wastewater appropriately treated. How to control saltwater intrusion in coastal zones implementing a MAR facility? The parameters required to answer that question include the selection of the most appropriate technology and the best location for MAR. The appropriate location must have good infiltration rates; enough space to store underground the recharged water; guarantee that the travel time of the recharged water in the aquifer is long enough, compatible with the expected frequency of drought periods; economic efficiency maximization; availability of areas for MAR; and positive impacts on the society. GABA-IFI model addresses those parameters allowing the selection of the most appropriate area for the location of MAR. Complementary, mathematical models are available to quantify MAR water injection rates required to recover groundwater depleted levels. Where should the injection be located? GALDIT is probably, today, the most used model worldwide to assess the vulnerability of saltwater intrusion in costal aquifers by a numerical calculation.

**Keywords**

Coastal zones • Saltwater intrusion • Mathematical models • GALDIT • GABA-IFI

J. P. Lobo-Ferreira (✉)
Laboratório Nacional de Engenharia Civil (LNEC), Lisbon, Portugal
e-mail: lferreira@lnec.pt

## 1 Introduction

The roles that managed artificial recharge (MAR) may play today, worldwide, within the framework of integrated water resources management are as follows: short- and long-term storage for later recovery during dry seasons, recovery of groundwater level of over-exploited aquifers, provision of barriers to seawater intrusion in coastal areas, improvement of water quality, use of the aquifer as a water distribution system for individual users, and flood prevention by deviating peak flows. In Portugal, the recent water supply crises caused by recent droughts, within the southern Algarve region, show the necessity to implement new measures for improved water resources management. The surface water reservoirs in the northern mountain areas of the Algarve suffer from significant water losses due to high evaporation rates and are not able to cover the water demand under drought conditions. During the rainy season, huge spillway losses have been observed from these reservoirs. In wet years more than 50hm$^3$. The river water flows to the sea. In drought years, a similar value of 50 hm$^3$ is overpumped from the aquifer causing seawater intrusion. MARSOL FP7 INNO-DEMO project Policy Brief clearly summarizes the legal framework of MAR (Schüth et al. 2019): The Water Framework Directive (2000/60/EC) considers "artificial recharge" of groundwater as one of the water management tools that can be used by EU Member States to achieve a good groundwater status. It has to be ensured, however, that the necessary regulatory controls are in place to warrant that such practices do not compromise quality objectives established for the recharged or augmented groundwater body. It is also acknowledged by the Groundwater Directive (2006/118/EC) that it is not technically feasible to prevent all input of hazardous substances into groundwater, in particular minor amounts which are considered to be environmentally insignificant and thus do not present a risk to groundwater quality. For such cases, the Groundwater Directive, under Article 6(3)(d), introduces a series of exemptions. Managed



M. Abrunhosa et al. (eds.), *Advances in Geoethics and Groundwater Management: Theory and Practice for a Sustainable Development*, Advances in Science, Technology & Innovation,
https://doi.org/10.1007/978-3-030-59320-9_44

artificial recharge is considered as one of these exemptions. MARSOL suggests a regulatory framework based on risk assessment, control mechanisms, and monitoring as a tool which can facilitate the application of the Water Framework and Groundwater Directives on MAR. It is the intention of such a regulatory framework to provide clear guidelines to Member States on the application of MAR techniques.

## 2 Materials, Methods, and Results

### 2.1 Groundwater Recharge Assessment Under Climate Change Conditions

The increase in extreme precipitation phenomena, even at the same annual volumes, can cause a decrease in groundwater recharge because soil infiltration capacity is exceeded more frequently, favoring runoff rather than recharge. The work carried out at LNEC on groundwater recharge (e.g., Lobo-Ferreira et al. 2012) demonstrates the influence of precipitation distribution series on groundwater recharge and the need to use daily sequential balance models (e.g., BALSEQ model, Lobo-Ferreira 1981) that take into account daily precipitation and evapotranspiration, as well as the area occupied by vegetation and its characteristics. The same authors conclude that, for selected analyzed scenarios, a reduction of the precipitation value down to 70% of the annual average value corresponds to annual recharge down to just 45% of the annual average.

Climatic conditions, such as rainfall, temperature, and atmospheric humidity, affect the volumes of water that are spent by the vegetation cover in evapotranspiration and those that are transferred to the recharge of the aquifers. Under climate change conditions, these climatic parameters are modified, having direct impacts on evapotranspiration and recharge. There is also an associated impact, which is due to the modification of the vegetation cover, which will affect the volumes of evapotranspiration, surface runoff, soil water content, and, consequently, recharge.

According to studies by Lobo-Ferreira et al. (2012), using BALSEQ model, we expect that the average recharge of the Torres Vedras aquifer system, in Central Portugal, will be between 84 and 98% of the recharge of the period 1979–2009, depending on the series of precipitation, temperatures, and reference evapotranspiration used. For the 2080 horizon, depending on the climatic series used, the average recharge will be 60–82% of the recharge of the period 1979–2009. It is impressive the expected potential reduction of aquifer recharge that can be observed in Fig. 1 (green, yellow, and orange areas correspond to losses greater than 50% in annual groundwater recharge).

### 2.2 GALDIT Method

The original development of GALDIT index was done in the framework of the EU-India INCO-DEV COASTIN project, proposed by Chachadi and Lobo-Ferreira (2007), aiming the assessment of aquifer vulnerability to seawater intrusion in coastal aquifers.

The most important factors controlling seawater intrusion were found to be the following: ***G***roundwater occurrence (aquifer type; unconfined, confined, and leaky confined); ***A***quifer hydraulic conductivity; depth to groundwater ***L***evel above the sea; ***D***istance from the shore (distance inland perpendicular from shoreline); ***I***mpact of existing status of seawater intrusion in the area; and ***T***hickness of the aquifer, which is being mapped. The acronym GALDIT is formed from the highlighted letters of the parameters for ease of reference.

In Fig. 2, a comparison between GALDIT scores for normal and raised sea levels in North Goa coast (left figure for normal sea level and right figure for raised sea level) is presented. One can easily see which areas will be more affected by sea level rise.

### 2.3 Managed Aquifer Recharge Response Strategy

Managed aquifer recharge (MAR) permits the managed conjunctive use of surface- and groundwater resources and, therefore, is an important component of integrated water resources management (IWRM). Decision support for MAR planning in the context of integrated water resources management, the implementation of a MAR system requires careful planning in terms of achieving efficient integration into the water resources system and the overall water resources management objectives.

Environmental and socioeconomic impacts of MAR planning options have been investigated in this context. The new H2020 MARSoluT project, i.e., a follow-up of FP7 INNO-DEMO MARSOL project, addresses MAR research via 12 PhD theses, one of them under development in LNEC on MAR GW quality in a large physical model.

Water availability during the dry period is nowadays the main constraint for the development of the Algarve region. Further exploitation of the already over-exploited Querença-Silves aquifer system is no sustainable response strategy to the water supply problem. The enormous losses of surface water resources in connection with over-exploitation of the local aquifer systems indicate the necessity of the conjunctive management of surface water and groundwater resources in order to guarantee water supply during

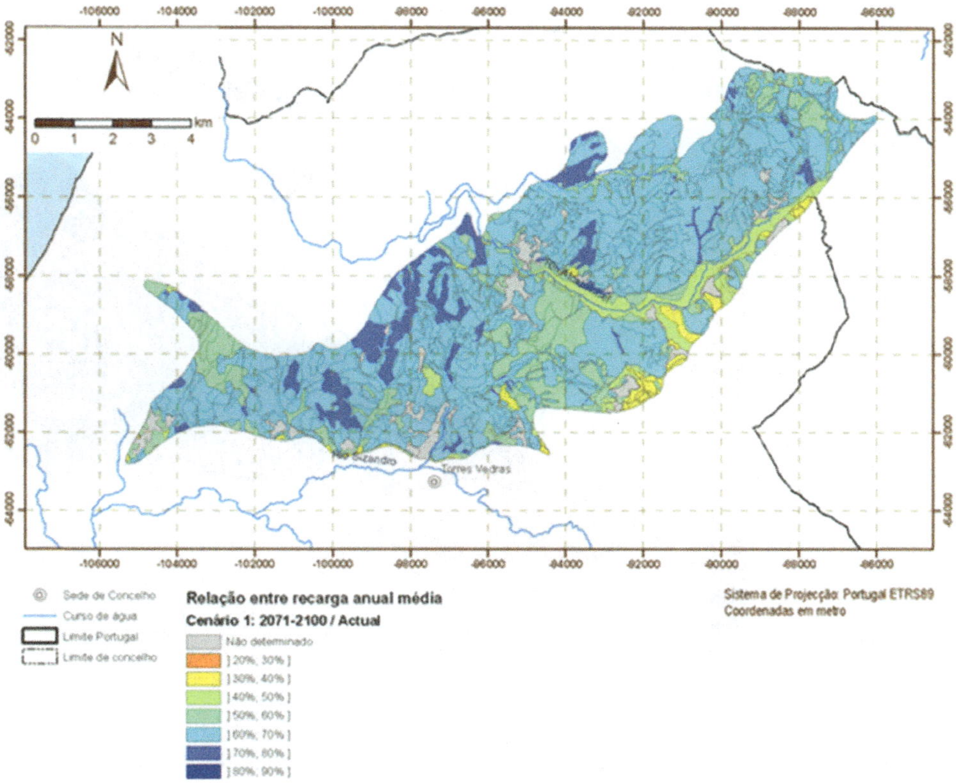


**Fig. 1** Relation between the current average and the expected annual recharge for Scenario 1 in 2071–2100 of the aquifer system of Torres Vedras, Central Portugal

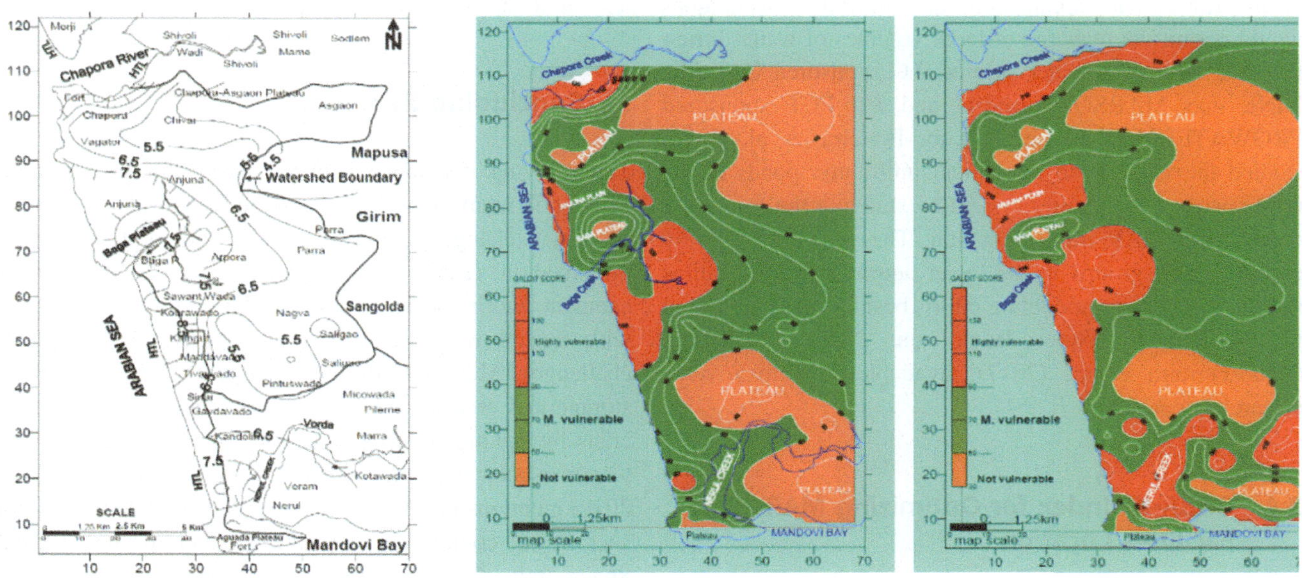


**Fig. 2** Comparison between GALDIT scores for normal and raised sea levels in North Goa coast (left and center figures for normal sea level and right figure for raised sea level; high vulnerable areas in red)

droughts, to mitigate their impacts and provide conditions for sustainable development of the Querença-Silves region.

Therefore, MAR is a potential response measure to the prevailing water management problems of the region, transferring surface water of the nearby Arade Dam to the upper part of the study area (cf. Fig. 3) and recharging this water to the Querença-Silves aquifer. The proposed water management response strategy uses the aquifer as a water

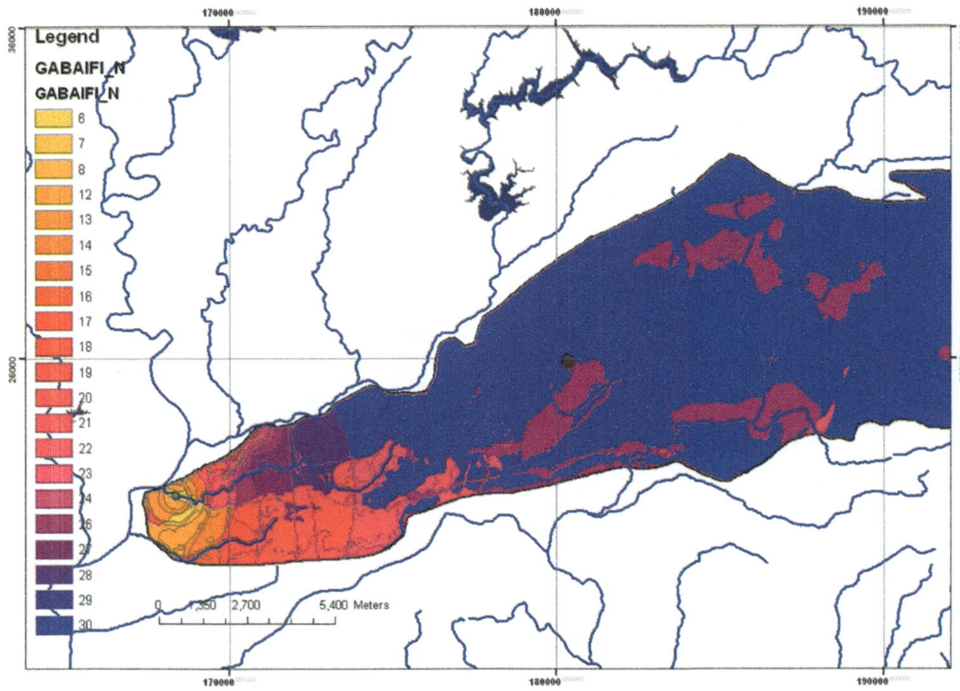


**Fig. 3** GABA-IFI_N index representing more favorable locations to the installation of MAR

transfer and storage system. The high residence times of the infiltrated water in the aquifer system would guarantee long-term storage of the recharged water in the underground with water recovery by wells in the lower part of the aquifer system located at strategic places, e.g., Vale da Vila near Alcantarilha. The application of infiltration techniques would permit the recharge of large volumes of water during the time of surface water surplus. EC sponsored GABARDINE and MARSOL projects and PT FCT sponsored PROWATERMAN projects (Lobo-Ferreira et al. 2011) studied the viability to implement infiltration ponds in the upper parts of the aquifer system by studying the suitability of areas for MAR implementation. The water recovery by pumping well groups in the lower aquifer system requires further groundwater modeling efforts besides those presented in MARSOL White Book on MAR Modelling (Lobo-Ferreira et al. (2017), available in https://www.researchgate.net).

## 2.4 GABA-IFI Index for Optimizing MAR Facilities Location

For the choice of sites favorable to the recharge, GABA-IFI index was developed in FP6 EU GABARDINE project (Oliveira et al. 2008) allowing the computation of the areas with more favorable natural conditions for MAR (in particular good recharge, good space for underground storage, and high residence time of the water in the aquifer system). The parameters considered relevant for the assessment of GABA-IFI index are the distance to the point of discharge of groundwater, the depth to groundwater level, the vertical transport time, and the horizontal hydraulic conductivity. Figure 3 shows the application of this index to the aquifer system of Querença-Silves aquifer in the Algarve (yellow less suitable; blue optimal place for MAR).

## 3 Concluding Remarks

The growing imbalance between water supply and water demand in many coastal areas around the world, exacerbated by climate change, population growth, agriculture needs, and urbanization, requires more efficient water resources management. Storing water in aquifers during times of excess or with treated wastewater can help address water scarcity challenges. Managed aquifer recharge (MAR) and aquifer storage and recovery (ASR) can be a key to solving water crisis by linking water reclamation, water reuse, and water resources management. MARSOL Policy Brief (Schüth et al. 2019) mentions that MARSOL operated eight demonstration sites in six countries around the Mediterranean (Portugal, Spain, Italy, Greece, Malta, Israel) applying various technologies, i.e., infiltration ponds, river bed infiltration, direct injection wells, canals, riverbank filtration, to infiltrate various water sources, i.e., river water, surface runoff, treated wastewater, and desalinated seawater. Now is the time to move on applying MAR, in a scientific-based way, to cope with expected dramatic climate change consequences, including those in ecosystems dependent on groundwater and seawater intrusion in coastal aquifers caused by sea level rise. This approach fulfills the hydrogeoethical approach as "… MAR is a sound, safe, and sustainable strategy that can be applied with great confidence", confirming uncertainty reduction in MAR modeling in eight MARSOL demo sites, improving the Campina de Faro, Algarve, groundwater quality.

**Acknowledgements** The MARSOL project received funding from the European Union's Seventh Framework Program for Research, Technological Development and Demonstration under grant agreement no 619120. MARSOL Policy Brief.

## References

Chachadi AG, Lobo-Ferreira JP (2007) Sea water intrusion vulnerability using GALDIT method: Part 2—GALDIT indicators description. In: Lobo Ferreira JP, Vieira J (eds) Water in Celtic countries: quantity, quality and climate variability, vol 310. IAHS Red Books, London, IAHS Publication, pp 172–180

Lobo-Ferreira JP (1981) Mathematical model for the evaluation of the recharge of aquifer in semiarid regions with lack of hydrological data. In: Proceedings of EUROMECH 143, flow and transport in porous media. Rotherdam, AA Balkema

Lobo-Ferreira JP, Oliveira LGS, Diamantino C (2011) Groundwater artificial recharge solutions for integrated management of watersheds and aquifer systems under extreme drought scenarios. In: Jones J (ed) Sustaining groundwater resources. International year of planet earth. Springer, Dordrecht

Lobo-Ferreira JP, Novo ME, Oliveira MM, Oliveira LGS (2012) Estudo do Impacto das Alterações Climáticas na Recarga do Sistema Aquífero de Torres Vedras. Lisboa. In: 11.º Congresso da Água, Associação Portuguesa dos Recursos Hídricos (APRH), ISEP, Porto

Lobo-Ferreira JP et al (2017) White book on MAR modelling: selected results from MARSOL PROJECT, Lisbon, LNEC. https://www.researchgate.net/publication/314957907_White_book_on_MAR_modelling_Selected_results_from_MARSOL_PROJECT/link/58c7e37fa6fdcc550caf5016/download

Oliveira L, Oliveira MM, Lobo-Ferreira JP (2008) Índice de suporte à escolha de áreas favoráveis à recarga artificial (Gaba - IFI). 9.º Congresso da Água. APRH, pp 1–12

Schüth C, Roehl KE, Fernández-Escalante E, Guttman J, Lobo-Ferreira JP (2019) MARSOL policy brief. Essentials on managed aquifer recharge for policy makers and water managers. In: Abstract, international symposium on managed aquifer recharge, ISMAR 10, Madrid. https://www.dina-mar.es/file.axd?file=2017/8/MARSOL+Policy+Brief_final.pdf

# Driving Factors of Karst Wetland Degradation from the Perspective of Hydrogeology: A Case Study in SW China

Liankai Zhang, Shengzhang Zou, Lina Shen, and Yi Zhao

**Abstract**

As the largest karst wetland in the low latitude region, Huixian Wetland is located in the watershed belt of Liujiang River and Lijiang River in SW China. It is of great significance on water resources regulation and ecological health maintenance for Pearl River. In recent decades, Huixian Wetland degraded and the water covering area gradually reduced. The insufficient understanding of the hydrogeological background weakened the restoration effort. In the view of karst hydrogeology, the driving factor of Huixian Wetland degradation was analyzed. Firstly, the rapid conversion of ground and surface water is one of the reasons. The short response time of groundwater to precipitation leads to the frequent occurrence of drought and waterlogging in wetland. Secondly, the underground karst pipes are narrow and easy to be blocked which resulting in the outflow of recharge water. It would reduce the storage of water resources in the core area of Huixian Wetland. Thirdly, the karst rock desertification makes a large amount of lost soil deposited in wetland. It obviously decreased the capacity of wetland regulation. With the aggravation of the rocky desertification, the sedimentation would be more intensified in the future.

**Keywords**

Karst wetland • Degradation • Drive factors • Hydrogeology

L. Zhang · S. Zou (✉) · L. Shen · Y. Zhao
Key Laboratory of Karst Dynamics, Ministry of Natural and Resources and Guangxi, Institute of Karst Geology, Chinese Academy of Geological Sciences, Guilin, 541004, China
e-mail: zshengzhang@mail.cgs.gov.cn

## 1 Introduction

The surface water in the special double-layer hydrogeological structure of the surface and underground in karst area is prone to leak to underground aquifer and then lead to the shortage of surface water and rare of wetland (Yuan 1994). Fortunately, many large lakes and wetlands exist in kart region of SW China, where the biodiversity is abundant, and the landscape is beautiful (Cai and Ma 2012). Huixian Wetland, located in Guilin, SW China, is the representative spot with the largest area in low latitude region. It is important for the Pearl River's hydro-ecological system maintaining. However, the water covering area in recent five decades degraded seriously (Wu et al. 2006) which threaten the water resources security around wetland. To save the wetlands, much effort had been done by governments and researchers, but with poor performance (Lin et al. 2008; Zhang et al. 2010, 2012). Lack of sufficient understanding of karst geological background is one of the crucial reasons. From the perspective of karst hydrogeology, the key factors driving the degradation of Huixian Wetland were analyzed.

## 2 Methods

### 2.1 Site Description and Water Table Monitoring

Huixian Karst Wetland locates on the watershed belt of Lijiang River and Liujiang River, two tributaries of the Pearl River. The catchment is 189 km$^2$, in which 42 km$^2$ is the core reserve area. It was seated at the center of the karst world heritage site of Guilin, China (UNESCO homepage). Since Devonian, a set of littoral-neritic clastic-carbonate sedimentary assemblages were deposited. The geomorphological types are solitary peak plain and peak cluster



M. Abrunhosa et al. (eds.), *Advances in Geoethics and Groundwater Management: Theory and Practice for a Sustainable Development*, Advances in Science, Technology & Innovation,
https://doi.org/10.1007/978-3-030-59320-9_45

depression, with an altitude of 150–300 m (Cai and Ma 2012). The climate here is typical subtropical monsoon, with an annual average temperature of 19.2 °C and an annual rainfall of 1894 mm. To monitor the water level, 11 monitoring stations on both surface and groundwater were set as shown in Fig. 1.

## 2.2 Remote Sensing Data

Landsat TM (NASA's Landsat Satellite, USA) images of 1973, 1991, 2000, 2004 and 2018 were selected. Combined with remote sensing interpretation and geographic information system technology, this paper analyzes the land use and rock desertification evolution in Huixian Wetland.

# 3 Results

## 3.1 Wetland Area Evolution

Remote sensing interpretation results suggested that, from 1973 to 2018, the water surface area in the core area of Huixian Wetland decreased continuously (from 7.24 $km^2$ to 5.5 $km^2$), with an average rate of decline of 0.04 $km^2$/a. But it was increased slightly from 2004 to 2018 with the area from 4.56 to 5.5 $km^2$. The seasonal flooded area (marshland) that impacts the wetland health significantly (Kobza et al. 2004) decreased from 3.57 $km^2$ in 1973 to 0.81 $km^2$ in 2018. The annual average decline rate was 0.06 $km^2$. The decreasing of water area in the core area is a sign of degradation.

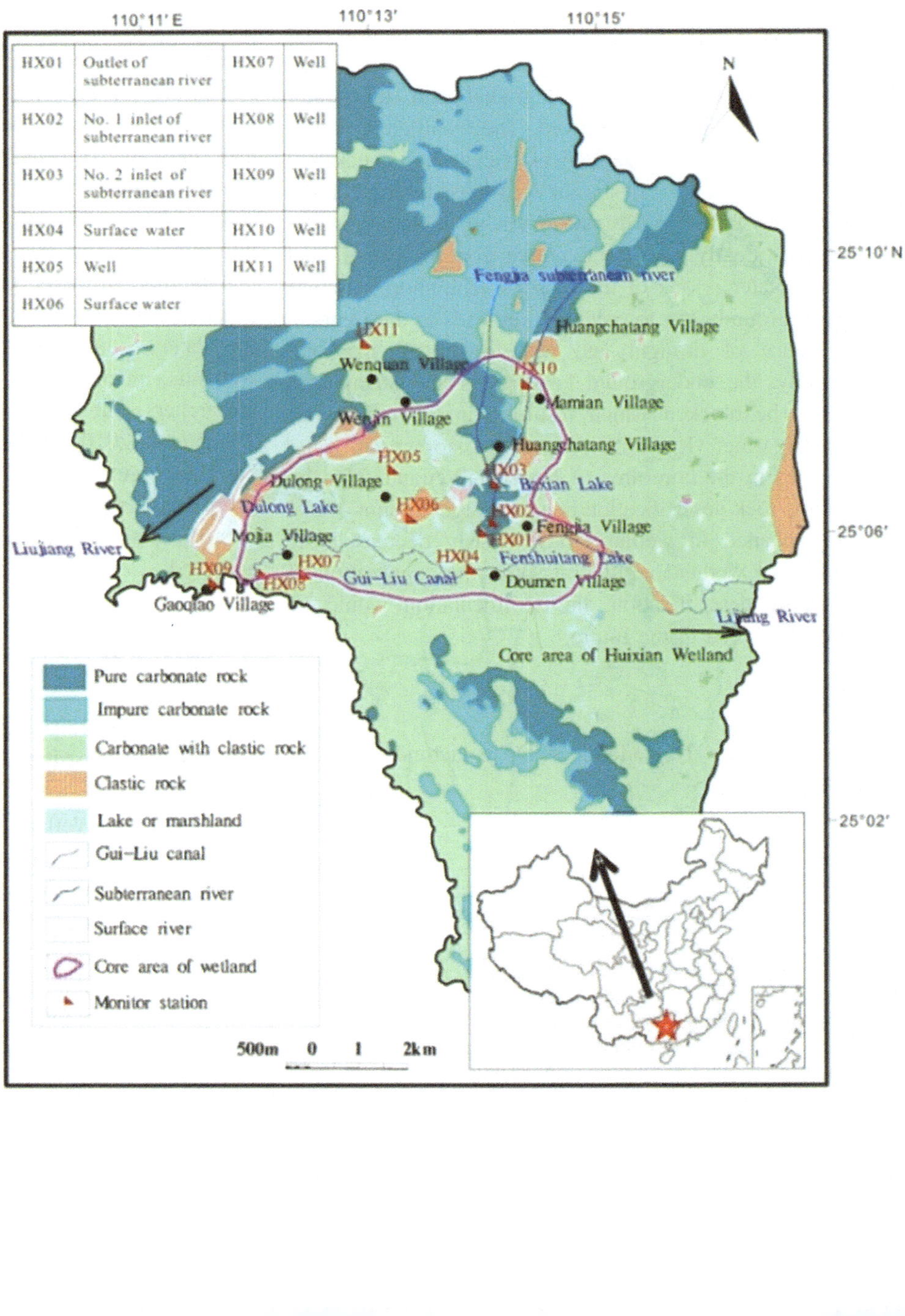


**Fig. 1** Hydrogeological map and the monitoring station of Huixian Karst Wetland

### 3.2 Variation of Water Level

Influenced by hydrogeological structure, the water level of both ground- and surface water at the runoff area of watershed has a significant response to rainfall (Fig. 2). The variation of water level at each monitoring point is characterized by multiple peaks. High water level peaks are always following closely to the heavy rainfall. The stagnation time of the water level peak is generally less than three days. For example, the variations of water level after rainfall in the runoff area reach 1.30 and 1.20 m in 24 and 32 h at HX10 and HX11 stations, respectively. It suggested that the hydrological process in karst system is much more rapid than non-karst area.

However, the variation of water level in the core of the wetland (drainage area) is not dramatically. When the change of groundwater table in the runoff area exceeds 1.0 m (HX10 and HX11), the value is less than 0.5 m as HX04, HX06, HX08 and HX09 shown. This indicates that the function of flood regulation is degraded.

### 3.3 Rocky Desertification in Wetland Catchment

The remote sensing interpretation shows that the rocky desertification was generally aggravated from 1973 to 2018. The area of moderate and severe rock desertification (MRD and SRD) gradually decreased, while the area of light rock desertification (LRD) increased. From 2004 to 2018, the area of these three kinds of rock desertification—LRD, MRD and SRD—increased (the classification of rocky desertification is based on Wang and Li 2005). Particularly, the area of MRD and SRD was increased by 2.4 times and 2.7 times, respectively. The rock desertification area was increased by 3.32 km$^2$, accounting for 1.24 times of 2004.

## 4 Discussion

### 4.1 Karst Development and Wetland Hydrological Characteristic

Around the Huixian Wetland Basin distribute pure carbonate rock of the upper Devonian. In the middle, there is argillaceous limestone with weak karst development. Under the argillaceous limestone is Devonian clastic rock that formed water-barrier floor (Cai et al. 2009). Though some karst forms such as caves, karst springs and ponor developed in the surrounding pure limestone strata, the amount of pipe-connected cave is few, and the karst development depth underground is shallow (with a buried depth of 5–20 m).

What is more, most of the underground river inlet is always narrow and flat. The drainage ability is poor in this condition. It tends to flood the upstream ecosystems during the rainy season. After the water receded, the land emerged again. This periodic flooding destroys the wetland ecosystem and difficulties its recovery. The rapid alternations of surface and groundwater aggravate the drought and flood disasters in the wetland, which are not conducive to the maintenance of wetland ecological health.

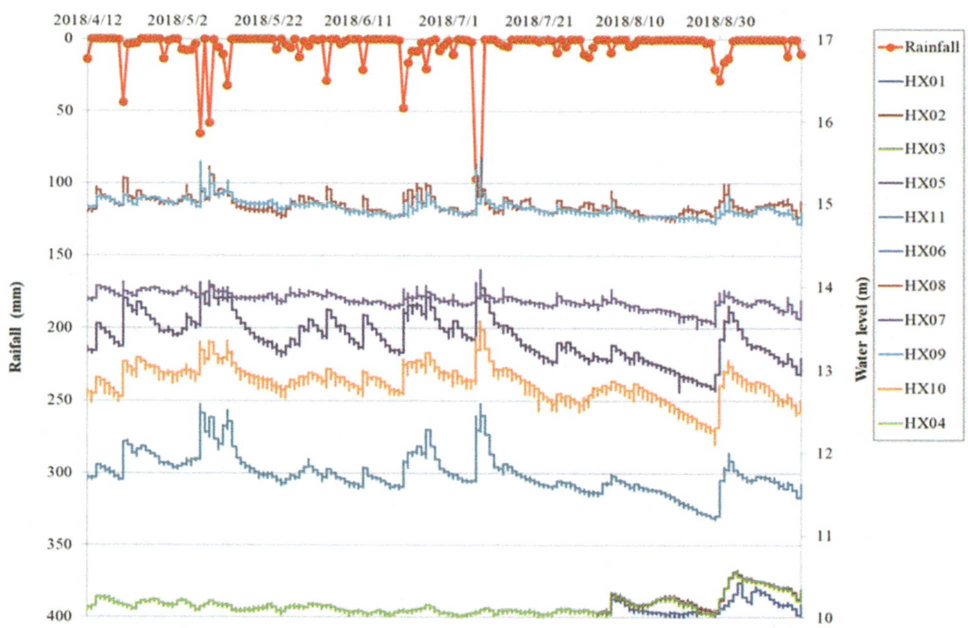


**Fig. 2** Response relationship between water level and rainfall

### 4.2 Pipe Blockage and Water Supply

The blockage of karst underground pipeline causes water outflow and reduce the water supply. Huixian Karst Wetland mainly consists of pipes and fissures. But the pipes and fissures are usually narrow and easily to be blocked. Then the groundwater will overflow to surface through the karst shaft, doline or blue hole. As mentioned above, Huixian Wetland is located on the watershed belt of Liujiang River and Lijiang River (Fig. 1). The altitude in the center is higher than the east and west. The surface water tends to flow into these two rivers, which results in the reduction of wetland water supply. It decreases the water collection and weakens the flood storage function.

In order to calculate the amount of outflowed drainage from wetland watershed accurately, the sub-system of Fengjia Subterranean River was selected, and the water balance calculation model (Eq. 1) was adopted according to the hydrogeological report (China Geological Survey 2016).

$$Q_{su} = Q_d + Q_e + Q_p + \Delta Q_{st} \tag{1}$$

where $Q_{su}$ is annual quantity of supply water; $Q_d$ is annual quantity of drainage water; $Q_e$ is annual quantity of evaporated water; $Q_p$ is phreatic water; $\Delta Q_{st}$ is the annual average variation of storage in wetland. The parameters are cited from the geological report (China Geological Survey 2016).

The results showed that the annual quantity of supply water, drainage water and the variation of storage in wetland is $30.71 \times 10^6$ t/year, $21.44 \times 10^6$ t/year and $22.24 \times 10^4$ t/year, respectively. The annual quantity that flows outside the watershed is $9.05 \times 10^6$ t/year, accounting for 29% of the total catchment supply. This outflow water would have great impact on the health maintenance of wetland.

### 4.3 Rock Desertification and the Regulation Capacity

There is an internal relationship between rocky desertification and wetland degradation. The lost soil caused by rocky desertification would be deposited at the bottom of the wetland and then reduce the storage capacity. The rocky desertification in Huixian Wetland presents a negative correlation with the wetland area (the correlation coefficient ranged from −0.215 to −0.652). In these three kinds of rocky desertification, LRD has a significant negative impact on the perennial water area (−0.652).

The relationship between rocky desertification and sediment in Huixian Wetland had been reported before. Shen et al. (2010) believed that before 1894, the lake was generally in a natural evolution stage, with no obvious interference from human activities and a low deposition rate. Since 1894, the influence of rocky desertification caused by human activities has been obviously increased. Rock desertification can accelerate sediment deposition in Huixian Wetland also confirmed by Li et al. (2009). His research showed that the average deposition rate of the sediment in Huixian Lake from 1952 to 2009 was 1.7 mm/year. However, in the period of large-scale steel refining policy in China from 1952 to 1963 during when a lot of vegetation on the mountain was cut down to make steel, the average deposition rate of the sediment reached 3.1 mm/a. It indicates that rock desertification and soil erosion caused by vegetation destruction have obvious influence on sediment of wetland.

## 5 Concluding Remarks

Under the influence of underground space development, karst wetland is a kind of rare resource especially in the karst area. To reveal underground space structure of karst wetland and the relationship between recharge, runoff and drainage of water resources is an important basis to reveal the evolution of wetland. The paper analyzes the degradation mechanism of karst wetland in Huixian Karst Wetland from the perspective of hydrogeology. It showed that the degradation of wetland is firstly due to the karst development underground which shorten the response time between hydrology and rainfall. Meanwhile, the poorly developed underground space in the recharge and runoff area causes the blockage of pipelines. The calculation results show that 29% of the recharge water is overflow to outside basin. Thirdly, rocky desertification aggravated the pipeline blockage and thickened the sediment deposition in the core area of wetland. It reduced the regulating capacity of wetland. The results of this study can provide support for the evaluation and management of wetland water resource as well as its sustainable development and utilization in karst region.

**Acknowledgements** This work was funded by National Key Research and Development Project of China (2017YFC0406104).

## References

Cai D, Ma Z (2012) Study on karst wetland ecosystem in Huixian. Geological Publishing House, Beijing

Cai D, Ma Z, Zhao X, Wang K (2009) Remote sensing supervision on spatio-temporal evolution of karst wetland in recent 40 years in Huixian District of Guilin, China. J Guangxi Normal Univ Nat Sci Ed 27(2):111–117

China Geological Survey (2016) Survey report of karst hydrogeology in Southwest China. In: Institute of karst geology, CAGS, Guilin

Kobza RM, Trexler JC, Loftus WF, Perry SA (2004) Community structure of fishes inhabiting aquatic refuges in a threatened Karst

wetland and its implications for ecosystem management. Biol Cons 116(2):153–165
Li S, Cai D, Zhang H, Shen D, Zhao X, Li C (2009) Environmental changes record derived from sediment cores in Huixian karst wetlands, Guilin, China. J Guangxi Normal Univ Nat Sci Ed 27 (2):94–100
Lin D, Wei Y, Zhang X, Chen T (2008) The restoration of Huixian wetland in Guilin from the perspective of karst hydrogeology. Resour Environ Eng 22(S1):221–222
Shen D, Li S, Cai D, Zhang H, Zhao X, Li C (2010) Element geochemistry of lake sediments from Sihu Lake in Guilin karst wetland region and its environmental effect. Geol J China Univ 16 (4):517–526
UNSECO homepage. https://whc.unesco.org/en/list/1248
Wang S, Li Y (2005) Discussion on the classification of karst rocky desertification in ecological constructions. Carsologica Sinica 24 (9):192–195
Wu Y, Mo Y, Zou S (2006) Ecologic problem and protection method of karst wetland in Huixian, Guilin. Carsologica Sinica 25(1):85–88
Yuan D (1994) Karst in China. Geological Publishing House, Beijing
Zhang L, Tan W, Wang H (2012) Wetland protection and restoration in Guilin Huixina Karst wetland park. Guangxi Forest Sci 41(3):303–305
Zhang M, Jin H, Cai D, Jiang C (2010) The comparative study on the ecological sensitivity analysis in Huixian karst wetland, China. Procedia Environ Sci 2(1):386–398

# Groundwater Vulnerability Assessment in the Naturtejo UNESCO Global Geopark, Portugal

Teresa Albuquerque, Natália Roque, Joana Rodrigues, Isabel Margarida Antunes, and Catarina Silva

**Abstract**

Nowadays, groundwater vulnerability assessment has become a useful tool for groundwater contamination prevention. Groundwater vulnerability maps provide useful data to protect groundwater resources and work as a tool for the improvement of changes in agricultural patterns and land use applications. The study area of this research survey is the Naturtejo UNESCO Global Geopark, located in central inland Portugal and corresponding to a mainly rural territory where intensive agricultural practices showed a rising tendency in the last decades. The most used method of vulnerability evaluation is the DRASTIC index. In this survey, a modified DRASTIC method, DRASTICAI, is introduced. A new attribute designated as *anthropogenic influence* is introduced. Map algebra in a GIS environment allowed the computation of two maps by overlaying the needed attributes. The Vila Velha de Rodão and Idanha-a-Nova municipalities show moderate to high vulnerability and, therefore, in need of monitoring, since intensive agricultural practices are the main economic activity. The algebraic subtraction of DRASTIC and DRASTICAI maps revealed a considerable increase in the risk of contamination, over the surveyed area, namely in Idanha-a-Nova where it is observed risk increase up to 45 points, changing from moderately vulnerable to highly vulnerable and, therefore, stressing the importance of anthropogenic activities.

**Keywords**

Naturtejo Geopark • Groundwater • DRASTIC • DRASTICAI • GIS

## 1 Introduction

Nowadays, groundwater vulnerability assessment has become a useful tool for groundwater contamination prevention. Groundwater vulnerability maps provide useful data to protect groundwater resources and to work as a tool for water management with changes in agricultural patterns and land use applications (Babiker et al. 2005; Albuquerque et al. 2013; Awawdeh et al. 2014; Singh et al. 2015). Several authors acknowledge two different types of groundwater vulnerability, the intrinsic vulnerability and extrinsic or specific vulnerability (Stiger et al. 2006): the first term as a function of hydrogeological factors and the second one defined through the potential anthropogenic influence. The most widely used method of vulnerability analysis is the DRASTIC index (Aller et al. 1987), as it is easy to compute with the minimum data requirement.

The main aim of this study was the evaluation of the groundwater vulnerability to contamination, in the Naturtejo Geopark (Fig. 1), using a modified DRASTIC method in a GIS environment. The modified DRASTIC index (DRASTICAI) was made by assigning a new attribute designated as anthropogenic influence.

## 2 Materials and Methods

The DRASTIC model is constructed using combined spatial datasets on depth to groundwater (*D*), aquifer recharge (*R*), aquifer media (*A*), soil media (*S*), topography (*T*), impact of

T. Albuquerque (✉) · N. Roque
CERNAS|QRural/Polytechnic Institute of Castelo Branco, Castelo Branco, Portugal
e-mail: teresal@ipcb.pt

T. Albuquerque
ICT/University of Évora, Évora, Portugal

J. Rodrigues
Geoparque Naturtejo, Idanha-a-Nova, Portugal

I. M. Antunes
ICT/University of Minho, Braga, Portugal

C. Silva
University of Lisbon/IDL, Lisboa, Portugal



M. Abrunhosa et al. (eds.), *Advances in Geoethics and Groundwater Management: Theory and Practice for a Sustainable Development*, Advances in Science, Technology & Innovation,
https://doi.org/10.1007/978-3-030-59320-9_46

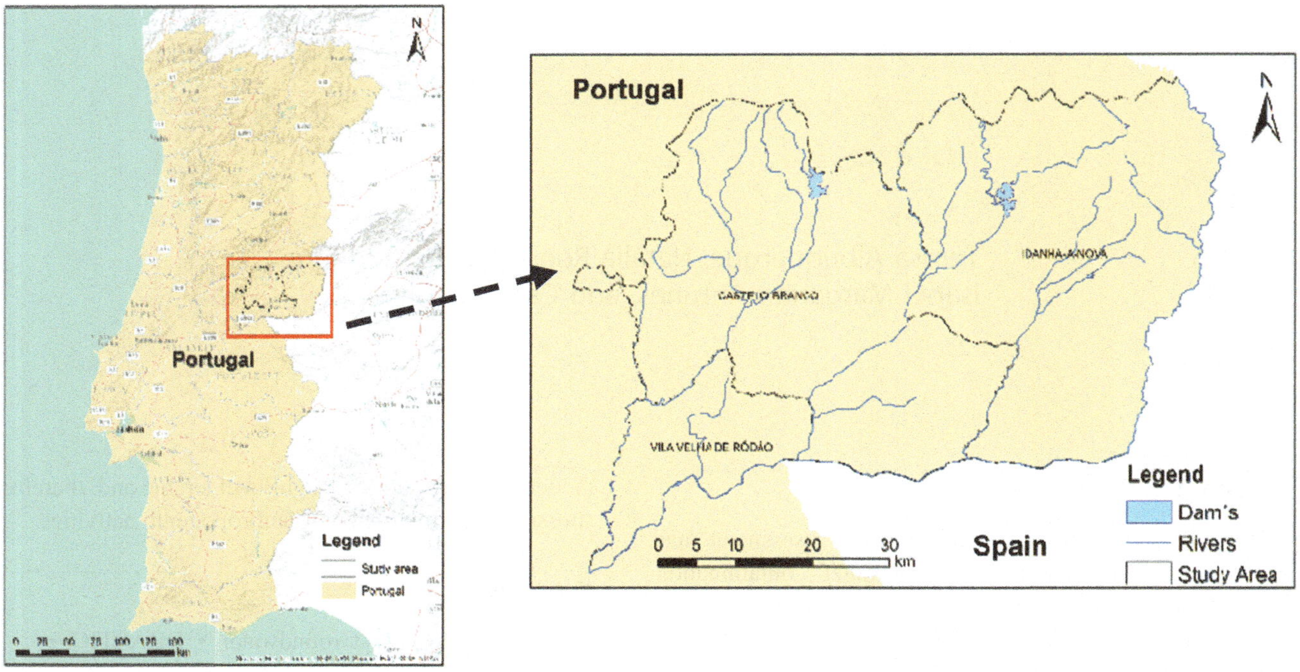


**Fig. 1** Study area: Vila Velha de Rodão, Castelo Branco and Idanha-a-Nova municipalities integrating the Naturtejo Geopark

the vadose zone (*I*) and hydraulic conductivity (*C*) of the aquifer (Aller et al. 1987). The purpose of the DRASTIC index implies multiplying each factor weight (Table 1) by its category rating (Table 2) as follows:

$$\text{DRASTIC} = D_r * D_w + R_r * R_w + A_r * A_w + S_r * S_w + T_r * T_w + I_r * I_w + C_r * C_w \quad (1)$$

In this study, one extra parameter was added to the DRASTIC model to map the groundwater vulnerability in the study area more accurately, including the *anthropogenic influence*. This new parameter, *anthropogenic influence* (AI), was assigned a weight value equal to 5, and the modified DRASTIC index, DRASTICAI, computed using the following equation:

$$\text{DRASTICAI} = D_r * D_w + R_r * R_w + A_r * A_w + S_r * S_w + T_r * T_w + I_r * I_w + C_r * C_w + AI_r * AI_w \quad (2)$$

where ***D*** is depth to groundwater, ***R*** is recharge rate (net), ***A*** is aquifer media, S is soil media, ***T*** is topography (slope), ***I*** is impact of the vadose zone, ***C*** is conductivity (hydraulic) of the aquifer, and ***AI*** is anthropogenic influence (Table 1):

ArcGIS 10 was used to process the datasets and to create the eight layers, corresponding to the eight considered attributes, and groundwater vulnerability maps by overlaying the available information (Fig. 2).

**Table 1** Assigned weights for DRASTIC parameters

| Parameters | Weight |
|---|---|
| Depth | 5 |
| Recharge | 4 |
| Aquifer media | 3 |
| Soil media | 2 |
| Topography | 1 |
| Impact of vadose zone | 5 |
| Hydraulic conductivity | 3 |
| **Anthropogenic influence** | **5** |

*Source* Aller et al. (1987) and DRASTICAI

**Table 2** DRASTIC

| DRASTIC/DRASTICAI | Range | Rating |
|---|---|---|
| Depth to groundwater | <1.5 m | 10 |
| Recharge rate (net) | 51–102 mm/year | 3 |
| Aquifer media | Metamorphic\|igneous rock<br>Stratified arenite and limestone | 3<br>6 |
| Soil media | Thin or absent | 10 |
| Topography (slope) | <2%<br>2–6%<br>6–12%<br>12–18%<br>>18% | 10<br>9<br>5<br>3<br>1 |
| Impact of the vadose zone | Shale and clay shale<br>Stratified arenite and limestone<br>Metamorphic \| Igneous rock | 3<br>6<br>4 |
| Hydraulic conductivity of the aquifer | <4.1 (m/d) | 1 |
| Anthropogenic index (*Source* Singh et al. 2015) | Built up with high density<br>Built up with medium density<br>Built up with low density<br>Built up with very low density<br>Agriculture<br>Forest<br>Water body<br>Shrub land<br>Waste land | 9<br>8<br>7<br>5<br>5<br>2<br>1<br>2<br>1 |

*Source* Aller et al. (1987) and DRATICAI parameters

## 3 Results

For aquifer vulnerability assessment of the study area, seven and eight thematic maps were prepared for the DRASTIC and the DRASTICAI indices computation, respectively (Fig. 3).

The DRASTIC risk map shows two different levels of vulnerability: low and moderate (Fig. 4a). The northern Idanha-a-Nova and Castelo Branco areas show low vulnerability (105–119) as the remaining territory is moderately vulnerable (120–138). However, when analysing the DRASTICAI map, it is possible to identify considerable changes in the spatial patterns of vulnerability (Fig. 4b). Indeed, five levels of growing vulnerability, from low to high, can be acknowledged. Idanha-a-Nova municipality is the most affected by the anthropogenic influence due to intensive farming activities.

## 4 Discussion

Land use parameters can significantly affect hydrogeological parameters. The properties of hydrogeological parameters can be changed by the use of pesticides, the addition of urban and industrial wastes, leakages from septic tanks and waste dumping sites. Land use classification of the study area showed that a major portion of the area is used for agriculture (Fig. 3). Groundwater is more vulnerable to nitrate concentration in agricultural fields. In groundwater systems, nitrate distribution principally depends upon the soil dynamics, recharge rate, groundwater movement and

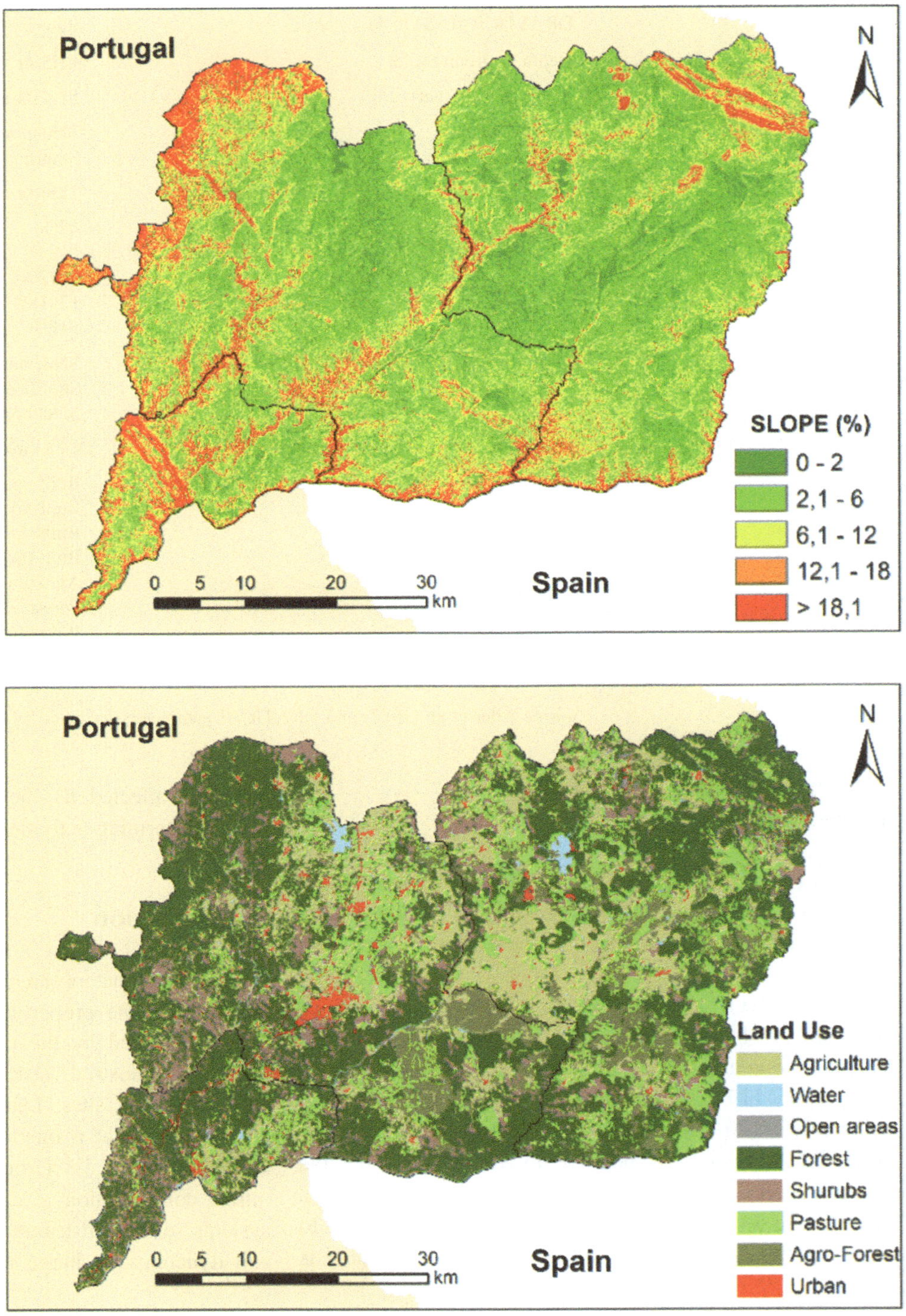


**Fig. 2** Maps corresponding to the topography and anthropogenic influence

on-ground nitrogen loading (Shirazi et al. 2013). The study area is significantly influenced by agricultural activities.

The algebraic subtraction between the DRASTIC and DRASTICAI maps shows an important contribution of the anthropogenic influence (Fig. 5). It is possible to mention all over the surveyed area, but specifically in Idanha-a-Nova municipality, a rousing effect from low/moderate vulnerability up to highly vulnerable.

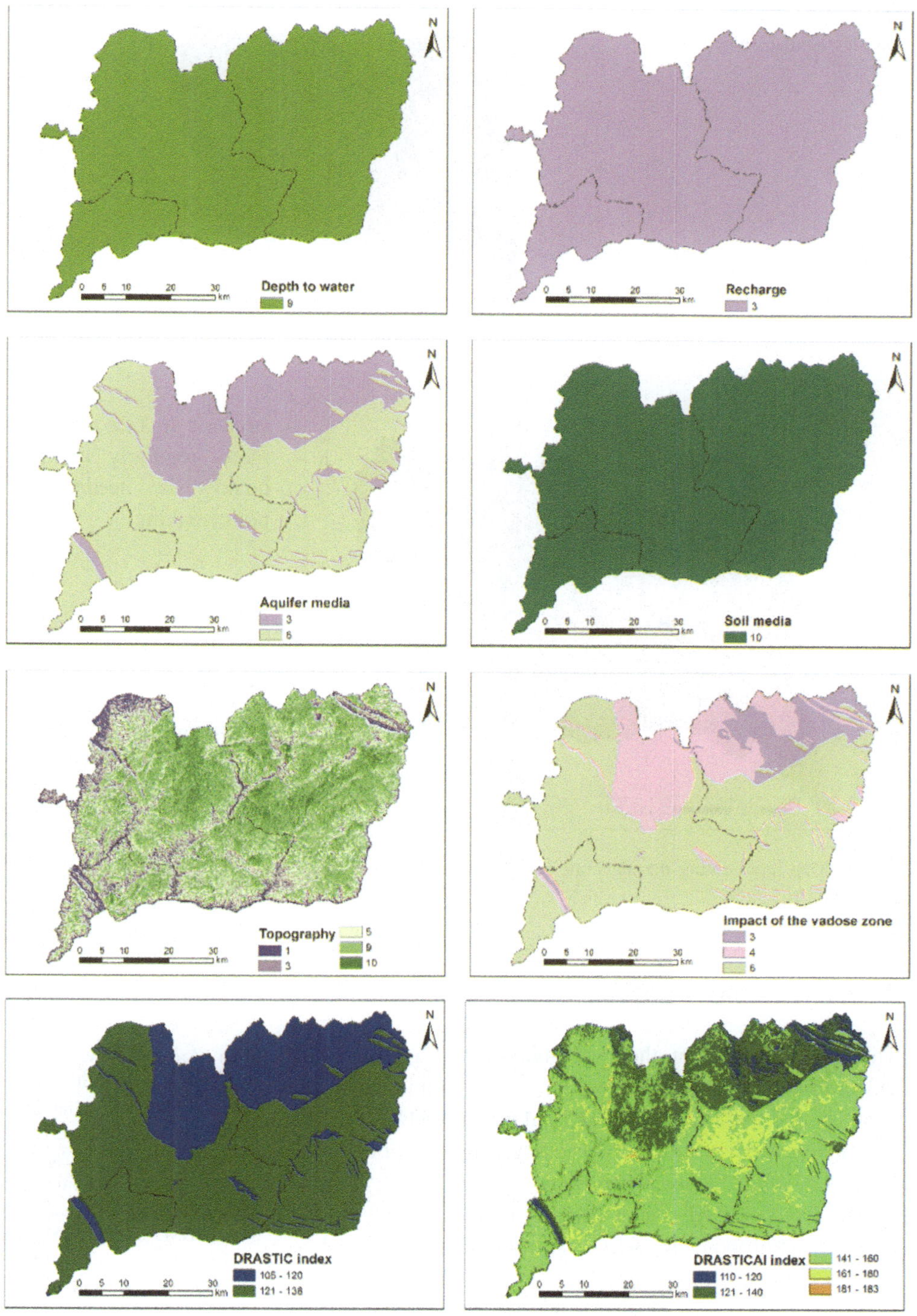


**Fig. 3** DRASTIC and DRASTICAI layer attributes

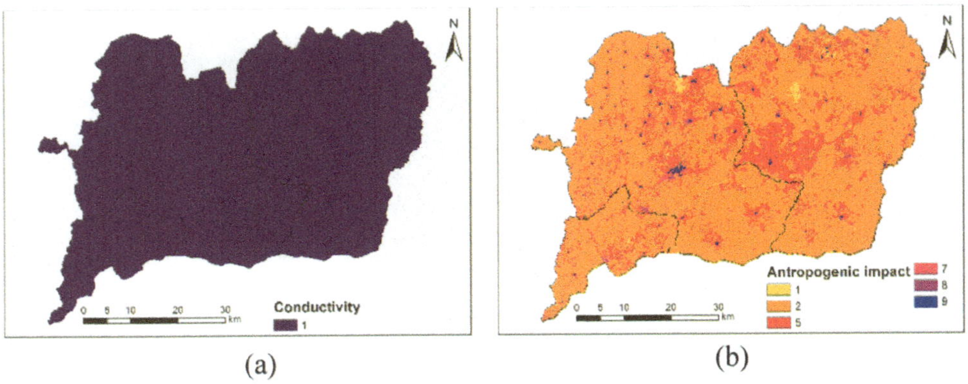


**Fig. 4** DRASTIC (**a**) and DRASTICAI (**b**) maps

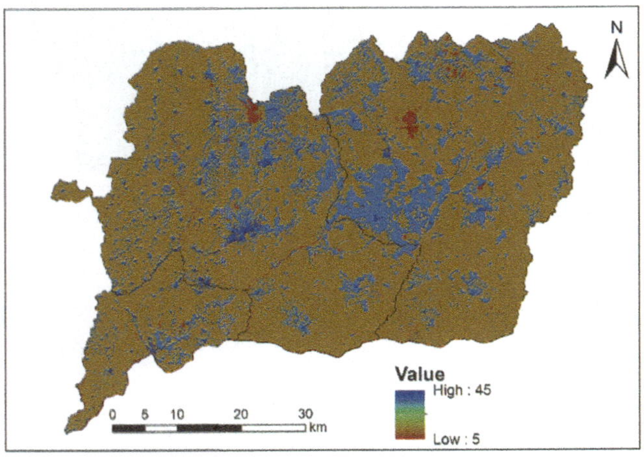


**Fig. 5** Map representing the algebraic subtraction of DRASTIC and DRASTICAI

## 5 Concluding Remarks

This survey aimed at the evaluation of the groundwater vulnerability to contamination, in the Naturtejo Geopark area, using a modified DRASTIC index, DRASTICAI. This new index was constructed by adding a new attribute designated as *anthropogenic influence.*

The DRASTICAI spatial patterns indicate a clear influence of anthropogenic activities, mainly in the Idanha-a-Nova municipality.

Water is one of the most strategic resources in the world. Portugal has important resources of groundwater that may be strategic to face the expected dry years to come. Furthermore, regularly monitoring and evaluating groundwater quality is needed for integrated management and policymaking.

## References

Albuquerque MTD, Lobón GS, Oliveira SF, Martínez-Alegría R, Antunes IMH (2013) Spatio-temporal groundwater vulnerability assessment: a coupled remote sensing and GIS approach for historical land cover reconstruction. Wat Res Manag 27(13):4509–4526

Aller L, Bennet T, Lehr JH, Petty RJ (1987) DRASTIC: a standardized system for evaluating groundwater contamination potential using hydrogeologic settings. J Geol Soc India 29:23–37

Awawdeh M, Obeidat M, Zaiter G (2014) Groundwater vulnerability assessment in the vicinity of Ramtha wastewater treatment plant, North Jordan. Appl Wat Sci 5:321–334

Babiker I, Mohammed M, Hiyama T, Kato K (2005) A GIS-based DRASTIC model for assessing aquifer vulnerability in Kakmigahara heights, Gifu Prefecture, central Japan. Sci Total Environ 345:127–140

Singh A, Srivastav SK, Kumar S, Chakrapani GJ (2015) A modified-DRASTIC model (DRASTICA) for assessment of groundwater vulnerability to pollution in an urbanized environment in Lucknow, India. Environ Earth Sci 74:5475–5490

Stigter T, Almeida P, Carvalho D, Ribeiro L (2006) Evaluation of an intrinsic and a specific vulnerability assessment method in comparison on a regional scale. Hydrogeol J 14(3):79-99A

Shirazi SM, Imran HM, Akib S, Yusop Z, Harun ZB (2013) Groundwater vulnerability assessment in the Melaka State of Malaysia using DRASTIC and GIS techniques. Environ Earth Sci 70:2293–2304

# Concentration of Chemical Species in Piezometers in an Agricultural Watershed, Santa Catarina, Southern Brazil

Mateus Melo, Adilson Pinheiro, Edson Torres, Gustavo Piazza, and Vander Kaufmann

**Abstract**

This study aims to analyze the concentration of chemical species in piezometers installed in riparian forests and in pasture, and in river, in Concórdia watershed, Lontras, SC, Brazil. Water samples were collected during both baseflow periods and storm events. Samples were analyzed to determine the concentration of chloride ($Cl^-$), nitrate ($NO_3^-$), phosphate ($PO_4^{3-}$), sulfate ($SO_4^{2-}$) and the carbon series of total organic carbon (TOC), inorganic carbon (IC) and total carbon (TC). The carbon series concentrations were bigger in summer. TOC concentrations were higher in piezometers than in river. Higher IC concentration in summer may be related to higher turbidity related to more intense rain events. The mean anion concentrations were higher in the piezometer located in the watershed outlet. Concentrations in river were more diluted when compared to piezometers.

**Keywords**

Agricultural watershed • Chemical species • Concentration

## 1 Introduction

One of the main drivers of chemical species is precipitation, which in turn affects groundwater quality through the infiltration process (Aubert et al. 2013). Know these impacts, it is an important step toward fostering geoethics, in view of sustainability of geosphere and humankind's activities (Peppoloni and Di Capua, 2015).

This work analyzed the concentration of chemical species in piezometers in riparian forest and in pasture, and in the river, in Concórdia watershed, Lontras, SC, Brazil. The study aims to generate knowledge of the local groundwater quality, seasonally and by monitoring point.

## 2 Materials and Methods

The study area was the Concórdia watershed, in the municipality of Lontras, Santa Catarina, Brazil. The watershed has an area of 30.93 $km^2$ and is located at latitude 27° 11′ 17.0″ S and longitude 49° 29′ 40.1″ W. Agricultural activities occupy approximately 15 $km^2$ (Piazza et al. 2014). Rural properties are characterized by small and medium sized (Lubitz et al. 2013). No-tillage, minimum tillage and conventional tillage are adopted as soil management (Pinheiro et al. 2009). The main crops produced are corn, beans, irrigated rice, tobacco, tomatoes and cabbage. Planted forests are composed of eucalyptus and pine, and riparian areas used for grazing with the constant presence of cattle.

53 water samples were collected in 3 piezometers every 2–3 weeks from January 01, 2012, to March 31, 2017. PZMC (depth of 2.6 m) is a piezometer located in area of riparian forest and PZ2127 (depth of 5.1) and PZ3 (depth of 3.8 m) area located in pasture areas. The PZ2127 is located near the watershed's outlet (Fig. 1). In the river section next to PZ2127 53, water samples were also sampled at the same sampling frequency.

A syringe connected to a silicone tube was used to collect water from the piezometers. Following suction, 100 mL samples were collected and stored in a polyethylene bottle followed the national standard—NBR 9898 (ABNT 1987).

Analyzed samples were used to determine the concentration of chloride ($Cl^-$), nitrate ($NO_3^-$), phosphate ($PO_4^{3-}$), sulfate ($SO_4^{2-}$) and the carbon series of total organic carbon (TOC), inorganic carbon (IC) and total carbon (TC).

M. Melo (✉) · A. Pinheiro · E. Torres · G. Piazza · V. Kaufmann
Environmental Engineering Postgraduate Program, Regional University of Blumenau, Blumenau, Brazil
e-mail: mateusnvm@gmail.com



M. Abrunhosa et al. (eds.), *Advances in Geoethics and Groundwater Management: Theory and Practice for a Sustainable Development*, Advances in Science, Technology & Innovation,
https://doi.org/10.1007/978-3-030-59320-9_47

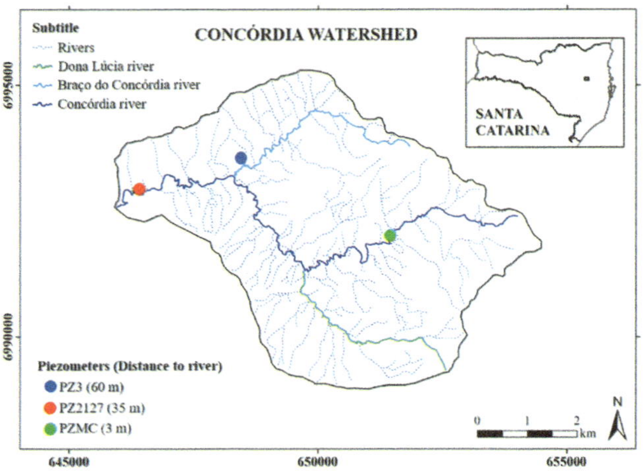


**Fig. 1** Location of PZMC, PZ2127 and PZ3 piezometers in the Concórdia watershed, Lontras, Santa Catarina, Southern Brazil

Analysis was performed using Dionex AG4A ion exchange chromatograph. Total carbon (TC), total organic carbon (TOC) and inorganic carbon (IC) were quantified using SHIMADZU model TOC—V CPH (Kaufmann et al. 2014).

The data analysis took place on seasonal scale and by monitoring point. The mean concentrations were submitted to ANOVA tests and the means compared by Tukey test, with a significance level of 5%. Data were organized in Excel software, and statistical tests were performed in PAST (version 3.25) software.

## 3 Results

The mean concentrations for river, PZMC, PZ2127 and PZ3 for seasons do not present statistical difference between monitoring points, according to ANOVA, post hoc Tukey (significance level = 5%). These results can be observed in Table 1 for river and PZMC and in Table 2 for PZ2127 and PZ3.

The mean, minimum and maximum concentrations of chemical species can be observed in Table 3.

## 4 Discussion

In seasonal analysis, no significant differences were found for anions, TOC, IC and TC. Results only indicated higher in the summer. Libânio et al. (2000) mention that although there are no legal parameters for the TOC limit, they recommend using the maximum value of 3 ppm, because within this universe may be found several toxic substances, such as pesticides. This value was exceeded especially in summer for the piezometers.

According to Nascimento and Barbosa (2005), the higher values of inorganic carbon may be related to the more intense rainfall events, associated with the season, in which higher runoff flows are generated, dragging by-products from these arable areas into the soil and river.

Nascimento and Barbosa (2005) reached a reference value for groundwater quality through the analysis of several studies and documents, including the Decree 518/2004 of the Brazilian Ministry of Health (Brasil 2004), World Health Organization (WHO) and United States Environmental Protection Agency (US/EPA).

In relation $Cl^-$, concentrations were below reference. $Cl^-$ can be a good indicator of pollution from landfills and dumps (Manzione 2015). To $NO_3^-$, Concórdias's concentrations were lower than the reference value, but some samples from PZ2127 and PZ3 exceeded the recommended value. It is noteworthy that these are mean values. According to Manzione (2015), values above 5 mg/L may be indicative of contamination (ditches, dumps, sewers, cemeteries, fertilization, animal remains). In the catchment, the main source of nitrogen is from fertilizers that are applied to the watershed during cropping season. The $PO_4^{3-}$ concentration values were below the reference limit, but some samples exceeded the recommended limit for all matches. Artificial sources P are the fertilizers used in agriculture and detergents. Phosphorus-based fertilizers are applied to the watershed.

It is noteworthy that Brazilian legislation does not recognize phosphorus as a soil contaminant, reforming the farmer's perception that phosphate fertilizer should obey critical and economic rather than environmental rivers

**Table 1** Mean concentrations (mg/L) of chloride ($Cl^-$), nitrate ($NO_3^-$), phosphate ($PO_4^{3-}$), sulfate ($SO_4^{2-}$) and mean concentrations (ppm) of total organic carbon (TOC), inorganic carbon (IC) and total carbon (TC) for river and PZMC

| CS | River | | | | PZMC | | | |
|---|---|---|---|---|---|---|---|---|
| | Su | Fa | Wi | Sp | Su | Fa | Wi | Sp |
| $Cl^-$ (mg/L) | 2.98 | 3.14 | 3.14 | 2.81 | 3.12 | 4.33 | 3.78 | 2.81 |
| $NO_3^-$ (mg/L) | 2.13 | 3.79 | 3.03 | 1.69 | 2.32 | 4.37 | 1.44 | 1.69 |
| $PO_4^{3-}$ (mg/L) | 0.11 | 0.09 | 0.06 | 0.07 | 0.16 | 0.60 | 0.04 | 0.07 |
| $SO_4^{2-}$ (mg/L) | 2.17 | 1.98 | 2.18 | 2.56 | 2.22 | 2.30 | 2.46 | 2.56 |
| TOC (ppm) | 4.07 | 3.94 | 2.79 | 1.77 | 6.00 | 2.73 | 1.47 | 1.77 |
| IC (ppm) | 9.83 | 9.45 | 7.13 | 11.71 | 15.87 | 14.27 | 11.10 | 11.71 |
| TC (ppm) | 6.37 | 5.52 | 4.28 | 9.94 | 11.01 | 11.37 | 9.59 | 9.94 |

Note *CS* means chemical species, *Su* means summer, *Fa* means fall, *Wi* means winter, and *Sp* means spring

**Table 2** Mean concentrations (mg/L) of chloride ($Cl^-$), nitrate ($NO_3^-$), phosphate ($PO_4^{3-}$), sulfate ($SO_4^{2-}$) and mean concentrations (ppm) of total organic carbon (TOC), inorganic carbon (IC) and total carbon (TC) for PZ2127 and PZ3

| CS | PZ2127 | | | | PZ3 | | | |
|---|---|---|---|---|---|---|---|---|
| | Su | Fa | Wi | Sp | Su | Fa | Wi | Sp |
| $Cl^-$ (mg/L) | 9.15 | 11.91 | 12.48 | 22.74 | 4.47 | 3.11 | 5.56 | 4.49 |
| $NO_3^-$ (mg/L) | 18.27 | 16.23 | 21.61 | 14.48 | 12.42 | 8.94 | 11.66 | 14.33 |
| $PO_4^{3-}$ (mg/L) | 0.07 | 1.46 | 0.03 | 0.12 | 0.17 | 0.23 | 0.04 | 0.07 |
| $SO_4^{2-}$ (mg/L) | 3.26 | 2.76 | 3.90 | 6.33 | 1.35 | 0.75 | 0.94 | 1.04 |
| TOC (ppm) | 8.27 | 5.93 | 2.10 | 4.03 | 6.40 | 2.28 | 1.87 | 1.30 |
| IC (ppm) | 20.75 | 17.24 | 11.15 | 10.75 | 13.62 | 14.65 | 11.02 | 9.74 |
| TC (ppm) | 14.44 | 11.30 | 8.90 | 6.72 | 7.26 | 12.37 | 9.01 | 8.44 |

Note *CS* means chemical species, *Su* means summer, *Fa* means fall, *Wi* means winter, and *Sp* means spring

**Table 3** Mean, minimum and maximum concentrations (mg/L) of chloride ($Cl^-$), nitrate ($NO_3^-$), phosphate ($PO_4^{3-}$), sulfate ($SO_4^{2-}$) and minimum, mean and maximum concentrations (ppm) of total organic carbon (TOC), inorganic carbon (IC) and total carbon (TC) for river, PZMC, PZ2127 and PZ3

| | River | | | PZMC | | | PZ2127 | | | PZ3 | | |
|---|---|---|---|---|---|---|---|---|---|---|---|---|
| | Me | Mi | Ma | Me | Mi | Ma | Me | Mi | Ma | Me | Mi | Ma |
| $Cl^-$ (mg/L) | 3.0a | 1.01 | 7.35 | 3.5a | 0.84 | 18.47 | 14.1b | 1.58 | 60.52 | 4.4a | 0.89 | 29.52 |
| $NO_3^-$ (mg/L) | 2.7a | 0.91 | 11.19 | 2.5a | 0.01 | 18.95 | 17.7b | 2.29 | 43.40 | 11.8c | 2.11 | 25.60 |
| $PO_4^{3-}$ (mg/L) | 0.1a | 0.00 | 0.47 | 0.1a | 0.00 | 6.34 | 0.1a | 0.00 | 15.11 | 0.1a | 0.00 | 1.39 |
| $SO_4^{2-}$ (mg/L) | 2.2a | 0.81 | 5.38 | 2.4ab | 0.73 | 6.14 | 4.1b | 0.22 | 20.34 | 1.0a | 0.12 | 4.97 |
| TOC (ppm) | 3.1a | 0.25 | 7.05 | 3.0a | 0.15 | 17.55 | 5.1a | 0.25 | 25.76 | 3.0a | 0.13 | 9.37 |
| IC (ppm) | 6.5a | 2.04 | 13.07 | 10.5a | 5.31 | 25.33 | 10.3a | 4.38 | 46.76 | 9.2a | 1.93 | 27.93 |
| TC (ppm) | 9.5a | 0.74 | 8.71 | 13.2a | 3.05 | 20.44 | 15.0a | 0.52 | 44.52 | 12.3a | 0.45 | 21.47 |

Note *Me* means mean values, *Mi* means minimum values, and *Ma* means maximum values

(Pinheiro et al. 2013). The $SO_4^{2-}$ concentration was below the reference limit and recommended limit. The increase in groundwater sulfate could be related to the application of fertilizers such as ammonium sulfate (Midões et al. 2001). The mean anion concentrations indicate that the outlet of the watershed PZ2127 presents higher concentrations of these chemical species, except for phosphate. Phosphate presented more similar concentrations in the different monitoring points compared to the other chemical species.

In general, the river presented lower concentrations of anions, TOC, IC and TC compared to piezometers. It can be explained by a dilution effect of this chemical species. TOC, IC and TC concentrations presented more similar concentrations in the different monitoring points compared to the anions.

## 5 Concluding Remarks

The carbon series species concentrations were higher in summer. The mean anion concentrations indicate a higher concentration in the piezometer located near the watershed outlet. The river presented a dilution effect.

In the geoethics context, for the efficient management of groundwater in agricultural watersheds, in view of their different land uses, impact and the demands of activities, it is necessary to carry out data monitoring and management and studies about the water quality to make recommendations to producers.

**Acknowledgements** This work was carried out with the support of the Coordination of Improvement of Higher Education Personnel (CAPES)—Brazil (Financing Code 001 and Process 88881.143991/2017-01). The authors thank CAPES for granting the master and doctoral scholarship to the first and third authors, respectively, and to National Council for Scientific and Technological Development (CNPq) (Process 309980/2017-8) for the second author's research productivity scholarship. The fourth author thanks Foundation of Support for Research of Santa Catarina (FAPESC) for granting a research scholarship.

## References

ABNT (1987) NBR 9898: preservação e técnicas de amostragem de efluentes líquidos e corpos receptores. Rio de Janeiro, Brasil

Aubert AH, Gascuel-Odoux C, Merot Ph (2013) Annual hysteresis of water quality: a method to analyse the effect of intra- and inter-annual climatic conditions. J Hydrol 478(25):29–39

Brasil (2004) Portaria nº 518. Estabelece os procedimentos e responsabilidades relativos ao controle e vigilância da qualidade da água para consumo humano e seu padrão de potabilidade e dá outras providências. Brasília, Brasil

Kaufmannm V, Pinheiro A, Castro NMR (2014) Transporte de compostos de carbono, nitrogênio e fósforo pelo escoamento da água em solos agrícolas na região Sul do Brasil. Rev Bras Rec Hídr 9(1):75–86

Libânio M, Libânio PAC, Costa BMP, von Sperling E (2000) Avaliação da relevância do carbono orgânico total como parâmetro de caracterização de águas de abastecimento. Revista Brasileira De Recursos Hídricos 5(4):41–55

Lubitz E, Pinheiro A, Kaufmann V (2013) Simulação do transporte de sedimentos, nitrogênio e fósforo na bacia do ribeirão Concórdia. SC Rev Bras Rec Hídr 18(2):39–54

Manzione RL (2015) Águas subterrâneas: conceitos e aplicações sob uma visão multidisciplinar. Jundiaí

Midões CF, Fernandes J, Costa CGJ (2001) Água subterrânea: conhecer para proteger e preservar. IGM, Lisboa. https://hdl.handle.net/10400.9/488

Nascimento SAM, Barbosa ISF (2005) Qualidade da água do aqüífero freático no Alto Cristalino de Salvador, Bacia do rio Lucaia, Salvador. Bahia Rev Bras Geoc 35(4):540–543

Peppoloni S, Di Capua G (2015) Geoethics: the role and responsibility of geoscientists. SP419, Geological Society of London, London. https://doi.org/10.1144/SP419

Piazza GA, Depine H, Pinheiro A, Kaufmann V, Gonçalves Júnior AC, Oliveira MHC (2014) Spatial distribution of soil attributes in the Concórdia river watershed in Southern Brazil. Environ Qualit Manag 24:1–12

Pinheiro A, Kaufmann V, Schneiders D, Gotardo R (2013) Transporte de sedimentos e espécies químicas em áreas de reflorestamentos e pastagem com base em chuva simulada. Revista Ambiente Água an Interdiscipl J Appl Sci 8(2):109–123

Pinheiro A, Teixeira LP, Kaufmann V (2009) Capacidade de infiltração de água em solos sob diferentes usos e práticas de manejo agrícola. Revista Ambiente Água Interdiscipl J Appl Sci 4 (2):188–199

# Effect of Subsurface Geological Conditions on Variation of Groundwater Quality in Part of Kurunegala, Sri Lanka

Ashvin Wickramasooriya, Stella Gunarathne, and Surangi Ekanayaka

**Abstract**

Majority of the people in rural areas in Sri Lanka still depend on surface water bodies, dug wells and deep tube wells for their water requirements. Most surface water bodies in dry zone and arid areas like Kurunegala district go dry during dry seasons. Therefore, people in Kurunegala district mainly depend on dug well and deep tube-well water during dry seasons. There are a few water quality issues recorded in this area. After preparing water quality distribution maps for calcium and total hardness, iron, fluoride and nitrate using groundwater quality analysis data of sixty-three wells, it was found that there is a different distribution pattern of the above parameters in southwestern part of the Kurunegala district. To understand whether there is a relationship between subsurface geological conditions and the variation of different water quality parameters, borehole log data of the area are considered. After correlate with these log data, it was found that high concentrated iron, calcium and total hardness, and fluoride and concentrated wells are located closer to laterite and biotite gneiss, marble and biotite gneiss rocks, respectively. Wells with high nitrate concentration are located in highly populated areas and are located closer to toilet drainage pits within high permeable subsurface soil profiles. Therefore, water quality parameters are influenced by the existing subsurface geological conditions of the area.

**Keywords**

Tube well • Borehole • Water quality • Distribution

A. Wickramasooriya (✉)
Department of Geography, University of Peradeniya, Kandy, Sri Lanka
e-mail: awickramasooriya@gmail.com

S. Gunarathne · S. Ekanayaka
Postgraduate Institute of Humanities and Social Sciences, University of Peradeniya, Kandy, Sri Lanka

## 1 Introduction

Water scarcity and water quality are two issues that the Kurunegala district of Sri Lanka faces with. Kurunegala district which is one of the main agricultural-based regions in the country belongs to dry zone and arid zones (Panabokke 2007). Dental fluorosis and water contaminated with hard water are two main water quality issues in the area (Dissanayake and Weerasooriya 1985a, b; Chandrajith et al. 2012). Most of the people living in the area do not have pipe-bone water system; thus, they have to depend on surface and groundwater for their domestic requirements. This study is focused on the determination of the distribution pattern of five important water quality parameters in groundwater in the southwestern part of the Kurunegala district and their relationship to subsurface geological conditions of the area. To achieve this task, five Divisional Secretarial Divisions in the southwestern part of Kurunegala district are selected (Fig. 1).

## 2 Materials and Methods

Groundwater quality analysis data during August 2018 (after the SW monsoon period) of sixty-three tube wells which are collected from the Wariyapola National Water Supply and Drainage Board are used to produce geochemical distribution maps of calcium hardness, total hardness, fluoride, iron and nitrate in the study area. Thereafter, lowest and highest concentrated areas of each chemical parameter in the area are demarcated. Then borehole logs closer to the lowest and the highest groundwater quality parameters mentioned in the above are used to understand the subsurface geological conditions such as available rock types, their fracture intensity, existing soil types, their thicknesses, porosity and permeability. Further, to understand the groundwater distribution pattern and flow directions, geomorphology map has produced using 1:10,000 contour maps of the Survey



M. Abrunhosa et al. (eds.), *Advances in Geoethics and Groundwater Management: Theory and Practice for a Sustainable Development*, Advances in Science, Technology & Innovation,
https://doi.org/10.1007/978-3-030-59320-9_48

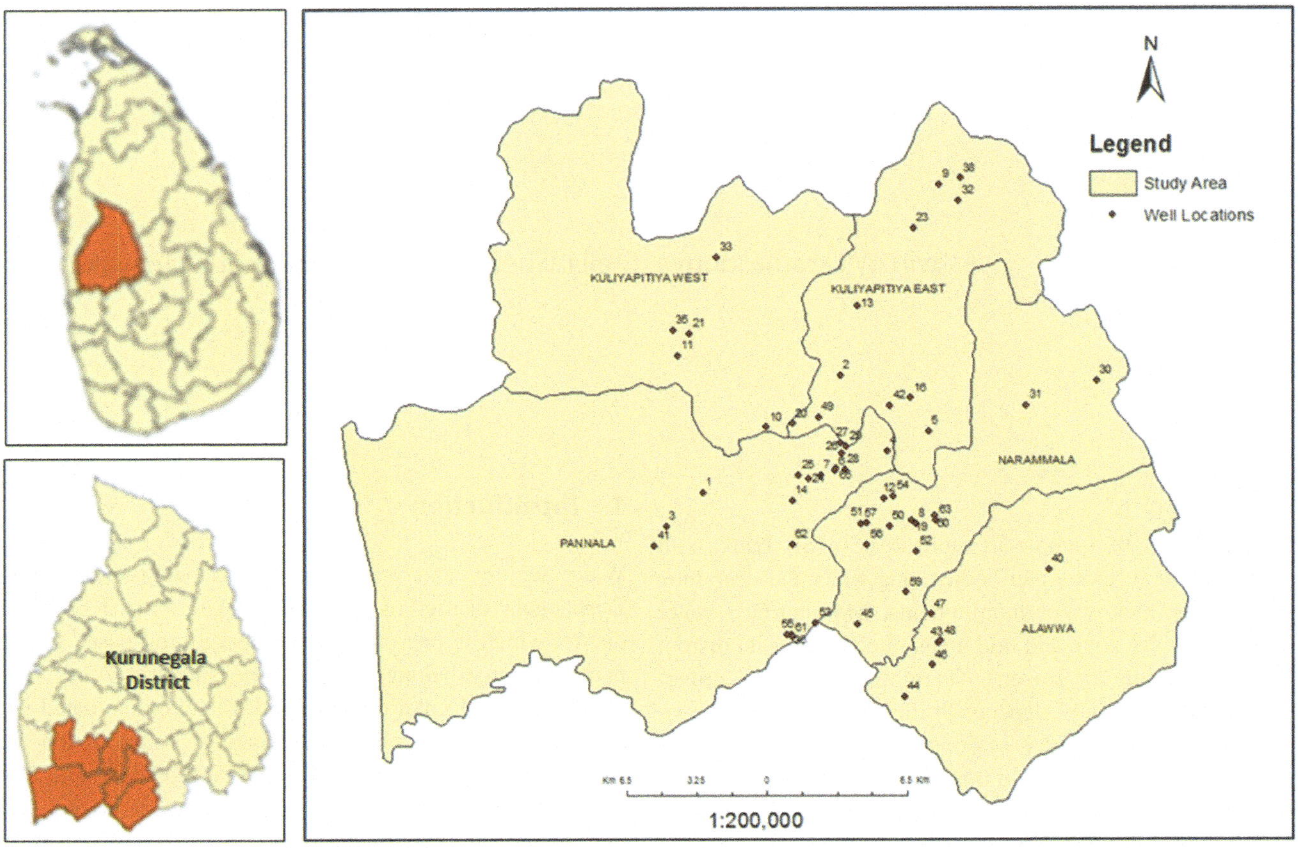


**Fig. 1** Study area (southwestern part of Kurunegala district, Sri Lanka)

Department of Sri Lanka. Groundwater distribution map has produced using depths to groundwater table of tube wells.

## 3 Results

### 3.1 Groundwater Quality Distribution in the Southwestern Part of Kurunegala District

According to the produced five geochemical distribution maps, it was found that their distribution patterns are different (Fig. 2). However, high concentration of calcium hardness and total hardness show in the northeastern part of the study area while fluoride and iron concentrations increase toward northern and southern part of the study area. The nitrate concentration increases toward southeastern direction.

### 3.2 Analysis of Borehole Log Data

Borehole log data are used to understand the subsurface geological conditions at selected sample locations. The sample locations of boreholes were selected closer to the highest and lowest concentrations of each groundwater quality parameter.

As the calcium hardness and the total hardness distribution patterns are almost similar, two borehole data logs each from closer to highest and lowest concentrations of calcium and total hardness were analyzed (Fig. 3) and compare the underground geological conditions at those locations. Similarly, two borehole log data each from closer to highest and lowest concentrations of fluoride and iron were considered and analyzed the status of the subsurface geological conditions of those locations as shown in Fig. 4 and Fig. 5, respectively. The nitrate concentration shows a different pattern of distribution (Fig. 3e). High concentrated nitrate in groundwater is found in highly populated southeastern part of the study area. However, it was found that two wells which are located very closer to each other (about 50 m apart), i.e., well no 49 and well no 26 (Fig. 6), are having very high and very low nitrate concentrate, respectively. Therefore, two borehole log data closer to those two wells are analyzed to identify the reason for such observation.

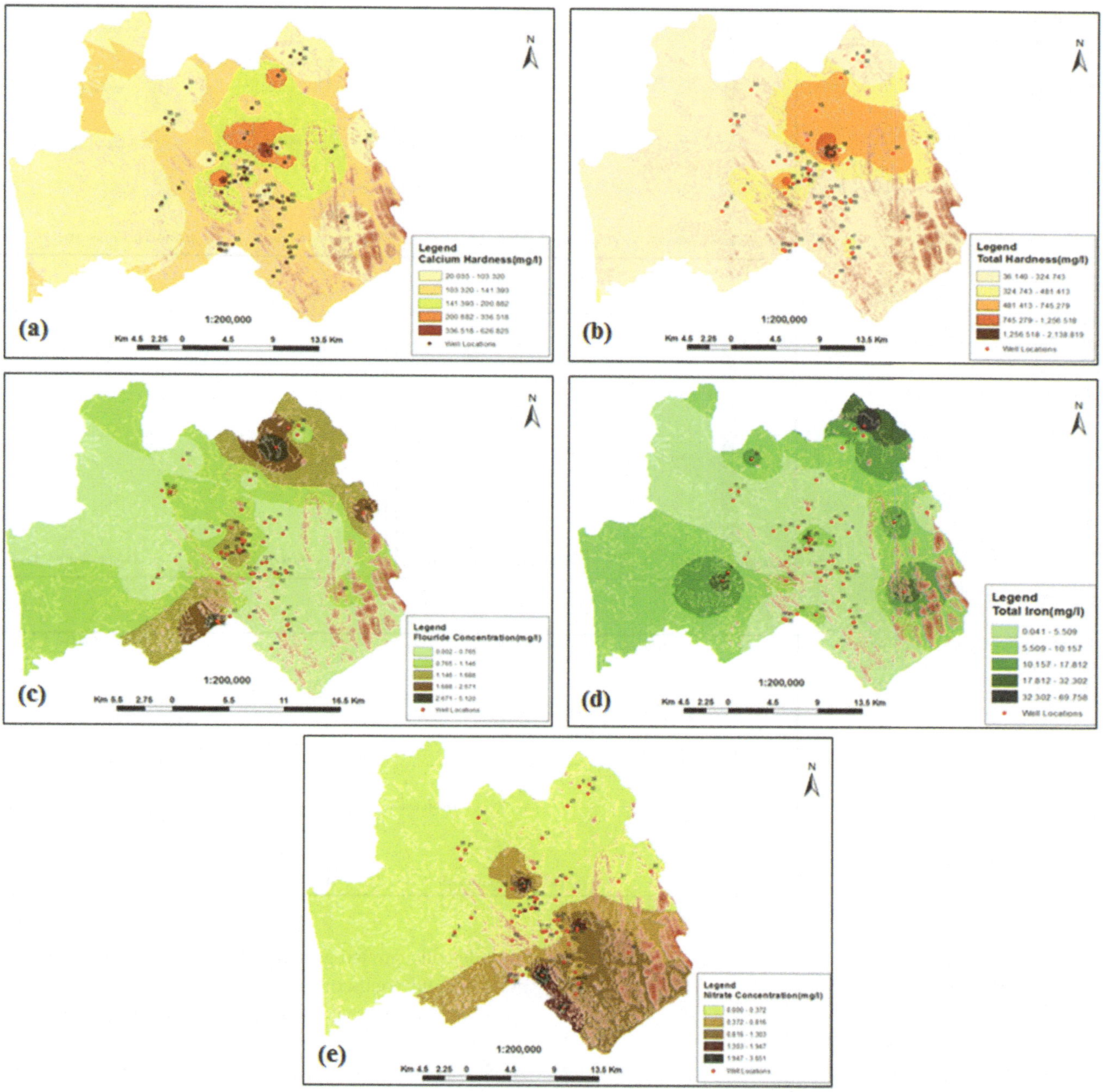


**Fig. 2** Distribution of chemical parameters in groundwater in southwestern part of Kurunegala district **a** calcium hardness, **b** total hardness, **c** fluoride, **d** iron, **e** nitrate

**Fig. 3** Borehole log data **a** closer to well no 42, **b** closer to well no 25, **c** closer to well no 41 and **d** closer to well no 33

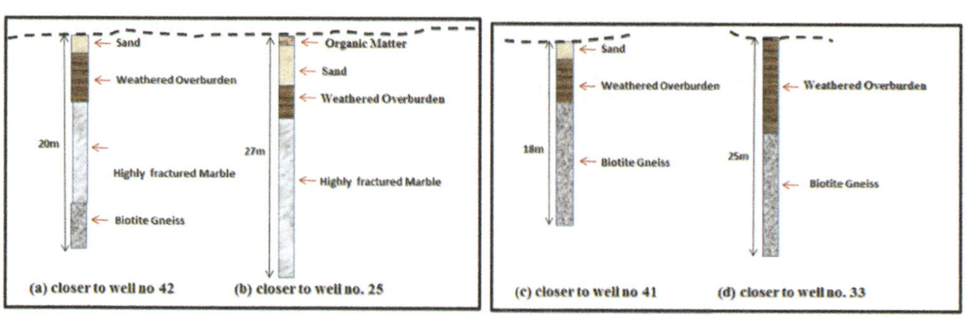

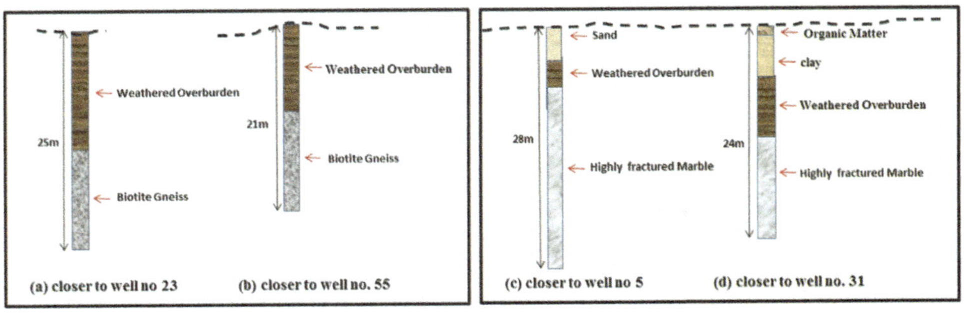


**Fig. 4** Borehole log data **a** closer to well no 23 **b** closer to well no 55, **c** closer to well no 5 and **d** closer to well no 31

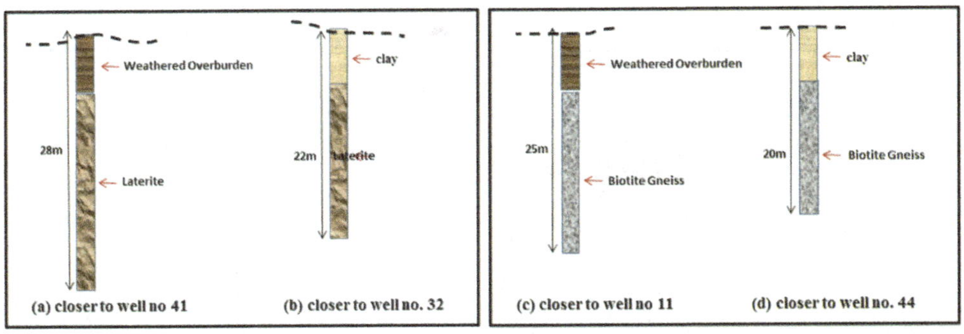


**Fig. 5** Borehole log data **a** closer to well no 41 **b** closer to well no 32, **c** closer to well no 11 and **d** closer to well no 44

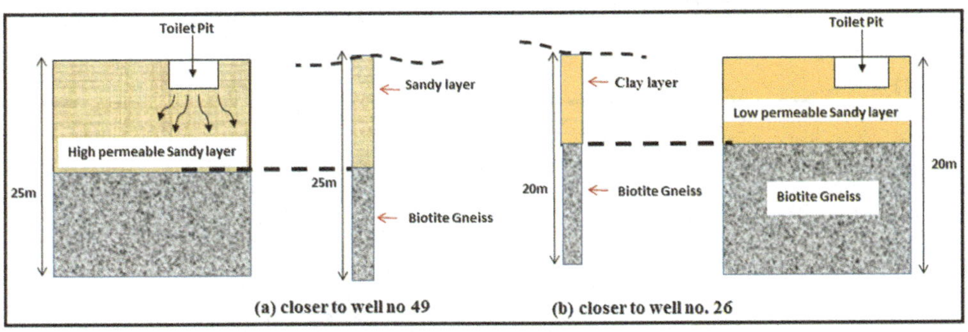


**Fig. 6** Borehole log data **a** closer to well no 49, **b** closer to well no 26

## 4 Discussion

According to prepared geochemical maps of the area, the calcium hardness and the total hardness concentrations increase toward the northeast of the study area. After comparing the borehole log data obtained from these areas and southwestern part of the study area, it was found that the northeast part consists of highly fractures marble layer about 5–8 m below the surface (Fig. 3a, b). The soil exists above the marble layer in this area is formed due to weathering of the marble layer. Therefore, $Ca^{+2}$ and $Mg^{+2}$ ions are contaminated with the groundwater which causes to increase the calcium hardness and total hardness in the northwestern region of the study area. In the southwestern part, the weathered overburden has formed due to weathering of biotite gneiss (Fig. 3c, d), and therefore, in this region, less amount of $Ca^{+2}$ and $Mg^{+2}$ ions are added to the groundwater and record less hardness in this area. It was noted that high fluoride concentrated groundwater has recorded in the part of northern and southern regions of the study area. Borehole log data showed that these areas consist of biotite gneiss underlain the soil profile. Biotite, i.e., $K_2(MgFe_2Al_2)Al_2Si_60_{20}(OH^-,F^-)$, contains $F^-$ and $Fe^{+2}$ ions, and these ions can be removed from the soil profile during chemical weathering process. Therefore, groundwater will be contaminated when these $F^-$ and $Fe^{+2}$ are leaching especially during rainy period. When examining the borehole log data in areas with low fluoride groundwater, areas consist of rock types which do not contain fluoride bearing minerals such as biotite. When comparing iron concentration of the groundwater and the borehole log data, it is obvious that high iron concentrated groundwater can be found in areas which consist of laterite which is composed of iron oxide minerals and biotite gneiss below the soil profile (Fig. 5). However, in some areas, fluoride and iron show similar pattern of distribution. When carefully analyzing borehole data at those locations, it was found that biotite can be released both $F^-$

and $F^{+2}$ to soil profile during chemical weathering process of biotite gneiss. Therefore, both biotite and laterite contributed to increase the iron concentration in groundwater in the study area.

Wells with high concentrated nitrate are associated with highly populated areas especially in the southeastern part of the study area (Fig. 3). However, when considering borehole log data, it was found that there is no significant relationship between available rock types with groundwater nitrate concentration. Even though there is no direct relationship between nitrate in groundwater and rock types, it was found that another two subsurface geological conditions, i.e., porosity and permeability of the available soil types, are influenced to change the groundwater nitrate concentration. Two wells, i.e., well no 49 and well no 26, are located within 50 m apart from each other (Fig. 1). However, well no 49 records high nitrate concentration, and well no 26 contains very low nitrate concentration (Fig. 6). After examining borehole log data closer to these two wells, it was found that a clay layer exists about 5 m below the surface closer to well no 26 while a sandy layer exists up to about 8 m below the surface closer to well no 49. Therefore, water seepage from toilet pits at two locations act differently as the clay layer and the sandy layer have different porosities and permeability. Therefore, it is possible that more nitrate removes from toilet pits can be contaminated with groundwater closer to well no 49 where there is a thick soil layer which exists below the surface while less seepage can be expected at well no 26 where there is an impermeable clay layer which exists below the surface (Fig. 6). According to groundwater contour map prepared for the area, it was noted that the groundwater flow is mainly toward north and northeastern directions where four chemical parameters other than nitrate are concentrated.

## 5 Concluding Remarks

It has been observed that concentrations of considered five chemical parameters, i.e., calcium hardness, total hardness, iron, fluoride and nitrate in groundwater of the study area, show a different pattern of distribution. Wells with high calcium and total hardness in groundwater are associated with highly fractures marble rock. Therefore, existing of marble which can be released $Ca^{+2}$ and $Mg^{+2}$ ions to groundwater is the main reason to record high calcium hardness and the total hardness in the area. Locations having biotite gneiss below the overburden record high fluoride concentrated in groundwater. Biotite in biotite gets weathered and releases $F^{-}$ to soil, and then groundwater can be contaminated with this $F^{-}$ which causes to have higher $F^{-}$ concentration than other areas. In some areas, distribution of $F^{-}$ and $Fe^{+2}$ shows a similar pattern of distribution. After analyzing the borehole log data closer to high concentrated $Fe^{+2}$ in the area, it was found that isolated laterite layers and biotite gneiss exist closer to those locations which cause to increase $Fe^{+2}$ concentration in groundwater. Even though there is no direct relationship between nitrate concentration in groundwater and existing rock types in the area, indirectly the subsurface conditions like porosity and permeability of soil are influenced to increase the nitrate concentration in groundwater at some locations.

According to the findings of this research, it can be concluded that the subsurface geological conditions such as mineral composition, fracture intensity of rocks, groundwater flow direction, porosity and permeability of soil types are influenced groundwater quality distribution in the southwestern part of the Kurunegala district.

**Acknowledgements** The manager, hydrogeologists and other staff members of the National Water Supply and Drainage Board at Wariyapola in Kurunegala district are highly acknowledged for their support to complete this research.

## References

Chandrajith R, Padmasiri JP, Dissanayake CB, Premathilaka KM (2012) Spatial distribution of fluoride in ground water of Sri Lanka. J Nat Sci Found Sri Lanka 40(4):303–309

Dissanayake CB, Weerasooriya SVR (1985a) A geochemical classification of groundwater of Sri Lanka. J Nat Sci Council Sri Lanka 13 (2):147–186

Dissanayake CB, Weerasooriya SVR (1985) The hydrochemical atlas of Sri Lanka. Natural Resources Energy and Science Authority of Sri Lanka, Sri Lanka

Panabokke CR (2007) Groundwater conditions in Sri Lanka: a geomorphic perspective. J Nat Sci Found Sri Lanka 36(1):113–114

# Nitrate Pollution in Groundwater of the Ouazi Basin: Case of Essaouira (Southwestern Morocco)

Otman El Mountassir, Mohammed Bahir, Driss Ouazar, and Paula M. Carreira

**Abstract**

Morocco is a semi-arid country with scarce and irregular precipitation; thus, surface water and groundwater resources are important for socioeconomic development. For this reason, aquifers are intensively exploited to satisfy growing agricultural, industrial and domestic needs. In addition to water scarcity, Morocco suffers from the degradation in groundwater quality, linked to the depletion of groundwater. In this paper, we zoom on the case of the Essaouira basin, where nitrate is one of the main pollutants that affect groundwater. The objective of this study is to obtain a better understanding of electrical conductivity and nitrate distributions in the aquifers during a period of 24 years. The results obtained indicate that the spatiotemporal distribution map of the electrical conductivity becomes very important from 1995 to 2018 and the $NO_3$ concentrations ranged from 0 to 375 mg/L based on 151 groundwater samples. The distribution of high nitrate content in groundwater is consistent with the urbanization and cultural practices, where it has sewage discharge. Groundwater extraction and rapid urbanization in the central of the study area accelerate the deterioration of groundwater quality. The results of this study could provide valuable information on the state of nitrate water concentrations in the Essaouira region.

**Keywords**

Pollution • Nitrate • Semi-arid • Groundwater • Essaouira

## 1 Introduction

Groundwater resources constitute significant water reserves, both quantitatively and qualitatively. Preserving the quality of groundwater is therefore a major concern for public health reasons (Bahir et al. 2018; Carreira et al. 2018). The deterioration of water quality generates significant health and environmental problems. Nitrate ($NO_3$) is a major pollutant in the world's groundwater. The study area is dominated by karst aquifers and is particularly vulnerable to contamination by nitrates of anthropogenic origin due to the rapid movement of water in their network of conduits (Bahir et al. 2019). The depletion of groundwater aquifers is a major issue in Morocco. Rapid declines in groundwater levels (0.5–2 m per year in average) are generally the result of (a) low groundwater recharge and (b) over-expansion of agricultural activities (Ait Kadi and Ziyad 2018). Concerning the hydrogeological aspect, the Ouazi sub-basin is a set of hydrogeological systems more or less independent, which corresponds to the various aquifer systems; in the Plio-Quaternary aquifer (shallow) in downstream part, the wall of this aquifer is formed by the Senonian gray marls. And in the Cenomanian–Turonian aquifer (deep) in upstream part, it is mainly formed by limestones and dolomitic-limestone layers (El Mountassir et al. 2020). The main objectives of this work were focused on two aspects: water quality assessment and analysis of temporal and spatial variations in groundwater nitrate.

O. El Mountassir (✉) · M. Bahir
High Energy and Astrophysics Laboratory, Faculty of Sciences Semlalia, Cadi Ayyad University, Marrakesh, Morocco
e-mail: otman.elmountassir@ced.uca.ma

M. Bahir · D. Ouazar
International Water Research Institute, Mohammed VI Polytechnic University, Ben Guerir, Morocco

P. M. Carreira
Centro de Ciências e Tecnologias Nucleares, C2TN/IST, Universidade de Lisboa, Lisbon, Portugal


M. Abrunhosa et al. (eds.), *Advances in Geoethics and Groundwater Management: Theory and Practice for a Sustainable Development*, Advances in Science, Technology & Innovation,
https://doi.org/10.1007/978-3-030-59320-9_49

## 2 Study Area and Methodology

The Ouazi basin is located on the Atlantic coast of Morocco, southeast of the city of Essaouira. It is characterized by a semi-arid climate with an average annual precipitation of around 300 mm and average temperatures of 20 °C. This basin occupies an area of 1196 km$^2$. 151 groundwater samples of the 4 campaigns distributed in the study area (Fig. 1) were collected in well-cleaned bottles during the 4 years (1995, 2007, 2016 and 2018). The water samples of the year 2018 were analyzed to test the water quality with special reference to nitrate problems. And the chemical parameters are compared with the World Health Organization (WHO) drinking water standards, which indicate the standard allowable limits.

## 3 Results and Discussion

### 3.1 Quality and Evolution of Groundwater

The problem of salinization of groundwater is a global phenomenon, particularly in areas suffering from water scarcity, such as semi-arid and arid environments. The quality of groundwater in the study area in the year 2018 was studied with the aim to evaluate the suitability of this groundwater for domestic. The analytical data revealed that all groundwater in the Ouazi basin is alkaline with pH values ranging from 7.2 to 8.8, except three samples, S50, S51 and S52, which are above 8.8. The electrical conductivity (EC) values in this zone range from 778 to 5845 µS/cm at 23.4 °C and are classified as "Saline" to "Brackish", according to the WHO standards. Also, it was found that all samples of EC exceed 500 µS/cm the standard values of WHO. Total dissolved solids (TDS) values ranged from 390 to 2920 mg/L, and most of the TDS values in the study area of Ouazi basin fall above the 1000 mg/L and are also classified as brackish water. Potassium (K) is in the range 0.3–36.5 mg/L, and 92% of the samples are below the standard values of WHO. In the study area, the spatio-temporal distribution of EC and nitrates was studied to assess the impact of climate change on the quality of the groundwater of Ouazi basin (Fig. 2, 1–4). For the 1995 campaign, the EC values vary between 350 and 3000 µS/cm with an average of 1200 µS/cm. In 2007, the CE values oscillated between 900 and 3880 µS/cm with an average of 2000 µS/cm. As for the 2016 campaign, the CE values vary between 540 and 6000 µS/cm with an average of 2264 µS/cm. For the 2018 campaign, they vary between 778 and 5845 µS/cm with an average of 2515 µS/cm. The spatio-temporal distribution maps of the EC (Fig. 2, 1–4) show that the values of the EC are very important from 1995 to 2018 and toward the Atlantic Ocean.

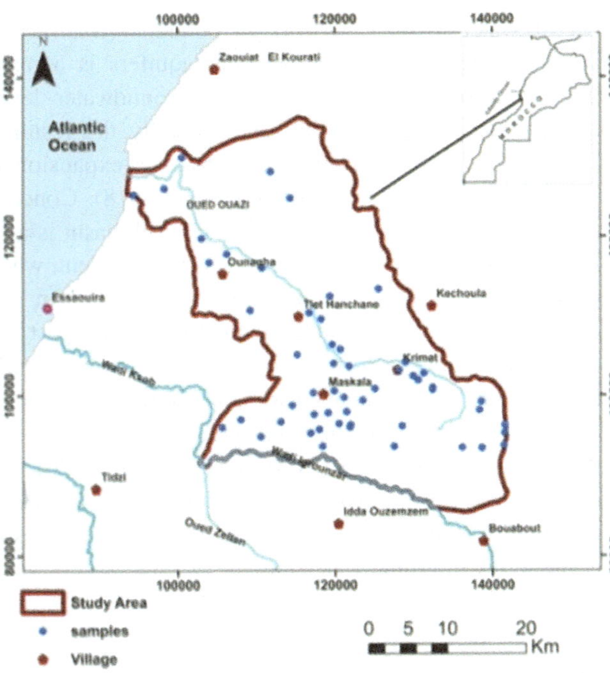


Fig. 1 Location of study area

### 3.2 Piezometric Surface Evolution

The piezometric maps of the study area, created with the data of the campaigns (1995, 2007, 2016 and 2018), show an overall direction of flow from SE to NW for the northern part and from NE to SW for the southern part of the study area (Fig. 3). This means that flow is conditioned by substratum of the Cenomanian–Turonian aquifer and the morphology. For the 24-year observation period, groundwater has the same direction of flow with a decline in the piezometric level. However, this situation is materialized comfortably, for example, by the shift of the iso-piezometric contours of 450 and 600 m a.s.l. more and more upstream, and this is observed on the 4 piezometric maps (Fig. 3, e–h). The groundwater level in the Essaouira basin shows significant fluctuations between periods of high and low water. These fluctuations are mainly related to precipitation, which thus controls the groundwater regime (Bahir et al. 2018; Ouhamdouch et al. 2020).

### 3.3 Potential Sources of Nitrate

To investigate the spatial changes in nitrate pollution in the study area of Ouazi basin, 4 campaigns are used (1995, 2007, 2016 and 2018), and the $NO_3$ concentration for all samples ranged from 0 to 375 mg/L (Fig. 4). For the water

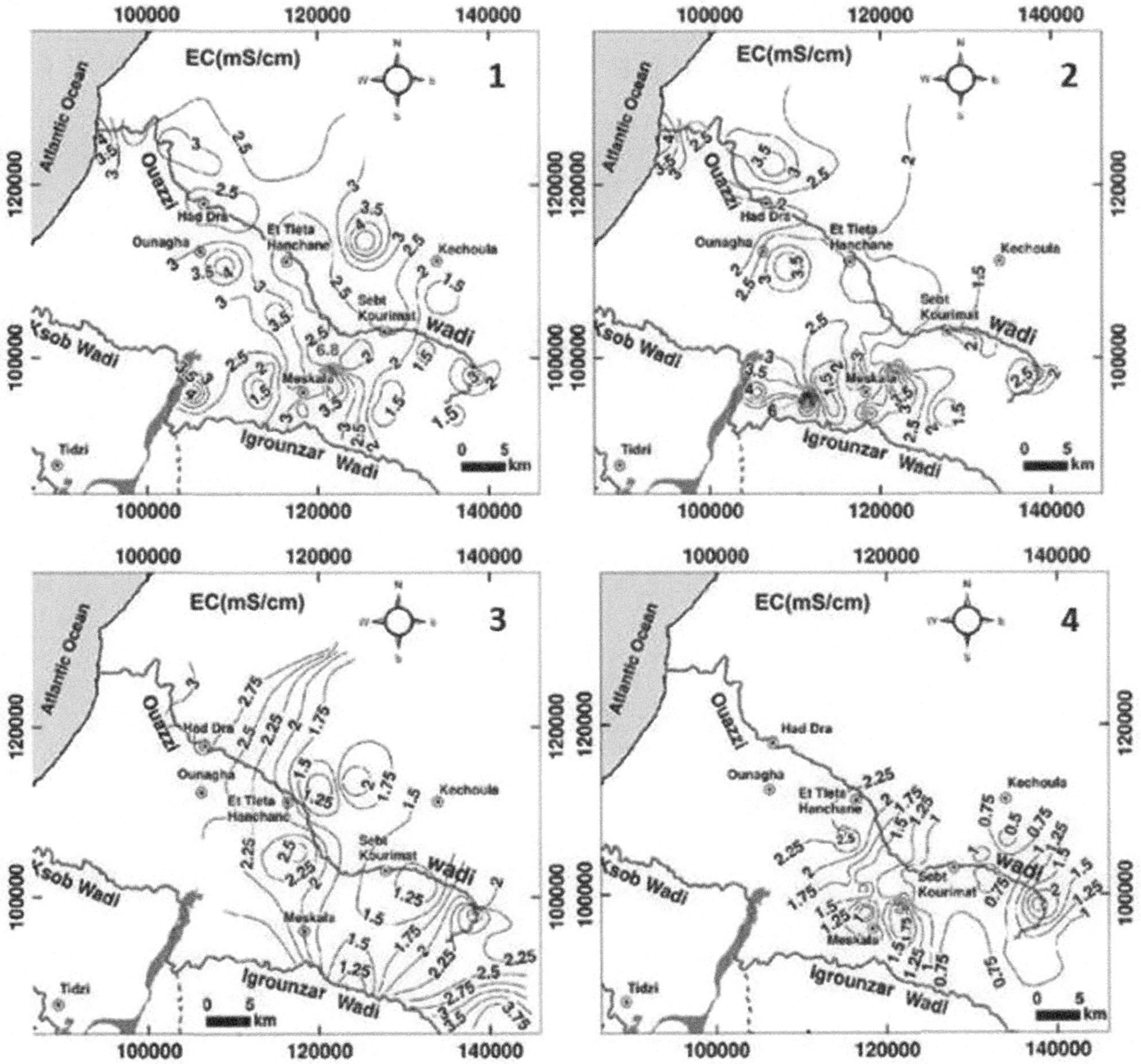


**Fig. 2** Spatio-temporal distribution of EC in the study area: campaign of 2018 (1), 2016 (2), 2007 (3) and 1995 (4)

points sampled in 1995, 70% of the samples had $NO_3$ concentrations above 50 mg/L (limit accepted by WHO 2011). For the 2007 campaign, 40% of samples exceed the threshold set by WHO (2011). As for the 2016 campaign, 36% of the water points have $NO_3$ levels above 50 mg/L. For 2018, 25% of the water samples exceeds the guideline for WHO (2011). These concentrations are clustered around the wells solicited following the waste of livestock rich in $NO_3$ which seep toward the aquifer during the extraction of water for the watering of livestock.

Spatial distribution of nitrate (Fig. 5a) showed that the highest concentrations had been observed in downstream and the upstream of study area where the highest concentrations had been observed at points S5, S9, S11, S12, S18, S25, S35, S40, S42, S44, S47 and S54. These highest concentrations could be explained by (1) the high concentration

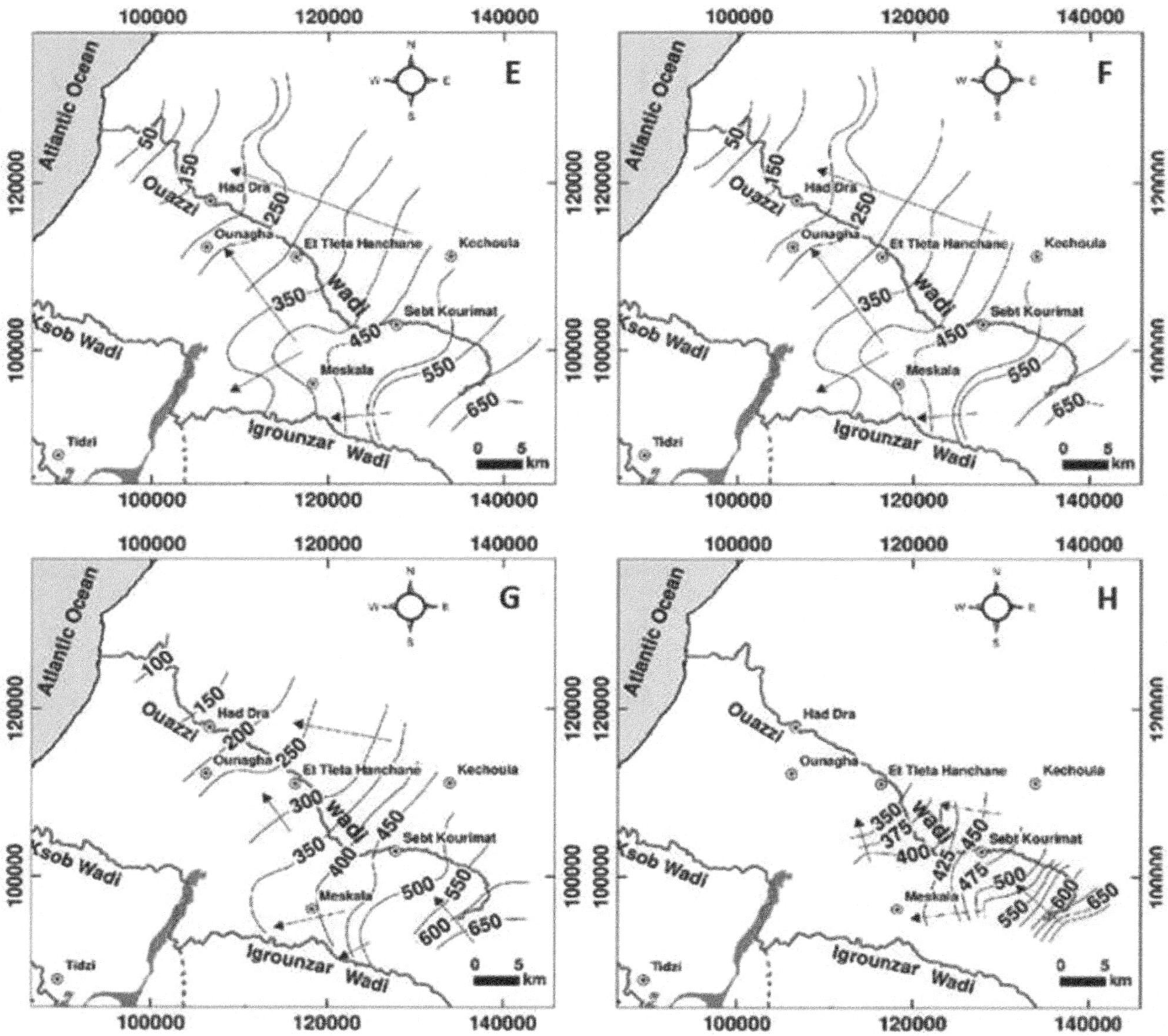


**Fig. 3** Piezometric maps of the study area: campaigns of 2018 (E), 2016 (F), 2007 (G) and 1995 (H)

of fertilizers use and (2) the gap of a sewage system and wastewater treatment plant (village of Sebt Krimat and Tlet Hanchane). Fig. 5b, showed there is no correlation between $NO_3$ concentrations and groundwater levels for most samples with ($R^2 = 0.029$). Therefore, the groundwater samples with high $NO_3$ concentrations are primarily concentrated in the downstream part of the oued Ouazi in the Plio-Quaternary aquifer (shallow), which suggests that it still constitutes a major threat to the quality of groundwater of the Ouazi basin. At the same time, an average nitrate concentration of less than 50 mg/L was found in the upstream part of oued Ouazi for the deep aquifer Cenonian-Turonian.

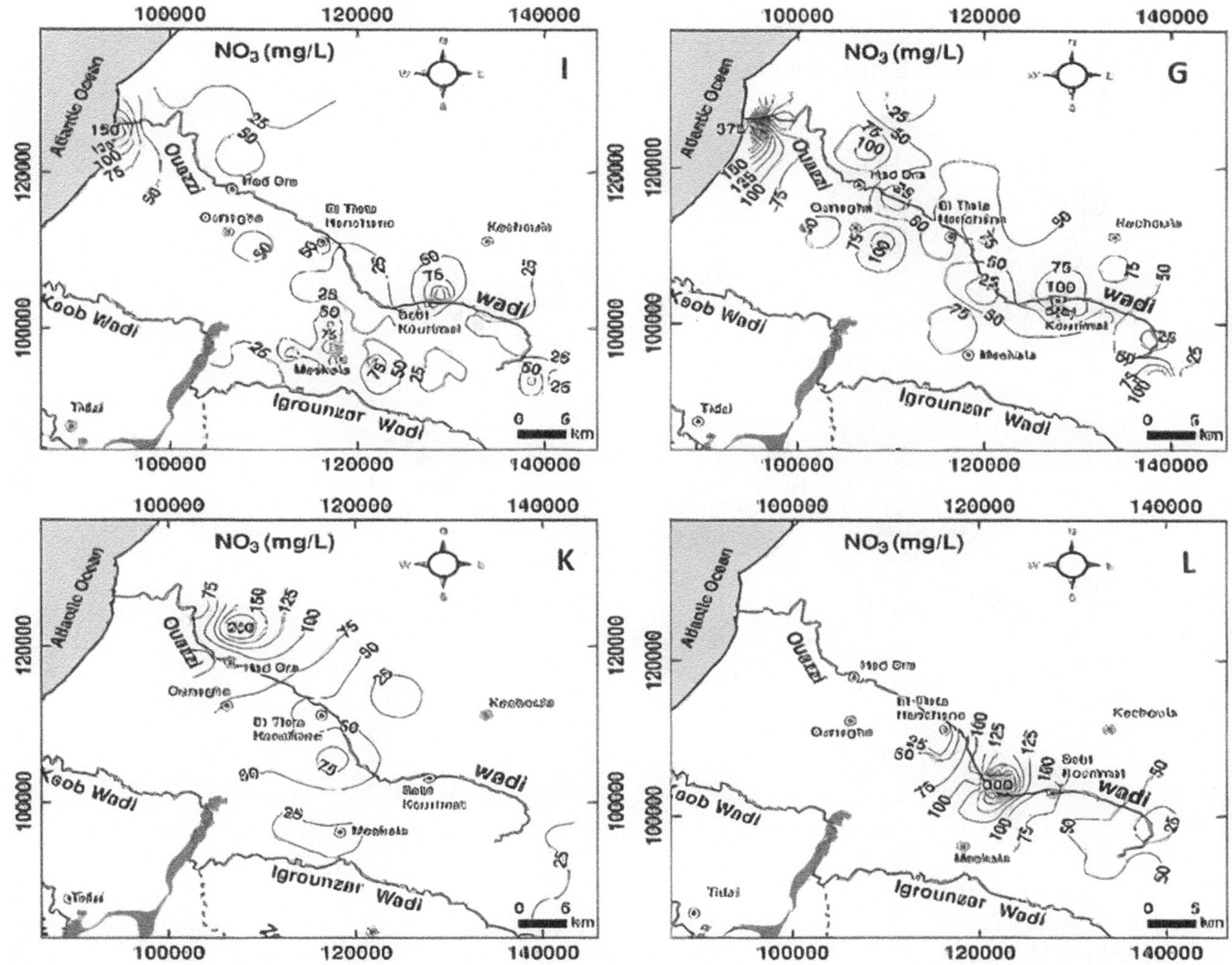


**Fig. 4** Spatio-temporal distribution of nitrates in the study area: campaigns of 2018 (I), 2016 (G), 2007 (K) and 1995 (L)

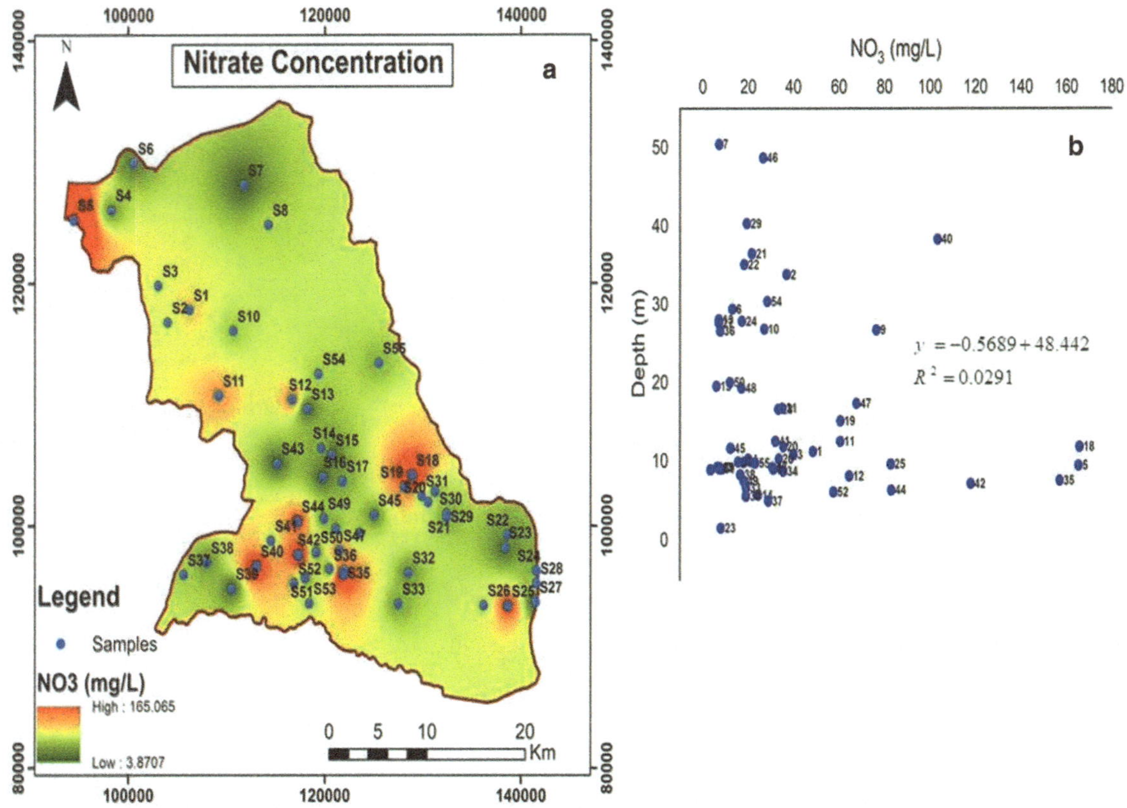


**Fig. 5** Spatial distribution of nitrate content of shallow and deep groundwater in 2018 (**a**), Plot of $NO_3$ concentrations versus depth of groundwater in 2018 (**b**)

## 4 Concluding Remarks

The threat to groundwater quality by nitrate pollution has always been of considerable concern in the Ouazi basin. In this paper, correspondence analysis, piezometric, groundwater quality evolution, identification of $NO_3$ pollution sources and its potential pollution paths based on the four years showed that a) the spatiotemporal distribution maps of the EC become very important from 1995 to 2018 and toward the Atlantic Ocean, b) the nitrate concentrations ranged from 0 to 375 mg/L, c) the climate change in the study area in recent years led to decreased rainfall, and the increasing the concentration of nitrate from 1995 to 2018, that can be explained by favoring the retention of nitrates in the superficial parts of the soil and by limiting their infiltration to the aquifer, and d) there is no correlation between $NO_3$ concentrations and groundwater levels. However, it depends on the thickness of the aquifer (shallow or deep aquifer). In addition, the demographic pressures, cultural practices and climate change make the Essaouira basin in alarming water situation shown by (1) drought which accentuates the desertification phenomenon and (2) enhancement of soil salinity and lowering groundwater levels groundwater. The complementarity of approaches hydroclimatic, hydrodynamic, hydrochemical and isotopic can lead to the diagnosis of aquifers vulnerable state of the Essaouira basin.

## References

Ait Kadi M, Ziyad A (2018) Integrated water resources management in Morocco. In: Global water security. Springer: Singapore, pp 143–163

Bahir M, Ouhamdouch S, Carreira PM (2018) Mineralization processes ruling the Cenomanian-Turonian and the Plio-Quaternary aquifer from semi-arid area, using isotopic and geochemical methods (Essaouira Basin, Morocco). Mar Freshwat Res 69:1290–1300

Bahir M, Ouazar D, Ouhamdouch S (2019) Hydrogeochemical mechanisms and recharge mode of the aquifers under semiarid climate from Morocco. Appl Water Sci 9:103

Carreira PM, Bahir M, Ouhamdouch S, Souhel A, Fernandes PG, Nunes D (2018) Tracing salinization processes by geochemical and isotopic features in two coastal aquifers: essaouira basin (W-Morocco) versus Lower Tagus-Lower Sado basin (Portugal). Hydrogeol J 26:2595–2615

El Mountassir O, Bahir M, Ouazar D, Ouhamdouch S, Chehbouni A, Ouarani M (2020) The Use of GIS and water quality index to assess

groundwater quality of krimat aquifer (Essaouira; Morocco). SN Appl Sci 2:871
Ouhamdouch S, Bahir M, Ouazar D (2020) Climate change impact assessment on a carbonate aquifer under semi-arid climate: example of the Cenomanian-Turonian aquifer within Meskala-Ouazzi region (Essaouira Basin, Morocco). Arab J Geosci 13:17
WHO (2011) Guidelines for drinking-water quality, 4th edn. World Health Organization, Geneva

# Hydrochemical Analysis and Evaluation of Groundwater Quality in Ouazi Basin (Essaouira, Morocco)

Mohammed Bahir, Otman El Mountassir, Driss Ouazar, and Paula M. Carreira

**Abstract**

Groundwater plays a dominant role in arid and semi-arid regions; it is among the most available water resources in Essaouira basin, where groundwater is the most important source of water supply. The aim of the study is to assess water quality with respect to agriculture and drinking for a better management of groundwater resources. To achieve such objectives, water analysis was carried out on 50 groundwater samples. The Piper plot reveals that the facies characterizing Ouazi basin was a combination of Ca-Mg-Cl type, Mg-Ca-$HCO_3$ type, $SO_4$-Ca-Mg and Na-Cl water type. The sustainability of groundwater for drinking and irrigation was assessed based on the World Health Organization (WHO), Wilcox and Richards's diagrams. The obtained results show that the consumption of groundwater in the study area requires a treatment before use as drinking water and most of the groundwater samples fall in C3S1–C4S1 indicating very high salinity and high to low sodium alkalinity hazard. Thus, groundwater quality is ranging between good to permissible and doubtful to unsuitable for irrigation uses under normal condition, and further action for salinity control is required in remediating such problem. The groundwater remains suitable for plants supporting high salinity.

**Keywords**

Essaouira • Irrigation • WHO • Groundwater quality • Wilcox

## 1 Introduction

Groundwater quality is equally important to its quantity owing to the suitability of water for various purposes. Water quality analysis is an important issue in groundwater studies. Variation of groundwater quality in an area is a function of physical and chemical parameters that are greatly influenced by geological formations and anthropogenic activities (El Mountassir et al. 2020). The hydrochemical study reveals the quality of water that is suitable for drinking, agriculture and industrial purposes and helps to understand the change in quality due to rock–water interaction or any type of anthropogenic influence (Carreira et al. 2018; Bahir et al. 2019a). Several studies have been realized in the Essaouira basin having partially identified and characterized some of the main mechanisms causing degradation of water quality in some coastal areas of this region, based on the application of geochemical and isotopic approaches (Bahir et al. 2019b; Ouhamdouch et al. 2019). These studies have demonstrated that (a) the groundwater recharge in Essaouira basin is supported by water runoff and precipitation; (b) important contribution of water–rock interaction in groundwater mineralization; (c) anthropogenic contamination was also identified.

The present study had the objective to understand the spatial distribution of hydrochemical constituents of groundwater related to its suitability for agriculture and domestic use, with application of cluster analysis.

M. Bahir · O. El Mountassir (✉)
High Energy and Astrophysics Laboratory, Faculty of Sciences Semlalia, Cadi Ayyad University, Marrakesh, Morocco
e-mail: otman.elmountassir@ced.uca.ma

M. Bahir · D. Ouazar
International Water Research Institute, Mohammed VI Polytechnic University, Ben Guerir, Morocco

P. M. Carreira
Centro de Ciências e Tecnologias Nucleares, C2TN/IST, Universidade de Lisboa, Lisbon, Portugal



M. Abrunhosa et al. (eds.), *Advances in Geoethics and Groundwater Management: Theory and Practice for a Sustainable Development*, Advances in Science, Technology & Innovation,
https://doi.org/10.1007/978-3-030-59320-9_50

## 2 Methodology

The Ouazzi basin is located in the north-eastern part of Essaouira city (Fig. 1). The basin is under semi-arid climate with irregular rainfall of about 300 mm year, and the temperature oscillates about 20 °C (El Mountassir et al. 2020). A set of 50 groundwater samples was analysed for 11 physical and chemical parameters comprising major ion concentrations. The samples were collected in 2019 from the Plio-Quaternary (downstream) and Cenomanian–Turonian (upstream) aquifers (depth 9.20–170 m) of Ouazzi basin. The physical parameters (temperature, pH, electrical conductivity) were measured in situ, and the major chemical parameters ($Ca^{2+}$, $HCO_3^-$, $Mg^{2+}$, $Na^+$, $K^+$, $Cl^-$, $SO_4^-$ and $NO_3^-$) were determined in the Laboratory of Geosciences and Environment—ENS (LGE–ENS) at 'École Normale Supérieure' of Marrakech (Cadi Ayyad University, Marrakech, Morocco). To identify possible groups and relationships among the samples analysed based on major chemical compositions, ion species ($Ca^{2+}$, $HCO_3^-$, $Mg^{2+}$, $Na^+$, $K^+$, $Cl^-$, $SO_4^-$ and $NO_3^-$) were considered as variables for application in Q-mode cluster analysis. The clustering procedure was performed by the Ward's linkage method with the Euclidean distance as a measure of similarity of samples using the XLSTAT statistical tool.

## 3 Results and Discussions

### 3.1 Hydrochemical Characterization and Water Type

The electrical conductivity of groundwater samples ranges from 615 to 5738 µS/cm with a mean value of 4479 µS/cm. The salinity increases in the direction of groundwater flow from north to south. The pH ranges between 7 and 8.4 with a mean value of 7.5, which shows that the groundwater of the study area is of alkaline nature. The mean temperature of groundwater samples was 20.4 °C. The maximum $Ca^{2+}$ and $Mg^{2+}$ concentrations of 769.5 and 297.4 mg/L, respectively, are however higher than their respective WHO standards of 200 and 30 mg/L. The mean sodium and potassium concentrations in the groundwater are 168 and 18.4 mg/L, respectively. The concentration of $HCO_3^-$ in the study area ranges between 244 and 898 mg/L. Most $Cl^-$ in the groundwater is from three sources including ancient sea water entrapped in sediment, solution of halite and related minerals in evaporate deposits in the region. The chloride value in the study area ranges between 113 and 1817 mg/L. The projection of 50 samples analysed on the Piper diagram (Fig. 2) shows four water types: Ca-Mg-Cl type, Mg-Ca-$HCO_3$ type, $SO_4$-Ca-Mg and Na-Cl water type. The dispersion of the points in the Piper diagram suggests that the groundwater mineralization is of several origins.

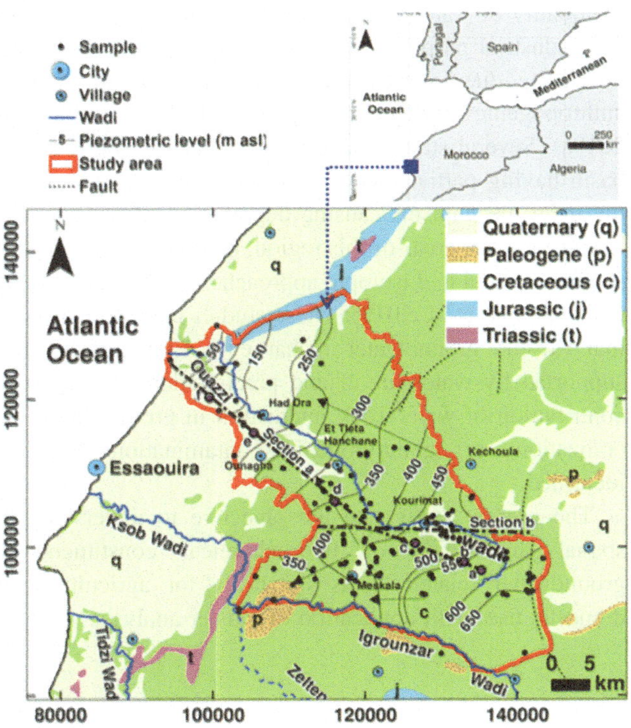


Fig. 1 Study area

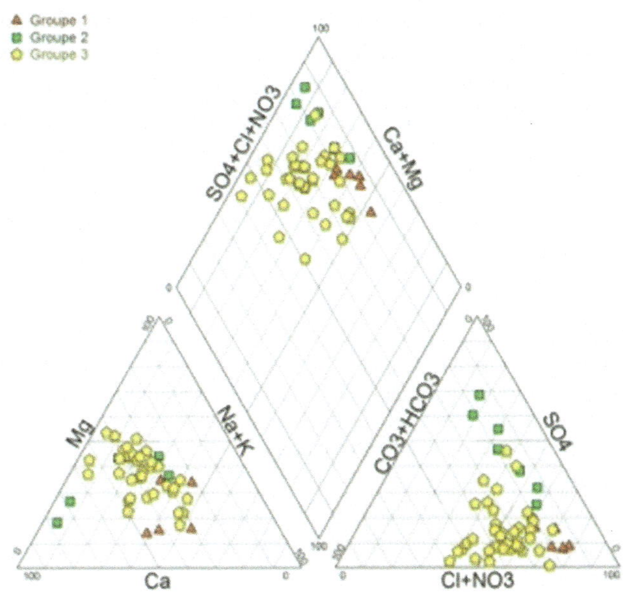


Fig. 2 Piper diagram of groundwater samples in the study area

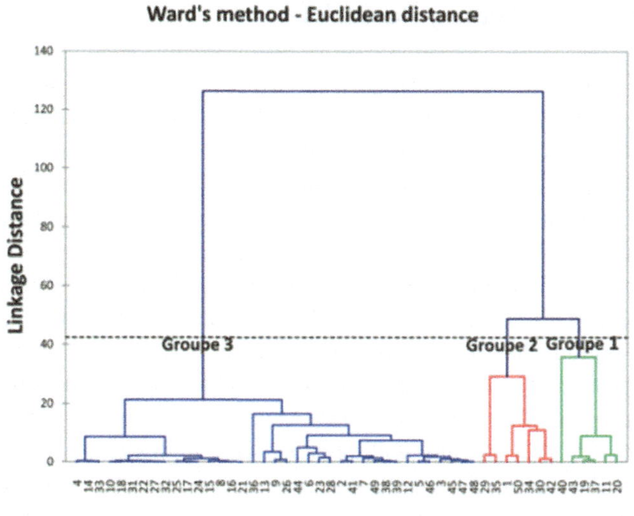


**Fig. 3** Q-mode cluster analysis. Dendrogram for 50 samples and eight chemical variables

## 3.2 Cluster Analysis

The output of the Q-mode cluster analysis is given as a dendrogram (Fig. 3). Three preliminary groups are selected based on visual examination of the dendrogram, each representing a hydrochemical facies (Table 1). The first group of waters, group 1, has high salinity (mean EC = 4275.67 μS/cm) and abundance orders (meq/L): $Na^+ > Ca^{2+} > Mg^{2+} > K^+$ and $Cl^- > HCO_3^- > SO_4^{2-} > NO_3^-$ (Table 1). These waters are classified as Ca-Mg-Cl and Na-Cl water type. Group 2 is made up of water samples wherein the anion composition was dominated by $SO_4^{2-}$ and $Cl^-$, with cation composition varying from dominantly $Ca^{2+}$ and $Mg^{2+}$ to dominantly $Na^+$ plus $K^+$ (Table 1). EC values (mean 4007.86 μS/cm) are significantly less than those of group 1, reflecting a more effective weathering process. These waters are classified as Ca-Cl and Ca-Mg-Cl. Group 3, made up of 37 water samples, has a salinity range (615 μS/cm < EC < 3842 μS/cm); mean 1857.9 μS/cm overlapping those of the former two groups (G1 and G2). On the basis of overall chemical composition, characterized by ion abundances $Ca^{2+} > Mg^{2+} > Na^+ > K^+$ and $Cl^- > HCO_3^- > SO_4^{2-} > NO_3^-$, these waters are classified as Ca-Mg-Cl, Ca-Cl and $SO_4$-Ca-Mg.

## 3.3 Irrigation Water Quality

The suitability of groundwater for agricultural purposes depends on the effect of mineral constituents of water on both plants and soil. Effects of salts on soils causing changes in soil structure, permeability and aeration indirectly affect plant growth. Wilcox and US Salinity Laboratory Staff proposed irrigational specifications for evaluating the suitability of water for irrigation use. There is a significant relationship between sodium adsorption ratio (SAR) values for irrigation water and the extent to which sodium is adsorbed by the soils. SAR was computed using the equation given below :

$$SAR = (Na^+)/\left[(Ca^{2+} + Mg^{2+})/2\right]^{1/2} \qquad (1)$$

where the concentrations are reported in meq/L.

The SAR values ranged from 0.22 to 5.83 with a mean varying from value of 1.95 to 4.8 in the three groups. Approximately, all samples fall in low SAR class (S1) and (S2) (Fig. 4). This implies that no alkali hazard is anticipated to the crops. If the SAR value is greater than 6–9, the irrigation water will cause permeability problems on shrinking and swelling of clayey soils types (Saleh et al. 1999).

The SAR and electrical conductivity values plotted on the US salinity richards diagram illustrate that most of the groundwater samples of the group 1 belong to the categories C4S2 (very high salinity and medium sodium) (Fig. 4). Most samples of the group 2 fall in the field of C4S1, indicating high salinity and low sodium water, and thus, they can be used for irrigation on almost all types of soil with little danger of exchangeable sodium (Fig. 4). The samples of the group 3 fall in the field of C4S2 and C4S1, indicating very high salinity and medium sodium hazard. These samples will be suitable for plants having good salt tolerance and hence restricted suitability for irrigation, especially in soils with limited drainage (Saleh et al. 1999).

The sodium percentage (Na %) is calculated using the formula given below, where all the concentrations are expressed in meq/L:

**Table 1** Mean concentrations of the major chemical parameters of three main water groups

| | pH | T (°C) | EC (μS/cm) | $Ca^{2+}$ | $Mg^{2+}$ | Na + | $K^+$ | $HCO_3^-$ | $Cl^-$ | $SO_4^{2-}$ | $NO_3^-$ |
|---|---|---|---|---|---|---|---|---|---|---|---|
| | | | | (meq/L) | | | | | | | |
| Group 1 | 7.26 | 20.87 | 4275.67 | 15.61 | 10.52 | 17.26 | 0.91 | 8.76 | 36.09 | 5.20 | 1.47 |
| Group 2 | 7.30 | 19.78 | 4007.86 | 21.98 | 18.55 | 9.08 | 0.64 | 9.76 | 21.40 | 27.11 | 0.68 |
| Group 3 | 7.54 | 20.51 | 1857.89 | 7.93 | 7.83 | 5.41 | 0.23 | 7.50 | 12.34 | 3.87 | 0.49 |

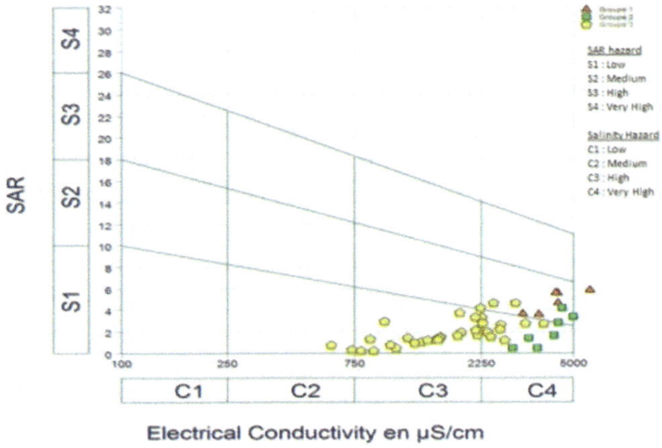


**Fig. 4** Salinity (µS/cm) versus alkalinity hazard in the US salinity diagram

$$Na\% = \left[(Na^+ + k^+)/(Ca^{2+} + Mg^{2+} + Na^+ + k^+)\right] \times 100 \quad (2)$$

The Na% of groundwater samples ranges from 4.3 to 53.18 with a mean of 40.92 for group 1, 18.76 in group 2 and 25.15 in group 3. The Wilcox diagram relating sodium percentage and electrical conductivity values shows that most of the groundwater samples fall in the field of good to permissible (group 3) except a few samples falling in the fields of doubtful and unsuitable category for irrigation (Fig. 5). The Na% indicates that the groundwater is unsuitable for irrigation in the group (1 and 2) (Fig. 5).

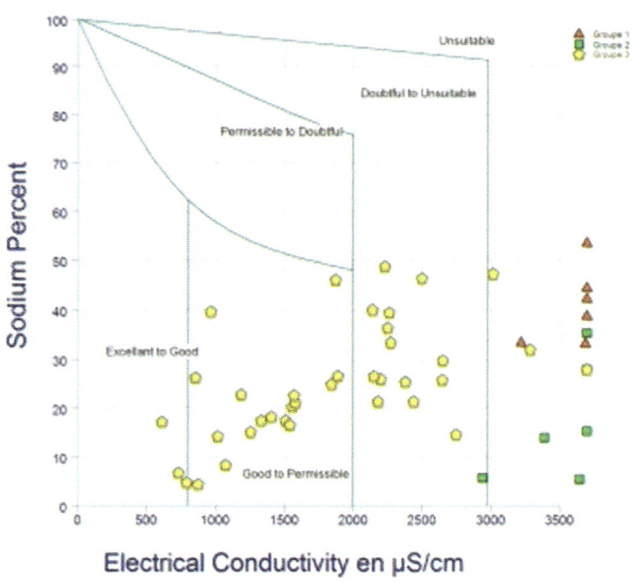


**Fig. 5** Wilcox diagram

## 4 Concluding Remarks

This study was conducted to evaluate factors regulating groundwater quality in an area with agriculture as main use. Fifty groundwater samples have been collected from Ouazi basin (Essaouira) for hydrochemical investigations, to understand the sources of dissolved ions and assess the chemical quality of the groundwater. Q-mode cluster analysis was applied to groundwater quality data sets and generated three clusters (groups 1, 2 and 3). The values of sodium absorption ratio and electrical conductivity of the ground water were plotted in the US salinity laboratory diagram for irrigation water. Most of the samples fall in C3S1–C4S1 quality with high and very high salinity hazard to low sodium hazard. However, the study has helped to improve understanding of water quality of the area for effective management and proper utilization of groundwater resources for better living conditions of the people. A continuous monitoring program of water quality is required to avoid further deterioration of the water quality of the study area. The hydroclimatic, hydrodynamic, hydrochemical and isotopic approaches have led to the diagnosis of the state of vulnerability of the aquifers of the Essaouira basin. Thus, the use of unconventional resources, such as desalinated sea water for drinking water supply or treated wastewater for agriculture, must be considered an immediate priority to avoid severe water shortage.

## References

Bahir M, Ouazar D, Goumih A, Ouhamdouch S (2019a) Evolution of the chemical and isotopic composition of groundwater under a semi-arid climate; the case of the Cenomano-Turonian aquifer within the Essaouira basin (Morocco). Environ Earth Sci 78:353

Bahir M, Ouazar D, Ouhamdouch S (2019b) Hydrogeochemical mechanisms and recharge mode of the aquifers under semiarid climate from Morocco. Appl Water Sci 9:103

Carreira PM, Bahir M, Ouhamdouch S, Fernandes PG, Nunes D (2018) Tracing salinization processes in coastal aquifers using an isotopic and geochemical approach: comparative studies in western Morocco and southwest Portugal. Hydrogeol J 26:2595–2615

El Mountassir O, Bahir M, Ouazar D, Ouhamdouch S, Chehbouni A, Ouarani M (2020) The use of GIS and water quality index to assess groundwater quality of Krimat aquifer (Essaouira, Morocco). SN Appl Sci 2:871

Ouhamdouch S, Bahir M, Ouazar D, Carreira PM, Zouari (2019) Evaluation of climate change impact on groundwater from semi-arid environment (Essaouira Basin, Morocco) using integrated approaches. Environ Earth Sci 78:449

Saleh A, Al-Ruwih F, Shehata M (1999) Hydrogeochemical processes operating within the main aquifers of Kuwait. J Arid Environ 42:195–209

# Use of WQI and Isotopes to Assess Groundwater Quality of Coastal Aquifers (Essaouira, Morocco)

Mohammed Bahir, Otman El Mountassir, Driss Ouazar, and Paula M. Carreira

**Abstract**

In the aquifers located along the Morocco coast, high mineralization of the groundwater is caused by dissolution processes of evaporitic rocks and carbonates and human impact from agriculture. The aim of the present work was to assess the origin and groundwater quality in the coastal aquifer of Essaouira synclinal basin. Therefore, 28 groundwater samples were used in the calculation of the Water Quality Index (WQI) during the period of March 2019. Eleven physico-chemical parameters were taken into account for the calculation of the WQI. The results obtained showed that the WQI values range from 110 to 890. The results obtained indicate that poor groundwater quality was specially found in the downstream part of the study area due to several factors such as the dissolution of evaporite minerals (e.g. gypsum, halite, and anhydrite), the use of pesticides and sea water intrusion. In addition, oxygen 18 ($^{18}O$) and deuterium ($^{2}H$) showed that the aquifer of study area is recharged by rapid infiltration (without evaporation) of meteoric waters ascribed to oceanic precipitation. Results obtained showed that this basin is sensitive and vulnerable to any climatic variation and therefore climate change.

**Keywords**

Groundwater quality • Essaouira • WQI • Semi-arid region • Climate change

M. Bahir · O. El Mountassir (✉)
High Energy and Astrophysics Laboratory, Faculty of Sciences Semlalia, Cadi Ayyad University, Marrakesh, Morocco
e-mail: otman.elmountassir@ced.uca.ma

M. Bahir · D. Ouazar
International Water Research Institute, Mohammed VI Polytechnic University, Ben Guerir, Morocco

P. M. Carreira
Centro de Ciências e Tecnologias Nucleares, C2TN/IST, Universidade de Lisboa, Lisbon, Portugal

## 1 Introduction

In semi-arid regions, given the scarcity of surface water, groundwater is the principal source of drinking, industrial and agricultural water. In recent decades, this resource has experienced qualitative and quantitative degradation due to overexploitation and the recurrence of drought episodes (Bahir et al. 2019a, b; Ouhamdouch et al. 2019). Groundwater salinization is a global problem. Several research studies have been focused on this problem under different geological and environmental approaches by using many tools such as isotope hydrology, hydrogeochemistry, statistics, geographic information system (GIS) and modelling (e.g. Carreira et al. 2018; El Mountassir et al. 2020). Coastal aquifers, in particular the Ksob subbasin, the downstream part of the Essaouira basin, may be vulnerable to sea water intrusion. However, there are other phenomena responsible for the groundwater resources degradation, for example, the dissolution of evaporite minerals (e.g. gypsum, halite and anhydrite), the use of pesticides and the effect of climate change (Bahir et al. 2019a). The main of this study was to investigate groundwater quality of the coastal zone within Essaouira basin (Morocco) based on 28 groundwater samples collected from the Plio-Quaternary aquifer, in 2019. To achieve this objective, the hydrogeochemical signatures and the Water Quality Index (WQI), an important tool for groundwater quality assessment, were determined. The isotopic signatures ($\delta^{18}O$ and $\delta^{2}H$) of the groundwaters were also determined to assess recharge zone. The WQI can be used by the local decision-makers to manage the groundwater resources and identification source of nitrate pollution in the coastal zone of Essaouira basin, especially the Plio-Quaternary aquifer.



M. Abrunhosa et al. (eds.), *Advances in Geoethics and Groundwater Management: Theory and Practice for a Sustainable Development*, Advances in Science, Technology & Innovation,
https://doi.org/10.1007/978-3-030-59320-9_51

## 2 Methodology

The studied aquifer system is the Plio-Quaternary alluvial aquifer of Essaouira basin which is characterized by a semi-arid climate with an average rainfall of 300 mm/year, and a mean temperature of 20 °C. The natural flow path of the system is SE to NW. This aquifer is limited to the east by Tidzi diapir and to the west by the Atlantic Ocean. (Fig. 1).

The calculation of WQI (1) was carried out in two steps:

$$\mathrm{WQI} = \frac{\sum_{i=1}^{n}(W_i \times Q_i)}{W_i} \quad (1)$$

Step 1: The relative weight ($W_i$) was assigned to each parameter analysed in the water samples. For this study, the unit weights of each of 11 physico-chemical parameters were assigned using the Formula (2):

$$W_i = \frac{K}{S_i} \quad (2)$$

With $S_i$ the permissible standard value of the water quality parameter and $k$ the proportionality constant calculated using the Formula (3):

$$k = \frac{1}{\sum(1/S_i = 1, 2, \ldots, i)} \quad (3)$$

Step 2: The quality index for each parameter was calculated according to Eq. (4):

$$Q_i = \frac{(V_a - V_i)}{(V_s - V_i)} \times 100 \quad (4)$$

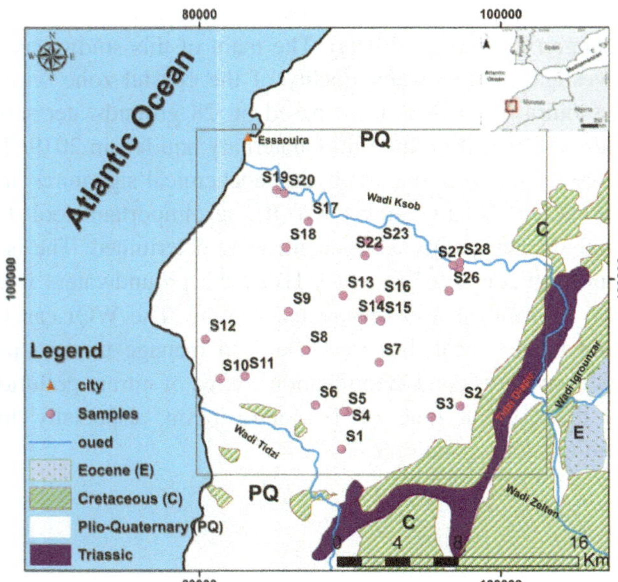


**Fig. 1** Study area of Essaouira synclinal basin

With $V_a$ the measured value of parameter $i$, $V_i$ the ideal value of parameter $i$ and $V_s$ the WHO recommended standard for the parameter $i$. The water quality can be evaluated using physico-chemical parameters compared to permissible limits prescribed at an international scale (WHO 2011). The best way to express the quality of water resources for drinking is the WQI, as it is one of the most effective tools by which water quality data is summarized and well presented (e.g. Bouteraa et al. 2019). The inverse distance weighting (IDW) technique was used in this study to delimit the geographic distribution of the WQI.

## 3 Results and Discussion

### 3.1 Groundwater Quality for Domestic Purposes

The geographic information systems (GIS) are important tools to study the spatiotemporal distribution of a given element. In this study, four parameters were processed ($NO_3$, electrical conductivity (EC), Cl and Na). The nitrate values in the studied groundwaters range between 0 and 400 mg/L with an average value of 40.68 mg/L. Most groundwater samples are not exceeding the maximum permissible limit of 45 mg/L according to WHO, except six samples: S10, S11, S12, S13, S15 and S19. This high concentration of the samples No (S10, S11 and S12) with a value 400, 95 and 135 mg/L, respectively, maybe it is due to a high concentration of tourism activity (village of Sidi Kaouki), b the gap of a sewage system and wastewater treatment plant and c waste from livestock during watering. The spatial nitrate concentrations distribution (Fig. 2) shown that the majority of high values are in the northwest of the study area. EC maximum values (12,250 µs/cm) were observed in sample No S12, and EC minimum values (880 µs/cm) were observed in sample No S3. All groundwater samples present EC values above 500 µs/cm, exceeding the standard values of WHO. EC values indicate that, 8% of the total groundwater samples (S2 and S3) fall under the WHO permissible type, 8% of the samples (S9 and S22) belong to brackish type, and 86% of the samples (S1, S4–S8, S10-S21, S23–S28) lie under the high enrichment of saline type. The spatial distribution map of the electrical conductivity in the study area is shown in (Fig. 2). Cl concentration ranges from 226 to 4800 mg/L with an average of 814 mg/L. The standard values set by the WHO for chloride concentration are 250 mg/L. It has been found that all chloride concentration in groundwater samples exceed the maximum allowable value approved by the WHO. These high Cl levels are mainly due to dissolution of evaporites, wastewater, industrial activity and infiltration of sea water (the overexploitation of the Plio-quaternary aquifer resource causing a deficit which leads to an intrusion of sea water). The spatial

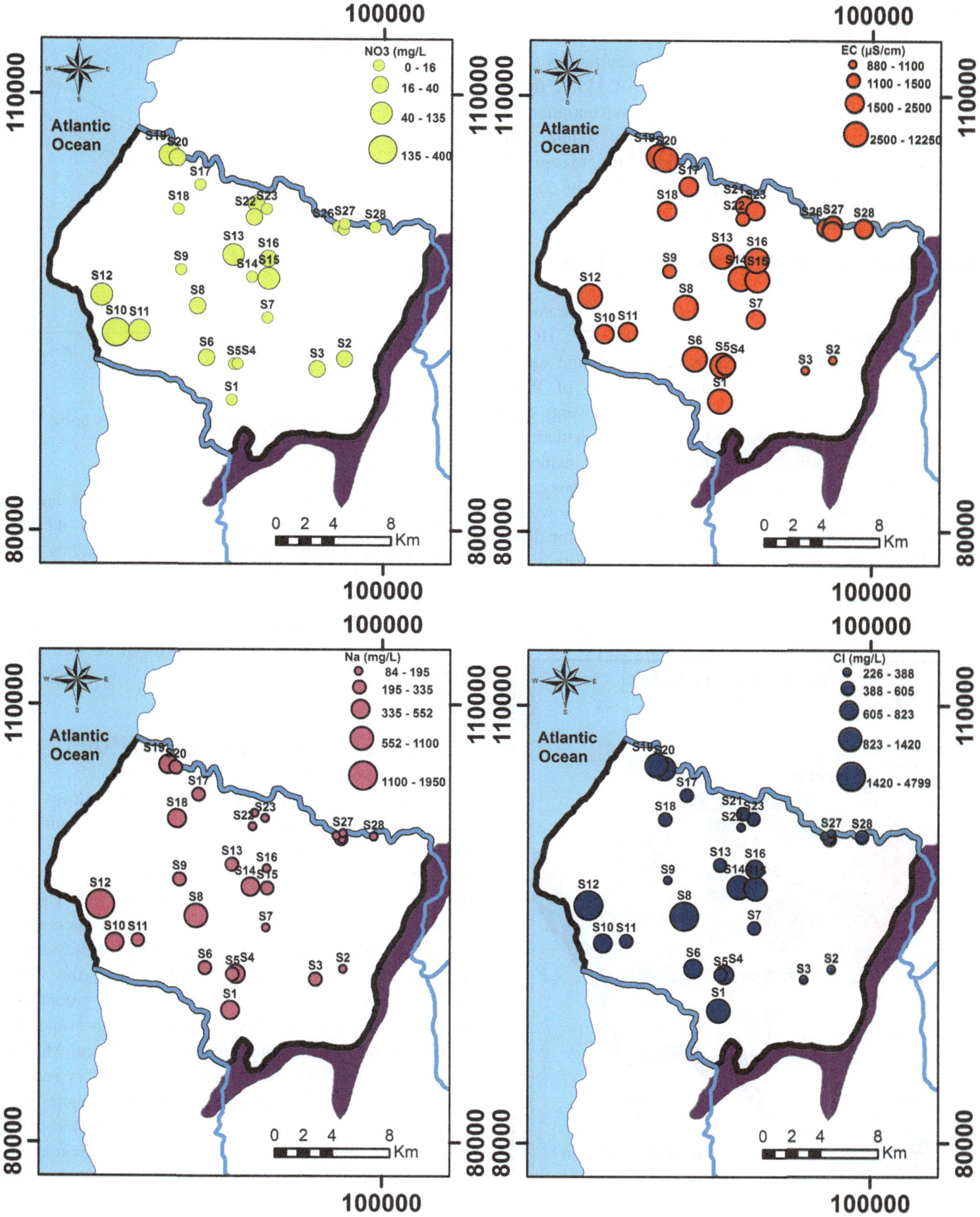


**Fig. 2** Spatial distribution of $NO_3$, EC, $Cl^-$, and $Na^+$ in the study area

distribution map of the chloride concentration shows that most of the high values are in the downstream part of the Essaouira synclinal basin (Fig. 2). $Na^+$ concentrations range from 84.3 to 1950 mg/L with an average of 359.72 mg/L, and the WHO standard values for sodium concentration are 200 mg/L. It was found that all samples of sodium concentration exceed the maximum allowable limit, except (S2, S3, S16, S21–S23, and S26–S28) is shown in (Fig. 2).

## 3.2 Estimation of the Water Quality Index

The average values of eleven physico-chemical parameters (pH, EC, TDS, $Na^+$, $K^+$, $Ca^{2+}$, $Mg^{2+}$, $NO_3^-$, $Cl^-$, $HCO_3^-$ and $SO_4^{2-}$) of groundwater Plio-Quaternary alluvial aquifer of Essaouira basin were used for calculation of Water Quality Index (WQI) during March 2019. Twenty-eight different sampling sites were used in the WQI calculation in order to assess the suitability of the groundwater quality for drinking. WQI values range from 110 to 890 and thus can be classified into three water classes. The results show that, 10.71% of the samples indicate water unsuitable for drinking, while 28.57% of groundwater sampled in the study area had very poor water quality. Regrettably, 60.71% are poor water quality. The spatial distribution map of the WQI shows that most of the groundwater samples belong to the very poor water (Fig. 3). However, this situation can be explained by the sea water intrusion, the dissolution of evaporite minerals (gypsum, halite and anhydrite), the use of pesticides and the effect of climate change in this coastal aquifer.

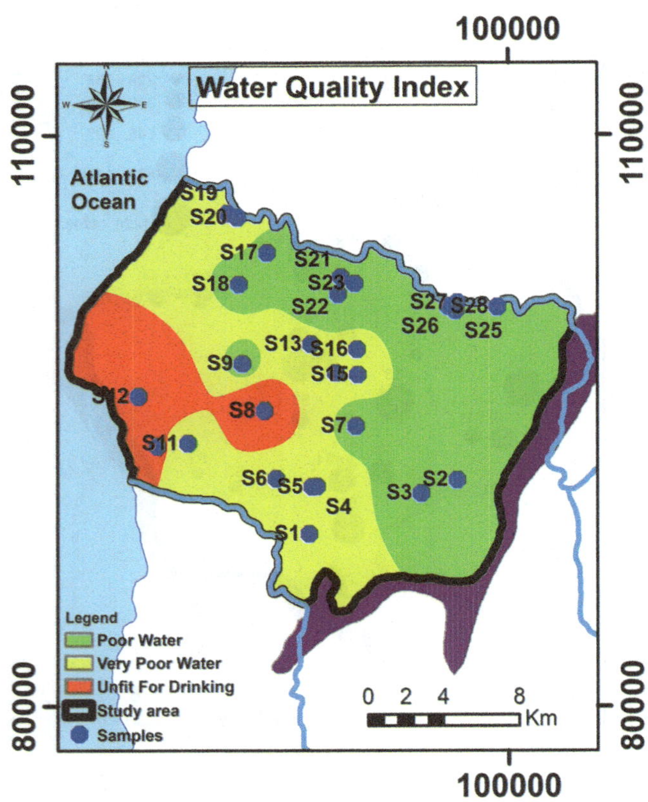


Fig. 3 Spatial distribution of water quality index in the study area

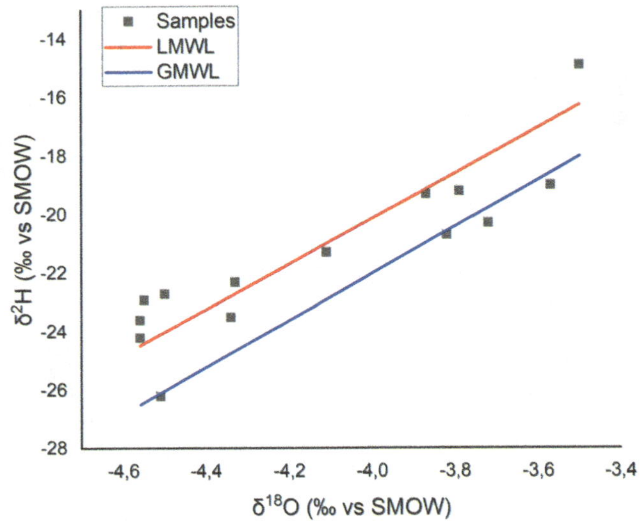


Fig. 4 $\delta^2H$ versus $\delta^{18}O$ relationship of groundwaters in the area of Essaouira synclinal

## 3.3 Isotope Hydrology ($\delta^2H$ and $\delta^{18}O$)

The stable isotopic signatures ($\delta^2H$ and $\delta^{18}O$ values) of the groundwaters of Plio-Quaternary alluvial aquifer of Essaouira basin (Fig. 4) indicated $\delta^2H$ values ranging between −38.37 and −20.69‰ and $\delta^{18}O$ values between −6.21 and −3.77‰ versus V-SMOW. The $\delta^2H$ versus $\delta^{18}O$ relationship of the studied groundwaters shown is the diagram of Fig. 4, shows that most of the sampled groundwaters of the study area plot above the Global Meteoric Water Line (GMWL). However, three groundwater points plot below to the GWWL indicating an evaporation trend. The other points are aligned on the Local Meteoric Water Line (LMWL). Nevertheless, the aquifers recharge within Essaouira basin is closely related to the precipitation, which makes Ksob basin sensitive and vulnerable to any climatic variation and in general of climate change. The staple isotopic data obtained indicates that aquifers recharge of study area is by infiltration without evaporation ensured by oceanic precipitation.

## 4 Concluding Remarks

In this study, the use of prominent tools to interpret water quality is demonstrated by applying them to data on the quality of groundwater of the Essaouira synclinal basin. The Water Quality Index (WQI) of the present study was calculated using 11 physico-chemical parameters from 28 groundwater sampling points distributed over the entire study area in order to assess and controlling the suitability of water for drinking purpose. The spatial distribution of the chemical elements shows that the highest concentrations are observed in the central and downstream part of the study area, following the dissolution of halite, gypsum and marine intrusion. The computed WQI for twenty-eight samples ranged from 102.60 to 879.77. In global, 10% of the samples indicate water unsuitable for drinking, while 27% had very poor water quality. Regrettably, 63% of groundwaters are poor water quality. In addition, the staple isotope shows that aquifers recharge of study area by rapid infiltration without evaporation ensured by oceanic precipitation. In general, the complementarity of approaches hydroclimatic, hydrodynamic, hydrochemical and isotopic can lead to the diagnosis of aquifers vulnerable state of the Essaouira synclinal basin.

## References


Bahir M, Ouhamdouch S, Carreira PM, Zouari K (2019) Relationship between hydrochemical variation and the seawater intrusion within coastal alluvial aquifer of Essaouira Basin (Morocco) Using HFE-Diagram. In: Chaminé HI, Barbieri M, Kisi O, Chen M, Merkel B (eds) Advances in sustainable and environmental hydrology, hydrogeology, hydrochemistry and water resources. Advances in science, technology and innovation series, Springer, Cham

Bahir M, Ouazar D, Ouhamdouch S (2019) Hydrogeochemical mechanisms and recharge mode of the aquifers under semiarid climate from Morocco. Appl Water Sci 9:103

Bouteraa O, Mebarki A, Bouaicha F, Nouaceur Z, Laignel B (2019) Groundwater quality assessment using multivariate analysis, geostatistical modeling, and water quality index (WQI): a case of study in the Boumerzoug-El Khroub valley of Northeast Algeria. Acta Geochem 38:796

Carreira PM, Bahir M, Salah O, Galego Fernandes PG, Nunes D (2018) Tracing salinization processes in coastal aquifers using an isotopic and geochemical approach: comparative studies in western Morocco and southwest Portugal. Hydrogeol J 26:2595–2615

El Mountassir O, Bahir M, Ouazar D, Ouhamdouch S, Chehbouni A, Ouarani M (2020) The use of GIS and water quality index to assess groundwater quality of Krimat aquifer (Essaouira; Morocco). SN Appl Sci 2:871

Ouhamdouch S, Bahir M, Ouazar D, Carreira PM, Zouari (2019) Evaluation of climate change impact on groundwater from semi-arid environment (Essaouira Basin, Morocco) using integrated approaches. Environ Earth Sci 78:449

WHO (2011) Guidelines for drinking-water quality, 4th edn. World Health Organization, Geneva

# Tritium and Carbon-14 Content as a Diagnostic Approach in Groundwater Resources Management and Protection

Paula M. Carreira, Dina Nunes, José Manuel Marques, Maria do Rosário Carvalho, Manuel Antunes da Silva, Augusto Costa, and António Monge Soares

**Abstract**

Groundwater isotopic composition is widely used nowadays in the management and protection of water resources. This work will focus not only on the temporal and spatial variations of tritium content in precipitation and their relationships with groundwater active recharge, but also on the radiocarbon groundwater dating. The $^{3}H$ levels determined in groundwater can be used as a complement to the carbon-14 dating of aquifer systems and in the identification of mixing between different water units or in the dynamics of water resources (transit time, definition of the main flow lines, for instance). Regional variations between the coast and the inlands of Continental Portugal will be analysed considering the influence of climatic and regional parameters in the isotopic composition of the groundwater. Two case studies will be discussed: (i) the study of Melgaço–Messagães aquifer, located in the NW of Portugal (granitic region), will be based on an isotopic approach that allowed the identification of the main origin of the dissolved carbon in the aqueous system; (ii) the Moura-Ficalho aquifer system, situated in SE of Portugal, where $^{3}H$ and $^{14}C$ measurements were applied to identify the active recharge of the system and in the characterization of two main flow paths with completely different velocities and apparent ages (2 and 17 ka BP).

**Keywords**

Tritium • Carbon-14 • Groundwater dating • Recharge area • Groundwater flow velocities

## 1 Introduction

The use of nuclear techniques in hydrogeological studies is built on the study of content variations of different isotopic species. These variations are a consequence of natural processes, not controllable by man, and their interpretation is made by comparison with absolute standards or with regional isotopic variations. The most used radioactive environmental isotopic species in hydrogeological studies are tritium ($^{3}H$) and carbon-14 ($^{14}C$).

The content of $^{3}H$ in precipitation is the result of two distinct processes. The first is of natural origin and results from the reaction of (thermal) neutrons, produced by the interaction of cosmic rays with the particles in the upper layers of the atmosphere, with nuclei of nitrogen atoms. The second has an artificial (anthropogenic) origin resulting from thermonuclear explosions in the atmosphere or from the nuclear industry (Mook 2000). $^{3}H$ is found in the environment in easily measurable quantities. $^{3}H$ has a half-life of 12.32 years (Lucas and Unterweger 2000). Both the $^{3}H$ produced in the atmosphere by natural processes and the $^{3}H$ resulting from the action of man (nuclear tests, nuclear power stations, etc.) are rapidly oxidized to atmospheric water vapour ($^{1}H^{3}HO$) and enter in the hydrological cycle through precipitation and isotopic exchange between air and ocean water bodies. Variations in the $^{3}H$ content are observed throughout the hydrological cycle (Araguas-Araguas et al. 1996). At present, $^{3}H$ levels in precipitation

P. M. Carreira (✉) · D. Nunes · A. M. Soares
Centro de Ciências e Tecnologias Nucleares (C2TN), Departamento de Engenharia e Ciências Nucleares (DECN), Instituto Superior Técnico, Lisbon University, Lisbon, Portugal
e-mail: carreira@ctn.tecnico.ulisboa.pt

J. M. Marques
Instituto Superior Tecnico (CERENA/IST), Lisbon University, Lisbon, Portugal

M. do RosárioCarvalho
Department of Geology, IDL, Faculdade de Lisboa, Lisbon University, Lisbon, Portugal

M. A. da Silva
Super Bock Group, S. Mamede de Infesta, Portugal

A. Costa
Geodiscover, Lda, Alcochete, Portugal



M. Abrunhosa et al. (eds.), *Advances in Geoethics and Groundwater Management: Theory and Practice for a Sustainable Development*, Advances in Science, Technology & Innovation,
https://doi.org/10.1007/978-3-030-59320-9_52

are close to pre-nuclear levels. The similarity of values between pre-nuclear tests and nowadays has led most applications of $^3H$ content in hydrogeology studies to be restricted to small-scale studies, both in space and time.

Carbon-14 present in the atmosphere has its origin also associated with two distinct processes: one natural, resulting from the interaction of cosmic radiation with nitrogen atoms in the upper layers of the atmosphere and an artificial origin related to nuclear activity developed by man (Mook 2000). The $^{14}C$ atoms after oxidation form carbon dioxide ($^{14}CO_2$) molecules that mix with non-radioactive atmospheric $CO_2$ and thereafter participate in the Global Carbon Cycle, i.e. the bio, litho and hydrosphere reservoirs. Carbon enters the hydrological cycle mainly through two mechanisms: by chemical processes associated with the dissolution in the aqueous system of atmospheric and soil $CO_2$ and/or through the dissolution of carbonated mineral and/or by biochemical production of $CO_2$ resulting from plant respiration. $^{14}C$ has a period (half-life) of 5730 years (Mook 2000).

When carbon in the aqueous system is primarily of biogenic origin (soil $CO_2$), the activity of $^{14}C$ in the total dissolved carbon of a water body represents exclusively the activity of this species of organic origin. However, when carbon is of organic and mineral origin in the aqueous system, the $^{14}C$ activity of biogenic $CO_2$ will be diluted by carbon resulting from the dissolution of "free" carbon-14 carbonated minerals, inducing an apparent ageing of the water.

Groundwater dating knowledge is an important tool under climate change scenario. Like many countries in the world, Portugal is facing a rise of the mean air temperature accompanied by a decrease of the precipitation amount, leading to decrease of the aquifers recharge. Understanding the mean residence time of the groundwater systems by dating techniques, allow to characterize an aquifer formation to be more or less vulnerable to anthropogenic actions, fundamental information for a proper management and protection of the groundwater resources. Besides, old groundwater systems can be considered as a strategic water resource to be preserved. In this work, examples how groundwater dating can help in the proper management of the aquifer systems are discussed.

## 2 Melgaço–Messegães Case Study (NW Portugal)

Melgaço–Messegães $CO_2$-rich cold mineral waters are located in the NW of Portugal mainland. Three types of granitic rocks can be recognized in the region based on their structural relationships and internal deformations, namely (i) syntectonic granites with muscovite and biotite inclusions strongly correlated with migmatitic rocks; (ii) tardi-tectonic granites (often associated with granodiorites), with a high presence of biotite; (iii) post-tectonic granites, generally characterized by the presence of K-feldspar and biotite mega crystals.

Major fault systems in the region have ENE-WSW, WNW-ESE, NNE-SSW and NNW-SSE directions. Geological-geophysical studies conducted in the region indicate that the ENE-WSW system is responsible for the morphology of the Minho river valley. Melgaço and Messegães $CO_2$-rich hydromineral systems emerge at the intersection of NNW-SSE structures (Fig. 1a).

Three fieldwork campaigns were carried out during February 2002 and February/July 2006. Groundwater samples were collected from boreholes (mineral system) and springs located at different altitude sites within the surroundings of the research region, and representative of the shallow dilute groundwater.

In the shallow aquifer systems, with low mineralization, $^3H$ content ranges from 2.1 to 5.2 ± 0.6 TU, while in the $CO_2$-rich mineral waters (Melgaço 1, 2 and Messegães), the $^3H$ content ranges from 0.0 to 2.2 ± 0.7 TU (Carreira et al. 2014).

Projecting $^3H$ content as a function of $\delta^{18}O$ and electrical conductivity, it is found that lower $^3H$ concentrations were recorded in samples with the higher mineralization, indicating a longer circulation path, i.e. more time for water–rock interaction processes. Mineral waters present recharge altitudes at higher altitudes, between 500 to 700 m asl, based on local isotopic fractionation data with altitude. The distribution pattern $^3H$ versus $\delta^{18}O$ seems to be associated with different recharge altitudes. Furthermore, within the Melgaço1 system, the $CO_2$-rich mineral waters show the lower mineralization along with the higher levels of $^3H$, indicating a shorter preferential pathway.

Considering the $^3H$ input in the same order of magnitude as observed at the Porto GNIP station, 4.5 TU—weighted arithmetic average (Carreira et al. 2004), the mean residence time should be over 40 years.

$^{14}C$ and $\delta^{13}C$ content was determined by AMS, performed on total dissolved inorganic carbon (TDIC) in Melgaço1 and Melgaço2 samples:

$$^{14}C = 2.33 \pm 0.07 \text{ pmC and } \delta^{13}C = 4.7\text{‰ (Melgaço 1)}$$

and

$$^{14}C = 1.01 \pm 0.04 \text{ pmC and } \delta^{13}C = 4.7\text{‰ (Melgaço 2)}.$$

The $\delta^{13}C$ enrichment found in the $CO_2$-rich mineral waters may be due to the upper mantle carbon mixture, although the hypothesis of methanogenesis should not be excluded, considering the positive $\delta^{13}C$ values. $^{14}C$ contents indicate a carbon source almost free of $^{14}C$.

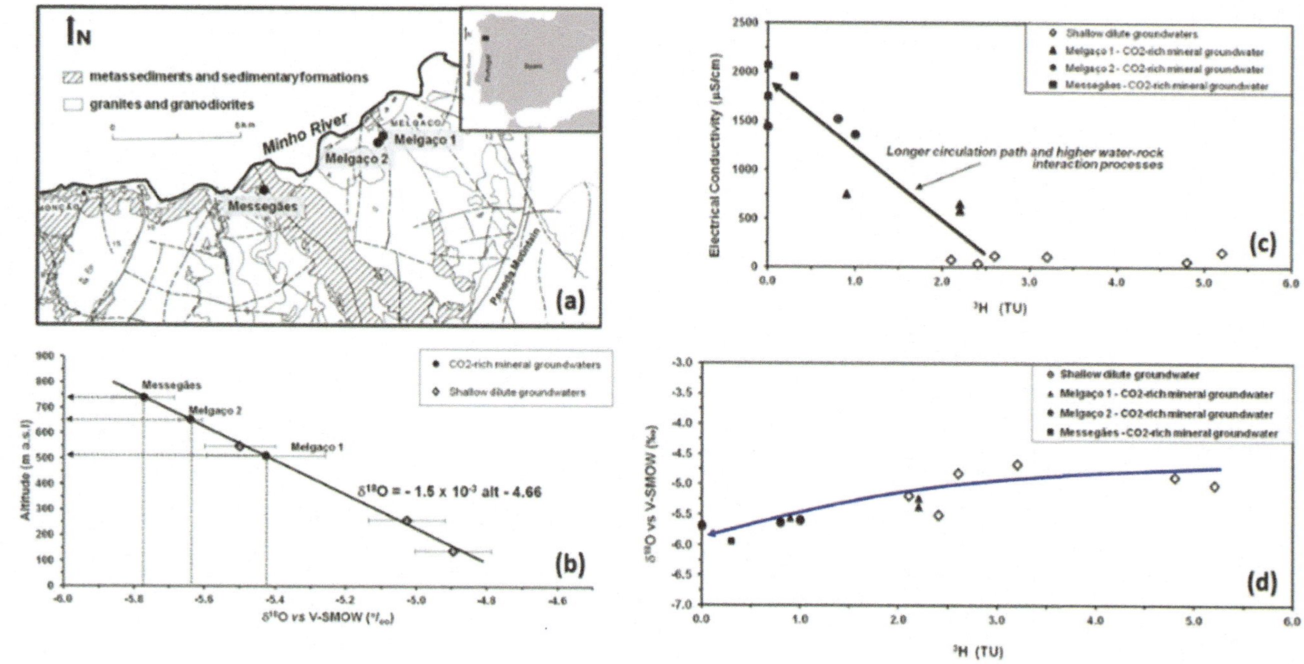


**Fig. 1** a Simplified geological map of the region. Filled circles (filled circle) represent the location of the mineral borehole waters. **b** Estimation of the recharge altitude of the $CO_2$-rich hydromineral systems. **c** $^3H$ versus electrical conductivity values. **d** $^3H$ versus $\delta^{18}O$ plot (Carreira et al. 2014)

## 3 Moura-Ficalho Case Study (SE Portugal)

Groundwater flow in the Baixo Ardila region is dominated by the existence of a karst-fissured aquifer system that develops between Vila Verde de Ficalho and Moura (Moura-Ficalho aquifer). The physical support of the aquifer consists essentially of Lower Cambrian dolomites formation (Costa 2008).

The mineral water host formations are carbonates. In such systems, significant delay of radiocarbon is expected with respect to water since losses of $^{14}C$ occur due to matrix diffusion, leading to high $^{14}C$ ages, although the $SI_{calcite}$ is not significant in all these waters. The apparent groundwater ages (Table 1) were calculated using Gonfiantini and Zuppi (2003) approach. The $\delta^{13}C$ content in the mineral waters is very similar, reflecting water–rock interaction processes ($^{13}C$ enrichment). However, the $^{14}C$ content in the sample of Casal St. André exhibits a very low $^{14}C$ content, which leads to a much higher groundwater apparent age.

The isotopic composition ($\delta^{18}O$ and $\delta^2H$) of the Casal de St. André water is also different, showing a shift (depletion) of 0.3‰ in $^{18}O$ and about 3‰ in $^2H$ (Fig. 2). The isotopic pattern ($^{18}O$ and $^{14}C$) supports different circulation times in Moura-Ficalho aquifer. The water in the deeper part of the aquifer (dark blue in Fig. 3) has longer path and residence time, and the Casal de St. André borehole, around 400 m deep, is the only outflow of the aquifer.

Três Bicas (spring and borehole) and Fonte da Telha represent the water circulating in the "upper" part of the aquifer—karst system, whereas Casal de St. André is preferentially exploiting the deeper part of the aquifer, a fissured medium, with slower flow velocities promoting higher water–rock interaction processes and leading to an increase of the apparent $^{14}C$ ages (Carreira et al. 2019).

## 4 Concluding Remarks

The great benefit of the simultaneous use of isotopic and geochemical characterization in hydrogeological studies, along with the use of traditional methodologies, is recognized in the conception of circulation models, since:

- stable isotopic data can play a fundamental role in the definition of preferential recharge altitudes, i.e. in the set-up of protection areas limits;
- $^3H$ becomes an ideal tracer in the characterization of the active recharge of systems and in determination of the average residence time of the water. The presence of this radioactive isotope points out to a modern water component that can also be interpreted in a qualitative way, i.e. in studies on the vulnerability of the water resources to pollution. The presence of $^3H$ in an aquifer should be regarded as indicatory of areas with relatively high risk to anthropic actions;

**Table 1** Carbon isotopes content and apparent $^{14}C$ groundwater ages

| References | $\delta^{13}C$ (‰) | $^{14}C \pm \sigma$ (pmC) | Apparent GW Age (BP) |
|---|---|---|---|
| B. Três Bicas | −15.22 | 79.17 ± 0.79 | 2182 |
| S. Três Bicas | −14.90 | 74.94 ± 0.33 | 2461 |
| Fonte Telha | −15.50 | 79.28 ± 0.34 | 2322 |
| C. St. André | −13.57 | 11.00 ± 0.12 | 17,550 |

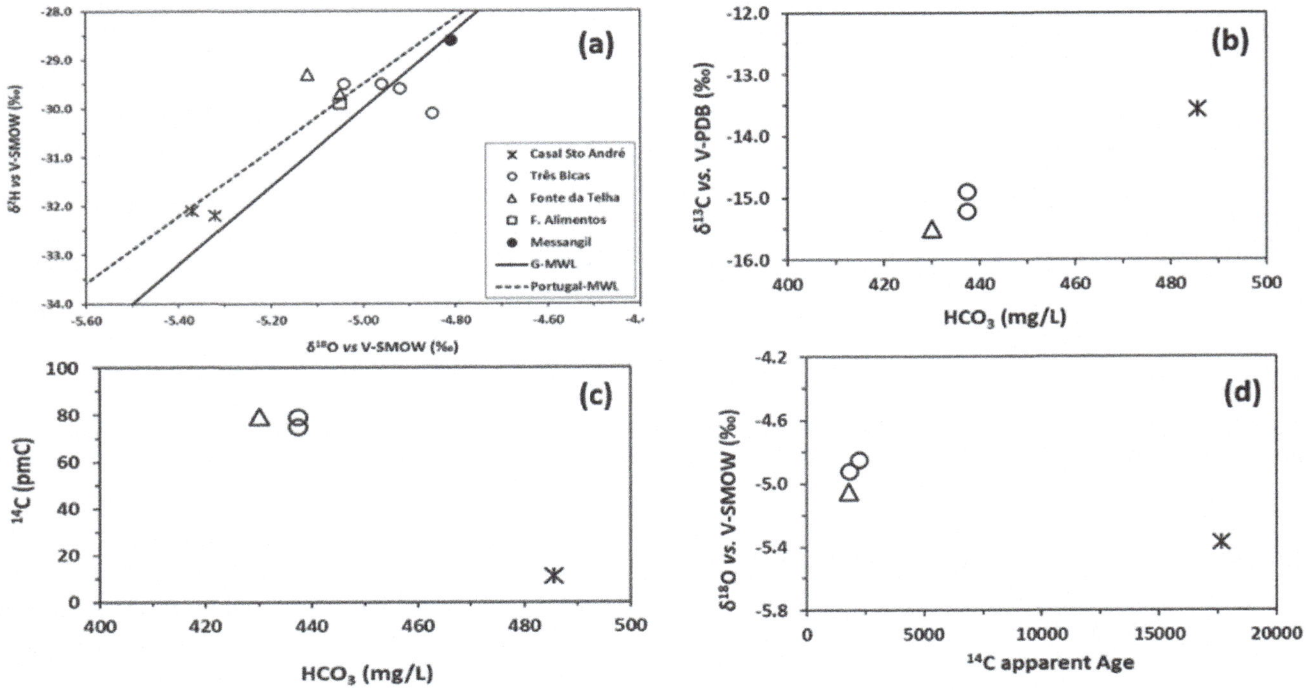


**Fig. 2** **a** $\delta^{18}O$ versus $\delta^{2}H$; **b** $\delta^{13}C$ versus $HCO_3$; **c** $^{14}C$ versus $HCO_3$; **d** $\delta^{18}O$ versus $^{14}C$ groundwater apparent age (Carreira et al. 2019)

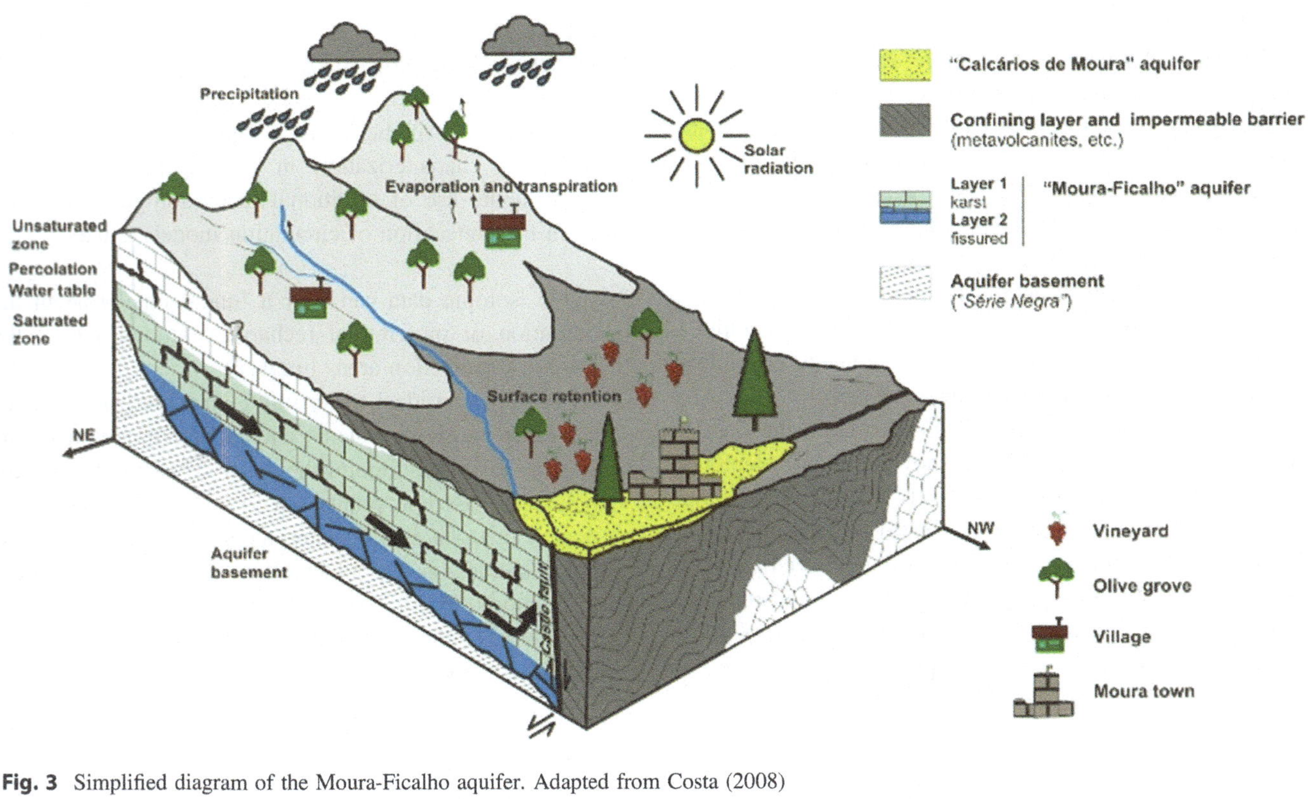


**Fig. 3** Simplified diagram of the Moura-Ficalho aquifer. Adapted from Costa (2008)

- the $^{14}C$ data combined with the $^{13}C$ content allows the (i) groundwater dating of old systems (ii) enables the identification and characterization and origin of dissolved carbon in aqueous systems, (iii) allows the definition of preferential pathways and (iv) estimation of mean water velocities.

**Acknowledgements** $C^2TN$ authors acknowledge the FCT support through the UIDB/04349/2020 Project. CERENA author acknowledges the FCT support through the UIDB/04028/2020 Project, and the author from Instituto Dom Luiz (IDL) acknowledges the FCT support through the UIDB/50019/2020 Project.

## References

Araguas Araguas L, Danesi P, Froehlich K, Rozanski K (1996) Global monitoring of the isotopic composition of precipitation. J Radioanal Nucl Chem 205(2):189–200

Carreira PM, Costa A, Soares AM, Nunes D, João M, Valadas A (2019) Carbon-14 content as a support for Moura mineral water aquifer conceptual model. Sust Wat Res Manag 5(4):1455–1468

Carreira PM, Marques JM, Carvalho MR, Nunes D, Antunes da Silva M (2014) Carbon isotopes and geochemical processes in $CO_2$-rich cold mineral water, N-Portugal. Environ Earth Sci 71:2941–2953

Carreira PM, Valério P, Nunes D, Araújo MF (2004) Temporal and seasonal variations of stable isotopes ($\delta^2H$ and $\delta^{18}O$) and tritium in precipitation over Portugal. In: Isotopes in environmental studies—aquatic forum 2004. conference and symposium papers, pp 370–373

Costa ATM (2008) Modelação matemática dos recursos hídricos subterrâneos da região de Moura. Technical University of Lisbon, Portugal (Ph.D. thesis)

Gonfiantini R, Zuppi GM (2003) Carbon isotopic exchange rate of DIC in karst groundwater. Chem Geol 197:319–336

Lucas LL, Unterweger MP (2000) Comprehensive review and critical evaluation of the half-life of tritium. J Res Natl Inst Technol 105:541–549

Mook WG (2000) Environmental isotopes in hydrological cycle. Principles and applications. IHP-V. Technical Documents in Hydrology, 39 (I), UNESCO and IAEA, Paris

# Impacts of Irrigated Cultures (Paddy-Rice) in Groundwater Quality in Tejo Alluvial River Basin, Portugal

D. Ferreira, M. Simões, F. Pessoa, F. Reboredo, A. Almeida, and F. Lidon

**Abstract**

Tejo River Plains, specially "Lezírias" from Vila Franca de Xira, are an important agricultural region where irrigated cultures prevail. Different water sources provide the demand fulfil for the agricultural practices. Surface waters, in particular those from Tejo River, are the most common in the region. Climate change future scenarios would affect hydric availability in quantity or quality. Irrigated cultures are introducing a stress in groundwater bodies where important nitrates concentrations have been already recognized. In addition, irrigation could increase fertilizers leaching to soils and, in depth, groundwater bodies. Ninety-eight water samples with different origins have been analysed and classified according to Piper and Wilcox diagrams. Piper classification was mostly chloride–sulphur–sodium type and Wilcox classification ranged from C1S1 to C4S3 with a great prevalence of C3S1 and C4S2 categories, meaning a salinity hazard that could pose a quality issue and compromising the soil fertility. Groundwater, as being a source where salinity is neglected, could be a valuable option as a water source. Considering it as an option has an increased pressure to groundwater bodies, where regulation and good agricultural practices are solutions to be considered.

**Keywords**

Groundwater quality • Agricultural practices • Paddy-rice production • Tejo alluvial plain

## 1 Introduction

Increasing demand for food products, in both quantity and quality, to fulfil nutrition scarcity is a challenge that must be considered in a population growth reality. To face this issue, it is required that agricultural lands become more efficient along with a balanced use of natural resources. Climate change is a driving force that could become a threat to this reality, especially in regions sensitive to this problem. Lower Tejo River Basin has been historically recognized as a prosper region and has been used for agriculture production since the eighteenth century. Non-irrigated crops, such as wheat, primarily occupied the region followed by maize and barley cultivation. However, in the beginning of the twentieth century, irrigated rice production starts to be gradually implemented. Usually located in river plains, this type of culture has benefited from the land extension and surface water availability, as occurred in other river basins in Portugal, such as Mondego and Sado. Tejo region, due to its location, could be severely affected by climate variations. These threats can include floods or droughts, affecting primarily water quantity and, in the long term, water quality. It is expected that Mediterranean regions suffer harsh precipitation irregularities along with decreasing rainfall (Ferreira et al. 2018). The major concern for rice production, as being an irrigated culture, is water availability. Surface water has been used as a supply source by its availability, but, as the agricultural production occurs in the area of the major aquifer in the Iberian Peninsula, Tejo–Sado basin could provide a source of this natural resource. The different users from Tejo–Sado aquifer include municipalities for water supply systems, industries and agricultural purposes.

The continuous usage of nitrogenous fertilizers associated with irrigated cultures increase the nitrogen leaching in the form of nitrates to the environment, both for runoff and percolation in soil, which in the long term will reach the groundwater bodies. Because nitrate is a conservative contaminant, it can pose a quality issue for the human health and

D. Ferreira (✉) · M. Simões · F. Pessoa · F. Reboredo · F. Lidon
Department of Earth Sciences and GeoBioTec, FCT-NOVA School of Sciences and Technology, University Nova de Lisboa, Lisbon, Portugal
e-mail: djo.ferreira@campus.fct.unl.pt

A. Almeida
National Institute for Agrarian and Veterinary Research (INIAV), Elvas, Portugal



M. Abrunhosa et al. (eds.), *Advances in Geoethics and Groundwater Management: Theory and Practice for a Sustainable Development*, Advances in Science, Technology & Innovation,
https://doi.org/10.1007/978-3-030-59320-9_53

for the environment leading to eutrophication phenomenon. Tejo Nitrates Vulnerable Area includes both aquifer systems, the Tejo-Sado/Left Bank aquifer and the Tejo Alluvium, which was implemented in Ordinance 164/2010 from 16 March where the area is considered a sensitive location for nitrates contamination from agriculture practices. The main agricultural area is described as "Lezírias" from Vila Franca de Xira region. Due to the agriculture implementation in this region, legislation builds the framework for preserving the groundwater and contaminations that could arise.

## 2 Sampling and Methods

Ninety-eight water samples were collected in HDPE bottles during the 2017 and 2018 agricultural campaigns (Fig. 1). Irrigation waters were superficial and groundwater. Superficial waters had different origins such as irrigation ditches, irrigation canal, water dam, supply systems, and river streams. Groundwater samples were collected in wells with different depths. Irrigation water from flooded paddy-rice parcels were collected as well.

Electric conductivity (EC), temperature (T) and pH were measured with multiparametric electrode. Bicarbonate was determined with titration method. Sulphate ($SO_4$), chloride (Cl) and nitrate ($NO_3$) were determined by photometry. Sodium (Na), potassium (K), calcium (Ca) and magnesium (Mg) were determined by ion chromatography.

## 3 Results

Piper diagram (Fig. 2) was made to classify the samples in their facies. Wilcox diagram (Fig. 3) indicates the adequation of the waters for irrigation purposes.

## 4 Discussion

According to Piper classification, irrigation waters are mainly chloride–sulphur–sodium type. In what concerns the Wilcox diagram classification, waters present a diverse irrigation ability. Most common classifications are C1S1 and C2S1 However, it is in irrigation ditch waters that the most

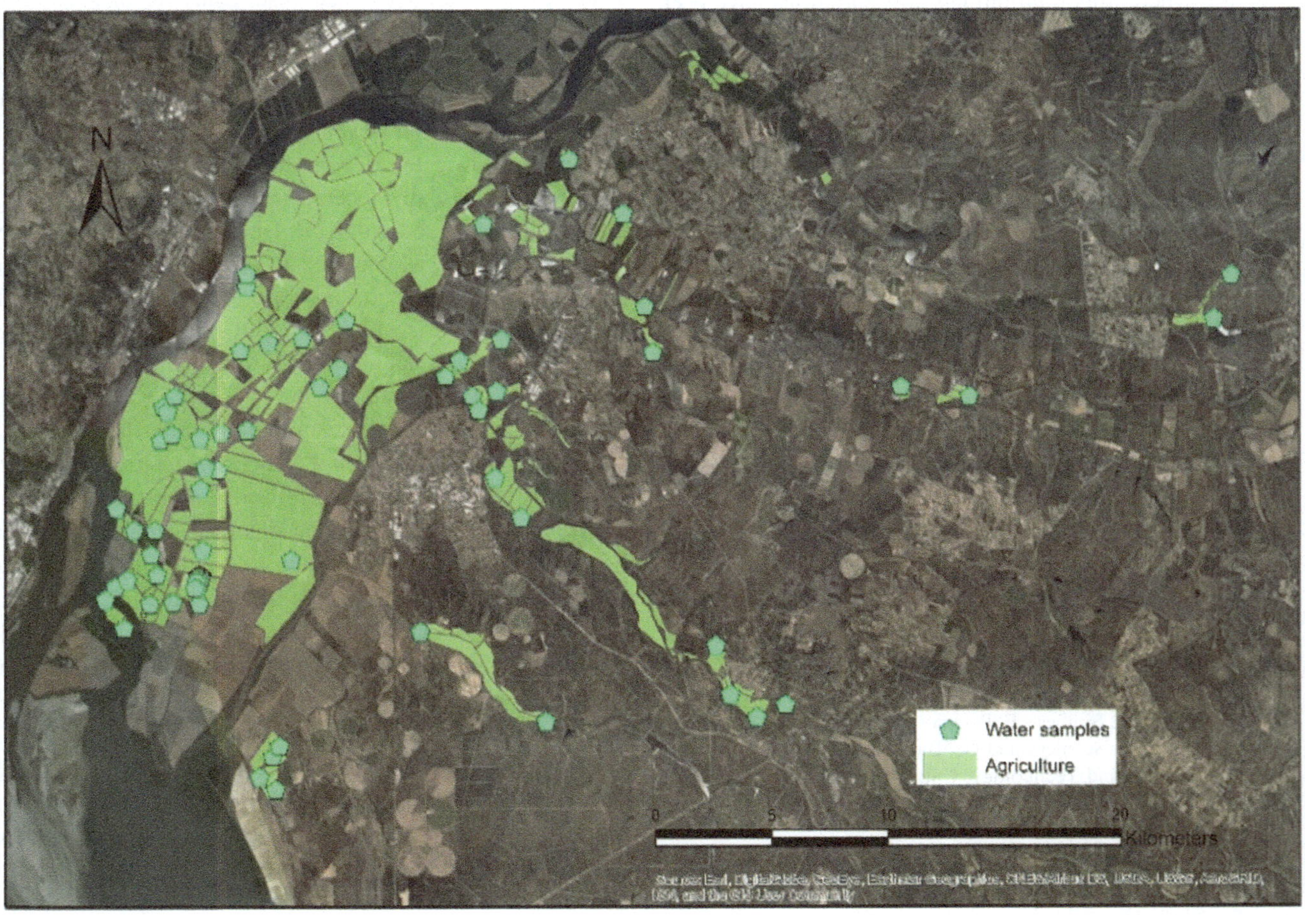


**Fig. 1** Agricultural fields and water samples location

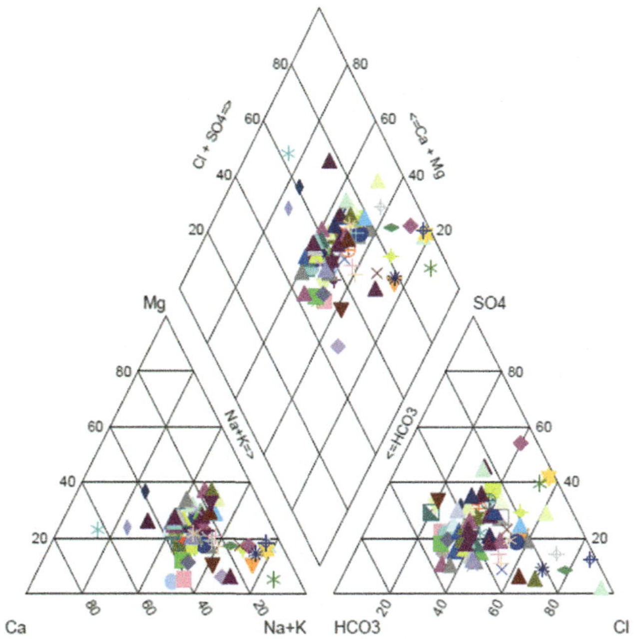


**Fig. 2** Chemical composition of water samples plotted in the Piper diagram

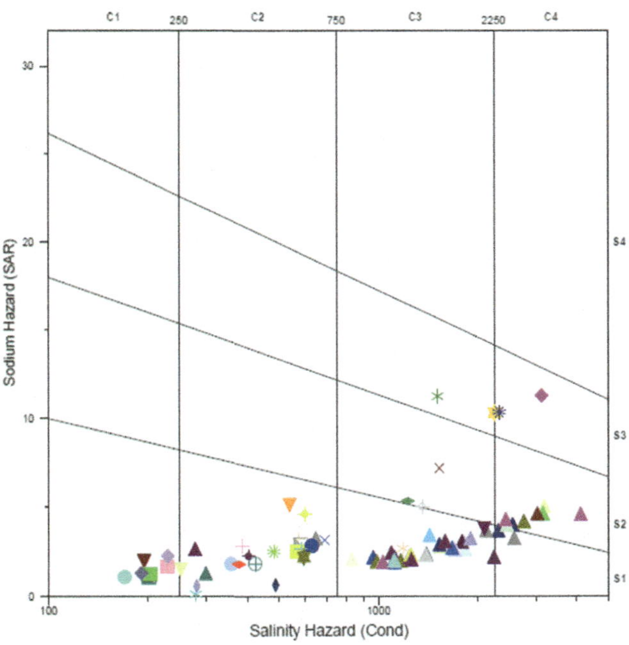


**Fig. 3** Composition of water samples plotted in Wilcox diagram

diverse water quality range is found. Irrigation waters from flooding paddy-rice parcels present C3S1, C4S1 and C4S2 classifications which could represent a salinity hazard for soils given their water conductivity.

Nitrates contamination has not been found in most of the sampled waters. However, in one groundwater sample, it has been detected an important nitrates contamination (70.8 mg $L^{-1}$) which shows that groundwater quality in that water body is already compromised due to agricultural practices. The other groundwater samples (103F1, 103F2 and 103 V) were collected from deeper wells (200 m) where, due to its depth and location, did not presented nitrates contamination.

## 5 Concluding Remarks

Agricultural practices could severely influence groundwater quality. The "Lezírias" from Vila Franca de Xira region suffered a shift of the type of agriculture; from non-irrigated cereal cultures to irrigated cultures, nitrogen fertilizers could be more easily leached to the environment. Water use and fertilizers efficiency, along with implemented nitrates measures, are described in Ordinance 259/2012 from 28 August which stablishes the rules for fertilizers application and good practices. Groundwater bodies are severely threatened by nitrogen contamination, even those considered confined, due to their hydraulic connection to upper bodies or even superficial water bodies. Climate change could increase groundwater stress in this region. As surface water could not be enough to fulfil irrigation demands, either in quantity or quality, an increased groundwater extraction will increase nitrates leaching, with severe groundwater quality deterioration.

Soils could be affected in the long term due to salinity hazard provided by high conductivity of some waters, mostly those that are related with paddy-rice irrigation parcels. Groundwater, having less conductivity, could be a valuable source to prevent soils spoiling in this highly productive area of Portugal.

Despite the preferred usage of surface water to irrigation, the implication of this use in groundwater quality in the Tejo alluvium aquifer has the major importance. Regarding some hydrogeoethical statements, such as preservation, and appropriate behaviour and practices (Limaye 2012), the reflection about the extent of the impact of agricultural practices must have a multidisciplinary approach. The efficient usage of fertilizers is a beginning for environmental consciousness, although, a transitory measure. Minimum impact agriculture with the goal of gathering the different stakeholders that merge into groundwater usage could provide a holistic approach for water resources preservation in a long term, achieving sustainability.

**Acknowledgements** This work was performed under the support of the research grant SFRH/BD/121892/2016 from funds through FCT—Portuguese Foundation for Science and Technology and with the financial support of GeoBioTec Centre (UIDB/04035/2020).

## References

Ferreira D, Simões M, Reboredo F, Pessoa F, Almeida AS, Daradzhanska I, Lidon F (2018) Effects of climate change (precipitation variations), on rice crop yields in alluvial plains of the Tejo and Sado rivers. Emirat J Food Agric 30(6):488–495

Limaye SD (2012) Observing geoethics in mining and in ground-water development: an Indian experience. Ann Geophys 55(3):379–381

# Sustainability Resource of the Hydrogeosphere to Anthropogenic Impacts with Urbanization

Viacheslav Iegupov and Genadiy Strizhelchik

**Abstract**

The main processes of changes in the upper part of the underground hydrosphere that occur in urban areas associated with engineering and construction activities of man are considered. An analysis was made of the causes of such changes and the consequences, which are often sharply negative. Some examples of incidents at urban infrastructure facilities are presented that are associated with deficiencies in forecasting and accounting for changes in the hydrosphere in urban areas as a result of building activity. Based on the indicator sustainable resource to external influences, a typification of Kharkiv city territory hydrogeological conditions stability was carried out with an assessment of the negative engineering and geological processes development possibility associated with building. This scheme can be used when choosing a development strategy and placement of construction projects. The concept of sustainable resource or resilience of the resource to external influences is quite universal; several examples of its application to various natural and natural-technogenic systems are given. The necessity of transition to a strategy of regulated interaction with the geological environment, in which ethical principles will be aimed at maintaining rational homeostasis of natural-technogenic systems, is emphasized.

**Keywords**

Sustainability • Resource • Hydrogeosphere • Impacts • Urbanization

V. Iegupov (✉) · G. Strizhelchik
Department of Geotechnics and Underground Structures, Kharkiv National University of Civil Engineering and Architecture, Kharkiv, Ukraine
e-mail: slavaegu@gmail.com

## 1 Introduction

The natural environment, including the hydrogeosphere (underground hydrosphere), is subjected to a powerful anthropogenic (man-made) impact on the territory of modern cities. The environmental response is often negative; changes occur in hydrogeological and geotechnical conditions that are unacceptable to humans. It should be understood that the anthropogenic load on natural ecosystems, including the hydrogeosphere, which exceeds a certain critical level, leads to consequences that will require dozens or even hundreds of years to eliminate and some of them are often irreversible, for example landslides, suffusion-karst dips, subsidence of the surface as a result of groundwater prolonged pumping, long-term groundwater pollution, etc. A detailed study of the factors of anthropogenic impact was carried out by many authors, for example Strizhelchik (1987), Foster (1990), Loucks and Gladwell (1999), Iegupov et al. (2006), ASCE (2010), Pain (2016) and others.

## 2 General Framework

In connection with industrial development and urbanization, the contradictions between real environmental management and the natural potential of the environment and its components (lithosphere, hydrosphere, atmosphere, and biosphere) to perceive and resist anthropogenic influences are becoming increasingly significant. This is especially noticeable in urban construction practice.

During the development of the territories, many changes occur in the natural hydrogeological systems functioning of under anthropogenic influence. Thus, the main anthropogenic impacts on the underground hydrosphere in the cities of Ukraine are as follows (Strizhelchik and Iegupov 2017):



M. Abrunhosa et al. (eds.), *Advances in Geoethics and Groundwater Management: Theory and Practice for a Sustainable Development*, Advances in Science, Technology & Innovation,
https://doi.org/10.1007/978-3-030-59320-9_54

- the construction of dams on the rivers, with the exception of a direct rise in the water level in the river bed and an increase in the groundwater level (GWL) near the river bed, leads to silt sedimentation on the bottom and banks, which causes a decrease in drainage capacity of the river as a whole;
- in the urban economic and drinking water circulation, there is a large proportion of water taken from the outside: sources and reservoirs located outside the city;
- in old cities, significant water losses from water mains, sewage networks, and technological cycles often occur;
- there is a volume decrease of the water extracted from local underground aquifers, due to their pollution and a corresponding water level increase in these aquifers;
- there are violations of the moisture regime in the aeration zone due to the shielding effect of asphalt pavements and buildings, which reduces evaporation and promotes the condensate formation under the building foundations and low-permeable coatings;
- often there is a barrage effect of blocking underground structures, causing an increase of the groundwater level in local areas;
- with insufficient quality of the territory planning, there is a violation of surface runoff in built-up areas and an increase in seepage compared with natural conditions.
- These effects lead to negative changes in the hydrogeological environment in urban areas, one of which is the process of flooding urban areas with groundwater. This process has the following negative effects:
- there is a decrease in strength characteristics and subsidence of soils, which cause buildings and structures deformation;
- landslide displacements of soils on slopes and escarpments occur;
- changes the chemical composition of groundwater and soil salinization occurs;
- degradation of woody vegetation is due to the excessive moisture of the roots;
- the area of impact and intensity of shock-vibration effects on buildings and structures increase;
- the soil and groundwater corrosivity to concrete and metals increases;
- the operational suitability of the buildings and structures buried parts is reduced due to the groundwater penetration into them;
- sanitary and hygienic conditions are deteriorating due to bacterial contamination and the infection transmission processes acceleration in the aquatic environment;
- the soil layers penetrating ability decreases, and the territory waterlogging occurs;
- suffosion and karst processes are intensified, and the earth's surface dips can occur.

Analysis of the main causes and consequences of settlements flooding clearly illustrates the complexity of the construction work problem interaction with the natural hydrogeological environment. The greatest danger is neglected by important factors and the wrong choice of the construction strategy and the construction objects placement. In such cases, there are prerequisites for major accidents and disasters.

For example, abnormally heavy rains in Kharkiv in June 1995 caused a sharp increase in waterlogging and led to an emergency—a slow "rise" of the shallow underground station "Central Market" began. The station, located in the floodplain, had been successfully operated for 25 years before. To eliminate this process, significant costs and reconstruction of the station were required (Iegupov et al. 2018).

## 3 Results

To prevent such situations, it is usually necessary to choose one of two strategies: consider the possibility of a hazard and protect yourself from it or the possibility of preventing a hazard. The second method, of course, is preferable.

It is worth emphasizing that each natural system is constantly changing, but at the same time, it has a certain sustainable resource (resilience) to external influences; in other words, it retains its main qualitative and functional characteristics. In our opinion, the concept and term "sustainable resource to external influences," first proposed by Strizhelchik (1987), is a fairly universal indicator and can be used to make strategic decisions on interaction with the geological environment.

The universality of the term sustainable resource to external influences (SR) does not change the known ideas about the uniqueness of each point of the geological space, and therefore, for typing and zoning of territories, a system of indicators will be required to highlight significant impacts and resonant factors. In actual practice, there are quite a few relevant examples. For example, in engineering hydrogeology, it is advisable to use the concept of sustainable resource for additional—anthropogenic infiltration and the possibility of hazardous processes. This means that, up to a certain value of this reserve, the groundwater level will not exceed the values characterizing the territory as flooded (usually less than two meters from the ground surface), and for large infiltration flows, the territory will be flooded. Also, the emergence of negative processes associated with the peculiarities of the relief, when even a temporary impact of water infiltration leads to dangerous consequences (e.g., in areas of prone to landslides river valleys, on the slopes of ravines).

As an example, we analyzed the hydrogeological conditions of the Kharkiv territory, typified and identified characteristic areas with different SR levels before additional infiltration and the possibility of hazardous processes (Fig. 1).

The basis was a geomorphological map and materials on the processes of groundwater flooding in the Kharkiv city, Iegupov et al. (2006). There are some types of territories:

I. territories with no SR due to the high position of the groundwater level (GWL). Additional infiltration due to construction causes flooding of the facilities and engineering protective measures is required (as a rule, various types of drainage, reinforced waterproofing, etc.). During heavy rains, intense infiltration in these areas can cause a sharp groundwater level rise, up to and including the release of water to the surface. In this case, the flooding by groundwater turns into a flood, which is accompanied by many negative consequences.
II. Areas with insufficient SR. Additional infiltration may lead to temporary, periodic, or permanent (less frequently) flooding. Measures to protect against flooding will depend on the deepening of the foundations, the responsibility of buildings.
III. Territories with medium (conditionally sufficient) SR. Local temporary flooding is possible here. Protective measures are preventative.
IV. Territories with high SR due to the large GWL depth. Flooding here is possible only because of a temporary perched aquifer.
V. Territories on the steep slopes of ravines and river valleys. The occurrence of flooding and, in some places, waterlogging is possible only near the thalweg. Erosion of soil and underlying grounds on steep slopes is possible.
VI. Areas with the possibility of landslides. They are located on the western slopes of the river valleys.

This scheme can be used when choosing a development strategy and placement of construction projects.

## 4 Discussion

The concept of sustainable resource to external influences (SR) is quite universal and can be applied to various natural and natural-technogenic systems. For example, in geotechnics, one of these SR indicators is the initial subsidence pressure, the value of which allows one to distinguish subsiding loess soils

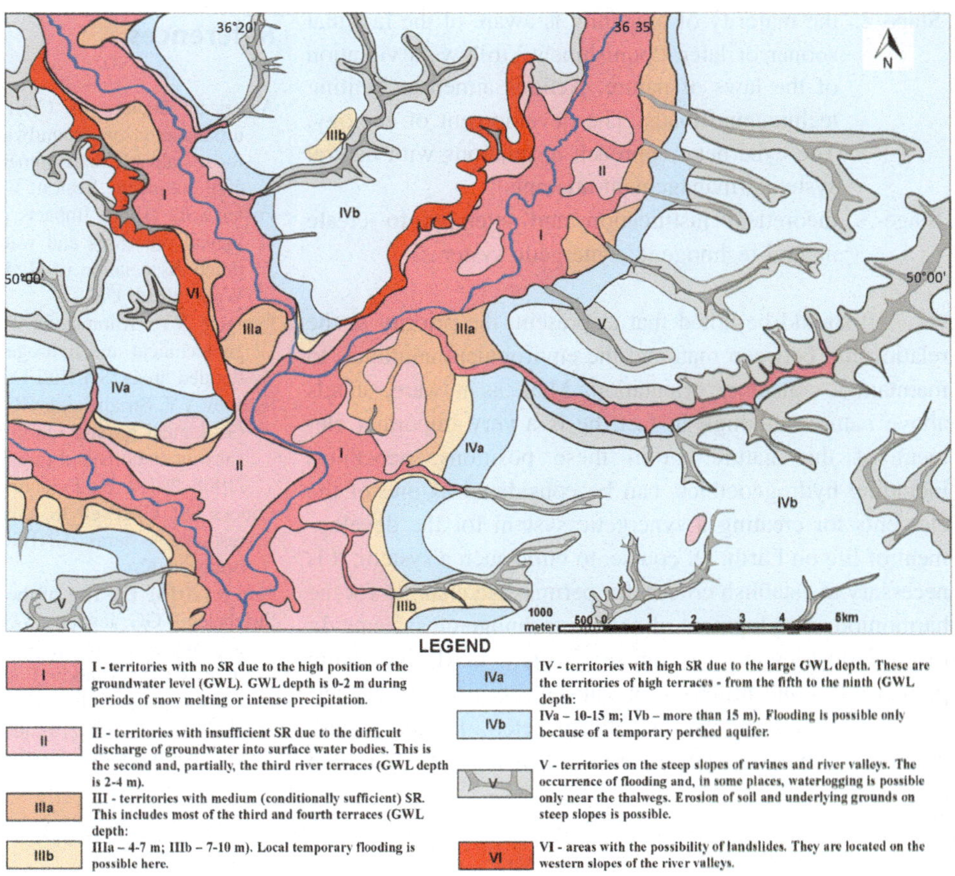


**Fig. 1** Hydrogeological conditions of the Kharkiv territory with different SR levels before additional infiltration and the possibility of hazardous processes

from practically non-subsiding during design and construction. In geotechnical practice, methods are used to increase the SR of soil layers by reducing the compressibility of the soils under load due to the technical restoration of soft soils, eliminating the subsidence properties of soils, etc.

The advantage of this approach we see in the unification of the principles of assessment, in the need to take into account the capabilities of the resource and to regulate both the intensity of the impacts and the increase in the natural resource, and this is also the basis for building harmonious relations with the environment.

Historically, mankind did not think about the possibility of the nature of the Earth to resist human activity (excessive consumption and pollution of water, extraction, and processing of minerals, etc.); i.e., SR natural systems were implied almost endless. Awareness of the limitations of SR appeared no more than a century ago. The following stages of civilizational time of human interaction with natural systems can be distinguished:

Stage 1. consumer attitude to nature. The lack of understanding that the excess of the anthropogenic load over the SR of natural systems can cause both a quick negative reaction (landslide, catastrophic destruction of the dam, etc.), and slow degradation with negative consequences (waterlogging, desertification of the earth, etc.);

Stage 2. the majority of humanity is aware of the fact that sooner or later, "punishment" follows a violation of the laws of nature. Actions aimed at limiting technogenic loads. The development of ecology. The cybernetic approach in relations with natural systems (management attempts);

Stage 3. theoretical justification and attempts to create natural-technogenic synergetic systems.

It should be noted that at present, the priority in the relationship between man and the environment continues to maintain a consumer orientation. Man, as it were, stands above nature, although in fact, he is a very important element of this nature. From these positions, geoethics, including hydrogeoethics, can be considered as one of the elements for creating a synergetic system for the development of life on Earth. Of course, to build such a system, it is necessary to establish criteria for permissible changes for the harmonious development of natural-technogenic systems. In many industries, this is already widely used (maximum permissible concentrations of substances in water bodies, maximum loads on various objects, etc.). In this regard, methods of system analysis are used, that is, the idea that each system is "live", that is, capable of interaction and development, if its phase coordinates do not exceed the permissible maximum or minimum values. Such an approach requires the development of ethical or philosophical standards that would ensure that all goals are subordinated to specific decisions and procedures of a common goal, i.e., the sustainable development of the "human–geological environment" system.

## 5 Concluding Remarks

Due to many variables and the high level of uncertainty in the dynamic system "human—geological environment," the process of harmonization of development is possible only if there are general concepts, such as sustainable resource to external influences. Moreover, this concept is universal and can be applied not only in geology but also in mechanics, biology, medicine, and other applied sciences.

We believe that the above indicates the need and the possibility of transition to a strategy of regulated interaction with the geological environment, where ethical principles will correspond to the synergistic level of development and preservation of rational homeostasis of natural-technogenic systems.

## References

American Society of Civil Engineers—ASCE (2010) Effects of urbanization on groundwater: an engineering case-based approach for sustainable development. In: Ni-Bin C (ed) American society of civil engineers, Reston, VA.

Foster SSD (1990) Impacts of urbanization on groundwater. Hydrological processes and water management in urban areas. International Association of Hydrological Sciences, Publication No. 198, Wallingford, UK

Iegupov VY, Khrapatova IV, Bondarenko AI (2006) Features of geotechnical and hydrogeological conditions of the Kharkov city flooded areas. Sci Bull Constr 39:276–280 ([in Russian])

Iegupov VY, Strizhelchik GG, Kupreychyk AY, Ubiyvovk AV (2018) Geological hazards during construction and operation of shallow subway stations and tunnels by the example of the Kharkiv Metro (1968–2018). Int J Geores Environ 4(4):187–200

Loucks DP, Gladwell JS (eds) (1999) Sustainability criteria for water resource systems. UNESCO, Cambridge University Press, Cambridge, UK

Pain S (2016) The rise of the urbanite. Nature 531:S50–S51

Strizhelchik GG, Iegupov VI (2017) Problems of subway construction in complex engineering and geological conditions by the example of the city of Kharkiv. Poltava National Technical Yuri Kondratyuk University, Ind Mach Build Civ Eng J 2(49):195–200

Strizhelchik GG (1987) Problems of engineering geology of cities and possible ways to solve them. Eng Geol 2:3–12 ([in Russian])

# Tiber Middle Valley: Hydraulic Risk Management and Urban Development of the Areas

Giuseppe Sappa, Stefania Vitale, and Flavia Ferranti

**Abstract**

This paper deals with different implications of hydraulic risk management in the Middle Valley of Tiber River, which is the Tiber River Basin part, starting from the Corbara dam and ending where this important river approaches and goes inside the large metropolitan area of Rome. It means that the hydraulic risk management in this area is strongly related to hydraulic defence of Rome town, and all its ancient and historically very important monuments protection. Here, they are presented the technical solutions, proposed for the hydraulic defence of the municipalities arising inside this area, without increasing the hydraulic risk in it, but also in the metropolitan area of Rome. This is quite a historical problem started at the end of seventies years of twentieth century, when the social and urban developments of large part of these areas have been frozen as they have been chosen to be the natural flooding areas in case of 100-year and 200-year return time meteoric and hydraulic extreme events. In the aim of respecting this target, this paper presents the hydraulic works plan, aimed to guarantee a sustainable urban development of these areas, without increasing hydraulic risks in the neighbouring areas and in the metropolitan area of Rome.

**Keywords**

Tiber river • Hydraulic risk management • Rome

G. Sappa (✉)
Department of Civil, Building and Environmental Engineering, Sapienza University of Rome, Rome, Italy
e-mail: giuseppe.sappa@uniroma1.it

S. Vitale
SPE Ingegneria Srl, Rome, Italy

F. Ferranti
Department of Environment, Rome, Italy

## 1 Introduction

In the aim of giving a sustainable development to municipality areas in the Tiber Middle Valley, flood management studies have been carried on, in the last three decades, by the Tiber River Basin Authority, now become Central Apennines District Authority. This target was included through national directives, and there are regional actions to implement the necessary measures. Legislative Decree no. 49/2010 transposed the EU Floods Directive (2007/60) into national law on assessment and management of flood risk. In this framework, according to Italian Law 183/89 on soil protection, the Tiber River Basin Authority had anticipated the EU directive contents developing a planning tool, the District River Basin Management Plan, for the identification of flood-prone areas and the level of risk exposure (Sappa and Ferranti 2020). This paper presents hydraulic works proposed as part of the Master Plan of the hydraulic risk conditions decrease in the Middle Valley of the Tiber, which has been, historically, part of the greater necessity giving hydraulic protection to the city of Rome. In other words, this very large area, located between Orte and Castel Giubileo, has been reserved for the natural expansion of the floods of the Tiber River in case of exceptional hydrological events, thus preventing damage to the city of Rome.

It is therefore possible to summarize that the aims of the project accord to this point:

- to maintain the current hydraulic safety conditions in the city of Rome unchanged,
- ensuring the lamination function of the flood waves;
- to satisfy the reasonable economic development needs of the communities living in the areas currently burdened by constraints (Municipalities of Capena, Castelnuovo di Porto, Fiano Romano, Riano, Settebagni and Monterotondo Scalo) through interventions suitable for a more equitable distribution of the same throughout the basin;



M. Abrunhosa et al. (eds.), *Advances in Geoethics and Groundwater Management: Theory and Practice for a Sustainable Development*, Advances in Science, Technology & Innovation,
https://doi.org/10.1007/978-3-030-59320-9_55

- to protect considerable river ecosystems in the Nazzano and Alviano oasis.

## 2 The Tiber Middle Valley

The Tiber River Basin, the largest river basin in Central Italy (third largest river in Italy), is located in the Central Apennines District and ends at the Mediterranean Sea, covering a land area of 17,500 km$^2$, crossing six administrative regions. The study area is the Tiber River section from Orte and Castel Giubileo, in the Middle Valley of the Tiber River (MTV), in Latium Region central part, few kilometres to North of Rome.

The MTV, structurally, corresponds to the "Paglia-Tevere Graben" (Funiciello and Parotto 1978), a NNW-SSE trending extensional basin, developed since the late Early Pliocene (Barberi et al. 1994). This basin is mostly filled by Plio-Pleistocene marine terrigenous deposits. However, coeval non-marine and transitional terrigenous and carbonatic deposits extensively outcrop along the western margin of the Mt. Peglia-Lucretili Mts. ridge (Mancini et al. 2004; Fiseha et al. 2013), while Pleistocene volcano-sedimentary successions are well exposed west of the Tiber River (Mazza and La Vigna 2011) (Fig. 1). Hydraulically speaking, the Middle Tiber Valley is the Tiber basin part, included between Corbara Dam, near Orte, a little town in the North part of Latium Region, and the Giubileo Castle Traverse, which is located just before Tiber River income the metropolitan area of Rome. This traverse is also considered, and it plays, as the hydraulic prevention work, against flooding events inside the metropolitan area of Rome.

## 3 Results

Hydraulic modelling and flood inundation mapping have been performed to provide important information about a flood event, including the level of inundation and water surface elevations, within the study area. Modelling simulations have been carried on, referring to different meteoric and hydraulic return times. According to aforementioned rules, they have been considered 100-year and 200-year return times. But only for this latter one, any hydraulic works had to be designed. Starting from these results, represented, as an example, in Fig. 2, they have been carried on many modelling simulations aimed to plan and design hydraulic works.

They are aimed to allow the involved municipalities, like Capena, Castelnuovo di Porto, Fiano Romano, Riano, Settebagni and Monterotondo Scalo to have an urban development in hydraulic safety conditions and, on the other side, to maintain, at the meantime, in the same hydraulic safety conditions also the Rome metropolitan area as the neighbouring areas. This target must be reached, without

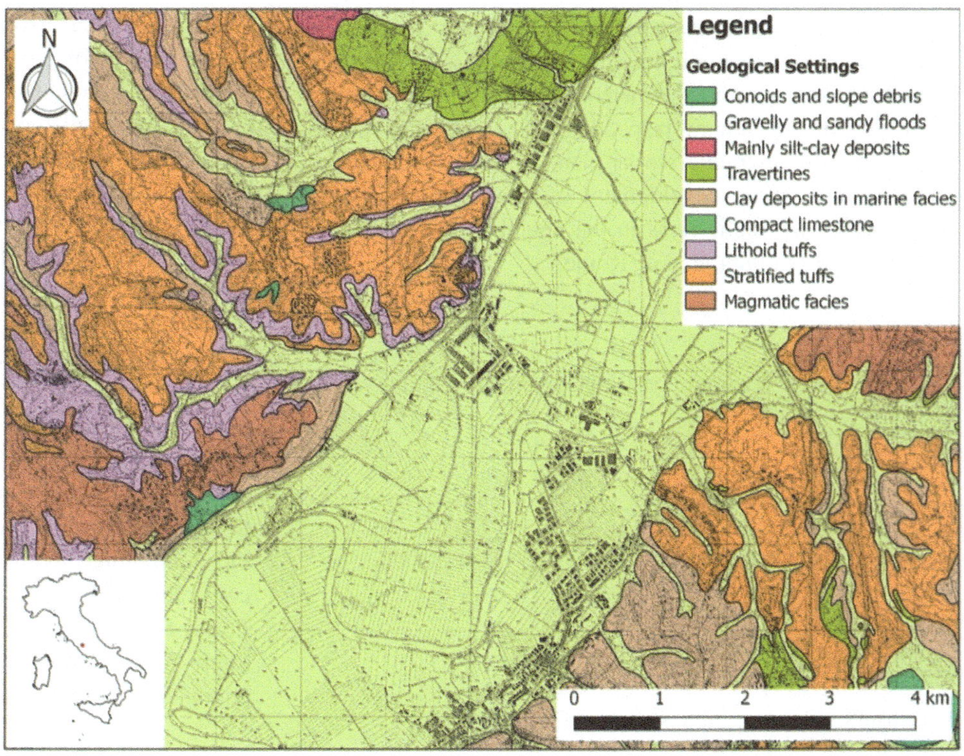


**Fig. 1** Geology, topography and drainage network in the Middle Tiber River Basin

**Fig. 2** Example of flooded area in the Tiber Middle Valley

increasing the hydraulic risk level in any other area of the Tiber Middle Valley.

Starting from these results, it has been designed a set of hydraulic works aimed to reach the aforementioned targets, and in the following, it will be described and presented (Fig. 3).

The reference constraints that addressed the original proposal, and that have been considered, to be reference, also for any verification's performance are the following:

- the flood flow rate at Castel Giubileo section of Tiber River must not exceed 2700 $m^3/s$
- the distribution of flooding volumes, subtracted from the areas, affected by the inhabited areas, which had to be defended, must be included in specially sized expansion tanks, located in areas, included in the same municipalities for which passive defence works have been proposed, or in contiguous areas;
- perimeter of passive defence works, related to areas, currently inhabited, and for which the writer is aware of urban development needs for the municipalities and inhabited areas mentioned above.

The designed hydraulic works set is made of passive defence interventions (conditioning flood levels in the alluvial valley between the Corbara reservoir and the Castel Giubileo traverse) and active defence (the recovered reservoir capacities must be located, in order to achieve the maximum hydraulic connection with areas to defend), as it is reported, as an example in Fig. 3. On the other hand, they have been suggested to introduce directives, framework prescriptions, regulation prescriptions, incentive measures and management guidelines in the aim of a better management of these areas.

## 4 Discussion

This paper presented the set of hydraulic works planned and designed in the aim of giving to Tiber Middle Valley municipalities some chances of a sustainable urban development, which means, in this case, to have it without increasing the hydraulic risk level neither in any other area in the same valley nor in the metropolitan area of Rome. In a previous study (Sappa and Ferranti 2020), it has been described the numerical model and the GIS applied for the updating the hydraulic risk assessment in this area, while here they have been presented the hydraulic risk management proposed and the constraints, they have been respected to set up a solution. This purpose contributes to give a different distribution of urban and agricultural areas in this Middle Valley, involving interests of different social and economic stakeholders, which had to be taken in account and which have been involved in the whole decision-making process.

## 5 Concluding Remarks

Hydraulic risk management in large areas, involved in urban development, requires choices referring to people living in those areas and classification of priority in land use and soil

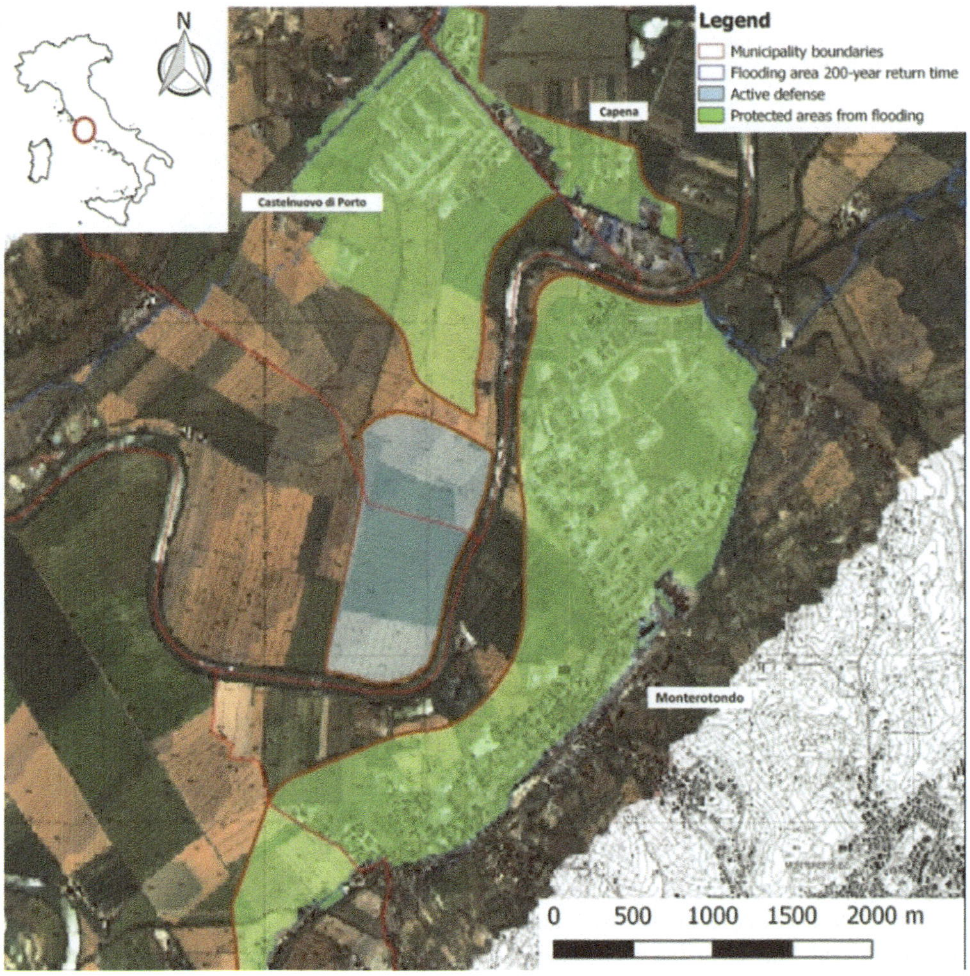


**Fig. 3** Example of passive defence works and expansion tanks in Tiber Middle Valley

consumption, which sometimes are quite complicate. This study presented a proposed solution for Tiber Middle Valley, which has been approved by the Tiber River Basin Authority, now become Central Apennines District Authority, in 2008, as from the methodological point of view as in the merit, which has been applying, step by step in the last ten years by the Administration of Latium Region.

## References

Barberi F, Buonasorte G, Cioni R, Fiordelisi A, Foresi L, Iaccarino S, Laurenzi MA, Sbrana A, Vernia L, Villa IM (1994) Plio-pleistocene geological evolution of the geothermal area of Tuscany and Latium. Memorie Descrittive Della Carta Geologa D'Italia 49:77–134

Fiseha BM, Setegn SG, Melesse AM, Volpi E, Fiori A (2013) Hydrological analysis of the Upper Tiber River Basin, Central Italy: a watershed modelling approach. Hydrol Process 27(16):2339–2351

Funiciello R, Parotto M (1978) Il substrato sedimentario nell'area dei Colli Albani: considerazioni geodinamiche e paleogeografiche sul margine tirrenico dell'Appennino centrale. Geol Romana 17:233–287

Mancini M, Girotti O, Cavinato GP (2004) Il Pliocene e il Quaternario della Media Valle del Tevere (Appennino Centrale). Geol Romana 37:175–236

Mazza R, La Vigna F (2011) Hydrogeology of the southern Middle Tiber Valley (Central Italy). AQUAmundi Am 04031:93–102

Sappa G, Ferranti F (2020) Flooding risk assessment in the Middle Tiber River Valley with remediation proposal. In: Fernandes F, Malheiro A, Chaminé HI (eds) Advances in natural hazards and hydrological risks: meeting the challenge. Advances in science, technology and innovation series, Springer, Cham, pp 111–115

# Risk Assessment for Groundwater: A Case Study from a Municipal Landfill in Southern Poland

Dominika Dabrowska and Wojciech Rykala

**Abstract**

Environmental Risk Assessment (ERA) is an analytical method suitable for assessing environmental impacts which uses historical data collection, identification of sources of regional risks, as well as a probability and impact estimation of signal risk type. It currently constitutes one of the practical aspects of geoethics. In this paper, EIA was performed with respect to groundwater quality in the region of a municipal landfills system in southern Poland. The consequences and causes analysis, the effect/probability matrix and SWOT method were used for assessing the impact of various factors on the quality of the groundwater in the region. In this paper, we indicated that the best method for a groundwater risk assessment in the region of the municipal landfills system was the SWOT analysis. It involved five strengths, six weaknesses, and opportunities and threats. The monitoring data of spatial and temporal variability of leachate quality and groundwater quality were considered, and a simulation of the longevity of both the top and bottom security system was performed. Moreover, the spatial planning was conducted, as well as an analysis of the impact of other objects on the groundwater and an examination of the terrain and climatic conditions were carried out. The total risk assessment for the groundwater in that region obtained using the SWOT method was determined as −4.

**Keywords**

Risk assessment • Landfills • Leachates • Tychy • Poland

D. Dabrowska (✉) · W. Rykala
Institute of Earth Sciences, University of Silesia, Katowice, Poland
e-mail: dominika.dabrowska@us.edu.pl

## 1 Introduction

One of the most popular methods of waste utilization is landfilling (Sobik 2007; Esmaeilian et al. 2018). In general, old landfills have no security from the ground, the technical construction of landfills and a security system from the ground appear to be of great importance (Dąbrowska et al. 2018a). It is not only the old landfills that can pose a threat to the groundwater quality (Mor et al. 2018), but also new landfills containing liner systems may be a source of groundwater pollution caused by their degradation (Sun et al. 2019).

Pollutants leached from the waste are dangerous for the groundwater in both physicochemical and bacteriological terms Taylor et al. (2004). Their size and chemical composition relate to the access of the waste to oxygen during storage. There are several indicators of groundwater pollution (Dąbrowska et al. 2018b) which should be measured in the groundwater and leachates, i.e., electrical conductivity (EC) and $Cl^-$, $SO_4^{2-}$, nitrogen, $B^+$, $Cr^{6+}$, $Cu^{2+}$, $Zn^{2+}$, $Pb^{2+}$, $Hg^{2+}$, $Co^{3+}$ and $Ni^{2+}$ (Bojakowska 1994).

Research and reflection on human activities influencing the geosphere are investigated by geoethics (Peppoloni and Capua 2012; Wyss and Peppoloni 2014 and references therein). In fact, geoethics encompasses both theoretical and practical aspects, including risk analysis and management (Wyss and Peppoloni 2014). One of the most popular methods for the proper management of the relationship between scientists and decision-makers is the Environmental Risk Assessment (ERA). ERA is an efficient analytical tool for analyzing environmental impacts which may also support the decision-making process to be conducted regarding the risk for the groundwater (Wu and Zhang 2013).

Risk management can be called a risk estimation process aimed at bringing it to an acceptable level. The ideal system of risk management should include the following phases: planning, acquisition, development, testing, proper deployment of information systems (Pritchard 2013). Risk analysis


M. Abrunhosa et al. (eds.), *Advances in Geoethics and Groundwater Management: Theory and Practice for a Sustainable Development*, Advances in Science, Technology & Innovation,
https://doi.org/10.1007/978-3-030-59320-9_56

is a process of risk recognition, understanding of its size and identification of areas that require protection. The aim of risk assessment is to create the largest list of risks and group the risks resulting from probable events in a representative manner (Wroblewski 2015).

In the case of risk assessment resulting from landfills, consideration should be given to the spatial and temporal variability of leachate quality and the simulation of the longevity of both the top and bottom security systems. Except for the chemical parameters of leachates and construction of the landfill, an assessment of the impact of external factors should also be performed. Spatial planning and climate conditions can be obtained based on historical data (Calamari and Zhang 2002).

In this paper, ERA was used for the assessment of groundwater quality and risk identification in the region of municipal landfills system in Tychy-Urbanowice (southern Poland). The determination of the impact of various internal and external factors on the groundwater quality was performed using risk analysis methods, such as the matrix and SWOT methods.

## 2 Settings

ERA was performed for the municipal landfill system in the eastern part of the city Tychy-Urbanowice (southern Poland) divided into a closed and reclaimed landfill and an active landfill with the surrounding infrastructure. The closure of part of the landfill was caused by its lack of an adequate liner system preventing the infiltration of pollutants into the groundwater. The active landfill consists of two adjacent quarters, and it is equipped with a modern security system and seals to prevent the entry of pollutants into the environment which include, among others, a triple drainage system: inter-soil drainage, drainage (over foliage) and sub-layer drainage, as well as a sealing system made up of two PEHD foils.

There are three aquifers in the region of the described landfills—Quaternary (Pleistocene), Triassic (locally) and Carboniferous. Locally, groundwater exists in the interbedding of sands or the sandstones of the Miocene clays. The Quaternary aquifer locally separated by poorly permeable sediments appears to be of the greatest importance. The lower aquifers are protected by very poorly permeable sediments (Czermiński et al. 1993).

## 3 Methods

An Environmental Risk Assessment for the complex of landfills under examination was performed using the following stages: planning of risk management, risk identification, estimation of the impact on operations, estimation of weaknesses and threats, implementation of the risk management plan, compliance measurements and review and monitoring. The risk analysis was designed to gain an in-depth understanding of each individual threat. Its results revealed which threat requires the application of appropriate methods of action. During the analysis, elements such as hazards, stressors, source of exposure, routes of exposure and receptors were considered.

The analysis of the groundwater risk assessment involved quantitative and qualitative methods (Korczowski 2009). In general, three methods were chosen: analysis of causes-consequences, effect/probability matrix and SWOT analysis (Phadermrod et al. 2019).

The analysis of causes and consequences is a combined method of the fault tree and event tree methods. Consideration is given to the causes and consequences of the situation under analysis. The initial part of the analysis involves the marking of the critical event, and then, logic gates yes or no are used for the analysis of the consequences of that situation (Bubbico 2018).

The consequence/probability (hazard) matrix is based on the classification of risk sources. It is used to determine which risks require immediate preventive action and which need further analysis. The determination of likelihood of an unfavorable situation provides information necessary for the assessment of the level of risk (Haddad et al. 2012).

The last selected method is among the most common methods applied to strategic management, also as regards hydrogeological aspects. The SWOT analysis was based on the following elements of the situation under analysis (Hajizadeh 2019): Strengths, Weaknesses, Opportunities and Threats.

## 4 Results

Several risk factors for the groundwater were found in the area subjected to analysis. The biggest facilities in the area presenting a risk are a system of municipal waste landfills, numerous car parts production plants, a sewage treatment plant located south of the landfills, a sewage treatment plant located at the car parts production plant and an animal shelter. There are a lot of possible factors that could pose a threat to the groundwater in the region, and they may be simulated. Additional events concerning the landfill that could possibly cause a risk for the groundwater in future include a failure of the leachate drainage system, leakage of the leachate tank and migration of leachate into the aquifer from the waste that has not yet been placed in the landfill.

Migration of pollutants to the groundwater could contaminate soils, surface and underground waters resulting in the extension of the cloud pollution zone and its migration in the groundwater flow direction. Such an event requires additional groundwater monitoring and remediation, and it will entail the imposition of financial penalties on the enterprises. Other possible factors affecting the groundwater quality could be human errors, installation failure or flood risk.

The second applied method was the effect/probability analysis. The analysis gives an example of two events that may occur in the region of the landfills under examination and have an impact on the groundwater. The first described situation involved a potential failure and leakage of sewage to the natural environment from the nearby sewage treatment plant located south of the landfill, while the second one related to the potential migration of leachate to the aquifer from the waste that has not yet been placed in the landfill.

The third used method was the SWOT analysis. The positive aspects were optionally divided into "strengths" and "opportunities," whereas the negative features were classified as "weaknesses" and "threats." The strengths included modern research monitoring, trained employee staff, modern security system, new jobs for the staff and the leveling of land. The identified opportunities were an increase in the financial inflow to the city budget, reduction of the amount of waste, development of waste management technologies and growth of the nearby area. In contrast, the revealed weaknesses and threats included among others: possibility of migration of pollutants into the soil and water environment, possibility of failure of security systems in the landfill, sewage treatment plant.

## 5 Concluding Remarks

Geoethics is beginning to play a valuable role in the field of the creation of new models of social and economic development, protection of the environment, at the same time supporting innovation in the management of environmental problems and natural resources. One of the effective methods which could help in a proper management of the risk for the environment and assessment of risk factors is risk analysis. It was used in this paper to carry out a structured assessment of the groundwater quality problem. The analysis facilitates the identification of weaknesses in the monitoring system, human work or errors in the procedures. Its important advantage consists of the improvement of the security system and the functioning of the area. It should also be noted that the occurrence of damage to nature is related, among others, to the factor dangerous for the environment, its quantity, intensity, the extent of contamination or its duration. The selected methods of risk analysis can help prepare a list of possible scenarios and situations that may occur in the described area.

## References


Bojakowska I (1994) Wpływ czynnika antropogenicznego na procesy geochemiczne w powierzchniowych warstwach litosfery (in Polish with English summary). Instructions and Methods of Geological Studies/Państwowy Instytut Geologiczny, Warszawa

Bubbico R (2018) A statistical analysis of causes and consequences of the release of hazardous materials from pipelines. The influence of layout. J Loss Prev Process Ind 56:458–466

Calamari D, Zhang L (2002) Environmental risk assessment of pesticides on aquatic life in Xiamen, China. Toxicol Lett 128:45–53

Czermiński P (1993) User manual of the groundwater monitoring network in the region of the municipal landfill in Tychy-Urbanowice (in Polish with English summary). EKO-SON, Tychy

Dąbrowska D, Witkowski A, Sołtysiak M (2018a) Application of pollution indices for the spatiotemporal assessment of the negative impact of a municipal landfill on groundwater (Tychy, southern Poland). Geol Q 62(3):496–508

Dąbrowska D, Witkowski A, Sołtysiak M (2018b) Representativeness of the groundwater monitoring results in the context of its methodology: case study of a municipal landfill complex in Poland. Environ Earth Sci 77(7)

Esmaeilian B, Wang B, Lewis K, Duarte F, Ratti C, Behdad S (2018) The future of waste management in smart and sustainable cities: a review and concept paper. Waste Manag 81:177–195

Haddad A, Galante E, Caldas R, Morgado C (2012) Hazard matrix application in health, safety and environmental management risk evaluation. In: Risk management for the future—theory and cases, pp 29–50

Hajizadeh Y (2019) Machine learning in oil and gas; a SWOT analysis approach. J Petrol Sci Eng 176:661–666

Korczowski A (2009) Zarządzanie ryzykiem w projektach informatycznych. Teoria i praktyka, Helion, Warszawa: 224

Mor S, Negi P, Ravindra K (2018) Assessment of groundwater pollution by landfills in India using leachate pollution index and estimation of error. Environ Nanotech Monit Manag 10:467–476

Phadermrod B, Crowder R, Wills G (2019) Importance-performance analysis based SWOT analysis. Int J Inf Manage 44:194–203

Peppoloni S, di Capua G (2012) Geoethics and geological culture: awareness, responsibility and challenges. Ann Geophys 55(3). https://doi.org/10.4401/ag-6099

Pritchard C (2013) Zarządzanie ryzykiem w projektach. WIG-Press, Łódź

Sobik K (2007) Badanie wpływu składowisk odpadów na środowisko gruntowo-wodne na przykładzie wybranych obiektów zlokalizowanych w obrębie zlewni Dunajca. AGH, Kraków (Ph.D. thesis)

Sun X, Xu Y, Liu Y, Nai Ch, Dong L, Liu J, Huang Q (2019) Evolution of geomembrane degradation and defects in a landfill: impacts on long-term leachate leakage and groundwater quality. J Clean Prod 224:335–345

Taylor R, Cronin A, Pedley S, Barker J, Atkinson T (2004) The implications of groundwater velocity variations on microbial transport and wellhead protection - review of field evidence. FEMS Microbio Eco 49(1):17–26

Wróblewski D (2015) Zarządzanie ryzykiem: przegląd wybranych metodyk. CNBOP-PIB, Józefów

Wu K, Zhang L (2013) Progress in the development of environmental risk assessment as a tool for the decision-making process. J Serv Sci Manag 7:131–143

Wyss M, Peppoloni S (2014) Geoethics. Elsevier

# Effects of Nanoplastics, Lithium, and Their Mixtures on *Corbicula fluminea*: Preliminary Findings

Rafaela S. Costa, Patrícia Oliveira, and Lúcia Guilhermino

**Abstract**

The effects of polystyrene nanoplastics (NP), lithium (Li), and their mixtures were investigated using the exotic invasive bivalve *Corbicula fluminea* as test organism for ethical reasons. In a 96 h laboratory bioassay, groups of bivalves were exposed to water (control), 0.8 mg/L NP, 3.2 mg/L NP, 1 mg/L Li, 4 mg/L Li, 0.8 mg/L NP + 1 mg/L Li (Mix1) or 3.2 mg/L NP + 4 mg/L Li (Mix2). Effect criteria were filtration rate (FR), lipid peroxidation levels (LPO), and the activity of the enzymes cholinesterases (ChE), isocitrate dehydrogenase (IDH), octopine dehydrogenase, glutathione *S*-transferases and glutathione reductase. Bivalves exposed to NP alone or in mixture had IDH activity inhibition (22%), and LPO reduced by 33%. Bivalves exposed to Li alone or in mixture had 22% of ChE inhibition. Bivalves exposed to Mix2 had FR induction (twofold). These results suggest that the NP tested may decrease the cellular energy production by anaerobic pathways despite providing some reduction of lipid oxidative damage that Li has anticholinergic effects, and that NP and Li may cause toxicological interactions in bivalves exposed simultaneously to the two substances. Further studies are needed to understand potential toxicological interactions between NP and Li under different abiotic conditions and exposure periods.

**Keywords**

Nanoplastics • Lithium • Bivalves • *Corbicula fluminea* • Biomarkers

## 1 Introduction

Microplastics (MP) are global pollutants of high concern. In the environment, MP are fragmented into smaller particles eventually reaching dimensions in the nanoscale known as nanoplastics (NP). NP used in several applications are also introduced in the environment. NP are considered an emerging environmental threat (Lehner et al. 2019). More knowledge on the effects of NP and their toxicological interactions with common contaminants is needed. Lithium (Li) is a widely used metal (e.g., medicine, several industries, batteries) and a common contaminant that is toxic to aquatic species at ecologically relevant concentrations. Therefore, investigating the toxicological effects of NP, Li, and their mixtures is of interest.

*Corbicula fluminea* is an exotic invasive species in Europe and other regions and is considered an adequate bioindicator (Oliveira et al. 2015a), including in relation to MP pollution (Su et al. 2018), and a good test organism (Oliveira et al. 2015b). Its use in environmental assessments helps to control bioinvasions and avoids the use of native species (Guilhermino et al. 2018). *C. fluminea* uptakes different types of MP (Li et al. 2019) and was used to assess the effects of some MP (Guilhermino et al. 2018; Oliveira et al. 2018) and NP (Rochman et al. 2017).

The goal of the present study was to conduct a preliminary assessment of the effects of polystyrene NP, Li, and their mixtures using *C. fluminea* as test organism to avoid the use of Portuguese native bivalves for ethical reasons.

R. S. Costa · P. Oliveira · L. Guilhermino (✉)
ICBAS—Institute of Biomedical Sciences of Abel Salazar, Department of Populations Studies, Laboratory of Ecotoxicology and Ecology (ECOTOX),, University of Porto, Porto, Portugal
e-mail: lguilher@icbas.up.pt

R. S. Costa · P. Oliveira · L. Guilhermino
CIIMAR—Interdisciplinary Centre of Marine and Environmental Research, Research Team of Ecotoxicology, Stress Ecology and Environmental Health (ECOTOX), University of Porto, Porto, Portugal



M. Abrunhosa et al. (eds.), *Advances in Geoethics and Groundwater Management: Theory and Practice for a Sustainable Development*, Advances in Science, Technology & Innovation,
https://doi.org/10.1007/978-3-030-59320-9_57

## 2 Methods

NP were fluorescent blue amine-modified polystyrene beads (mean diameter 50 nm, 1.04–1.06 g/mL density, excitation and emission wavelengths: 360 and 420 nm, respectively) in aqueous suspension purchased from Sigma Aldrich (USA). Lithium chloride (Merck, Germany) was used as Li source.

*C. fluminea* specimens were collected in the Minho River estuary (NW Iberian Peninsula). They were transported to the laboratory in water from the collection site, nine animals were used for biomarkers determination, and the others were gradually acclimatized to experimental conditions: dechlorinated tap water for human consumption (water), 16 h light: 8 h dark, 16 ± 1 °C, continuous air supply, and feed with *Chlorella vulgaris* ($8 \times 10^5$ cells/day/bivalve), other conditions as in Oliveira et al. (2018). After 14 days of acclimatization and 78 h without food, 9 bivalves were used for biomarkers determinations, and the others were used in the bioassay. Specimens used in the bioassay had a mean and standard deviation (SD) of maximal shell length (2.3 ± 0.1 cm).

In the 96-h bioassay, bivalves were exposed individually through the water to the following treatments (nominal concentrations): control (water only), 1 mg/L NP, 4 mg/L NP, 1 mg/L Li, 4 mg/L Li, 1 mg/L NP + 1 mg/L Li (Mix1), and 4 mg/L NP + 4 mg/L Li (Mix2). Abiotic conditions were as in the acclimatization period but without food. Three bivalves per treatment were used. Water was renewed at each 24 h. The NP actual concentrations were determined in new and old water by spectrophotometry from a calibration curve (fluorescence *versus* NP concentration) as indicated in Guilhermino et al. (2018) but using 360 nm and 420 nm as excitation and emission wavelength, respectively. No significant ($p > 0.05$) differences in NP concentrations between NP-L and Mix1 either in new and old water were found, and a similar result was obtained for NP-H and Mix2 in both new and old water. The actual concentrations of NP in the water decreased over 24 h by more than 20%. Thus, the estimated exposure concentrations along the bioassay were determined (Guilhermino et al. 2018), and they were 0.8 mg/L for the lowest NP concentration and 3.2 mg/L for the highest NP concentration tested. The biological results were expressed in relation to these concentrations.

At the end of the bioassay, biomarkers were determined as in Oliveira et al. (2018): filtration rate (FR), lipid peroxidation levels (LPO) and the activity of the enzymes cholinesterases (ChE), octopine dehydrogenase (ODH), isocitrate dehydrogenase (IDH), glutathione S-transferases (GST), and glutathione reductase (GR). FR was expressed in mL/h/bivalve, enzymatic activities in nanomoles per min per mg of protein (nmol/min/mg prot), and LPO levels in thiobarbituric acid reactive species per mg of protein (TBARS/mg prot).

NP concentrations were analyzed by the Student's *t* test. Each biomarker dataset was analyzed through the analysis of variance (ANOVA), one-way ANOVA (1-ANOVA) to compare different treatments followed by the Tukey test, and two-way ANOVA (2-ANOVA) with interaction (fixed factors: NP and Li). The SPSS (version 25) was used for data analyses. Significance level was 0.05.

## 3 Results

The results of the biomarkers per treatment are shown in Table 1. Significant ($p \leq 0.05$) differences in FR among treatments were found, with bivalves exposed to Mix2 having significantly higher FR than those of the control group. Regarding the other biomarkers, no significant ($p > 0.05$) differences among treatments were found.

The data analysis of each biomarker by 2-ANOVA indicated significant differences in ChE activity between treatments with and without Li and significant differences in IDH activity and LPO levels between treatments with and without NP (Table 2). No other significant differences or significant interactions were found ($p > 0.05$). In relation to the control group and based on mean values, bivalves exposed to Li (alone or in mixture with NP) had 22% of ChE inhibition. Bivalves exposed to NP (alone or in mixture with

**Table 1** Mean (± standard error of the mean) of the biomarkers determined in *Corbicula fluminea* in different treatments

| Biom | Control | NP-L | NP-H | Li-L | Li-H | Mix1 | Mix 2 |
|---|---|---|---|---|---|---|---|
| FR | 29 ± 2a,b | 21 ± 10a | 50 ± 2b,c | 27 ± 3a | 12 ± 3a | 22 ± 3a | 59 ± 3c |
| ChE | 35 ± 5 | 35 ± 3 | 39 ± 4 | 32 ± 4 | 27 ± 4 | 28 ± 2 | 25 ± 3 |
| IDH | 11 ± 2 | 5.7 ± 0.6 | 8 ± 1 | 8.4 ± 0.5 | 8 ± 1 | 8 ± 1 | 8 ± 1 |
| ODH | 9.1 ± 0.2 | 14 ± 5 | 10.7 ± 0.6 | 11.5 ± 0.3 | 11 ± 3 | 15 ± 2 | 9 ± 2 |
| GST | 14 ± 3 | 13 ± 2 | 27 ± 10 | 64 ± 7 | 13 ± 2 | 14 ± 2 | 13 ± 3 |
| GR | 5 ± 1 | 5 ± 1 | 10 ± 3 | 19.5 ± 0.7 | 6 ± 1 | 5 ± 2 | 7 ± 1 |
| LPO | 1.03 ± 0.04 | 0.4 ± 0.1 | 0.7 ± 0.1 | 0.8 ± 0.2 | 0.9 ± 0.2 | 0.8 ± 0.1 | 0.64 ± 0.04 |

Different letters indicate significant differences (ANOVA, Tukey test, $p \leq 0.05$). NP-L—0.8 mg/L; NP-H—3.2 mg/L; Li-L—1 mg/L; Li-H—4 mg/L; Mix1—0.8 mg/L NP + 1 mg/L Li; Mix2—3.2 mg/L NP + 4 mg/L Li

**Table 2** Significant differences detected by 2-ANOVA ($p \leq 0.05$)

| Biom | Factor | Level | N | Mean ± SD | F | P |
|---|---|---|---|---|---|---|
| ChE | Li | No | 9 | 36 ± 6 | $F_{1, 17} = 6.968$ | 0.017 |
| | | Yes | 12 | 28 ± 6 | | |
| IDH | NP | No | 9 | 9 ± 2 | $F_{1, 17} = 6.537$ | 0.020 |
| | | Yes | 12 | 7 ± 2 | | |
| LPO | NP | No | 9 | 0.9 ± 0.3 | $F_{1, 17} = 7.193$ | 0.016 |
| | | Yes | 12 | 0.6 ± 0.2 | | |

Li) had 22% of IDH inhibition and LPO levels reduced by 33%.

## 4 Discussion

The biomarkers determined immediately after field collection, after the acclimatized period (data not shown), and in the control group (Table 1) are in the range of corresponding values determined in *C. fluminea* from the Minho estuary (Oliveira et al. 2015b, 2018; Guilhermino et al. 2018).

The 22% inhibition of ChE by Li indicates that this metal has anticholinesterase effects in *C. fluminea* in agreement with findings in the fish *Danio rerio* (Oliveira et al. 2011). The inhibition of ChE activity may lead to excess of acetylcholine in the nervous system and neuromuscular junctions, potentially negatively affecting several functions that are important for the survival and performance of bivalves, such as valve opening and closing, coordination, feeding, among several others.

The inhibition of *C. fluminea* IDH activity by NP suggests that bivalves exposed to this contaminant may get less energy through aerobic pathways of energy production a situation that may decrease the efficiency of several processes including facing chemical and other stress. As IDH is also involved in antioxidant defenses, this function may also be altered in bivalves exposed to NP. However, no lipid damage occurred. On the contrary, the tested NP seem to provide some protection against lipid oxidative damage as indicated by the lower LPO levels in bivalves exposed to NP in relation to organisms not exposed to NP (Table 2). No anticholinesterase effects of NP were observed. In *Mytilus galloprovincialis*, other polystyrene NP increased LPO and inhibited AChE activity (Brandts et al. 2018). These distinct findings may be due to differences in species sensitivity, NP properties, methodological procedures, among other potential causes. Despite a higher FR mean in bivalves exposed to the highest concentration of NP alone, no significant effects were found in agreement with findings in *C. fluminea* exposed to other type of NP (Baudrimont et al. 2020).

Mix2 caused a significant increase of *C. fluminea* FR, whereas Li and NP alone did not (Table 1). Although the difference in relation to the control group was not significant, the FR was higher in bivalves exposed to 3.2 mg/L of NP and not significantly different from those exposed to Mix2. These results suggest that NP played a major role on Mix2, but that toxicological interactions between Li and NP may have occurred too. Considering the results obtained in this preliminary study and the interest of NP and Li as environmental contaminants, the effects of NP and Li mixtures deserve deeper investigation, with higher number organisms and long-term exposures.

## 5 Concluding Remarks

The effects of NP, Li, and their mixtures were investigated using the Asian clam *Corbicula fluminea*, an exotic invasive species in Europe and other regions of the world. The use of *C. fluminea* in environmental studies avoids the use of native bivalves and helps in the control of its bioinvasions and therefore has ethical and conservation interest. Moreover, this type of studies is in line with the hydrogeoethical approach. In *C. fluminea*, Li inhibited ChE activity indicating anticholinesterase effects of this metal. NP inhibited IDH suggesting that exposed bivalves may get less cellular energy through aerobic pathways. NP decreased LPO levels suggesting that somehow they provided some protective effect in relation to lipid peroxidation damage. The mixture containing 3.2 mg/L NP + 4 mg/L significantly increased the FR. Although NP seem to be a major contributor to increased FR, toxicological interactions may have occurred too. Although the number of bivalves used was relatively low, these evidences stress the need of further studies on the combined effects of NP and Li, including in relation to long-term effects and climate changes.

**Acknowledgements** This study was funded by national funds from "Fundação para a Ciência e a Tecnologia/Orçamento do Estado," PTDC/BIA-BMA/30922/2017), Portugal, and ERDF European funds ("Programa Operacional Competitividade e Internacionalização") through the project EMERGEMIX (project number 30922).

## References


Baudrimont M, Arini A, Guégan C, Venel Z, Gigault J, Pedrono B, Prunier J, Maurice L, Ter Halle A, Feurtet-Mazel A (2020) Ecotoxicity of polyethylene nanoplastics from the North Atlantic oceanic gyre on freshwater and marine organisms (microalgae and filter-feeding bivalves). Environ Sci Pollut Res 27:3746–3755

Brandts I, Teles M, Gonçalves AP, Barreto A, Franco-Martinez L, Tvarijonaviciute A, Martins MA, Soares AMVM, Tort L, Oliveira M (2018) Effects of nanoplastics on *Mytilus galloprovincialis* after individual and combined exposure with carbamazepine. Sci Tot Environ 643:775–784

Guilhermino L, Vieira LR, Ribeiro D, Tavares AS, Cardoso V, Alves A, Almeida JM (2018) Uptake and effects of the antimicrobial florfenicol, microplastics and their mixtures on freshwater exotic bivalve *Corbicula fluminea*. Sci Tot Environ 622–623:1131–1142

Lehner R, Weder C, Petri-Fink A, Rothen-Rutishauser B (2019) Emergence of nanoplastic in the environment and possible impact on human health. Environ Sci Techn 53:81748–81765

Li L, Su L, Cai H, Rochman CM, Kolandhasamy P, Peng J, Shi H (2019) The uptake of microfibers by freshwater Asian clams (*Corbicula fluminea*) varies upon physicochemical properties. Chemosphere 221:107–114

Oliveira RL, Seibt KJ, Rico EP, Bogo MR, Bonan CD (2011) Inhibitory effect of lithium on nucleotide hydrolysis and acetylcholinesterase activity in zebrafish (*Danio rerio*) brain. Neurotox Terat 33:651–657

Oliveira C, Vilares P, Guilhermino L (2015a) Integrated biomarker responses of the invasive species *Corbicula fluminea* in relation to environmental abiotic conditions: a potential indicator of the likelihood of clam's summer mortality syndrome. Comparat Bioch Physiol A-Mol Integr Physiol 82:27–37

Oliveira P, Lopes-Lima M, Machado J, Guilhermino L (2015b) Comparative sensitivity of European native (*Anodonta anatina*) and exotic (*Corbicula fluminea*) bivalves to mercury. Estuar Coast Shelf Sci 167:191–198

Oliveira P, Barboza LGA, Branco V, Figueiredo N, Carvalho C, Guilhermino L (2018) Effects of microplastics and mercury in the freshwater bivalve *Corbicula fluminea* (Müller, 1774): filtration rate, biochemical biomarkers and mercury bioconcentration. Ecotox Environ Saf 164:156–163

Rochman CM, Parnis JM, Browne MA, Serrato S, Reiner EJ, Robson M, Young T, Diamond ML, The SJ (2017) Direct and indirect effects of different types of microplastics on freshwater prey (*Corbicula fluminea*) and their predator (*Acipenser transmontanus*). PLoS ONE 12:e0187664

Su L, Cai H, Kolandhasamy P, Wu C, Rochman CM, Shi H (2018) Using the Asian clam as an indicator of microplastic pollution in freshwater ecosystems. Environ Pollut 234:347–355

# Reuse of Treated Effluents in a Food-Processing Industry

Aías Lima, Tiago Abreu, and Sónia Figueiredo

**Abstract**

In view of the constant population growth, the unrestrained economic development and the consequent demand for water resources in different sectors (e.g., industrial, agricultural and domestic), the water stress index became alarming in several countries, and the scarcity of water is an increasingly frequent and worrying phenomenon. International policies to mitigate water consumption, as well as the reuse of water resources, emerged with the aim of reducing imminent risks for the coming decades. The sectors of agriculture and industry are those with the highest consumption of fresh water, and it is expected that the annual global use of water in industry will increase. The circular economy emerges as a very useful tool in the reuse of treated wastewater in the industrial sector, contributing to the sustainability of water resources. The present work deals with a case study in a food-processing industry, where the implementation of a wastewater treatment process is explored in a concept of circular economy. New stages in the treatment process are proposed, which allow to reuse treated wastewaters.

A. Lima · T. Abreu (✉)
Civil Engineering Department, School of Engineering (ISEP), Polytechnic of Porto, Porto, Portugal
e-mail: taa@isep.ipp.pt

T. Abreu
CESAM-Centre for Environmental and Marine Studies, University of Aveiro, Aveiro, Portugal

S. Figueiredo
Chemical Engineering Department, School of Engineering (ISEP), Polytechnic of Porto, Porto, Portugal

S. Figueiredo
REQUIMTE-LAQV, School of Engineering (ISEP), Polytechnic of Porto, Porto, Portugal

**Keywords**

Circular economy • Sustainability • Water reuse • Wastewater treatment • Food industry

## 1 Introduction

We are facing a world crisis, where almost two-thirds of the world population face shortage of water resources for at least one month a year (Mekonnen and Hoekstra 2016). According to the United Nations World Water Assessment Program (WWAP 2017), the industry is in second place in the ranking of consumption of water and consumes 19% of the bulk (agriculture represents 69% of water consumption). Every year since 1980, water consumption rates have increased by 1%, and this tendency is expected to continue until 2025. United Nations (UN) studies indicate that the planet may face a 40% water supply deficit by 2030 (UN 2015).

Regarding the quality of water for its different uses, Feitosa and Filho (1997) reported the concept of polluted water, defining it as a change in its physicochemical characteristics, which interfere in the pre-established standards for a given purpose. More than 80% of wastewater is discharged in the environment without treatment (WWAP 2017).

Facing the needs to create strategies to mitigate water consumption by the most representative sectors in the demand for water resources, the circular economy appears as a model of sustainable development that allows materials to be returned to the production cycle through treatment and/or reuse/recovery/recycling, ensuring greater efficiency in the use and management of resources (COTEC Portugal 2016). According to Ellen MacArthur Foundation (2012), this model is based on three principles: (i) preservation and increase of natural capital through the control of finite stocks; (ii) optimization of yields and resources through the


M. Abrunhosa et al. (eds.), *Advances in Geoethics and Groundwater Management: Theory and Practice for a Sustainable Development*, Advances in Science, Technology & Innovation,
https://doi.org/10.1007/978-3-030-59320-9_58

circulation of products and materials in technical and biological cycles; (iii) stimulating the system's effectiveness through the identification and elimination of negative externalities.

The present work deals with a case study of a food-processing industry producing frozen pre-fried potatoes and dehydrated potato flakes, with an annual production of 200 thousand tons. Water is used in all stages of the manufacturing process, and consequently, large volumes of wastewater are generated for treatment. Considering quality requirements of the water used in the production process and the lack of a wastewater treatment that guarantees adequate characteristics for its reuse, there is a great financial expense both in capturing and treating the water used in the process, which is then launched downstream into the nearest watercourse, without any reuse. This work explores the possibility of reusing treated water in the production cycle, in a context of circular economy, contributing to the sustainability of water resources.

## 2 Methods and Data

The research, development and evaluation of this case study aim to present a solution for conscious use and reuse of water. The main purpose is associated with the increase of the potential use of treated water within the industrial sector of the food-processing industry, which is one of the sectors that demands large quantities of water in its manufacturing processes.

This research is characterized as an exploratory method, as it aims at improving mutual knowledge and understanding of this water reuse question, through the analysis of the productive process of the case study and through bibliographic research addressing this subject.

Regarding bibliographic research, data were collected on successful cases of reuse of industrial wastewater, complementary processes of secondary and tertiary treatments of wastewater and their economic impacts.

Data collection on the production process was performed, as for example, the volume of water consumption, the current processes for the treatment of water before it gets into the production process, the following treatment of the wastewater and its later rejection. Subsequently, a bibliographic search work was done in order to gather information on existing solutions that allow the reduction of costs related to water consumption and wastewater discharge, which minimize environmental impacts.

In the data collection of this case study, all stages of the raw material processing were characterized, as well as all the existing stages of the wastewater treatment process. In particular, the volume of water consumed in each production cycle and the volume of treated water were further detailed and evaluated, for subsequent analysis envisaging the possible reuse of that water. Potential financial gains associated with the capture water and subsequent rejection of treated wastewaters were also assessed.

The company has a system for capturing water to feed the boilers from artesian wells and a water treatment station, which includes coagulation/flocculation followed by flotation, filtration and disinfection by chlorination.

The current wastewater treatment consists of a preliminary, primary and secondary conventional treatments, in which the biological treatment is done by lagoons (anaerobic lagoons followed by aerobic lagoons).

The characterization of the treated wastewater was carried out to assess the possibility of its reuse in the industrial process; the parameters monitored at each stage of the treatment process were: dissolved solids (mg/L), settleable solids (mg/L), total suspended solids (mg/L), total solids (mg/L), chemical oxygen demand (COD, mg/L), biochemical oxygen demand at 5 days ($BOD_5$, mg/L), pH, temperature (°C), oils and fats (mg/L), electrical conductivity (μS/cm). Table 1 shows the annual average values for each parameter and respective standard deviation (SD). Standard methods from American Public Health Association (APHA 2012) were followed in the analysis of these parameters, and their respective references are provided in Table 1. The limits presented in this table were defined by the enterprise and are related to the quality standards demanded for a treated industrial wastewater that is going to be further treated prior to its reuse, following the same treatment as raw captured water does.

## 3 Results

The characterization of the treated effluents revealed high levels of electrical conductivity. This was found to be the only impediment to the reuse of water within the production process, which requires conductivity levels below 400 μS/cm for water use. Since potato, the raw material of the manufacturing process, is a great electrical conductor, all stages of the industrial processing will end up increasing the levels of electrical conductivity, with the potassium cation ($K^+$) being the cause of this conductivity increase.

The research pointed to treatments by phytoremediation, with roots of floating plants, such as *Eichhornia Crassipes*, commonly known as water hyacinth. It has demonstrated a 40% ion reduction capacity, with a retention time of 15 days. Its implementation would be more precisely in the aerated basins that already exists at the wastewater treatment plant. However, this change was not enough to reduce the levels of electrical conductivity to the desired values, as the retention time in these basins was lower. Figure 1 shows the prototype simulation results for a couple of weeks of the

**Table 1** Analysis of treated raw effluent—2018

| Test | Method | Limit | Annual average | SD |
|---|---|---|---|---|
| Thermotolerant coliforms (CFU/100 mL) | 9222 B | – | $25 \times 10^3$ | $42 \times 10^3$ |
| Total coliforms (CFU/100 mL) | 9222 B | – | $167 \times 10^3$ | $334 \times 10^3$ |
| $BOD_5$ (mg $O_2$/L) | 5520 D | (>75% removal) | 71 (98% removal) | 23 |
| COD (mg $O_2$/L) | 5210 D | (>70% removal) | 186 (98% removal) | 46 |
| Oils and Greases (mg/L) | 5520 B | 50 | 10 | 1 |
| Dissolved oxygen (mg $O_2$/L) | 4500—O G | – | 7.5 | 2.7 |
| pH | 4500-$H^+$ B | 6.0–9.0 | 7.9 | 0.1 |
| Total dissolved solids (mg/L) | 2540 C | – | 1.4 | 0.2 |
| Settleable solids (mL/L) | 2540 F | 1.0 | 0.14 | 0.07 |
| Total suspended solids (mg/L) | 2540 D | 150.0 | 38 | 16 |
| Anionic surfactants (mg/L) | 5540 C | 2.00 | 0.06 | 0.03 |
| Temperature (°C) | 2550 B | – | 25 | 3 |
| Turbidity (NTU) | 2130 B | – | 45 | 26 |
| Electric conductivity (μS/cm) | 2510 B | <400 | 1.720 | 42 |

implementation of rhizofiltration in aerated lagoons, within the existing retention time. There is a clear reduction of the water conductivity values, but the desirable threshold of 400 μS/cm is not achieved.

Dilution tests were also carried out on the treated effluent with drinking water and a reduction in the levels of electrical conductivity from 1600 to 400 μS/cm was obtained, by means of a 20% dilution (in volume) of wastewater treated in drinking water.

A third hypothesis involved the application of a filter containing a mixture of both cationic and anionic ion-exchange resins, which removes $K^+$ in a more efficient way. Moreover, the anionic resins can remove $CO_3^{2-}$ and $HCO_3^-$ ions that are one of the main causes of scale deposits in the tubes and equipments in industrial processes when hot water is used. Figure 2 shows the result of experimentation with ion exchange of the wastewater under study. The experiment was carried out through a filtration column with a volume of 7.5 ml of resin.

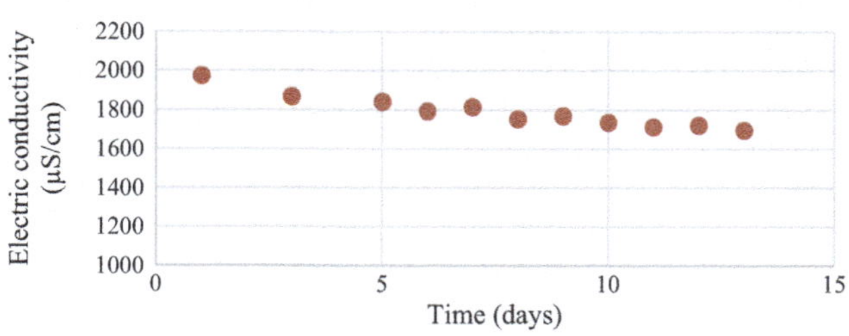


**Fig. 1** Electric conductivity *versus* retention time, using *Eichhornia Crassipes*

## 4 Discussion

Several treatment options were studied in order to propose a readjustment of the wastewater treatment process within the perspective of a circular economy, envisaging the reduction of existing costs and of environmental impacts. The set of solutions to be integrated in the production process could propose a considerable reuse of the treated water volume. Figure 3 outlines the processes for reusing treated wastewater and the respective percentage of its reuse in this food-processing industry.

The proposed use of rhizofiltration with floating plants is a very interesting solution from an economic and environmental point of view. The chosen plant has self-replicating characteristics, as long as there are nutrients for consumption, and after its useful life has been completed, its residues can be transformed into biomass for energy generation. This implementation would require only an initial investment and basic maintenance of the plants; however, it was demonstrated that this treatment alone was not enough to achieve the desired values of electric conductivity.

Only the use of ion-exchange resins was able to achieve the desirable values of conductivity. In addition, these resins can be reused after regeneration by an appropriate chemical treatment.

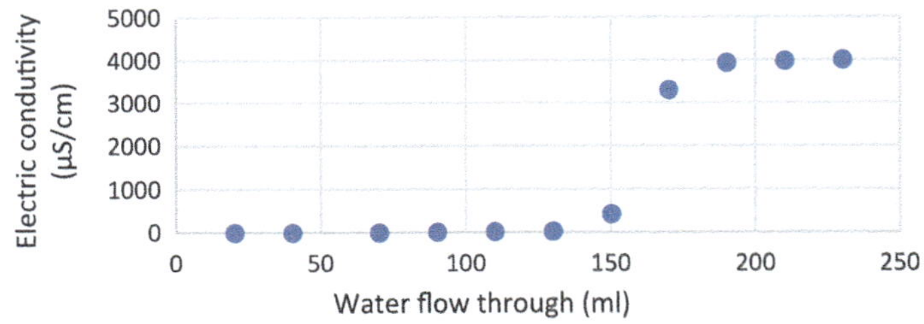


**Fig. 2** Electric conductivity through ionic exchange resin filter

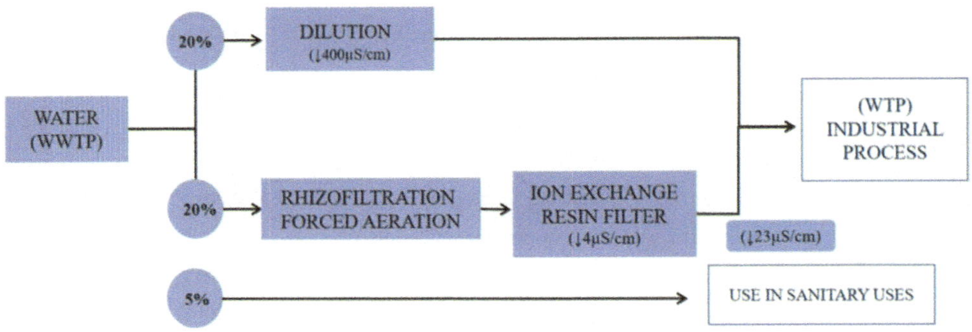


**Fig. 3** Process template for reusing treated wastewater

## 5 Concluding Remarks

This work discusses the reuse of treated effluents of a potato-processing industry. The treated waters have a high concentration of potassium cations and, consequently, increase the levels of electrical conductivity, preventing the reuse of treated water within the production process. Different solutions for this problem were studied as an essential component of a circular economy. In short, the importance of making use of more than one process to obtain a considerable volume of water to be reused is recognized, reducing the volume of groundwater abstracted. This leads to a better and sustainable management of water resources and consequent hydrogeoethical approach, contributing to the company's financial economy in a cost-effective manner. Economic analysis of large-scale implementation is being developed to measure financial savings and benefits.

**Acknowledgements** Thanks are due for the financial support to CESAM (UID/AMB/50017/2019, UIDB/50017/2020, UIDP/50017/2020), to FCT/MEC through national funds, and the co-funding by the FEDER, within the PT2020 Partnership Agreement and Compete 2020. Likewise, it is acknowledged the financial support to REQUIMTE, the Associate Laboratory for Green Chemistry-LAQV, through national funds from FCT/MCTES (UIDB/50006/2020), co-financed by the ERDF under the PT2020 Partnership Agreement (POCI-01-0145-FEDER-007265).

## References

APHA (2012) Standard methods for the examination of water and wastewater. In: Rice EW, Baird RB, Eaton AD, Clesceri LS (eds) American Public Health Association (APHA), 22nd edn. American Water Works Association (AWWA) and Water Environment Federation (WEF), Washington, D.C., USA

COTEC Portugal (2016) Economia circular: preservar, otimizar e assegurar recursos essenciais para o nosso futuro. COTEC Portugal https://www.cotecportugal.pt/imagem/20161122_EC_Booklet_Exposição.pdf. Last accessed 09 Oct 2019

Ellen MacArthur Foundation (2012) Towards the circular economy Vol.1: an economic and business rationale for an accelerated transition. Ellen MacArthur Foundation, 96 p

Feitosa ACF, Manoel-Filho J (1997) Hidrogeologia: conceitos e aplicações. CPRM-LABHID/UFPE, Brasil, Fortaleza

Mekonnen MM, Hoekstra AY (2016) Four billion people facing severe water scarcity. Sci Adv 2(2):e1500323–e1500323

UN (2015) Até 2030 planeta pode enfrentar déficit de água de até 40%, alerta relatório da ONU. https://nacoesunidas.org/ate-2030-planeta-pode-enfrentar-deficit-de-agua-de-ate-40-alerta-relatorio-da-onu/. Last accessed 05 May 2019

WWAP (2017) The United Nations World Water Development Report 2017. Wastewater: The Untapped Resource. UNESCO, Paris

# Diagnosis of a Conventional Water Treatment Station: Qualitative Analysis of Treatment Capacity

Anne Louise de Melo Dores, Allan Thiago de Oliveira, and Felipe Corrêa Veloso dos Santos

**Abstract**

Analysis and proposal of low filtration rate filter readjustment of a Conventional Water Treatment Station in order to increase the treatment capacity and agree the parameters of water quality index are recommended by the Ministry of Health. The case study is the Water Treatment Station of Nova Crixás, city located in the interior of the State of Goiás (Brazil), which was analysed in the last semester of 2018. In Nova Crixás, the expansion and concomitant population growth had resulted in the impairment of the treatment components of the station, making them possible spots of water contamination, and consequently in its inefficiency (in the case of the filter). In order to perform the efficiency analysis of the system as a whole, the authors based in projects and calculation memorials to verify the design parameters. According to the technical norms and values proposed by bibliography, it is evaluated that the Treatment Station operates within the normative parameters, but within the limit of its operating capacity. According to the results, it is clear that the Decanter's Treatment Capacity is within its operational limit, and the filtering capacity influences the water quality index.

**Keywords**

Health • Water quality • System efficiency • Parameterization • Water system upgrade

## 1 Introduction

The city of Nova Crixás, as well as several cities in the interior of the State of Goiás, has been undergoing a process of exponential population growth in recent decades, directly affecting the city's sanitary arrangement, especially in relation to water supply. This growth is due to the migratory flow of people directly encouraged by the construction of the GO-164 highway and the intense agricultural activity in the region IBGE (2017), which places the city first as the holder of the largest cattle herd in the State of Goiás, a herd of 12 thousand head of cattle.

This growth leads to the need for expansion of water treatment components, as presented in this publication, characterized by the analysis of treatment capacity and its compliance with normative parameters. The main objective of the improvement proposal is to evaluate the operating conditions of the Conventional Water Treatment Station, when, in a situation of non-compliance with the quality parameters determined by the Ministry of Health and the Brazilian standards, the ABNT norms propose the readjustment of the components of the water treatment process.

In order to analyse qualitatively the capacity of the Water Treatment Station installed, it established parameters such as potability standards, flow rate, detention time, decanter and filter which are analysed to decide if the station needs readjustment in their units of treatment.

## 2 Materials and Methods

### 2.1 Potability Standard: Water Quality Index

The main source of water caption in the State of Goiás, Brazil, is through surface water. Those commonly does not express potability quality, demanding a better treatment before being ready for public consume. Therefore, the water quality index is one of the most important qualitative

A. L. de Melo Dores (✉) · A. T. de Oliveira · F. C. V. dos Santos
Department of Civil Engineering, Pontifical Catholic University of Goiás, Goiânia, Brazil
e-mail: annelouisedemelo@gmail.com



M. Abrunhosa et al. (eds.), *Advances in Geoethics and Groundwater Management: Theory and Practice for a Sustainable Development*, Advances in Science, Technology & Innovation,
https://doi.org/10.1007/978-3-030-59320-9_59

potability standards, setting acceptable levels for a minimum of quality for water to public consummation. The city of Nova Crixás satisfactorily corresponds to a water quality index of 97.04%.

However, it is noteworthy that some analyses were non-standard. The sample analysis performed during the treatment process and at the entrance of the distribution network showed an amount in percentage of non-standard parameters such as turbidity, total coliform (CT) and Escherichia Coli (EC) that are parameters of analysis recommend by the Ministry of Health (2011) in the Ordinance No 2.914/2011.

Therefore, Table 1 shows the percentage for each non-conforming analysis obtained through the Operating System of the Laboratory of Water Quality (LQA083-2018) provided by the water treatment company of the State of Goiás, SANEAGO.

## 2.2 Facility Treatment Capacity: Flow Rate

At the Water Treatment Station a monthly data record measures the amount of water produced and ready for consume. With the assistance of the OP030 (2018), which corresponds to the compilation of data registered from the station operating system, values of water flow for the last 6 months of 2017 and the first 7 months of 2018 were obtained in order to determine the maximum and average treatment capacity of the station in this period. As illustrated in Table 2, the result for the average flow is 24 L/s and the maximum value corresponding to 31 L/s.

## 2.3 Flocculation Evaluation: Detention Time

According to the Brazilian norm ABNT NBR 12,216/1992 concerning the elaboration of projects for the Public Water Treatment Station, flocculation units are "used to promote the aggregation of particles formed in the rapid-mix", which may be mechanical or hydraulic. Two Alabama-type hydraulic flocculators are in operation at the Nova Crixás Water Treatment Station, illustrated by Casanova et al. (2019) in Fig. 1.

The flocculation capacity is determined by the flocculation unit hold time parameter. According to the norm ABNT NBR 12,216, an Alabama-type hydraulic flocculation operates at a satisfactory level for detention time limits ranging from 20 to 30 min for conventional treatment.

To determine the installed capacity of the flocculation, the maximum filtered flow of 31 L/s is used as the worst case, representing the largest volume of water treated by the unit. According to Netto et al. (1998), the hydraulic detention time is express by

$$t = \frac{V}{Q} \tag{1}$$

$t$ = hydraulic detention time; $V$ = volume of water in $m^3$; $Q$ = maximum flow treated by the filter in $m^3/s$.

In order to determinate the installed capacity of the flocculation unit, the project specifications such as superficial area, volume and flow rate were considered.

Substituting the projects parameters shown at Table 3 in the Eq. (1), the time of detention for this flocculation unity is determined as 1850s, 31 min.

## 2.4 Decanter Evaluation

According to Netto et al. (1998), the surface flow rate (speed at which sedimentation is processed) is adopted as a factor to characterize the treatment efficiency of the Decanter. The Brazilian norm ABNT NBR 12,216, in sub-item 5.10.3 states that this surface run-off rate is expressed by the relationship between the treatment flow and the surface area of the settling unit.

For the evaluation of the decanter, using the relation between flow rate and superficial area, the flow rate expressed on Table 2, for a maximum flow rate of 31 L/s is considered as the worst case, therefore, used for the evaluation. The transformation to $m^3$/day is required by the average of the operating time in hours/day taken from Table 2 and transformed to $m^3$/day ($Q$ = 1815.73 $m^3$/day).

If laboratory testing is not possible, the standard provides that the run-off rate shall be determined according to the

**Table 1** Summary of irregularities detected in the sample analysis

| Treatment Station | Turbidity | Colour | pH | F | Cl | Fe | Al | Mn | CT | EC |
|---|---|---|---|---|---|---|---|---|---|---|
| % Non-standard | 2.380 | 1.785 | 0.000 | 2.976 | 1.785 | 0.000 | 5.000 | 0.000 | 0.595 | 0.000 |
| Total | **1.384** | | | | | | | | | |
| Distribution network | Turbidity | Colour | pH | F | Cl | Fe | Al | Mn | CT | EC |
| % Non-standard | 3.389 | 3.428 | 0.000 | 0.000 | 3.389 | 0.000 | 0.000 | 4.000 | 0.564 | 0.000 |
| Total | **2.056** | | | | | | | | | |

**Table 2** Produced water volume and operating time in the last 12 months

| Month/Year | 07/2017 | 08/2017 | 09/2017 | 10/2017 | 11/2017 | 12/2017 | – |
|---|---|---|---|---|---|---|---|
| Flow rate ($m^3$/month) | 52,930.00 | 62,490 | 55,280 | 37,774 | 35,910 | 37,677 | – |
| Time (hours/month) | 480.34 | 488.17 | 578.56 | 694.55 | 487.45 | 434.27 | – |
| Flow rate (L/s) | 31 | 31 | 18 | 14 | 20 | 24 | – |
| **Month/Year** | **01/2018** | **02/2018** | **03/2018** | **04/2018** | **05/2018** | **06/2018** | **07/2018** |
| Flow rate ($m^3$/month) | 38,210 | 39,720.14 | 49,270 | 47,370 | 49,802 | 40,200 | 42,310 |
| Time (hours/month) | 489.38 | 415.16 | 446.04 | 475.54 | 591 | 528.11 | 567.57 |
| Flow rate (L/s) | 22 | 27 | 31 | 28 | 23 | 21 | 21 |
| **Average Flow Rate: 24 L/s** | | | | **Max. Flow Rate: 31 L/s** | | | |

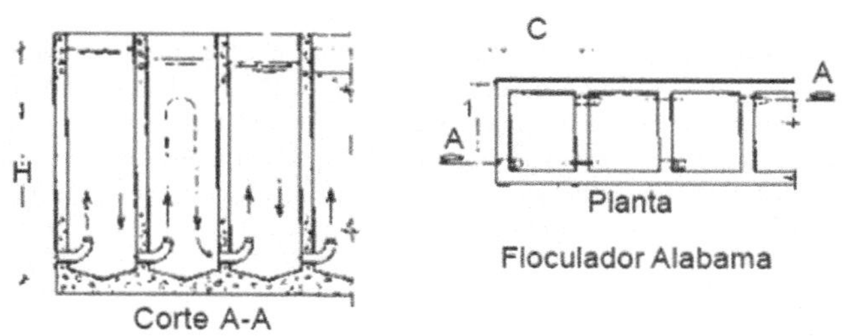


**Fig. 1** Alabama-type flocculation's project (Casanova et al. 2019)

**Table 3** Detention time: project specifications

| Superficial area $m^2$ | Volume $m^3$ | Flow Rate 'Q' $m^3/s$ |
|---|---|---|
| 20.125 | 57.35 | 0.031 |

*Source* From the author (2019)

Treatment Station Capacity. For Water Treatment Stations with capacities ranging between 1000 and 10,000 $m^3$/day, in order to ensure good operational control, the surface run-off rate must be 25 $m^3$/ ($m^2$ day). Therefore

$$\frac{Q}{As} = \frac{1815.73}{2 \times 3.9 \times 11} = 21.16\,\mathrm{m}^3/(\mathrm{m}^2\,\mathrm{day}) \tag{2}$$

## 2.5 Slow Filter Evaluation

Filters are components of the Water Treatment Station, which have the function of trapping particles within the filter matrix by adhesion and mechanical retention forces, having large particle retention capacity. In Nova Crixás WTS, the installed filter units consist of a simple filter with a slow filtration sand layer.

According to Seckler (2017), the slow filter operates satisfactorily with a filtration rate in between 2 and 6 $m^3$/($m^2$ day). From the relation between flow rate and area, the filtration rate '$q$' is determined by Eq. (3) as 333.13 $m^3$/ ($m^2$ day).

$$q = \frac{Q}{A} = \frac{31 \times 86.4}{8.04} = 333.13\mathrm{m}^3/(\mathrm{m}^2\,.\,\mathrm{dia}) \tag{3}$$

# 3 Results

The filtration rate calculated in the last topic of the previous chapter shows that under the current conditions, the filter is overloaded and therefore does not fit the standards recommended by the Ministry of Health for drinking potable water demonstrated by the data analyses of the percentage of non-conformity samples. It is possible to realize that this lack of filtration capacity interferes in the quality of the water.

In order to maintain the dimensions of the filter in design, the same flow rate suggested the replacement of the slow filtration filters by a fast filter—also named high rate filters, with a double layer of Sand + Anthracite, as recommended by Abreu (2009).

The filtration rate remains the same as previously calculated but considering the filtering limit recommended by the norm ABNT NBR 12,216 for double layer filters, such as 360 $m^3$/($m^2$.day), the filter overload is considerably relieved. Therefore, the water treated in the filters can assure quality for human consumption.

Following the determinations of ABNT NBR 12,216 for high up flow rate filters, the sand and anthracite layers to be implemented on existing filters will have a thickness of 0.40m (divided into 0.20 m of thin sand and 0.20 m of thick sand) and 0.50 m, respectively. Table 4 represents the equivalent quantity of materials in order to implement the adjustment suggest.

**Table 4** Volume of materials required for five filters

| Filtering Material | Volume $m^3$ | Quantity (per bag of 25 kg) |
|---|---|---|
| Thick Sand | 8.05 | – |
| Thin Sand | 8.05 | – |
| Anthracite | 20.01 | 885 |

*Source* From the Author (2019)

To comply with ABNT NBR 12,216/1999, the filtering layer plus the backing layer must necessarily be at least 40 cm below the pipe gutter, meeting the 30% expansion of the layer provided by the norm. The filters will have a capacity of 33 L/s.

## 4 Discussion

It is understood that due to the development and population growth of Brazilian cities, the diagnosis of operational and technical activities is of utmost importance for the improvement of the water supply system, quality of drinking water supply and consequent adaptation of the component units of the Water Treatment Stations. Those responsible for the system operation, in this case the state Treatment Company Saneago, are able to report all non-conformities to the Municipal Health Secretariat and to make the necessary changes to ensure good water quality (State of Goiás, MP, 2016).

Table 1 showed several non-conformity in the water quality index samples, in juxtaposition with the diagnosis of filtration unit efficiency. These values correlate in cause and effect, as a low filtration rate leads to loss of water quality, making the filtering unit a possible focus of system contamination.

If increased the capacity of filtration units to 33 L/s, by changing to a high filtration rate as proposed, the efficiency of the treatment system will be within the theoretical and normative limits present in the literature. In case of no future capacity expansions of the Water Treatment Station, the system will continue to be in disability according to the parameters established in Ministry of Health Ordinance No. 2.914 (2011), as on-site operational capacity is limit by the capacity of treatment.

## 5 Concluding Remarks

High rate filtering is required to meet acceptable water quality index parameters for drinking water intended for human consumption. This measure should be combined with the expansion of the Water Treatment Station units, as it currently operates at the operating capacity limit.

**Acknowledgements** The analysis of the Water Treatment Station of Nova Crixás was developed within the scope of the Operational Technical Development Management (P-GDT) of the sanitation company of the State of Goiás, Saneago, from which maps and technical data essential for the development of research and application of quality improvement were consulted.

## References

ABNT NBR 12216 (1992) Projeto de Estação de Tratamento de Água Para Abastecimento Público. ABNT, Rio de Janeiro

Abreu SB (2009) Comportamento de Filtros Rápidos de Camada Profunda no Tratamento de Águas de Abastecimento Mediante o Emprego de Polímeros como Auxiliares de Filtração. Polytechnic School of the University of São Paulo, São Paulo (PhD thesis)

Casanova JM, Breytmann MS, Wulf MR (2009) Trabajo de investigación: planta de tratamiento de aguas servidas y agua potable para rahue bajo- Osorno. Universidad Tecnica Federico Santa Maria, Valparaiso, Chile

IBGE–Nova Crixás Online Catalog, ID 3127. https://www.ibge.gov.br/ . Last accessed 04 2019

Ministry of Health (2011) Ordinance No 2.914/2011. For the control and surveillance procedures of drinking water quality and its potability standard. Ministry Cabinet, Brasília

Netto A, Fernández MF, Araújo R, Ito AE (1998) Manual de Hidráulica, 8th edn. Blucher, Rio de Janeiro

Operational System of SANEAGO, for Water Quality Index and Flow Rate (2017)

Seckler SF (2017) Tratamento de Água – Concepção, Projeto e Operação de Estações de Tratamento. Elsevier, Rio de Janeiro

# Combination of Adsorption in Natural Clays and Photo-Catalytic Processes for Winery Wastewater Treatment

Nuno Jorge, Ana R. Teixeira, Marco S. Lucas, and José A. Peres

**Abstract**

The winery wastewater (WW) in study holds high organic contaminant levels expressed in chemical oxygen demand (COD, 2145 mg $O_2$/L), total organic carbon (TOC, 400 mg C/L) and total polyphenols (22.6 mg gallic acid/L), which are very hard to degrade. With the objective of treating the WW, it was combined adsorption with heterogeneous photocatalytic oxidation. The adsorption was carried out by addition of a sodium montmorillonite natural clay (Na-Mt), which achieved a COD and TOC removal of 68.1 and 52.6%, respectively. The Na-Mt was then converted into a heterogeneous catalyst (Fe-Mt), and it performed heterogeneous photocatalysis. With the optimization of operational conditions, it achieved a TOC removal of 75.4% (heterogeneous photocatalysis); however, when both processes were combined, it achieved a COD and TOC removal of 93.0 and 88.3%, with an $Fe^{2+}$ leaching value of 6.8 mg/L. In conclusion, the use of Na-Mt for adsorption and photocatalytic processes can be adapted to WW treatment.

**Keywords**

Winery wastewater • Sodium montmorillonite • Adsorption • Heterogeneous photocatalysis

## 1 Introduction

Winery wastewater is a major waste stream resulting from several activities that include cleaning of tanks, washing of floors and equipment (Petruccioli et al. 2000; Malandra et al. 2003).

In order to meet the increasing stringent environmental regulations, many technologies such as chemical oxidation, biological degradation, solvent extraction and adsorption were developed (Yan et al. 2007). The application of clays, especially bentonites have attracted more and more interest because of their high cation exchange capacity, surface area and wide availability. The adsorption of neutral organic contaminants can be greatly improved by replacing the inorganic interlayer cations by organic cations (Boyd et al. 1988; Yan et al. 2007).

A possible mean of treatment could be the application of advanced oxidation processes (AOPs), in order to reach the degradation of organic (Mosteo et al. 2008; Sarria et al. 2003). Fenton and Fenton-like systems have prevailed for many years for the production of HO· (Herney-Ramirez et al. 2011). However, there are several critical disadvantages for these systems, such as a narrow working pH range (2–4), instability of $H_2O_2$ during storage and transportation and massive consumption of $H_2O_2$ during application (Moreira et al. 2017). To overcome the disadvantages of homogeneous photo-Fenton, it studied the application of heterogeneous photo-Fenton for the treatment of WW, by conversion of sodium bentonite (montmorillonite) used in wine clarification process (Lambri et al. 2012). Therefore, the objectives of this study were (1) study of the efficiency of sodium montmorillonite in adsorption process; (2) development of a heterogeneous catalyst based on sodium montmorillonite; (3) treatment of a red winery wastewater (RWW) by adsorption—heterogeneous photocatalysis oxidation system.

N. Jorge (✉)
Campus da Auga, Campus Universitário de Ourense, Universidad de Vigo, Ourense, España
e-mail: njorge@uvigo.es

N. Jorge · A. R. Teixeira · M. S. Lucas · J. A. Peres
Departamento de Química, Centro de Química de Vila Real (CQVR), Universidade de Trás-Os-Montes E Alto Douro (UTAD), Vila Real, Portugal



M. Abrunhosa et al. (eds.), *Advances in Geoethics and Groundwater Management: Theory and Practice for a Sustainable Development*, Advances in Science, Technology & Innovation,
https://doi.org/10.1007/978-3-030-59320-9_60

**Table 1** Winery wastewater characterization

| Parameters | Decree Law nº 236/98 | RWW |
|---|---|---|
| pH | 6.0–9.0 | 4.0 |
| Biochemical oxygen demand—$BOD_5$ (mg $O_2$ $L^{-1}$) | 40 | 550 |
| Chemical oxygen demand—COD (mg $O_2$ $L^{-1}$) | 150 | 2145 |
| Biodegradability—$BOD_5$/COD | | 0.26 |
| Total organic carbon (mg C $L^{-1}$) | | 400 |
| Turbidity (NTU) | | 296 |
| Total suspended solids (mg $L^{-1}$) | 60 | 750 |
| Electrical conductivity (μS $cm^{-1}$) | | 62.5 |
| Total polyphenols (mg gallic acid $L^{-1}$) | 0.5 | 22.6 |
| Iron (mg $L^{-1}$) | 2.0 | 0.05 |

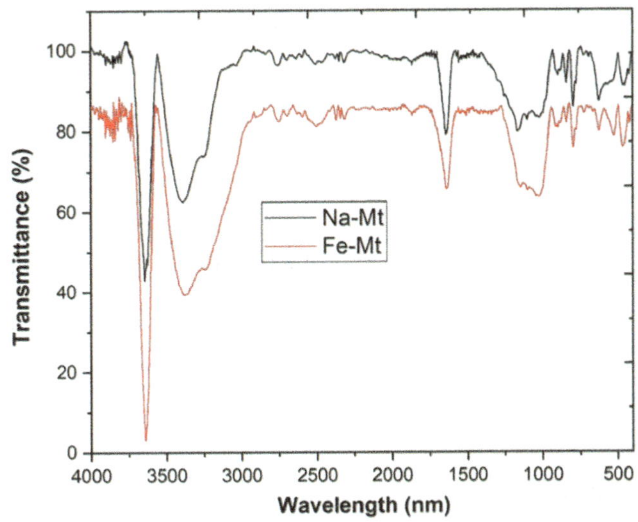


**Fig. 1** Infrared spectra obtained before and after iron catalyst preparation (Na-Mt and Fe-Mt)

## 2 Materials and Methods

### 2.1 Reagents and Winery Wastewater Sampling

RWW was collected from a cellar located in the Douro Region from Portugal. After collecting the samples in plastic containers to be transported to the laboratory, they were stored at −40 °C. Montmorillonite clay was supplied by Angelo Coimbra & Ca., Lda, $H_2O_2$ (30% w/v) by Sigma-Aldrich, $FeSO_4{\cdot}7H_2O$ by Panreac, NaOH and $H_2SO_4$ (95%) were both obtained from Analar Normapur. Deionized water was used to prepare the respective solutions (Table 1).

### 2.2 Catalyst Preparation

The catalyst Fe-Mt was prepared by the adsorption of $Fe^{2+}$ on montmorillonite, promoting the adsorbent saturation by adding 10 g of clay to 500 mL of 0.10 M $FeSO_4{\cdot}7H_2O$ solution, during 24 h with vigorous stirring. After saturation process, both samples were washed several times with deionized water and dried at 60 °C during 72 h. The structural characterization of the Fe-Mt catalyst was obtained using Fourier transform infrared spectroscopy (FTIR) measurements.

## 3 Results

### 3.1 Structural Characterization of the Heterogeneous Catalyst

In Fig. 1, activated sodium bentonite shows the stretching vibration of structural O–H groups at 3645 $cm^{-1}$, structural Si–O groups at 1103, 999 and 789 $cm^{-1}$, structural Al–Al–OH groups at 902 $cm^{-1}$, structural Al–Fe–OH groups at 883 $cm^{-1}$, the free and interlayer water in bond stretching vibration at 3396 $cm^{-1}$ and adsorbed water yielded bending at 1643 $cm^{-1}$ (Schuttlefield et al. 2007; Servim et al. 2012). The FTIR spectra of iron immobilized montmorillonite only exhibit one significant structural alteration at 530 $cm^{-1}$ comparing to the original spectrum, due to the appearance of this vibration band, which could be linked to the isomorphic substitution of the predominant $Fe^{3+}$ by $Fe^{2+}$, suggesting the retention of $Fe^{2+}$ not only in the interlayer surface of Mt, but also in the internal structure (Guimarães et al. 2019a, b).

### 3.2 Adsorption Optimization

The adsorption efficiency of Na-Mt for the removal of organic contaminants was observed by its capacity to remove TOC from WW. Initially, the adsorption process was tested under different pH conditions (3.0–9.0). After pH optimization, it optimized Na-Mt dosage (1.5–10 g $L^{-1}$). It was observed that using the best operational conditions ($[TOC]_0$ = 400 mg C $L^{-1}$, [Na-Mt] = 1.5 g $L^{-1}$, pH 4.0, agitation 350 rpm/24 h, temperature 298 K), there was a 52.6% TOC removal, which corresponds to 140 mg C/g. The amount of organic carbon was calculated in accordance with Eq. (1) (Guimarães et al. 2016):

$$q_e = \frac{(TOC_0 - TOC_e)*V}{m} \tag{1}$$

where $q_e$ (mol $kg^{-1}$) is the equilibrium concentration, $V(L)$ is the solution volume used during batch experiments, and $m$ ($g$) is the mass of Fe-Mt used.

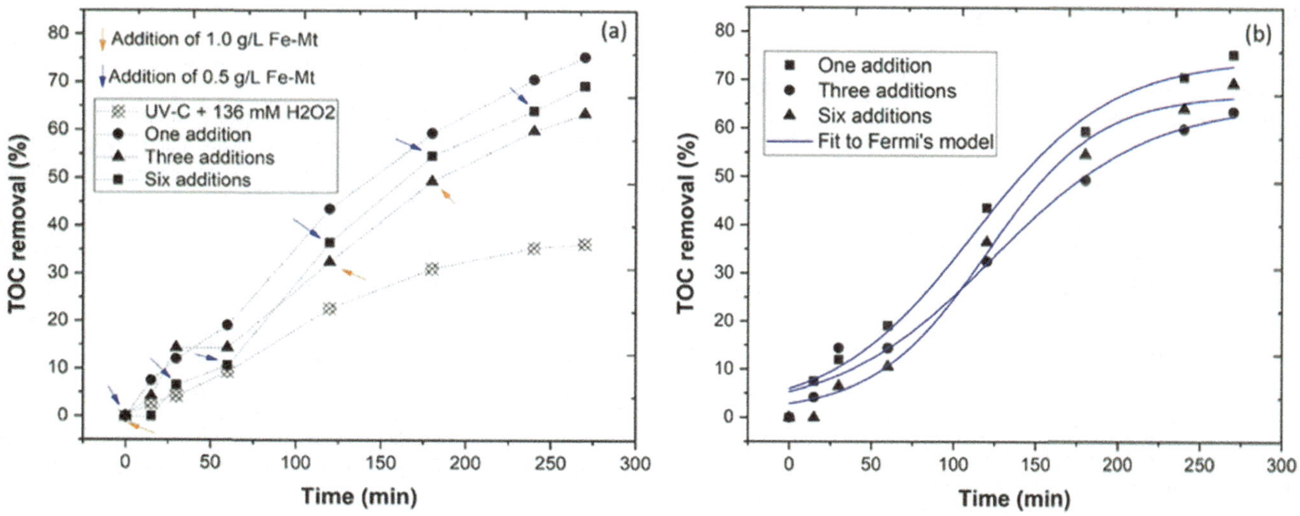


**Fig. 2** **a** Influence of Fe-Mt multiple addition in TOC removal. **b** Fit of Fermi's based lumped kinetic model to the experimental data

## 3.3 Heterogeneous Photocatalysis Optimization

After optimization of adsorption process, TOC degradation was further enhanced by performance of heterogeneous photo-Fenton. After pH optimization (3.0–7.0), $H_2O_2$ dosage (68–272 mM), Fe-Mt dosage (1.5–6.0 g $L^{-1}$) and Fe-Mt multiple addition (1–6 times), it observed a high TOC removal (75.4%) with operational conditions (Fig. 2a): [Fe-Mt] = 3.0 g $L^{-1}$—single addition, [$H_2O_2$] = 136 mM, pH 3.0, agitation 350 rpm, reaction time 270 min, radiation UV-C mercury lamp (254 nm). It observed a high $H_2O_2$ consumption (89.9%) and low $Fe^{2+}$ leaching value (6.8 mg $L^{-1}$). A kinetic study was performance by fit of Fermi's model (Eq. 2) (Guimarães et al. 2019a, b):

$$\frac{TOC}{TOC_0} = \frac{1 - x_{TOC}}{1 + \exp[k_{TOC}(t - t^*_{TOC})]} + x_{TOC} \qquad (2)$$

where $k_{TOC}$ corresponds to the apparent reaction rate constant; $t^*_{TOC}$ represents the transition time related to the TOC content curve's inflection point, and $x_{TOC}$ corresponds to the fraction of non-oxidizable compounds that are formed during the reaction. By observation of Fig. 2b, six times addition of Fe-Mt catalyst had the highest $k_{TOC}$ ($2.95 \times 10^{-2}$ $min^{-1}$) regarding one addition step ($2.36 \times 10^{-2}$ $min^{-1}$); however, when conjugated the induction time and TOC removal, one single addition becomes the best choice parameter for this catalyst.

## 4 Discussion

By increasing the pH, a significant fraction of $Fe^{2+}$ precipitated, contributing to the lower photocatalytic efficiency, thus pH 3.0 was the best selection. With increasing $H_2O_2$ concentration from 136 to 272 mM, it observed a lower TOC removal (62.4%), which could be attributed to a scavenger activity, due to the competition of $H_2O_2$ for additional HO· radicals (Peres and Lucas, 2007). When the Fe-Mt dosage was increased from 3.0 to 6.0 g $L^{-1}$, TOC removal decreased from 75.4 to 65.1%, respectively. Two different justifications can be understood from these results: (1) an excess of catalyst concentration may have contributed to increase the solution turbidity, decreasing the UV-C radiation from penetrating the wastewater (Guimarães et al. 2019a, b due to the higher concentration of iron present in solution, there could be present scavenging activities by the $Fe^{2+}$, which competed with organic matter for HO· radicals, decreasing the efficiency of the reaction (Lucas and Peres 2006).

Both adsorption and heterogeneous photo-Fenton were successfully applied in the treatment of WW. It achieved a total TOC removal of 52.5 and 88.3%, respectively. After determination of COD, it was observed a reduction of 68.1 and 93.0% (685 and 150 mg $O_2$ $L^{-1}$, respectively). This value is compliant with Portuguese legislation for discharge of wastewaters.

## 5 Concluding Remarks

Regarding the objectives established for this work, it was concluded as follows:

1. Sodium montmorillonite is economically viable for adsorption processes with pH 4.0, dosage 1.5 g $L^{-1}$ with 52.5 and 68.1% TOC and COD removal;
2. Sodium montmorillonite can successively impregnate $Fe^{2+}$ and slowly release it into solution during photocatalysis process;
3. The application of heterogeneous photo-Fenton best operational conditions has a TOC removal of 75.4% with a $k_{TOC} = 2.36 \times 10^{-2}$ $min^{-1}$;
4. The combination of adsorption/heterogeneous photo-Fenton has a final TOC and COD removal of 88.3 and 93.0%, which complies with Portuguese legislation for WW discharge (<150 mg $O_2$ $L^{-1}$);
5. Combination of adsorption in natural clays and photocatalytic processes could be a very interesting proposal for RWW treatment.

**Acknowledgements** The authors are grateful for the financial support to Project INTERACT—Integrative Research in Environment, Agro-Chains and Technology—NORTE-01-0145-FEDER-000017; Project INNOVINE & WINE—Innovation Platform of Vine and Wine—NORTE-01-0145-FEDER-000038; Centro de Química de Vila Real (CQVR) through UID/QUI/00616/2019; FCT project I&D Projecto Exploratório Investigador 2015—action number IF/00802/2015 managed by Marco S. Lucas.

## References

Boyd SA, Lee JF, Mortland MM (1988) Attenuating organic contaminant mobility by soil modification. Nature 333(6171):345–347

Guimarães V, Lucas MS, Peres JA (2019a) Combination of adsorption and heterogeneous photo-Fenton processes for the treatment of winery wastewater. Environ Sci Pollut Res 26(30):31000–31013

Guimarães V, Teixeira AR, Lucas MS, Silva AM, Peres JA (2019b) Pillared interlayered natural clays as heterogeneous photocatalysts for $H_2O_2$-assisted treatment of a winery wastewater. Sep Purif Technol 228:115768

Guimarães V, Rodríguez-Castellón E, Algarra M, Rocha F, Bobos I (2016) Influence of pH, layer charge location and crystal thickness distribution on U (VI) sorption onto heterogeneous dioctahedral smectite. J Hazard Mater 317:246–258

Herney-Ramirez J, Silva AM, Vicente MA, Costa CA, Madeira LM (2011) Degradation of Acid Orange 7 using a saponite-based catalyst in wet hydrogen peroxide oxidation: Kinetic study with the Fermi's equation. Appl Catal B 101(3–4):197–205

Lambri M, Dordoni R, Silva A, De Faveri DM (2012) Comparing the impact of bentonite addition for both must clarification and wine fining on the chemical profile of wine from Chambave Muscat grapes. Int J Food Sci Technol 47(1):1–12

Lucas MS, Peres JA (2006) Decolorization of the azo dye Reactive Black 5 by Fenton and photo-Fenton oxidation. Dyes Pigm 71 (3):236–244

Lucas MS, Peres JA (2007) Degradation of reactive black 5 by Fenton/UV-C and ferrioxalate/H2O2/solar light processes. Dyes Pigm 74(3):622–629

Malandra L, Wolfaardt G, Zietsman A, Viljoen-Bloom M (2003) Microbiology of a biological contactor for winery wastewater treatment. Water Res 37(17):4125–4134

Moreira FC, Boaventura RA, Brillas E, Vilar VJ (2017) Electrochemical advanced oxidation processes: a review on their application to synthetic and real wastewaters. Appl Catal B 202:217–261

Mosteo R, Sarasa J, Ormad MP, Ovelleiro JL (2008) Sequential solar photo-Fenton-biological system for the treatment of winery wastewaters. J Agric Food Chem 56(16):7333–7338

Petruccioli M, Duarte J, Federici F (2000) High-rate aerobic treatment of winery wastewater using bioreactors with free and immobilized activated sludge. J Biosci Bioeng 90(4):381–386

Sarria V, Kenfack S, Guillod O, Pulgarin C (2003) An innovative coupled solar-biological system at field pilot scale for the treatment of biorecalcitrant pollutants. J Photochem Photobiol, A 159(1):89–99

Schuttlefield JD, Cox D, Grassian VH (2007) An investigation of water uptake on clays minerals using ATR-FTIR spectroscopy coupled with quartz crystal microbalance measurements. J Geophys Res Atmosp 112(D21)

Sevim AM, Hojiyev R, Gül A, Celik MS (2012) Preparation of novel heterogeneous catalysts by adsorption of a cationic tetrapyrrole on to bentonite: equilibrium, kinetics, and thermodynamics. Monatshefte Für Chemie-Chemical Monthly 143(3):385–400

Yan LG, Wang J, Yu HQ, Wei Q, Du B, Shan XQ (2007) Adsorption of benzoic acid by CTAB exchanged montmorillonite. Appl Clay Sci 37(3–4):226–230

# Scientific and Humanistic Components of Hydrogeoethics in Groundwater Education and Professional Training

**Introductory Note**

Maurizio Barbieri (La Sapienza University, Rome, Italy), Tibor Stigter (IHE, Delft, The Netherlands), Rosário Carvalho (FCUL, Lisbon, Portugal), and Susana Neto (APRH and UNL, Lisbon, Portugal)

Geoethics should be a fundamental aspect of groundwater education and training. Meanwhile for many people, its concept remains vague and is rarely addressed explicitly. The aim of this part is to launch the theme on the subject at several education levels and forms in Portugal and within an international perspective.

A number of case studies are presented from Spain, Portugal, Ireland, and Brazil about the groundwater management practices and use of some tools, even isotopes, to diagnose the current conditions of water in the actual water crisis scenario and the risk to increase contamination under extreme weather events. A Canadian regional initiative stresses groundwater data transfer and information exchange as an essential tool facilitating the sustainable management of groundwater resources. A real-life application from Portugal promotes knowledge about geoethics and groundwater sustainability among the younger students, using a project-based teaching approach.

Citizen science, in particular for groundwater management, is increasingly seen as an approach to gather information with unprecedented time and territorial resolution, especially after the introduction of new IT solutions. Citizen science bears the promise of a higher level of involvement and commitment of the commoners into subjects relevant to society. A proposed study from Portugal provides the context for higher levels of citizenship and democracy-building that we need to pursue if we aim to create a better society.

Geoethics is intrinsically linked to sustainability, as it provides tools that allow reflection before action, in the earth systems' best interest. Finally, geoethics concerning the professional approach involves:

- the moral decisions to be made regarding the undertaking of fundamental and applied hydrogeology in practice;
- the appropriate professional use of the scientific advancements in groundwater science, technology, and management.

**Highlights**

- Although geoethics is not formally integrated in the analyzed curricula of Portuguese secondary education, the principles, and moral values about the importance of preserving the planet are clearly explored, providing pupils with geoethical awareness and responsibility.
- On a higher education, a more explicit integration of geoethics into the training of groundwater experts is needed, to promote scientific accuracy, impartiality, and capacity to assess different social, cultural, and political contexts. This can be actively promoted by explicit surveys and discussions with students on the importance and integration of geoethics.
- Groundwater education is not merely a transfer of knowledge and skills; it is an interactive process of exchange of ideas, experiences, and wisdom, in which all come out with enriched knowledge.
- Co-construction of knowledge, whether through education, training, or research, is a process that requires time, mutual respect and strong investment by all parties involved.
- The growing availability of groundwater expertise in the world is essential to support groundwater resources management as well as the debate on its geoethical implications.
- Continued and improved communication on these geoethical implications through different channels and platforms will promote lifelong learning and awareness.
- Stimulating the debate on groundwater and geoethics within the context of international cooperation is crucial to promote the conservation and adequate use of this precious resource for global development.

# Over Fifty Years of Hydrogeological Practice and Geoethics: An Intergenerational View of a Changing World

José Martins Carvalho and Helder I. Chaminé

**Abstract**

The exercise of professional hydrogeology is a privilege that requires sound scientific and technical knowledge, field training and interpersonal skills. A skilled hydrogeologist is required to have ethics, deontology, integrity, eco-responsibility, leadership, and behaviour but also the acceptance of the high-standards and codes of ethics and boundaries to avoid discrimination and harassment. In addition, the hydrogeology practice must encompass the geoethical approach. Water supply and infrastructure design and construction for groundwater monitoring are not popular nowadays around the scientific and technical community. However, thousands and thousands of wells and boreholes are drilled worldwide annually, without appropriate hydrogeological support. A professional hydrogeologist cannot support, for example, policies of abandon of fieldwork, leaving the aquifer exploitation restrained to high levels of management and governance. Those practices must be developed through proper conceptual site modelling based on field and laboratory data, complemented by GIS mapping, geovisualization techniques, numerical tools to predict scenarios and climate change understanding. On the ethical point of view, this systematic methodology underpins personal scientific integrity but also a comprehensive understanding of the problem to solve. That includes the moral decisions to be made regarding the undertake of the pure and applied hydrogeology in the practice, as well as the professional appropriate use of the scientific advancements in groundwater science, technology, and management.

J. M. Carvalho (✉) · H. I. Chaminé
Laboratory of Cartography and Applied Geology (Labcarga), School of Engineering (ISEP), Polytechnic of Porto, Porto, Portugal
e-mail: jmc@tarh.pt

J. M. Carvalho · H. I. Chaminé
Centre GeoBioTec|UA, Aveiro, Portugal

**Keywords**

Hydrogeology • Groundwater • Conceptual modelling • Military hydrogeology • Geoethics

## 1 Introduction

This essay, based in selected examples of an intergenerational field practice, aims to emphasize the role of micro-ethics as a key value that is part of a process framework (D'Anselmi and Di Bitetto 2013), i.e. ethics related to personal values and professional integrity. In a broad context related with Society and Stewardship of Earth, emerge the macro-ethics (Herkert 2003). The motto "*water is a unique resource: a drop can save a life, in war and peace*" is still topical (details in Chaminé and Carvalho 2015).

Regarding the diverse personal experience of the leaded author (J. M. Carvalho) in the last 50 years and the long common grounds partnership mixing academia and industry shaped with H. I. Chaminé, since 1997 (details in Chaminé and Carvalho 2015; Chaminé et al. 2015b) this work highlights the importance of ethics on groundwater engineering and applied hydrogeology in practice. In particular, the field of groundwater exploration and exploitation based on comprehensive and multidisciplinary knowledge of the hydrogeological conceptual site model must be created with ethics, eco-responsibility, and scientific integrity. To figure out, some key sites in hydrogeological African complexity (e.g., Eales et al. 1984; Wright 1992; MacDonald et al. 2008) were revisited in order to illustrate that approach (Fig. 1). The impressive thoughts of the Portuguese Explorer Major Alexandre Serpa Pinto highlights so well that African reality: "*Water continued to be scarce*" (Serpa Pinto 1881; p. 55); or the amazing African groundwater projects ("Chercheur d'eau: en Afrique et ailleurs") reported by Bourguet (2017).



M. Abrunhosa et al. (eds.), *Advances in Geoethics and Groundwater Management: Theory and Practice for a Sustainable Development*, Advances in Science, Technology & Innovation,
https://doi.org/10.1007/978-3-030-59320-9_61

**Fig. 1** The selected professional scenarios in a changing world: **a** selected studied sites in Africa; **b** J.M. Carvalho with the "water platoon" in Mozambique warfare during hydrogeological surveys in Mapé Mountain area (southern edge of Macondes Plateau) (1968–1970); **c** hydrogeological surveys in Sahara desert, Aozou Strip, Libya–Chad border for ACavaco company (1981); **d** hydrogeological mapping in Namaacha area (Southern Mozambique) with the Geologist H. I. Chaminé for groundwater studies (2012); **e** keywords cloud based on introductory text and outlined concluding remarks (generated using https://www.wordle.net/)

Since the landmark concept '*land ethic*' stated by Leopold (1949) to the need to a sustainable water ethos in society, nature and Earth systems (e.g., Wittfogel 1956; Carson 1962; Biswas 1970; Leopold 1974, 1990; Llamas 1975, 2004; Datta 2005; Custodio 2000, 2002; Llamas et al. 2009; Chaminé 2015; Gundersen 2017; Abrunhosa et al. 2018; Groenfeldt 2019, and references therein). In addition, it must also incorporate insights on geoethics, eco-responsibility and scientific integrity in geosciences, as well as getting a better understanding of resilience of Earth systems design with nature and geo-hazards (e.g., McHarg 1992; González de Vallejo 2012; Kresik and Mikszewski 2013; Chaminé et al. 2013, 2015a; Chaminé 2015; Wyss and Peppoloni 2015; Bobrowsky et al. 2017; Koepsell 2017; Chaminé and Gómez-Gesteira 2019). The characterization, assessment, development, and protection of groundwater systems involve high standards for practitioners, including ethical dimensions (Chaminé et al. 2013), and the professional hydrogeologists and groundwater engineers are uniquely skilled on this thematic (Petitta 2013). Last but not least, conceptual site model for groundwater systems should be grounded on Earth-based models, mathematical modelling to outline predicting scenarios and a reliable comprehension of climate change and multi-hazard systems. Consequently, going back to basics is important to shape a reliable conceptualisation on groundwater systems established on cartographic reasoning (Chaminé et al. 2015a).

## 2 Turning Back 50 years in Selected Scenarios in a Changing World

> To a European, and generally to any man who has never travelled in the wilds of Africa, what explorers have to endure in penetrating into that continent, what difficulties they have at every instant to overcome, and what iron labour they have to go through, will be well-nigh incomprehensible. (Serpa Pinto, 1881; p. xvii).

### 2.1 Groundwater Supply to Military Forces in the Context of Guerrilla Warfare at Cabo Delgado, Northern Mozambique (1968–1970)

In areas of military conflicts or humanitarian crisis, access to drinking water sources is critical for the military personnel involved in the operations and or for the civilian populations (Mather and Rose 2012). Negligence and less proper judgments and assessments can jeopardize many lives. Without discussing the context of a fair war, the core values of all military hydrogeology reflect honour, courage, integrity, and a commitment of sound ideals of service and mission.

In 1968 Portugal led a colonial military conflict against insurgents fighting for the independence of those territories on three fronts in Africa (Guinea-Bissau, Angola, and Mozambique). This conflict was designated by the Portuguese political and military authorities, and Portuguese people as *Overseas War ("Guerra do Ultramar")*, or *African War* and by *Liberation* or *Independence War* by the insurgents. From 1961 until the end of operations, dictated by the carnation revolution of April 25, 1974, about one million young Portuguese were involved in the conflict that left tens of thousands of civilian and military casualties on both sides. This conflict was part of the context of the Cold War (USA *vs.* Soviet Union) and of the liberation movement born after World War II and promoted by the Bandung Conference (1955) that led to the end of colonial rule in all Non-Self-Governing Territories worldwide, in line with the United Nations Charter (article 73).

Portugal, the first European power to settle in Africa in the fifteenth century, was the last to withdraw after 13 years of a relatively low-intensity military struggle that, however, at the end of operations in particular at Guinea (nowadays, Guinea-Bissau) was close to a medium-intensity conflict. In the case of Mozambique the fighting was located, since 1964, mainly in the far north of the territory in the Cabo Delgado region (on the border with Tanzania), although it later spread to Niassa, Tete (border with Zambia) and at the end also to the corridor Beira-South Rhodesia (present-day Zimbabwe).

In a counter-insurgency struggle, more important than the control of the territory is the conquest of the local populations: in this sense the Portuguese military effort consisted in the maintenance of small garrisons (battalion or company level) of few hectares, provided with some defensive works, most of them near old small villages and native communities scattered on the subtropical bush. Near some of these military posts, disseminated in a squared system ("modelo de quadrícula"), new settlements with health and social support, water and agricultural land were built to install local populations, voluntarily or coercively, and subtract them from the insurgent's influence. Water was needed for all this military effort and the Portuguese Army Corps of Engineers created a structure, the Water Platoon of the AEM (Mozambican Military Engineering Group) based in Nampula, whose function was exclusively to provide water, usually groundwater, to the troops and to the population under Portuguese control.

In a war that, at the time, for a large part of the Portuguese population was fair, based on the centennial historic Portuguese presence in Africa, J. M. Carvalho (JMC) as a young conscripted officer in command of the AEM Water Platoon, started his professional career as a military hydrogeologist in this framework. The tasks that were committed to this military unit consisted of coordinating, based on the hydrogeological knowledge of the area, field campaigns for groundwater exploration and exploitation. These campaigns were then carried out on the ground by Portuguese private companies and/or official institutions under the control of the AEM. Normally the location of military bases and settlements was selected by strategical or tactical military reasons. However, in later stages, particularly in Tete area, groundwater exploration preceded the installation of several barracks' settlements.

At Cabo Delgado district two main approaches were carried out, determined by the hydrogeological conditions, as follows: (i) In the Precambrian basement geophysical surveys were carried out to map hydrogeological traps (faults, veins, weathered bands or lithological boundaries) and try to get drinkable water, followed by drilling with conventional percussion rigs, and (ii) in the Macondes Plateau and on the coastal aquifer direct drilling also with conventional percussion or mud base rotary rigs.

To groundwater exploration in the Precambrian basement, the geophysical approach proposed by "Geotécnica e Minas" from Lourenço Marques (now Maputo) comprised of sophisticated electrical surveys—that anticipated the current electrical tomographies—and geomagnetic profiles. The geomagnetic surveys carried out with highly portable equipment, were unbeatable to delimit lateral contacts and proved especially useful in the definition of brackish waters circulation along mafic bands. The cross interpretation of these techniques with aerial photographs (on a scale of 1/50,000) proved very effective. Obviously, the exploration means were consistent with the capacity of the available percussion drilling machines in metamorphic and igneous rock (30 or 40 m depth, depending on the rate of weathering and faulting). In fact, only the weathered zones were drilled and screened leaving out the spectrum of the deepest fractured areas. So, they only were made available, with adequate diameters, for groundwater extraction in Mozambique after 1973.

The sustainable flow rates, designed with pumping tests carried out with piston pumps, were generally relatively modest, and it was normal to complete wells with flows of 0.6 $m^3$/h, or about 0.17 L/s. Under the conditions of that conflict, for military barracks of less than 150 men (Company level), this was a considerable amount. The alternative supply of these men was the use of springs or surface water of poor quality and whose use was often shared with the enemy. Those attempts led to often lethal encounters (land mines activation and/or ambushes) situation analogous to that experienced in the other Portuguese theatres of operation at Angola and Guinea.

Water well extraction systems were based on submersible electric pumps and in some cases on self-priming pumps with hydro-injector powered by petroleum engines. The installation of the extraction equipment was the main problem of the operation as we do not found a way to ensure the immediate exploitation of the wells soon after the completion of the pumping tests. This inefficiency took some flavour to the hydrogeological successes achieved.

In the Macondes plateau, traditionally known for its poverty in water resources (Bourguet 1961), and in its extension to the south, the Serra Mapé, several wells were drilled in places as unlikely as Nambude, Miteda and Omar, among others. The most emblematic was a 300 m deep well drilled by ACavaco for the Portuguese Air Force (FAP) aerodrome AM-45 in Mueda. This well, drilled till the top of the Precambrian gneisses deserved electrical logs (self-potential and single point resistivity) and it was a real technological adventure. In addition, to the avatars of drilling and designing the casing in a war area, at the end it has been necessary to install an extraction system with two electric submersible pumps to ensure the appropriate total head to feed the surface storage reservoirs. At the end, an enormous logistic problem was solved and, furthermore, FAP pilots could enjoy a comfortable bath after their latest daily missions, a particularly important war issue on a psychological point of view.

Much insight, team spirit, competence, dedication… and luck… allowed the success achieved by the Water Platoon operations in Mozambique. During those times there was still no mention to the hydrogeological models: the emphasis was all placed in the geological model (represented as geological cross-sections) and its implications in the circulation of water, the geology of groundwater. In the case of crystalline rocks, the joint interpretation (remember, no GPS was available), very precise mapping with compass and tape of all possible favourability factors to the circulation of groundwater was considered vital…the whole job done under machine gun protection. But and this is a main issue, at the time, FAP and the Portuguese Army highly valuated the technical capacity of the drillers, geologists, and groundwater engineers: water well was considered a real engineering matter. No place for adventurers, dowsers, and opportunists.

Nowadays, since 2017, Cabo Delgado province after the important discovery of gas reserves (Mozambique is now quoted as the new Qatar) is the stage of cruel Jihadist insurgents attacks, apparently out of control of the Mozambican Army.

## 2.2 Hydrogeological Survey at Aozou Strip, Libyan/Chad Border (1981)

The conflict around the Aozou strip dates to colonial times, French and Italians running for a territory supposedly rich in uranium and other strategic minerals north of the Tibesti Mountains. Libya under the rule of Muammar Gaddafi (1942–2011) engaged in military operations in the Aozou Strip to gain access to the geological resources and to use it as a base of influence in Chadian politics. This ultimately resulted in the Chadian–Libyan conflict ended with the so-called Great Toyota War in 1987.

In the 70's of the last century, some people in Europe look with curiosity the Libyan leader Muammar Gaddafi as a revolutionary. His "Green Book", first published in 1975, was intended to be read by all people in the world and it rejected both capitalism and communism, as well as representative democracy. Instead, it suggested a type of direct democracy overseen by the General People's Committee which should allow direct political participation for all adult citizens. In this context, in 1979, JMC have been in Libya in a Portuguese mission supported by Partex company (at the time the armed branch of Fundação Calouste Gulbenkian for non-cultural activities) to develop groundwater in the country. Some months later, following that first mission, at ACavaco firm, JMC received a strange request: a certain Peter Meyer from Germany invited us to participate in the groundwater supply for the Uzu military airfield, already located inside the Aozou Strip. The airfield had a 3500 m length runway and was fully equipped to accept B 747 Jumbo aircrafts.

After a first Libya logistical and commercial visit by Mr. Parreira, one of our operational experts, it was decided that

JMC will go there in February 1981 to make the field reconnaissance and promote the adequate further exploration and exploitation operations. From my personal point of view this work was fascinating because it allowed me to redeem my participation as a member of a so-called colonial army, working, now, for a revolutionary movement as the Jamahiriya Army was considered! Also, the action would be in the Sahara Desert that I did not visit before.

I prepared accurately this intervention as I did before in military operations in Mozambique, trying to collect all the available information that should be relevant for hydrogeological purposes. A week before I went to Montpellier (Southern France) were in the CNRS shared aerial photos and geological mapping of northern Chad and to Paris where my friends Dr. Pierre Yves Berthou and Dr. Jacques Lauverjat (University Pierre and Marie Curie, France) provided additional technical information.

Afterwards I flight to Tripoli where I acquired additional information namely a geological map of Libya (scale 1:2000.000) from USGS that basically was my only topographical source of the site! The next step to arrive at the site was Sabha, an oasis capital of the Fezzan region, deep in the Libyan Desert at the confluence of several migration routes from sub-Saharan Africa, around 600 km NNW of Uzu airfield. For two or three days, Peter Meyer and I look forward waiting at the Djebel Hotel in Sabha to get transport to Uzu. Finally, a Piper Cherokee single engine plane appeared with a friendly Libyan pilot who quickly set out on his way to Uzu, following roughly the course of the dirt track leading to Chad. As it was my practice in those conditions, I opened the geological map to follow the flight and determine the position from the formations. Additionally, I decided to control the time with a small digital chronometer, great novelty I acquired a few weeks before in London.

The trip was very educational, it was taking place normally until, near the destination, where a thin and persistent fog at low altitude made it difficult to observe the ground. And it was then, when by my control we were flying over the Aozou Strip, south of Uzu airport, probably on the trail to N'Djamena that the very excited pilot informed us that he did not know where he was and wanted to make an emergency landing, for lack of fuel. We later learned that he had taken a wrong radio frequency from the Uzu airport and that prevented him from communicating with the ground.

Our astonishment was total. I have decided to take control of the operation and drive the plane North along the track to Shaba. My inexperience of the desert was evident, I could not identify the trail and about 40 km to NNW of Uzu and 30 km to W of that trail the pilot was terrified and decided to land for good in the Sarir Tibesti.

Total astonishment: no, it was not a movie. We were really landed, without fuel, in the desert without water and with half a dozen oranges, about 50 km out of the only human settlement, the Uzu airfield. The next morning after a frozen night barely slept under a wonderful sky, we decided to review our position and convince the pilot to fly about fifteen minutes to West. Not too convinced, our pilot, after controlling the fuel levels, wanted to fly to NW towards Sabha. As I suggested he accepted to manage to take off direction West and thirteen minutes later there appeared the famous track to N'Djamena, signalled kilometrically with black drums. We were rescued two days later, quite dehydrated, by a caravan of Mercedes trucks. We then learned that the Libyan Air Force was trying to find us, to the south, in the Tibesti Mountains, a recurring point of unsuccessful landings. But no: we were somewhere in the Sarir Tibesti because an unlikely Portuguese geologist intervened in the pilot's decisions.

On the fourth day I did my job, escorted by the Libyan army, as my mentor Dr. Georges Zbyszewski (1909–1999) always advised to pursuit: *"the main point is that you have to perform your duty always accurately"*. The Cambrian-Ordovician sandstone were very easy to distinguish from the metamorphic formations in the desert and thus in a work of geology of groundwater of a few hours it was possible to implant favourable drilling sites. I then confirmed that the job was practically done thanks to the cartographic elements collected in France. On the fifth day, inside a Twin Otter plane, loaded with soldiers, the Libyan Air Force returned me to Bengazi and a few hours later I was inside the comfort of the Meridien Hotel in Athens. In a local bar at night, an ouzo (what an ironic homophony!) with water calmed me down and prepared me for the return to civilization.

I did not come back to Libya, since 1981, but my dreams still often pass through the Djebel Hotel of Sabha and the immensity of the Sarir Tibesti. Some spherical quartz samples from that magical place are on my bookshelf and help me to tell grandchildren what wind abrasion is like in the Sahara Desert...

## 2.3 Groundwater Exploration and Exploitation for the Bottled Water Industry in Namaacha, Southern Mozambique (2011–2020)

The small town of Namaacha is located about 75 km west of the Mozambican capital, Maputo, just in the border with the Kingdom of Swaziland and the Republic of South Africa. The area is located in the Lebombo Mountains that locally reach elevations of about 500 m and are aligned approximately to N and NNE standing out clearly in the savanna that descends gently to the Indian Ocean.

The geological framework of Lebombo Mountains is a monocline series, gently dipping to East and composed of Jurassic (Karoo) volcanic rocks, essentially basaltic lavas, and rhyolitic flows, interbedded with tuffs and volcanic breccias (Freire de Andrade 1929; Torre de Assunção et al. 1962). The alternating resistant rhyolite and the easily eroded basalts produce a series of parallel cuesta ridges separated by savanna plains. The volcanic sequence is overlain by cretaceous sandstone, marl, and clay and tertiary and the recent sedimentary cover. Since the colonial times, the beauty of the mountain, cool climate, and freshwater abundance (the climate being tropical humid to temperate with an average annual rainfall of 750 mm and temperature of 20ºC) made Namaacha a popular tourist destination. Such a rich natural environment favours the occurrence of the industry of bottled water. This is the case of the 'Água da Namaacha' whose bottling plant, former springs and new wells are located in Germantine, about 10 km far from Namaacha in the road to Maputo. The exploitation of groundwater for the bottling water industry started, in the early days, with the springs Ferrão 1 and Ferrão 2, located over the rhyolitic series in the Impaputo river basin. These rocks and the nearby basaltic lavas are typical fractured rocks, groundwater flowing through the hydrogeological traps, largely referred in the literature.

In 2011, TARH–Terra, Ambiente & Recursos Hídricos and Labcarga|ISEP consortium was invited by 'Sociedade das Águas de Moçambique' (SAM) to support technically the development of the resource 'Água da Namaacha'. Since then we have carried out several hydrogeological programs that included desk and field studies, such as: remote sensing, geomorphological and geological surveys (with field volcanological support of J.C. Nunes; groundwater inventory by T. Carvalho, and field hydrogeological mapping by H. I. Chaminé, J. Teixeira, and J. M. Carvalho), geophysics (electrical tomographies and high-resolution seismic reflection by Ocsa Consultancy, Spain) and drilling with down-the-hole hammer rigs with local contractors under TARH supervision. That permitted developed a keen hydrogeological conceptual site model.

The first surveys demonstrates that the overall transmissivity in the rhyolitic series were quite low so all drilling efforts were focused in the weathered/fractured basaltic lavas up to 120 m depth (nearby Germantine, elevation 400 m) and in a new well field, designated by Mountain (depths up to 150 m, elevation 300 m) located about 2 km north of the bottling plant. In this area the productive sectors are clearly associated with the intersection of the basaltic lava "bedding" with WNW-ESE lineaments. In both areas the rate of success is very low. All wells are properly completed with artificial gravel packing and adequate sealing in the non-productive sections and are equipped with submersible pumps. Mostly wells are fitted with flowmeter and water level monitoring devices, and additionally in each well field, one piezometer monitor the long-term water level evolution. Transmissivity in the productive sectors are in the range 0.2–1.4 m$^2$/day with equivalent hydraulic conductivity of $3 \times 10^{-2}$ to $5 \times 10^{-3}$ m/day. Piezometric levels answer well the rainfall episodes and the total abstraction are of the same order of magnitude of the recharge. Bacteriological and physico-chemical water quality (electrical conductivity of 200 μScm$^{-1}$ at Germantine and 900 μScm$^{-1}$ in Mountain sector) is controlled daily in each well.

Presently, the total water production is of 130 m$^3$/day in Germantine (with 14 wells) and of 53 m$^3$/day in the Mountain sector with 4 wells, that meaning about $67 \times 10^6$ L/year, a remarkable figure on the industrial point of view, and quite surprising considering the existing hydraulic parameters of the occurring basalts.

## 3 Concluding Remarks

The impressive words of the Roman military engineer Marcus Vitruvius (33BC) still topical: *"we should also consider the nature of the place when we search for water"*. In fact, the models must incorporate the intrinsic geological ground variability and uncertainty of earth-based systems and numeric modelling to shape predicting scenarios of hydrological systems under a climate change framework, design with natural hazards, societal pressures, or conflicts. The characterization, design, planning, monitoring, evaluation, and management of groundwater shall include integrative sustainable methodologies, nature-based solutions and a hydrogeoethical approach. It is indispensable to shape a balanced multidisciplinary and transdisciplinary network in both the academic and the practitioners to preserve the mutual interchange of learned lessons.

Hydrogeological conceptual site modelling—grounded in a reliably data, such as, climatology, geology/morphotectonics, hydrogeological inventory and mapping, hydrodynamics, water chemistry and isotopic hydrology—must be embraced as the key tool to support a balanced Integrated Water Resources Management, sustainable groundwater policies, understanding about climate change in Earth systems and geoethical manner for the post-COVID-19 pandemic world. And, furthermore, Leonardo da Vinci quote must be always present in the conceptual site models: "*Simplicity is the Ultimate Sophistication*".

**Acknowledgements** Our warmest thanks to: (i) Portuguese Army, where both authors served in different time and realities, but where the mission spirit and field training were always present and, undoubtedly, contribute to the alma mater of a Geologist; (ii) Clients, Private Companies, Laboratories, and Agencies and to all colleagues, managers, operators, and technical teams who carried out fieldwork, surveys, drilling jobs and desk studies. Lastly, our appreciation to the retired Major General Manuel de Campos Almeida for the detailed comments, which helped to improve the clarity of the manuscript and also to the colleague Hydrogeologist Lucien Bourguet for confirming some Mozambican bibliography.

## References


Abrunhosa MJ, Peppoloni S, Chambel A (2018) Introducing geoethics in groundwater science, technology, and management. In: Abstracts Book, Session T1.1 socioeconomic aspects of groundwater resources management, 45th IAH Congress, Daejeon, Korea, p 26

Biswas AK (1970) History of hydrology. North-Holland Publishing Company, Amsterdam

Bobrowsky P, Cronin VS, Di Capua G, Kieffer SW, Peppoloni S (2017) The emerging field of geoethics. In: Gundersen LC (ed) Scientific integrity and ethics with applications to the geosciences. Special Publication American Geophysical Union, John Wiley and Sons, Inc, vol. 73, pp. 175–212

Bourguet L (1961) Rapport de mission dans le district de Cabo Delgado. BURGÉAP, Direcção Dos Serviços De Geologia E Minas, Lourenço Marques 288:1–19

Bourguet L (2017) Chercheur d'eau: en Afrique et ailleurs – quarante années au Burgéap 1956–1996. EdiLivre, Paris

Carson R (1962) Silent spring. Houghton Mifflin Company, Boston

Chaminé HI (2015) Water resources meet sustainability: new trends in environmental hydrogeology and groundwater engineering. Environ Eart Sci 73(6):2513–2520

Chaminé HI, Carvalho JM (2015) Meeting sustainability and water resources: a dialogue about groundwater science and hydrogeology practice. Environ Eart Sci 73(6):2531–2542

Chaminé HI, Carvalho JM, Afonso MJ, Teixeira J, Freitas L (2013) On a dialogue between hard-rock aquifer mapping and hydrogeological conceptual models: insights into groundwater exploration. Eur Geol J 35:26–31

Chaminé HI, Carvalho JM, Teixeira J, Freitas L (2015) Role of hydrogeological mapping in groundwater practice: back to basics. Eur Geol J 40:34–42

Chaminé HI, Fernández-Rubio R, Simões Cortez JA, Oliveira Silva M, Custodio E, Llamas MR, Chambel A, Lydon JW, Fox RA, Jones GL, Cotelo Neiva JM, Gama Pereira LC, Borges FS, Oliveira R, Costa Pereira A, Gomes Coelho A, Baptista R, Nunes JC, Senos Matias M, Rocha F (2015b) Hydrogeologist J. Martins Carvalho: professional, professor and gentleman. Environ Eart Sci 73(6):2521–2529

Chaminé HI, Gómez-Gesteira M (2019) Sustainable resource management: water practice issues. Sustain Water Resour Manag 5:3–9

Custodio E (2000) Some relevant ethical issues in relation to freshwater resources and groundwater. Bol Geol Min, Madrid 111(6):121–130

Custodio E (2002) Overexploitation, what does it mean? Hydrogeol J 10(2):254–277

D'Anselmi P, Di Bitetto M (2013) Micro-ethics. In: Idowu SO, Capaldi N, Zu L, Gupta AD (eds.), Encyclopedia of Corporate Social Responsibility. Springer, Berlin, Heidelberg

Datta PS (2005) Groundwater ethics for its sustainability. Curr Sci 89 (5):1–6

Eales HV, Marsh JS, Cox KG (1984) The Karoo igneous province: an introduction, Petrogenesis of the volcanic rocks of the Karoo Province. Spec Publ Geol Soc South Afr 13:1–26

Freire de Andrade C (1929) Esbôço geológico da província de Moçambique. Imprensa Nacional de Lisboa, Lisboa

González de Vallejo LI (2012) Design with geo-hazards: an integrated approach from engineering geological methods. Soils & Rocks—J Geotech Geoenviron Eng 35(1):3–28

Groenfeldt D (2019) Water ethics: a values approach to solving the water crisis, 2nd edn. Earthscan Water Text, Routledge, New York

Gundersen L (2017) Scientific integrity and ethics in the geosciences. American Geophysical Union, Wiley, New York

Herkert JR (2003) Microethics, macroethics, and professional engineering societies. Emerging technologies and ethical issues in engineering papers from a workshop (October 14–15, 2003). The National Academies Press, Washington DC, pp 107–114

Kresik N, Mikszewski A (2013) Hydrogeological conceptual site models: data analysis and visualization. CRC Press, Boca Raton

Koepsell D (2017) Scientific integrity and research ethics - an approach from the ethos of science. SpringerBriefs in Ethics, Springer, Cham

Leopold A (1949) A sand county almanac: and sketches here and there. Oxford University Press, Oxford

Leopold LB (1974) Water: a primer. W.H. Freeman Co., San Francisco

Leopold LB (1990) Ethos, equity and the water resources. Environ Sci Policy Sustain 32(2):16–41

Llamas MR (1975) Non economic motivations in ground water use: hydroschizophrenia. Groundwater 13(3):296–300

Llamas MR (2004) Water and ethics: use of groundwater. UNESCO series on water and ethics, essay 7. UNESCO, New York

Llamas MR, Martínez-Cortina L, Mukherji A (Eds) (2009) Water ethics. 3rd Marcelino Botín Foundation Water Workshop. CRC Press, Boca Raton

MacDonald A, Davies J, Callow RC (2008) African hydrogeology and rural water supply. In: Adelana S, MacDonald A (eds) Applied groundwater studies in Africa. CRC, Boca Raton, pp 127–148

Mather JD, Rose EPF (2012) Military aspects of hydrogeology: an introduction and overview. In: Rose EPF, Mather JD (eds) Military aspects of hydrogeology, geological society, vol 362. Special Publications, London, pp 1–17

McHarg IL (1992) Design with nature. Wiley series in sustainable design, 25th edn. Wiley, New York

Petitta M (2013) New challenges for hydrogeologists. Eur Geol J 35:6–8

Serpa Pinto A (1881) How I crossed Africa: from the Atlantic to the Indian ocean, through unknown countries: discovery of the great Zambesi affluents, 2 vols, Sampson Low, Marston, Searle & Rivington, London

Torre de Assunção CF, Pinto Coelho AVT, Tavares Rocha A (1962) Petrologia das lavas dos Libombos (Moçambique): subsídios para o seu conhecimento. Estudos Ensaios E Documentos, Série Da Junta De Investigações Do Ultramar, Lisboa 99:1–74

Vitruvius M [33 BC] (1960) De Architectura (The Ten Books on Architecture—Book VIII (Water Supply); translated by Morgan MH, Dover Publications)

Wittfogel KA (1956) The hydraulic civilizations. In: Thomas WL (ed) Man's role in changing the face of the earth. The University of Chicago Press, Illinois, pp 152–164

Wright EP (1992) The hydrogeology of crystalline basement aquifers in Africa. In: Wright EP, Burgess WG (eds) The hydrogeology of crystalline basement aquifers. Special Publication nº 66, The Geological Society London, London

Wyss M, Peppoloni S (2015) Geoethics: ethical challenges and case studies in earth sciences. Elsevier, Amsterdam

# Ethical Issues on the Use of Citizen Science Approaches

António Dinis Ferreira, Eduardo Barai, Inês Leitão, António Massena Ferreira, Anne-Karine Boulet, Paulo Pereira, Maria de Fátima Oliveira, and Carla Santos Ferreira

**Abstract**

Citizen science is increasingly seen as an approach to gather information with an unprecedented time and territorial resolution, especially after the introduction of new IT solutions. This allows citizens to interact with researchers and provide detailed information about the scientific questions under appraisal in real time, and documented with extra information (e.g. photos, sound, videos), and be of high value for groundwater management. Nevertheless, these new approaches are flaw with problems and profound ethical significance—to start with, the Orwellian type of problems. By adhering to a citizen science activity, one may lose some of his/her privacy that becomes public domain. In addition, there is the risk of misuse, due to poor preparation or to evil behaviour. Also, there are ethical problems related to data property, and with the feedback of the results to the citizens, in a format that they can understand. Yet, citizen science bears the promise of a higher level of involvement and commitment of the commoners into subjects relevant to the society. This provides the context for higher levels of citizenship and democracy building that we need to pursue if we are aimed to create a better society.

**Keywords**

Citizen science • IT solutions • Participation • Citizenship building • Ethical aspects

A. D. Ferreira (✉) · E. Barai · I. Leitão · A.-K. Boulet · M. de Fátima Oliveira · C. S. Ferreira
CERNAS, Instituto Politécnico de Coimbra, Escola Superior Agrária, Bencanta, Portugal
e-mail: aferreira@esac.pt

A. M. Ferreira
ISCAA, Universidade de Aveiro, Aveiro, Portugal

P. Pereira
Environmental Management Laboratory, Mykolas Romeris University, Vilnius, Lithuania

## 1 Introduction

Citizen science bridges science and society, involving the common individuals in scientific discovery across disciplines (Bonney et al. 2014). Citizen science has recently evolved to unforeseeable levels of implementation and proficiency, making monitoring at a territorial and time resolution hardly imaginable a few decades ago. IT has a fundamental role in the process, allowing a stronger interaction between scientists and citizens, as highlighted by Hand (2010). The new IT technologies such as the advent of smartphones fostered the involvement of citizens in scientific endeavour. This process generates new knowledge and understanding, making citizens to act as contributors, collaborators or even as project leaders, thus performing meaningful roles in the acquisition and processing of scientific data.

Citizen science is often seen as a pathway towards a higher level of citizenship, by increasing participation, self-awareness and commitment. Nevertheless, as for a number of other IT-based tools, such as social media and social networks, citizen science can be tainted with Orwellian-type problems that ultimately may hamper the potential impacts of this approach and lead to ethical problems.

Among the potential problems we can list: (a) the Orwellian type of problems, namely the "big brother is watching you" threat, that is often associated with the social media. By adhering to a citizen science activity, the citizen may lose some of his/her privacy that becomes of the public domain, and subject to criticism and nudge bullying; (b) risk of misuse, due to poor preparation, evil behaviour, or greed in the cases where some kind of tangible or intangible advantage can be extracted from participating in citizen science actions (see, for instance, Hand 2010).

In addition, there are ethical problems related to the property of the data acquired, and with the feedback of the results to the citizens, in a way that they can understand.



M. Abrunhosa et al. (eds.), *Advances in Geoethics and Groundwater Management: Theory and Practice for a Sustainable Development*, Advances in Science, Technology & Innovation,
https://doi.org/10.1007/978-3-030-59320-9_62

This work aims to address the ethical implications of citizen science and explores possible solutions to overcome the increment of exposure, and the new situations that a large number of contributors create.

## 2 Methods

This work derives from the mental process of implementing an app to rank the ecosystem services provided by the Urban Green and Blue Infrastructure (MapNat) under the project URBANGAIA.

## 3 Results and Discussion

A commonly definition of citizen science is that provided by the Oxford English dictionary: "scientific work undertaken by members of the general public often in collaboration with or under the direction of professional scientists and scientific institutions" (Hecker et al. 2018), predominantly collecting or analysing data (Fast and Haworth 2020). Citizen science provides geographical discriminated data and may represent an opportunity to integrate different layers of scientific information and local knowledge and perspectives into a scientific study.

Citizen science is an increasingly appealing tool to increase the territorial and time resolutions of data acquisition and to involve citizens as manpower in research experiments, for instance, in the environment and earth sciences areas. In addition, citizens can play a major role in complex data processing, as observed by Hand (2010).

Turrini et al. (2018) argue that citizen science is an excellence tool to generate new knowledge, increase awareness, raising and facilitate in-depth learning, and enhance civic participation.

Internet and communication technologies evolution fostered the evolution of citizen science, as has the global tendency to perfecting the democracy in the western societies. They claim that there is a need for increasing civic participation in the governance of assets relevant to the public good. This path to a higher involvement, commitment and citizenship is essential to the strengthening of democracies.

As Fast and Haworth (2020) pinpoint, there are important considerations to bear in mind such as issues of uneven societal access to participation, geographic and other sampling biases, data quality, and participant motivation that may hamper the scientific outcomes. Despite the fine territorial and temporal scale and broad scope of citizen science data, they may also provide sensitive personal information, or sensitive information about the processes of issues that are the study object of analysis (Anhalt-Depies et al. 2019). For instance, the location of endangered species, if turned public, may increase the visitation to the area and the pressure, contributing to the disappearance of these individuals. Cappa et al. (2016) found that the limited contact of citizen researchers with professional researchers (especially in online participation) may limit participation and decrease data quality.

To be effective, citizen science approaches have to be considered a benefit not only to individuals that may gain tangible of more frequently intangible satisfaction/benefits from their participation but has to be meaningful for the communities. This may be achieved, among other issues, by supporting the co-creation of relevant knowledge by making community-based monitoring part and parcel of citizen science activities that add value to the available information. In addition, they must strength the capacity of local communities to organize via connective action, a new form of collective mobilization, less reliant on formal organizational coordination (Leeuwis et al. 2018).

The relevance to groundwater management is evident, with citizens able to provide detailed time and territorial information on several visual parameters of water quality and quantity in wells and other water bodies.

In what concerns citizen science, we argue that attention should be given to key aspects related to ethics, when designing and implementing actions based on citizen science, to reduce the likelihood of conflicts, misinterpretations and poor or ill-use of the platforms set for data acquisition.

To reduce the problems with the ill-use of citizen science tools, we defend that a set of principles has to be thought and implemented to maximize their performance and minimize the risks. This includes a particular attention to the participatory processes that should be directed to different target populations, according with the particular aspects of the scientific area/study to be addressed. We can start by preaching to the converted. Since this would not lead to the potential full commitment of the entire population, a strategy can be drawn to expand participation beyond the converted. Openness is fundamental to strength the entire process. Feedback must be given to the citizens participating in science projects, in a format that they are able to understand. This implies an effort to produce contents in a format other than scientific papers. This aspect is of particular importance to improve the confidence in complex institutions/scientists. Relevance is a touchstone of the citizen science process. Citizens will be more committed to their tasks if the outcomes are relevant for them or their community. This requires feedback. From a motivational design perspective, the provision of feedback is considered as an effective design principle to promote users' experience and performance in many contexts (Zhou et al. 2020). The follow-up is effectiveness, citizen science outcomes must be effective and timely, delivering what is needed on the basis of clear objectives, an evaluation of future impact and, where available, of past experience. Effectiveness also depends on implementing policies in a proportionate manner and on taking decisions at the most appropriate level. Acknowledgement, citizen scientists must be identified and praised by their work in the project interim and final outputs.

## 4 Concluding Remarks

Citizen science is a research approach with multiple benefits, namely the capacity to provide information with a fine geographical and temporal resolution, that would not be able to acquire with any other approach, and the capacity to empower and commit citizens, a process viewed as fundamental to increase citizenship, that is at the basis of strengthen democracy. The application to groundwater management is evident, allowing for high time and territory resolution of water amount and visual assessment of quality by the stakeholders. Yet, citizen science is flawed with ethical problems, some of which derived from the use of IT solutions. The Orwellian problems, the evil doing, the lack of preparedness or reckless behaviour, can only be supplanted if it takes into consideration the interests of individuals and communities, the interaction between scientist citizens and professional scientists, and the observance of some basic principles of openness, feedback, effectiveness and acknowledgement.

**Acknowledgements** This work was performed under project URBANGAIA Managing Urban Biodiversity and green infrastructure to increase city resilience. JPI BiodivERsA3-2015-143/FCT-BIODIVERSA/0008/2015.

## References

Anhalt-Depies C, Stenglein JL, Zuckerberg B, Townsend PA, Rissman AR (2019) Tradeoffs and tools for data quality, privacy, transparency, and trust in citizen science. Biol Cons 238:108–195

Bonney R, Shirk JL, Phillips TB, Wiggins A, Ballard HL, Rushing AJM, Parrish JK (2014) Next steps for citizen science. Sci 343 (6178):1436–1437. https://doi.org/10.1126/science.1251554

Cappa F, Laut J, Nov O, Giustiniano L, Porfiri M (2016) Activating social strategies: face-to-face interaction in technology-mediated citizen science. J Environ Manage 182:374–384

Fast V, Haworth BT (2020) Citizen science. In: International encyclopedia of human geography, 2nd edn, pp 209–214

Hand E (2010) People power. networks of human minds are taking citizen science to a new level. Nature 466:685–687

Hecker S, Haklay M, Bowser A, Makuch Z, Vogel J, Bonn A (2018) Innovation in open science, society and policy—setting the agenda for citizen science. In: Hecker S, Haklay M, Bowser A, Makuch Z, Vogel J, Bonn A (eds) Citizen science innovation in open science, society and policy. UCL Press, London, pp 1–23

Leeuwis C, Cieslik KJ, Aarts MNC, Dewulf ARPJ, Ludwig F, Werners WE, Struik PC (2018) Reflections on the potential of virtual citizen science platforms to address collective action challenges: lessons and implications for future research. NJAS - Wageningen J Life Sci 86–87:146–157

Turrini T, Dorler D, Richter A, Heigl F, Bonn A (2018) The threefold potential of environmental citizen science – Generating knowledge, creating learning opportunities and enabling civic participation. Biol Cons 225:176–186

Zhou X, Tang J, Zhao Y Wang T (2020) Effects of feedback design and dispositional goal orientations on volunteer performance in citizen science projects. Comput Hum Behav (in press)

# Expedition Piracicaba: For a Resilient and Sustainable Hydro-future of a Watershed

José Gonçalves, Paulo Rodrigues, Geraldo Gonçalves, and Amanda Primola

**Abstract**

Expedition means that a group of people go on a journey with the particular purpose of exploring and investigating a fact, object or region, usually on a scientific basis. The objectives of the Expedition "For the Life of the River" were to diagnose the current conditions of water in the Piracicaba Basin in a "water crisis" scenario; to measure the flow capacity; to reiterate the need for water preservation; discuss the importance of environmental appreciation; and to foster debates with a socio-environmental and cultural scope, promoting the mobilization of urban and rural populations. Currently, by large mining enterprises and reckless groundwater extraction in the Piracicaba basin could reduce the basic river discharge, drain wetlands and springs, reduce soil moisture that sustains the natural or cultivated biomass, or cause other forms of environmental impacts. For the use of groundwater, it is fundamental to differentiate a well from a simple hole from which water is extracted. To this end, there must be a federal, state or municipal control of well drilling as well as groundwater abstraction, to ensure that the constructed, operated and abandoned wells are adequately based on the tripod: Ethics, Sustainability and Economy.

**Keywords**

Expedition • Piracicaba • Water crisis • Groundwater • Brazil

J. Gonçalves (✉) · A. Primola
Institute of Applied and Pure Sciences, Federal University of Itajubá, Itabira, Brazil
e-mail: jaucosta@gmail.com

P. Rodrigues
Development Centre of Nuclear Technology, Belo Horizonte, Brazil

G. Gonçalves
Piracicaba Tribune Newspaper, João Monlevade, Brazil

## 1 Introduction

At the time when Portugal discovered Brazil, in April 1500, Europe arose and developed. In this period, the world seemed limitless and the natural goods inexhaustible, in particular, water. Thus, when Pêro Vaz de Caminha wrote his first letter to the king of Portugal about the territories he had just discovered, he coined the famous description: "Waters are many; endless, graceful in such a way that anything that is planted will thrive" (Dias 1992).

The Piracicaba Expedition, For the Life of the River, 2019, is a movement of all "hydro-actors" of the Piracicaba Basin, without any dependence on a political party or interest group, which has, however, gained an immensely potentiated influence in society. Such an environmental expedition into a watershed strongly affected by mining, the steel industry and forestry activities, is a powerful tool to require adherence to the ethical concepts of managers and users of surface and groundwater. On-site information is compiled, and the society is mobilized, in an absolutely participatory and decentralized integration with power to influence, create or recreate a new environmental order, with a view to implementing a new model of regulation and control of water, based on the foundations of ethics and citizenship.

Nowadays, not only in the Piracicaba basin but in most of Brazil, pessimists and optimists are invited to debates about the "water crisis", as observed in the decreasing flow capacity and intense siltation of the water bodies, in the water contamination, sinking groundwater levels and reduction of the local and regional well production flow. In this controversy, the pessimists emphasize the hydro-problem and promote a "water scarcity strategy" based on rainfall reduction, without however considering the inefficiency of the water sector, the culture of water waste, the environmental degradation by different economic activities, the technic-scientific unawareness of the groundwater potential and the coexistence with and dumping of our



M. Abrunhosa et al. (eds.), *Advances in Geoethics and Groundwater Management: Theory and Practice for a Sustainable Development*, Advances in Science, Technology & Innovation,
https://doi.org/10.1007/978-3-030-59320-9_63

garbage into the rivers, turning them into open sewage channels. On the other hand, if we believe that the pessimists are mistaken, this does not mean that the optimists are right.

Experts and prophets of the "water crisis" generally do not think about using groundwater sources, although this is comparatively the largest volume of freshwater on Earth. This reservoir is the best-protected against contamination agents by the unsaturated soil layer and overlying rocks and whose discharge sustains the rivers during periods of rainfall deficit.

The Brazilian territory is part of a dense hydrographic network, whose river discharge is in the order of 183.000 $m^3/s$, ANA (2002), i.e. nearly 20% of the total river discharge in the world. On over 90% of the Brazilian territory, more than 1.000 mm rain falls annually, and the rivers never run dry. This means that the groundwater flows in the respective watersheds are enough to not run dry even in rainless periods. With such an abundance of water in rivers that never cease to flow, it is not surprising that the groundwater is so poorly esteemed and maintained.

In Brazil, the river base-flow indicates that the recharge rate of our aquifers is around 3.400 $km^3$/year, Rebouças et al. (1999). Undeniably, to supply domestic consumption, the alternative of using groundwater would be cheaper. This is just one example of how a new understanding of the hydro-environment could entail a revolution of the hydro-concepts that would contribute to solve the water crisis, moving away from the "scarcity strategy" supported by a small mislead part of our society, which is in fact a way of getting subventions or prime interest investments to construct ever more expensive exceptional hydraulic structures, as if that were the only way to solve problems with water.

The renewable groundwater reserves in Brazil are estimated at 42.289$m^3$ $s^{-1}$, corresponding to 24% of the river flow of the national territory, and to 49% of the drought flow (drought flow considered to have 95% of hydrologic permanence). The 27 main sedimentary aquifers alone, which cover 32% of the national territory, have an output of 20.473$m^3$ $s^{-1}$. In simple words, this large water volume is distributed into two major domains: one with aquifers of rock and sedimentary material and the other with fractured-rock aquifers, ANA (2005) and Hirata et al. (2006).

On the other hand, until 1990, groundwater in Brazil was considered a private good, which could be used to supply hotels and luxury condominiums, clubs, hospitals and industries in metropolitan areas.

The water crisis in Minas Gerais and in most of Brazil is all about distribution and tradition. However, the first point to take into account is that no one may be excluded from the right to have access to clean drinking water. The use of groundwater in the water matrix of the state of Minas Gerais is not being duly appreciated, and the opportunity is lost that would allow great savings by reducing investments in the construction of large water supply system. The latter model was stimulated in Brazil by governments that made decisions on behalf of society, preferring the construction of extraordinary structures that consume huge sums of public money. If the groundwater were duly appreciated, it would certainly come to be better protected against anthropogenic contamination and adverse natural events.

The main pending issues are the regulation and governance of groundwater use, to define and prioritize underground reservoir protection areas, the dangers of intensive extraction and an integrated and sustainable use of surface and groundwater. The increasingly efficient use of different aquifer functions—production, natural or artificial storage of water protected from contamination agents and large evaporation losses, supply regularization, filter function, biogeochemical self-purification of reused water, among others—are only a few examples of how to trigger a true ethical revolution that can lead the country and its people to the place and standard of living we truly deserve.

## 2 Materials and Methods

The Piracicaba River headwaters lie in the municipality of Ouro Preto, at 1.680 m asl, and flow eastwards for 241 km, they flow into the Rio Doce in the urban area of Ipatinga, at 210 m asl. In this region, the landscape is composed of the watersheds of the rivers Conceição, São João, Peixe, Santa Bárbara, Maquiné, Una and Prata (Fig. 1). The total area of the Piracicaba River basin is 5465.38 $km^2$ and includes 21 municipalities, more precisely, 21 hydro-administrative units, wholly or partially within their limits.

The Piracicaba Expedition lasted 11 days, from late May to early June 2019, during which 21 municipalities were visited, along a route of 241 km travelled on the river and 1600 km on the road. One team of the Expedition travelled on the river, wherever navigable, collecting water samples, sediments and general information.

Another travelled along the riverside and banks, visiting the riverine populations and hearing their experiences, historical records and information about the way of life of the populations in partnership with and dependence on the natural cycle of the river. In the communities, social mobilization actions, meetings, debates, discussion forums, awareness workshops and public performances of the local culture and traditions were promoted (Fig. 2).

## 3 Results

For all campaigners and users of the basin, the main results regarding aspects of mobilization and public announcement of the expedition can be resumed as:

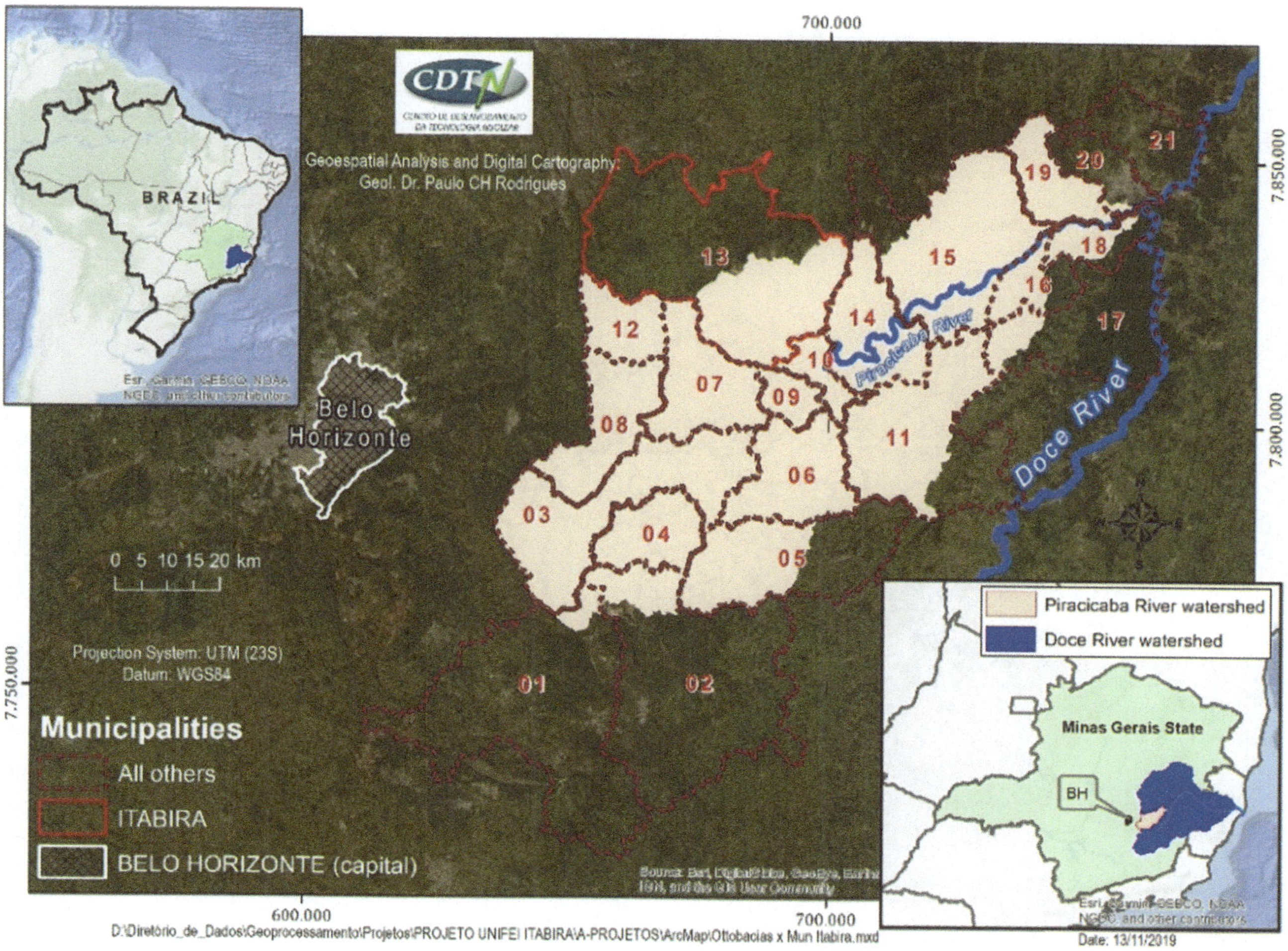


**Fig. 1** Piracicaba expedition site

**Fig. 2** Images of the actions developed during the Piracicaba expedition

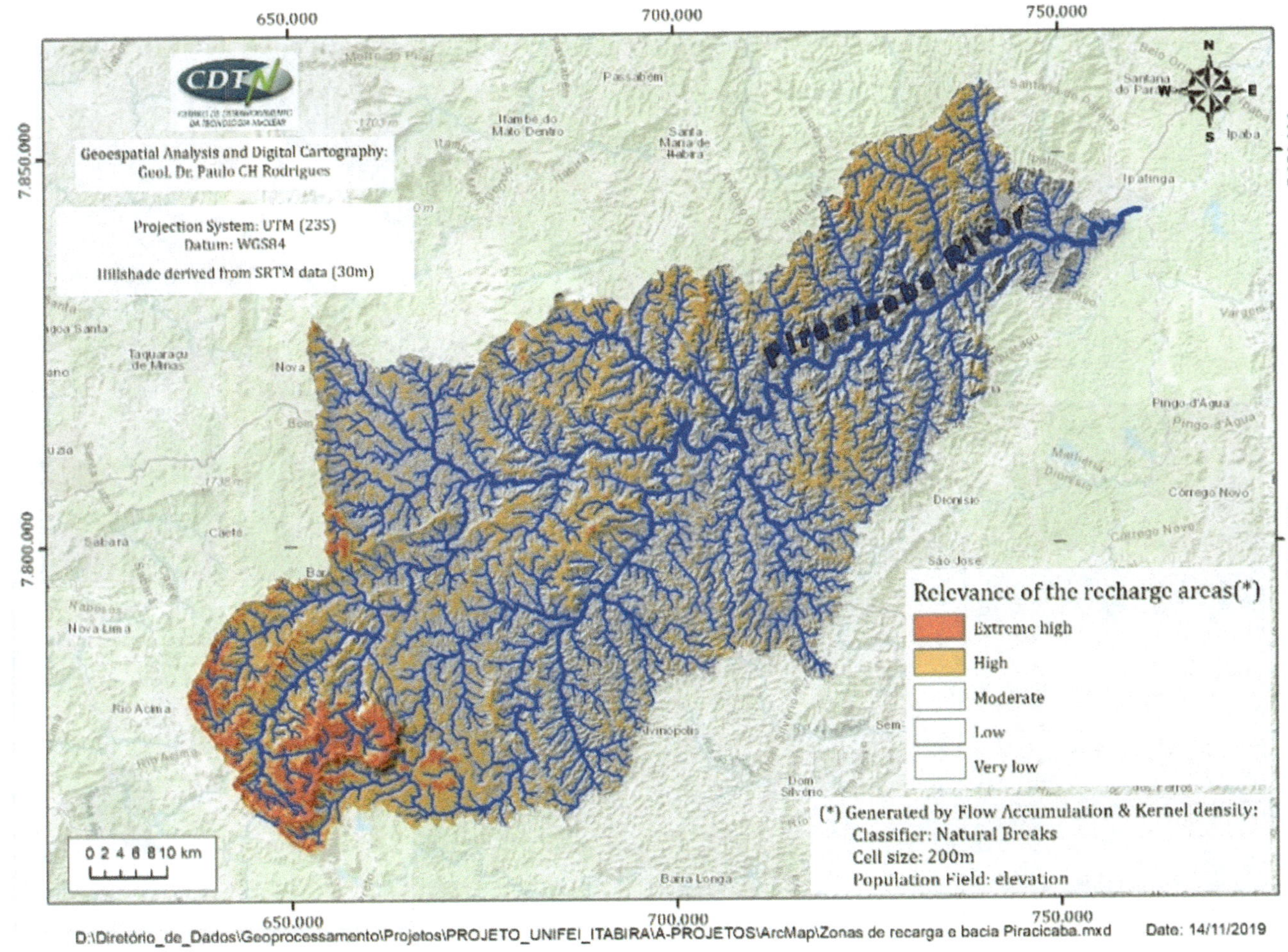


**Fig. 3** Watershed potential recharge areas

1. The Piracicaba Expedition gained widespread media coverage, with articles in newspapers, radios, and broadcasts on websites and television channels.
2. The social networks and website of the Expedition created a dynamic environment for the daily coverage of the journey step by step, which fuelled the audience's interest and engagement in the project.
3. Spontaneous media was fundamental for the success of the mobilization events in the 21 cities visited. Over the course of the journey, the Expedition increased its legitimacy and recognition, as indicated by the presence of authorities such as mayors, legislature representatives, renowned environmental experts and a considerable number of residents who attended the meetings.
4. It is worth considering the image gains for the partners of the Expedition, who were mentioned in events, interviews, website content and press releases.
5. The media coverage was unanimously positive. All published material highlighted the key views defined by the communication strategy: the technic-scientific character of the expedition and the parallel objectives of awareness raising and integration for the causes of the watershed.

An important result regarding the scientific aspect of the expedition are new studies to evaluate the watershed recharge areas, Fig. 3.

## 4 Concluding Remarks

More than eight hundred thousand people live in the Piracicaba basin, and although most of them are unaware of it, their lives depend directly or indirectly on the life of the river.

The Piracicaba Expedition found that there is the possibility and desire for a great mobilization of all active campaigners and dependents on the waters of the Piracicaba River towards a necessary intervention in the regional environment.

Changes in attitude and behaviour of all segments (civil society and public and private institutions) are essential for a new order in the environmental scenario of the Piracicaba River basin, to redirect the current trajectory towards a new awareness of the hydro-environment, raising protagonists of a new hydro-citizenship who participate in the co-creation of a new hydro-environment, based on new hydro-ethical values.

**Acknowledgements** Thanks to the support of the institutions below for the expedition: Federal University of Itajubá—Itabira Campus, to the Postgraduate Program in the National Network on Water Resources Management and Regulation—PROFÁGUA CAPES/ANA AUXPE Project Nº. 2717/2015, and the Coordination for the Improvement of Higher Education Personnel—Brazil (CAPES)—Financing Code 001/Piracicaba Tribune/CBH Piracicaba.

## References

ANA (2005) Disponibilidade e demandas de recursos hídricos no Brasil. Agência Nacional de Águas, Brasília, Brasil. https://www.ana.gov.br/sprtew/recursoshidricos.asp. Last accessed 27 Sept 2019

APHA (2005) Standard methods for the examination of water and wastewater, 21st edn. American Public Health Association/American Water Works Association/Water Environment Federation, Washington DC

Dias EM (1992) Brazil's birth certificate: the letter of Pero Vaz de Caminha. Pacific Coast Philology 27(1/2):10–15

Hirata R, Zobbi J, Fernandes A, Bertolo R (2006) Hidrogeología del Brasil: Una breve crónica de las potencialidades, problemática y perspectivas. Boletín Geológico Y Minero. 217(1):25–36

Rebouças A, Braga B, Tundisi J (1999) Águas Doces no Brasil: capital ecológico, uso e conservação. Escrituras, São Paulo, Brasil

# Exploring Groundwater Management in La Galera and Tortosa Aquifers: A Geoethics Approach

Francesc Bellaubi and Alvaro Arasa

**Abstract**

The irrigated citrus farming around the Ebro River and the olive trees at the foothills of the Tortosa-Beseit mountain range shape the distinct landscape of the lowlands in the Ebro River Basin. The inhabitants of this region express a strong connection with their land through a hydro-social network of "Canals i Sequias" (channels and ditches) whilst the "pagesia" (farmers) holds strong identity values. Since the 1990, extensive use of groundwater for citrus trees was boosted in the region resulting in a water table decline developing a culture of water hoarding contrasting with a well-established traditional surface water management. The work explores the groundwater management dilemma around La Galera aquifer and provides some insights into the values that underpin groundwater management and governance in the region.

**Keywords**

Groundwater • Management • Governance • Geoethics • La Galera aquifer

## 1 Introduction

La Galera and Tortosa aquifers (henceforth La Galera aquifer) extend from the left bank of the Sénia River to the right bank of the Ebro River. Considered as a single hydrogeological unit, the stratigraphy of the aquifer is characterized by quaternary detritic materials with thickness ranging from 30 m in the distal areas to 200 m in the proximity of the Tortosa-Beseit mountain range. Sealed by Miocene clay materials, a second deep aquifer level within cretaceous carbonates ranges from 70 m in the distal areas to 300 m depths in the nearby mountains (Fig. 1).

La Galera aquifer groundwater management is co-shared with the Agencia Catalana de l'Aigua—ACA (Catalan Water Agency) and the Confederación Hidrográfica del Ebro—CHE (Ebro River Basin Authority), but in practical terms the CHE is responsible for reporting on existing boreholes, extending licences and reporting on complaints.

It is estimated that there are around 1900 wells in La Galera aquifer, which exploit both the surface and the deep aquifer (CGRXS 2004). The total extraction is estimated close to 60 $hm^3$/year. The recharge, in general, is carried out by direct infiltration and lateral contributions from the Tortosa-Beseit mountain range feeding the surface and deep aquifer with resources in the region of 200 $hm^3$/year (CHE 1991; Espinosa 2014) The predicted drawdowns are stimated at 46.8 m in the south-west of the Plana de La Galera aquifer (Samper et al. 2014).

Together with the Ebro right side irrigation channel, running parallel to the Ebro River, groundwater contributes to the agricultural economic development of the Low Ebro lands. Both the citrus trees in the intensively irrigated areas close to the Ebro River and the olives trees at the foothills of the Tortosa-Beseit mountain range shape the distinct landscape of the lowlands in the Ebro River Basin. Due to the current low market price of citrus fruits, farmers incur losses and have decided to remove 20–30 year old citrus trees and substitute them with fodder, corn, which have higher water needs, or highly intensive olive tree farming. In spite of this traditional economic activity, nowadays most farmers have other work that provides them with their main income; meanwhile they look for higher productivity of their plots by increasing the water demand.

F. Bellaubi (✉)
Department of Philosophy, South Ural State University, Chelyabinsk, Russia
e-mail: bellaubi@hotmail.com

A. Arasa
Grup EbreRecerca, Barcelona, Spain


M. Abrunhosa et al. (eds.), *Advances in Geoethics and Groundwater Management: Theory and Practice for a Sustainable Development*, Advances in Science, Technology & Innovation,
https://doi.org/10.1007/978-3-030-59320-9_64

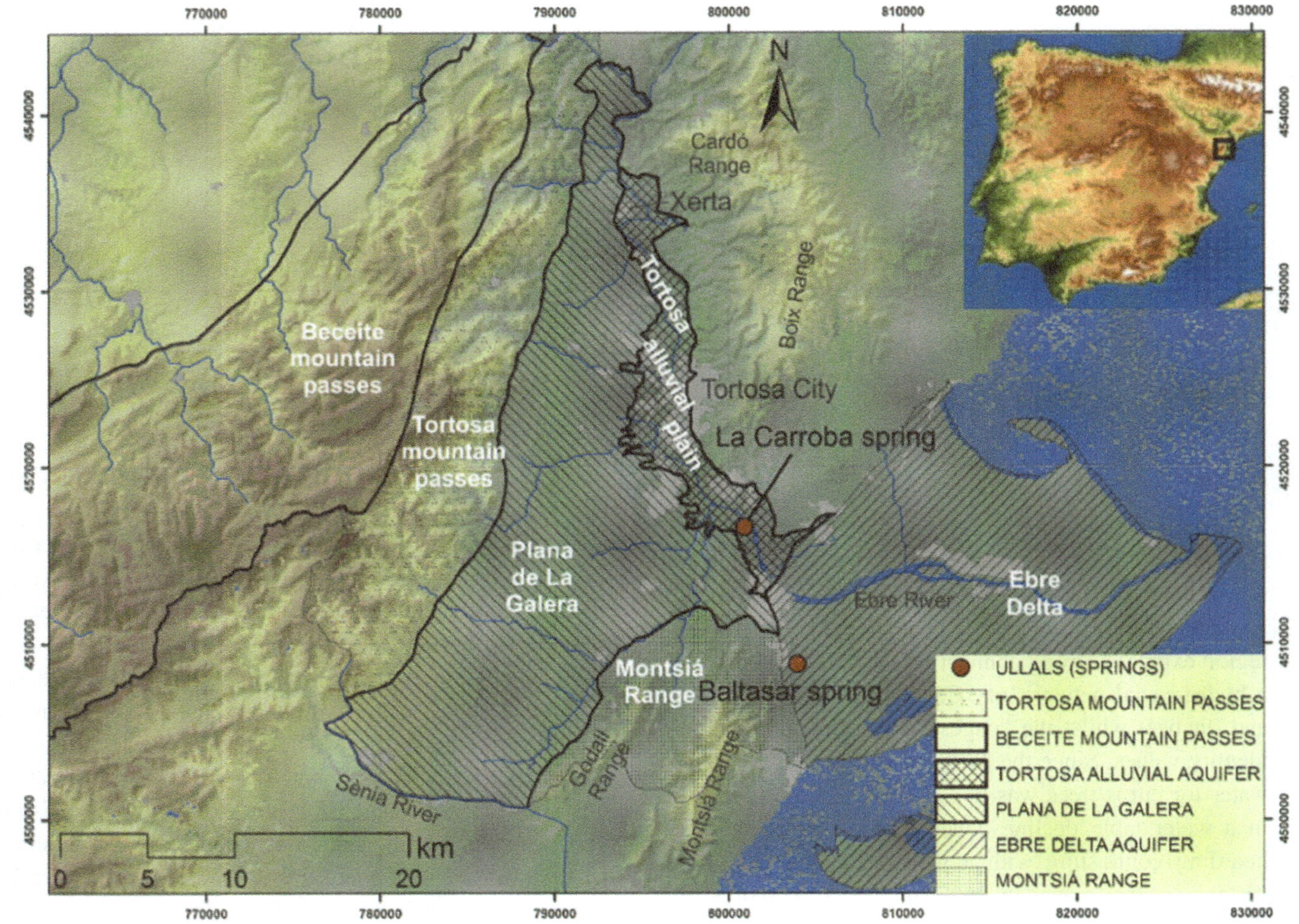


**Fig. 1** Location of La Plana de La Galera and Tortosa alluvial aquifers (Samper et al. 2014)

**Table 1** Crop's water demand

| $m^3$/ha/a | Period 1983–93 | Period 2009–19 | Increase % | Current demand |
|---|---|---|---|---|
| Citrus tree farming (400 units/ha) | 5006 | 5381 | 7.48 | 4600 |
| Olive tree farming (100 units/ha) | 972 | 1228 | 26.30 | 1000 |
| Olive trees—intensive farming (200/ha units) | | | | 3500 |

Water demand of each crop is based on the climatic characteristics and the irrigation coefficients (Table 1), and rainfall variations have a clear impact increasing water demand. At the same time, farmers practise water-saving strategies.

Groundwater costs range from € 0.07/$m^3$ in boreholes with a water table less than 30 m depth, to € 0.15–0.25/$m^3$ in boreholes with water tables between 100 and 200 m. This contrasts with surface water costs from the Ebro right side irrigation channel where an initial contribution is required for the amortization of works of € 5.480/ha that must be paid for 5 years. Subsequently, the annual expenses are set at € 219/ha/year for the use and maintenance of the irrigation infrastructure.

Current shifts in crop preferences between the farmers and the low cost of groundwater pose a challenge in aquifer management. The paper explores what values underline the current groundwater management in order to overcome social and environmental challenges faced in the management of La Galera aquifer.

## 2 Materials and Method

Addressing values is key in solving social and environmental-related challenges Glenna (2010). Conflicting values in the relation of the human with the geosphere (or their constitutive lithosphere, hydrosphere and atmosphere cycles) (Williams and Ferrigno 2012) may be explored using geoethical dilemmas (Marone and Peppoloni 2017). Geoethical dilemmas are defined as the ethical attitude when explicit values in normative rules or social norms "confront" the individual or community values, affecting the human–geosphere relationship. The existing values held by different stakeholders interacting in the groundwater management of La Galera aquifer, expressed in terms of agricultural itineraries, may be framed as a geoethical dilemma. This dilemma is developed through a case study approach. A case study brings together quantitative and qualitative data from different sources, such as interviews, field observations and literature review. Data collection stops once information becomes largely redundant (Dougherty 2019).

## 3 Results

The geoethical dilemma is presented as a double-entry matrix representing the alternative values of two of the main stakeholders involved in the dilemma: the farmers located on the Tortosa plain near to the Ebro right side irrigation channel as a main source of water but also using groundwater through boreholes and the CHE. Farmers located far from the channel on La Galera plain and only with access to groundwater are observers of the dilemma.

Table 2 is showing different scenarios when each alternative is crossed with the other stakeholder alternative. These scenarios are expressed in terms of impact (I) on the geosphere (hereby La Galera aquifer), and, oppositely, there is a derived socio-economic vulnerability on human activities (V). Whilst scenario 1 represents the current trend in the La Galera aquifer management, scenario 4 would present a more socially and environmentally fair option.

## 4 Discussion

The geoethical dilemma presents different scenarios as a result of divergent values the stakeholders may adopt regarding the groundwater management. Some scenarios show a higher impact on the aquifer in terms of water table drawdown affecting cultures and farming systems' vulnerability.

To interpret the dilemma, it is necessary to explore how the local farmers have traditionally related to the different sources of water, either through the main man-made infrastructure on the Ebro right side irrigation channel or from the groundwater from La Galera aquifer in terms of inherited values (Groenfeldt 2016). Whilst farmers located in the Tortosa plain mainly rely on the irrigation channel, although they may also have access to groundwater, they participate in community maintenance works and are involved in water users associations that date from 1864, thus creating a solid hydro-social network (Wesselink 2016). However, when it refers to groundwater, such organizational structures are non-existent, cooperative values are less developed and competition for groundwater prevails. Under this situation, it is a matter of fact farmers choose not to have a cooperative culture rather than competing amongst themselves driven by self-interests.

In the light of the interpretation of the geoethical dilemma, it would seem reasonable to promote the development of groundwater associations in line with the efforts promoted by the CHE. In that way, it is possible to endure a traditional farming system that preserves the local cultural

**Table 2** Geoethical dilemma—La Galera aquifer

| Observers: Farmers on La Galera plain | Alternative 1<br>Farmers on Tortosa alluvial plain change to more profitable but higher water-consuming crops | Alternative 2<br>Farmers on Tortosa alluvial plain introduce new crops and diversify production to counter low citrus prices |
|---|---|---|
| Alternative 1<br>CHE exerts strict control over the farmers' borehole pumped volumes | Scenario 1<br>I: water table drawdown<br>V: increase groundwater demand opens confrontation with local authorities | Scenario 2<br>I: stabilization of water table<br>V: new farming initiatives are abandoned because of lack of financial support |
| Alternative 2<br>CHE promotes agricultural plan to strategically switch to less water-consuming farming | Scenario 3<br>I: water table drawdown<br>V: political cost, lost confidence of farming sector and lack of private investment | Scenario 4<br>I: stabilization of water table<br>V: preservation of farming culture identity, investment backed-up by subsidies |

identity: The farmers become holders of such values, and their durable management of the aquifer is broadly recognized by the regional water authorities and the citizenship, which allows moving forward to scenario 4.

## 5 Concluding Remarks

The geoethical dilemma of La Galera aquifer presented in this work allows a better understanding of the values at stake in the management and governance of the groundwater in relation to the different stakeholders. The method formalizes the analysis of geoethical dilemmas to explore future scenarios with less impact and vulnerability in the relationship between humans and the geosphere. At the same time, understanding of geoethical dilemmas follows a pedagogic approach in helping to set up a geoethical dialogue as a politically contested process in groundwater management.

## References

CHE (1991) Estudio de los recursoshídricos subterráneos de los acuíferos de la margen derecha del Ebro. Zona III-Acuíferos de la zona baja. Informe General MOPT, Oficina de Planificación Hidrológica de la Confederación Hidrográfica del Ebro

Comnitat General Regants Xerta-Senia (CGRXS) (2004) Estudi tècnic i jurídic d'alternatives de la zona regable del canal Xerta-Sénia. Regs de Catalunya S.A., Generalitat de Catalunya, Departament d'Agricultura, Ramaderia i Pesca, Direcció General de Desenvolupament Rural

Dougherty ML (2019) Boom times for technocrats? How environmental consulting companies shape mining governance. Extr Indus Soc 6:443–453

Espinosa S (2014) Estimación de la recarga media anual de acuíferos: aplicación en el Baix Ebre. Universidad Politécnica de Cataluña, Barcelona (Phd thesis)

Glenna L (2010) Value-laden technocratic management and environmental conflicts: the case of the New York City watershed controversy. Sci Technol Hum Values 35(1):81–112

Groenfeldt D (2016) Toward a new water ethic. Leopold Outlook 16:12–16

Marone E, Peppoloni S (2017) Ethical Dilemmas in Geosciences. We can ask, but, can we answer? Ann Geophys 60, FAST TRACK 7, doi: https://doi.org/10.4401/AG-7445

Samper J, Li Y, Pisani B (2014) An evaluation of climate change impacts on groundwater flow in the Plana de La Galera and Tortosa alluvial aquifers (Spain). Environ Earth Sci 73:2595–2608

Wesselink A, Kooy K, Warner J (2016) Socio-hydrology and hydrosocial analysis: toward dialogues across disciplines. WIREs Water 4:e1196. https://doi.org/10.1002/wat2.1196

Williams RS Jr, Ferrigno JG (eds) (2012) State of the Earth's cryosphere at the beginning of the 21st century–Glaciers, global snow cover, floating ice, and permafrost and periglacial environments. US Geological Survey Professional Paper 1386–A

# During and Post-COVID-19: Challenges in Water Sector and Ethical Issues

Gopal Krishan and Umesh Kulshrestha

**Abstract**

During and post-COVID-19, the main challenges in water sector are (i) access to safe water, and (ii) sanitation. A huge amount of clean water is being consumed for hand washing as a protective measure against the spread of the COVID-19 which goes as wastes. Increased consumption of drinkable water at the times of depleting groundwater levels will aggravate the adversity in this sector. This situation will further deteriorate in countries having water stress and water scarcity conditions. Presently, efforts should be made on monitoring of the water usage, augmentation of depleting water levels, conservation of water resources, finding the sustainable ways for treating the wastewater and reuse of this treated water. If needed, new regulations and control mechanisms must be practiced. Finally, a hydrogeoethical approach shall be persecuted by all in the sustainable management of water resources.

**Keywords**

COVID-19 • Water sector • Challenges • Rural and urban areas • Hydrogeoethical approach

## 1 Introduction

COVID-19 (coronavirus disease) pandemic has a profound effect globally, where a total of 4,429,235 positive cases and 298,165 deaths have been reported till date, i.e., May 14, 2020 (Worldometer, 14-05-2020). This is one of the severest adversities which have impacted the global economy and has a discernible impact on water resources. On earth surface, the usable water resource is only 1% out of total 3% freshwater resources (Pacific water-13-05-2020). Global freshwater use has increased from 500 to 3000 BCM in last 100 years (IGB) out of this 70% is used for irrigation and 30% is used for domestic and industrial use. This clearly shows that the majority of water is used for producing food. Demand of water from surface water resources and groundwater resources for rising population has been increasing consistently (grew by 4 times during 1900 to 2000 century) which resulted in rapid increase in groundwater withdrawals during last 3 decades (Wada et al. 2014; MacDonald et al. 2016). Besides water availability, pollution of water resources surface as well as groundwater in rural and urban areas is becoming a serious concern. Increasing geogenic and anthropogenic contamination is a major threat to the people depending on these resources.

The COVID-19 scenario due to lockdowns has rejuvenated the nature and this can be used as a boon for preventing our water resources for future (Gopal 2020). In the light of above, this work briefly discusses the impacts of COVID-19 on water resources and hydrogeoethical approaches for safeguarding the precious but vulnerable resource.

## 2 Impact of COVID-19 on Water Resources

Human greed over need has always been a problem for natural resource conservation. Mankind started overexploitation of water in agriculture, domestic and industrial sectors with heavy use of borewells. In rural areas, the water stress conditioned was already experienced due to overexploitation in agriculture irrigation due to unethical practice of electricity consumption and unlimited water usage because of it people migrated to the urban areas to find out the jobs (Sheikh and Kulshrestha 2016). The COVID-19 situation is forcing all the migrated laborers to go back to their villages which will force them to try agriculture again with the

G. Krishan (✉)
Groundwater Hydrology Division, National Institute of Hydrology, Roorkee, Uttarakhand, India
e-mail: drgopal.krishan@gmail.com

U. Kulshrestha
School of Environmental Sciences, Jawahar Lal University, New Delhi, India


M. Abrunhosa et al. (eds.), *Advances in Geoethics and Groundwater Management: Theory and Practice for a Sustainable Development*, Advances in Science, Technology & Innovation,
https://doi.org/10.1007/978-3-030-59320-9_65

**Table 1** Water usage for hand washing as protection tool against COVID-19

| Period | Water used in one hand washing activity (L) | Water wasted if tap running (L) | Wastewater generated (L) |
|---|---|---|---|
| Daily | 20 | 30 | 45 |
| Monthly | 620 | 930 | 1395 |

available reduced quantity of groundwater. It may work for some time, but ultimately, people have to leave agriculture and find out some alternate occupation for their survival. We guess that such water stressed conditions once again force the people of villages to migrate to the nearby cities. The conditions might trigger a new development in socio-economics of the region.

The water crisis in urban areas is foreseen different than in rural areas. Urban areas will be facing a water stress due to more water usage for vegetable washing and cleaning of various items brought from the outdoor in order to decontaminate from the COVID-19 infection. Frequent bathing due to virus contamination will also increase water consumption. The precaution and prevention of infection from the toilet sheets will also be increasing the consumption of water in lavatories at various public and commercial places. Drinking water used for car washing and gardening are important factors too which we need to monitored strictly as per the regulations.

Similarly, unethical practice of water overconsumption in the industries and the water supply to the illegal colonies, generally which are not parts of distribution policy, affect the water availability. In addition to this, water is being used as a protection tool for hand washing where it is believed that washing hands with soaps for 20 s is effective against COVID-19. The water usage in this act of hand washing is calculated per person as (details in Table 1).

Assumptions are that in 1 time event of hand washing 2 L of water is required and if a person wash hands 10 times 20 L will be required and assumed that 4.5 L when tap is running during applying soap, i.e., around 45 L of wastewater will be generated during 10 hand washes. Wastewater generation is based on 90 percent of water used, including the volume wasted when taps are on (Downtoearth 2020). All this water will come under category of wastewater and additional water will be required to treat this water. All these will put a lot of pressure on the supply side to meet out the water demand.

## 3 Suggested Measures: Toward a Hydrogeoethical Approach

### 3.1 A Holistic Approach is Needed

The water crisis can be solved by opting a holistic approach. Both quality, quantity and ethical management of water have been the challenges for the civil bodies. Some of the measures for improvement in quality and augmentation of the resources are discussed below.

### 3.2 Augmentation of Water Resources

Source of water in surface water resources is precipitation while groundwater resources are recharged through infiltration/percolation of water from surface water resources and precipitation. Practice of water conservation techniques, rainwater harvesting and construction of recharge structures can be very much useful in augmentation of water resources. In addition, forestation or intensive plantation is also useful in improving soil conditions hence more infiltration/percolation and more recharge.

### 3.3 Water Quality Testing

The quality of water can be increased by capturing the runoff water during monsoon by creating artificial huge lakes in the urban areas. These lakes may serve multiple purpose such as drinking water supply, groundwater recharge, moisture in air, biodiversity increase, greenery increase, less erosion, tourism, and occasional rain. The quality of water needs less polluted river water which can be processed with easy steps.

### 3.4 Control of Industrial Effluent During Post-COVID-19 Lockdown

The COVID-19 scenario has rejuvenated the nature. Water streams are seen cleaner. Considering these new normal values of water streams, the operation of industrial units can be monitored in order to make sure that the industries are not pouring unprocessed effluent to the rivers. This will be helpful in water purification with easy steps and lesser chemicals and at faster rate.

## 4 Concluding Remarks

Monitoring and regulating of water resource is required at this time of water crisis. The rainwater harvesting which is mandatory in each house (in India) must be implemented effectively in the urban and rural areas. The untreated effluents discharge in the rivers streams from various industries must be regulated through stringent measures, including the correct protection of the water sources, springs, and ecosystems.

In fact, there is an urgent need of changing our lifestyle under a climate change and eco-responsibility framework, as well as a hydrogeoethical approach is essential to a sustainable water management, particularly, understood by academia, practitioners, authorities, and society.

## References


Gopal K (2020) Scientific imperatives for COVID-19: water sector. Current World Environ 15(1):5–7

https://www.downtoearth.org.in/blog/water/covid-19-outbreak-more-hand-washing-can-increase-india-s-water-woes-69900. Accessed in 13 May 2020

https://www.worldometers.info/coronavirus/. Accessed in 13 May 2020

MacDonald AM, Bonsor HC, Ahmed KM, Burgess WG, Basharat M, Calow RC, Dixit A, Foster SSD, Gopal K, Lapworth DJ, Lark RM, Moench M, Mukherjee A, Rao MS, Shamsudduha M, Smith L, Taylor RG, Tucker J, van Steenbergen F, Yadav K (2016) Groundwater depletion and quality in the Indo-Gangetic Basin mapped from in situ observations. Nat Geosc 9:762–766

Pacific Water: SPC water, sanitation and hygiene: water distribution. Accessed in 13 May 2020

Seikh AA, Kulshrestha U (2016) Climate change and demographic shift. In: Kulshrestha U (ed) Air pollution and climate change in South Asia, Athena Publishers, UK

Wada Y, Wisser D, Bierkens MFPT (2014) Global modeling of withdrawal, allocation and consumptive use of surface water and groundwater resources. Earth Syst Dynam 5:15–40

# Engaging Communities and Extending the Lives of Water Systems with Technology

Sarah Evans, Mary Hingst, and Kathryn Bergmann

**Abstract**

It takes more than just good intentions and money to provide a community with a reliable source of clean water. Most water systems installed in Africa will fail within the first few years because of insufficient technical planning, lack of community investment and improper maintenance. We have developed a customizable and easy-to-use app to help train and assist community members to perform regular maintenance and diagnosis issues with their water systems. Additionally, the app is equipped with tools for tracking and reporting usage measurement and system evaluation. This information will provide our technical team with data about the frequency and commonality of issues which can teach us how to mitigate such issues in the future.

**Keywords**

Water systems • Technology • Community engagement

## 1 Introduction

Well Beyond is a woman-owned small business that grew from a passion for addressing water issues in the developing world. We are tackling global water scarcity by sharing our successful model with other organizations and launching a patent-pending app to provide water system technical support anywhere in the world. Through our decade of experience, we have learned we can minimize problems when community operators fix systems on site. When a problem occurs, the Well Beyond App navigates the user—on or offline—to identify the issue and repair the system. If the fix cannot be solved through the app, a fix-it ticket is generated and Well Beyond provides real-time support. The app will first be available in English and the official language of Kenya, Swahili additional supporting images and videos will also help to navigate user through diagnostics, and future languages will be integrated into the app after beta. Well Beyond's technical expertise, along with the app, places even more ownership of the water system into the hands of the communities. While also saving time, money and extending the overall life of the water system.

## 2 Settings

As of 2017, almost 30% of the global population still did not have access to safely managed water supplies, and in sub-Saharan Africa, this number increases to 73% (UNICEF 2019). Despite numerous organizations trying to reduce this number, 30–60% of rural water systems have failed in African countries (Harvey and Reed 2006). The International Institute for Environment and Development reported that approximately 50,000 water points in rural communities are no longer functional, which equates to a loss of about $360 million USD (Skinner 2009), while the World Bank estimates that failure of hand pumps alone in sub-Saharan Africa represents a loss of over $1 billion USD (World Bank 2011). Throughout our work over the past ten years, we at Well Beyond have found that many of these projects fail because of poor technical planning, lack of community involvement, and improperly maintained and malfunctioning equipment. Our findings agree with those from Hoko and Hertle (2006), Foster (2013), and Fisher et al. (2015) that active community involvement and proper training can significantly extend the life and sustainability of a water project. Of the top five water-centered charity organizations around the world, only two of them emphasize the need for dedicated technical staff, and less than five percent of water projects are revisited after implementation by NGOs (World Bank 2011).

S. Evans · M. Hingst · K. Bergmann (✉)
Well Beyond, Austin, TX, USA
e-mail: kathryn@wellbeyondwater.com


M. Abrunhosa et al. (eds.), *Advances in Geoethics and Groundwater Management: Theory and Practice for a Sustainable Development*, Advances in Science, Technology & Innovation,
https://doi.org/10.1007/978-3-030-59320-9_66

## 3 Results

Without measurement and evaluation there is no accountability. Without records, it is difficult to understand how and why a water system failed. Rosenstock et al. (2017) discussed the need for reliable monitoring and reporting in order to track progress of sustainable development goals. Despite this call for evaluation, less than one percent of projects receive any sort of long-term monitoring (World Bank 2011). The Well Beyond App includes reporting tools to track and record water usage and conduct community surveys. The data collected on usage will allow us to track and better understand both the usage and impact of the system. Data from the issues reported will be aggregated and analyzed to common points of failure in systems. A better understanding of the frequency and commonality of failures is necessary to determine underlying causes and how to address such issues prior to implementation of the water systems.

## 4 Discussion

As with any system that has moving parts, regular maintenance is required to prolong the life of water systems, especially pumped groundwater systems. Once water systems have been installed in rural communities, many implementing organizations do not sufficiently train the communities how to properly maintain the systems nor do they provide long-term support to repair any issues that may arise. When water systems are not working, community members may be forced to get water from distant and unclean sources, which increases the risk of disease and takes time away from other activities such as schooling for children and income opportunities.

To reduce ongoing technical problems, we have developed a beta version of an app that helps train community members to perform regular maintenance of water systems, fix common, simple problems and track usage to measure and evaluate impact of the water project. Despite the lack of access to clean water, the Pew Research Center reported that a median of 45% of adults in developing countries has a smartphone (Pew 2019), making an app an easily accessible way to disseminate information to people around the world. Unlike the very few apps that exist which are for specific, patented systems, our app supports hand pump, submersible

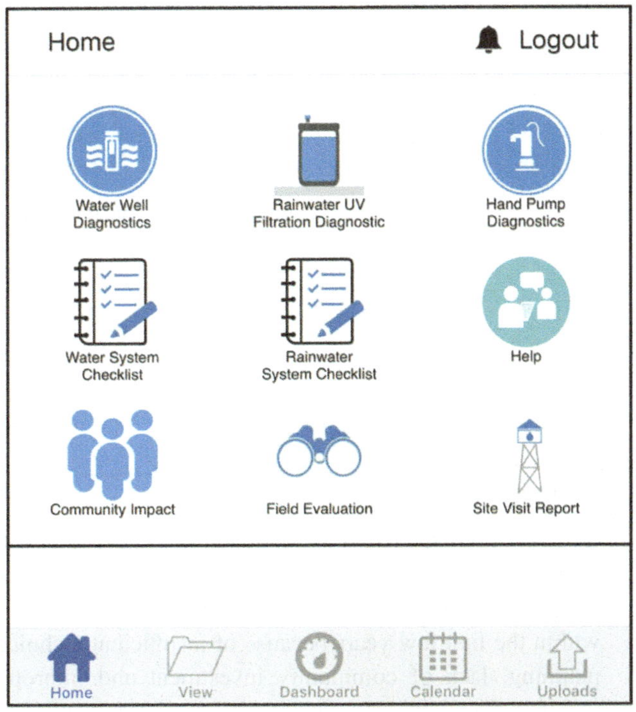


**Fig. 1** Home screen of the Well Beyond App displaying the customizable tools

pumps and rainwater harvesting systems and will be customized to each community's system setup. The app provides reminders for performing regular maintenance and a decision tree of yes/no questions to diagnose common issues (Fig. 1). When a problem arises, the Well Beyond App navigates the user, on or offline, uses step-by-step instructions and visual aids to help the user identify the issue and repair the system. If the problem cannot be solved through the app, a fix-it ticket is generated and Well Beyond provides real-time support (Fig. 2).

## 5 Concluding Remarks

By enabling communities to diagnose and service water system issues, this app will not only reduce the downtime of water systems, but will also reduce communities' dependence on external organizations, increasing ownership of the system. The data collected will be used to better inform organizations on how to mitigate common issues and increase the impact of their water projects.

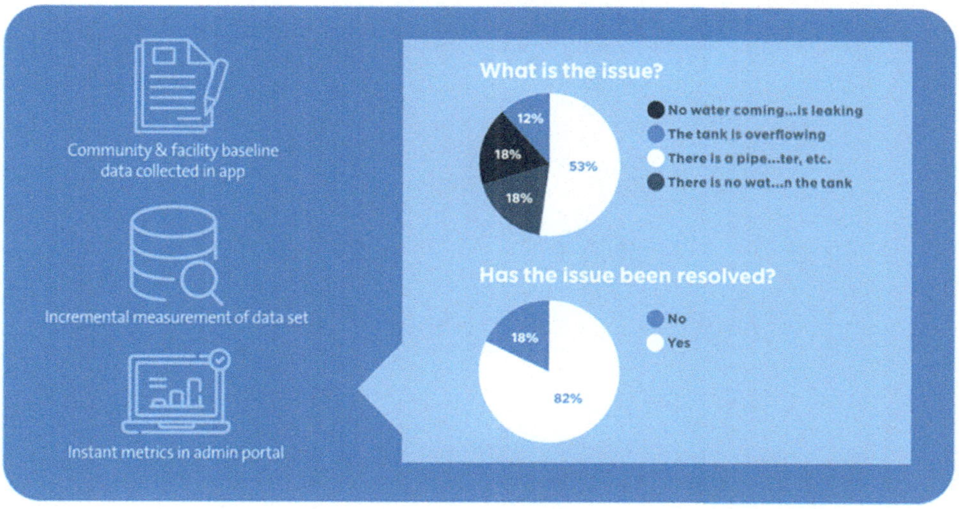


**Fig. 2** Example of the type of data to be collected and aggregated

## References


Fisher MB, Shields KF, Chan TU, Christenson E, Cronk RD, Leker H, Samani D, Apoya P, Lutz A, Bartram J (2015) Understanding handpump sustainability: determinants of rural water source functionality in the Greater Afram Plains region of Ghana. Wat Resour Res 51(10):8431–8449

Foster T (2013) Predictors of sustainability for community-managed handpumps in sub-Saharan Africa: evidence from Liberia, Sierra Leone, and Uganda. Env Sci Tech 47(21):12037–12046

Harvey PA, Reed RA (2006) Community-managed water supplies in Africa: sustainable or dispensable? Comm Develop J 42(3):365–378

Hoko Z, Hertle J (2006) An evaluation of the sustainability of a rural water rehabilitation project in Zimbabwe. Phys Chem Eart Parts a/B/C 31(15–16):699–706

Pew (2019) Smartphone ownership is growing rapidly around the world, but not always equally. Pew Research Center, Washington DC

Rosenstock TS, Lamanna C, Chesterman S, Hammond J, Kadiyala S, Luedeling E, Shepherd K, DeRenzi B, van Wijk MT (2017) When less is more: innovations for tracking progress toward global targets. Current Opin Environ Sustain 26:54–61

Skinner J (2009) Where every drop counts: tackling rural Africa's water crisis. International Institute for Environment and Development (IIED), London

UNICEF (2019) Progress on household drinking water, sanitation and hygiene 2000–2017. Special focus on inequalities. United Nations Children's Fund (UNICEF) and World Health Organization, New York

World Bank (2011) WASH sustainability forum summary report. World Bank, Washington D.C.

# A Regional Initiative for the Efficient Transfer of Groundwater Knowledge Between Experts and Stakeholders

Julien Walter, Alain Rouleau, Melanie Lambert, Romain Chesnaux, Anouck Ferroud, and Laura-Pier Perron-Desmeules

**Abstract**

Recent hydrogeological mapping throughout the Saguenay-Lac-Saint-Jean region in Quebec, Canada, has yielded a very comprehensive geodatabase (GDB) on groundwater and aquifer systems, designed to fulfil the requirements of groundwater management at the regional level. The high density of observed hydrogeological data points also makes the GDB useful for many local or sub-regional applications. Recently acquired data from continuing geotechnical or groundwater-related activities must be integrated into the GDB in order to maintain its relevance for a variety of problems concerning groundwater quality, its management and its protection. When it is consistently updated, such a hydrogeological (HG) GDB also stands as a critical and constantly evolving link between the scientific community, government agencies, regional and municipal authorities, and ultimately the population who draw upon the groundwater resource. The HG-GDB represents a central hub of effective collaboration between scientists and data users, and the intent is that it may lead to more appropriate decision-making when faced with different issues and/or when taking actions that may include applied groundwater research projects. Groundwater data transfer becomes an information exchange process, an essential tool facilitating the sustainable management of groundwater resources.

**Keywords**

Regional hydrogeology • Knowledge transfer • Land-use management

## 1 Introduction

The transfer of scientific knowledge to large user communities has long been recognized as a necessity in the health sciences (e.g. Lemire et al. 2009), but it is also applicable to all disciplines of science. The transfer of knowledge to stakeholders is particularly important for groundwater, as this underground component of the water cycle, although ubiquitous, remains invisible. Hydrogeologists must decipher for the benefit of stakeholders a number of phenomena affecting groundwater, many of which are based on a geological description of the subsurface; such knowledge is essentially based on interpretation.

The efficient use of groundwater knowledge is hampered by numerous obstacles in many states and provinces in North America (Lavoie et al., 2013, 2014). Pelchat (2015) has identified important barriers to the implementation of groundwater protection measures, including poor groundwater knowledge by land managers and the lack of resources to implement the required measures.

In Quebec, great advancements have been made since 2008 in groundwater and aquifer characterization, thanks to regional and provincial initiatives. Recent hydrogeological knowledge acquisition projects have produced extensive hydrogeological geodatabases (HG-GDB) that are made available to public and private users. This short paper presents an overview of an ongoing project for the transfer of groundwater knowledge contained in a regional HG-GDB to potential users in the Saguenay-Lac-Saint-Jean (SLSJ) region (275,000 inhabitants), particularly to land-use planners.

J. Walter · A. Rouleau (✉) · M. Lambert · R. Chesnaux
A. Ferroud · L.-P. Perron-Desmeules
Department of Applied Sciences, Université du Québec à Chicoutimi, Saguenay, Canada
e-mail: arouleau@uqac.ca



M. Abrunhosa et al. (eds.), *Advances in Geoethics and Groundwater Management: Theory and Practice for a Sustainable Development*, Advances in Science, Technology & Innovation,
https://doi.org/10.1007/978-3-030-59320-9_67

## 2 Regional Context Regarding Hydrogeological Knowledge and Land-Use Planning

Over the last three decades, researchers in hydrogeology at the *Université du Québec à Chicoutimi* (UQAC) have carried out numerous research projects in the SLSJ region, mostly in collaboration with municipalities and industries. These projects made use of available data on the subsurface that had been acquired by public or private organizations, particularly for geotechnical, environmental or hydrogeological purposes. Although extremely useful, these data were owned and hosted by a diversity of data owners and presented a variety of formats and quality levels. Gathering and integrating these existing data required much time and effort.

In 2008, UQAC entered into an agreement with several administrative entities in the SLSJ region, including the City of Saguenay, the Mashteuiatsh First Nations Community, the four regional county municipalities (*Municipalités régionales de comté*; MRC), and the regional offices of four government ministries: municipal affairs, natural resources, agriculture, and health and social services. The objective was to integrate all available data related to groundwater into a single, comprehensive HG-GDB.

This agreement was followed by a major program launched in 2009 by the Quebec Government: the Groundwater Knowledge Acquisition Program (*Programme d'acquisition de connaissances sur les eaux souterraines*; PACES; https://www.environnement.gouv.qc.ca/eau/souterraines/programmes/acquisition-connaissance.htm). Managed by the Ministry of Environment, this province-wide program was given the objective of mapping the hydrogeological resources of the "municipalized" southern and more densely populated portion of the province of Quebec. The PACES program is subdivided into several regional projects, each managed by an expert team including hydrogeologists and GIS experts operating out of one of several universities in Quebec. These PACES projects involve three main steps: the integration of available data on the subsurface into a geodatabase, the acquisition of new groundwater data, and a general description of regional hydrogeological systems.

One of these regional PACES projects was conducted from 2009 to 2013 across the municipalized territory (13.210 $km^2$) of the SLSJ region (Fig. 1). It required the partnership of all the municipal entities in the region as well as the Mashteuiatsh First Nations Community. The overall results of the PACES-SLSJ project are based on more than 12,000 data points covering a variety of information types, including: stratigraphic logs from boreholes, descriptions of rock outcrops, groundwater geochemistry analyses, hydraulic properties of aquifers, grain size distributions of superficial deposits, piezometry data, and surface and borehole geophysical data (Fig. 1; CERM-PACES 2013; Walter et al. 2018). These data were used to develop the PACES project deliverables, including approximately 30 maps on a 1/225 000 scale and a regional geodatabase (Chesnaux et al., 2011) integrating all of the collected data. The results were delivered to the Quebec Ministry of Environment and to all regional partners in the PACES-SLSJ project. The scientific value of these deliverables has been leveraged and enhanced by numerous research projects conducted by professors and graduate students at UQAC.

### 2.1 A Pilot Project on Hydrogeological Knowledge Transfer to Stakeholders

In 2015, a series of one-day workshops were organized by the Quebec Groundwater Network (*Réseau québécois sur les eaux souterraines*; RQES) in all regions covered by the PACES projects completed at that date. The purpose of these workshops was to teach land planners how to access and use the regional geodatabases developed by PACES project scientists (Ruiz et al. 2015, 2016). Although these workshops were well organized and useful, they had several important limitations: (1) they were limited over time to one day; (2) the level of groundwater knowledge remains limited for most workshop participants, who probably face multiple issues other than groundwater in their day-to-day functions; (3) the number of workshop spaces is limited, so an entire organization is usually represented by a single participant; (4) a participant may not be anymore the actual person in charge when his/her organization is later on facing problems related to groundwater.

Meanwhile, an informal post-PACES work committee was created in SLSJ, with the aim of developing a more efficient use of the new, comprehensive HG-GDB; the committee brought together researchers in hydrogeology from UQAC as well as representatives of the MRCs and the Mashteuiatsh First Nations Community. The meetings of this committee gave rise in 2017 to a new pilot project called RISUp, to respond to the needs of municipalities in relation to groundwater, through research (R), implementation of the HG-GDB in every MRC (I), support (S), and updating (Up) of PACES-SLSJ hydrogeological data. The RISUp project has been self-funded by the partners over the last three years; it pays the salaries of two part-time research associates at UQAC. Specifically, the main objectives of this pilot project constitute a circular data exchange process (Fig. 2) involving: (1) the acquisition of hydrogeological knowledge and data by experts and researchers, (2) updating the HG-GDB and adapting it to the needs of every MRC, (3) the transfer of this knowledge to regional stakeholders, (4) providing ad hoc scientific and technical support to users

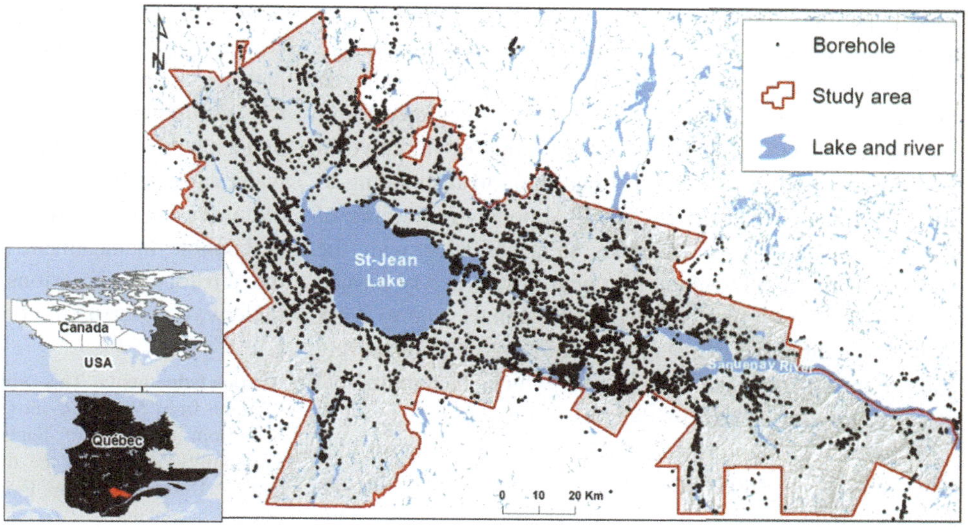


**Fig. 1** Municipal territory of the Saguenay-Lac-Saint-Jean region in Quebec, showing locations of the boreholes providing data to the hydrogeological geodatabase

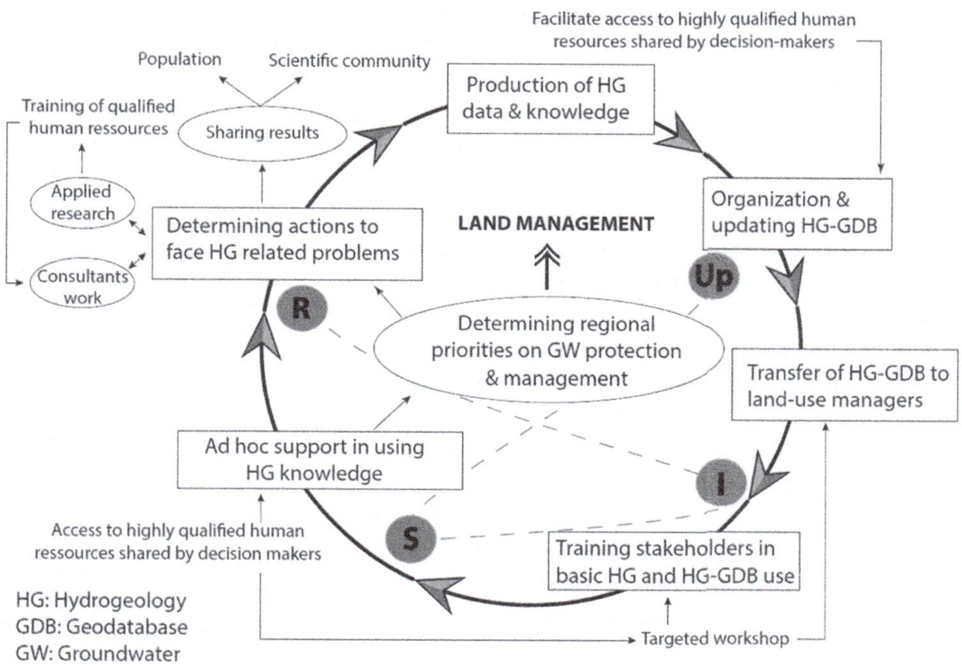


**Fig. 2** Circular transfer of groundwater knowledge resulting from the RISUp project

on groundwater-related issues; (5) helping to define actions to be taken to address identified groundwater-related problems. These actions may involve hydrogeological researchers or consultants, or both; they often result in new data being gathered and entered into the regional HG-GDB in order to keep the circle turning.

## 3 Discussion

The RISUp project strengthens and leverages the university's role as a center of knowledge and expertise in hydrogeology and creates a climate of trust that favours the development of applied research projects. The RISUp project also makes it possible to address critical regional issues such as: groundwater contamination by septic systems, 3D modeling of regional aquifers and chemical variability of groundwater, development of hydrodynamic conceptual models of specific aquifers. These customized projects generate new and relevant data that are quickly integrated into the regional HG-GDB and are made readily accessible to users. These recorded data can then be used for a much wider variety of purposes than what had been initially planned during their acquisition; as a geoethics result, data produced is more likely to be used than wasted.

The projects with a research component are mostly carried out by students under the supervision of professors. The issues to be addressed by these projects are first identified by

RISUp partners, then translated into research problems by researchers or professors. The resulting research projects are then assigned to students, or in rare cases, directly managed by UQAC researchers and professors. The integrated training aspect of RISUp makes it a very valuable initiative in geoethics for the entire region. In addition to responding to specific regional issues, RISUp will support the training of the next generation of scientists in accordance with the realities of the municipal environment. These newly formed experts become highly qualified professionals able to work in the community, for instance, within private engineering and environmental consulting, for which the municipal sector represents a large market share in Quebec.

Another important aspect of RISUp is that it can provide local support to those in charge of map production and database management in municipal organizations. This support takes the form of individual visits by RISUp professionals, directly to the offices of partner organizations. During these visits, various exercises (mapping, SQL queries, etc.) are carried out using the data contained in the regional HG-GDB. As a contribution in geoethics, these activities ensure an efficient transfer of knowledge using specific tools developed for RISUp partners.

However, in some cases, the university can be seen as a direct competitor to hydrogeological consulting firms. Indeed, university researchers accompany regional decision-makers in groundwater-related projects, as some consultants would do. This competition, at least in appearance, raises the fundamental question of the role of the university in a region that comprises private, governmental, and municipal entities.

Meetings of the RISUp partners are scheduled at least twice a year. Their purpose is (1) to communicate key results to partners, (2) to present future projects, and (3) to define priorities for action. Representatives from the two watershed organizations active in the SLSJ region are invited to participate in the RISUp management meetings.

## 4 Concluding Remarks

Effective collaboration between scientists and data users gives rise to the development of applied groundwater research projects that are based on actual problems. The resulting groundwater information exchange process facilitates the sustainable management of groundwater as a resource and a component of ecosystems.

A significant change has occurred over the last ten years among regional decision-makers in the SLSJ region. Although relatively new to the specialized field of groundwater, regional stakeholders now understand the general hydraulic and chemical aspects of groundwater dynamics. Some partners now see a potential for economic development (water supply in remote areas, water bottling, etc.), and all the partners are now well-informed concerning the fragility of the groundwater resource. All stakeholders now fully understand the importance of adequately protecting groundwater and also have a better understanding of the importance of continuing research and developing knowledge for a more sustainable management of the groundwater resource over the territory. Discussions among the RISUp partners are underway to pursue the project for an additional three years.

**Acknowledgements** This project was made possible by in-kind contribution from *Université du Québec à Chicoutimi*, and financial support from the City of Saguenay, the Mashteuiatsh First Nations Community, as well as the following regional county municipalities (MRC): Domaine-du-Roy, Fjord-du-Saguenay, Lac-Saint-Jean-Est and Maria-Chapdelaine.

## References

CERM-PACES (2013) Résultats du programme d'acquisition de connaissances sur les eaux souterraines de la région Saguenay-Lac-Saint-Jean. Centre d'études sur les ressources minérales, UQAC, 308 p. https://www.environnement.gouv.qc.ca/_PACES/rapports-projets/SaguenayLacStJean/SLSJ-scientif-UQAC-2013.pdf. Accessed 20 Nov 2019

Chesnaux R, Lambert M, Fillastre U, Walter J, Hay M, Rouleau A, Daigneault R, Germaneau D, Moisan A (2011) Building a geodatabase for mapping hydrogeological features and 3D modeling of groundwater systems: application to the Saguenay-Lac-St-Jean region, Canada. Comp Geosc 37:1870–1882

Lavoie R, Lebel A, Joerin F, Rodriguez MJ (2013) Integration of groundwater information into decision making for regional planning: a portrait for North America. J Environ Manag 114:496–504

Lavoie R, Joerin F, Rodriguez MJ (2014) Incorporating groundwater issues into regional planning in the Province of Québec. J Environ Plann Manag 57(4):516–537

Lemire N, Souffez K, Laurendeau M-C (2009) Animer un processus de transfert des connaissances. Bilan des connaissances et outils d'animation. Institut National de Santé publique, Gouvernement du Québec, 46 pp. https://www.inspq.qc.ca/publications/1012

Pelchat J (2015) La protection de la ressource en eau souterraine au Québec : portée des outils d'aménagement du territoire et de l'information hydrogéologique régionale. Université du Québec à Trois-Rivières, Mémoire de maîtrise, p 105

Ruiz J, Tremblay Y, Larocque M, Cloutier V, Ferlatte M (2015) Eaux souterraines et aménagement du territoire: Défis et stratégie pour un transfert des connaissances réussi. Vecteur Environnement, Septembre 2015:46–51

Ruiz J, Larocque M, Cloutier V, Tremblay Y, Decelles A-M, Ferlatte M (2016) Les ateliers de transfert des connaissances sur les eaux souterraines. Réseau québécois sur les eaux souterraines, 11 pp. https://rqes.ca/wp-content/uploads/2016/08/AteliersRQES2016.pdf. Accessed 20 Nov 2019

Walter J, Rouleau A, Chesnaux R, Lambert M, Daigneault R (2018) Characterization of general and singular features of major aquifer systems in the Saguenay-Lac-Saint-Jean region. Canad Wat Res J 43(2):75–91

# Groundwater Contamination and Extreme Weather Events: Perception-Based Clusters of Irish Well Users

Simon Mooney, Jean O'Dwyer, and Paul Hynds

**Abstract**

Extreme weather events (EWEs) may significantly increase pathogenic contamination of private (unregulated) groundwater supplies. Given the current paucity of guidance in this context, well owners may be ill-equipped to undertake responsive actions. With rising cases of waterborne illness documented in groundwater-dependent rural regions, a better understanding of well user risk perception is required to develop appropriate interventions. To this end, a survey was developed to identify current risk perceptions among Irish private well owners concerning EWEs and groundwater contamination. Respondents were surveyed on actions taken in the aftermath of five recent national weather events including drought and flood and clustered based on perceived consequences. Results suggest that well owners are not inclined to undertake systematic well maintenance in the wake of an event, with only 13.1% ($n$ = 37) of affected respondents subsequently testing their water. Two-step cluster analysis identified three distinct respondent groups based on perception of climate change impacts on well contamination, with perception level significantly related to respondent age ($p$ = 0.003), education ($p$ = 0.000) and gender ($p$ = 0.002). These findings may enable development of customized risk information for private well owners and reduce the risk of human illness associated with groundwater.

**Keywords**

Climate change • Extreme weather • Groundwater contamination • Private wells • Risk perception

## 1 Introduction

Groundwater is the most extracted resource in the world, with the security of existing reserves under increased pressure due to climate change (Green 2016). As aquifer recharge is regulated by precipitation and interaction with surface water bodies, the rising incidence of extreme weather events (EWEs) such as drought and pluvial flooding may have a deleterious effect on groundwater quality (Taylor et al. 2013). Increases in event frequency, combined with warmer temperatures favoring bacterial growth and transport, are projected to escalate rates of groundwater contamination and, by extension, waterborne illness (Forzieri et al. 2017).

The growing occurrence of climate extremes in high groundwater reliant, rural regions such as the Republic of Ireland (ROI) will necessitate regular, systematic source maintenance to mitigate against contamination and subsequent infection risk. Approximately 16% of Irish residents receive daily drinking water from unregulated private wells, with many supplies susceptible to microbial contamination originating from adjacent farmyards and domestic wastewater treatment systems (CSO 2017). Linked to a rising number of livestock and septic tanks (Naughton and Hynds 2014), private well contamination has become widespread and contributed to the growing national incidence of gastrointestinal illnesses such as Verotoxigenic *E.Coli* (VTEC) —currently >10 times the EU average (ECDC/EFSA 2018). With intensified rainfall and drought events becoming increasingly prevalent, mobilization of viral and bacterial

S. Mooney (✉) · P. Hynds
Environmental Sustainability and Health Institute, Technological University Dublin, Dublin, Ireland
e-mail: d17125411@mytudublin.ie

J. O'Dwyer · P. Hynds
Irish Centre for Research in Applied Geosciences (iCRAG), University College Dublin, Dublin, Ireland

J. O'Dwyer
School of Biological, Earth and Environmental Sciences, University College Cork, Cork, Ireland

J. O'Dwyer
Environmental Research Institute, University of Cork, Cork, Ireland


M. Abrunhosa et al. (eds.), *Advances in Geoethics and Groundwater Management: Theory and Practice for a Sustainable Development*, Advances in Science, Technology & Innovation,
https://doi.org/10.1007/978-3-030-59320-9_68

pathogens into private groundwater may be further expedited by additional surface water runoff and preferential flow paths (O'Dwyer et al. 2016).

As the onus to safeguard private groundwater lies with well owners themselves, the requirement for enhanced awareness via risk communication is evident. However, there is limited guidance available to Irish well owners in the context of climate change and EWEs. Current risk perceptions of groundwater contamination in relation to climate change are unknown. The current study responds to this deficit via a national survey of private well users to establish current actions, perceptions and audience clusters relating to groundwater contamination and EWEs.

## 2 Methods

The survey was undertaken in the ROI, which is characterized by high private groundwater reliance and significant rates of microbial contamination (Naughton and Hynds 2014). Accordingly, the majority of questions concerning groundwater contamination focused on microbial (as opposed to geogenic) contaminants. The survey followed a structured, standardized format and was adapted from the knowledge, attitudes and practice (KAP) model (Médecins du Monde 2011). Question types included numerical, multiple response and Likert-scale questions to establish household composition and respondent perceptions. Additionally, a scoring protocol was developed to compare respondent behavior, knowledge and perceptions. Questions employed both dichotomous (0/1) and ordinal scoring (0–2).

A total of 41 questions were devised, with three filter questions employed to delineate responses based on well maintenance, experience with EWEs and information seeking behaviors. The survey comprised four principal sections: demographics, source maintenance, climate risk perception and information preferences. Respondents were surveyed based on knowledge of groundwater contamination, climate change impacts and actions taken during five recent national weather events: *Storm Deirdre* (December 2018 rainfall/floods), *summer drought 2018*, *Beast from the East* (March 2018 snowstorm), *Storm Ophelia* (October 2017 rainfall/floods) and *winter 2013/2014 floods*. Questions pertaining to climate change examined perceived likelihood, severity and consequence of impacts and both actual and conjectural responses.

Following a pilot study in August 2019 ($n$ = 39), the finalized survey was distributed electronically between September 17 and November 26, 2019. The survey was hosted on SurveyMonkey and distributed via a series of rural interest groups, government bodies and primary/post-primary schools. Eligible respondents were required to be 18 years old and avail of either a private well or private group water scheme as a primary drinking water source. After closure of the survey, data were imported to IBM SPSS Statistics 26 software for statistical analyses. Two-step cluster analysis was undertaken to explore respondent perceptions regarding effects of extreme weather on groundwater quality and identify discrete audience cohorts for future risk communication activities.

## 3 Results

### 3.1 Descriptive Results

The final number of surveys included for analysis was 515, with respondents originating from all 26 counties in the ROI. A total of 449 (87.2%) respondents availed of a private well while 66 (12.8%) respondents availed of a private group water scheme. Prior testing of supply was reported by 68% ($n$ = 350) of respondents, of whom 10.6% ($n$ = 37) reported testing more than once per year. The most frequently cited reasons for resting were 'to determine water quality' (77.4%, $n$ = 274), 'peace of mind' (32.9%, $n$ = 115) and 'as part of regular maintenance' (24.9%, $n$ = 87). 'Occurrence of an extreme weather event' was cited by just 6.9% ($n$ = 24) of respondents.

Over half of surveyed respondents (54.8%, $n$ = 282) cited experience with $\geq 1$ specified EWE (Fig. 1). While 55.3% ($n$ = 156) of these respondents took some form of responsive action to mitigate impacts on well water quality, only 13.1% ($n$ = 37) had their water tested for contamination in the aftermath of an event. Conversely, over one-third of respondents (36.5%, $n$ = 85) who did not experience a recent weather event stated they would submit a water sample for testing were such an event to occur.

### 3.2 Cluster Analysis

The elbow method was adopted to optimize optimum number of clusters and two-step clustering used to identify final clusters. Clusters were developed based on four 'climate statement' variables (i.e., perceived consequences of EWES on groundwater quality) outlined in Table 1. A total of three clusters was deemed appropriate for classification (silhouette measure = 0.6).

Final clusters (presented in Table 2) were categorized based on overall agreement with climate statements. Respondents in cluster 1 ('high perception') demonstrated agreement with all statements. Respondents in cluster 2 ('uncertain') demonstrated apathy toward effects of drought and warmer temperatures while respondents in cluster 3 ('high to moderate perception') demonstrated high uncertainty concerning drought.

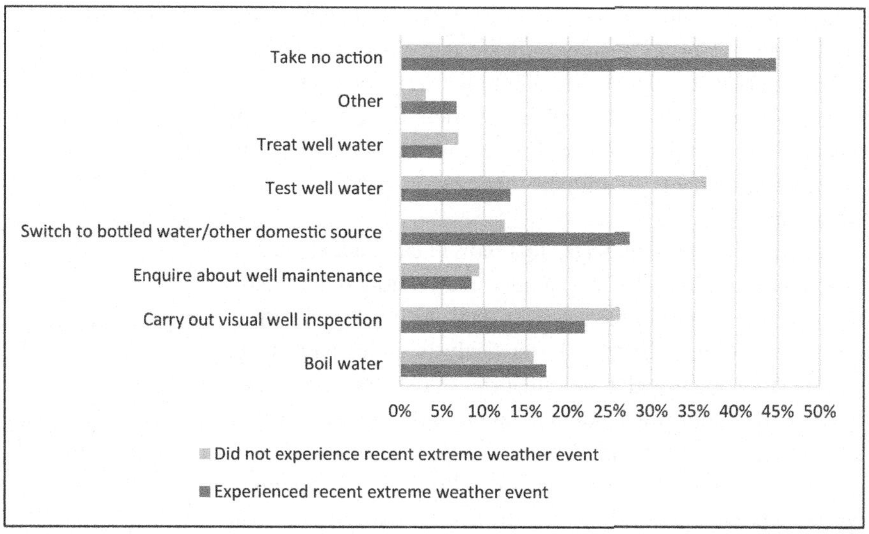


**Fig. 1** Actual versus hypothetical response to extreme weather events

**Table 1** Respondent agreement with climate statements

| Climate statement | Agree | Neither agree nor disagree | Disagree |
|---|---|---|---|
| 1. Well water may remain contaminated after flood and heavy rainfall | 381 (74%) | 103 (20%) | 31 (6%) |
| 2. Occurrence of drought may lead to increased contaminant levels in well water | 304 (59%) | 160 (31.1%) | 51 (9.9%) |
| 3. Increased rainfall may impact transport of contaminants to well water | 417 (80.1%) | 82 (15.9%) | 16 (3.1%) |
| 4. Warmer temperatures may affect survival of pathogens in well water | 386 (75%) | 103 (20%) | 26 (5%) |

**Table 2** Cluster profiles for climate statements by most common perception category

| Climate statement | Cluster 1: high perception ($n$ = 220) | Cluster 2: uncertain ($n$ = 189) | Cluster 3: high to moderate perception ($n$ = 106) |
|---|---|---|---|
| 1 | Agree (100%) | Agree (53.4%) | Agree (56.6%) |
| 2 | Agree (100%) | Neither agree nor disagree (42.3%) | Neither agree nor disagree (75.5%) |
| 3 | Agree (100%) | Agree (48.1%) | Agree (100%) |
| 4 | Agree (100%) | Neither agree nor disagree (54.5%) | Agree (100%) |

Cluster membership was significantly related to householder gender ($\chi^2 = 12.825$, $p = 0.002$), education ($\chi^2 = 26.449$, $p = 0.000$) and age ($\chi^2 = 26.883$, $p = 0.003$) but displayed no such association with experience of recent weather event ($\chi^2 = 0.861$, $p = 0.650$). Over half of respondents in cluster 1 were female (55.8%, $n$ = 121) while the majority of respondents in cluster 3 comprised males (61.5%, $n$ = 115). With respect to educational attainment, almost 80% of respondents ($n$ = 170) in cluster 1 had a third level degree. Respondents aged 18–24 constituted the highest proportion of respondents in cluster 2 (36%, $n$ = 68) and cluster 3 (33%, $n$ = 35).

## 4 Discussion

With approximately 35–59 million global illnesses annually attributable to microbial groundwater contamination alone, groundwater quality constitutes a significant public health concern (Murphy et al. 2017). In contrast to many public groundwater end-users, rural private well owners are disproportionately vulnerable to source contamination due to geographic location and climate vulnerability and responsible for ensuring supply integrity (Andrade et al. 2018). The current study represents the first attempt to quantify levels of risk perception among private well owners concerning climate change. Survey results indicate that there is a marked difference between actual and theoretical actions among well owners in the wake of an EWE. As such, there is a need to promote improved, visible well user guidance and elucidate the myriad impacts of climate change on private groundwater supplies.

Existing literature indicates that individual, context-specific experiences with well water quality and demographic factors may exert a strong influence over well user risk perceptions (Munene and Hall 2019). Cluster analysis undertaken in the current study identified three categories of risk perception level relating to groundwater contamination and extreme weather, with cluster membership demonstrating a relationship with respondent gender, education and age. Educational attainment has been previously noted as the strongest predictor of climate change risk perceptions globally (Lee et al. 2015). As such, engagement with younger and more disadvantaged populations may be key toward encouraging future private well risk awareness and risk mitigation behaviors. In the absence of clear policy and regulation surrounding private groundwater, such information may enable focused tailoring and audience segmentation of risk information. More broadly, study findings support the emerging sustainability science of socio-hydrogeology (the dynamic feedbacks and interactions between social and hydrogeological systems) and provide new impetus for emerging citizen science and rural engagement initiatives promoting groundwater awareness (Re 2015).

## 5 Concluding Remarks

Availability of supply maintenance information for private well owners has long been an issue for groundwater policy-makers and end-users alike. The quality of existing guidance is given comparatively less attention—particularly, in relation to EWEs and climate change impacts on groundwater contamination. Results from this study may contribute toward closing this gap in an era where increased groundwater education and citizen stewardship are urgently required to ensure pristine groundwater quality in rural areas. Framed against the emerging scientific field of *(hydro)geoethics*—defined as 'the study and promotion of the evaluation and protection of the geosphere' (Peppoloni and Di Capua 2012)—the presented research highlights the importance of public/private equity in groundwater quality management and the social dimension of groundwater contamination. Study findings underscore the responsibility of hydrogeologists and state authorities to impart risk information to vulnerable groundwater end-users strategically and acknowledge the role of climate change within future groundwater education initiatives.

**Acknowledgements** The authors would like to thank all anonymous reviewers for taking the time to consider this abstract and for providing constructive suggestions to help position the work within the conference theme. The authors also wish to note that the current study constitutes a work package forming part of an overarching Ph.D. project co-funded by Technological University Dublin (TUD) and the Irish Health Service Executive (HSE). Ethical approval was granted in principle prior to commencement of the survey and was formally ratified by TUD's Research Ethics Committee on February 7, 2020.

## References

Andrade L, O'Dwyer J, O'Neill E, Hynds P (2018) Surface water flooding, groundwater contamination, and enteric disease in developed countries: a scoping review of connections and consequences. Environ Pollut 236:540–549

CSO (2017) Census 2016: small area population statistics. Central Statistics Office, Dublin

ECDC, EFSA (2018) The European Union summary report on trends and sources of zoonoses, zoonotic agents and food-borne outbreaks in 2017. EFSA J 16(12):5500

Forzieri G, Cescatti A, Silva FB, Feyen L (2017) Increasing risk over time of weather-related hazards to the European population: a data-driven prognostic study. Lancet Planetary Health 1(5):200–208

Green TR (2016) Linking climate change and groundwater. In: Jakeman AJ, Barreteau O, Hunt RJ, Rinaudo JD, Ross A (eds) Integrated groundwater management: concepts, approaches and challenges. Springer Nature, Basel

Lee TM, Markowitz EM, Howe PD, Ko CY, Leiserowitz A (2015) Predictors of public climate change awareness and risk perception around the world. Nat Clim Change 5:1014–1023

Médecins du Monde (2011) Data collection. Quantitative methods. The KAP survey model (knowledge, attitudes and practices). Available at: https://www.medecinsdumonde.org/6c27001736f069d23fab6b06b30ee3a1-6.pdf

Munene A, Hall DC (2019) Factors influencing perceptions of private water quality in North America: a systematic review. Syst Rev 8:111

Murphy HM, Prioleau MD, Borchardt MA, Hynds PD (2017) Review: epidemiological evidence of groundwater contribution to global enteric disease, 1948–2015. Hydrogeol J 25(4):981–1001

Naughton O, Hynds PD (2014) Public awareness, behaviours and attitudes towards domestic wastewater treatment systems in the Republic of Ireland. J Hydrol 518:108–119

O'Dwyer J, Downes M, Adley CC (2016) The impact of meteorology on the occurrence of waterborne outbreaks of vero cytotoxin-producing Escherichia coli (VTEC): a logistic regression approach. J Water Health 14(1):39–46

Peppoloni S, Di Capua G (2012) Geoethics and geological culture: awareness, responsibility and challenges. Ann Geophys 55(3):335–341

Re V (2015) Incorporating the social dimension into hydrogeochemical investigations for rural development: the Bir Al-Nas approach for socio-hydrogeology. Hydrogeol J 23(7):1293–1304

Taylor RG, Scanlon B, Döll P, Rodell M, van Beek R, Wada Y, Longuevergne L, Leblanc M, Famiglietti JS, Edmunds M, Konikow L, Green TR, Chen J, Taniguchi M, Bierkens MFP, MacDonald A, Fan Y, Maxwell RM, Yechieli Y, Gurdak JJ, Allen DM, Shamsudduha M, Hiscock K, Yeh PJF, Holman I, Treidel H (2013) Ground water and climate change. Nat Clim Change 3:322–329

# Environmental Isotopes as Tools in Sustainable Groundwater Management: Essential or Dispensable?

José Manuel Marques and Paula M. Carreira

**Abstract**

In this work, a special emphasis is given to the role that environmental isotopic techniques can play in sustainable groundwater management practices. A brief review of two case studies is made, in which isotopic techniques combined with other tools of Geosciences (e.g. Geology and Geochemistry) have proved to be fundamental in the sustainable management of different types of groundwater resources. In the case of a study on Caldas da Rainha thermal waters, the multidisciplinary approach involving conventional geochemical and isotopic techniques (e.g. $\delta^2H$, $\delta^{18}O$, $^3H$, $\delta^{34}S_{(SO4)}$ and $\delta^{18}O_{(SO4)}$) led to the conclusion that the Caldas da Rainha thermal waters are "geologically protected" from anthropogenic contamination ascribed to intensive agriculture practices in the soils of the Upper Jurassic geological formations. In the case of the study of the Santa Margarida Military Camp, the use of environmental isotopes ($\delta^2H$ and $\delta^{18}O$) clearly showed no mixing between the waters from the ponds (projectile impact zones associated with real fire training) and the groundwaters of the region.

**Keywords**

Environmental isotopes • Groundwaters • Surface waters • Sustainable management • Case studies

## 1 Introduction

Groundwater availability is one of the major issues facing mankind today, as the problems associated with it affect the lives of many millions of people around the world. We are therefore facing a problem that in recent decades has attracted the attention of a large number of International Agencies (e.g. UN, UNESCO, IAEA, WMO, etc.) and governmental and non-governmental organizations (at regional/international level) related to the "groundwater issue".

The main objective of this paper is to provide a brief review of the basic theoretical concepts as well as the potential practical applications of Isotopic Hydrology. Attention is also drawn to the importance of expanding Isotope Hydrology techniques throughout the curricula of Universities around the World, as already defined in 2002 by the UNESCO Joint International Isotopes in Hydrology Program (JIIHP). The two case studies reviewed are distinct, but at the same time paradigmatic, and contain reference material for groundwater scientists, engineers, policymakers, managers, organizations, and professionals, who may be interested/involved in practical applications of Isotope Hydrology related to the assessment, development, protection and sustainable management of groundwater resources.

## 2 Methodology

Environmental isotopes play a key role in the development of conceptual hydrogeological models, where the use of stable isotopes [oxygen-18 ($^{18}O$) and deuterium ($^2H$)] as groundwater flow tracers from recharge areas to where they emerge, resides in that these isotopes are an integral part of the water molecule, and are conservative after rainwater infiltration (e.g. IAEA 1981). We are thus using water itself to write "*the book of its history*".

J. M. Marques (✉)
Centro de Recursos Naturais e Ambiente, Instituto Superior Técnico, Universidade de Lisboa, Lisbon, Portugal
e-mail: jose.marques@tecnico.ulisboa.pt

P. M. Carreira
Centro de Ciências e Tecnologias Nucleares, Departamento de Engenharia e Ciências Nucleares, Instituto Superior Técnico, Universidade de Lisboa, Lisbon, Portugal


M. Abrunhosa et al. (eds.), *Advances in Geoethics and Groundwater Management: Theory and Practice for a Sustainable Development*, Advances in Science, Technology & Innovation,
https://doi.org/10.1007/978-3-030-59320-9_69

In a $\delta^2H$ versus $\delta^{18}O$ diagram, the isotopic composition of atmospheric precipitation is aligned along the so-called Global Meteoric Water Line (GMWL) $\delta^2H = 8\ \delta^{18}O + 10$ defined by Craig (1961), and later improved by Rozanski et al. (1993) and Terzer et al. (2013). Thus, water from atmospheric precipitation, which has not been subject to variation in isotopic composition along the underground flow system, will have $\delta^2H$ and $\delta^{18}O$ values that are distributed along and/or near GMWL (see Fig. 1).

Tritium ($^3H$), the radioactive hydrogen isotope, has a half-life of 12.32 years, being its concentration in natural waters expressed in terms of tritium units (TU) (Lucas and Unterweger 2000). The use of tritium content present in groundwater provides primarily qualitative information on: (i) active recharge of the system, (ii) the "age" of groundwater, (iii) the duration of underground pathways and (iv) the existence of a mixture of recent infiltration water and older groundwater.

Analysis of the $\delta^{34}S_{(SO4)}$ and $\delta^{18}O_{(SO4)}$ values in groundwater sulphate is a very important hydrogeological tool for determining sulphate origin (e.g. interaction with evaporites, sulphate from both atmosphere and soil, sulphate from oxidation of reduced inorganic sulphur species) and is also useful in assessing surface water/groundwater interaction and, consequently, in analysing vulnerability to anthropogenic groundwater pollution (Krouse and Mayer 2001).

Lakes/ponds are complex dynamic systems, interacting with the local environment and strongly associated with the hydrological cycle through surface and ground "inlet" and "outflow" flows, as well as through atmospheric precipitation/evaporation phenomena (Mook 2000). In a diagram $\delta^2H$ versus $\delta^{18}O$ (see Fig. 1), water resulting from evaporation processes is evidenced along a line projected below the GMWL and with a slope between 4 and 6 (Mook 2000).

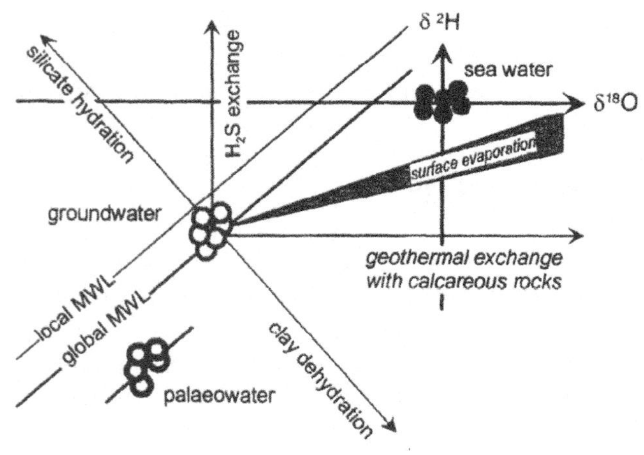


**Fig. 1** Scheme of the various processes causing a deviation in the $\delta^2H$ versus $\delta^{18}O$ of precipitation and/or groundwater relative to GMWL – global MWL ( adapted from Mook 2000)

Water sampling procedures and storage (for later chemical analyses), type of waters physico-chemical parameters measured in situ, the techniques used for surface water and groundwater chemical analysis and isotopic determinations ($\delta^2H$, $\delta^{18}O$, $\delta^{34}S_{(SO4)}$, $\delta^{18}O_{(SO4)}$ and $^3H$) were those described in detail in Matias et al. (2008) and Marques et al. (2019).

## 3 Results

### 3.1 The Case Study of Caldas da Rainha Thermal Waters

The region where Caldas da Rainha thermal waters emerge (Fig. 2) is dominated by a vast synclinal structure responsible for the regional flow of thermal waters.

The thermal waters emerge through springs and boreholes near a N-S oriented oblique fault (60° E), at the contact between the Upper Jurassic detrital rocks (e.g. D – clayey sandstone) and the Dagorda marls (A) that function as a hydrogeological barrier. The regional diapiric structures were responsible for the elevation and consequent folding of the Jurassic formations. At greater depth, we find mainly limestone rocks (C and B), later covered by sedimentary deposits (E) (details in Marques et al. 2019).

The Caldas da Rainha thermal waters ($T \approx 33$ °C) belong to the group of sulphur waters of $Cl\text{-}SO_4\text{-}HCO_3\text{-}Na\text{-}Ca$ facies (with the presence of $H_2S$, $HS^-$ and $S_2^-$). According to Marques et al. (2019), the Caldas da Rainha thermal waters sampled at AC1-B, AC2 and JK1 boreholes indicate significant $^3H$ concentrations: 2.8; 2.5 and $1.1 \pm 0.6$ TU, respectively. On the other hand, the thermal waters sampled in the Piscina da Rainha spring are characterized by the absence of $^3H$ or by $^3H$ values close to the detection limit. The obtained results indicate the existence of two groups of waters with different geochemical signatures and with

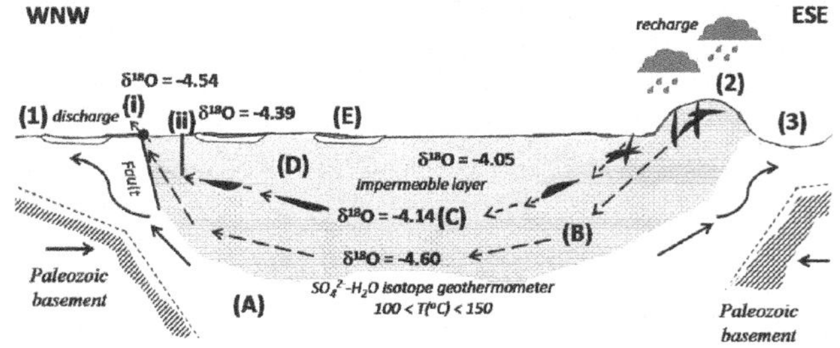


**Fig. 2** Schematic cross section of the research region. Geological formations: **(A)** Hetangian-Retian; **(B)** Lower and **(C)** Middle Jurassic (thermal aquifer formations); **(D)** Upper Jurassic; **(E)** Plio-Plistocene. Sampling sites: (1) Caldas da Rainha diapir; (2) Candeeiros Mountain; (3) Fonte da Bica diapir. (i) thermal spring at Caldas da Rainha Spa; (ii) exploitation borehole of thermal water. $\delta^{18}O$ values in ‰ versus V-SMOW (after Marques et al. 2019)

equally different $\delta^{34}S_{(SO4)}$ and $\delta^{18}O_{(SO4)}$ values (Marques et al. 2019): (i) the thermal waters of Caldas da Rainha with $\delta^{34}S_{(SO4)}$ and $\delta^{18}O_{(SO4)}$ between 14.9 and 19.1 ‰ and 11.1 and 16.2 ‰, respectively, and (ii) groundwater associated with geological formation (D), with a less deep circulation and sampled in wells, exhibiting $\delta^{34}S_{(SO4)}$ and $\delta^{18}O_{(SO4)}$values between 1.5 and 4.1 ‰ and 8.6 and 9.3 ‰, respectively.

## 3.2 The Case Study of Santa Margarida Military Camp

The main objective of this case study was to evaluate the environmental impact caused by the Santa Margarida Military Camp (CMSM) after 50 years of use, with special emphasis on the characterization of surface waters and groundwaters (see Matias et al. 2008). The CMSM covers an area of about 62 km$^2$ and is divided into two sectors with distinct characteristics: an "urban" zone where all the organic units of the field are located, supporting infrastructures, residential areas, etc., and a military training zone.

From the geological point of view, the CMSM is located in the Tertiary Basin of the Tagus River, with geological formations ranging from the Precambrian to the Pleistocene. It is important to consider the presence of coarse detrital formations (sandstones and gravel) interspersed with upper Pliocene and Upper Miocene clay formations, which strongly influence soil porosity and permeability (see Matias et al. 2008 and references therein).

In this case, an evaporation tendency is strongly evidenced by the water samples collected in the CMSM ponds. These ponds represent the impact zones of the real fire training projectiles, carried out by the mechanized brigades (Matias et al. 2008).

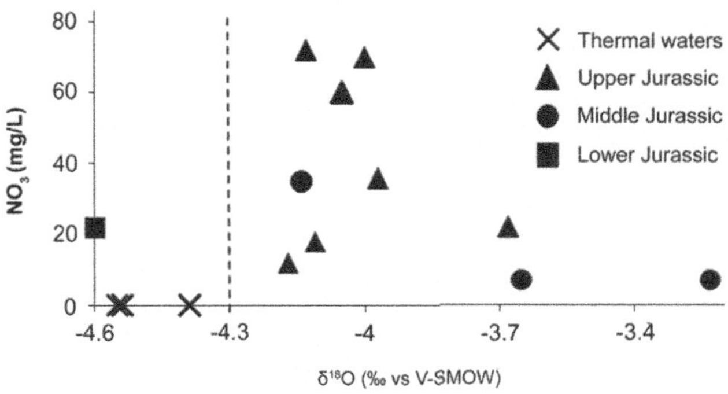


**Fig. 3** $NO_3^-$ versus $\delta^{18}O$ for groundwaters sampled in different geological formations. Caldas da Rainha case study (Marques et al. 2019)

# 4 Discussion

The results indicate that Caldas da Rainha thermal waters are "isolated" from any processes of anthropogenic contamination. Such contamination was only detected in the shallow groundwater associated with geological formation (D – Upper Jurassic), due to the high concentrations in $SO_4$ and $NO_3$ (Marques et al. 2019).

The $\delta^{18}O$ values of Caldas da Rainha thermal waters and groundwater associated with geological formation (D – Upper Jurassic) reflect different recharge altitudes associated with different underground flow systems without any evidence of mixing (Fig. 3).

The absence of $^3H$ in Caldas da Rainha thermal waters seem to indicate that these thermal waters should be regarded as the most representative of the deep hydrothermal system (Marques et al. 2019).

$\delta^{34}S_{(SO4)}$ and $\delta^{18}O_{(SO4)}$ values seem to indicate that in the thermal waters of Caldas da Rainha sulphate is clearly the result of water–rock interaction with evaporitic formations (e.g. gypsum and anhydrite) associated with regional synclinal structure. However, in the Upper Jurassic (D) groundwater samples, the $SO_4$ origin may be attributed to: (i) atmospheric deposition which is incorporated into the soil area and subsequently assimilated by groundwater; and (ii) local anthropogenic influence (agricultural practices), taking into account both the $\delta^{34}S_{(SO4)}$ values close to $0°/_{oo}$, and the increased concentrations of $SO_4$, Cl and $NO_3$ in these waters (Marques et al. 2019).

According to Matias et al. (2008), the projection of the isotopic composition of surface waters (streams, ponds and a dam) and groundwaters (springs, wells, boreholes and a piezometer) from Santa Margarida Military Camp (CMSM) in a $\delta^2H$ versus $\delta^{18}O$ diagram (Fig. 4) reveals the strong evaporation of the CMSM pond waters.

On the other hand, it does not appear to exist any mixing trend between pond waters and local groundwater systems (Fig. 3). As reported by Matias et al. (2008), the projection of local groundwater near the line of world meteoric waters allows them to be classified as normal meteoric waters.

# 5 Concluding Remarks

The widespread application of environmental isotopic techniques in hydrogeology has proven to be a valuable tool for improving our understanding on the behaviour of groundwater systems, essential for their sustainable management. The case studies reviewed here are well illustrative of the fact that environmental isotopic techniques can provide an independent approach to solve a particular hydrogeological problem, as isotopic groundwater signatures (whether

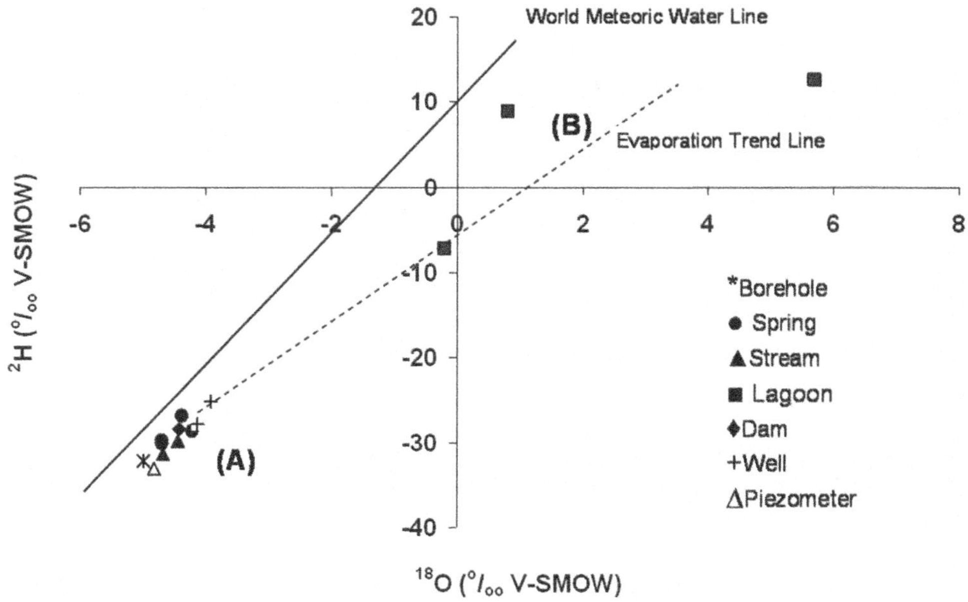


**Fig. 4** Isotopic composition ($\delta^2H$ and $\delta^{18}O$ values) of surface water and groundwater of the Santa Margarida Military. (adapted from Matias et al. 2008)

thermal or not) are authentic archives that hold the memory of its evolution: recharge area altitude, underground flow direction and mean velocities, residence time, etc. However, we should be aware that these isotopic techniques are just one of the many available tools that can be applied in hydrogeology surveys, and that their use in parallel with other disciplines of geoscience (geology, geochemistry and geophysics) will certainly be much more efficient.

**Acknowledgements** CERENA/IST thankfully acknowledges the FCT support through the UIDB/04028/2020 Project, and $C^2TN$/IST gratefully acknowledges the FCT support through the UIDB/04349/2020 Project.

## References

IAEA [International Atomic Energy Agency] (1981) Stable isotope hydrology. Deuterium and oxygen-18 in the water cycle, technical reports series 210. IAEA. Vienna

Craig H (1961) Isotopic variations in meteoric waters. Science 133:1703

Rozanski K, Araguás-Araguás L, Gonfiantini R (1993) Isotopic patterns in modern global precipitation. In: Swart PK et al (eds) Climate change in continental isotopic records, Geoph. Monog. Series, AGU, Washington DC, vol 78, pp 1–36

Terzer S, Wassenaar LI, Araguás-Araguás LJ, Aggarwal PK (2013) Global isoscapes for $\delta^{18}O$ and $\delta^2H$ in precipitation: improved prediction using regionalized climatic regression models. Hydrol Earth Syst Sci 17:4713–4728

Lucas LL, Unterweger MP (2000) Comprehensive review and critical evaluation of the half-life of tritium. J Res Natl Inst Stand Technol 105(4):541–549

Mook WG (2000) Environmental isotopes in hydrological cycle. Principles and applications. Technical Documents in Hydrology, 39 (I), IHP-V UNESCO & IAEA, Paris

Krouse HR, Mayer B (2001) Sulphur and oxygen isotopes in sulphate. In: Cook P, Herczeg AL (eds) Environmental tracers in subsurface hydrogeology, 2nd edn. Kluwer Academic Publishers, Boston, pp 195–231

Matias MJ, Marques JM, Figueiredo P, Basto MJ, Abreu MM, Carreira PM, Ribeiro C, Flambó A, Feliciano J, Vicente EM (2008) Assessment of pollution risk ascribed to Santa Margarida Military Camp activities (Portugal). Environ Geol 56:1227–1235

Marques JM, Matos C, Carreira PM, Neves MO (2019) Isotopes and geochemistry to assess shallow/thermal groundwater interaction in a karst/fissured-porous environment (Portugal): a review and reinterpretation. Sust Wat Res Manag 5:1525–1536

# The Role of International Cooperation in Sustainable Groundwater Development

Raquel Sousa and Fabio Fussi

**Abstract**

Groundwater plays a key role in the achievement of the Sustainable Development Goals contributing to many of them, such as ensure availability and sustainable management of water and sanitation for all, poverty eradication, food security, gender equality, sustainability of cities and human settlement, combating climate change and protecting terrestrial ecosystems. But, groundwater, along with water resources, is very vulnerable to population growth and climate change and groundwater stress is a present and increasing issue worldwide that needs to be addresses in the decision-making processes leading water resources management. In the context of International Cooperation, it is a priority that programs directly or indirectly dealing with groundwater address groundwater sustainable management and conservative approaches based in scientific knowledge, invest in adequate boreholes constructions with proper planning and supervision and enhance investment in preliminary research, training and implementation of participatory approaches.

**Keywords**

Sustainable development goals • Cooperation • WASH • Boreholes • Scientific knowledge

## 1 Introduction

Groundwater is an essential resource for human life: it makes up to 25–40% of the world's drinking water, being also essential for food production. Even if in small amounts, in some region's groundwater is the only option during dry seasons and droughts (Vrba and van der Gun 2004). Being the main non-frozen freshwater resource on earth, corresponding to 96% of earth's unfrozen freshwater (Shiklomanov et al. 2003), it is a fragile resource: less than 6% of the water contained in the uppermost portion of the continental crust is less than 50 years old (Gleeson et al. 2016); therefore, only a small part of groundwater is renewable. Furthermore, contamination of groundwater makes it unsafe for human use and decrease its availability for our life.

In the context of International Cooperation and WASH (Water Sanitation and Hygiene) programs, groundwater plays an important role and two issues arise: a) the importance of sustainable and conservative approaches in groundwater exploitation based on scientific knowledge; b) the importance of adequate boreholes constructions and implementation of adequate management models for water systems.

## 2 (Ground)Water, WASH and the SDGs—Background and Development

In January 1992, a group of experts in water issues signed the Dublin Statement on Water and Sustainable Development, during the International Conference on Water and the Environment, setting four prevailing principles: freshwater is a finite and vulnerable resource, essential to sustain life, development and environment; water development and management should be based on a participatory approach, involving users, planners and policy-makers at all levels; women play a central part in the provision, management and safeguarding of water; water has an economic value in all its competing uses and should be recognized as an economic good, The Dublin Statement (1992).

By that time, the first steps towards Water and Sanitation programs had already taken off in the 70s. From there on, the lack of an UN agency in charge of the water sector resulted

R. Sousa (✉)
International Groundwater Resources Assessment Centre (IGRAC), Delft, The Netherlands
e-mail: raquel.sousa@un-igrac.org

F. Fussi
University Milano Bicocca, Milano, Italy



M. Abrunhosa et al. (eds.), *Advances in Geoethics and Groundwater Management: Theory and Practice for a Sustainable Development*, Advances in Science, Technology & Innovation,
https://doi.org/10.1007/978-3-030-59320-9_70

in an attempt to create different collaborative councils, and in the 90s the "WASH Sector" *was in disarray*, without a shared focus and *with too many "mothers"*, (Cross 2013). Finally, in 2003 the UN-Water was created as a coordinating mechanism among UN agencies and other international stakeholders.

In the meanwhile, in 2000, at the UN Millennium Summit, the eight Millennium Development Goals (MDG) were developed, aiming to reduce poverty by 2015. In 2015, at the UN Sustainable Development Summit, the negotiation process on the post-2015 development agenda concluded with the adoption of the 2030 Agenda for Sustainable Development with its core at 17 Sustainable Development Goals (SDG), UN (2019).

Groundwater is a key resource in the achievement of the UN Sustainable Development Agenda for 2030. Besides SDG-6, *ensure availability and sustainable management of water and sanitation for all,* groundwater contributes to other goals like poverty eradication (Goal 1), food security (Goal 2), gender equality (Goal 5), sustainability of cities and human settlement (Goal 11), combating climate change (Goal 13) and protecting terrestrial ecosystems (Goal 15). (IGRAC 2018).

As for the Agenda for 2030 SDG-6 is currently off-track to solve the global water crisis. Today, 2.1 billion people lack safely managed drinking water, 4.5 billion lack access to safely managed sanitation, water governance structures remain weak and water-related ecosystems continue to decline (UN-WATER 2018). As for a *sustainable management* of water, there is still a long way to go and in the particular case of groundwater, it *is still weakly conceptualized in the SDG-6 indicators or insufficiently known to provide a reliable indicator value* (IGRAC 2018).

## 3 Groundwater Development and Scientific Knowledge in International Cooperation

Over the year's population growth and expansion of human activities have increased their impact on natural resources and environment. In the specific case of groundwater, the impact has resulted in increased groundwater stress, which in turn has negative repercussions on water access, social stability, human health and environmental degradation, in a continuous cause–effect relationship.

With an increasing water demand by the population and several billions people still lacking access to safe water, it is a priority to better know water resources, their natural dynamics and equilibrium and the effect of human actions on it in order to prevent from unexpected results. Scientific understanding of aquifers' dynamics and related environmental processes can and must undertake the impact of groundwater development aiming the implementation of sustainable and safe solutions.

While addressing SDG-6, international cooperation can play an essential role in sustainable groundwater management solutions through:

(a) a scientific engagement in the development of strategies for water supply;
(b) an increased effort for data collection and information sharing at global level.

An example of insufficient groundwater knowledge leading to large consequences is the case of Bangladesh. References of arsenic contamination in groundwater in West Bengal, India, date from 1983. Groundwater contamination was first attributed to boreholes' pipes, pesticides and insecticides, but studies revealed to be geological (from groundwater captured between 20 and 60 m). Based on the same geological, climate and hydrologic conditions, investigations were carried out in the early 90s, financed by WHO (World Health Organization) and arsenic contamination in Bangladesh soon became a national issue. Until the mid-1970s, most people in rural areas drank bacteria-infested water from surface sources and dug wells leading to waterborne diseases, especially in children. To address this situation, boreholes, financed by UNICEF and World Bank, were constructed in the whole country and in 2000 around 96% of the population was drinking water from shallow tube wells. At that moment in the past, arsenic was unfortunately not recognized as a problem and it was not tracked in the standard quality testing procedures. What was a groundwater revolution for agriculture and an important decrease in deaths caused by waterborne diseases became a danger for human health, and an estimation of 120 million people were exposed to arsenic, a toxic and carcinogenic chemical, in "potable" water (Alam et al. 2002).

## 4 Technical Challenges

The exploitation of groundwater has a long history, starting millennia ago. Although human settlements were preferably located near permanent and abundant surface water sources, in some contexts (arid regions or during dry periods) shallow groundwater and rainfall collection were the only source of exploitable water. In this situation, population used to dig medium–large diameter wells by hand, or in some cases applied rudimentary drilling techniques to exploit shallow groundwater, especially in geological contexts with thick unconsolidated sediments in alluvial deposits, it is the case in India and Nepal, where manual drilling is a common practice (Danert 2015). The techniques for shallow groundwater exploitation were part of the traditional culture and its management part of the social dynamics.

At the beginning of the nineteenth century mechanized drilling appeared, powered by steam engines. With this technical revolution, the exploitation of deep groundwater and drilling in hard rocks became possible. On the one hand, mechanic groundwater exploitation improved the access to safe water but on the other hand this new approach increased the distance to community-based water management strategies. Relying on more sophisticated and expensive techniques brought the local population to have little participation in the definition of the strategies to obtain water. Population became then dependent on private drilling companies, external funding policies and governmental priorities. External support and funding for boreholes construction are at first glance the key factor to improve access to safe water, but in the meantime most of communities are unable to maintain their water supply systems operational and expand them to in need areas. The risk is that communities, losing their traditional coping mechanisms to access water, have become more vulnerable.

The approach used in the delivery of external support is thus crucial for the success of strategies aiming to ensure efficient, safe and sustainable water supply. The experience shows that what has been delivered, in terms of infrastructures for groundwater exploration, has often failed. An analysis of the efficiency of boreholes after their construction in four African Countries (Malawi, Liberia, Sierra Leone and Tanzania) shows that more than 20% of the water points are not functioning after the first year of operation. After 6 years this rate increases to 30–40% (Carter and Ross 2016). According to a report from UNICEF 2018, the overall percentage of waterpoints failing during the first year after construction is 15 to 30%, mainly as a result of poor planning, siting, and repercussions from bad construction. After the first year and within the first 5 years fails are mainly due to poor water points' management models, UNICEF (2018) since boreholes' maintenance and operationality fail. The rate of operational wells after the first year of construction is however quite different from country to country (Foster 2020), and it depends from a combination of different factors. Some of them are related to the quality of construction and installation, while others related to the lack of proper management and maintenance after construction (Fig. 1).

Concerning the quality of the infrastructures, construction must be done under close supervision in each phase of work. Furthermore, it is important to take into consideration the real needs and capacity of population. Many water practitioners consider that the private sector could ensure a higher quality in the boreholes construction than public institutions or Non-Governmental Organizations. But on the other hand, there are social aspects that must be considered in the planning phase (e.g. the involvement of local communities, the economic support to small traditional drilling groups, the promotion of local expertise as even cheaper and more efficient technical solutions can be imported, the mandate to provide an adequate service to marginal or discriminated groups), although they can be partially in contrast with a fully "business-oriented" approach as private contractors often do.

To improve the sustainability of water systems, the following aspects need to be covered:

(a) resources for effective maintenance of the systems when specific weak technical aspects appear as a common problem (e.g. the problem of corrosion in hand pumps).
(b) adequate training of technicians in charge of the infrastructures, from local Water Users Committees (Bonsor et al. 2015) to National Water Authorities.
(c) stronger sectorial strategies addressing the maintenance of water supply systems (commercialization of spare parts and technical training)

In a recent e-discussion promoted by the Rural Water Supply network (RWSN 2019), specific aspects were pointed out as receiving insufficient funds: collecting and analysing data, understanding communities' capacity, boreholes siting, capacity building to stakeholders, embedding the conditions of WASH access for all in national and subnational policies, sector guidance; the cost of corruption and the cost of including people with disabilities.

Funding in water supply projects is often too focused on the infrastructures. There is limited investment in preliminary research and planning, training, and participatory approaches in the definition of actions and elaboration of fully inclusive solutions. If these aspects are not adequately considered and financed, the investments delivered in the WASH sector will often fail in providing an adequate and sustainable water supply to the population.

## 5 Concluding Remarks

In many countries, a sustainable groundwater development is the most adequate strategy to improve the living conditions of large sectors of population who have limited access to safe water. A specialized assessment of groundwater status, trends and risks is then required to understand groundwater resources, whose sustainability is essential for achievement of SDG-6 Targets, IGRAC (2018). Groundwater development must take into consideration the knowledge of aquifers and groundwater resources, capacity building, understanding of the social structures and be consistent with the overall development of national, legal and political frameworks.

Better cooperation between stakeholders and a clear definition of their roles, can give a great contribution

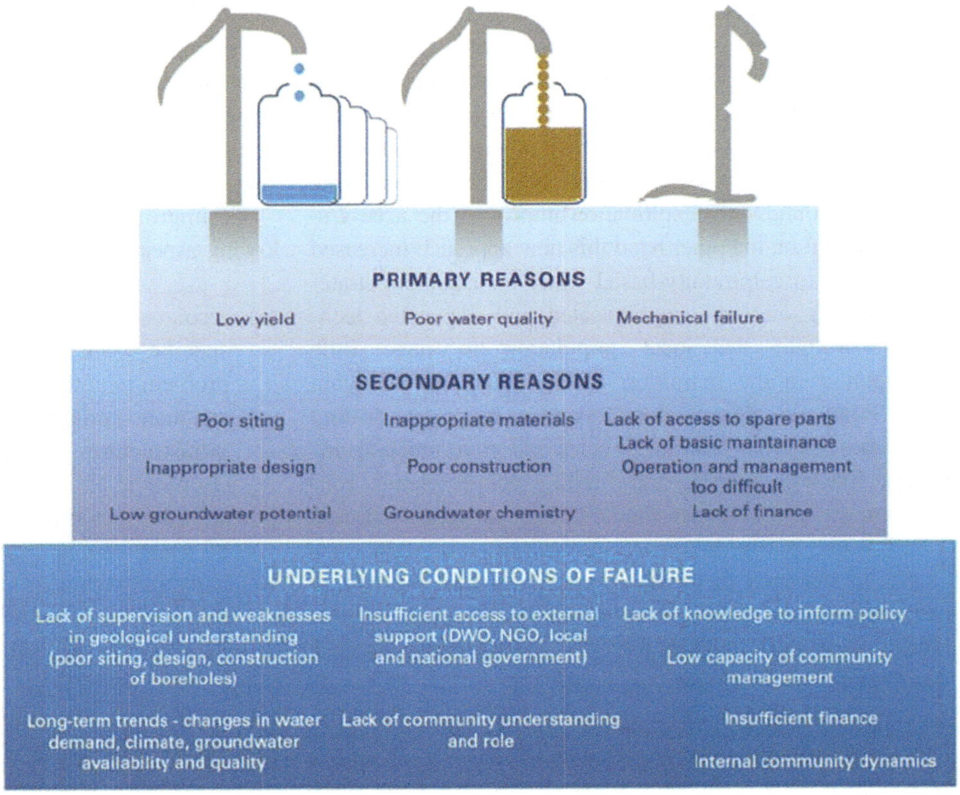


**Fig. 1** Some of the causal factors and underlying conditions which lead to water failure (Bonsor et al. 2015)

towards the achievement of the UN Sustainable Development Agenda for 2030, in what depends on safe and sustainable groundwater resources management.

Governmental authorities must define national and local strategies and guidelines for the implementation of water supply systems, international organizations give technical support, promote capacity building and intermediate dialogue between local stakeholders, donors give financial support and the private sector assure the technical means for groundwater exploitation considering the interest of the population instead of a solely business-oriented policy.

**Acknowledgements** The identification of crucial aspects in the role of international cooperation to promote an adequate and fair groundwater management strategy has been supported by the opinion of a large group of experts within the groundwater discussion group of the Rural Water Supply Network, and we want to acknowledge the important contribution they have provided.

## References

Alam MGM, Allinson G, Stagnitti F, Tanaka A, Westbrooke M (2002) Arsenic contamination in Bangladesh groundwater: a major environmental and social disaster. Intern J Environ Healt Res 3 (12):235–253

Bonsor HC, Oates N, Chilton PJ, Carter RC, Casey V, MacDonald AM, Calow R, Alowo R, Wilson P, Tumutungire M, Bennie M (2015) A hidden crisis: strengthening the evidence base on the sustainability of rural groundwater supplies: results from a pilot study in Uganda. BGS report reference: OR/15/019

Carter C, Ross I (2016) Beyond 'functionality' of handpump supplied rural water services in developing countries. Waterlines 35(1):94–110

Cross P (2013) International WASH cooperation: past and future: an insider's view. IRC. https://www.ircwash.org. Last accessed 29 Nov 2019

Danert K (2015) Manual drilling compendium. RWSN Publication 2015-2

Foster S (2020) Global policy overview of groundwater in urban development—a tale of 10 cities!. Water 12:456. https://doi.org/10.3390/w12020456

Gleeson T, Befus K, Jasechko S, Luijendijk E, Bayani-Cardenas M (2016) The global volume and distribution of modern groundwater. Nature Geosci 9:161–167

IGRAC (2018) Groundwater overview: making the invisible visible. UN-WATER/IGRAC, New York

RWSN (2019) Summary of RWSN E-discussion: cost effective ways to leave no-one behind in rural water and sanitation. RWSN, The Netherlands

Shiklomanov IA, Rodda JC (2003) World water resources at the beginning of the 21st century. Cambridge University Press, UNESCO/IHP International Hydrology Series

Sustainable Development Goals (SDG): knowledge platform homepage https://sustainabledevelopment.un.org/sdgs. Last accessed 29 Nov 2019

Sustainable Development UN Homepage: https://sustainabledevelopment.un.org/. Last accessed 25 Nov 2019

UNICEF (2018) Programming for sustainability in water services: a framework. UNICEF, Paris

UN-WATER (2018) Annual Report 2018.UN-Water, Genève, Switzerland

Vrba J, van der Gun J (2004) The world's groundwater resources. UNESCO & IGRAC, Utrecht. https://www.ircwash.org/sites/default/files/212.0-04WO-18539.pdf

# Training in Groundwater Science and Technology in Portuguese Higher Education (2018–2019)

Ana Isabel Andrade and Manuel Abrunhosa

**Abstract**

In order to develop a review on the current higher academic training in groundwater science in Portugal, a comprehensive survey of the national offer of undergraduate, master and postgraduate specialization courses was carried out. This survey consisted of an extensive research on institutional websites and a survey. This allowed the identification of curricular units (common name: discipline) that include significant groundwater topics. An analysis of the obtained data made it possible to recognize that the training offer included a considerable number of courses from science, engineering, technology, health and humanities. Taken globally, the identified curricular units include subjects from probably all scientific and technological areas in which groundwater is subject of study and intervention. Ethics topics have some relevance in the training. It is thus possible to conclude that there is higher education offer in Portugal covering the diverse areas of scientific and professional knowledge on groundwater. Factors other than the unavailability of qualified professionals and scientists with a higher education degree obtained in Portugal with at least some significant knowledge on groundwater, may explain the existing perception that there is a current deficit in the capacity to intervene in relevant matters in which groundwater plays an essential role.

**Keywords**

Higher education • Groundwater • Training offer • Portugal • Survey

A. I. Andrade · M. Abrunhosa (✉)
Centre for Earth and Space Research of the University of Coimbra (CITEUC), Observatório Geofísico e Astronómico, Almas de Freire, 3040-004, University of Coimbra, Coimbra, Portugal
e-mail: manuel.abrunhosa@gmail.com

A. I. Andrade · M. Abrunhosa
Portuguese Chapter of the International Association of Hydrogeologists (AIH-GP), Porto, Portugal

## 1 Introduction

When liquid freshwater came to consideration immediately rivers, reservoirs and lakes are mentioned and deserve the bulk of attention from scientists, managers and the public for their visible, diverse and undoubtedly important roles in nature and human uses.

However, groundwater that continuously flows within underground aquifers and account for about 99% of the planet's liquid freshwater (Shiklomanov and Rodda 2003) does not receive a proportional consideration. Groundwater is an inseparable part of the global water cycle, which in turn is one of the most important global mechanisms acting at the surface of the Earth, being intimately interconnected with the climate, the ecological, and the geological systems. Water, namely groundwater, since always, has been a pillar of the human system (Cuthbert and Ashley 2014; Manning 2020). It is mainly used to support human life, health and the world economy, through its contribution to food production, domestic and industrial supply, mining and energy. Its availability and quality are determining factors for human, economic and social development. It is by far the most exploited geological resource, about $1 \times 10^{12}$ tonnes per year in 2010 (Margat and van der Gun 2013) with the prospect of continuing to grow for decades (Conti et al. 2016; FAO 2016; Leyronas et al. 2016). This means for the water cycle an increased competition and trade-offs mediated by the human factor in its intervention in the natural hydrological processes, in the surface geological processes, in the maintenance of ecosystems, the biogeochemical processes, in the climate, in the geotechnical processes, and the human development needs. Groundwater is also an essential aesthetic element of the landscape and relates to material and immaterial cultural values and heritage. Human intervention in the water cycle and in particular in the underground cycle has been insufficiently considered (Abbott et al. 2019). However, it has taken such magnitude that the human factor is currently on par with natural factors with an equivalent



M. Abrunhosa et al. (eds.), *Advances in Geoethics and Groundwater Management: Theory and Practice for a Sustainable Development*, Advances in Science, Technology & Innovation,
https://doi.org/10.1007/978-3-030-59320-9_71

capacity to alter dynamics, mass and biogeochemical balances, local and worldwide.

In these circumstances, due to the increased demand for extraction, diversion, consumptive use or rejection and the new attributes and values, groundwater acquired the ability to significantly interact with the dynamics of society, submitting, in addition to the hydrological laws, to human laws.

This happens in a world that is undergoing rapid transformation as a result of population growth and urbanization, increased global levels of development, procurement for geological and other natural resources, climate change and its consequences, and also an aspiration for economic development, equality and safe environment.

Hence, the long lasting and still insufficiently filled need to increase knowledge about groundwater in all its aspects to gain power on its complexity in order to support good practices at all levels, to cooperate in its transdisciplinarity, to regulate, to manage, to protect, to educate the public and eventually litigate on its uses to cope with the daily needs and the challenges of the future (UNESCO 1974; Nash et al. 1990; Wagener et al. 2007; Seibert et al. 2013; Gleeson et al. 2012; Conti et al. 2016).

The modern society responds to those needs by (1) the provision of professionals with sound training in groundwater hydrology and hydrogeology, as is the particular case under consideration in this work, and, concurrently, (2) with the creation and maintenance of jobs where professionals proficient in relevant groundwater issues are able to interact with other professionals and stakeholders, including the public to apply their skills for progress.

In this work, we deal only with training of groundwater science, technology, law, and culture, achieved through teaching at the university, polytechnic or technical levels in Portugal.

The main objective of this study was to assess the situation of higher education courses offer in Portugal that include curricular units (CUs) (formerly designated "disciplines") that address in their syllabus any of the different aspects of the "groundwater" topics. This issue was previously addressed in Barradas and Silva (1985). To achieve this objective, an investigation was designed, including an extensive research on websites and a survey, that allowed us to ascertain facts regarding to: (1) which undergraduate, master and postgraduate specialization courses include groundwater training, (2) the intensity of training in each course and the importance given to geoethics and related topics in this training and (3) the academic training of teachers responsible for the CUs. This will allow a better understanding of the available skills and the potential of response capabilities, with a focus on the professional.

## 2 Materials and Methods

The basic information for this work was obtained from a research carried out on (1) a web-portal of the "Direção-Geral do Ensino Superior" (General Directorate of High Education), where Portuguese higher education institutions are listed, and on (2) the website of each one of the listed institutions where the pages of the undergraduate, master, integrated master (integrating undergraduate and master courses) and postgraduate specialization courses were accessed to identify which courses include CUs with some groundwater training. The criteria for selecting CUs (mandatory or optional) were: (1) explicit terms relating to groundwater in the name of CUs (e.g. hydrogeology, aquifers and groundwater) and (2) the detection of pre-selected keywords in the syllabus of CUs, when available, that indicate the inclusion of topics on groundwater (e.g. hydrogeology, flow in the saturated zone and flow in porous media). The search on the websites took place between 7 December 2018 and 31 January 2019, with 2018–2019 as the adopted reference period.

Following this research, a survey was carried out in order to obtain data not available or only partially available on the websites (e.g. on teacher's academic training and syllabus content, namely the presence of topics on Geoethics, Ethics or Professional Deontology). The link to this survey was sent by e-mail, in March 2019, and resent in May 2019, to all potential CUs teachers/course coordinators identified in the previous websites research; responses were received between March and December 2019.

The information compiled and verified for validation was included in a database in which several markers were broken down allowing the statistical exploration and the generation of representative graphs in order to better achieve the proposed objectives.

## 3 Results and Discussion

The websites search allowed the identification of 669 potential CUs containing groundwater topics in their syllabus (exclusively or with a significant focus). These CUs are divided between public (university: 76.2%, polytechnic: 17.6% and military: 0.4%) and private (university: 4.9% and polytechnic: 0.9%) educational institutions, with greater prevalence in university public education.

Considering only the survey responses, of the total of 669 potentially identified CUs in the website search, responses were obtained for 49 CUs (7 only identified in the survey, increasing the number of potential CUs to 676). As some of the CUs are taught in more than one course, this results in a total of 66 CUs considered for this study, corresponding to 9.8% of the potential CUs. The margin of error of the answers to the survey is 11% with a confidence level of 95%.

It was found that there is a good correlation ($r^2 = 0.9987$) between the number of the above potential CUs and the number of CUs with responses in the survey in each type of institutions, although there were no survey responses for any CU from polytechnic private and military public institutions (that add up to a total of only 9 potential CUs).

The analysis of all data collected shows that during the reference period, a total of 187 CUs were identified in operation, fulfilling the defined requirements, comprising a total of 113 courses divided by 58 undergraduate degrees, 38 master degrees, 13 integrated master degrees and 4 post-graduate specializations; each one of these courses includes at least one curricular unit (CU) that meets the defined criteria (see Fig. 1).

The exhaustive procedure of the methodology allowed to detect CUs integrating topics related to groundwater in courses in which common sense would point out as unlikely (e.g. Undergraduate Degree in Cultural Heritage and Archaeology—Archaeology Branch at the University of Algarve). Some CUs may also not have been included because they do not have their syllabus accessible to the public outside the academic community, and there were no data available from the survey responses. According to Fig. 2, many of the 187 CUs identified in operation are part of the curriculum of courses in the Sciences area, mostly in Geology, Geosciences and Earth Sciences, and in the area of Technology (e.g. Environmental Engineering and Civil Engineering).

The scientific and technological areas in groundwater subjects included in the training offer, at first glance, are very diverse (see Fig. 2), reflecting the adaptation to the training objective at the course level and the recognized great diversity of topics that are considered important and associated with this resource and in general with groundwater hydrology and hydrogeology (Gleeson et al. 2012).

It was found that 93.8% of the CUs teachers that respond to the survey have a Ph.D. degree. Regarding the presence of topics on Geoethics, Ethics or Professional Deontology in the syllabus of the CUs it was obtained from the survey that in 34.8% of the CUs with response there is a concern with these topics, with 1.4–10.0% of the teaching time dedicated to it. This is a very promising result demonstrating the high qualification of the instructors and the interest in including Geoethics, Ethics or Professional Deontology in the academic training of the students.

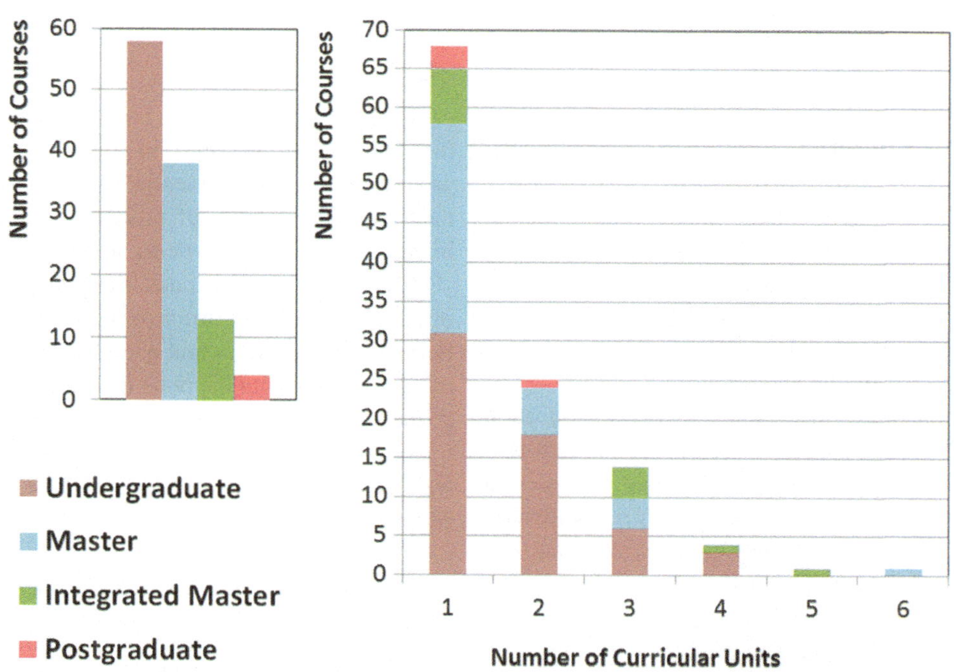


**Fig. 1** Number of courses by different degree (left) and with a discrimination of the total number of selected curricular units included in their syllabus (right)

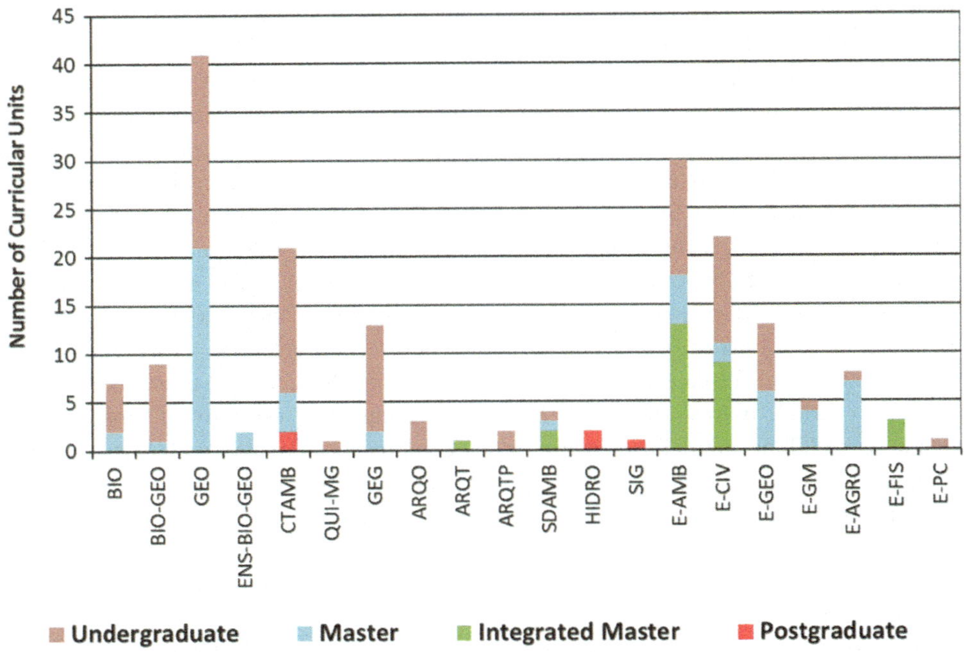


**Fig. 2** Number of curricular units by level of education and area of the courses that they integrate

**Scientific Area of the Courses:**

BIO - Biology
BIO-GEO - Biology/Geology
GEO - Geology
ENS-BIO-GEO - Biology/Geology Teaching
CTAMB - Environmental Sciences and Technologies/Sea/Natural Resources
QUI-MG - Chemistry with minor in Geology
GEG - Geography
ARQO - Archeology
ARQT - Architecture
ARQTP - Landscape Architecture
SDAM - Health and Environment
HIDRO - Hidrogeology
SIG - Geographic Information Systems
E-AMB - Environmental Engineering
E-CIV - Civil/Geotechnics/Mining an Geo-environment Engineering
E-GEO - Geological Engineering
E-GM - Geological and Mining Engineering
E-AGRO - Agricultural/Rural Engineering
E-FIS - Physical Engineering
E-PC - Civil Protection Engineering

## 4 Concluding Remarks

There is a need for CUs syllabus to be available on the institutions' websites in order to allow the identification of their contents.

Groundwater literacy is available to more undergraduate and master degree professionals in science and technology than initially expected by the authors.

This study concludes that there is higher education instituted in the most diverse areas of knowledge and professionals trained to cover a very high range of topics in groundwater subjects. It also extends to other disciplinary areas in which knowledge about groundwater was not commonly expected to be represented, such as some courses in Health, Archaeology and Architecture.

There is a promising result demonstrating the interest in including Geoethics, Ethics or Professional Deontology during the academic training.

The perception that there is in Portugal a current deficit in the ability to intervene in relevant matters in which groundwater plays an essential role stems from factors other than the unavailability of professionals and scientists with higher education acknowledgeable in groundwater.

This study demonstrates that groundwater topics are offered in a wide spectrum of higher education curricula in Portugal. However, several questions were raised during this research, motivating the potential for its future development, as the subsequent inclusion of the university or polytechnic training offer in not graduated and of doctoral courses, as well as the detailing of topics of the curricular subjects adopted in the CUs.

**Acknowledgements** We thank all the CUs teachers/course coordinators that kindly answered our survey. CITEUC is supported by the Portuguese Government through the Foundation for Science and Technology—FCT (Projects UIDB/00611/2020 and UIDP/00611/2020).

## References

Abbott BW, Bishop K, Zarnetske JP, Hannah DM, Frei RJ, Minaudo C, Chapin III FS, Krause S, Conner L, Ellison D, Godsey SE, Plont S, Marçais J, Kolbe T, Huebner A, Hampton T, Gu S, Buhman M, Sayedi SS, Ursache O, Chapin M, Henderson KD, Pinay G (2019) A water cycle for the anthropocene. Hydrol Process 33:3046–3052

Barradas JJM, Silva MAM (1985) O ensino da investigação no domínio da hidrogeologia em Portugal: estado actual e perspectivas. Geonovas Rev Assoc Portg Geol 8(9):151–156

Conti KI, Velis M, Antoniou A, Nijsten G-J (2016) Groundwater in the context of the sustainable development goals: fundamental policy considerations. Brief for GSDR—2016 Update

Cuthbert MO, Ashley GM (2014) A spring forward for hominin evolution in East Africa. PLoS ONE 9(9):e107358

Direção-Geral do Ensino Superior, Pesquisa de Cursos e Instituições. https://www.dges.gov.pt/pt/pesquisa_cursos_instituicoes?plid=372. Last accessed on 31 Jan 2019

FAO (2016) Global diagnostic on groundwater governance. Food and Agriculture Organization of the United Nations, Rome

Gleeson T, Allen DM, Ferguson G (2012) Teaching hydrogeology: a review of current practice. Hydrol Earth Syst Sci 16:2159–2168

Leyronas S, Rojat D, Maurel F, Giraud G (2016) Un cadre d'analyse pour la gouvernance des ressources naturelles. Le cas des eaux souterraines. Rev Econ Dévelop 24:129–148

Manning P (2020) A History of humanity: the evolution of the human system. Cambridge University Press, Cambridge

Margat J, van der Gun J (2013) Groundwater around the world: a geographic synopsis. CRC Press/Balkema

Nash JE, Eagleson PS, Philip JR, van der Molen WH, Klemeš V (1990) The education of hydrologists (Report of an IAHS/UNESCO Panel on hydrological education). Hydrol Sci J 35:597–607

Seibert J, Uhlenbrook S, Wagener T (2013) Preface "Hydrology education in a changing world". Hydrol Earth Syst Sci 17:1393–1399

Shiklomanov IA, Rodda JC (2003) World water resources at the beginning of the twenty-first century. International hydrology series. Cambridge University Press

UNESCO (1974) The teaching of hydrology. UNESCO. Technical papers in hydrology 13. The UNESCO Press, Paris

Wagener T, Weiler M, McGlynn B, Gooseff M, Meixner T, Marshall L, McGuire K, McHale M (2007) Taking the pulse of hydrology education. Hydrol Process 21:1789–1792

# Geoethics in Higher Education of Hydrogeology

José Manuel Azevedo

**Abstract**

The education and training of geosciences (where hydrogeology is included) should be marked out by: (1a) a total scientific accuracy with adequate detail, depth and scope; (1b) the use of appropriate and updated teaching methodologies and didactic means; (2a) a full respect for the trainees' scientific and cultural background, as well as for the socioeconomic reality of the local (continent, country or region) of education or future practice; (2b) the regional, national and supranational legislation about groundwater and water resources; and, fundamentally, (3) with a maximum respect for the Earth natural processes and ecosystems. The teaching and training of hydrogeology with ethics prepare geosciences experts, whose activity is crucial to promote the social and ecological balancing, and consequently to avoid local and regional conflicts.

**Keywords**

Hydrogeology • Higher education • Didactic tools • Geoethics

## 1 Introduction

Here, it is understood by hydrogeology teaching the set of actions, activities and methodologies that are focused to transfer and instill in the trainee (mostly university students of first, second and third cycles) knowledge and skills to teach, investigate and act in all processes directly or indirectly related to groundwater bodies and their natural reservoirs (aquifers).

On the other hand, here it is understood by geo-education with ethics (Wyss and Peppoloni 2014; Di Capua and Peppoloni 2019), the teaching and training of the different fields of geosciences (where hydrogeology is included, Gleeson et al. 2012) supported and marked out by (Fig. 1):

1a. Total scientific accuracy with adequate detail, depth and scope;
1b. The use of appropriate and updated teaching methodologies and didactic means;
2a. A full respect for the trainees' scientific and cultural background, as well as for the socioeconomic reality of the place (continent, country or region) of education or future practice;
2b. The regional, national and supranational legislation about groundwater and water resources;
and, fundamentally,
3. A maximum respect for the Earth natural processes and ecosystems.

This communication tries to express in a synthetic way the transposition of the three educative principals and pillars above described (1a, b; 2a, b and 3) to hydrogeology graduate teaching. This attempt is strongly based on the author acquired experience in hydrogeology teaching for more than 30 years, to students of very different origins (Portuguese, island and continental students, European-Erasmus, students from African countries, Brazil and East-Timor, etc.) and in different sociocultural and continental contexts, such as Europe, Africa and SE Asia.

J. M. Azevedo (✉)
Departamento de Ciências da Terra, Faculdade de Ciências e Tecnologia da, Universidade de Coimbra—FCTUC, Coimbra, Portugal
e-mail: jazevedo@dct.uc.pt

J. M. Azevedo
Centro de Investigação da Terra e do Espaço da Universidade de Coimbra—CITEUC, Coimbra, Portugal


M. Abrunhosa et al. (eds.), *Advances in Geoethics and Groundwater Management: Theory and Practice for a Sustainable Development*, Advances in Science, Technology & Innovation,
https://doi.org/10.1007/978-3-030-59320-9_72

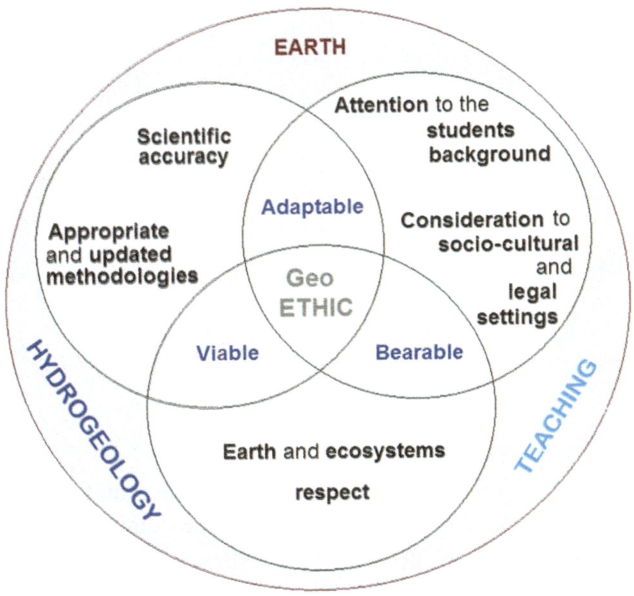


**Fig. 1** Schematic representation of the foundations and behaviors for teaching hydrogeology in the light of geoethics

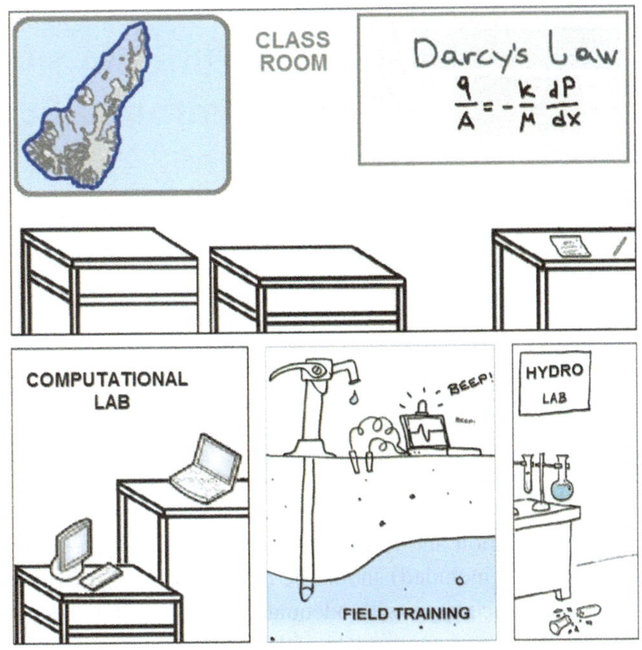


**Fig. 2** Major stages and methodologies for a correct and effective teaching of hydrogeology

## 2 Scientific Accuracy and Appropriated Methodologies

The hydrogeology education and training with ethics necessarily implies a rigorous, profound and updated presentation of the various theoretical and theoretical-practical contents. This assumes a permanent and continuous updating of the teacher-instructor.

On the other hand, the best didactic methodologies and techniques should be used. This implies teaching and training in different stages: classroom, field, chemical hydraulic and computational laboratories (Fig. 2). Obviously, nowadays the field and computational training are crucial on teaching hydrogeology for advanced graduation levels (M.Sc. and Ph.D.).

## 3 Students' Scientific Background and Local, Regional or National Socioeconomic Reality

The success of students' training with geoethics imposes to the teacher-trainer the need to take into account the educational level and skills of the students, as well as knowledge of the social, economic and cultural environment of the region where future hydrogeologists will work. For example, the practice of hydrogeology on a volcanic island requires the kind of knowledge and skills that are considerably different from the same practice in a continental domain, or a European student expresses a susceptibility to the exploration of groundwater bodies very different from a student from Africa or from SE Asia.

## 4 National and Supranational Water Legislation

Along the recent decades, there has been a large convergence between national laws about water and about different activities, directly and indirectly, related to groundwater management. However, there are still significant disparities, particularly between the European legislation and the laws of other countries. These disparities are mainly found within the law application. Thus, the teaching of hydrogeology should include the knowledge and interpretation of national legislations. The usual non-fulfillment of the law guidelines in some regions recommends to alert students for groundwater impacts and remediation techniques.

## 5 Earth and Ecosystems Respect

The continuous increase on human population and societies development leads to a greater water demand on quantity and quality. As a result of this, the pressure on groundwater bodies is growing almost everywhere, and the intervention of hydrogeologists is crucial nowadays. Therefore, the graduate teaching and training in hydrogeology should be guided by considering issues of interest for geoethics such as (1) sustainable development practices, which are based on balanced exploitation and consumption of natural resources (including groundwater bodies); (2) recovery and remediation techniques for stressed aquifers and contaminated groundwater bodies; (3) mitigation of natural risks; (4) the preservation of geodiversity and geoheritage (for instance, the protection and valorization of water springs) and (5) the dissemination and social education in geosciences with emphasis on matters related to natural water bodies.

## 6 Concluding Remarks

Ethical behavior is important in all human activities, including teaching practices. Otherwise, ethics in higher education is crucial for the future of the geosciences, including hydrogeology (Gundersen 2017). Ethical teaching and training in universities is essential to mold hydrogeologists capable to face problems that are not only focused on technical and professional aspects, but also related to societal issues connected to groundwater management. In addition, over-exploitation and increasing contamination of many water bodies have led to important environmental problems and to local, interregional and international conflicts. In this context, the intervention of hydrogeologists assumes a great importance nowadays, and consequently, teaching and training hydrogeology with ethics is fundamental to promote social and ecological balancing.

## References

Di Capua G, Peppoloni S (2019) Defining geoethics. Website of the IAPG, International Association for Promoting Geoethics. https://www.geoethics.org/definition. Accessed on Jan 2020

Gleeson T, Allen DM, Ferguson G (2012) Teaching hydrogeology: a review of current practice. Hydrol Earth Syst Sci 16:2159–2168

Gundersen L (ed) (2017) Scientific integrity and ethics in the geosciences. American Geophysical Union and Wiley, NY

Wyss M, Peppoloni S (2014) Geoethics, ethical challenges and case studies in earth sciences. Elsevier

# And if the Spring that Provides the Farm with Water Should Run Dry?—A Geoethical Case Applied in Higher Education

Alexandra Cardoso, Nir Orion, Cristina Calheiros, and Clara Vasconcelos

**Abstract**

Geoethics is a scientific discipline in development whose primary concern is to instill a better relationship between citizens and the planet they inhabit. With a multiplicity of ethical, social, and cultural values, the significant input geoethics can have on decision-making worldwide is recognized. Knowing how the planet Earth works is the basis for all geoscientific knowledge and essential for the resolution of the great challenges that humankind faces every day. The Geoethics Outcomes and Awareness Learning (GOAL) Project is the first to focus on taking geoethics to higher education curricula, to spread a geoethical perspective across our society. One of the educational resources, developed by the GOAL team of experts, was implemented on a higher education course. The results showed that students ($n$ = 10) are aware of the strict relation between human actions and Earth system dynamics. Students also expressed the concern to educate citizens on the global problems faced, as for daily decisions. They take a preventive position when it concerns human action impacting Earth's balance. This study is important to develop the students' geoethical background, as well as to look for ways to improve it, hopefully contributing to a better future on Earth.

**Keywords**

Geoethics • Higher education • Geoheritage • Earth system

## 1 Introduction

Geoethics is a thematic that has been arising over the last decade on the scientific community. In the past years, geoethics has been topic of discussion on important scientific reunions like the International Geological Congresses, EGU General Assemblies, The Geological Society of America Annual Meetings, International Union of Geological Sciences Executive Committee Meetings, Geological Society of London Events, International Conference on Geoscience for Society, among others. However, geoethical thinking and knowledge are absent among the broader public. Consequently, geoethics is not a subject considered in the daily life decisions of most citizens. This issue creates a great need to educate and raise awareness among citizens about this scientific field, especially the future geoscientists, an issue already identified by the International Association for Promoting Geoethics (IAPG)—the leading organization fostering geoethics (Bobrowsky et al. 2017; IAPG 2020; Mogk et al. 2017).

Aware of the importance and urgency of dissemination and practice of geoethics, and recognizing that education is the best way to change future decisions (Mogk et al. 2017; Almeida and Vasconcelos 2015; Vasconcelos et al. 2016), an international and interdisciplinary consortium, coordinated by a Portuguese team and with the supporting partnership of IAPG, designed the Erasmus + project GOAL—Geoethics Outcomes and Awareness Learning (2017-1-PTO1-KA203-035790). An educational geoethical framework was defined, covering a wide range of geosciences topics, such as sustainable resources exploitation and water management, by combining GOAL members' multiple

A. Cardoso (✉)
Interdisciplinary Centre of Marine and Environmental Research (CIIMAR) and Science Teaching Unit, Faculty of Sciences, University of Porto (FCUP), Porto, Portugal
e-mail: alexandra.cardoso@fc.up.pt

N. Orion
Weizmann Institute of Science, Rehovot, Israel

C. Calheiros
Interdisciplinary Centre of Marine and Environmental Research (CIIMAR), University of Porto, Porto, Portugal

C. Vasconcelos
Interdisciplinary Centre of Marine and Environmental Research (CIIMAR), Science Teaching Unit & Department of Geosciences, Environment and Spatial Planning, Faculty of Sciences, University of Porto (FCUP), Porto, Portugal


M. Abrunhosa et al. (eds.), *Advances in Geoethics and Groundwater Management: Theory and Practice for a Sustainable Development*, Advances in Science, Technology & Innovation,
https://doi.org/10.1007/978-3-030-59320-9_73

expertise. The project aims to develop a geoethical syllabus and several educational resources to make possible the teaching of geoethics in higher education (HE) (GOAL 2020). Afterward, the syllabus will be to implement on different higher institutions during geosciences courses. The purpose is to start to develop a geoethical culture from aspiring geoscientists, passing through the geoscience's scientific community, and reaching the society. In the long term, it is expected that geoethics can become a must-take approach for the future of society and, more importantly, of the planet Earth. A geoethical approach contemplates a series of ethical, social, and cultural values as a work basis. These values must be considered for good decision-making concerning human actions on the planet. Also, they express a wide span of points, from values related to geosciences practice—like honesty, integrity, and transparency—to values associated with society's scientific literacy (Bobrowsky et al. 2017; Mogk et al. 2017).

In this work, the educational resource (ER) "Earth system nexus human interaction: a geoethical perspective" was implemented in a HE class. The latter intends to promote "geoethical values (ethical, cultural, and social) related to human interaction with the Earth system (ES) through reflection on georesources, geoheritage and the need for geoscientists to raise public awareness of their work" (Calheiros et al. 2019, p. 1)—available here. To achieve this, Paço de Calheiros, a seventeenth-century farm located in the north of Portugal, was taken as an example. The authors explain the different relationships between the farm activities and the ES, contextualizing them within the local geological particularities. Once the water is of high importance to the farm's subsistence, the possibility of jeopardizing its supply is the focus of the geoethical dilemma raised in this ER. The disappearance or shift of the natural watercourse upstream might implicate severe changes in the Earth subsystems downstream, where the farm is located. Through a hypothetical dialogue, between the Count of Calheiros and his grandson, several geoethical questions were placed, concerning this scenario.

## 2 Methodology

During a geosciences education class, on a HE in biology and geology science teaching's master course, the referred case was implemented. The participants were divided into three groups: two groups with three and one group with four students. The groups analyzed the case and answered to the proposed questions, following the procedure described in the ER. All further information presented in the ER was available and open access. The tasks lasted for one hour and a half. The research team assured the intergroup homogeneity and intragroup heterogeneity.

This qualitative research was based on content analysis of the students' answers, to detect answer patterns and trends, as well as the frequency of responses to objectives. The count of frequency should be noted that the frequency unit is the group itself and not individual responses. For each question, there is a variable number of objectives associated. Also, only the objectives correctly identified, for each question, were considered. The content of the responses was cited whenever the results needed additional information. In all cases was guaranteed confidentiality.

The content analysis made on this stage considered the five questions on the case, as well as the respective correspondence to the objectives set. For question 1—"What would happen if the spring that provides the house with water should run dry?"—were set the following objectives: (a) to defend geoethical values to preserve the ES and (b) to enhance the geological landscape by raising human aesthetic values like, for example, respect for the land that sustains our lives. To the question 2—"If there was a decision to divert the part of the stream the farm has a right to (e.g., to support building a factory that would bring jobs to the village inhabitants), what would be the impact on wine production and organic farming?"—were established, besides the objective (a), the following objectives: (c) to understand the need to strictly respect the natural systems and dynamics when designing interventions on the environment and (e) to develop sustainable activities in order to ensure energy supply and natural resources for future generations. Concerning question 3—"If a road were to be built across the farm (e.g., to improve access to the main village), to what extent would we be able to preserve the geological heritage?"—the objectives attributed were (a), (b), and (c). In the case of question 4—"What are the consequences of not informing the owners of these houses and the village inhabitants about the area's geology?"—the authors corresponded it, beyond the objectives (a) and (b), with the objectives: (g) to bolster citizens' awareness of the work of geoscientists; (h) to boost geoethical education in schools and HE; and (i) to raise citizens' awareness of geoethics. Finally, to the question 5 "How can we avoid the risk of not preserving this geoheritage site?", in addition to the objectives (a), (b), and (g), it corresponded with the objectives: (d) to discover ways of protecting and enhancing geo-diversity for the sustainable development of communities and (f) to explain the work of geoscientists to preserve the ES better.

The convenience sample comprised ten students ($n$ = 10) that participated voluntarily. All the students were enrolled in the same master's degree in biology and geology science teaching, but they had different backgrounds—having first degrees in biology, geosciences, environmental sciences, and landscape architecture. Despite the diversity of the participants' initial training, the prerequisites, necessary for understanding the resource, were ensured, once all respondents had completed geosciences curricular units.

## 3 Results

The results obtained are systematized in Table 1. For each question, the objectives achieved per group are identified, and the correspondent frequencies are listed.

Regarding the first question, all groups reached objective (c), but the same did not occur for the objective (a). The following question had three objectives and one of the groups achieved them all. Groups 1 and 3 only identified the objective (c). The objectives of question 3 were fully achieved by groups 1 and 2. However, group 3 did not answer concerning the objective (a). Concerning the fourth question, group 1 achieved all objectives, while group 2 only answered to objective (b). Group 3 did not complete the objective (h). Finally, the objectives of question 5 were achieved by groups 2 and 3, but group 1 failed the objectives (f) and (g). Overall, the results were satisfactory. In total, each group could have identified 17 objectives, being group 2 reached 13 (76.5%) and groups 1 and 3 achieved 12 (70.6%).

## 4 Discussion

All groups showed that they understood the water spring is essential to the maintenance of the house and inherent farm activities, and if it would run dry the dynamics of the natural system would be negatively affected. Nevertheless, they all seem not to invoke explicitly geoethical values in the answers. Maybe this happened because the students could not relate the possibility of the spring run dry with the disrespect of several values associated with the ES. All groups predicted the impact on wine production and organic farming, but only one of them (group 2) answered focusing on all objectives. The other two groups only based their answers on the perturbation of ES dynamics. Only one group (group 2) suggested some ways to prevent or minimize the impact of the stream diversion from the farm and enunciated the need to respect and value the heritage. The other groups centered their answer on the negative consequences of this possible change, maybe because it was the most direct answer considering the formulated question. The geological heritage preservation in case of road construction across the farm was the subject of thought of the third question, and two of the groups (groups 1 and 2) considered all the objectives on their answers. However, one group did not make a direct relation between road construction and the jeopardizing of important geoethical values. Concerning the importance of geosciences literacy addressed on the fourth question, only one group (group 1) was capable of completing all objectives, defending that "[…] knowing what surrounds us is the first step in heritage protection" (group 1). However, one of the groups (group 2) only mentioned the consequences of a non-informed community, not reflecting on the communication needed between geoscientists and society. The third group only failed in not referring to the importance of teaching geoethics in schools and HE. Maybe the question was not formulated clearly and leads to a directed answer instead of focusing on the consequences of lack of geoscientific literacy. The last question's objectives were all achieved by two of the three groups, but the first group did not consider the need to explain the geosciences

**Table 1** Results obtained regarding each question per group

| Question | Group 1 | Group 2 | Group 3 | Frequencies |
|---|---|---|---|---|
| 1 | (c) | (c) | (c) | (a) ($n$ = 0; 0.0%)<br>(c) ($n$ = 3; 100.0%) |
| 2 | (c) | (a), (c), (e) | (c) | (a) ($n$ = 1; 33.3%)<br>(c) ($n$ = 3; 100.0%)<br>(e) ($n$ = 1; 33.3%) |
| 3 | (a), b), (c) | (a), (b), (c) | (b), (c) | (a) ($n$ = 2; 66.6%)<br>(b) ($n$ = 3; 100.0%)<br>(c) ($n$ = 3; 100.0%) |
| 4 | (a), (b), (h), (i) | (b) | (a), (b), (i) | (a) ($n$ = 2; 66.6%)<br>(b) ($n$ = 3; 100.0%)<br>(h) ($n$ = 1; 33.3%)<br>(i) ($n$ = 2; 66.6%) |
| 5 | (a), (b), (d) | (a), (b), (d), (f), (g) | (a), (b), (d). f), (g) | (a) ($n$ = 3; 100.0%)<br>(b) ($n$ = 3; 100.0%)<br>(d) ($n$ = 3; 100.0%)<br>(f) ($n$ = 2; 66.6%)<br>(g) ($n$ = 2; 66.6%) |
| Total | 12/17 (70.6%) | 13/17 (76.5%) | 12/17 (70.6%) | 17 objectives |

practices for ES preservation. One hypothesis is that students may not recognize or have the needed knowledge to relate the geoheritage conservation with geosciences practice. Likewise, the group may have focused the reflection and consequent answer only on possible ways to conserve the geoheritage of the region.

Additionally, the students may not have achieved some objectives because the expecting ones for each question were not evident. For future case applications is recommended that the direct connection between questions and objectives is explained clearly to the students. This way, probably even better results will be achieved.

Some previous studies showed that there is a lack of understanding about geoethics and its connections with the ES dynamics, but after elucidated about the purpose of geoethics, citizens acknowledge its importance and relevance for the future society (Almeida and Vasconcelos 2015; Vasconcelos et al. 2016). It is noteworthy that research concerning the application of geoethics-based educational resources on science classes is scarce. This fact might be related to geoethics being a recent field, so the existence of more geoscience's educators and researchers conscious of these ethical aspects is needed. However, two recent works revealed that participant students show an identic way of applying geoethical principles to cases presented through different scenarios. Both applied cases focused on georesources exploitation and presented a series of dilemmas students had to reflect about, having in mind a geoethical perspective. As has happened in the present work, it is clear that students are capable of predicting the pros and cons of several human actions and take a preventive approach for planet Earth, considering some essential geoethical values, even though sometimes indirectly (Vasconcelos 2018; Ribeiro et al. 2020).

## 5 Concluding Remarks

The introduction of geoethics in geosciences education curricula, namely HE, is of utmost importance considering the current planetary panorama. Students need to understand the responsibility of decision-making for the society's future, and it can be done through reflection upon geoethical cases and dilemmas. In this work, the implementation of the ER showed students the need to respect the ES dynamics, and all participating groups generally recognized this. Furthermore, participants predicted consequences concerning possible changes in the farm and its surroundings that may come from non-informed decision-making and can jeopardize a balanced human–Earth interaction. However, it is also relevant to consider that some of the students did not achieve all the objectives. This fact can be due to a multiplicity of factors, but for future practices, the teacher can try to surpass this by better clarifying the objectives expected for each question. Nevertheless, the teachers can try to transcend this by promoting a discussion of the geoethics' values and principles before groups starting to work.

**Acknowledgements** This work was supported by: (1) Erasmus+ Programme thought GOAL project-ref. 2017-1-PTO1-KA203-035790; (2) FCT through the PhD scholarship—"Geoethics in Sustainable Development: implementation and evaluation of a syllabus in higher education"—ref. SFRH/BD/137852/2018; (3) FCT through ref. UIDB/04423/2020 and UIDP/04423/2020; and (4) Science Teaching Unit, FCUP.

## References

Almeida A, Vasconcelos C (2015) Geoethics: master's students knowledge and perception of its importance. Res Sci Educ 45 (6):889–906

Bobrowsky P, Cronin VC, Di Capua G, Kieffer SW, Peppoloni S (2017) The emerging field of geoethics. In: Gundersen LC (ed) Scientific integrity and ethics with applications to the geosciences. Wiley, Hoboken, pp 175–212

Calheiros C, Orion N, Vasconcelos C (2019) Earth system nexus human interaction: a geoethical perspective. In: https://tinyurl.com/y3cqnmey. Last accessed on 01 Feb 2020

GOAL. https://goal-erasmus.eu/. Last accessed on 28 Feb 2020

IAPG. https://www.geoethics.org/. Last accessed on 03 Feb 2020

Mogk DW, Geissman JW, Bruckner MZ (2017) Teaching geoethics across the geoscience curriculum: why, when, what, how, and where? In: Gundersen LC (ed) Scientific integrity and ethics in the geosciences. Wiley, Hoboken, pp 231–265

Ribeiro T, Cardoso A, Pinto T, Vasconcelos C (2020) Can we dare say modern society doesn't need raw material? Reflecting about the increasing demand for teaching geoethics. In: Pixel (ed) NPSE conference proceedings 2020. Filodiritto Editore, Bologne, pp 299–302. https://conference.pixel-online.net/NPSE/virtual_presentation_scheda.php?id_abs=4410

Vasconcelos C (2018) The experience of the University of Porto with cases of education for sustainable development. In: Iwińska K, Jones M, Kraszewska M (eds) Widening interdisciplinary sustainability education. Collegium Civitas, Warsaw, pp 241–242

Vasconcelos C, Torres J, Vasconcelos L, Moutinho S (2016) Sustainable development and its connection to teaching geoethics. Episodes 39(3):509–517

# Teaching Groundwater Resources and Geoethics in Portuguese Secondary Schools

Gina P. Correia and Hélder Pereira

**Abstract**

The growing importance of geoethics justifies its inclusion in formal education, particularly, when related to essential themes to the sustainability on Earth. In Portugal, the study of groundwater resources is part of the curriculum of subject areas, such as geology and geography. In order to evaluate the possibility of address geoethics while teaching about groundwater resources in secondary education, we analysed the curriculum reference documents of biology and geology (11th grade - 16th years old students), geography A (10th grade - 15th years old students), geography C (12th grade - 17th years old students) and geology (12th grade). The data obtained show that, although the concept of geoethics is not formally integrated into the analysed curricula, it is possible to explore the principles and moral values about the importance of preserving the Planet, throughout formal activities. This is important as we consider it to be fundamental that, by the end of upper secondary education, students should have acquired a responsible and geoethical awareness that could influence their daily life and actions. In our opinion, this should not only be a competence of Earth science professionals but, should involve all professions and all citizens, to which the promotion of non-formal activities can also contribute.

**Keywords**

Curriculum • Education • Geosciences

G. P. Correia (✉)
University of Coimbra, Centre for Earth and Space Research (CITEUC), Coimbra, Portugal
e-mail: gina_maria@sapo.pt

H. Pereira
Escola Secundária de Loulé, Loulé, Portugal

## 1 Introduction

The concept of geoethics has evolved, as its use in the contemporary geoscientific panorama has increased. If its initial use was related to the sustainable exploitation of non-renewable resources, currently, the definition of geoethics is much broader, as can be seen from the content of the "Cape Town Statement on Geoethics" (Di Capua et al. 2017). According to several authors (e.g. Bobrowsky et al. 2017), it is geosciences and geoscientists who seem to have the proper competence for addressing and resolving issues on sustainability and protection of the planet, including those related to the exploration and maintenance of the quality of groundwater resources.

In the Portuguese secondary education system, the teaching of groundwater resources is carried out within the geoscience content of science/humanities courses (one of secondary education offers that are mainly aimed at students wishing to continue their studies in higher education) in the subject areas of biology and geology (11th grade - 16th years old students) (BG11), geography A (10th grade - 15th years old students) (GeoA10), geography C (12th grade - 17th years old students) (GeoC12) and Geology (12th grade) (G12). With this study, we intended to evaluate the possibility of address geoethics through the scope of the teaching of groundwater resources in secondary education in Portugal.

## 2 Methods

This study focused on the analysis of current curriculum reference documents: curriculum programmes and core learning from subject areas such as BG11, GeoA10, GeoC11 and G12, on the contents/domains/themes in which groundwater resources are taught (Table 1).

With the analysis of the curriculum reference documents, we wanted to identify the objectives, competences and core


M. Abrunhosa et al. (eds.), *Advances in Geoethics and Groundwater Management: Theory and Practice for a Sustainable Development*, Advances in Science, Technology & Innovation,
https://doi.org/10.1007/978-3-030-59320-9_74

**Table 1** Groundwater resources in the curriculum reference documents of scientific-humanistic courses in secondary education (Alves et al. 2001; Amador et al. 2003; Amador and Silva 2004; Martins et al. 2002; ME 2018a, b, c, d)

| Level of education | Subject areas | Contents[a]/Domain[b]/Theme[c] |
|---|---|---|
| 10th grade | Geography A | The natural resources available to the population uses, limits and potentialities |
| 11th grade | Biology and geology | Sustained exploitation of geological resources |
| 12th grade | Geography C | A world of contrasts |
| | Geology | The Earth yesterday, today and tomorrow |

[a]Designation on the curriculum programmes
[b]Designation on the BG11/G12 core learning
[c]Designation on the geography core learning

learning that allow students to acquire knowledge and develop skills in geoethics. For convenience and curriculum likeness, the subject areas were divided into two analysis groups: BG11 and G12; GeoA10 and GeoC11.

## 3 Results

The data obtained from curriculum reference documents analysis are presented in Tables 2 and 3.

## 4 Discussion

The data obtained, in the content and subject areas under consideration, allows us to affirm that in the teaching of this curriculum there is an obvious approach of ethical, social and cultural values that form an individual ethically committed to the preservation of the Geo-Planet.

Thus, in the BG11 and G12 curriculum reference documents (Table 2) and in those of GeoA11 and

**Table 2** Results of the analysis of the curriculum programmes and the core learning in the subject areas of biology and geology (11th grade) and geology (12th grade) in the contents/domains/themes in which groundwater resources are taught

| | Biology and geology 11th grade | Geology 12th grade |
|---|---|---|
| Curriculum | | |
| Contents | Theme IV—Geology, problems and everyday materials<br>Subtheme 3—Sustained exploitation of geological resources | Theme III—The Earth yesterday, today and tomorrow<br>Subtheme 3.1—Water exploitation and contamination |
| Main goals | • To identify geological resources and their applicability in a Science, Technology, Society and Environment (STSE) perspective<br>• To develop attitudes of geological heritage valorization (memory of the Earth) | • To recognize the role of mankind in water exploration and contamination<br>• To synthesize, within the framework of geological knowledge, the main environmental problems at the beginning of the twenty-first century |
| Skills | • To recognize the contributions of geology in the areas of geological risk prevention, spatial planning, environmental resource management and environmental education<br>• To assume opinions supported by a scientifically based environmental awareness<br>• To assume attitudes in defence of the geological heritage | |
| Core learning | | |
| Domain | Sustained exploitation of geological resources | The Earth yesterday, today, and tomorrow |
| Core learning | • To relate the geological characteristics of a region to the aquifer formation conditions (unconfined and confined)<br>• To analyse data and formulate critical judgments, scientifically based, on the sustainable exploitation of geological resources in Portugal | • To explain experimental data related to contamination of geological resources, appreciating knowledge from other disciplines (e.g. mathematics, biology, computer applications B)<br>• To infer on possible scenarios for the twenty-first century, because of global warming and environmental changes |
| Transversal core learning | • To formulate and communicate critical opinions, scientifically grounded and related to Science, Technology, Society and Environment (STSE)<br>• To communicate knowledge from different subjects to deepen geological contents | |

**Table 3** Results of the analysis of the curriculum programmes and the core learning in the subject areas of geography A (10th grade) and geography C (12th grade) in the contents/domains/themes in which groundwater resources are taught

| | Geography A 10th grade | Geography C 12th grade |
|---|---|---|
| Curriculum | | |
| Contents | Theme 2—Natural resources available to the population: uses, limits and potentialities<br>Subtheme 2.3—Water resources | Theme 4—A world of contrasts<br>Subtheme 4.3—Environmental problems, different human impacts? |
| Main goals | • To be interested in reconciling economic growth and improving the quality of life of the people, associating them with the enhancement of the natural and cultural heritage<br>• To speak about the existence of conflicts in the use of space and resource management with situations of unequal development at local and/or regional level | |
| Skills | • To recognize the role of the water cycle in the Earth's balance preserving<br>• To relate water availability with the amount and type of precipitation<br>• To know the factors that condition aquifer productivity<br>• To recognize that human activities interfere with water quantity and quality<br>• To assess risks in water resources management<br>• To infer the need to establish international agreements in the management of water resources<br>• To discuss measures leading to the control of water quantity and quality | • To understand the global dimension of some environmental problems<br>• To understand the need for international cooperation to solve global problems<br>• To discuss measures proposed at international conferences to address global environmental problems<br>• To relate the different impacts caused by environmental degradation to the degree of development of countries<br>• To discuss the sustainability capacity of large urban agglomerations |
| Core learning | | |
| Theme | Natural resources | A world of contrasts |
| Core learning | • To relate the distribution of the main features of the subsoil with the geomorphological units<br>• To compare the distribution of major energy resources and energy distribution and consumption networks with hydrography, solar radiation and underground resources<br>• To relate water availability to energy production, agricultural use, population water supply or other uses<br>• To identify the main river basins and their relationship with water availability | • Issue an opinion on concrete actions that enhance the appropriate use of essential resources at the global level.<br>• To discuss and express an opinion on the measures proposed at international conferences to solve global environmental problems, considering the sustainability of the planet |

GeoC12 curriculum reference documents (Table 3), it is clear the intention to develop in the geoscience student, a future conscious and ethically responsible citizen, skills that will allow him to make concrete decisions, in his daily actions and as a professional, for the conservation of the geoenvironment or to deal with the global economic and environmental challenges (e.g. Bobrowsky et al. 2017); and, at the same time, to face one of the challenges of Geoethics that of having a greater social intervention (Almeida and Vasconcelos 2014). In these curriculum reference documents, the following main goals/core learning become more relevant: 'to develop attitudes of geological heritage valorization' (curriculum programme, BG11); 'to recognize the role of mankind in water exploration and contamination' (curriculum programme, G12); 'to analyse data and formulate critical judgments, scientifically based, on the sustainable exploitation of geological resources in Portugal' (core learning, BG11); 'to infer on possible scenarios for the twenty-first century, because of global warming and environmental changes' (core learning, G12); 'to speak about the existence of conflicts in the use of space and resource management with situations of unequal development at local and/or regional level' (curriculum programme, GeoA11 and GeoC12); 'issue an opinion on concrete actions that enhance the appropriate use of essential resources at the global level' (core learning, GeoC12). To achieve all these skills the effectiveness of using a didactic methodology incorporating hands on skills should be explored.

## 5 Concluding Remarks

Studying geoethics in geosciences can contribute to broadening the applications of geology to everyday life; to achieving a greater perception of the characteristics of geological knowledge and the potential and limits of this knowledge in society; to clarify the role of the geoscience professional in the search for a path for the development of a humanity that is more just and more respectful of the environment (Almeida and Vasconcelos 2014). In this context, we believe that the secondary education curriculum in which groundwater resources are taught, enables the training of citizens who, at the end of compulsory education and regardless of whether they undertake a professional activity in the geosciences area, demonstrates an ethical awareness and a capacity to act responsibly on the abiotic environment. Therefore, these citizens should be able to evidence a personal awareness and capacity for autonomy and responsible actions in daily life as well.

This study focused on the approach to geoethics through formal education. However, the training of an individual is not confined to what he learns only in the curriculum. His full education includes non-formal learning in which the above-mentioned skills can also be developed. Examples of these non-formal learning activities include: the Side-Event to the Geoethics and Groundwater Management Congress, an educational programme that promoted a national competition/contest (GEOETH&GWM'20) and the International Geoethics Day promoted by the IAPG (2019), the first for pre-university school communities and the second for society in general.

**Acknowledgements** CITEUC is funded by National Funds through FCT—Foundation for Science and Technology (project: UID/Multi/00611/2020) and FEDER—European Regional Development Fund through COMPETE 2020—Operational Programmer Competitiveness and Internationalization (project: POCI-01-0145-FEDER-006922).

## Reference

Almeida A, Vasconcelos C (2014) Dilemas da Geoética: suas potencialidades na perceção das características do conhecimento geológico. Comunicações Geológicas 101(III):1189–1192

Alves ML, Brazão MM, Martins OS (2001) Programa de Geografia A 10.° e 11.° anos – Curso Científico-Humanísticos de Ciências Socioeconómicas e de Ciências Sociais e Humanas. DES, Ministério da Educação, Lisboa

Amador F, Silva CP, Batista JP, Valente RA (2003) Programa de Biologia e Geologia. Componente de Geologia. 11.° Ano. DES, Ministério da Educação, Lisboa, pp 15–36

Amador F, Silva M (2004) Programa de Geologia 12.° ano – Curso Científico-Humanístico de Ciências e Tecnologias. DGIDC, Ministério da Educação, Lisboa

Bobrowsky P, Cronin VS, Di Capua G, Kieffer SW, Peppoloni S (2017) The emerging field of geoethics. In: Gundersen LC (ed) Scientific Integrity and ethics with applications to the geosciences. Special Publication American Geophysical Union, Wiley, Washington, DC, USA, pp 175–212

Di Capua G, Peppoloni, S, Bobrowsky, PT (2017) The Cape Town statement on geoethics. Ann Geophys 60, Fast Track 7. https://doi.org/10.4401/ag-7553

GEOETH&GWM'20—Geoethics & groundwater management congress homepage. https://geoeth-gwm2019.wixsite.com/porto. Last accessed on Jan 2020

IAPG—International Association for Promoting Geoethics (2019) Homepage. https://www.geoethics.org/. Last accessed on Jan 2020

Martins OS, Brazão MM, Alves ML (2002) Programa de Geografia C 12.° ano – Curso Científico-Humanísticos de Ciências Socioeconómicas e de Ciências Sociais e Humanas. DES, Ministério da Educação, Lisboa

ME—Ministério da Educação (2018a) Aprendizagens Essenciais de Biologia e Geologia 11.° Ano Ministério da Educação, Lisboa

ME—Ministério da Educação (2018b) Aprendizagens Essenciais de Geografia A 10.° Ano Ministério da Educação, Lisboa

ME—Ministério da Educação (2018c) Aprendizagens Essenciais de Geografia C 12.° Ano Ministério da Educação, Lisboa

ME—Ministério da Educação (2018d) Aprendizagens Essenciais de Geologia 12.° Ano Ministério da Educação, Lisboa

# Geoethics Calls for Action: An Interactive Module to Communicate Geosciences

Tiago Ribeiro, Alexandra Cardoso, Joana Silva, Dulce Lima, and Clara Vasconcelos

**Abstract**

In a society that is facing problems related to the planet's sustainability for humankind's survival and where science is part of the daily life of any citizen (in a more prominent or hidden way), it is increasingly essential to establish a correct and coherent geosciences' communication. This latter emergency is particularly significant in the geoethics' scope, since it leads to dilemmas that affect humanity and the planet itself, being essential to promote public awareness about their actions. However, we live in a society where conscious effects of human dependence on the Earth are not fully understood, and it is still extremely required. The change in our daily behaviours is urgent and necessary. In geoethics' field of study, human interaction with the Earth' subsystems—geosphere, biosphere, atmosphere, hydrosphere, and cryosphere—is the primary concern, but there is yet a lack of proposals for communicating geoethics principles and values among the citizens. Within this context, an interactive geoethics communication module—"Geoethics calls for action"—which may be applied in a museum or science centre, was created to reach a broad public and to provide a reflection on human actions and their consequences on planet Earth. The citizens' engagement with the geoethical dilemmas, present in the module related to the human interactions with the Earth, may develop a geoethical thinking and more conscious actions. Finally, it could raise people's awareness regarding geoethics and its importance. Underlying these intents are not only the geoethics communication but also the endorsement of higher respect for planet Earth and its subsystems.

**Keywords**

Geoethics • Geosciences communication • Interactive module

T. Ribeiro (✉) · A. Cardoso · J. Silva · D. Lima · C. Vasconcelos
Science Teaching Unit (UEC), Faculty of Sciences (FCUP), University of Porto, Porto, Portugal
e-mail: tiago.ribeiro@fc.up.pt

T. Ribeiro · A. Cardoso · D. Lima
Institute of Earth Sciences (ICT), Porto, Portugal

C. Vasconcelos
Interdisciplinary Centre of Marine and Environmental Research (CIIMAR), Matosinhos, Portugal

## 1 Introduction

Science is an important aspect present in the daily life of any citizen, so it is essential to have sensitivity and understanding of certain scientific concepts to be capable of taking advantage of their potential to solve global and even everyday problems. The need to bring science closer to society—to motivate the development of a scientific culture where the reality of science is recognized, and the communication between its protagonists and the citizens is transparent—becomes unavoidable. The vehicles to achieve it can be diverse: from science centres, museums, and media to a simple website. Science communication (SC) is a scientific field in full development and is of supreme importance in modern times that today's society faces. With approximately half a century of discussion, analyses, results, SC practice and investigation are becoming central at the science-society interface (Burns et al. 2003; Guenther and Joubert 2017; Trench et al. 2014). Naturally, there have been various interpretations of what SC is, for a science communicator, in the present context, it is defined as "the use of appropriate skills, media, activities, and dialogue to produce one or more of the following personal responses to science (the AEIOU vowel analogy): Awareness, Enjoyment, Interest, Opinion-forming, and Understanding" (Burns et al. 2003, p. 183).

Knowledge and understanding of the geosciences by society is necessary for greater awareness of the direct relationship that all human beings have with the Earth



M. Abrunhosa et al. (eds.), *Advances in Geoethics and Groundwater Management: Theory and Practice for a Sustainable Development*, Advances in Science, Technology & Innovation,
https://doi.org/10.1007/978-3-030-59320-9_75

system (Mogk et al. 2018; Peppoloni and Di Capua 2016). A society aware of its daily dependence on the natural environment is a society more receptive to the (re)education of its current behaviours and the adoption of geoethical values. This awareness may create new and more sustainable and respectful behaviours and, consequently, create a society capable of improving its relationship with the natural systems. Geosciences communicators and educators, as well as geoscientists, are the most suitable professionals to promote geosciences communication (by developing awareness and public engagement about geoethical issues) and education (by teaching scientific concepts and skills to understand the geoethical dilemmas better), based on geoethical values. A geoethical approach should be not only accessible to school-age citizens but also extendable to the rest of society, promoting lifelong education and constant contact with the most recent geoethical approaches in facing dilemmas that also interest society as a whole (Bobrowsky et al. 2018; Peppoloni and Di Capua 2016). Since the human being can be considered a "geological agent"—capable of drastically altering the dynamics of the planet—changes of natural environments have to be carried out more consciously. Today, more than ever, it is essential that all citizens understand their role in the actions they take against the Earth and its consequences.

Since geoethics is a recent and emerging field, it is essential to create new and different ways to communicate it among all citizens. Geoethics lends itself to research and reflection on a set of values which scientific practice, as well as other human activities, has to take into account in its interaction with the Earth system (Bobrowsky et al. 2017; Di Capua and Peppoloni 2019). Geoethics studies the ethical, cultural, and social implications of geosciences, both in their theoretical and practical dimensions, as well as in their educational and communicative ones, with a particular focus on the social role and responsibilities of geoscientists in conducting their activities (Di Capua and Peppoloni 2019). From a social point of view, geoethics has the mission of making geoscientists and society, in general, more actively aware of humanity's duties towards the planet (Di Capua and Peppoloni 2019). In this sense, geoethics makes use of the geosciences knowledge to ensure an informed and conscious way of solving problems resulting from the interaction between human beings and natural systems (Di Capua et al. 2017). Its importance increases as the dilemmas arising from the human–geosphere interaction intensify.

In this sense, an interactive module to deal with the human-Earth system interactions and its interface with geoethics was developed, making these less abstract and more tangible to citizens, whatever their socio-cultural context.

## 2 An Interactive Module on Geoethics Communication—"*Geoethics Calls for Action*"

Under the scientific theoretical framework previously explained, and following the mentioned ambitions, the interactive module "Geoethics calls for action" was conceived. The developed module (Fig. 1) consists of a digital interface (which can assume the form of a website, a mobile application format, or even a physical element with a touch screen) that displays case studies by using photographs. The problems/scenarios are related to a range of topics explored from a geoethical point of view. These topics include: climate and environmental changes, geodiversity, and geoconservation, sustainable use of natural resources, pollution, and its impacts on human health (medical geology), the management of natural and anthropogenic risks, the management of coastal areas and marine life, geosciences literacy and education, and finally, science ethics and professional codes of conduct.

These different topics, linked to photographs presenting a case or a location before and after a specific action, are initially described, representing the impact that results directly or indirectly from human intervention. Through this approach, it is possible to induce a reflection on the responsibilities of each individual. The goal is to transform common "bad" behaviours towards ecosystems into more sustainable ones. In each case included in the module, the central question and the photographic animation confront the message receiver—the animation changes by crossing the cursor or finger over it. After the animation, the user is encouraged to change some daily life behaviours.

Furthermore, all the module pages have an associated audio file (that can be activated or not by the user, according to personal preferences), in which several Earth elements—for example, the water or atmosphere—establish a "conversation" with the user. This audio is related to the short video clips that appear at the bottom of each page (Fig. 2).

Additionally, to contribute to the geoethics communication and its importance, there is an upper horizontal bar, containing a tab with the following items: (1) What is geoethics; (2) GOAL (Geoethics Outcomes and Awareness Learning), an international project dedicated to geoethics education, coordinated by a Portuguese team—available in https://goal-erasmus.eu; (3) IAPG (International Association for Promoting Geoethics)—the leading organization developing and promoting geoethics worldwide (allowing the recipient to become an IAPG member, without paying any money fee)—available in https://www.geoethics.org; (4) Help: operating instructions.

**Fig. 1** Home page of the interactive module "Geoethics calls for action". Available in https://bit.ly/371QmFv

**Fig. 2** One of the pages of the module. In the upper left corner is the audio widget. Available in https://bit.ly/3kQDFCw

The module targets can be quite diverse since it covers topics ranging from those that are part of the school curricula, to those related to the most advanced and complex scientific practice, as science ethics and professional codes of conduct. The user self-regulate interaction with the module is intended, having total freedom to explore it in any way wished. Additionally, there is the intention of creating two interfaces: one in Portuguese and another in English, with translations of the audio content in the respective sign languages, making it more inclusive. It is also our intention to continually update this platform, whether with the inclusion of new cases or other relevant content, making it dynamic and versatile.

Regarding possible places of application of the module proposed, it does not require a specific physical space, since any mobile device can access it. However, it is desirable to include it in a broad and interdisciplinary communication action on natural environments, such as, for example, its integration in science centres or geoparks.

A more immersive 360° experience adapted to the present module can be included through the edition and capture of images using specific software and hardware to the effect (e.g. the capture of images through a 360° camera and the use of virtual reality glasses).

## 3 Concluding Remarks

As explained, a more reliable and distinct geosciences communication is necessary and fundamental in the field of geoethics. Besides the work already carried out by the IAPG and its national sections, it is essential to extend the scientific communication initiatives development to all types of public. This action may bring this scientific area closer to society, allowing citizens to adopt more conscious and consistent practices with geoethical values.

The engagement of the citizens with geoethical issues may create a productive environment to develop projects with a profound social component in geoethics, leading to citizen science projects—bringing more people towards the geoethical thinking. Underlying these intents are not only the geoethics communication but also the endorsement of higher respect for planet Earth.

**Acknowledgements** This work was supported by: (1) the *Fundação para a Ciência e a Tecnologia* (FCT), through the Ph.D. Scholarship —"*An approach to the Earth System from a Geoethics perspective: from Science Education to Citizen Science*" with reference SFRH/BD/143306/2019; (2) the Science Teaching Unit, Faculty of Science of the University of Porto (FCUP).

## References

Bobrowsky P, Cronin V, Di Capua G, Kieffer S, Peppoloni S (2017) The emerging field of geoethics. In: Scientific integrity and ethics: with applications to the geosciences special publications, vol 73. American Geophysical Union: Washington DC, pp 175–212

Bucchi M, Trench B (2014) Science communication research: themes and challenges. In: Bucchi M, Trench B (eds) Routledge handbook of public communication of science and technology. Routledge, London, pp 1–14

Burns TW, O'Connor DJ, Stocklmayer SM (2003) Science communication: a contemporary definition. Publ Underst Sci 12(2):183–202

Di Capua G, Peppoloni S (2019) Defining geoethics. IAPG—International Association for Promoting Geoethics, Rome

Guenther L, Joubert M (2017) Science communication as a field of research: identifying trends, challenges and gaps by analysing research papers. J Sci Commun 16(2):1–19

Mogk DW, Geissman JW, Bruckner MZ (2018) Teaching geoethics across the geoscience curriculum: why, when, what, how, and where? In: Gundersen LC (ed) Scientific integrity and ethics in the geosciences. Wiley, Hoboken, pp 175–212

Peppoloni S, Di Capua G (2016) Geoethics: ethical, social, and cultural values in geosciences research, practice, and education. In: Geoscience for the Public Good and Global Development: toward a sustainable future. Geological Society of America, Colorado, pp 17–21

Trench B, Bucchi M, Amin L, Cakmakci G, Falade BA, Olesk A, Polino C (2014) Global spread of science communication: Institutions and practices across continents. In: Bucchi M, Trench B (eds) Routledge handbook of public communication of science and technology. Routledge, London, pp 214–230

# Teaching Geoethics and Groundwater Sustainability Through a Project-Based Approach

Marta Paz, Maria Lurdes Abrunhosa, and Clara Vasconcelos

**Abstract**

Groundwater is a natural resource of great importance. In addition to the demographic growth and the consequent rise in global demand to meet people's water needs, the vital role of groundwater is undermined by significant threats such as pollution, urbanization, or climate change. Combined with misuse and ineffective management systems, these factors determine its scarcity, quality degradation and reduction of replacement rates, enhancing the challenges of its sustainability. Geoethics is intrinsically linked to sustainability, as it provides tools that allow reflection prior to action, in the Earth systems best interest. The aim of this investigation was to check and to promote knowledge about geoethics and groundwater sustainability among the youngest, using a project-based teaching approach. The sample consisted of a group of sixteen 11th grade students, from a school in the city of Porto (north of Portugal). The final aim was the collaborative development of a video related to the theme. The study relied on qualitative methods and some educational research techniques. The results demonstrate the increase of students' content knowledge, as well as their higher awareness of the importance of groundwater sustainability. These results support the potential of this approach in promoting geoethics awareness.

**Keywords**

Groundwater • Sustainability • Geoethics • Project-based teaching

M. Paz (✉) · M. L. Abrunhosa · C. Vasconcelos
Science Teaching Unit, Faculty of Sciences, University of Porto, Porto, Portugal
e-mail: martacpaz@gmail.com

C. Vasconcelos
Interdisciplinary Centre of Marine and Environmental Research, University of Porto, Porto, Portugal

## 1 Introduction

Groundwater is the invisible component of the water in the hydrosphere cycle. It accounts for over 97% of all usable freshwater on the planet and contributes to a large percentage of all human uses of it, whether public or domestic, agricultural, livestock or industry production (Margat and Van der Gun 2013; Fetter 2018; Vadiati, Adamowski and Beynaghi 2018). It is also the main source of supply in water-scarce regions and acts as a buffer against extreme weather events (Chambel and Abrunhosa 2017).

As a result of rising demographics and world consumption, global groundwater extraction has grown more than four times in the last 50 years (Margat and Van der Gun 2013; Chambel and Abrunhosa 2017). Intensive groundwater extraction can break this aquifer systems' balance, resulting in a progressive reduction in groundwater storage (Margat and Van der Gun 2013; Chambel and Abrunhosa 2017; Fetter 2018). Furthermore, agricultural and industrial pollution, increasing urbanization or climate change combined with misuse and ineffective management systems, also determine its scarcity, degradation of quality and reduction of replacement rates (Vadiati et al. 2018).

The need for more sustainable societies was unanimously recognized by the United Nations in September 2015, when 17 Sustainable Development Goals were set in the so-called Agenda 2030 (United Nations 2015). Issues related to education (4th goal) and drinking water (6th goal) are identified as important challenges. Thereby, the education of future generations to change behaviors and achieve those goals is urgent.

Following the growing awareness of environmental problems, a new geoscientific field emerged in the 1990s—geoethics (Martínez-Frías et al. 2011; Almeida and Vasconcelos 2015; Bobrowsky et al. 2017). Within its scope is a reflection on the values that should underlie behaviors and practices that interfere with the way the human being relates to the Earth system, embodying a point of intersection


M. Abrunhosa et al. (eds.), *Advances in Geoethics and Groundwater Management: Theory and Practice for a Sustainable Development*, Advances in Science, Technology & Innovation,
https://doi.org/10.1007/978-3-030-59320-9_76

between geosciences, sociology and philosophy (Peppoloni and Di Capua 2015, 2017; Bobrowsky et al. 2017). Additionally, it's related to the responsibility and social role of geoscientists in conducting their activities, presenting itself as a tool to influence society's awareness of geological resources and the environment (Peppoloni and Di Capua 2015, 2017; Bobrowsky et al. 2017; Vasconcelos et al. 2016). Geoethics is closely linked to the sustainability of geological resources, namely underground water resources. This theme is aligned within the Portuguese 11th grade curriculum of the discipline of Biology and Geology.

The presented study was conceived with the aim of checking and promoting knowledge and awareness about geoethics and groundwater sustainability in Secondary Education. To achieve this goal, we used project-based teaching (PjBT) approach, which follows a socio-constructivist perspective of learning. This teaching methodology intends to develop learning through organized projects, to promote collaborative work and students have to attempt to answer problems and challenging questions in an integrated way (Blumenfeld et al. 1991; Bell 2010; Dobber et al. 2017). The final product of the project consisted of a 3min video-pill, done by students and teachers, to disseminate the importance of a geoethical perspective about groundwater management.

## 2 Methods

Within the presented framework, a summative evaluation study was done. Our purpose was to evaluate the quality and merit of the applied project, as well as its final product (the video-pill), at the end of the developed project.

A convenience sample of sixteen 11th grade students, from a public school in the north of Portugal (Porto) participated in the study voluntarily. To achieve our goal, this research relied on qualitative techniques. As such, the researchers used observations, a test and interviews.

The final product, a 3 min' video-pill, consisted of the adaptation of the lyric of a song well known to the general public, to the scientific theme of geoethics and groundwater sustainability. The lyric was written by the students with the help of teachers of the Biology and Geology curricular unit.

Before starting the work, students participated in a test phase, which aimed to gauge their degree of knowledge about the addressed topic. Following, students were gathered into three groups according to their personal characteristics: choosing the music and writing the lyric, choosing images and producing the video, or research relevant data within the topic. The collaborative work was undertaken during four classes of 50 min.

After collecting data about the importance of groundwater and written the lyric, students recorded the music, in a special room of the school library. The final editing phase of the video was made up by the teachers. Following, the authors tested its impact by presenting it to other students and also other members of the school community.

To finish the research, the teachers gathered students in three focal groups of 5 or 6 elements. An interview was done to evaluate the impact of the project developed. Some interviews were also done with seven teachers of the Natural Sciences disciplinary group and the school principal. The triangulation of the data collected was used to validate and support the reliability of the research.

## 3 Results

### 3.1 Video

The background music selected to produce the video-pill was the 2008 theme "Viva la Vida" from the popular band Coldplay. Both lyrics and the visual art were developed by students with the aim of raising public awareness of geoethics and groundwater sustainability. In addition, some technical data were introduced by the teachers to highlight scientific concepts. In the end, the video was submitted to participate in a video contest promoted by the International Geoethics and Groundwater Management Congress. This motivational competition was a target since the beginning of the project.

### 3.2 Diagnostic Test and Students' Interviews

Regarding the diagnostic test and the final focus group interviews, students answered to several questions, namely about aquifers, sustainability and geoethical values. Questions ranged from general concepts to specific scientific knowledge on the topics.

The initial test revealed heterogeneity in the results, but also low knowledge and some scientific misconceptions about the topics addressed. Data collected with the interviews made in the end of the project showed that students increased their comprehension of the subjects.

One of the questions of the diagnostic test was also included in the interviews. Students had to mention words or concepts related to geoethics and groundwater sustainability. Results are presented in the form of a "word cloud", where larger words mean higher response frequency (Fig. 1).

### 3.3 Teachers' and School Principal Interviews

The individual interviews started with a previous visualization of the video-pill. Teachers of the Natural Sciences disciplinary group, as well as the school principal, were

**Fig. 1** Word cloud made up from students' answers to the question: "Indicate some words/concepts that you associate with geoethics and groundwater sustainability". Left image from the diagnostic test and right image from final interviews

**Table 1** Results from the individual interviews

| Video quality regarding the theme | Relevance as teaching resource | Major disadvantage of PjBT |
|---|---|---|
| Excellent<br>$n$ = 8; 100% | Excellent<br>$n$ = 8; 100% | Time consumed<br>$n$ = 5; 62.5% |

asked about the quality of the video-pill produced, its relevance as an educational resource and the importance of the PjBT methodology. The results are presented in Table 1.

As almost any methodology that is not based on the traditional way of teaching (lectures and textbook reading), the PjBT methodology was classified as a time-consuming approach, although its final result was unanimously classified as excellent.

## 4 Discussion

Comparing the results obtained from the initial diagnostic test to the ones recorded from the focus group interviews made to the students, a clear difference is revealed. Starting from a low level of knowledge on this subject, students showed a substantial increase in understanding the theme.

In the initial test, only six students ($n$ = 6; 37.5%) achieved a score equal or superior to 50% and the average score was 39.6%. The results of the final student interviews revealed that their knowledge grew, with an increase in the percentage of correct and complete answers, while neither incorrect answers nor scientific misconceptions were verified.

Additionally, as shown in Fig. 1, there was a noticeable change in the perspective of students. In the left word cloud (initial test), students' answers repeated the words in the question. Thus, when we asked them words or concepts regarding geoethics and groundwater sustainability, they replayed by saying "geology", "ethics", "sustainability" or "groundwater". After the end of the project, by answering the same question, the more frequent answers were "responsibility", education", "citizenship", "sustainability", "values" and "pollution" (see the word cloud on the right in Fig. 1). This reveals students became more capable to relate broader and seemingly distant concepts, with a clear appeal to action and collective responsibility, thereby showing their increased sensibility to the theme.

Furthermore, the audiovisual was considered, by Natural Science Department teachers and school principal, to be well constructed and a useful pedagogical tool while teaching about this topic, either in the beginning or at the end of teaching activities ($n$ = 8; 100%). Moreover, it was unanimously recognized the importance and advantages to students' participation in this kind of projects ($n$ = 8; 100%), namely because "it promotes their autonomy, critical thinking, engages them to the subject and links them to the real world" and it provides "(…) meaningful knowledge, students never forget it for life, they can study for a test and the next year they don't remember it, but this stays recorded in memory". These competences referred by teachers, related to greater autonomy, motivation, responsibility, critical thinking and increased knowledge, are also widely referred in the literature as flags of project-based education (Blumenfeld et al. 1991; Bell 2010; Dobber et al. 2017).

However, a considerable number of teachers ($n$ = 5; 62.5%) listed disadvantages in using projects as a teaching methodology, due to the larger time needed to develop this kind of projects. In Portugal, high school teachers face long curriculums and a lot of bureaucracy, which are very time-consuming, and thereby reducing their willingness to adopt different teaching methodologies, like PjBT.

## 5 Concluding Remarks

Quality in education and sustainable managements in water are part of the 2030 Agenda of the United Nations Sustainable Development Goals. As such, raising awareness and improving knowledge about these topics are part of the geoscience teacher role in society.

This study aimed to show that, even with a simple project, it is possible to increase conceptual knowledge and develop attitudes to change young students' perspectives, enhancing their own active role in society.

This investigation was developed with a convenient sample of few students and therefore does not allow to generalize its results. Nevertheless, the outcomes obtained seem to be good indicators that support the use of PjBT methodology, to promote awareness and to highlight the need to teach geoethical values in schools, in order to form conscious, responsible and participating citizens, as well as to spread a culture of Earth Science Literacy in society (Mogk et al. 2015; Bobrowsky et al. 2017; Peppoloni and Di Capua 2017; Peppoloni et al. 2019). In our plans for future studies, we would like to analyze if the developed video can act as a tool to raise public awareness about geoethics and groundwater sustainability. More than promoting pro-environmental behaviors, we need to develop pro-innovative actions.

## References


Almeida A, Vasconcelos C (2015) Geoethics: master's students knowledge and perception of its importance. Res Sci Educ 45 (6):889–906

Bell S (2010) Project-based learning for the 21st century: skills for the future. Clearing House 83(2):39–43

Blumenfeld PC, Soloway E, Marx RW, Krajcik JS, Guzdial M, Palincsar A (1991) Motivating project-based learning: sustaining the doing, supporting the learning. Educ Psychol 26(3–4):369–398

Bobrowsky P, Cronin VS, Di Capua G, Kieffer SW, Peppoloni S (2017) The emerging field of geoethics. In: Gundersen LC (ed). Scientific integrity and ethics with applications to the geosciences. Special Publication American Geophysical Union, Wiley

Chambel A, Abrunhosa M (2017) Governança de águas subterrâneas: uma prioridade a nível internacional. In: Livro de atas do 11° Seminário sobre Águas Subterrâneas. APRH, IAH-GP & ISEP, Porto

Dobber M, Zwart R, Tanis M, van Oers B (2017) The role of the teacher in inquiry-based education. Educ Res Rev 22:194214

Fetter CW (2018) Applied hydrogeology. Waveland Press

Margat J, Van der Gun J (2013) Groundwater around the world: a geographic synopsis. CRC Press

Martínez-Frías J, González JL, Pérez FR (2011) Geoethics and deontology: from fundamentals to applications in planetary protection. Episodes 34(4):257–262

Mogk D, Bruckner M, Kieffer S, Geissman J, Reidy M, Taylor S, Vallero D (2015) Teaching geoethics across the geoscience curriculum. In: EGU general assembly conference abstracts, vol 17

Peppoloni S, Bilham N, Di Capua G (2019) Contemporary geoethics within the geosciences. In: Exploring geoethics. Palgrave Pivot, Cham, pp 25–70

Peppoloni S, Di Capua G (2015) The Meaning of geoethics. In: Wyss M, Peppoloni S (eds) Geoethics: ethical challenges and case studies in Earth Science. Elsevier, Waltham, Massachusetts

Peppoloni S, Di Capua G (2017) Geoethics: ethical, social and cultural implications in geosciences. Ann Geophys 60. https://doi.org/10.4401/ag-7473

United Nations (2015). Transforming our world: the 2030 agenda for sustainable development. General Assembly 70 session

Vadiati M, Adamowski J, Beynaghi A (2018) A brief overview of trends in groundwater research: progress towards sustainability? J Envir Manag 223:849

Vasconcelos C, Torres J, Vasconcelos L, Moutinho S (2016) Sustainable development and its connection to teaching geoethics. Episodes 39(3):509–517

# Socio-hydrogeology and Ethical Groundwater Management

**Introductory Note**

Viviana Re (University of Pisa, Pisa, Italy)

Socio-hydrogeology is recently developed discipline fostering the systematic inclusion of the social dimension in hydrogeological assessments targeted to support groundwater governance. The key aspects of socio-hydrogeology are: (1) assessing the impact of human activities on groundwater resources; (2) evaluating the (socioeconomic) impact of groundwater resources (and its changes in terms of both quality and quantity) on human life and well-being; (3) unveiling how social relationships (their patterns and structures) influence attitudes and behaviors toward groundwater use and its governance; (4) promoting better use of the outcomes of a hydrogeological investigation, thus ensuring their use in the promotion of science-based management practices for groundwater protection; (5) contributing to bridge the gap between science and practice; and (6) demystifying science and scientists.

Overall, the discussion would permit to highlight the potential of socio-hydrogeology to: support ethical groundwater governance, incorporate geoethical values in hydrogeology, foresee conflicts of interests, and ensure an effective interaction with water users and all relevant stakeholders. Eventually, the session would permit to pave the way for promoting a more conscious engagement of groundwater scientists willing to contribute to the long-term management of global groundwater resources.

**Highlights**

- Specific examples of integrated socio-hydrogeological assessments in different cultural and geographical contexts;
- Importance of participatory approaches, stakeholder's analysis, and public engagement for effective ground-water management;
- Innovative and multidisciplinary frameworks to tackle groundwater issues;
- Challenges and opportunities raising from the integration of hydrogeology and social sciences;
- The role of hydrogeologists and groundwater scientists in fostering geoethics in scientific research and daily life actions.

# Socio-hydrogeology and Geoethics—State of the Art and Future Challenges

Viviana Re

**Abstract**

This contribution aims at exploring the nexus between the recently developed discipline of socio-hydrogeology and geoethics. Socio-hydrogeology targets the inclusion of social sciences into hydrogeological assessment, with the overall goal of assessing the reciprocity between people and groundwater. For doing so, it promotes the integration of hydrogeological assessments with tools and activities typical of the social sciences (e.g. social network analysis, social impact assessment, public engagement). A closer look at the social implications of any hydrogeological assessment would not only permit to better frame the research and propose more adequate science-based managed practice, but also to ensure that (still existing) gap between science and society is effectively bridged. Indeed, socio-hydrogeology fosters the development of mutual trust between scientists and concerned stakeholders: a fundamental requisite for the sustainability of any action targeted to the long-term protection of water resources. Socio-hydrogeology hence encourages hydrogeology goes beyond classical sectorial approaches, by looking at ethical, social, and cultural implications of hydrogeological assessments. As a result, they will be capable of promoting new management strategies that will not only based on sound scientific knowledge, but also on the adequate assessment of the social dimension of groundwater.

**Keywords**

Socio-hydrogeology • Science • Society • Groundwater • Governance

## 1 Introduction

Socio-hydrogeology is a recently developed discipline promoting the structured incorporation of the social dimension into hydrogeological assessments Re (2015).

Barthel and Siedl (2017) highlighted that, even when collaborations between natural and social sciences targeted to groundwater research are implemented, description of applied interdisciplinary methodology is scarcely found. Moreover, Re (2015), and Hynds et al. (2018) pointed out that participatory approaches are often included in studies about groundwater resources, however, a lack of structured approach is often a common factor in these cases. As a result, hydrogeologists often miss a great opportunity to engage first hand with the stakeholders involved in studied groundwater-related issue.

Therefore, understanding the social structure of the studied groundwater system should be a fundamental step when hydrogeological investigations are targeted to propose new e-science-based management practices respectful of the real needs of local populations, thus reducing the likelihood of creating conflicts among stakeholders. The latter is indeed a key issue, especially when competition for a shared good may occur (The Water Channel 2019).

The goal of this contribution is to discuss how the rapidly evolving discipline of socio-hydrogeology can underpin groundwater management and the key issues to face when geoethics is at stake.

Emphasis will be put in discussing the need for a more active (and structured) engagement from the international community of hydrogeologists, and the potentials of socio-hydrogeology in supporting ethical groundwater management.

V. Re (✉)
Department of Earth Science, University of Pisa, Pisa, Italy
e-mail: viviana.re@unipi.it





M. Abrunhosa et al. (eds.), *Advances in Geoethics and Groundwater Management: Theory and Practice for a Sustainable Development*, Advances in Science, Technology & Innovation,
https://doi.org/10.1007/978-3-030-59320-9_77

## 2 Understanding the Reciprocity Between People and Groundwater

The socio-hydrogeological approach is centred on the identification of the cause and effect relationship between people and groundwater. The key is therefore to go beyond classical hydrogeological and hydrogeochemical assessments, by not only looking at how human activities can affect groundwater properties (i.e. quality and quantity), but also at the impacts of inadequate groundwater quality and quantity on human wellbeing Re (2015). Disentangling the complexity of human–groundwater interaction also means to both understand how the aforementioned interactions happen but also to understand why they happen.

To do so, a stakeholder analysis performed at the early stage of a hydrogeological assessment would allow hydrogeologists to better understand the relationships among all the actors involved (directly or indirectly) in the groundwater issue under investigation. This would permit avoiding conflicts due to the competition over a shared resource, and to ensure that all interests, including those of marginal groups, are considered. In particular, the integration of Social Network Analysis (SNA) in hydrogeological assessment favours the understanding of the power relations among concerned stakeholders, a necessary requisite to facilitate collaboration and trust in view of the challenges eventually imposed by the implementation of new science-based management practices (Tringali et al. 2017). Additionally, as demonstrated by Musacchio et al. (2020), SNA can also permit to unveil the social factors underpinning groundwater governance. Therefore, the adoption of a social relational approach (Bodin et al. 2011) in hydrogeology would represent a benefit in terms of assessing the values which underpin appropriate behaviours and practices when human activities interact with groundwater resources.

Furthermore, the inclusion of structured interviews with well's owners and farmers involved in the monitoring network in the groundwater assessment procedure to would allow hydrogeologists to better constraint research challenges and to retrieve reliable information on local know-how, useful to support hydrogeological and hydrogeochemical findings (Re et al. 2017). Indeed, by strengthening the interaction and confrontation with (ground)water end-users, it will be possible to foster capacity building and results dissemination. In this framework, it is important to underline that public engagement (PE) should be not only targeted to information collection, but also to develop a network of mutual trust with local water users. Therefore, the same time and energies spent to questionnaires administration should be dedicated to bringing results "back to the farmers" (Re et al. 2017), and to find adequate tools for effective communication and outreach strategies (Re and Misstear 2017).

## 3 Ethical, Cultural, and Social Implications of Hydrogeology

Socio-hydrogeology requires an effort for hydrogeologists to open up to new disciplines and go beyond the "silo mentality" (Daly et al. 2016), by looking at ethical, social, and cultural implications of hydrogeological assessments.

This clearly means to ensure that a research does not cause any conflict or involuntarily marginalize some stakeholders or minorities. Getting acquainted with the social factors influencing water use and management also implies understanding the social value of groundwater for a given community. To this end, the integration of groundwater quantity and quality assessments with social impact assessment (SIA) would also be an asset. In addition, hydrogeologists should also take care of project suitability, thus ensure that project impacts continue also once all the activities are completed.

For this reason, specific attention should be paid to capacity building and knowledge transfer to local communities and authorities. The result will be an increased and broader awareness of groundwater resources issues for both hydrogeologists and groundwater users, thus resulting in more effective research activities, education, and communication. In fact, as stressed by Limaye (2017) hydrogeologists should orientate their research work towards solving practical problems and share knowledge to local communities so as to ensure stakeholders participation in groundwater management.

In addition, the assessment of the social impact of any hydrogeological assessment should be a fundamental requisite to avoid conflicts generated by the lack of adequate knowledge of the relations between local communities and water in any given cultural background. This has been recently as demonstrated by "Pani Check" and "Pani Doctors", a documentary film project about a socio-hydrogeological project in Jaipur (Frommen and Ambrus 2019).

## 4 Social Responsibilities of Hydrogeologists and Groundwater Scientists

In line with the overall description of geoethics (IAPG 2019), socio-hydrogeology can represent an excellent opportunity for hydrogeologists to become more conscious of their social role and responsibilities.

These include (but are not limited to):

- To ensure an adequate transfer of research outcome and ultimately the support of good groundwater governance.
- To ensure that participation and outreach are not just a unilateral dissemination of knowledge or a mere "ticked box" in project's deliverables. Instead, hydrogeologists should use socio-hydrogeology to create the opportunity for testing new communication tools, and interactive learning, thus paving the way for citizen science and participatory groundwater monitoring assessment. Ultimately, co-management through local collective action should be explored as a way of effectively target critical groundwater management issues (Shalsi et al. 2019).
- To foster the inclusion of all visions and needs of concerned water users, including minorities and marginal groups. In addition, although groundwater and gender are intrinsically linked, groundwater research generally does not have a gendered approach. To this end, the collection of sex-disaggregated data (e.g. by adopting the UNESCO-WWAP Toolkit on Sex-disaggregated Water Data in PE activities; UNESCO-WWAP 2019) would facilitate the understanding of the possibly unequal power relations between men and women when groundwater issues are at stake.
- To walk the talk, hence bringing science (and scientific knowledge related to groundwater protection and conservation) in our daily life, thus making hydrogeologists advocate of (ground)water protection and management (Responsible Water Scientists 2019).

## 5 Future Challenges

Socio-hydrogeology works at the interface between geosciences and social sciences. It must be stressed, however, that by encouraging hydrogeologists to get familiar with tools and techniques typical of the social sciences, this approach does not aim to replace social scientists, nor does it aim at substituting interdisciplinary projects. Instead, it paves the way for holistic assessments, which are a starting point for projects where a full integrated assessment cannot be implemented (Re 2015).

Socio-hydrogeology can therefore represent a powerful tool to influence the awareness of society regarding problems related to geo-resources and geo-environment. In this framework, future development and challenges include the application of such an approach in different case studies worldwide and to identify the most suitable tools for adequately integrate geosciences and social sciences.

To support these challenges, the new Socio-Hydrogeology Network of the International Association of Hydrogeologists (IAH 2019) has been recently launched. The overall goal of the network is to create a platform for hydrogeologists who work at the interface between society and groundwater for discussing the challenges, opportunities, and advances within such an emerging research field.

**Acknowledgements** This review and forward look of socio-hydrogeology benefited from the interaction and discussion with many colleagues and friends who are contributing to the development of this new discipline. The author wishes to express her sincere thanks to Theresa Frommen for the critical review of the text.

## References

Barthel R, Seidl R (2017) Interdisciplinary collaboration between natural and social sciences – status and trends exemplified in groundwater research. PLoS ONE 12(1):e0170754

Bodin Ö, Ramirez-Sanchez S, Ernstson H, Prell C (2011) A social relational approach to natural resource governance. In: Bodin Ö, Prell C (eds) Social networks and natural resource management: uncovering the social fabric of environmental governance. Cambridge University Press, Cambridge, UK, pp 1–54

Daly D, Archbold M, Deakin J (2016) Progress and challenges in managing our catchments effectively. Biol Environ: Proc R Irish Acad 116B(3):157–166

Frommen T, Ambrus K (2019) https://www.avbstiftung.de/projekte/artikel/news/pani-check-pani-doctors/. Last accessed on 09 Nov 2019

Hynds P, Regan S, Andrade L, Mooney S, O'Malley K, DiPelino S, O'Dwyer J (2018) Muddy waters: refining the way forward for the "sustainability science" of socio-hydrogeology. Water 10(9):1111

IAH (2019) https://iah.org/groups/commissions-networks. Last accessed on 09 Nov 2019

IAPG (2019) https://www.geoethics.org/geoethics. Last accessed on 03 Nov 2019

Limaye SD (2017) Socio-hydrogeology and low-income countries: taking science to rural society. Hydrogeol J 25(7):1927–1930

Musacchio A, Re V, Mas-Pla J, Sacchi E (2020) EU Nitrates Directive, from theory to practice: environmental effectiveness and influence of regional governance on its performance. Ambio 49:504–516

Re V (2015) Incorporating the social dimension into hydrogeochemical investigations for rural development: the Bir Al-Nas approach for socio-hydrogeology. Hydrogeol J 23(7):1293–1304

Re V, Misstear B (2017) Education and capacity development for groundwater resources management. In: Villholth KG, Lopez-Gunn E, Conti K, Garrido A, Van Der Gun J (eds) Advances in groundwater governance. CRC Press, London, UK, pp 212–228

Re V, Sacchi E, Kammoun S, Tringali C, Trabelsi R, Zouari K, Daniele S (2017) Integrated socio-hydrogeological approach to tackle nitrate contamination in groundwater resources. The case of Grombalia Basin (Tunisia). Sci Total Environ 593–594:664–676

Responsible Water Scientists (2019) https://responsiblewaterscientists.wordpress.com/the-sci-nexus/

Shalsi S, Ordens CM, Curtis A, Simmons CT (2019) Can collective action address the "tragedy of the commons" in groundwater management? Insights from an Australian case study. Hydrogeol J 27(7):2471–2483

The Water Channel (2019) Theatre of the absurd: Is there a happy ending to the tragedy (of the Commons)? https://www.thewaterchannel.tv/media-gallery/5473-theatre-of-the-absurd-is-there-a-happy-ending-to-the-tragedy-of-the-commons. Last accessed on 09 Nov 2019

Tringali C, Re V, Siciliano G, Chkir N, Tuci C, Zouari K (2017) Insights and participatory actions driven by a socio-hydrogeological approach for groundwater management: the Grombalia Basin case study (Tunisia). Hydrogeol J 25(5):1241–1255

UNESCO-WWAP (2019) https://www.unesco.org/new/en/natural-sciences/environment/water/wwap/water-and-gender/. Last accessed on 03 Nov 2019/11/03

# Responsible Water Scientists: Bringing Socio-hydrogeology in Our Daily Life

Viviana Re, Raquel Sousa, Vincent Post, and Chiara Tringali

**Abstract**

Responsible Water Scientists (RWSci) was created with the ambition of encouraging a broader discussion on the lifestyle changes that can make scientists real advocates of (ground)water protection for a more sustainable world. Created in 2016 by two early career hydrogeologists, the Responsible Water Scientists blog is currently a platform for international debate on the social responsibility of water scientists, providing practical examples on how to "walk the talk". In the era of globalization, dominated by communication and social media, engaging firsthand in water saving and pollution reduction is definitely something that can contribute maintaining the credibility of researchers in the eyes of the general public. Indeed, too often researchers and academics are still perceived as being isolated from the real world, and "the theory without direct commitment" could actually be seen as a lack of interest. To avoid this, RWSci gives the chance to water scientist to dismantle this stereotype, not only by sharing practical experiences of their direct engagement, but also providing practical tips for those interested in taking a new path towards sustainability "one drop at a time".

**Keywords**

Social responsibility • Engagement • Socio-hydrogeology • Groundwater • Science

## 1 Introduction

Responsible Water Scientists (RWSci) is a project that aims at encouraging water scientists around the world to live more sustainably and to help drive the changes they aim to inspire with their research (Responsible Water Scientists 2019).

The project took its firsts steps in 2014 when two young hydrogeologists, with many personal and professional interests in common, started to work together in the Early Career Hydrogeologists Network (ECHN) within the International Association for Hydrogeology (IAH). While doing an assessment on educational videos on online hydrogeology platforms, they started to wonder on the missing link between the professional community's message about the protection of groundwater resources, and the actions in daily life of researchers sending out that message. This insight was reinforced by Viviana Re's work and research on socio-hydrogeology (Re 2015) and by the then arising zero waste movements.

The motto was set, and the pieces were brought together with a main ambition: encourage a broader discussion on the little changes in the daily routine that can make scientists real advocates of groundwater protection for a more sustainable world (Responsible Water Scientists 2019). Therefore, RWSci provides a scientific and grounded perspective on the impact's sustainable practices in our daily life may have in terms of water resources protection. Since we can only protect what we know, RWSci advocates for a strongest consciousness within the scientific community on the need for a more holistic approach to water resources issues. This means, on the one hand to perform sound investigations to address the global water crisis, and on the other to apply the

V. Re (✉)
Department of Earth Sciences, University of Pisa, Pisa, Italy
e-mail: viviana.re@unipi.it

R. Sousa
International Groundwater Resources Assessment Centre (IGRAC), Delft, The Netherlands

V. Post
Amsterdam, The Netherlands

C. Tringali
Trieste, Italy



M. Abrunhosa et al. (eds.), *Advances in Geoethics and Groundwater Management: Theory and Practice for a Sustainable Development*, Advances in Science, Technology & Innovation,
https://doi.org/10.1007/978-3-030-59320-9_78

knowledge acquired also to our lifestyle. By giving the example, we encourage others and our message is stronger and in consonance with our professional activities in the protection of water and especially groundwater resources.

## 2 The Social Responsibilities of Water Scientists

The social responsibility of scientists is a concept that has been debated for a long time (Brunner and Ascher 1992 and references therein; Peppoloni et al. 2019).

By opposing the common idea that scientists should not care about the use of the results of their work, in Russel (1960) argued that "the scientist is also a citizen, and citizens, who have any special skill have a public duty to see, as far as they can, that their skill is utilized in accordance with the public interest". With a specific reference to the "intimate and sinister connection with war" he also stresses that scientists should engage in individual actions against the misleading use of scientific outcomes. Quite a topical question indeed, even in the new millennium, as also highlighted by Bird (2014).

More recently Pain (2013) highlighted how it is important for scientists not only to take care of their professional responsibility (i.e. how research should be conducted with respect to international standards), but also to start considering their "external responsibility", i.e. the impact of their research toward the larger community (Post et al. 2018).

With specific reference to environmental issues, scientists should get out of the ivory tower of academia and engage firsthand for nature conservation, instead of calling for the society to do so (Fig. 1). Therefore, as far as water resources go, this implies that besides performing innovative and outstanding research, scientists should also make sure that the outcome of their investigations will be translated into concrete actions for water protection in the long term.

Being a responsible water scientist also means to contribute towards "a cultural shift in society that advocates for more responsible interaction with the Earth system" as promoted by geoethics (Bohle and di Capua 2019). Eventually, engaging firsthand for water protection would become as important as writing a paper in a high impact peer review journal.

## 3 The Scientific Nexus

A point that gets raised a lot during climate summits like the most recent one in Spain is the fact that the climate scientists themselves fly an awful lot. This does not sit well with the general public, and this perception may be one reason why

**Fig. 1** Academics in the ivory tower (Ahlefeldt 2019)

so many people choose to ignore warnings by science that urgent action is required to ward off the most dangerous consequences of global warming. The accusation is somewhat unfair, because holidaymaking and other personal reasons are responsible for most of the global air traffic. Nevertheless, during this critical time of transition in humankind's history, science needs to take an inward look.

Science is not a machine but instead consists of millions of human beings, who through their daily choices can have an enormous impact on the environmental and social wellbeing of the planet. At RWSci we document how (ground) water scientists make such choices, given that we are acutely aware of the water problems and the effects that climate change will have on water security and the health of ecosystems. RWSci's main goal is to bring a new consciousness to both the scientific community and water end-users. For the scientific community, our aim is to remind our colleagues that they are also citizens like everyone else, whose behaviour to protect the environment and fight climate change should logically align with the latest research. For the broader public, we aim to give the right example, advocate for water protection, and educate by sharing knowledge and information.

RWSci asks questions such as: What is our responsibility towards society? Is there something that we can do to bring the scientific knowledge in our everyday life? What if we could better engage to inspire people, our friends and local communities, and act as guardians of water resources and thus the environment at all levels, both professionally and privately? How far should we go in taking the ethical and social responsibility of doing our part and giving a good example firsthand?

The contributions to the RWSci blog reflect the thinking and actions of a small group of individuals. Many scientists around the world share similar values and are driven by their expert knowledge to take action in their everyday lives, each of them in their own way. Nonetheless, we believe there is a need of the scientific community to become even more aware. Aware of our own daily impacts on the environment, on water resources in particular, and our role as advocates for sustainability.

Therefore, we argue that scientists have an obligation to further lift their game and take actions on a daily basis that contribute to the sustainability of this planet. Recently, the British band Coldplay announced that they will not go on tour until they could do it in a $CO_2$ neutral way. The fashion magazine Vogue published their January edition in Italy without photographs to make a statement about sustainability. Scientists have no choice but to follow suit. After all, if those who are warning about humankind's impact on our planet aren't leading the charge in abandoning our old wasteful ways, how do they expect others to follow? People are right to expect the same level of commitment from academia that is expected from industry, governments, rock bands and fashion magazines.

Starting with the workplace, there is of course a myriad of everyday things individual scientists could do. One reason why it is particularly important to do this is to inspire students. They need to feel inspired by the staff rather than develop the resentment that is now prompting high school students around the world to organize gatherings like Fridays for Future.

At the institutional level, there need to be mechanisms that steer an organization toward impact reduction. Approval of travel applications can be made conditional on the mode of transport, with flights only allowed when alternatives are unfeasible (and remembering that the train is often the perfect workplace for finishing journal articles). New technologies and smart timetabling should be put to use to reduce student travel to and from campus. Environmental literacy should become part of all academic programs, the physical sciences and arts alike. And so on….

Many things could be said as well about the environmental footprint of experimental work, including the materials used and the waste streams generated. Such criteria are currently not routinely assessed by funding bodies, but they deserve to be valued equal to ethical selection criteria that apply to the use of human subject or laboratory animals. As such there is a role to play by science councils by implementing policies that favour research proposals that score well on environmental criteria. Laboratory staff and suppliers of equipment should investigate ways to reduce the use of disposables and identify opportunities for recycling programs.

As life on our planet finds itself in one of the worst crises in its entire history, not allowing environmental thinking to permeate all levels of science would be a fatal miscalculation. Not only is it a moral obligation, society expects this as those who warned of climate change will be held accountable for setting in motion the enormous transformations that will affect the lives of everyone on the planet.

## 4 Walk the Talk

Being knowledgeable about the potential issues and factors that can trigger water overexploitation and contamination, water scientists should become the main reference for their local communities, by inspiring new actions for water protection.

To do so, a shift in mindset is required to start applying the knowledge acquired to everyday life (and choices). As mentioned in the previous section, there is a large amount of possibilities and actions to undertake, that can be summarized as follows:

- **Research**. Broaden the spectrum of the scientific interest also to the understanding of the socio-economic implications of research on communities relying upon the studied water system.
- **Communicate**. Outreach and capacity building should become part of the basic activities to bring science out of academia and research centres. Acquiring new communication and social media skills would therefore be an asset.
- **Inspire**. Sharing new approaches and new ideas with colleagues in order to lead them to walk the talk in their own contexts and through their actions.
- **Promote**. Any time is a good time to foster more sustainable choices and to put into practice the knowledge gained about water issues. In addition, any activity (i.e., a congress, a course, a meeting, etc.) should be organized taking into account its environmental impact and trying to reduce it as much as possible.
- **Be an example**. As scientists we have to be even more trustworthy and, in order to avoid loss of credibility in front of society, it is our responsibility to "practice what we preach". For example, at a congress, it may appear contradictory to see a speaker talking about the impact of microplastic pollution on water resources, while drinking from a plastic bottle.

Some actions may not seem straightforward at the beginning, but the main idea behind RWSci is actually that, "one drop at the time", the change is possible.

**Acknowledgements** We would like to acknowledge the help and support of all guest blog contributors, in particular, Smita Jairam, Georgios Xenakis, Tomaso Fortibuoni and Nadir Faria. We are also thankful to all our colleague's hydrogeologists whose comments and inputs inspired many of our posts.

## References

Bhole M, Di Capua G (2019) Setting the scene. In: Bohle M (ed) Exploring geoethics. Ethical implications, societal contexts, and professional obligations of the geosciences. Palgrave Pivot, Switzerland, pp 1–24

Bird SJ (2014) Socially responsible science is more than "Good Science." J Microbiol Biol Educ 15(2):169–172

Brunner RD, Ascher W (1992) Science and social responsibility. Policy Sci 25:295–331

Ahlefeldt F (2019) https://fritsahlefeldt.com/2011/09/22/academic-avoid-reality/book-tower-expertsby-hikingartist/ . Last accessed on 15 Nov 2019

Pain F (2013) The social responsibilities of scientists. Science. https://www.sciencemag.org/careers/2013/02/social-responsibilities-scientists. Last accessed on 15 Nov 2019

Peppoloni S, Bilham N, Di Capua G (2019) Contemporary geoethics within the geosciences. In: Bohle M (ed) Exploring geoethics. Ethical implications, societal contexts, and professional obligations of the geosciences. Palgrave Pivot, Switzerland, pp 25–60

Post V, Re V, Sousa R (2018) Responsible but objectives. Reflections about science and environmental activism. https://responsiblewaterscientists.wordpress.com/2018/02/08/responsible-but-objectives-reflections-about-science-and-environmental-activism/. Last accessed on 14 Jan 2020

Re V (2015) Incorporating the social dimension into hydrogeochemical investigations for rural development: the Bir Al-Nas approach for socio-hydrogeology. Hydrogeol J 23:1293–1304

Responsible Water Scientists (2019) https://responsiblewaterscientists.wordpress.com/. Last accessed on 14 Nov 2019

Russell B (1960) The social responsibilities of scientists. Science 131 (3398):391–392

# Gender-Responsive Indicators to Close the Sex-Disaggregated Water Data Gap

Michela Miletto, Vasudha Pangare, Laurens Thuy, and Paola Piccione

**Abstract**

The first official recognition of the role of women in water management was made as early as 1977 during the UN Water Conference at Mar del Plata, and now, over 40 years later, the gap between policy and practice remains to be bridged, as gender inequality in the water sector continues to be the norm. Sex-disaggregated water data are essential for understanding and addressing gender inequalities in the water realm. However, the scarcity of water and gender data is hampering this process at the national and regional level. The UNESCO World Water Assessment Programme (WWAP) released in 2019 a new toolkit on sex-disaggregated water data. The publications contain a set of gender-responsive indicators, along with a methodology, guidelines, and questionnaire for the collection of sex-disaggregated water data on a variety of water resources-related topics, including groundwater resources. It comes as a response to the wide international recognition of the UNESCO WWAP 2015 methodology and takes into account the results of field testing in Southern Africa and Central America, as well as of the 2030 Agenda for Sustainable Development with its interlinked goals and targets.

**Keywords**

Gender • Water • Data • 2030 agenda • Methodology

## 1 Introduction

Interest in the gendered nature of water has long been on the international institutional agendas, from as early as 1977 with the UN Water Conference at Mar Del Plata, followed by the Dublin commitments in 1992 and the Beijing Declaration and Platform for Action in 1995. These milestones represent some of the most progressive declarations ever for the advancement of women's rights, referring to the central role played by women in the provision, management and safeguarding of water.

Nowadays, almost all the key global frameworks, commitments, declarations and action plans on water mention the importance of including gender considerations in their overall field of vision, and the most essential areas for action have been identified. However, the progress made over the last two decades in the field of gender equality has been uneven and unacceptably slow (UN Women 2015). In addition, evidence has revealed a clear gap between policy and practice, due to insufficient funding, limited awareness of gender-related concepts, and monitoring and evaluation processes that come short to reveal the gendered power dynamics occurring within a water resources management context (GWP 2017).

One of the major obstacles to the achievement of gender equality is the lack of data to track progress and inform evidence-based policies. Relevant gender statistics are still not produced regularly, or no historic data are available for the observation of trends. According to a global survey of 126 countries conducted by the UN Statistics Division (UNSD) in 2011, only 37% of the countries regularly produce sex-disaggregated statistics on access to clean water and/or sanitation (and on other topics like informal employment, unpaid work, entrepreneurship, agriculture, etc.). Further, among this group of less-covered topics, most of the remaining 60–70% of countries tend not to collect data at all or only infrequently.

M. Miletto
World Water Assessment Programme (WWAP), United Nations Educational, Scientific and Cultural Organization's (UNESCO), Perugia, Italy

V. Pangare · L. Thuy · P. Piccione (✉)
UNESCO WWAP, Perugia, Italy
e-mail: p.piccione@unesco.org


M. Abrunhosa et al. (eds.), *Advances in Geoethics and Groundwater Management: Theory and Practice for a Sustainable Development*, Advances in Science, Technology & Innovation,
https://doi.org/10.1007/978-3-030-59320-9_79

## 2 Methods

Data are the lifeblood of decision-making; however, there is evidence that sex-disaggregated water data are among the least available across the national-level indicators. The lack of sex-disaggregated water data is a major obstacle to the production of scientific evidence on gender inequalities related to water and to the formulation of gender-transformative policies. Investing in 'engendering' water contributes to the strengthening of social inclusion, eradicating poverty and advancing environmental sustainability. In 2014, UNESCO WWAP started a groundbreaking initiative on gender-responsive water assessment, monitoring and reporting, which included the identification of key indicators, the creation of a methodology, as well as guidelines and questionnaires to collect sex-disaggregated water data. In line with the results of field testing of the methodology and taking into consideration the interlinked character of the 2030 Agenda for Sustainable Development, UNESCO WWAP has fine-tuned its indicators and Toolkit (Miletto et al. 2019). The 2019 Toolkit for sex-disaggregated water data contains 105 updated or new key indicators—grouped in ten priority topics (Tool 1)—as well as relevant updates in the methodology (Tool 2), and a new edition of the guidelines (Tool 3) and questionnaire (Tool 4). A summary of results of the field testing is presented below.

UNESCO WWAP leads the gender component of the Groundwater Governance in Transboundary Aquifers (GGRETA) project, funded by the Swiss Agency for Development and Cooperation (SDC) and led by the UNESCO International Hydrological Programme (UNESCO-IHP). The first edition of the Toolkit has been applied within the framework of the GGRETA project that incorporated the first time ever in the assessment of transboundary aquifers elements related to gender equality and the role of women in the use, management, and decision-making of groundwater resources.

Two gender surveys were conducted in 2017 in key sites within the Stampriet Transboundary Aquifer in Botswana and Namibia (STAS), as well in key sites within the Ocotepeque-Citalá Transboundary Aquifer in El Salvador and Honduras (ATOC). By applying the UNESCO WWAP Toolkit on sex-disaggregated water data, the surveys evolved around key WWAP gender-responsive indicators selected according to the contextual needs. The selected gender-responsive indicators for the STAS Aquifer included access to safe drinking water, sanitation and hygiene, water governance, decision-making, water for agriculture, and income generation, while the selected gender-responsive indicators for the ATOC Aquifer included water, sanitation and hygiene, water for agricultural uses, transboundary water resources management, human rights-based water resources management, water education and training.

## 3 Discussion

Over the last two decades, there has been uneven progress towards gender equality, as discrimination in laws has been gradually removed and laws for the promotion of gender equality have been adopted. Overall progress, however, has been unacceptably slow with stagnation and even regress in some contexts. Change towards gender equality has not been deep enough, nor has it been irreversible. Women's enjoyment of human rights in the field of water access and sanitation continues to be limited in several areas. Women's already limited presence in decision-making at all levels is frequently subject to setbacks, and women remain significantly underrepresented at the highest levels of political leadership. Moreover, it can be stated that gender inequalities are worsened by the lack of universal access to improved water sources (UN Women 2015).

Overall evidence has revealed a clear gap between policies and practice: gender strategies are seldom funded adequately, gender mainstreaming is not well understood, and monitoring and evaluation processes are not sophisticated enough to reveal the true gender and inclusion power dynamics occurring within the water resources management context. There is a need to better understand and account for a broader range of factors that can lead to exclusion and marginalization, such as age, disability, ethnicity, caste, and sexuality, if we are to ensure that no one is left behind. Key is the development of appropriate governance frameworks, and the empowering of both women and men to contribute to and make decisions about water management issues that affect them is central. Access to water resources and safely managed water and sanitation underpins economic resilience, and inclusive practice will therefore ensure that the benefits of water management are shared equally among people; reducing inequalities, contributing to economic growth and increasing social cohesion (GWP 2017; Thuy et al. 2019).

In light of the Agenda 2030 for Sustainable Development, tracking of progress is critical to ensure its successful implementation. If monitoring and evaluation are taken seriously, then disaggregated data should be easily available and accessible. However, to date, data are lacking or incomplete for many indicators, providing an incomplete picture for women and girls. In addition, many of the most critical indicators for gender equality do not have agreed methodology yet and, therefore, no globally comparable data. This lack of data weakens the ability to monitor SDG progress. There exists the need for data to track progress at a highly disaggregated level to show disparities by age, sex, location, ethnicity, disability, income groups, and other categories. Women's Major Group (2017).

The review of gender-related policies in the field of water resources suggests a discrepancy between the national water policy's goal of empowerment and full inclusion of women in water- and sustainability-related issues and decisions-making on one hand and the fact that men still emerge as primary decision-makers at the household level, on the other hand. More scientific evidence is needed to monitor the effectiveness of national and regional policies through gender analysis and the regular collection and analysis of sex-disaggregated water data. The lack of the latter, however, may exacerbate existing inequalities between women and men in the use, management, and decision-making related to groundwater resources.

The findings of the field testing in the STAS Aquifer and in the ATOC Aquifer suggest, among other things, that men still emerge as primary decision-makers in water-related decision-making at the household level, in contrast with the national water policy's goal of empowerment and full inclusion of women in issues and decisions relating to sustainable development and management of water resources. More scientific evidence is needed to monitor the effectiveness of national and regional policies through gender analysis and the regular collection and analysis of sex-disaggregated water data.

## 4 Concluding Remarks

Despite the growing global recognition of gender equality and related issues, the international community is making a slow and uneven progress in the water field. This slow progress is mainly due to a lack of attention to the social root causes of inequality, a disabling economic context and insufficient financial resources, topped by a lack of political support and accountability mechanisms.

Up to date there exists a pronounced lack of timely and reliable sex-disaggregated water data, which is necessary to detect trends, strengths, and weaknesses of water policies and institutions dealing with groundwater on all levels. This lack of data on virtually all topics related to water and gender makes it difficult to assess and address progress in the field and therefore also impedes the timely achievement of the 2030 Agenda. It could overall be stated that there is the need for a concerted effort in order to accelerate the implementation of historical international commitments and agreements.

The UNESCO WWAP provides indicators that allow the collection of such sex-disaggregated water data through its toolkit that captures quantitative and qualitative data on water and gender issues (UNESCO WWAP 2015, 2019). Field surveys have been carried out based on the WWAP Toolkit (2015 edition).

The findings of the field analyses carried out in Botswana and Namibia suggest that shortcomings in water-related policies disproportionately burden women, while men emerge as primary decision-makers in household and also farm water-related decision-making. The observed lack of gender responsiveness in groundwater governance frameworks at all levels stands in contrast to the pivotal role of women in the use, management, and decision-making for sustainable management of groundwater resources.

UNESCO WWAP is committed to advancing women's empowerment and gender equality in the water realm. Its newly developed indicators, methodology, and tools aim to promote the integration of gender responsiveness within groundwater governance frameworks at all levels and to generate insights and understanding of the role of women and men in the management of water resources. The creation of clear-cut benchmarks of sex-disaggregated water data is crucial to inform regional, national, and transboundary water policies and to set up gender-transformative strategies in water governance.

## References

Global Water Partnership—GWP (2017) Gender equality and inclusion in water resources management. https://www.gwp.org/globalassets/global/about-gwp/publications/gender/gender-action-piece.pdf. Accessed on Jan 2020

Globescan and Sustainability (2017) Survey—Evaluating progress towards, the sustainable development goals. https://globescan.com/evaluating-progress-towards-the-sdgs/. Accessed on Jan 2020

https://globescan.com/wp-content/uploads/2017/07/GlobeScan-SustainAbility-Survey-Evaluating-Progress-Towards-the-Sustainable-Development-Goals-March2017.pdf. Accessed on Jan 2020

Miletto M, Pangare V, Thuy L (2019) Tool 1—Gender-responsive indicators for water assessment, monitoring and reporting. UNESCO WWAP toolkit on sex-disaggregated water data. UNESCO, Paris

Thuy L, Miletto M, Pangare V (2019) Tool 2—Methodology for the collection of sex-disaggregated water data. UNESCO WWAP toolkit on sex-disaggregated water data. UNESCO, Paris

UN Women (2015) Summary report—The Beijing declaration and platform for action turns 20. https://www.unwomen.org/-/media/headquarters/attachments/sections/library/publications/2015/sg%20report_synthesis-en_web.pdf?la=en&vs=5547. Accessed on Jan 2020

UNESCO WWAP (2015) The United Nations world water development report 2015: water for a sustainable world. United Nations world water assessment programme, UNESCO, Paris

UNESCO WWAP (2019) The United Nations world water development report 2019: leaving no one behind. United Nations world water assessment programme, UNESCO, Paris

Women's Major Group (2017) High level political forum—Position paper. https://sustainabledevelopment.un.org/content/documents/14774HLPF_WMG_Paper_2017_final_UPDATED.pdf. Accessed on Jan 2020

# "Pani Check—The Sisterhood of Water": A Transdisciplinary Documentary Film Project

Theresa Frommen and Katalin Ambrus

**Abstract**

The transdisciplinary cooperation between a hydrogeologist and a freelance filmmaker resulted in two films about a participatory and interdisciplinary socio-hydrogeological project of Freie Universität Berlin, Germany, which took place in India between 2016 and 2019: Rekha Devi and Zeenat Begum have a problem: the little water in their slum is of poor quality. Hydrogeologist Theresa Frommen is asked to help. But can illiterate women learn the scientist's complicated methods? And is the hydrogeologist prepared to get involved in the social structures of an Indian slum? A story of women empowerment in slums of India, where water supply is a challenge every day. It shows a contrasting picture of scientific research and its impacts on the social structures. "Pani Check – The Sisterhood of Water" is a 52-min documentary about the project, where the camera itself served as a research instrument and thus made new insights possible. The conclusion: the more different the ways of thinking and working, the broader a team is positioned in terms of its resources to solve a problem in a complementary and creative way.

**Keywords**

Socio-hydrogeology • India • Women • Peri-urban • Film

T. Frommen (✉)
IRI THESys, Humboldt-Universität zu Berlin, Berlin, Germany
e-mail: theresa.frommen@hu-berlin.de

T. Frommen
Hydrogeology Group, Freie Universität Berlin, Berlin, Germany

K. Ambrus
Freelance Filmmaker, -, Berlin, Germany

## 1 Introduction

From 2016–2019, the Hydrogeology Group at the Geological Institute of Freie Universität Berlin (FUB), Germany, conducted a participatory and interdisciplinary socio-hydrogeological project in Jaipur, India. This research was part of the project "Women's Action towards Climate Resilience for Urban Poor in South Asia" under the lead of the Indian organization Mahila Housing SEWA Trust. The project took place in the frame of the Global Resilience Partnership funded by Rockefeller Foundation, USAID and SIDA. The aim of the FUB project was to establish a participatory groundwater management system in two peri-urban low-income communities. Research has been conducted on how the groundwater component can be integrated into local water projects and how socio-economic aspects can be given greater importance in hydrogeological research.

The idea of filming the project arose shortly after the funding was approved, even before the official project start. The basic thought was to ensure the transferability and sustainability of the project results by finding a form that is also accessible to lay people and, with regard to India, to illiterate people. Film, through its visual and emotional qualities, has the unique opportunity to reach many people who would never read a scientific publication or project report. This is particularly important in relation to hydrogeology, as groundwater-related work and research always has a target group or "affected group" that does not belong to the scientific world (Frommen and Ambrus 2019).

Filmic methods are being used in science in various ways, e.g. documentary, educational film, filming very fast processes which otherwise would not be seen, filming long-term processes. To use filmic methods in a scientific way, several preconditions have to be fulfilled. The film must be embedded in a scientific discourse, the transparency of the methods and sources has to be given, and the access to the sources used has to be guaranteed (Schaedler 2010).


M. Abrunhosa et al. (eds.), *Advances in Geoethics and Groundwater Management: Theory and Practice for a Sustainable Development*, Advances in Science, Technology & Innovation,
https://doi.org/10.1007/978-3-030-59320-9_80

The main questions for this film are how to:

– increase the participating women's commitment;

– increase the project impact and achieve transferability;

– achieve awareness and increased knowledge about groundwater among the local population as well as local NGOs; and

– achieve valuable insights about socio-hydrogeological (research) projects for hydrogeologists and/or water-related professionals.

## 2 Approach and Methods

In order to achieve the above-mentioned goals, the general approach was to make a project documentary (45–60 min). The target groups of the documentary are the science community, NGOs working in the water sector as well as local communities.

During the project period, the filmmaker Katalin Ambrus visited the project three times for 3–4 weeks to record the progress, to document challenges and solution strategies and to portray the everyday life within the low-income communities. The visits were distributed over the whole duration of the project: the first visit took place two months and the second ten months after the field work has started, whereas the third visit took place nine months after the field work has ended.

Methods applied by the filmmaker for the documentary were in-depth unstructured interviews with all important stakeholders (local women assistants, NGO staff, water supplier, hydrogeologists), participatory observation at the main protagonists' communities and houses (three women assistants), observing and filming the hydrogeological and the participatory work of the hydrogeological team, documenting the water supply system and the wastewater situation of the city, shooting general pictures of the city and, finally, developing and shooting the moderation of the film.

## 3 Results and Discussion

### 3.1 Commitment of the Participants

The widespread interest in India in movies, filming and in being filmed generated a positive atmosphere in general. The protagonists were open to participate in the film and felt appreciated for their thoughts about the project. This was also due to the layout of the documentary with in-depth interviews, where the women had time to explain in detail their current and past lives as well as their hopes and thoughts concerning the project. In addition, the fact that the filmmaker was coming a second and a third time, as was promised at the beginning, increased the trust into the whole project.

### 3.2 Impact and Transferability

First screenings of the documentary at national (test screening and premiere in Berlin) and international level (IAH 2019 in Spain) made it clear that there is great interest in the films, especially from a hydrogeological but also from a humanities perspective. With the film, the experiences from the socio-hydrogeological project could already be transferred to many scientists. The planned film tour in India at the beginning of 2020 will show if the transfer of the project results to other Indian communities will work as well.

### 3.3 Insights for Hydrogeologists

#### 3.3.1 Expectations and Communication

The in-depth interviews with the participating women provided remarkable insights into the expectations they had on the project at the beginning. Their expectation that the project would literally solve all water problems led the hydrogeologist to reflect critically on her communication strategy with the local partner NGO and the local assistants regarding the project objectives. Communication seems to be obvious, but even if one is aware of communication barriers and prepares oneself, one will often underestimate them. Being aware of the great discrepancy between the planned and communicated project goals on the one hand and the expectations of the people involved on the other hand can help to adapt the communication strategy and prevent conflicts in the further course of a project.

#### 3.3.2 Visualization of the Development

Another finding from the documentary was the visualization of the development in the relationship with the local assistants as well as of the behaviour of them. In one of the scenes at the beginning of the documentary, the first groundwater sampling with the participating women community leader is shown. The body language of both, the hydrogeologist as well as the women leader is reserved and uncertain. In contrast to that, the pictures shot at the second visit of the filmmaker eight months later show the women leader taking a groundwater sample by herself, completely naturally and self-confident. That is a visible development which otherwise would not have been captured. Furthermore, the observation of one's own work and behaviour by the camera, which leads to the fact that one can look at oneself afterwards from an external perspective, helps to critically reflect one's own appearance, e.g. how different and thus conspicuous one looks among the locals, even if an adaptation to the local customs was tried.

### 3.3.3 Social, Cultural and Ethical Implications

The most compelling evidence of how the filmic approach changed the project and enabled an additional evaluation of the participatory work is the history of one young local woman assistant in one of the study areas, whose family received real threats from the village community, after working on the hydrogeological project. What the hydrogeologist did not know, when she started working with her, was that the young woman had left her husband and returned to her family. Since then, she has been an outsider within the Muslim community. The natural scientist's work thus had a direct influence on the social structure of the community, which was only fully reflected and came to light through the interviews with the filmmaker.

## 4 Concluding Remarks

Regarding the transdisciplinary cooperation and the choice of film as a new method, it can be concluded that this approach has influenced the project itself and generated knowledge which otherwise would not have been detected. Therefore, it definitely enabled a new way of evaluation of the project. The subjective view of the world and the aesthetic and artistic perception of it can contribute to an understanding of reality as can critical scientific research and thinking (Oester 2010).

In regard to the outcomes of the participatory socio-hydrogeological project, it can be concluded that even if natural scientists, geoscientists or, like in this case, hydrogeologists, are motivated to include the social perspective into their work, this does not mean that the project will automatically be more inclusive and sustainable. The problem is that natural scientists are normally not trained to work with communities and people and therefore tend to underestimate social, cultural and ethical implications. In the frame of the future of geoethics and socio-hydrogeology, we therefore highlight the importance of including these aspects into higher education.

## References

Frommen T, Ambrus K (2019) Pani Check & Pani Doctors: a documentary film project about an interdisciplinary hydrogeological project in Jaipur, India. Lernpapier, Andrea von Braun Stiftung - voneinander wissen, München

Schaedler L (2010) Über die Wissenschaftlichkeit von Dokumentarfilmen. In: Schöbi S, Bader Egloff L, Christen G (eds) research@film - Forschung zwischen Kunst und Wissenschaft. Alexander Verlag, Berlin, pp 46–55

Oester K (2010) Filmische Bilder als Erkenntnismittel. In: Schöbi S, Bader Egloff L, Christen G (eds) research@film - Forschung zwischen Kunst und Wissenschaft. Alexander Verlag, Berlin, pp 22–30

# "Pani Doctors—Join the Sisterhood of Water": A Participatory Film Project and an Educational Musical

Theresa Frommen and Katalin Ambrus

**Abstract**

The transdisciplinary cooperation between a hydrogeologist and a freelance film-maker resulted in two films about a participatory and interdisciplinary socio-hydrogeological project of Freie Universität Berlin, Germany, which took place in Jaipur, India, between 2016 and 2019: The documentary "Pani Check – The Sisterhood of Water" and the educational short film "Pani Doctors – Join the Sisterhood of Water". In the latter, the peri-urban low-income area Khara Kuaa transforms into a colourful theatre stage as Rekha Devi, Zeenat Begum and their neighbours unpack the scientific water testing kit and explain it to other women. "Pani Doctors – Join the Sisterhood of Water" is a 10-min short film and educational musical about groundwater monitoring which is developed in a participatory way by the director, the scientist and the Indian women, who took part in the socio-hydrogeological model project, together. New ways of science communication and education are followed with this approach. On the one hand, the film design was adapted to local viewing habits, and on the other hand, the target groups – in this case poor urban women in India – were included into the creation process itself.

**Keywords**

Socio-hydrogeology • Participation • Education • Science communication • Film • India

T. Frommen (✉)
IRI THESys, Humboldt-Universität zu Berlin, Berlin, Germany
e-mail: theresa.frommen@hu-berlin.de

T. Frommen
Hydrogeology Group, Freie Universität Berlin, Berlin, Germany

K. Ambrus
Freelance Filmmaker, Berlin, Germany

## 1 Introduction

From 2016–2019, the Hydrogeology Group at the Geological Institute of Freie Universität Berlin (FUB), Germany, conducted a participatory and interdisciplinary socio-hydrogeological project in Jaipur, India. The research was part of the project "Women's Action towards Climate Resilience for Urban Poor in South Asia" under the lead of the Indian organization Mahila Housing SEWA Trust. The project took place in the frame of the Global Resilience Partnership funded by Rockefeller Foundation, USAID and SIDA. The aim of the FUB project was to establish a participatory groundwater management system in two peri-urban low-income communities. Research has been conducted on how the groundwater component can be integrated into local water projects and how socio-economic aspects can be given greater importance in hydrogeological research.

The idea of filming the project arose shortly after the funding was approved, prior to the official start of the project. One of the big questions in planning the dissemination of the study results was: Which form is suitable to achieve the greatest possible understanding with the greatest possible influence among the largest possible group of people who have no access to scientific literature and can neither read nor write? The medium of film was chosen because its barrier-free nature and its entertaining and emotional qualities make it the most suitable medium for conveying content in a sustainable way. In the educational film, the methods and conclusions of the hydrogeological research project should be presented in an easily understandable and descriptive way. The film should be a practical support for future projects. The main target group is Indian women community groups.

To sum up, the main goals of the educational film are as follows:

—to generate a support for similar projects in the future,
—to be able to transfer the knowledge and experience of this project and


M. Abrunhosa et al. (eds.), *Advances in Geoethics and Groundwater Management: Theory and Practice for a Sustainable Development*, Advances in Science, Technology & Innovation,
https://doi.org/10.1007/978-3-030-59320-9_81

—to motivate women with similar backgrounds to participate in groundwater projects.

## 2 Approach and Methods

The original idea of a classical educational film, where the water testing is explained step by step, changed during the project. People in India have different viewing habits than people in Europe. The film industry in India is characterized by Bollywood productions that live from music and dance, and Indian viewers are used to interpreting, working out and dealing with the film during the screening (Srinivas 1998; Tirumala 2009). It was therefore necessary to understand the conventions of representation and to apply the codes of language and culture that Indians can read (Hall 1997). Since this would not have been possible for non-cultural outsiders within the given short time frame, the idea developed by the film-maker was to involve the women, who participated in the hydrogeological project, in the making of the film and to develop a script together. It was also assumed that it would be more effective for the project to tell essential parts of the educational film from the point of view of the local assistants, as they had already gone through the learning process at that time and would therefore know best how to communicate the content to other Indian women. Another aspect would be that in this way the women would learn something new, that is about film-making, and they would get even more motivation to share their knowledge (Frommen and Ambrus 2019).

## 3 Results and Discussion

### 3.1 Preparation

The chosen approach required a time-intensive preparation. In a first step, the main topics which had to be included in the educational film were defined by the hydrogeologist based on the structure and content of the overall socio-hydrogeological project. Three themes evolved out of this: the water cycle, water infrastructure mapping and (ground-)water monitoring. The hydrogeologist summarized all important aspects of each topic. In a next step, the film-maker transferred the given scientific content into song texts which were then, after another review by the scientist, translated into Hindi. This process demands an open mind and willingness to compromise by both the scientist and the film-maker as storytelling and scientific knowledge transfer had to take place at the same time (Frommen and Ambrus 2019).

The second part of the preparation happened on the ground. First, the content of the planned educational movie was discussed together with the women. After this, the key participatory component started where we examined together how to bring the songs into life and how to make the knowledge visible. The women took lead during this framing and presentation part by developing storylines and building scenarios. This required several intense meetings in the study area Khara Kuaa where the shooting should take place later. Various challenges arose during this participatory film-making which are similar to challenges in participatory approaches in general, e.g. confidence of the participants to fulfil this (new) task, inclusion of all participants, defining the level of participation and based on this, reflecting on the own role as facilitator (Arnstein 1969).

The third task was to find musicians who are ready to participate in the planned project. This again required a lot of effort, especially on the side of the film-maker. After finding local musicians, the challenge was how to actually transform the content and the lyrics of the songs into the traditional Rajasthani way of musical storytelling. These songs have a long-established structure and history. They talk about the beauty of the nature or love, and they are usually very short and follow a typical rhythm. The project songs did not fit into these principles at all. So, the challenge was to bring these two approaches—the hydrogeological and cinematographic storytelling approach and the traditional Rajasthani music approach—together. The film-maker had to make compromises regarding content, and the musicians had to make poetic and then later presentation-related compromises.

### 3.2 Shooting

The shooting, which took place on only one day, was a demanding and at the same time exiting practice for all participants as the film-maker was the only professional at the whole film set. She had to motivate people when scenes had to be repeated several times, had to hold the team together on the set, even if some of them had to wait and could only watch other scenes, had to be patient despite numerous interruptions from outside and had to explain things again and again to all non-experts, including the scientist.

Due to unpredictable incidences, the script had to be changed continuously, which required from the main participants, that is the film-maker, the scientist and the leading local women, a flexible performance and a lot of communication between all of them.

### 3.3 Reflection

The employed method falls under the category of participatory film-making, which is increasingly used by researchers, activists as well as artists and film-makers (Sudbury 2018; Hermans Amir 2019). The main motivation behind this is to address power inequalities and to create new forms of knowledge. In the present case, the primary motivation was different. Most Indian films live from music and dance and to be able to apply these codes of language and culture, the Indian protagonists took part in the film-making process. The cooperation could be described as an exchange in a certain way. The Indian protagonists provided the language of how to transmit the knowledge and the Western film-maker and hydrogeologist the necessary means. The goal was the same: to pass the project on to other women in India. Thus, it was an artistic decision to include the Indian women which allowed insights into other perspectives and aesthetics.

In general, the actual level of participation varies significantly in participatory film-making projects. In the present project, the scientist set the theme and the film-maker had the last word on aesthetics and narratives, which is also reflected in the fact that the editing took place at a later point in Germany without the Indian participants. To sum up, the power was still with the western facilitators and not with the local communities. Kapoor (2005) heavily criticizes participatory approaches in the development sector as they appear to be inclusive but only helps to mask the power of the mainly western educated facilitators. In the same way as Hermans Amir (2019) participatory film-making project in Nigeria and Cameroon, this presented case of participatory film-making is not able to overcome Kapoor (2005) points. But as discussed above, that was also not the main aim of this project. It was an artistic decision, and the slight shift of power from the western facilitators to the local women was an additional effect. Therefore, it can only be emphasized that the facilitators are aware of their own role and that they have taken a first step, which should be seen as a starting point rather than an endpoint.

In the end, the educational film became more a motivational film with many dance and music scenes, which meander around the topics of water cycle, well mapping and groundwater sampling. It is not a step-by-step instruction on how to do the water testing and monitoring as this is not the way knowledge transfer within this Indian target group would happen. To motivate the people to learn more about groundwater and take part in a groundwater related project is much more important.

## 4 Concluding Remarks

The applied approach resulted in a short film, which is entertaining, colourful and at the same time still transfers knowledge. This new way of science communication and education through an educational film is characterized on the one hand by a film design which is adapted to local viewing habits and on the other hand, by an inclusion of the target group into the creation process itself. In the field of science communication and education, it is important to take the respective social and cultural conditions into account in order to achieve a meaningful output.

## References

Arnstein SR (1969) A ladder of citizen participation. J Amer Inst Plann 35(4):216–224

Frommen T, Ambrus K (2019) Pani Check & Pani Doctors: a documentary film project about an interdisciplinary hydrogeological project in Jaipur, India. Lernpapier, Andrea von Braun Stiftung - voneinander wissen, München

Hall S (1997) Representation: cultural representations and signifying practices, vol 2. Sage Publications, Thousands Oaks, California

Hermans Amir A (2019) Folk filmmaking: a participatory method for engaging indigenous ethics and improving understanding. Conserv Soc 17(2):123–134

Kapoor I (2005) Participatory development, complicity and desire. Third World Quart 26(8):1203–1220

Srinivas L (1998) Active viewing: an ethnography of the Indian film audience. Vis Anthropol 11(4):323–353

Sudbury S (2018) Visualising the everyday: participatory filmmaking in rural India. Vis Ethnogr 7(2):6–23

Tirumala LN (2009) Bollywood movies and identity construction amongst second generation Indian Americans. Texas Tech University, Lubbock, Texas

# Transformative Art Applied to the Social Hydrogeology of the Cape Flats, South Africa

Rowena Hay, Anni Snyman, and Christopher J. H. Hartnady

**Abstract**

The Cape Flats Aquifer (CFA) is a primary coastal aquifer under the low-lying eastern suburbs and townships in the City of Cape Town (South Africa). It was identified as a major groundwater source since the 1970s but only intermittently explored and monitored until 2018, when Cape Town initiated an emergency programme in response to a prolonged drought. The objectives in this programme—still in progress—include detailed mapping of aquifer extent and properties, establishment and maintenance of a monitoring programme, and design and implementation of a Managed Aquifer Recharge abstraction scheme. Sustaining the potential of the CFA requires a holistic approach, considerate of the social, cultural and economic context. Support from the local communities is essential if this natural resource is to be harnessed for the benefit of all, as a supply and a means to store water for reuse from wastewater treatment works situated above the aquifer. CFA development can potentially help heal many of the historic societal wounds arising from the apartheid system of segregation and discrimination. The practices of transformative art may provide a practical and useful way to consider and structure the necessary "edges of engagement" in the socio-hydrogeological and environmental context.

**Keywords**

Transformative arts practice • Adaptive aquifer management • Societal engagement • Water-sensitive urban design

R. Hay (✉) · C. J. H. Hartnady
Umvoto Africa (Pty) Ltd., Muizenberg, South Africa
e-mail: rowena@umvoto.com

A. Snyman
Site Specific Land and Nature Art Collective, Johannesburg, South Africa

## 1 Introduction

South Africa's Mother City, Cape Town, experiences a profound challenge in meeting the demand for water from its growing population, as well as from agriculture, industry and business. Its response to climate change and the optimization of scarce resources provides valuable lessons for other cities, around the country and beyond its borders (Taing et al. 2019 and references therein). While there are other large aquifers in the area of the Western Cape Water Supply System (WCWSS), the present focus is the Cape Flats Aquifer (CFA), underlain by sedimentary units between the mountainous Cape Peninsula and the mainland ranges (Fig. 1). The present contribution deals with the broader hydro-sociological issues of CFA developments in the current drought-emergency response (McGibbon et al. 2021).

The Cape Flats is a geomorphological province dominated by Aeolian processes in which siliciclastic sediment from rivers and carbonate shell fragments from marine molluscs are mixed in the surf zone and beach environments of the False Bay coastline and are driven onshore by powerful SE trade winds. The original, mobile dune scape mapped in the late nineteenth century is now largely gone, replaced by extensive and dense human settlements (formal and informal).

Unlocking the potential of the CFA requires a holistic approach that addresses the underlying groundwater architecture, the above-ground surface water and transport infrastructure, demographic patterns and processes, and the natural and urban water cycles that draw in large volumes of water from distant sources and store and release these via the aquifer and rivers into the near-shore environment of False Bay (Fig. 2). It requires collaboration between numerous different stakeholders, such as scientists, engineers, small-scale farmers, municipal officials and many others, including communities who live above the aquifer.


M. Abrunhosa et al. (eds.), *Advances in Geoethics and Groundwater Management: Theory and Practice for a Sustainable Development*, Advances in Science, Technology & Innovation,
https://doi.org/10.1007/978-3-030-59320-9_82

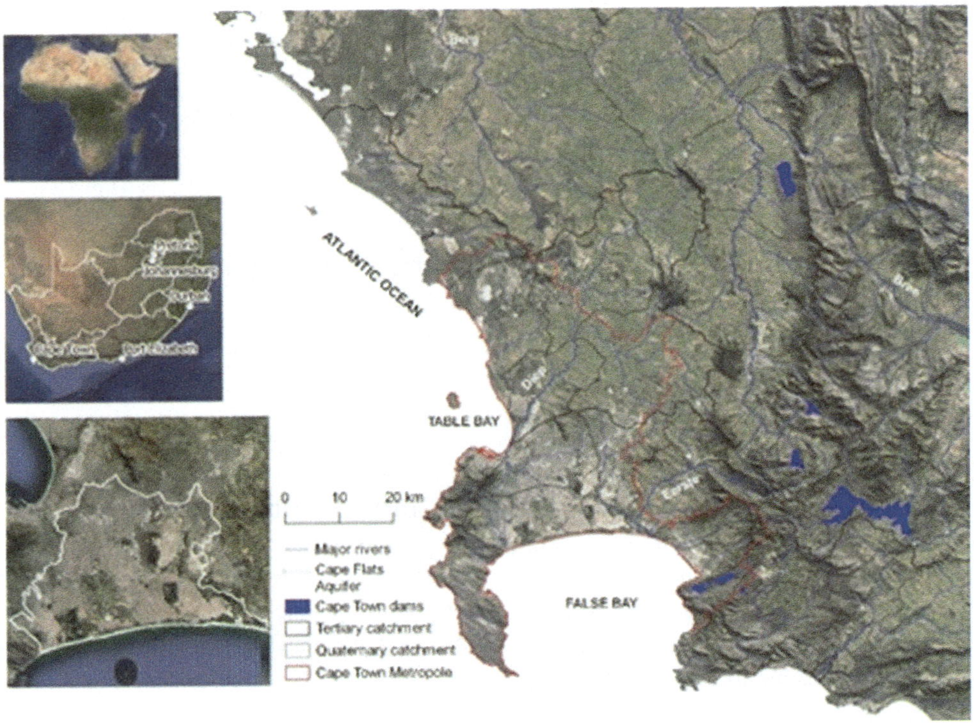


**Fig. 1** The Cape Town metropolitan area (red line boundary) at the southern tip of African continent (upper left inset), in the Western Cape Province of South Africa (middle left inset), contains the Cape Flats Aquifer (CFA), bounded on the west by the mountainous Cape Peninsula extending from Table Bay to Cape Point, on the south by the False Bay coastline and on the east by the foothills and mountains of the hinterland Hottentots Holland range. Major dams of the Western Cape Water Supply System (WCWSS; blue areas) are located within the hinterland mountains, mostly outside the metropolitan edge. The CFA edge (white border in lower left inset) contains topography in which slope is generally less than 2 degrees (except for localized areas of relict dune field) and sedimentary thickness is >5 m

**Fig. 2** Present-day land-use patterns in the CFA region (shadowed black edge) between Table Bay and False Bay, showing watercourses (generally canalized), water bodies of four kinds including Waste Water Treatment Works (WWTW; purple areas), and four generalized categories of land use including agriculture (olive green area) within the intra-urban Philippi Horticultural Area (PHA). The Big Lotus Canal (BLC)—a focus of the present initiative (red rectangle locating Fig. 3)—channels the stormwater run-off from the Cape Town International Airport (CTIA) via the northern end of the PHA to the Lotus River and to Zeekoei Vlei (ZV) lake and False Bay

## 2 Water, Wind and People

In the course of urban expansion, the natural waterways and local endorheic, interdune wetlands have been transformed. Most of the streams have been canalized, generally as stormwater drains. During development of the Cape Town International Airport (CTIA; Fig. 2) in the 1950s, the Big Lotus Canal (BLC; Fig. 2) was constructed to carry stormwater from runways on the divide between the Vygekraal and Kuils Rivers, and transport it obliquely across a gentle westerly topographic slope to the lower (~20 m amsl) divide between the Vygekraal and Lotus Rivers.

The Great Lotus Canal (Gorgens et al. 2001) or BLC (Fig. 2) flows through formal and informal settlements, receiving variable amounts of grey and black water. After crossing informal squatter settlements erected upon an elevated former municipal landfill site (Fig. 3, upper right), the BLC flows between the formal townships of Nyanga and Gugulethu, passing local schools and a sports stadium (cf. photo insets on Fig. 3). Turning southwards past the SW section of Nyanga and the suburban railway line, it flows WNW parallel to Govan Mbeki Drive, between the Edith Stephens Nature Reserve and the Consol Glass Sand Mine, before turning sharply southwards and passing through a new industrial area to enter PHA farmlands (Fig. 3).

For weeks after rainfall, the BLC in the Nyanga–Gugulethu area continues to flow and is often foul smelling. Water quality tests indicate high ammonia, and significant amounts of faecal matter and litter are visible, with abundant large and small plastic pieces that are not easily contained by litter traps. At a culvert within the PHA, BLC waters are diverted into a sewage main during low-flow stages, when the canal is effectively an open sewer. After storms or substantial winter rains, however, the BLC overflows to join the main canal of the (Great) Lotus River, which runs southwards to enter the NE corner of Zeekoevlei (ZV lake on Fig. 2), a large freshwater body and wetland area of conservation and recreational significance.

The BLC between the CTIA, on Vygekraal–Kuils River divide, and the PHA, on the Vygekraal–Lotus River divide, is a priority target for urgent intervention to restore a healthy surface water environment and protect a substantial sector of the underlying CFA from toxic pollutants of both industrial and urban-domestic origins. Initiatives to prevent or at least inhibit the inflow of human urine, excrement, rubbish and litter into the BLC will require both formal intervention by the municipal authorities and profound behavioural change by citizens who inhabit the BLC environs.

Pollution accumulation in the BLC headwater is a formidable problem, precisely because it is inadequately addressed in current municipal policy and practice. Water management in this area "… should be integrated with management of other activities such as waste handling, industrial production, and transportation. The most important action must be "prevention at source". Tools for implementation of solutions are legislative and administrative actions coupled with education programmes on all societal levels and increasing public awareness. The goal is sustainability of the society by resource recovery and re-use" (Gorgens et al. 2001, p. 212). Gorgens et al (2001) define "best management practices" (BMPs): structural BMPs include run-off minimization design, infiltration

**Fig. 3** The Great Lotus Canal (or BLC in Fig. 2; blue line with red triangle markers) collects stormwater from the Cape Town airport and its industrial area (upper right) and transports it to the north-western corner of the PHA (lower left), where—just outside of the image edge—it is intercepted during low-flow periods by a weir connected to a main sewage line

enhancement and (hydrochemical) treatment, generally grouped as "stormwater treatment measures" (pp. 214–216); Non-structural BMPs include educational, regulatory and enforcement measures (pp. 213–214).

## 3 Adaptive Management and Transformative Art: Final Remarks

While developing the CFA using structural BMPs and an adaptive management approach (McGibbon et al. 2021), authorities can support creation of collaborative platforms through transformative art practices and projects (LeBaron and Sarra 2018) and the education of a network and generation of Citizen Scientists (Heigl et al. 2019) to support resource restoration and protection. Through conversation with and learning from other stakeholders and disciplines, the officials, professional consultants and communities can play a vital role in addressing a variety of social and environmental challenges. Development of a sustainable CFA resource can contribute to the medium- and long-term behavioural shifts within institutions, professions and communities necessary for the achievement of the vision of water-sensitive urban design (Armitage et al 2014; Taing et al. 2019).

Our initial approach to BLC rehabilitation is of a non-structural kind, focussed on collaboration and community building through transformative arts practices (LeBaron and Sarra 2018), which refers to the creative process by which practitioners engage participants in artistic activities to promote change in the lives of all who participate, as well as those who witness what is created. These activities are integrated into education and advocacy outreach initiatives such as a recent Fynbos and Water Stewardship course held for residents who live on or near the BLC. Collaboration between artists, scientists and other disciplines is integral in the engagement of the community and raising awareness of deep collaboration needed to rehabilitate and protect our city's resources and acknowledging individual and shared trauma and injustice.

**Acknowledgements** The ongoing initiatives to engage with BLC local communities has benefited from discussions and collaboration with management of Edith Stephens Nature Reserve (Luzann Isaacs, Stacey-Ann Michaels), CommuniTree NGO (Frances Taylor, Paul Hoekman) and Paula Hay, with whom we co-designed and ran a Fynbos and Water Stewardship workshop with participants who live on or nearby the BLC and during which artwork (Fig. 4) was produced. Participants are acknowledged for their input and feedback, and for initiatives some have taken forward. Umvoto funded the workshop and artwork, providing staff who presented and contributed to discussions. Mr Mike Ryder and Prof Neil Armitage also contributed their time and expertise. This work is ongoing and acknowledges lessons learned from all persons, whether named above or not.

**Fig. 4** Photograph mosaic illustrating the setting of the Great Lotus Canal (or BLC in Fig. 2) where it flows past the Edith Stephens Nature Reserve (ESNR) alongside Govan Mbeki Drive, with artwork along a wall directly opposite ESNR

## References

Armitage NP, Fisher-Jeffes L, Carden K, Winter K, Naidoo V, Spiegel A, Mauck B (2014) Water sensitive urban design (WSUD) for South Africa: framework and guidelines. Report No. TT 588/14. Water Research Commission, Pretoria, 154 pp. https://doi.org/10.13140/2.1.3042.5922

Gorgens A, Pegram G, Uys M, Grobicki A, Loots L, Tanner A, Bengu R (2001) Integrated catchment management in an urban context: The Great and Little Lotus rivers, Cape Town. Report No. 864/1/01. Water Research Commission, Pretoria

Heigl F, Kieslinger B, Paul KT, Uhlik J, Dörler D (2019) Toward an international definition of citizen science. Proc Nat Acad Sci 116 (17):8089–8092

LeBaron M, Sarra J (eds) (2018) Changing our worlds: art as transformative practice. African SUN MeDIA, Stellenbosch

McGibbon D, Hugman R, Towers L, Riemann K, Hay R, Hartnady C (2021) Long-term planning during emergency response—geoethics of a no-regrets emergency supply scheme with a long-term vision for the development of the Cape Flats Aquifer (Cape Town). In: Abrunhosa A, Chambel A, Peppoloni S, Chaminé HI (eds), Advances in geoethics and groundwater management: theory and practice for a sustainable development, Proceedings of the 1st congress on geoethics and groundwater management (GEOETH&GWM'20), Porto, Portugal 2020. Springer ASTI, Cham (this volume)

Taing L, Chang CC, Pan S, Armitage NP (2019) Towards a water secure future: reflections on Cape Town's Day Zero crisis. Urb Wat J 16(7):530–536

# Multisector Collaborative Groundwater-Surface Water Modelling Approach to Improve Resilience to Hydrological Extremes in the Limpopo River Basin

Syed Md. Touhidul Mustafa, Anne Van Loon, Luis Artur, Zareen Bharucha, Annatoria Chinyama, Farisse Chirindja, Rosie Day, Fulvio Franchi, Josie Geris, Stephen Hussey, Edward Nesamvuni, Alcino Nhacume, Alfred Petros, Hanne Roden, Melanie Rohse, Sithabile Tirivarombo, and Jean-Christophe Comte

**Abstract**

It is necessary to combine the understanding of physical environmental drivers with social, economic, cultural and political perspectives and information to build resilience to future flood and drought hazards. We present a flexible collaborative modelling approach to improve resilience to hydrological extremes in large basins with application to the Limpopo River Basin (LRB). It uses an iterative, knowledge co-production process to strengthen crucial bridges between scientists and water management stakeholders on the appropriate scale(s). In the proposed collaborative modelling approach, the integrated hydro (geo)logical model is combined with regional to transboundary people's knowledges and policies. We analyse the effect and importance of stakeholders' feedback on the numerical model prediction. The proposed methodology is applied in the Limpopo River Basin (LRB) where floods and droughts are recurrent events. Through this iterative multisector collaborative modelling approach, we aim to develop a reliable and feasible management instrument to help reduce the impact of alternating droughts and floods and increase the resilience to hydrological extremes.

**Keywords**

Collaborative modelling approach • Drought–flood cycles • Limpopo river basin • Knowledge co-production • Resilience

## 1 Introduction

Floods and droughts are recurrent events in the Limpopo River Basin (LRB) and cause serious water and food insecurities. Both may have considerable adverse effects on socio-economic, agricultural and environmental conditions. Traditionally, floods and droughts are studied separately. However, impacts of floods and droughts are influenced by each other, whereby floods following severe drought are often more damageable. Future predictions emphasize increases in the frequency and magnitude of both climatic extremes globally, with southern Africa among the most

S. Md.T. Mustafa (✉) · J. Geris · J.-C. Comte
University of Aberdeen, Aberdeen, UK
e-mail: syed.mustafa@abdn.ac.uk

A. Van Loon · R. Day
University of Birmingham, Birmingham, UK

L. Artur · F. Chirindja
Eduardo Mondlane University, Maputo, Mozambique

Z. Bharucha · M. Rohse
Anglia Ruskin University, Cambridge, UK

A. Chinyama
National University of Science and Technology, Bulawayo, Zimbabwe

F. Franchi · S. Tirivarombo
Botswana International University of Science & Technology, Palapye, Botswana

S. Hussey
Dabane Trust, Bulawayo, Zimbabwe

E. Nesamvuni
University of Free State, Bloemfontein, South Africa

A. Nhacume
Directorate of Water Supply and Sanitation, Government of Mozambique, Maputo, Mozambique

A. Petros
Department of Water and Sanitation, Government of Botswana, Gaborone, Botswana

H. Roden
German Red Cross, Changanine, Mozambique



M. Abrunhosa et al. (eds.), *Advances in Geoethics and Groundwater Management: Theory and Practice for a Sustainable Development*, Advances in Science, Technology & Innovation,
https://doi.org/10.1007/978-3-030-59320-9_83

exposed and vulnerable regions (Kusangaya et al. 2014; Thornton et al. 2014).

In order to build resilience to future flood and drought hazards, it is necessary to combine understanding of physical environmental drivers with social, economic, cultural and political perspectives and information (Kahinda et al. 2016; Basco-Carrera et al. 2017). Integrated Water Resources Management (IWRM) principles have completely changed the water resources management by introducing the stakeholder participation in planning and decision-making processes (GWP 2000). Tortajada (2010) suggested that stakeholder participation is mandatory to ensure good water governance.

People's vulnerability to flood and drought events depends on both the availability and development of physically based predictive tools and society's coping strategies. Risks related to both climatic extremes could be reduced by using interdisciplinary methods and prediction tools. However, implementation of new strategies and more generally, the capacity of water end-users and stakeholders to adapt their management in an efficient and timely manner, relies on accurate predictions of groundwater-surface water systems and good connections/dialogues across management levels.

Accurate predictions of groundwater-surface water systems, as well as sustainable water management practices, are essential for policy-making. Numerical models are used to understand and forecast water flow systems under anthropogenic and climatic influences. They provide primary information for decision-making and risk analysis. However, water governance frameworks as well as local people's hydrogeological knowledge and needs require to be better integrated as part of the physical modelling design and process.

Although much research has been conducted on stakeholder's engagement in water resources planning and decision-making processes (Basco-Carrera et al. 2017), in contrast, very little research has been conducted in developing management scenarios using the numerical hydrological models within these participatory planning and decision-making processes.

In this study, we present a flexible collaborative modelling approach to co-create management solutions to reduce impacts and increase benefits of drought–flood cycles throughout the LRB. In the proposed collaborative modelling approach, the integrated hydro(geo)logical model is combined with local to transnational knowledges and policies. The proposed methodology is applied in the LRB where hydrological and groundwater data are fragmented and highly uncertain.

## 2 Study Area

The study area is located in southern Africa as shown in Fig. 1. The transboundary Limpopo River Basin (LRB) spans Botswana, South Africa, Zimbabwe and Mozambique over an area of 415,000 km$^2$. The elevation of the study area varies between sea level to 2331 m. Because of its long-term water scarcity and multidimensional water problems, the LRB is an ideal basin to apply our multisector collaborative groundwater-surface water modelling approach to improve resilience to hydrological extremes.

## 3 Methodology

The general approach integrating local people's hydrogeological knowledge within numerical modelling to improve resilience to hydrological extremes in the Limpopo River Basin is summarized in Fig. 2.

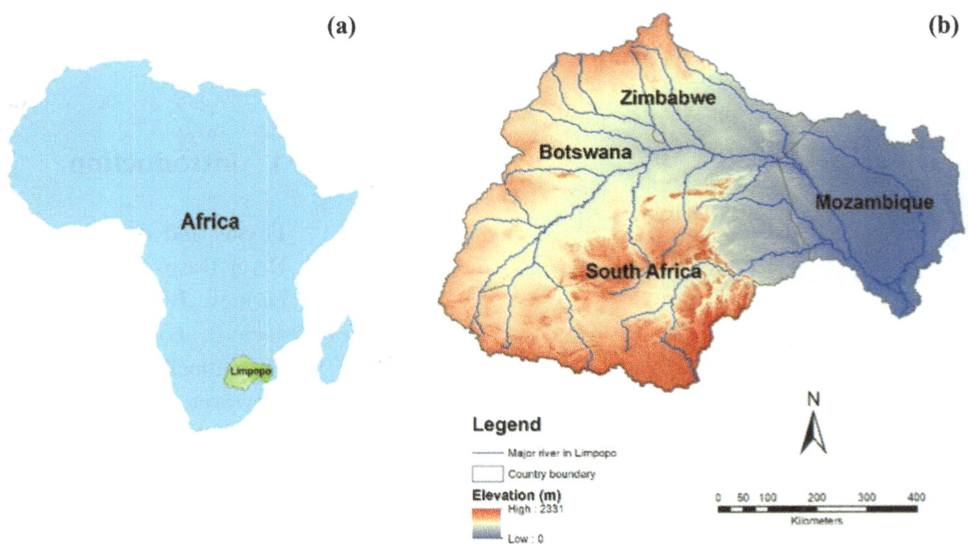


**Fig. 1** Limpopo River Basin (LRB): **a** the location of the LRB catchment area within the African continent, **b** LRB topographical features and major river network

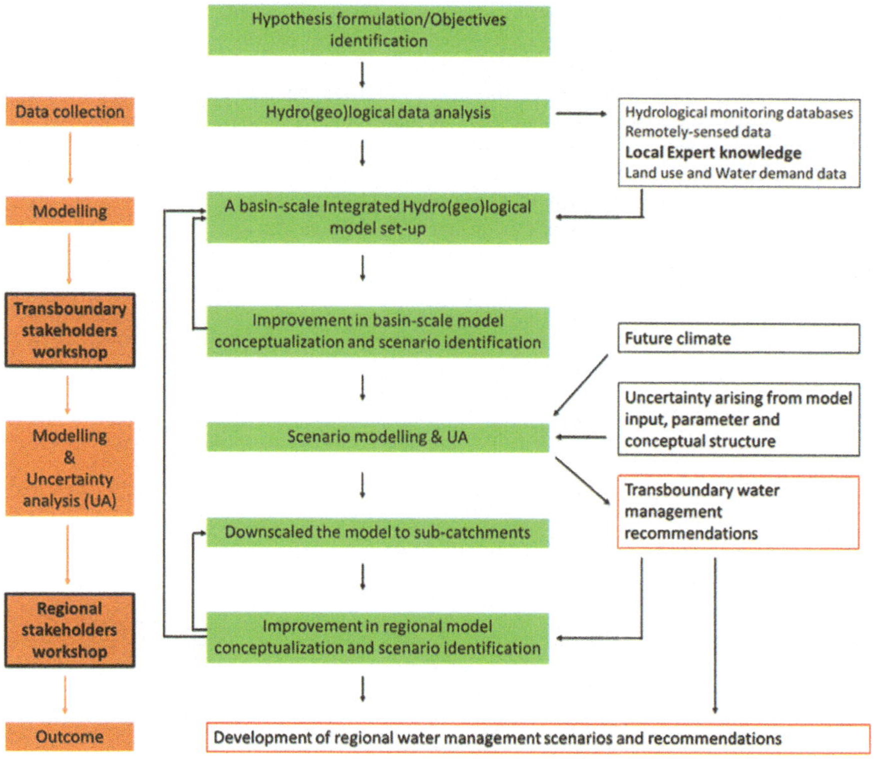


**Fig. 2** Proposed modelling methodological approach integrating regional to transboundary people's hydrogeological knowledge

In a first step, the available hydrogeological data of the Limpopo River Basin are characterized in detail. Based on existing climate data, flow measurements, remote sensing, reanalysis and literature data, the spatial, temporal, and vertical variability of the hydrogeological parameters are identified. Next, a basin-scale groundwater-surface water integrated model is set-up using the existing datasets. Then, through transboundary stakeholders' dialogues, model conceptualization will be improved, and management scenarios will be co-developed and further explored with the hydrogeological model. Subsequently, through running the model with identified scenarios, transboundary water management recommendations will be produced for further decision-making. Finally, the basin-scale (transboundary levels) results will be discussed with the stakeholders at lower regional level to improve the model conceptualization and feasible regional management scenarios. Thereby a multiscale, robust, and more reliable management instrument will be developed, which will help improve communication between managers/policy-makers and local water users towards reducing negative impacts of, and increasing resilience to, drought–flood cycles. The uncertainty arising from model input, parameter and conceptual model structure will be quantified following Mustafa et al. (2018, 2020).

## 4 Concluding Remarks

The proposed collaborative modelling approach will be able to generate more accurate and reliable management scenarios both at regional and transboundary levels. Through the iterative process, we aim to develop a reliable and feasible management instrument to help reduce the impact of alternating droughts and floods and increase the resilience to future flood and drought hazards.

**Acknowledgements** The research is supported by the Natural Environment Research Council, UK, under the framework of Connect4 Water Resilience project (Project Reference: NE/S005943/1).

## References

Basco-Carrera L, Warren A, van Beek E, Jonoski A, Giardino A (2017) Collaborative modelling or participatory modelling? A framework for water resources management. Environ Model Softw 91:95–110

GWP (2000) Integrated water resources management, TAC background papers. Global Water Partnership (GWP)

Kahinda JM, Meissner R, Engelbrecht FA (2016) Implementing integrated catchment management in the upper Limpopo River basin: a situational assessment. Phys Chem Earth Parts A/B/C 93:104–118

Kusangaya S, Warburton ML, van Garderen EA, Jewitt GP (2014) Impacts of climate change on water resources in southern Africa: a review. Phys Chem Earth Parts A/B/C 67:47–54

Mustafa SMT, Nossent J, Ghysels G, Huysmans M (2018) Estimation and impact assessment of input and parameter uncertainty in predicting groundwater flow with a fully distributed model. Water Res Res 54(9):6585–6608

Mustafa SMT, Nossent J, Ghysels G, Huysmans M (2020) Integrated Bayes-ian Multi-model approach to quantify input, parameter and conceptual model structure uncertainty in groundwater modeling. Environ Modell Soft 126:104654

Thornton PK, Ericksen PJ, Herrero M, Challinor AJ (2014) Climate variability and vulnerability to climate change: a review. Glob Change Biol 20(11):3313–3328

Tortajada C (2010) Water governance: some critical issues. Int J Water Resour Dev 26:297–307

# Socio-hydrological Analysis Protocol Adaptation to NW São Paulo State, Brazil

Adriana Sanches Borges, Rodrigo Lilla Manzione, and Viviana Re

**Abstract**

Groundwater is an economically viable resource for the supply of agricultural, industrial, public and small rural areas. However, there is a lack of knowledge about groundwater quantity and quality issues, being a passive resource for depletion and pollution. Given that the agricultural sector is one of the largest consumers of water and that it is an essential and strategic component of food security, proper control and administration enables balanced use management. At present, effective procedures for bringing water users and technicians closer together are not structured to ensure the application of practices that lead to sustainable groundwater use in agriculture. Therefore, this work aimed to create ways to bring farmers and rural communities closer to the theme of groundwater, to ensure its preservation and good use, by using socio-hydrogeology approaches. These approaches targeted to bridge the gap between technicians and water users by disseminating information and discussions on groundwater management, and this work intends to develop and implement a groundwater communication protocol in rural communities based on previous studies. Modification and adaptations of the analysis protocol are proposed to the northwester part of São Paulo State, Brazil.

A. S. Borges
Bunge Açúcar e Bioenergia, Orindiuva, SP, Brazil

A. S. Borges
Professional Master Course in Water Resources Management and Regulation (PROFÁGUA), São Paulo State University (Unesp), School of Engineering (FEIS), Ilha Solteira, São Paulo, Brazil

R. L. Manzione (✉)
Biosystems Engineering Department, School of Sciences and Engineering (DEB/FCE), São Paulo State University (UNESP), Tupã, SP, Brazil
e-mail: lilla.manzione@unesp.br

V. Re
Earth Sciences Department, University of Pisa, Pisa, Italy

**Keywords**

Water users • Socio-hydrology • Rural communities

## 1 Introduction

Groundwater is essential for life, not only for supplying cities, rural communities, and as inputs for a variety of economic activities, but also for sustaining various aquatic systems such as rivers, lakes, mangroves and marshes. Without groundwater, vegetation in dry or tropical climates would not stand, nor would aquatic environments exist or fulfil their environmental functions (Tuinstra and van Wensem 2014, Griebler and Avramov 2015). Due to the growing demand for groundwater use, overexploitation, that is, the extraction of water in a larger volume than the one replenished by nature, can cause a reduction in the amount of water that supplies watersheds, the drought of natural springs, among many other negative impacts, as recharges flow slowly.

In Brazil, groundwater is extracted through tubular wells (popularly known as artesian or semi-artesian) or even excavated wells. The actual number of wells in the country is currently unknown, although the requirement has been established since 1997 through Law 9433, where it established the National Water Resources Policy, the number of regular wells is still very low. Given that the agricultural sector presents large water demands and that it is an essential and strategic component for food security, proper groundwater control and management enables fair and balanced management, allowing allocation to other uses (Re 2015). It should be analysed from different perspectives to ensure that it is adequately included in the water management system, and for this purpose, a joint analysis with the other components of the water cycle (taking into account the social perspective and interaction between land use and water demands) is needed.



M. Abrunhosa et al. (eds.), *Advances in Geoethics and Groundwater Management: Theory and Practice for a Sustainable Development*, Advances in Science, Technology & Innovation,
https://doi.org/10.1007/978-3-030-59320-9_84

Nowadays, effective procedures for bringing water users and technicians closer together are not structured to ensure the application of practices that lead to sustainable groundwater use in agriculture. Therefore, the socio-hydrogeological approach promotes the structured interaction between members of society, especially rural communities and water technicians (Re 2015; Limaye 2017). In many cases, water resource management is recognized as an objective and is therefore not a tool for solving a relevant issue. For example, many projects that are funded are based on water management principles but do not consider the reality itself in the field.

In fact, as one of the key aspects of socio-hydrogeology is engaging with water users/polluters in order to favour knowledge sharing, it can also provide useful information for the study of human–aquifer co-evolution dynamics, as promoted by socio-hydrology for river basins (Di Baldassarre et al. 2013). Therefore, with this information it will be possible to propose new groundwater management plans that are not only based on sound scientific knowledge, but also take into account the socio-economic relevance of a given aquifer. In this framework, the present work aims at contributing to the emerging field of socio-hydrogeology by adapting the proposed protocol to the reality of farmers and rural communities in the north-western part of São Paulo State, Brazil, in order to evaluate their perceptions on groundwater issues. This application will therefore permit to bridge the gap between scientists/hydrogeologists and farmers and rural communities when groundwater protection and preservation is at stake. Additionally, this paper can serve as an example for possible adaptation of socio-hydrogeological surveys to different geographical and cultural contexts.

## 2 Materials and Methods

### 2.1 Study Area

The municipality of Populina is located in the northwest of the State of São Paulo (19°57′14″ south and 50°32′16″ west), in the hydrological context of two watersheds, the Cascavel/Cã-Cã sub-basin and the Ribeirão Santa Rita sub-basin, both belonging to the Water Resources Management Unit (UGRHI) 15-Turvo/Grande (Fig. 1). Populina has an area of 315.94 km$^2$, with average altitude of 443 m a.s.l., inserted in the São José do Rio Preto Administrative Region and Fernandópolis Government Region. In population terms, it is considered a small municipality, with a population of 4015 inhabitants, representing 3.63% of the total population of the Government Region of Fernandópolis. Its territorial extension imposes a demographic density of 12.71 inhabitants/km$^2$. In the population evolution phase, Populina has a negative annual geometric growth rate of −0.56% per year (2010–2019). The largest participation in the municipality is in the service sector, followed by the agricultural sector and the industrial sector. Having a GDP per capita in the 2016 Population of R$ 17,997.47 per inhabitants/year. Finally, the Human Development Index (HDI) was 0.714 in 2010 (IBGE 2011).

The municipality is inserted in the geological context of the Paraná Sedimentary Basin, located in the northeast portion of the Bauru Basin. The Bauru Basin is characterized as a predominantly sandy sedimentary sequence, in the order of metres, composed of three larger units: São Bento Group, Bauru Group and Caiuá Group (CPRM 2006). The area is located in the geomorphological context of the Paulista Western Plateau. This unit is represented by degradation relief forms on dissected plateaus with soft hills relief, smoothed hills and localized residual mounds (Ross and Moroz 1996). The area covered by the municipality lies in an NW–SE direction interfluvium, with the south-central region as the topographically highest portion of the land and the northern region as the lowest portion. The eastern limit of the municipality coincides with the low course of Ribeirão Santa Rita, the eastern limit coincides with the middle and low course of Ribeirão do Arrancado and the northern limit coincides with the middle course of Rio Grande. The topographic range of the municipality is approximately 160 m, with heights ranging from approximately 320–480 m. The diversity of relief and geology of the municipality of Populina gives rise to a low variety of soils. According to the Oliveira et al. (1999), the predominant soils identified in the municipality are: Leptsols, Acrisols, Ferralsols and Gleysols. According to CEPAGRI (2019), Populina is located in the tropical climate region, designated as Aw in the Köppen classification, characterized by the occurrence of dry winter with rainfall less than 30 mm. The average temperature of the warmest month is over 25 °C, and the coldest month is over 20 °C, with an average annual rainfall of 1.200 mm.

### 2.2 Socio-hydrology Analysis Protocol

The survey scheme proposed by Re (2015) is analysed in this paper, and modifications are provided in order to adapt it the reality of the northwest part of São Paulo State. This adaptation would permit to administer the questionnaire to the rural population of the region. In particular, in order to better understand the local perception of groundwater resources, specific interviews will be organized with members of the rural association of dairy farmers in the municipality of Populina.

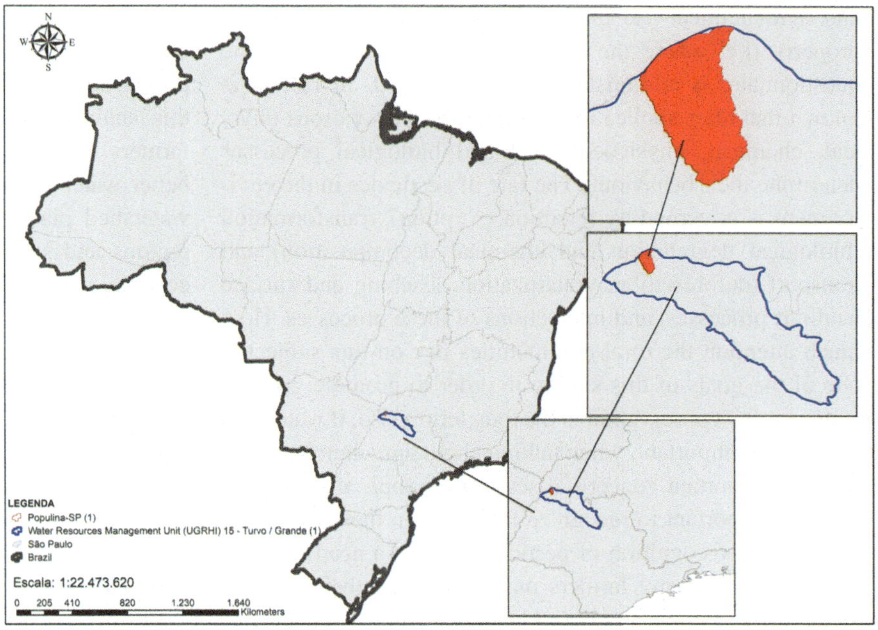


**Fig. 1** Location of the municipality of Populina, UGRHI 15—São Paulo State, Brazil

## 3 Results

Originally, the survey proposed by Re (2015) contains questions about four main topics: water use, use of water for irrigation purposes, awareness of water issues and potential for participation. The list of modifications performed to adapt the questionnaire to the local context is provided in this section.

As concerns water use, the main original questions concerns (1) the main water sources used in the household (among tap water/public water supply, rainwater harvesting, surface water, groundwater and bottled water); (2) the description of the main issues (if any) faced per water source (in both quality and quantity); (3) information of water use rates; (4) the person responsible for water management in the household. In this topic, a new question was incorporated, in order to understand if additional treatment is performed after groundwater withdrawal, as filtering and chloride application after pumping are a common practice in many regions of São Paulo State and Brazil. The second section of the survey refers to the use of water for irrigation purposes and is targeted to assess the rates and type of irrigation associated to the different is cultivated in the area and the purpose of agricultural production. In the awareness of water issues part, the original questionnaire is focused on understanding the awareness of water scarcity and climate changes and to assess the perception of the local population about these issues, which are the core of many hydrogeological and hydrogeochemical assessments worldwide. Finally, the last section targets the evaluation of the willingness of local farmers and well's owners to be included in participatory actions for water management and to assess the potential for the creation of new citizen sciences actions.

For the north-western part of São Paulo State, the main modification of this survey is the inclusion of a topic about land use and cover in the surroundings of the rural properties. Cash crops cultivated with sugarcane, citrus and reforestation with Eucalyptus are common and occupy extensive areas that can be potential disrupters of the water cycle in long term. On this topic, it is asked (1) what kind of crop is cultivated in the surroundings of the property; (2) if there is agrochemicals application in these crops; (3) what kind of sprayers are used among terrestrial and aerial; (4) if there is legal vegetation reserve in the property; (5) if the springs and water courses are protected and vegetated in the property; and (6) if the property is part of the rural cadastre required by the govern.

## 4 Discussion

Some characteristic features of the region have been added to adapt the questionnaire to the local peculiarities of the study area, such as groundwater treatment by filtration and/or chlorine dosing. The inclusion of this question is recommended also in other areas where this practice can occur, as it will provide useful information to avoid biased interpretation of hydrogeochemical data.

Another important new point addressed in the questionnaire refers to the use of fertilizers and/or pesticides in crops not only directly in the property, but also in the surrounding areas. To this goal, a specific question about crops production and agrochemical use in the area surrounding the studied property (i.e. where the sampled well is located, and the questionnaire is administered), was included. In fact, it is known that after application of these products various physical, chemical, physicochemical and biological processes determine their behaviour. The fate of pesticides in the environment is governed by retention (sorption), transformation (biological degradation and chemical decomposition) and transport (deforestation, volatilization, leaching and surface hauling) processes, and interactions of these processes. How much attention the rural communities pay on this subject is one of the goals of this survey in order to promote groundwater quality conservation in the long term. Also, if water was not enough important, when talking about agrochemicals in a form an important concern arises in the scope of geoethics, and it is important to recognize if and when these questions appear. The recognition of pesticides use as a need for crop production for many farmers make then take their chances using these products without enough criteria, creating a vicious cycle of environmental contamination and health problems that is difficult to break, but geoscientists should identify an ethical criterion that can orient them, on which to base technical decisions, building a framework of common values to be adopted by the geoscience community and society as a whole (Peppoloni and Di Capua 2017).

As concerns the expected results, overall, the engagement of farmers thanks to the questionnaire administration will not only result in a more conscious use of water resources but may eventually lead to long-term bottom-up participatory management actions. As a next step, we may have a methodology that brings rural producers closer to water technicians, creating a two-way flow of information that promotes rational use and protection of groundwater on their properties. This would be the way to empowering the rural producers to maintaining and preserve water springs, creeks and rivers buffering zones and terrace farming. The latter are practices established in the Brazilian forest code and techniques easy to implement in order to groundwater quantity conservation.

## 5 Concluding Remarks

The proposed socio-hydrological protocol was adapted to the Brazilian conditions in the NW of São Paulo State. Despite the evident applicability for the purposes of socio-hydrogeology, modifications regarding the land use and cover were necessary for the reality of the region's rural producers to be respected and addressed in the survey. From this analysis protocol, it will be possible to improve the farmers' perception about the groundwater issues, aiming a better water resources management and its inclusion in the watershed plans. This protocol may be replicated to other regions and states of Brazil where agriculture is the main economic activity.

**Acknowledgements** The authors are grateful to PROFÁGUA (Professional Master Course on Water Resources Management and Regulation) — UNESP unit, supported by CAPES (Coordination for the Improvement of Higher Education Personnel — financial code 001) and ANA (Brazilian National Water Agency) — CAPES/ANA AUXPE n° 2717/2015.

## References

CEPAGRI – Centro de Pesquisas Meteorológicas e Climáticas Aplicadas a Agricultura (2019) Clima dos Municípios Paulistas. https://www.cpa.unicamp.br/outras-informacoes/clima-dos-municipios-paulistas.html. Accessed on 16 Nov 2019

CPRM – Serviço Geológico do Brasil (2006) Domínio geológico litoestratográfico do Estado de São Paulo. 1:750,000. CPRM, São Paul, Brasil

Di Baldassarre G, Viglione A, Carr G, Kuil L, Salinas JL, Blöschl G (2013) Socio-hydrology: conceptualising human-flood interactions. Hydrol Earth Syst Sci 10(4):4515–4536

Griebler C, Avramov M (2015) Groundwater ecosystem services: a review. Freshwater Sci 34:355–367

IBGE – Instituto Brasileiro de Geografia e Estatística (2011) Censo Demográfico. IBGE, Rio de Janeiro. https://www.ibge.gov.br/home/estatistica/populacao/censo2010/caracteristicas_da_populacao/resultados_universo.pdf. Accessed on 16 June 2019

Limaye S (2017) Socio-hydrogeology and low-income countries: taking science to-rural society. Hydrogeol J 25(7):1927–1930

Oliveira JB, Camargo MN, Rossi M, Calderano Filho B (1999) Mapa Pedológico do Estado de São Paulo.1:500,000. IAC/FAPESP, Campinas (In Portuguese)

Peppoloni S, Di Capua G (2017) Geoethics: ethical, social and cultural implications in geosciences. Ann Geophys 60: Fast Track 7. https://doi.org/10.4401/ag-7473

Re V (2015) Incorporating the social dimension into hydrogeochemical investigations for rural development: the Bir Al-Nas approach for socio-hydrogeology. Hydrogeol J 23(7):1293–1304

Ross JLS, Moroz IC (1996) Mapa geomorfológico do Estado de São Paulo. Revista do Departamento De Geografia 10:41–56 (In Portuguese)

Tuinstra J, van Wensem J (2014) Ecosystem services in sustainable groundwater management. Sci Total Environ 485–486:798–803

# A Comprehensive, Up-To-Date Evidence Base to Inform Public, Planning and Policy for Australia's Great Artesian Basin

Carlos Miraldo Ordens, Neil McIntyre, Jim Undershultz, and Phil Hayes

**Abstract**

The Great Artesian Basin (GAB) (Australia) is one of the world's most iconic groundwater basins and the lifeblood of much of Australia's interior with high eco-hydro-socio-economic importance. Nonetheless, the GAB is currently mostly perceived and managed based on outdated hydrogeological conceptual models, which give rise to over-simplistic and incorrect public understanding of the basin. Capitalizing on 10 years of research, mainly in areas targeted by gas and mining industries (e.g. Surat Basin), this project aims to contribute to updated conceptual understandings of the GAB that can lead to more informed discussions and management of the basin. It also explores how the knowledge gained from intensive work conducted in the Surat Basin can inform research and management in other parts of the GAB. This is firstly achieved through (i) the 2020 interdisciplinary Hydrogeology Journal Special Issue "Advances in hydrogeologic understanding of Australia's Great Artesian Basin". Secondly through a science-communication and public engagement program, targeting a range of publics, aiming at societal education on GAB-related groundwater science and management, which includes a dedicated Website (temporarily https://natural-gas.centre.uq.edu.au/gab) with all produced scientific and communication material (informational videos, other educational material and links to the Special Issue articles).

C. M. Ordens (✉) · N. McIntyre
Centre for Water in the Minerals Industry, Sustainable Minerals Institute, The University of Queensland, Brisbane, QLD, Australia
e-mail: c.ordens@uq.edu.au

J. Undershultz · P. Hayes
Centre for Natural Gas, The University of Queensland, Brisbane, QLD, Australia

**Keywords**

Special issue • Communication • Public engagement • Great Artesian Basin • Australia

## 1 Introduction

Australia's Great Artesian Basin (GAB) is one of the world's most iconic groundwater basins and is the lifeblood of much of Australia's interior for Aboriginal Peoples, rural towns, agriculture and industry. Its annual economical value is estimated at AU$12.8 billion (Frontier Economics 2016). The GAB aquifers support a vast number of springs and associated groundwater dependent ecosystems (GDEs), which are closely related to Aboriginal subsistence, cultural values, traditions and spirituality (e.g. Ah Chee 2002; White and Lewis 2011; Moggridge et al. 2019). The Australian Aboriginal culture is the world's oldest living culture and has been thriving for over 60,000 years on the driest inhabited continent, partly because of accessing water and food from artesian springs and associated GDEs. Aboriginal peoples have a deep spiritual and cultural connection to water, which is part of a single connected system that includes people, land and all living beings. This system is protected by lore, the customs and stories learnt from the Dreamtime, present in songs, dance, dreaming stories and art.

Despite its eco-hydro-socio-economic importance, as documented in Ordens et al. (2020), the GAB is still mostly perceived, and to some degree managed, based on outdated hydrogeological conceptual models, and not according to recent scientific findings. These concepts have been given rise to over-simplistic and incorrect public understanding of the basin and therefore led to ill-informed perceived risks to groundwater (e.g. interconnectivity across aquifers and sub-basins, role of faults on local and regional hydrogeology, recharge processes, impacts of climate change, springs



M. Abrunhosa et al. (eds.), *Advances in Geoethics and Groundwater Management: Theory and Practice for a Sustainable Development*, Advances in Science, Technology & Innovation,
https://doi.org/10.1007/978-3-030-59320-9_85

hydrology and human-activity impacts). The last 10 years have seen a large increase in available data and a corresponding large investment in improving scientific understanding of the GAB's hydrogeology and socio-hydrogeological connections, including in areas that have undergone intensive monitoring and assessment prior to and during coal seam gas (CSG) extraction or mining particularly in the Surat Basin (Towler et al. 2016). Documenting the acceleration in understanding forms the basis of this project, which aims to collect and disseminate the most recent science on the GAB.

The overarching objectives of the project are to (i) develop on-line, freely available publications that update the evidence base on the hydrological functioning of the GAB aquifers and associated aquifer management; and (ii) disseminate the content of the publications to the public through a program of engagement activities and production of easily accessible educational material. More specifically, this project aims to contribute to updated conceptual understandings of the GAB that can lead to more informed discussions and, eventually, to improved management of the basin. It also explores how the knowledge gained from intensive work conducted in the Surat Cumulative Management Area (CMA) can inform research and management in other parts of the GAB.

## 2 Settings and Methods

The GAB (Fig. 1) covers over 1.7 million km$^2$ ($\sim$22% of Australia) and is a sedimentary complex groundwater basin composed of four main basins: Eromanga, Carpentaria, Surat and Clarence-Moreton (Smerdon et al. 2012). It has been deposited 200–65 million years ago (Jurassic and Cretaceous periods). It is largely formed by alternating sandstone, siltstone and mudstone beds that constitute aquifer and aquitard units, which reaches a thickness of $\sim$3000 m in the central part of the Eromanga Basin (Ransley et al. 2015). The GAB overlays deeper, older geological basins (e.g. the Bowen and Galilee basins) and is overlaid by newer surface drainage divisions and associated alluvium (e.g. the Lake Eyre and Murray-Darling River basins) (Smerdon et al. 2012).

The project is composed of two components. The first includes organizing a Special Issue (SI) of Hydrogeology Journal (HJ) collating the latest science on the basin. The second includes a public engagement and communication program, which follows on the SI, aiming at communicating the science of the GAB to the general public, including a Website, roadshows and the delivery of a range of educational material.

## 3 Results

### 3.1 Hydrogeology Journal Special Issue on the GAB

The first project delivery is a HJ's SI "Advances in hydrogeologic understanding of Australia's Great Artesian Basin", which will the journal's first issue of 2020. It compiles the advances in knowledge of the GAB and overlapping groundwater basins, covering pure and applied research contributions from process to management scales. It delivers up-to-date, robust scientific knowledge to inform improved future management of the GAB, providing insights into novel methodologies for the study and management of geologically complex, continental-scale aquifer systems subject to intense human activities.

The SI has 26 articles, including a review paper on the history of the basin (Habermehl 2020), and an Eminent Researcher paper on the career of Justin Costelloe (Western et al. 2020). The remaining articles cover groundwater flow processes and groundwater governance and management tools. The SI includes a preface that provides a summary of each paper and a synthesis that offers the Guest Editors' collective view of the scientific and practical significance of the papers within and across topics (Ordens et al. 2020).

### 3.2 Public Engagement and Scientific Dissemination

The dissemination to the science community started with a special session dedicated to the SI content at the IAH/NCGRT Australasian Groundwater Conference in Brisbane in November 2019 and an official launch during the conference dinner. The project team is currently working on producing videos that will be used to reach a range of audiences on (i) introducing the GAB, (ii) GAB springs and GDEs, (iii) threats to and pressures on the GAB and (iv) water extraction from the GAB. The team will also produce informational material to be delivered during roadshows in the GAB region, which will be done in conjunction with existent government citizen-science engagement initiatives that are outlined by Jamieson et al. (2020). Additional dissemination activities include meetings with politicians through parliamentary briefings and the publication of two summary articles in The Conversation. All material, including links to the SI articles, will be shared through a Website, which is currently "https://natural-gas.centre.uq.edu.au/gab" but likely to change in the future.

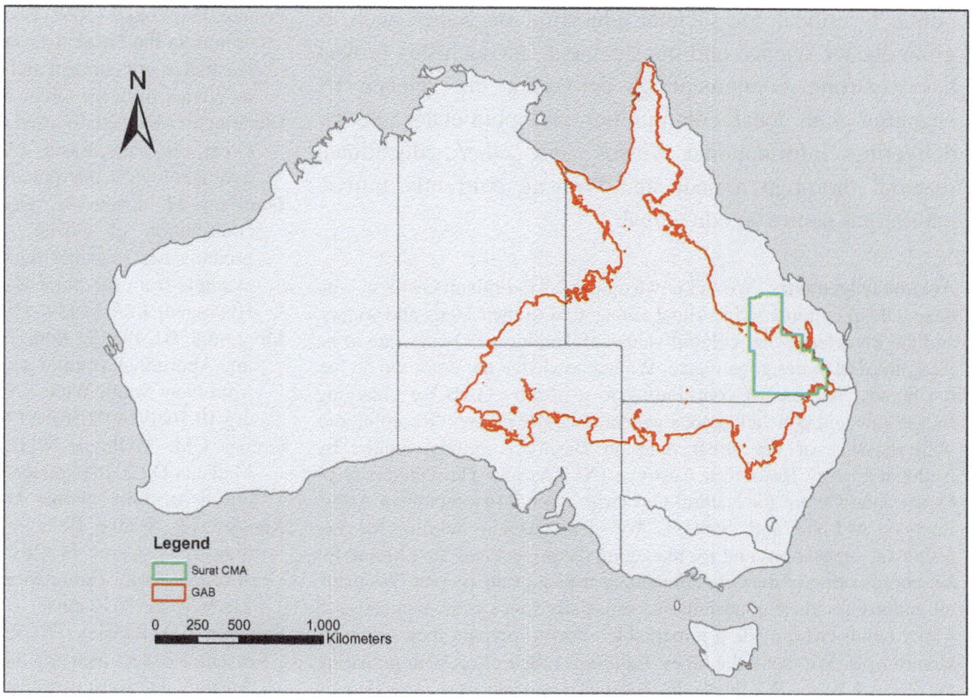


**Fig. 1** Location of the GAB and Surat CMA within Australia

## 4 Discussion

A key aspect of the SI is that it is a highly collaborative initiative and combines research done across institutions, such as Universities, research agencies, government departments and CSG and consultant companies, often with manuscripts being co-authored across institutions. As well as encouraging the publication of peer-reviewed science from organizations that are not accustomed to doing so, it forges multi-institutional collaborations to continuing study and better manage the GAB into the future. This inclusivity has its challenges but is important for the project to ensure that valuable data sets and insights are made public and are peer-reviewed. More than half of the articles (14) focus on the Surat CMA, which saw a large expansion of the CSG industry over the past decade that led to the need for research related to the impacts of this industry on groundwater. This spatially uneven distribution of contributions reflects the priority interest for research funders. There are many important messages coming out of the SI. Brief examples are as follows, and the reader is invited to access the SI preface for a complete discussion of lessons learnt. The heterogeneity and complexity of the basin are exposed at different scales, for example, new understanding of sub-unit heterogeneity and inter-unit connectivity, its importance for local to regional assessments and new ways of characterizing it. The articles collectively force us to recognize the exponentially increasing volumes of data available, including satellite and other geophysical data, and also on the increasing role of big data management and visualization, and initiatives to turn big data into better conceptual and numerical models. There are papers addressing climate change, land and water use dynamics, such as pumping and re-injection. These papers address the interactions between stressors and begin addressing the question of "how can we better manage the basin in a changing world?". Three social research papers discuss effective participatory models for better management, including water-governance analyses, perception and acceptance of new technologies and citizen science.

The public engagement and scientific dissemination component of the project is just starting and as such there is not sufficient material for a discussion yet. Nevertheless, there are currently many unfortunate examples around the world reminding us that simplistic attitudes to sustainability, which are not well-informed by science, can have a strong voice and lead to socio-environmental disasters. As such, we believe it is an important part of our work as hydrogeologists to contribute to a better societal understanding of groundwater.

## 5 Concluding Remarks

The project "Resetting our understanding of the Great Artesian Basin" delivers the most updated science on Australia's GAB, one of the world's biggest groundwater basins, through an interdisciplinary HJ SI titled "Advances in hydrogeologic understanding of Australia's Great Artesian Basin". To communicate science and to engage with the public is critical for societal education on issues such as groundwater science and management. As such this project has a strong component of delivering the science by engaging with local communities and politicians and by delivering informational videos and other educational material thorough a dedicated Website (currently https://natural-gas.centre.uq.edu.au/gab).

**Acknowledgements** We acknowledge the Traditional Owners, who have a deep cultural and spiritual connection to their lands and waters, and the knowledge and cultural values of Traditional Owners are to be recognized in water governance. We acknowledge the work we do has the highest purpose of contributing to a healthy GAB for emerging Indigenous and non-Indigenous generations.The project "Resetting our understanding of the Great Artesian Basin" is jointly funded by National Energy Resources Australia (NERA) and The University of Queensland Centre for Natural Gas (with its industry members Arrow Energy, APLNG and Santos). We thank Helen Schultz for her insightful suggestions and for managing the project and Suzi Moore for her coordination of the communications aspect of the project. We thank all authors for their contributions, especially those from non-research backgrounds, who made a remarkable effort to prepare their scientific manuscripts. We thank the Guest Editors for their effort with preparing the Special Issue. We thank the assistance and guidance provided by Hydrogeology Journal's executive editor Clifford Voss, editorial office manager Susanne Schemann and technical editorial advisor Sue Duncan. We thank all agencies and companies assisting us with the public engagement and communication aspect of the project.

## References

Ah Chee D (2002) Kwatye, Indigenous peoples connection with kwatye (water) in the Great Artesian Basin. Environ SA 9:20–21

Frontier Economics (2016) Economic output of groundwater dependent sectors in the Great Artesian Basin. A report commissioned by the Australian government and Great Artesian Basin jurisdictions based on advice from the Great Artesian Basin Coordinating Committee

Habermehl MA (2020) Review: the evolving understanding of the Great Artesian Basin (Australia), from discovery to current hydrogeological interpretations. Hydrogeol J 28(1):13–36

Jamieson M, Elson M, Carruthers R, Ordens CM (2020) The contribution of citizen science in managing and monitoring groundwater systems impacted by coal seam gas production: an example from the Surat Basin in Australia's Great Artesian Basin. Hydrogeol J 28(1):439–459

Moggridge BJ, Lyndal Betterridge L, Thompson RM (2019) Integrating Aboriginal cultural values into water planning: a case study from New South Wales, Australia. Aust J Env Manag. https://doi.org/10.1080/14486563.2019.1650837

Ordens CM, McIntyre N, Underschultz JR, Ransley T, Moore C, Mallants D (2020) Preface: advances in hydrogeologic understanding of Australia's Great Artesian Basin. Hydrogeol J 28(1):1–11

Ransley TR, Radke BM, Feitz AJ, Kellett JR, Owens R, Bell J, Stewart G, Carey H (2015) Hydrogeological Atlas of the Great Artesian Basin. Geoscience Australia, Canberra. https://doi.org/10.11636/9781925124668

Smerdon BD, Ransley TR, Radke BM and Kellett JR (2012) Water resource assessment for the Great Artesian Basin. A report to the Australian Government from the CSIRO Great Artesian Basin water resource assessment. CSIRO Water for a Healthy Country Flagship, Australia

Towler B, Firouzi M, Underschultz J, Rifkin W, Garnett A, Schultz H, Witt K (2016) An overview of the coal seam gas developments in Queensland. J Nat Gas Sci Eng 31:249–271

Western A, Matic V, Peel M (2020) Justin Costelloe: a champion of arid zone water research. Hydrogeol J 28(1):37–41

White DC, Lewis MM (2011) A new approach to monitoring spatial distribution and dynamics of wetlands and associated flows of Australian Great Artesian Basin springs using QuickBird satellite imagery. J Hydrol 408:140–152

# Private Groundwater Supply Management as a Response to Flooding Events: Perceptions of Irish Well Owners

Luisa A. de Andrade, Cillian P. McDowell, Jean O'Dwyer, Eoin O'Neill, Simon Mooney, and Paul D. Hynds

**Abstract**

Over 720,000 people in the Republic of Ireland rely on private groundwater resources (i.e. private wells) for daily consumption, and as these extractions are unregulated, users are solely responsible for managing/mitigating contamination risks to their supplies. However, low levels of exposure to appropriate guidance on well water protection and ongoing maintenance are not uncommon, particularly regarding responses to sporadic environmental threats, such as significant flooding. Despite this, very little is known regarding the factors leading to (or inhibiting) preparedness among groundwater-reliant individuals in the context of health threats triggered by flooding events. Accordingly, the purpose of this study is to bridge this knowledge gap and explore current behaviours, knowledge, risk perception, and experience relating to this issue in the Irish context. This was attempted via a combination of quantitative and qualitative methodologies, including a nation-wide online survey with 405 Irish well owners and six localized focus group meetings. Results show the need to go beyond knowledge-based interventions, and use socio-hydrogeological and/or socio-epidemiological approaches to target risk perception and potential structural constraints as a mean to turn protective intentions into protective actions when dealing with adverse effects of sporadic natural events, particularly in a changing climate.

**Keywords**

Private wells • Flooding • Survey • Focus group • Preparedness • Health behaviour

## 1 Introduction

Flooding events have been shown to increase the risk of human exposure and subsequent outbreaks of waterborne infection, particularly in communities that rely on groundwater supplies for daily consumption (Andrade et al. 2018). Compounding this, both the frequency and intensity of flooding events are predicted to increase over the coming years as a result of anthropogenic climate change (Arnell and Gosling 2016). As such, incentives to protect groundwaters against flood contamination are paramount in communities that rely on groundwater supplies for daily consumption. However, a barrier often faced by intervention strategies is motivating people to undertake desired behaviours.

Individual responses (protective or otherwise) to potentially hazardous circumstances are complex and can be influenced by myriad factors, such as risk perception (Bradford et al. 2012) and previous experiences. With regard to well stewardship, structural constraints and personal motivations, or lack thereof, also represent significant barriers to protective action, at times even more so than knowledge gaps (Kreutzwiser et al. 2011; Chappells et al. 2015). Additionally, low levels of exposure to appropriate guidance on well water protection and ongoing maintenance are not uncommon. This, coupled with widespread lack of

L. A. de Andrade · J. O'Dwyer
School of Biological, Earth and Environmental Sciences, University College Cork, Cork, Ireland

L. A. de Andrade · J. O'Dwyer · P. D. Hynds (✉)
Irish Centre for Research in Applied Geosciences, University College Dublin, Dublin, Ireland
e-mail: hyndsp@tcd.ie

C. P. McDowell · E. O'Neill
Planning and Environmental Policy, School of Architecture Planning and Environmental Policy, University College Dublin, Dublin, Ireland

J. O'Dwyer
Environmental Research Institute, University College Cork, Cork, Ireland

S. Mooney · P. D. Hynds
Environmental Health and Sustainability Institute, Dublin Institute of Technology, Dublin, Ireland



M. Abrunhosa et al. (eds.), *Advances in Geoethics and Groundwater Management: Theory and Practice for a Sustainable Development*, Advances in Science, Technology & Innovation,
https://doi.org/10.1007/978-3-030-59320-9_86

well stewardship (i.e. water treatment, continuous maintenance, and well water testing) among private well owners even under normal circumstances (Kreutzwiser et al. 2011; Hynds et al. 2013), could make well users ill equipped to prevent and manage the consequences of flood-triggered contamination. However, to date, no studies have identified the factors leading to (or inhibiting) preparedness among groundwater-reliant individuals in the context of the health threats triggered by flooding events. Thus, the purpose of this study is to bridge this knowledge gap and explore current experience, risk perception, and behaviours relating to this issue among Irish groundwater supply users.

## 2 Methods

The current study was undertaken in the Republic of Ireland (RoI) where over 720,000 residents (>15% of national population) are supplied by unregulated groundwater sources, which are exempted from compliance with the European Commission (Drinking Water) Directive 98/83/EC (EU 2018). Both quantitative and qualitative approaches were employed in order to gage groundwater users' perspectives on the issue at hand.

Quantitative data were collected via an online survey conducted between November 2017 and February 2018 and disseminated among well users in the RoI through local organizations, local authorities, governmental bodies, and social media platforms. The implemented questionnaire comprised 39 questions, and included multiple-choice, checkbox, numerical, forced preference ranking, and Likert-scale style questions. Questions included in the survey belonged to four distinct sections, namely (i) respondent and household-related information, (ii) groundwater supply characteristics, (iii) previous experience with well water testing, and (iv) flood-related risk perception, experience, knowledge, and behaviours.

Additionally, focus group meetings were hosted in specific locations in RoI as a qualitative methodology for gathering data. These were conducted from April to July 2018, with five sessions meeting the minimum number ($\geq 4$) of participants typically required to be considered successful (Krueger 1997). One further focus group was successfully completed despite lower participant numbers, and no data were collected for the remaining three sessions due to insufficient participant numbers. Participants had to be 18 years of age and characterized as being a private well user/owner. Recruitment was undertaken via invitation letters to between sixty and one hundred households in chosen groundwater-reliant areas or via local institutions (e.g. local parish, school, or community centre). Focus groups were steered by a topic guide based upon the Health Belief Model (HBM) framework (Rosenstock 1974). Primary topics covered included: (i) general thoughts on water quality locally and nationally; (ii) general thoughts on their own water; (iii) previous flood experience, perceived vulnerability, and perceived severity; (iv) general thoughts and attitudes regarding protective behaviours; (v) thoughts on social acceptance and one's own ability to take protective action; and (iv) trusted/preferred information sources. Focus group participants also completed a short survey comprised of twelve multiple-choice questions. Questions examined ages of participants and all household members, gender, residential ownership, number of years living in current household, and household location. Participants were also asked to state whether they drank from their private household well and if they used any microbial treatment before consumption (e.g. chlorination, UV treatment, or reverse osmosis), and if they had (or new others who had) experienced flooding events near ($\leq 100$ m) their groundwater supplies.

In order to determine respondents' main characteristics, knowledge gaps, risk perception, awareness levels, and engagement or intended engagement with desired behaviours, descriptive statistics were used to analyse the quantitative data. With regards to the analyses of focus group data, all meeting recordings were transcribed, and manually analysed/coded in order to identify obvious patterns, which were interpreted in the context of the HBM.

## 3 Results

### 3.1 Online Survey

All 26 counties in RoI were represented by a total of 405 survey respondents. The majority were male (58.5%) and of all respondents, 40.5% were in the 35–49 age range (13.8% below and 42.7% above). Overall, 94.8% of respondents lived in a rural area, with 5.2% from villages, towns, or another (peri)urban settlement. Young children (<5 years) and persons over 65 years of age resided in 20 and 39.5% of the surveyed households, respectively. In all, 81.7% of study participants were supplied by a private groundwater source (i.e. household well), with the remainder self-reporting as members of private group water schemes. Of the latter participants, 59.5% reported use of a microbial water treatment system prior to consumption (i.e. chlorination, UV lights, etc.), compared to just 14.2% of respondents that owned a household well.

One in five survey respondents (19.8%) reported previous experience with flooding near their groundwater supplies (direct experience), 18.8% reported no direct flood experience but knew others that have experienced flooding in the past (indirect experience), 55.0% declared no experience with flooding, and 6.4% abstained from answering the question.

**Table 1** Focus groups description, including code, date, recruitment strategy, number of participants, and the level of flood experience in the area

| Focus group | Date (2018) | Recruitment strategy | Number of participants | Level of flood experience in the area (in the past 10 years) |
|---|---|---|---|---|
| FG1 | 28/04 | Letters | 9 | No history of floods in the area |
| FG2 | 28/04 | Letters | 2 | No history of floods in the area |
| FG3 | 05/05 | Letters | 6 | Large-scale nearby floods in the past 5 years |
| FG4 | 05/05 | Letters | 1 | Large-scale nearby floods in the past 5 years |
| FG5 | 07/06 | Local recruiter | 6 | Incidences of small-scale localized flooding |
| FG6 | 07/06 | Local recruiter | 0 | Incidences of small-scale localized flooding |
| FG7 | 23/06 | Local recruiter | 4 | No history of floods in the area |
| FG8 | 18/07 | Letters | 8 | Incidences of small-scale localized flooding |

With regards to risk perception, 69.1% of all participants believed it to be unlikely or very unlikely that future flooding could negatively impact their supplies (low objective risk perception) and 57.5% of participants reported not being at all worried that their wells could become contaminated by future floods (low affective risk perception). Even among respondents that have directly experienced floods in the past, low levels of objective and affective risk perception were exhibited, equating to 41.3% and 26.3% of them, respectively. Overall, just 0.7% of respondents suspected that an illness to a member of their household was caused by their well being contaminated. Moreover, Fig. 1 shows actual versus conjectural responses to flooding by respondents with and without flood experience.

### 3.2 Focus Groups

In total, 35 private household well owners participated across the six successfully conducted focus groups (FG1, FG2, FG3, FG5, FG7, FG8; Table 1). Participation was balanced between genders with a total male to female ratio of 1.19 (i.e. 19 male and 16 female participants). Most participants reported drinking the well water (91.43%), but few (14.28%) made use of a microbial water treatment (i.e. UV, reverse osmosis, or chlorination) before consumption. Of all participants, 22.71% have experienced flooding events near their own well in the past (direct experience) and a further 25.71% knew others that have experienced this issue (indirect experience).

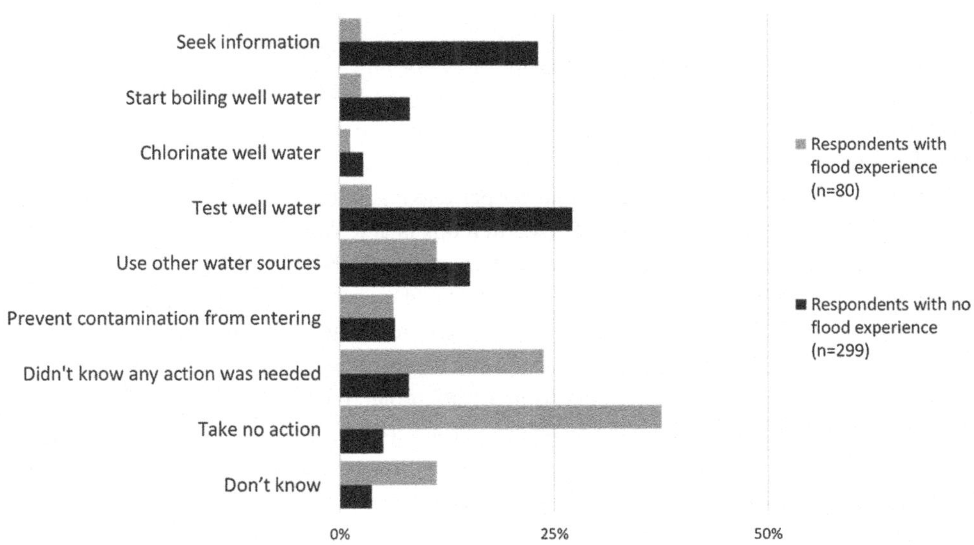


**Fig. 1** Composition of actual response to flooding by survey respondents that have directly experienced it ($n$ = 80) versus indented responses by respondents that have not experienced floods near their supplies in the past ($n$ = 299)

In general, participants voiced satisfaction with their own well water, mentioning good taste, lack of issues and praise from visitors, despite very infrequent well water testing habits. The vast majority of participants did not consider themselves at risk of flooding. The most commonly cited reason was that their well was elevated, while participants that experienced small-scale floods tended to associate flooding and flood-related contamination to large-scale events only. Nevertheless, most participants agreed that their wells would be at risk of contamination if flooding did occur, particularly if the well got inundated. In the case of floods nearby (<100 m), some participants reported that they would not be concerned about contamination. When asked about the reasons for concern, the vast majority raised the issue of nearby agriculture, including chemical and animal contaminations, only acknowledging human sources (e.g. septic tanks) when probed.

Regarding post-flood action, most participants stated that they would get the well water "checked"/"tested", with some also reporting that they would boil the water before drinking. However, among the few that have experienced flooding, only one participant pursued protective action, i.e. installing a treatment system, other participants attributed lack of action to "putting on the long finger". The following statement by one of the participants at the end of a meeting succinctly summarizes the primary findings of this study:

> FG5: "The highest risk to all wells is lack of information and lack of knowledge. People not even understanding how, where the water comes from, let alone what to do with it. If people understand the risk of flooding, 90% of them would get a remedy fairly quick, it's just that they're not aware of the risks".

## 4 Discussion

The research described in this paper addresses the knowledge gap concerning the experiences, risk perception, and behaviours of well owners in RoI with regard to the adverse health effects of flooding near their supplies. Results identified low levels of protective behaviour by participants with previous flood experience. It is considered that this is at least partially attributable to low levels of risk perception exhibited by these respondents regarding the effects of nearby flooding to their own well water quality. According to Bradford et al. (2012), risk perception is a primary driver of health behaviours (i.e. behaviours aimed at protecting health), along with well stewardship (Morris et al. 2016).

Findings also indicate a significant disparity between intended and actual engagement with protective actions, i.e. actions performed by respondents that have previously dealt with the issue versus conjectural actions intended by respondents with no flood experience. Previous work by Sandman and Weinstein (1993) explains that there are practical and psychological barriers when transforming intention into action in the context of well stewardship. However, in the Irish context, these barriers could be further exacerbated by the lack of publicly available evidence-based guidelines on what to do if flooding happens near groundwater supplies, as well as lack of socio-hydrogeological and/or socio-epidemiological methodologies to engage with these at-risk populations.

## 5 Concluding Remarks

Results from this study reinforce the need to go beyond knowledge-based interventions, and employ socio-hydrogeological and/or socio-epidemiological approaches to target risk perception and potential structural constraints as a mean to turn protective intentions into protective actions when dealing with adverse effects of sporadic natural events, particularly in a changing climate.

## References


Andrade L, O'Dwyer J, O'Neill E, Hynds P (2018) Surface water flooding, groundwater contamination, and enteric disease in developed countries: a scoping review of connections and consequences. Environ Pollut 236:540–549

Arnell NW, Gosling SN (2016) The impacts of climate change on river flood risk at the global scale. Clim Change 134(3):387–401

Bradford RA, O'Sullivan JJ, Van der Craats IM et al (2012) Risk perception: issues for flood management in Europe. Nat Hazards Earth Syst Sci 12(7):2299–2309

Chappells H, Campbell N, Drage J, Fernandez CV, Parker L, Dummer TJ (2015) Understanding the translation of scientific knowledge about arsenic risk exposure among private well water users in Nova Scotia. Sci Total Environ 505:1259–1273

European Commission (2018) Drinking water legislation. Available at: https://ec.europa.eu/environment/water/water-drink/legislation_en.html. Accessed 21 Sept 2019

Hynds PD, Misstear BD, Gill LW (2013) Unregulated private wells in the Republic of Ireland: consumer awareness, source susceptibility and protective actions. J Environ Manage 127:278–288

Krueger RA (1997) Analyzing and reporting focus group results. Sages Publications. https://us.sagepub.com/en-us/nam/analyzing-and-reporting-focus-group-results/book6630

Kreutzwiser R, de Loë R, Imgrund K, Conboy MJ, Simpson H, Plummer R (2011) Understanding stewardship behaviour: factors facilitating and constraining private water well stewardship. J Environ Manage 92(4):1104–1114

Morris L, Wilson S, Kelly W (2016) Methods of conducting effective outreach to private well owners—a literature review and model approach. J Water Health 14(2):167–182

Rosenstock IM (1974) Historical origins of the health belief model. Health Educ Monogr 2(4):328–335

Sandman PM, Weinstein ND (1993) Predictors of home radon testing and implications for testing promotion programs. Health Educ Q 20 (4):471–487

# Geoethics of Decision Making Under Uncertainty and Ethical Issues in Neglecting Groundwater Functioning

**Introductory Note**

Rui Hugman (Umvoto Africa, Cape Town, South Africa), José Joel Carrillo-Rivera (UNAM, Ciudad de México, Mexico)

Uncertainty is inherent in groundwater management. Yet decisions must be made which may have long-lasting impacts. As practitioners, we strive to reduce or define that uncertainty when informing about a decision so that decision-makers are informed of the risks involved.

With growing pressure on water sources and the expected increase in climate volatility, what are the ethics of our management decisions? What are the ethical implications of implementing management decisions when uncertainty is high? And on the other hand, what are the ethical implications of inaction?

In this part, it is aimed to explore the ethics of taking (or avoiding) decisions affecting groundwater, society, and other linked systems, as well as how tools at our disposal may be geoethically used to guide our decisions. This aim searches to develop insights into how groundwater professionals, regulators, and society in general address (or fail to) uncertainty in their decision-making process, and what implications this has on society's resilience to improve their water sources in a rapidly changing world.

The topic collates contributions that outline an overview of experience on both ethical and unethical responses by government institutions, private holdings, and practitioners towards groundwater management. These include reports on unethical (and ethical) decision-making and their impact on society and related groundwater sources, describing actions (or lack thereof) such as:

- The development of country or region in terms of industry, urban, and agriculture where actions are directed to the dispossession of groundwater rights from socially vulnerable groups of the population;
- The creation of public opinion by neglecting scientific facts and results based on well-established myths to develop strategies to bring society away from proper land planning and social justice;
- The needed education, teaching, and application of knowledge on the actual functioning of groundwater seeing its integrative actual conditions from a local to a regional scale rather than biasing results meeting a predetermined improper model away from field conditions;
- Government activities related to permits and groundwater concessions made against local or international groundwater pacts where its managing responds to political corruption criteria, where laws and regulations are hardly applied to water and environmental protection;
- Proper groundwater functioning definition requires a correct review from basic sampling methods to data analyses avoiding estimates erasing "interference" of selected data by standardizing samples presenting "anomalies", which could mean valuable end members;
- Report water policy showing corruption via flaws on international obligations or authoritarian use of water policymaking reinforced by unethical use of words and concepts that need consistency and transfer to a proper scientific application of related groundwater jargon.

**Highlights**

- An exploration of how ethics could be brought into the decision-making process to assist practitioners to better understand and resolve water governance issues in groundwater conflicts.

- Several case studies of responsible decision taking under uncertainty during extreme events.
- Demonstrated implications of inaction and an exploration of the ethical responsibility of geoscientists to contest the lack of action.
- The rise in computational tools and capacity has opened new ways to assimilate information and inform decisions. But what new ethical dilemmas accompany these new technologies?
- The reality of unethical decision-making and potential long-lasting impact on the environment and society.
- The need for correct and clear communication between stakeholders, decision-makers, practitioners, and society grounded in scientific facts and data.
- The key importance of education, training, and application related to the functioning of the groundwater flow systems to facilitate communication and decision taking.
- Groundwater functioning definition requires a correct review and application from basic sampling methods to data analysis.
- The need to acknowledge the spirit of local and international treaties and related knowledge to be acquired and implemented in decision-making.

The unethical use of words and/or concepts from some players needs consistency and transfers to a proper scientific application of related legal groundwater jargon.

# Geoethical Issues Around Water-Security for Cape Town and Groundwater Resilience in Uncertain Circumstances

Christopher J. H. Hartnady and Rowena Hay

**Abstract**

In early 2018, Cape Town faced "Day Zero," the date when it was expected to run out of water, and when all municipal supply would be rerouted to emergency collection points. A three-year drought, considered ~1-in-400-year hydrological event, had brought the level of its largest storage reservoir to around 11% of full capacity. Day Zero was averted by relatively good rainfall early in 2018. Two years on, water-security remains precarious and uncertain in the face of rapid urban expansion, slow environmental degradation, and long-term climate change. The wider catchment region contains two important subsurface resources: the Palaeozoic Table Mountain Group (TMG) Aquifer System, and the Cenozoic Cape Flats Aquifer (CFA). The development of these groundwater options is confronted by challenges related to environmental and societal impacts. In the case of the TMG resource, which underlies mountain biosphere reserves of the extraordinary Cape Floral Kingdom, geoethical concerns arise in the context of uncertainties related to anticipated impacts on stream flow and floral biodiversity. In the CFA case, geoethical issues revolve around uncertainties related to impacts on the coastal environment, the groundwater–seawater interface, and the power-generation costs of groundwater management and treatment under constraints imposed by South Africa's fossil-fuel dependence.

**Keywords**

Cape Town • Water-security • Groundwater resilience • Environmental impact • Water-energy nexus

## 1 Introduction

At the south-western tip of Africa, the City of Cape Town (4.004 million inhabitants in 2016 census) depends on the Western Cape Water Supply System (WCWSS) for domestic and industrial usages within the metropolitan municipality and surrounding towns in the water-management area of the Berg River. A substantial proportion of the WCWSS also supports an economically important agricultural demand. This supply comes from a surface-water resource impounded by large dams in a mountainous hinterland, connected to the city by a system of tunnels and pipelines (Fig. 1). After good winter rainfall in 2013 and 2014, the WCWSS entered a drought crisis with declining rainfall and dam levels from 2015 through 2017. The crisis culminated in early 2018 with the declaration of a disaster emergency and the anticipation of 'Day Zero'—the day when all the taps run dry. The latter was averted by a combination of municipal water supply restriction with citizen co-operation (Wolski 2018) and the relative improvement of winter rainfall in 2018 (Fig. 2). Although groundwater augmentation options had been introduced into water-resource planning in the 2001–2009 period, their implementation had been postponed after 2012 in an atmosphere of complacency. They were, however, re-introduced in an emergency context in early 2018, as a 'crash programme' under severe pressures of time and budget.

The planning and rapid implementation of groundwater schemes under such extreme circumstances, compounded by the hydrological uncertainties associated with contemporary climate change in a region considered highly susceptible to future recurrence of extreme drought conditions, posed several geoethical issues. Because of the great local inequalities of Cape Town society, in which the poorest communities generally suffered most under the drought crisis, a 'no-regrets approach' (Siegel and Jorgensen 2011), drawn from existing instruments and innovative ideas from

C. J. H. Hartnady (✉) · R. Hay
Umvoto Africa (Pty) Ltd., Muizenberg, Cape Town, South Africa
e-mail: chris@umvoto.com



M. Abrunhosa et al. (eds.), *Advances in Geoethics and Groundwater Management: Theory and Practice for a Sustainable Development*, Advances in Science, Technology & Innovation,
https://doi.org/10.1007/978-3-030-59320-9_87

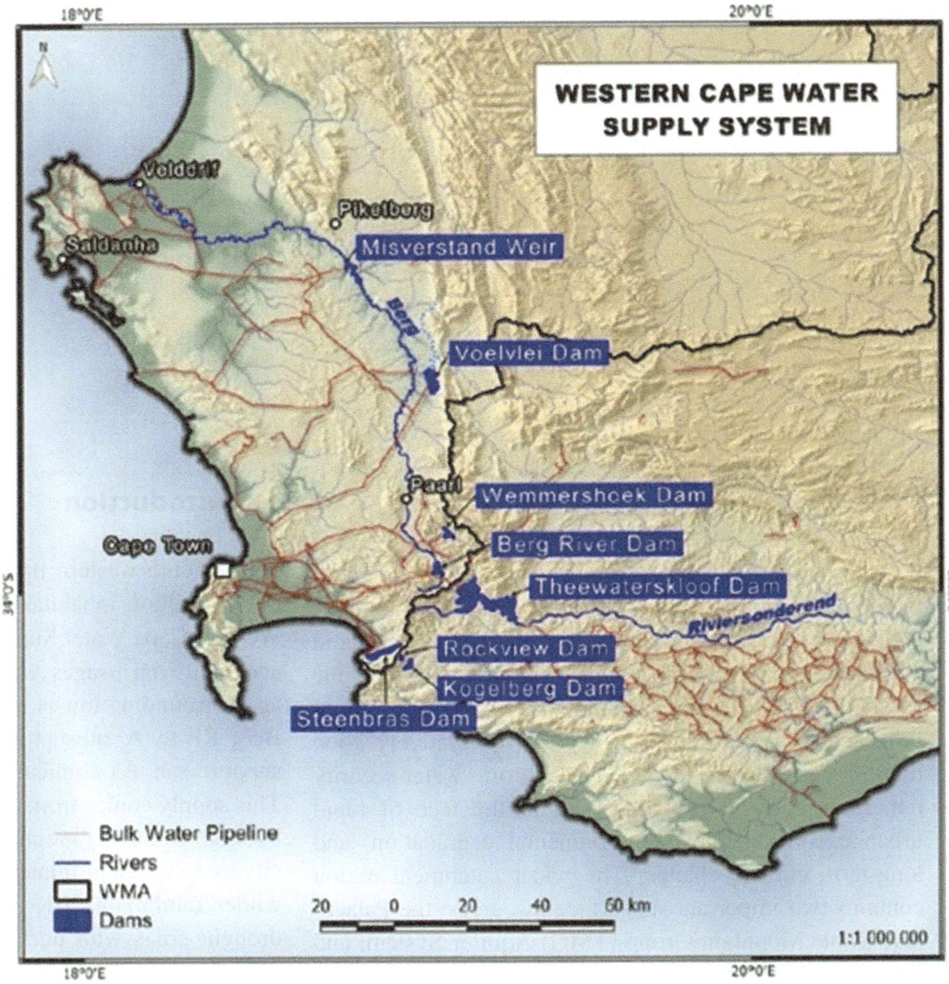


**Fig. 1** Main rivers (blue lines), large dams (blue areas), and bulk-water pipelines (red lines) of the WCWSS on an earth-toned, artificially sun-shaded, elevation basemap. The 'big six' dams are Theewaterskloof, Berg River, Voelvlei, Wemmershoek and Steenbras (Lower and Upper) (from DWS 2014)

the disciplines of disaster risk management (DRM), climate change adaptation (CCA) and social protection (SP), was adopted. Considering the 2015–2017 drought as possible evidence of climate change and the substantial uncertainties associated therewith, a *no-regrets approach* involves the taking of "actions that are justifiable from economic, social, and environmental perspectives whether climate change takes place or not—that can help increase human resilience to multiple hazards whether caused by weather, economic imbalances, food shortages etc." (Siegel and Jorgensen 2011, p. 2).

## 2 Hydrogeological Setting

The vulnerability of the surface-water resource that underpins the WCWSS had been long recognized, and the option of large-scale augmentation from the Table Mountain Group (TMG) Aquifer System (Fig. 2) had been explored long before onset of the severe drought (Hartnady et al. 2004; Blake et al. 2010). The development of the intra-urban Cape Flats Aquifer (CFA) had also come under serious consideration (DWS 2014) prior to the water supply emergency.

The existing surface-water, treatment-plant, and pipeline infrastructure of the WCWSS is near an undeveloped groundwater treasure in the TMG Aquifer System, an exceptional resource of giant storage volume, characterized by a unique, non-volcanic, geothermal association with fault- and fracture-controlled hot springs. To counter present and future urban water scarcity—with its negative impacts on human health, social well-being and economic growth in the face of high population growth—and build resilience against future transient drought crises and long-term aridification linked to climate-change, it is essential to develop access to the TMG groundwater storage.

The shallow CFA provides an opportunity to better manage the treatment and recycling of urban wastewater, generated from freshwater supply sources in the mountainous hinterland. Abstraction of CFA groundwater, initially for non-potable, recreational and agricultural usage contributing to urban food security, can be combined with Managed Aquifer Recharge (MAR) with treated and/or desalinated

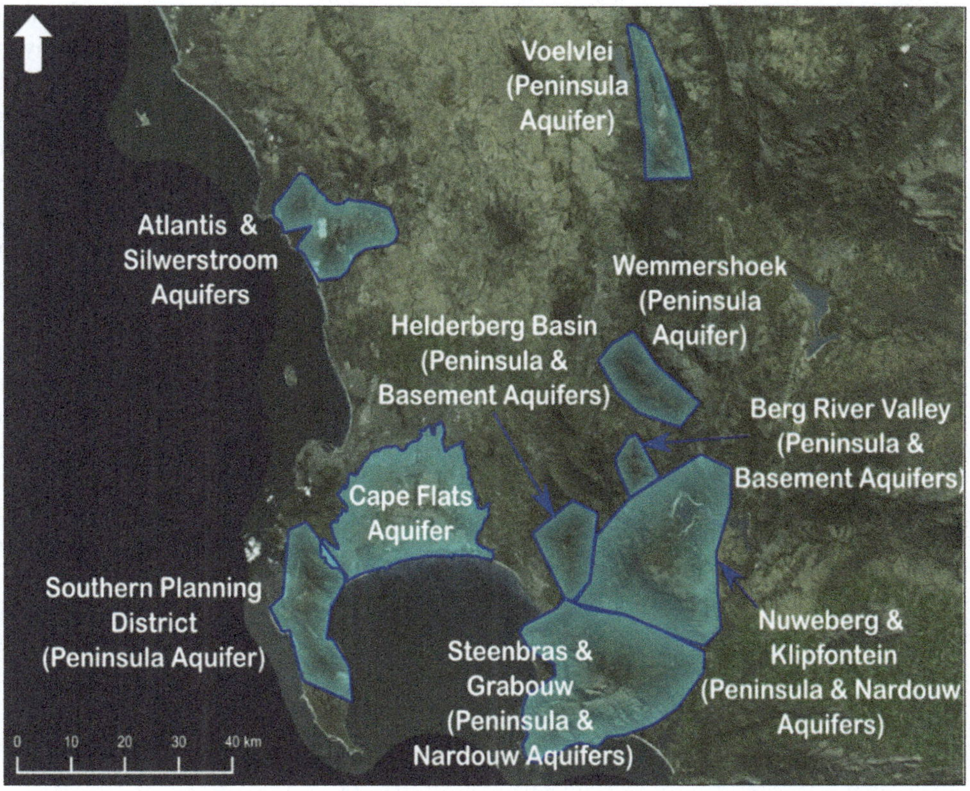


**Fig. 2** Major aquifer systems within and in the vicinity of the City of Cape Town. The Peninsula and Nardouw Aquifers are hydrostratigraphic units with the Paleozoic Table Mountain Group (TMG) Aquifer System (or Super-Aquifer). The Cape Flats Aquifer (CFA) and Atlantis and Silwerstroom Aquifer are shallow coastal Cenozoic systems

wastewater, to conserve TMG and other resources for potable use.

The TMG (Steenbras) development confronts challenges posed by its mountain biosphere reserve setting (Blake et al. 2021), whereas the CFA development (McGibbon et al. 2021) is challenged by its coastal setting on the northern shoreline of False Bay (cf. Fig. 2). In each case, there has been a vigorous, activist resistance to the development of the groundwater options, and the advocacy (generally in the popular media) of alternative supply schemes. The TMG development has been opposed by the idea of streamflow increase via clearing of invasive alien plant (IAP) infestations in the mountain catchments. Desalination of seawater at coastal sites along the CFA extent is proposed as a potential solution to the drought crisis and the longer-term challenge of climate change.

The realism of these alternative options is questionable. From a quantitative perspective on "green-water" stocks, flows and residence time, the restoration of a pre-IAP environmental state in the mountain catchments of the Western Cape cannot alone resolve the current urban water supply crisis and build long-term resilience. Given high power costs and the current vulnerability of South African electricity generation to disruption, the desalination of seawater remains an expensive last resort.

## 3 Geoethical Considerations

### 3.1 Professional Obligations

In South Africa, the profession of Earth Science is governed by the South African Council for Natural Science Professionals (SACNASP) Code of Conduct, which at its outset states that the practitioner must:

1. Have due regard to public safety, public health, and public interest generally
2. Have due regard to harmful practices against the environment.

During the 35th International Geological Congress (27 August–4 September 2016), the International Association for Promoting Geoethics (IAPG) prepared the "Cape Town Statement on Geoethics" (Di Capua et al. 2017), which defines the discipline of Geoethics and lists as its first Fundamental Value the "Honesty, integrity, transparency and reliability of the geoscientist, including strict adherence to scientific methods." It advocates a "... Hippocratic-like oath (the 'Geoethical Promise')," in which the geoscientist promises that:

> … I will practice geosciences being fully aware of the societal implications, and I will do my best for the protection of the Earth system for the benefit of humankind.
> … I understand my responsibilities towards society, future generations, and the Earth for sustainable development.
> … I will put the interest of society foremost in my work.

Geoscientists always have an ethical duty to be mindful of potentially harmful impacts on the local or global environment and, whenever these emerge, to promote the interests of society, public safety, public health, and the protection of the environment above any narrower interests. In pursuit of geoethical integrity and transparency, they must also inform and educate the public, including their employers and/or clients, about wider societal implications concerning protection of the Earth system.

### 3.2 Specific Groundwater-Resource Issues with Geoethical Implications

Among the perceived negative impacts of *TMG groundwater* schemes are the following:

- Land-use change and physical encroachment/deleterious visual impact of extraction wellfields on mountain biosphere reserves;
- Reduction of streamflow in mountain catchments affecting quantity of the surface-water resource;
- Water-table declines threatening floral biodiversity;
- Potential for earthquakes triggered by large-scale groundwater abstraction.

Among the perceived negative impacts of *CFA groundwater* development are:

- Seawater intrusion due to groundwater over-abstraction;
- Degradation/loss of aeolian and coastal-zone geodiversity and associated ecosystem services;
- Competition with local agriculture for access to limited resource around proposed extraction wellfields;
- Collateral flooding in areas around MAR injection wells;
- Deficient aquifer monitoring and protection;
- Deleterious changes in quantity and quality of submarine groundwater discharge to False Bay.

## 4 Concluding Remarks

These environmental challenges are exacerbated by a widespread public ignorance about groundwater systems within the WCWSS, including their diversity between shallow, porous sand types (CFA), and deep fractured-rock systems (TMG), and the inherent complexity of their flow regimes under natural hydraulic gradients. It necessitates the devotion of more resources to aquifer monitoring and modeling, and the integration of such technical guidance and information into the political processes in the City, preferably in a format and language that empowers community leaders and politicians to act appropriately and timeously.

The way forward presupposes collaboration "with experts from the social sciences and humanities, notably economics, policy and law, to develop water-management tools that decision-makers and the public can understand and use" (Muller 2018, p. 176). It also presupposes a radically new perception of the Earth, and of humankind's natural environment, reflected in the recent emergence of the "critical zone" concept (Lin et al. 2011) and an developing ambition to realize a local Critical Zone Observatory in this region of the Western Cape, South Africa.

**Acknowledgements** We thank our colleagues Dylan Blake, Tyrel Flugel, Rui Hugman, David McGibbon, Kornelius Riemann and Luke Towers for discussions and co-operation.

## References

Blake D, Mlisa A, Hartnady CJH (2010) Large scale quantification of aquifer storage and volumes from the Peninsula and Skurweberg Formations in the southwestern Cape. Water SA 36(2):177–184

Blake D, Hartnady CJH, Hay ER, Riemann K (2021) Geoethics of bulk groundwater abstraction in ecologically sensitive areas (Steenbras Wellfield, City of Cape Town). In: Abrunhosa A, Chambel A, Peppoloni S, Chaminé HI (eds) Advances in geoethics and groundwater management: theory and practice for a sustainable development, proceedings of the 1st congress on geoethics and groundwater management (GEOETH&GWM'20), Porto—Portugal 2020, Springer ASTI, Cham (this volume)

Department of Water and Sanitation (2014) Support to the continuation of the water reconciliation strategy for the Western Cape Water Supply System—Cape Flats Aquifer: situation assessment & potential for successful urban groundwater development and management. Prepared by Umvoto Africa (Pty) for the Department of Water Affairs, South Africa

Di Capua G, Peppoloni S, Bobrowsky PT (2017) The Cape Town statement on geoethics. Ann Geophys 60, Fast Track 7. https://doi.org/10.4401/ag-7553

Hartnady CJH, Hay ER, Riemann K (2004) Table mountain group aquifer feasibility study and pilot project: preliminary phase—hydrogeological report. TMG Aquifer Alliance/CMC Administration, Report no. 2, Oct 2004, 138pp

Lin H, Hopmans JW, Richter DdeB (2011) Interdisciplinary sciences in a global network of critical zone observatories. Vadose Zone J 10:781–785

McGibbon D, Hugman R, Towers L, Riemann K, Hay R, Hartnady C (2021) Long-term planning during emergency response—geo-ethics of a no-regrets emergency supply scheme with a long-term vision for the development of the Cape Flats Aquifer (Cape Town). In: Abrunhosa A, Chambel A, Peppoloni S, Chaminé HI (eds) Advances in geoethics and groundwater management: theory and practice for a sustainable development, proceedings of the 1st congress on geoethics and groundwater management

(GEOETH&GWM'20), Porto—Portugal 2020, Springer ASTI, Cham (this volume)

Muller M (2018) Lessons from Cape Town's drought. Nature 559: 174–176

Siegel PB, Jorgensen S (2011) No-regrets approach to increased resilience and climate change justice: toward a risk adjusted social protection floor. In: Conference on "social protection for social justice". Institute for Development Studies (IDS) University of Sussex, Brighton, UK, 13–15 Apr 2011, 70pp.

Wolski P (2018) What Cape Town learned from its drought. In: Bulletin of the atomic scientists, vol 74, 16 Apr 2018. https://thebulletin.org/2018/04/what-cape-town-learned-from-its-drought/. Last accessed 13 Nov 2019

# Cycles of Uncertainty: An Exploration of 40 Years of the Atlantis-Managed Aquifer Recharge Scheme Through a Geoethical Lens

Luke Towers and Rui Hugman

**Abstract**

The Atlantis Water Resource Management Scheme (AWRMS) in the Municipality of Cape Town (South Africa) is recognized as one of the first large-scale-managed aquifer recharge schemes in the world. It was implemented during 1970s and ran (apparently) successfully for almost 30 years. However, in the early 2000s, it was essentially abandoned and replaced with surface water supply. Most of the infrastructure was allowed to deteriorate and monitoring of the aquifer system was carried on without proper scientific purpose. Due to the recent drought crisis, in which Cape Town almost ran out of water to supply over 3 million people, the need for groundwater as an additional source of supply was once again recognized. As one of several projects to increase the City's resilience to drought, the AWRMS was targeted for refurbishment and re-design. Although there have been many technical challenges, perhaps the most complex to solve has been coordinating the various disjointed, and often opposed, institutions and entities that affect the functioning of the scheme. The fundamental geoethical principles of scientific practice, communication, and the education of all those who affect and are affected by the scheme are paramount to ensure the long-term sustainability of the system.

**Keywords**

Managed aquifer recharge • Decision-takers • Education • Communication • Risk

## 1 Introduction

The Atlantis Water Resource Management Scheme (AWRMS) has operated since the 1970s and is renowned within Southern Africa (see Fig. 1). The scheme's design demonstrates cost effective, responsible, and wise water use through visionary town planning and water recycling via managed aquifer recharge (MAR). AWRMS produces up to 20 ML/d and has the potential to offer water security to both the low-income residential and noxious trade industrial sectors of the town of Atlantis near Cape Town, South Africa.

## 2 In the Past

For AWRMS to be a success, it required the integration of its water supply, wastewater, and storm water systems, each of which require a multidisciplinary management approach. In previous years, this proved an easier task than it does today where the numerous components are distributed between several different departments of the larger City of Cape Town Municipality (see Fig. 2). Adding to challenges of inter-departmental co-operation and communication on an upper management level is the lack of investment into a skilled operation and maintenance work force resulting in a reactive rather than proactive response to risk identification and mitigation.

Further complications arise when considering the in-situ heterogeneity and vulnerability of the primary sand aquifer. Biofouling, vandalism, and readily available surplus surface water were negative drivers for the AWRMS which saw production decrease from ~15–5 ML/d. Atlantis was connected to a surface water supply system in the year 2000, after which the scheme fell into disrepair with elevated groundwater levels and contamination risk due to continued uncontrolled MAR of ~7500 $m^3$/annum (see Fig. 3).

L. Towers (✉) · R. Hugman
Umvoto Africa (Pty) Ltd., Cape Town, South Africa
e-mail: luke.t@umvoto.com





M. Abrunhosa et al. (eds.), *Advances in Geoethics and Groundwater Management: Theory and Practice for a Sustainable Development*, Advances in Science, Technology & Innovation,
https://doi.org/10.1007/978-3-030-59320-9_88

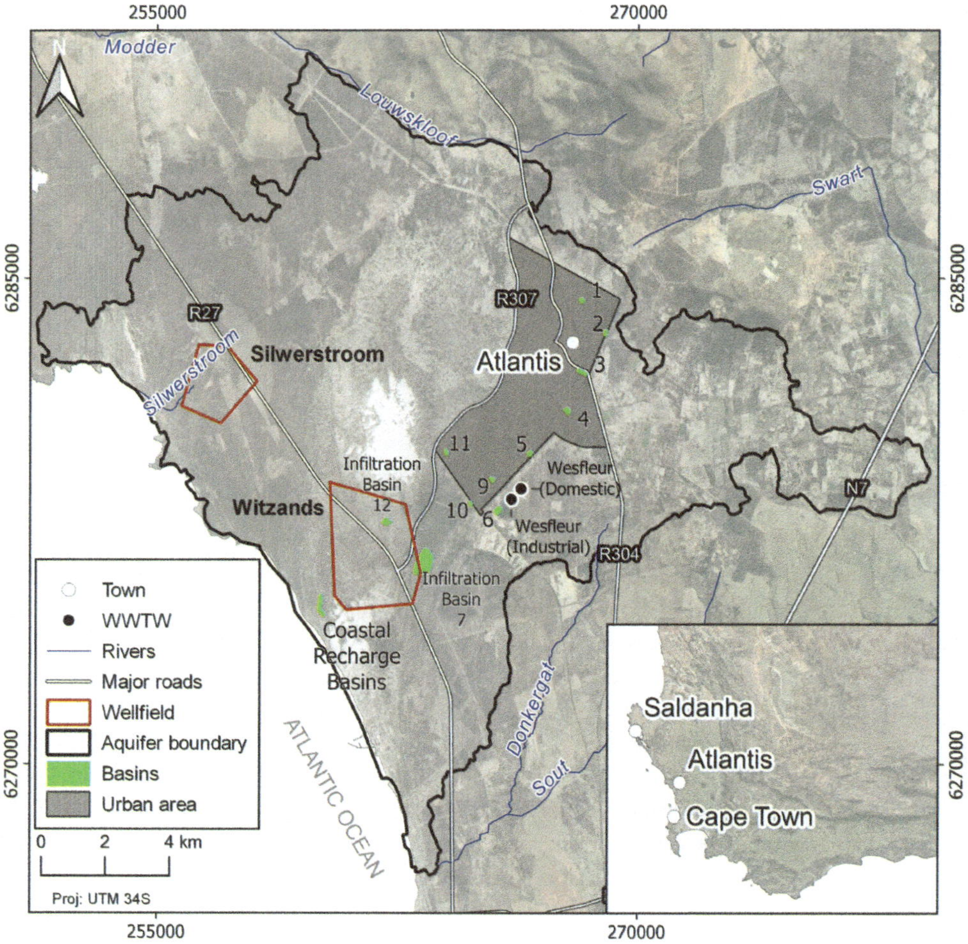


**Fig. 1** Locality map and general schematic layout of the AWRMS, highlighting the locations of stormwater and wastewater collection network, managed aquifer recharge basins, and abstraction well-fields

These factors pose significant geoethical questions not only of the governance authorities and professionals involved in the design and management of the scheme but also of the day-to-day operational workforce and importantly the water and land users around which the scheme was built. There lies an ethical obligation with geoscientists to ensure that governing and management entities, operational sectors, and end users are aware of one another and are encouraged to practice effective communication to identify and overcome the risks and sensitivities posed to the scheme. This becomes increasingly complex and uncomfortable for geoscientists when dealing with large politically driven municipalities and a predominantly unskilled work force. It is imperative that geoscientists bridge the gap between these role players and in so doing see the contributions of an unskilled workforce be validated through performance related incentives and recognized training. To create and sustain awareness and a culture of accountability within governing and managing role players as well as end users for whose benefit the scheme was designed is the responsibility of the professionals advising and overseeing the large-scale development of this natural resource.

## 3 Going Forward and Concluding Remarks

The emergency response to the drought experienced in the Western Cape of South Africa has driven the refurbishment of the system and since early 2018 groundwater has once again been positioned to replace surface water as the sole source of supply to the town of Atlantis. However, there remains the need to establish co-operative monitoring and user committees and educate all levels of people from the top down or from the bottom up; only then will assurance of supply from the scheme improve through enhanced understanding of groundwater flow paths, aquifer characteristics, response to long-term abstraction and MAR. Accountability and acknowledgement of the water suppliers geoethical responsibility are key to ensure the scheme is not abused and over exploited in the future.

Geophysical investigations, exploratory drilling, and wellfield optimization efforts are reducing uncertainties in stratigraphy, paleontography, and potential interflows between basement rocks and the overlying aquifer, but resilience is only assured by institutions, individuals and

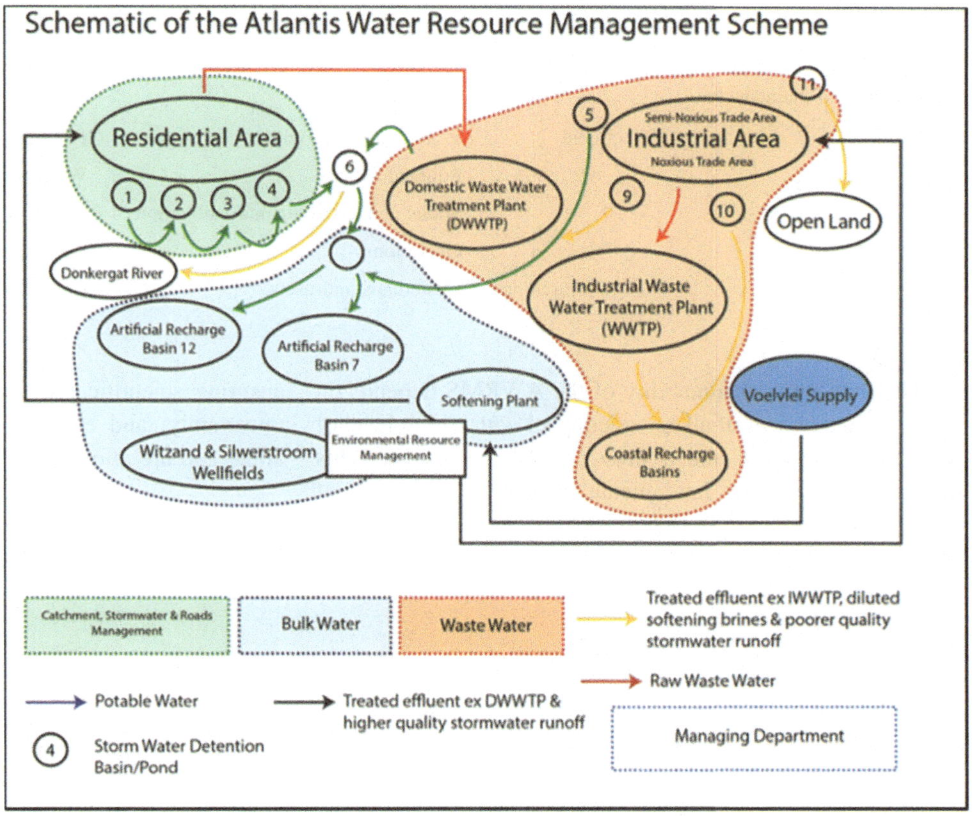


**Fig. 2** Disjointed and uncoordinated management of interconnected entities of the AWARMS scheme

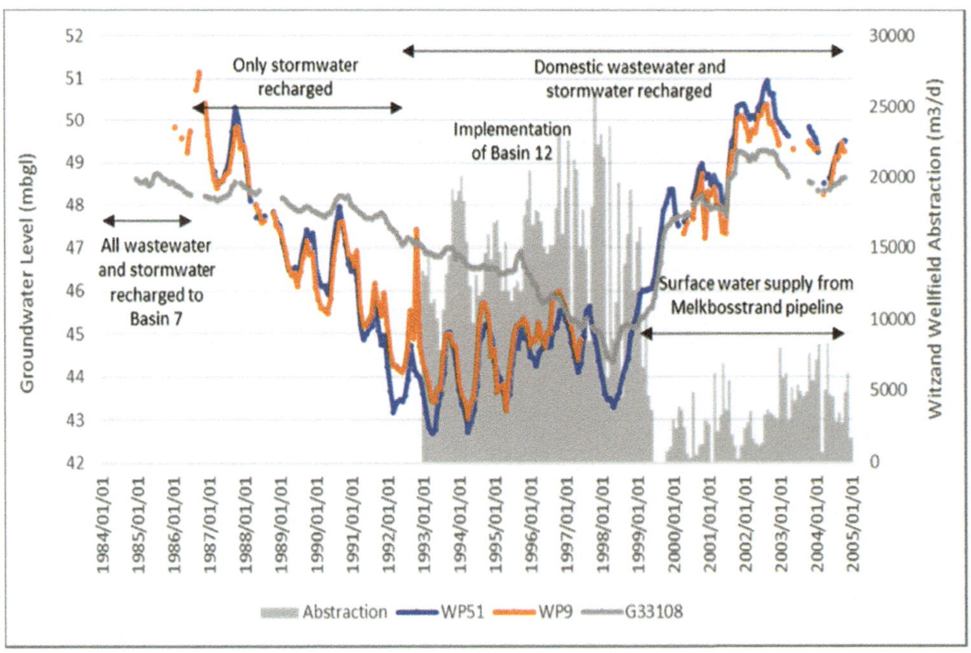


**Fig. 3** History of the good, poor, and non-management of the AWRMS illustrated through hydraulic head timeseries

communities taking timely and appropriate decisions. Initiating or facilitating inter-departmental communication and upgrading knowledge and insight through improved investigative techniques, monitoring, modelling and adaptive management are the first and foremost steps to eliminating the risks to the AWRMS (Table 1).

**Table 1** Levels of resource and institutional interaction and management required to coordinate and maintain the AWRMS

| Scale | Resource management | People management |
|---|---|---|
| Regional | Regulator | National government |
| | | Catchment management agencies |
| Local | Water service provider | Catchment management agencies |
| | Political | Ward councilors, planned developments |
| | Social and environmental obligations | NPO, NGO, corporates |
| Direct | Operations | Local employees |
| Overarching | Custodians and users | Community, consumers, land users and industry |

The long-term sustainability of the AWRMS depends on proper management of all actors coupled with a high level of scientific confidence and a strong geoethical resolve. This entails abiding to the fundamental geoethical principles of ensuring scientific practice, communicating understanding, uncertainty, and ensuring the education of all those who affect and are affected by the scheme.

# Geoethical Groundwater Modelling: Aligning Decision-Support Models with the Scientific Method

Rui Hugman and John Doherty

**Abstract**

Management of groundwater resources is increasingly reliant on numerical simulation. Unfortunately, decision-support modelling is often conducted under the premise that predictive reliability increases with modelling complexity. In truth, while modelling complexity can support quantification of predictive uncertainty, the latter is a function of data availability. Excessive complexity can often erode, rather than enhance, a model's ability to quantify and reduce the uncertainties of decision-critical predictions by reducing its capacity to assimilate prediction-salient information. We submit that a groundwater model is more productively viewed as a data assimilator for decision-pertinent information than as simulating subsurface processes, even though the latter role (though imperfect) underpins the former. Assimilated data may, or may not, allow rejection of the hypothesis that a certain course of management action will have adverse consequences. Either way, the decision-making process requires that this hypothesis be tested. In the following document, we outline how decision-support environmental modelling can be implemented with the scientific method, and discuss how uncertainties of decision-salient predictions can be addressed with appropriate model complexity so that stakeholder expectations are better aligned with what models can and cannot deliver to the decision-making process.

**Keywords**

Uncertainty • Complexity • Decision support • Modelling • Scientific method

R. Hugman (✉)
Umvoto Africa (Pty) Ltd., Cape Town, South Africa
e-mail: rthugman@gmail.com

J. Doherty
Watermark Numerical Computing, Brisbane, Australia

## 1 Introduction

Decision-makers and stakeholders are often reliant on models when planning and implementing interventions in the environment. The outcome of these decisions can often carry a large cost, whether economic, social or to the environment. Yet, sadly, there is a frequent and expensive misuse of what is fundamentally a very useful technology, with resources wasted on expensive yet fruitless modelling exercises that do not provide the insights into the future that were promised as justification for their construction (Doherty 2010a). This has led to a growing realization that models are not quite the magical crystal balls that they are often made out to be. Stakeholders and end-users of model outputs, perhaps too often burnt by wildly inaccurate predictions, are turning their attention to model uncertainty and the implications on expensive or controversial decisions.

Uncertainty is a term that any modeler is familiar with (we hope!) but few are well versed in. In the groundwater industry, uncertainty has in the past tended to be a relatively niche and often dispersed topic in academia and rarely at the forefront of 'real-world' applications. This is likely due to the perception that uncertainty is complicated, in conjunction with the associated cost of carrying out uncertainty analysis in both human resources and computation time. However, as an industry, we are going through a paradigm shift in how we develop and use models to support decisions. The intersection of growing recognition of how wrong models can be (in particular amongst those paying for them), with the maturation of software and computational capacity that allows us to tackle the problem, has brought us to a point in which addressing the potential for wrongness in a model's prediction must become the norm. This paradigm shift creates an inherent change in the process of how to think about and develop models for decision support: away from '*how well does the model match available data?*', and towards '*can the model reject a hypothesis and still match available data?*'.



M. Abrunhosa et al. (eds.), *Advances in Geoethics and Groundwater Management: Theory and Practice for a Sustainable Development*, Advances in Science, Technology & Innovation,
https://doi.org/10.1007/978-3-030-59320-9_89

The following chapters present a very summarized overview of how decision-support environmental modelling can be implemented geoethically with the scientific method. We highlight that the focus here is on models for 'decision-making', as modelling can have other purposes to which some of these concepts may not be fundamental (i.e. exploring process interactions or conceptual ideas of a system). It is our hope to contribute both to current modelling practice, as well as better align stakeholders and end-users' expectations with what models can (and cannot) deliver to the decision-making process. These concepts and practical applications are developed in greater detail in several publications which we hope the reader will be inspired to delve into (Doherty 2003; James et al. 2009; Doherty 2010a, b; Herckenrath et al. 2011; Doherty 2015; Burrows and Doherty 2015).

## 2 The Scientific Method and Environmental Modelling

The Cape Town Statement on Geoethics expresses an international consensus of geoscientists and suggests several fundamental values of Geoethics (Di Capua et al. 2017). Amongst these are the following:

- Honesty, integrity, transparency, and reliability of the geoscientist, including strict adherence to *scientific methods*.

The hypothetic-deductive model of the scientific method is an empirical method of acquiring knowledge through hypothesis rejection. It follows the general steps of: (1) a hypothesis is proposed and (2) an appropriate experiment is designed to test the hypothesis. Should the outcome of the experiment disagree with the hypothesis, then it can be invalidated. However, even if the data agrees with the hypothesis—it cannot be validated. Thus, uncertainty is reduced by increasing the number of invalidated hypothesis.

Classical groundwater modelling rarely follows the scientific method. Instead of finding the hypotheses which are not rejected by available data, model development tends to focus on finding 'a' hypothesis which 'best' fits historical data, which is then used to make predictions. Effectively, this validates a hypothesis of how the system works but does not necessarily explore all the others which may also be valid. Additionally, the quest to find an 'adequate' fit can lead to unnecessary model complexity—which may, or may not, be relevant for the prediction of concern—and can hinder any subsequent exploration of predictive uncertainty due to increased numerical burden.

As geoethical practitioners, how can our environmental modelling for decision support '*adhere to scientific methods*'? Well, in practice, an unwanted event can be considered a scientific hypothesis. Say, for example, that a decision must be made that could impact the flow to a spring that is the sole source of supply for a town. The hypothesis in this case would be 'the decision will cause the spring to run dry'. The purpose of a modelling exercise in such a setting must be to provide a basis to reject the hypothesis that the event can occur if a certain management option is taken. This is achieved through assimilating all available information into the model—including 'hard' measurement data as well as 'soft' expert knowledge. If processing all this data leads to rejection of the hypothesis that the unwanted event can occur, then model-based environmental data analysis has provided support to the decision-making process in a defensible manner. Alternatively, while it may not be possible to entirely reject the possibility that an event can occur, it may be possible to ascribe a probability that society is willing to risk in order to receive the benefits of the proposed development. These concepts provide the philosophical basis for model development, which can be summarized as follows (Doherty 2010a):

*A model is deployed specifically to test the hypothesis that an unwanted event will occur. This is done by including system states corresponding to that event in the model's calibration dataset, along with historical measurements of system state. The model is then calibrated against this composite dataset. The hypothesis that the proposed state may occur can be rejected if:*

- *the model cannot replicate the occurrence of the event in the future while simultaneously respecting historical system behaviour;*
- *the model can accommodate the simultaneous occurrence of the proposed and historical system states only through use of parameters that are unrealistic.*

It follows from the above that a model prediction from a single 'calibrated' model cannot provide this evidence. A hypothesis can only be considered invalidated (to a degree of confidence given uncertainty in the data/information) if it does not occur for any of the parameter combinations that can explain available/historical data and information.

## 3 Making a Decision with an Uncertain Model

Reality is messy. And in this messy reality, the future cannot be perfectly known. We are all inherently aware of this to some degree and on a day-to-day basis make decisions based on a probabilistic outcome. We take decisions based on imperfect and incomplete data, combined with inadequate and simplified understandings of the systems around us in order to maximize the probability of something good/bad happening/not happening. If the outcome is uncertain, we make the decision based on how willing we are to risk the possibility of the opposite event occurring.

One of the fundamental values in the Cape Town Statement on Geoethics states the following (Di Capua et al. 2017): "Sharing knowledge at all levels as a valuable activity, which implies communicating science and results, *while taking into account intrinsic limitations such as probabilities and uncertainties*".

Decision-making that affects groundwater, and environmental systems in general, rely on quantifying the risk associated to an event. Given the complexity of environmental systems, models are the best tools available for this. It follows that, purposeful (and inherently geoethical) modelling must define the confidence interval associated with the prediction of interest so that risk can be considered when taking the decision.

Models are simplification of reality. And modelers are all well versed in the concept of Occam's Razor, which can be paraphrased as 'the simplest solution is usually right'. However, in environmental modelling, this idea of parsimony can result in models with poor predictive capacity and, perhaps more importantly, severely hinder the ability to quantify predictive error.

## 3.1 The Need for Many Parameters

In groundwater modelling, the model domain is typically divided into 'as few as possible' piece-wise constant zones to allow the model to fit the data according to an—arbitrary—measure of what is acceptable. This approach assumes that the 'zoning' is based on a firm understanding of parameter variability. However, constraining the model's capacity for heterogeneity (1) reduces information gain during the calibration process and (2) ignores *possible* property complexity beyond that which can be inferred from available data. This inherently weakens the fundamental motivation for model development—testing hypothesis—by reducing the ability to capture complexity which a prediction may depend on. Doherty et al. (2010) describe these concepts in greater detail and propose an alternative solution for model parameterization and representation of system complexity: *A model should include parameter and process detail to the extent to which model predictions depend on such detail, since omission of this detail contravenes the original motivation for building the model.*

Capacity to represent heterogeneity can be increased by using highly parameterized models (i.e. pilot-points). However, the use of large amounts of parameters brings other technical challenges (Doherty et al. 2010). Essentially, the issue of computation time becomes an important driver in the practical application of highly parameterized models. The more parameters, the greater the number of models runs required to 'history match' a model, as well as to explore the range of predictive uncertainty. Although there are methods which reduce this cost (i.e. singular value decomposition and ensemble methods), effectively if a model takes too long to run at least several thousand times in a practical time frame, the ability to judge how wrong it may be decreases or is simply not possible.

## 3.2 The Need for Simple Models

So, we have seen that models that simplify parameter heterogeneity will misrepresent the range of possible predictions, but on the other hand, models that are too complex are too expensive to characterize when using many parameters. Which leaves us with the unpleasant conundrum of *how complicated is complicated enough?* One could argue that the answer could be: *As complicated as necessary to quantify the level of predictive error to within a (problem specific) confidence range.* Or, if not possible given available data, then: *As complicated as the budget allows, whilst still being able to provide a quantification of predictive error.* It follows from the above that this level of complexity is specific both to the problem/decision as well as the context in which the model is being developed (i.e. dependent on computation capacity, timeframe and budget).

The tools and computational capacity to cope with the numerical burden of thousands of model runs are readily available. Cloud computing and cheap clusters of distributed off-the-shelf computers make the parallelization of model runs a cheap and trivial task. Software to extract information from data and quantify the implications of unknowns on model predictions is still maturing but does exist (i.e. the PEST suite of tools provides several options). Hopefully as the recognition within our industry for the need for such software tools becomes more widespread and accepted, software developers will respond by building more.

# 4 Concluding Remarks

Models are simplifications of reality built from imperfect data by people with imperfect knowledge. To expect certainty from a model prediction is unreasonable. To expect a quantification of how wrong the prediction could be is. If used properly, models are tools to assimilate all available data—including expert knowledge, point measurements of system properties, and historical measurements of system state. Perhaps, more importantly, they can then be used to quantify the potential for error that remains given the available information.

Unfortunately, the latter is rarely done. Nevertheless, barriers to developing models able to incorporate these concepts are rapidly decreasing, alongside a growing recognition by end-users for the need to consider them. It is our opinion that for modelling to support decision making in a geoethical manner, any prediction must be accompanied by the quantification of predictive uncertainty. Only by doing so can a decision be informed by an assessment of the associated risk. This requires finding a balance between model simplicity and complexity so that prediction specific parameter and process complexity is captured, whilst also maintaining low computational burden to allow for uncertainty to be quantified through numerous models runs. Retaining this nimbleness may require prediction specific models, rather than the common drive for a single complicated model to answer all questions.

We hope that the overview presented here will raise awareness amongst model practitioners and (perhaps more importantly) end-users to what should and should not be expected from model predictions and help move our industry towards a more scientifically sound practice when providing input to the decision-making process.

## References


Burrows W, Doherty J (2015) Efficient calibration/uncertainty analysis using paired complex/surrogate models. Groundwater 1:1–11

Di Capua G, Peppoloni S, Bobrowsky P (2017) The Cape Town statement on geoethics. Ann Geophys 60, Fast Track 7. https://doi.org/10.4401/ag-7553

Doherty J (2003) Ground water model calibration using pilot points and regularization. Groundwater 41:170–177

Doherty J (2010a) Model-based environmental decision-making. Watermark Numerical Computing, Brisbane, Australia

Doherty J (2010b) Methodologies and software for PEST-based model predictive uncertainty analysis. Watermark Numerical Computing, Brisbane, Australia

Doherty J (2015) Calibration and uncertainty analysis for complex environmental models. Watermark Numerical Computing, Brisbane, Australia

Doherty J, Hunt R, Tonkin M (2010) Approaches to highly parameterized inversion: a guide to using PEST for model-parameter and predictive-uncertainty analysis: U.S. geological survey scientific investigations report 2010-5211, 71p

Herckenrath D, Langevin CD, Doherty J (2011) Predictive uncertainty analysis of a saltwater intrusion model using null-space Monte Carlo. Water Resour Res 47:W05504. https://doi.org/10.1029/2010WR009342

James S, Doherty J, Eddebbarh A (2009) Practical postcalibration uncertainty analysis: Yucca Mountain, Nevada. Ground Water 47 (6):851–869

# Geoethics of Bulk Groundwater Abstraction in an Ecologically Sensitive Area: Steenbras Wellfield (Cape Town)

Dylan Blake, Christopher Hartnady, Rowena Hay, and Kornelius Riemann

**Abstract**

The City of Cape Town initiated its "New Water Programme" in 2017 to diversify its bulk water supply, thereby improving long-term water security and resilience against future droughts—this includes bulk groundwater abstraction of potentially ~140–400 Ml/day from the major fractured Peninsula and Nardouw Aquifers of the Table Mountain Group (TMG; in the mountain catchments to the east of the city). Current TMG groundwater exploration and wellfield development are taking place in the vicinity of Steenbras Dam, in the form of the ~15–20 Ml/day Steenbras Wellfield scheme. The TMG aquifers are also essential in sustaining groundwater-dependent ecosystems associated with the Cape Floral Kingdom—a global biodiversity hotspot with exceptional endemic diversity, but also a global extinction hotspot. A strong geoethical, "no-regrets" approach is, therefore, required to develop TMG wellfield schemes for the City of Cape Town (and other towns/cities in the Western/Eastern Cape), in order to reduce the risk of any negative ecological and environmental impacts, while still enhancing the drought resilience of the city, providing water for future urban growth, and meeting Sustainable Development Goals 6 and 11.

**Keywords**

Fractured aquifer • Wellfield development • Bulk groundwater supply • Groundwater-dependent ecosystem

## 1 Introduction

The City of Cape Town (CCT; within the Western Cape Province of South Africa) is the most-populated coastal city, and second most-populated city within South Africa (~4 million people). The city is one of the most multi-cultural in the world, and is a major economic, transport, tourist, design, and agricultural (in association with surrounding farm regions) hub in South Africa and Africa. The Cape Floristic Kingdom (CFK) also occurs within the city and surrounding mountain catchments, and despite it being the smallest of the six floral kingdoms of the world, is a biodiversity hotspot with high economic and ecological value. The CCT is currently supplied by surface water stored in the six major dams of the Western Cape Water Supply System (WCWSS), which has a total storage of ~900 million $m^3$, and an original 98% assurance of supply of 600 million $m^3$ per annum (~50–60% of which is used by the CCT, and the remaining ~40–50% used by surrounding agriculture/other urban users).

The Western Cape is a relatively water-scarce area as a result of the Mediterranean climate experienced, and this scarcity is likely to be exacerbated as a result of climate change. An extreme, extended 1:590 year drought (CCT 2019) in the Western Cape from 2015 to 2017 (although continuing through to present in certain areas) has put severe strain on the WCWSS to such an extent that the supply system came close to failing in early 2018 (total dam storage volumes of <20%), even with severe water restrictions and water conservation/demand measures in place (reducing the city's water demand from ~1000–1200 Ml/day to ~500–600 Ml/day) (CCT 2019). Consequently, the CCT was at risk of becoming the first modern day city on Earth to run out of water (as extensively reported nationally and internationally as "Day Zero"), which would have had enormous societal and economic impacts.

The CCT initiated its "New Water Programme" (NWP) in earnest to diversify its water supply to improve long-term

D. Blake (✉) · C. Hartnady · R. Hay · K. Riemann
Umvoto Africa, Cape Town, South Africa
e-mail: dylan@umvoto.com



M. Abrunhosa et al. (eds.), *Advances in Geoethics and Groundwater Management: Theory and Practice for a Sustainable Development*, Advances in Science, Technology & Innovation,
https://doi.org/10.1007/978-3-030-59320-9_90

water security and resilience against future droughts by implementing alternative bulk water supply options (CCT 2019). The NWP also aims to meet the demand by an ever growing urban population, improving the standard of living of approximately half of the city's population through meeting Sustainable Development Goals (SDGs) 6 and 11. Identified alternative bulk water supply options include desalination, water re-use, and the abstraction of groundwater from three major aquifer systems that the city has access to, namely:

- the Atlantis Aquifer and Cape Flats Aquifer, both primary sand aquifers within the city's municipal boundaries; and
- the Table Mountain Group (TMG) aquifers (largest of the three major aquifer systems; sub-divided into the fractured quartzitic sandstone Peninsula and Nardouw Aquifers), which occurs within the mountain catchments that surround the city and incorporate the dams of the WCWSS.

## 2 Project Implementation

The vast potential of the globally unique, deep (up to 1–2 km thick), evaporation free, extensive ($\sim$11,000 km$^2$), voluminous (saturated aquifer volume of $\sim$400,000 km$^3$) and high yielding (individual borehole yields of $\sim$10–70 l/s, potential total yield of $\sim$140–400 Ml/day or $\sim$50–150 million m$^3$/a) Palaeozoic TMG aquifers was identified through pioneering groundwater exploration and supply projects by Umvoto Africa and others since the 1990s (Hartnady and Hay 2002, 2008; Hartnady et al. 2005, 2009; Blake et al. 2010).

The CCT has been investigating the feasibility of groundwater abstraction from the TMG aquifers since the early 2000s (Hartnady et al. 2004). Current exploration (including additional exploration via detailed geological mapping and heliborne geophysics) and wellfield development is occurring in the vicinity of the WCWSS Steenbras Dam east of the CCT, in the form an $\sim$15–20 Ml/day wellfield scheme (expected to be operational in early-mid 2020) (Blake et al. 2019) (see Fig. 1).

The planned "Steenbras Wellfield" targets both TMG aquifers along a high-yielding major crustal strike-slip structure/hydrotect termed the "Steenbras-Brandvlei Mega-fault Zone" (see Fig. 2). Current drilling activities have included shallow to deep ($\sim$100–400 m, Nardouw Aquifer) and ultra-deep (up to 1000 m, Peninsula Aquifer), wide diameter abstraction and core exploratory boreholes, with tested abstraction borehole yields ranging between 10–70 l/s. Further CCT TMG groundwater exploration and wellfield scheme development (potential total combined supply of $\sim$50–150 million m$^3$/a or $\sim$140–400 Ml/day) is planned along major TMG structures within the Grabouw-Eikenhof and Theewaterskloof basins, Wemmershoek, Voelvlei, Berg River and the CCT South Peninsula region. These areas are in vicinity of the major WCWSS dams (which are fortuitously near large fault systems and hydrotects), as the infrastructure costs of transporting large volumes of groundwater from extremely rugged, ecologically sensitive mountain terranes can be excessive.

## 3 Geoethical Considerations

The "Cape Town Statement on Geoethics" (Di Capua et al. 2017), prepared during the 35th International Geological Congress in Cape Town (2016), describes the purpose of geoethical thinking as: "to improve both the quality of professional work and the credibility of geoscientists, to foster excellence in geosciences, to assure sustainable benefits for communities, as well as to protect local and global environments; all with the aim of creating and maintaining the conditions for the healthy and prosperous development of future generations".

The TMG aquifers are essential in sustaining groundwater-dependent ecosystems (GDEs; both floral and faunal components) associated with the CFK (see Fig. 2), through spring/seep discharge of high quality, low nutrient, acidic groundwater along geological (specifically aquifer-aquitard) contacts and structural features (e.g., faults, fractures and dolerite dykes), and the provision of mountain stream baseflow (Colvin et al. 2009). The understanding of any potential changes in groundwater quality and quantity within the TMG aquifers as a result of bulk groundwater abstraction, through monitoring of GDEs and monitoring-modeling-management of surface–groundwater interactions, is a critical geoethical approach in determining the long-term sustainable viability of the groundwater resource (while still supplying water to the city).

Planned bulk TMG groundwater abstraction has raised both valid concerns and perceived claims by numerous botanists, ecologists, and environmentalists, especially in the Western Cape where there is a strong (and sometimes over-zealous) ecological conservation culture and green environmental lobby. There is valid concern in protecting the CFK, which is a global biodiversity hotspot with exceptional endemic diversity ($\sim$3% of all species on Earth, covering only 0.06% of Earth's land surface) (Myers et al. 2000), but also a global extinction hotspot—37 seed-bearing plants (13 within CCT municipal boundaries; Rebelo et al. 2011) in the CFK have gone extinct since 1900 (2nd most to only Hawaii with 79 extinctions) (Humphreys et al. 2019),

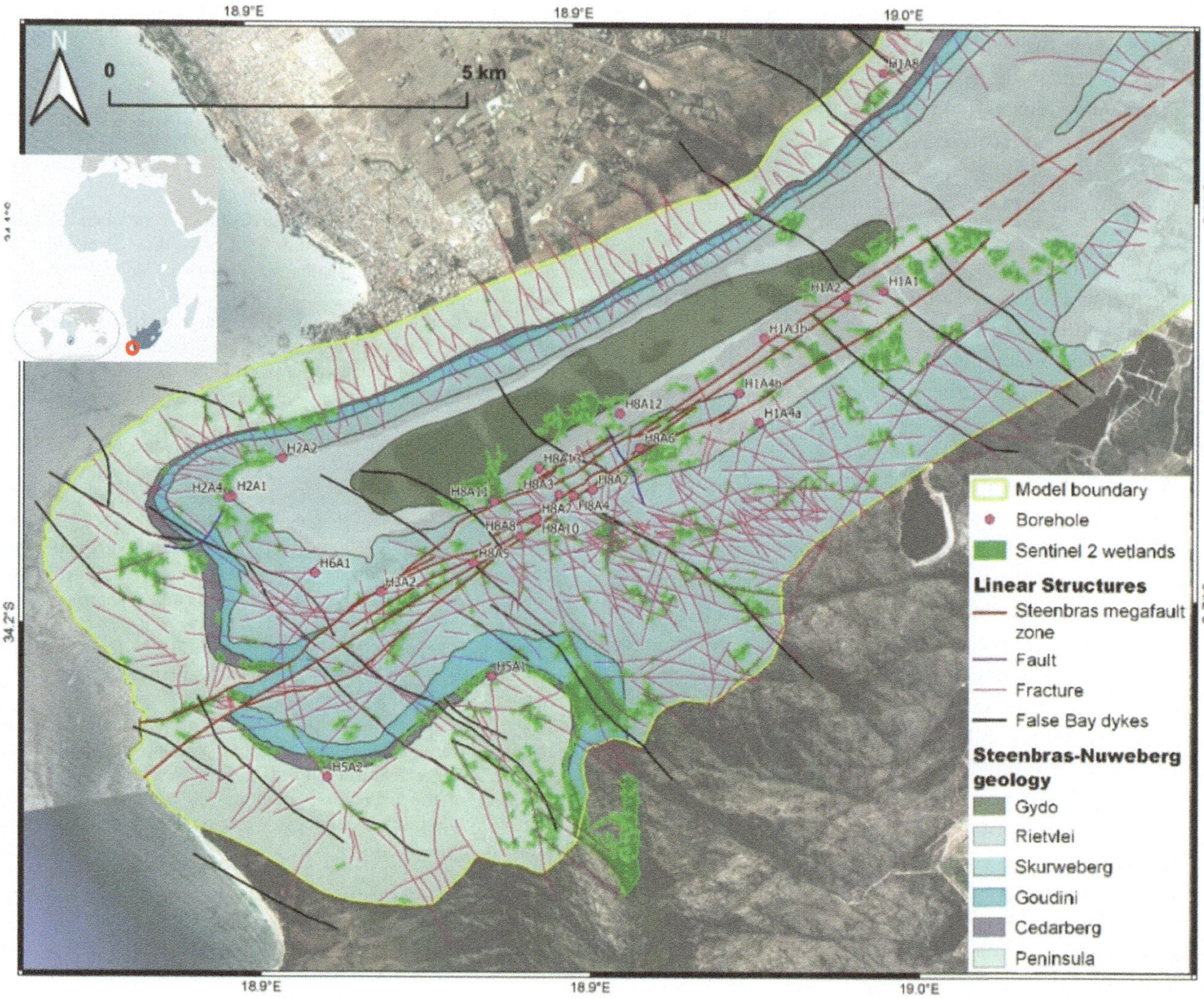


**Fig. 1** Steenbras Wellfield lithology, structure, borehole positions, and wetlands/GDEs identified using Sentinel 2 imagery. Skurweberg and Rietvlei Formations of the TMG form the Nardouw Aquifer, whereas the Peninsula Formation (also TMG) forms the Peninsula Aquifer

**Fig. 2** Example of likely groundwater-related wetland/seep/channel ecosystems in the Steenbras Wellfield. SBMZ (red dashed line)—Steenbras-Brandvlei Megafault Zone strand, Ss— Skurweberg Fm., Dr—Rietvlei Fm. (both Nardouw Aquifer, contact line in solid blue)

with ~13% of all threatened plant species on the planet occurring within the CFK (with an additional ~1500 species considered rare or critically rare). Main ecological concerns identified by botanists/ecologists include:

- direct habitat and biodiversity loss due to drilling/wellfield construction;
- indirect impacts such as changes in hydrology and fire regimes due to wellfield infrastructure, which can potentially lead to increased alien species invasion potential, and soil erosion (depending on the terrane);
- impacts of bulk abstraction on ecosystems (incorporating both fauna and flora) that are directly (GDEs) and indirectly dependent on groundwater.

The CCT has taken a "no-regrets" (and in essence geoethical) approach in the development of TMG wellfield schemes to reduce the risk of any negative ecological and environmental impacts (sometimes at additional excessive costs to reduce perceived ecological fears), while still enhancing the drought resilience of the city, providing water for future urban growth, and meeting SDGs 6 and 11:

- restricting drilling target sites and wellfield infrastructure to existing access roads/tracks/firebreaks to reduce direct habitat and biodiversity loss;
- if natural area clearing is required for drill rig setup or other wellfield infrastructure, then search-and-rescue of endangered flora species (stored in a temporary on-site botanical nursery) and topsoil removal/storage is undertaken for subsequent site rehabilitation to the prior natural state;
- containment of all groundwater expelled during the drilling process (which can be very large volumes, with some boreholes having blowyields of $\sim$100–300 l/s), as a sometimes over-cautious approach to ensure no groundwater and biodegradable drilling foam enters the receiving environment (even in plantation forested areas earmarked for rehabilitation);
- transfer of all groundwater expelled during test-pumping to Steenbras Dam via temporary PVC pipelines, to ensure the characteristic of streams and drainage lines are not altered;
- field and remotely sensed mapping/identification and monitoring (see Figs. 1 and 2) of GDEs relating to the TMG aquifers;
- detailed geological/hydrogeological mapping (especially aquifer-aquitard contacts and structural features associated with GDES) and numerical groundwater modeling, to determine potential groundwater abstraction-related impacts to at-risk GDEs;
- development of detailed hydrogeological, hydrological and ecological monitoring protocols for wellfield operation, in line with (and in addition to) the water use license requirements detailed by government regulators;
- establishment of an aquifer monitoring committee and ecological focus groups with relevant stakeholders, in order to present monitoring results, educate on earth/groundwater science principles, capture/develop relevant hydrogeological-ecological concerns/solutions, and holistically manage the various planned TMG wellfield schemes.

This "no-regrets" geoethical approach should be applied (within reason) across future TMG wellfield scheme developments in the Western and Eastern Cape of South Africa.

**Acknowledgements** The City of Cape Town: Water and Sanitation Department: Bulk Water Branch is acknowledged for the funding and undertaking of the TMG wellfield developments.

## References

Blake D, Mlisa A, Hartnady CJH (2010) Large scale quantification of aquifer storage and volumes from the Peninsula and Skurweberg formations in the southwestern Cape. Water SA 36 (2):177–184

Blake D, Riemann K, Rademan Z, Geldenhuys P (2019) Groundwater development of the Table Mountain Group Aquifers as part of the City of Cape Town's new water programme. In: 16th South African groundwater division conference, Port Elizabeth, South Africa, Oct 2019

City of Cape Town (2019) Cape Town water strategy—our shared water future

Colvin C, Riemann K, Brown C, Le Maitre D, Mlisa A, Blake D, Aston T, Maherry A, Engelbrecht J, Pemberton C, Magoba R, Soltau L, Prinsloo E (2009) Ecological and environmental impacts of large-scale groundwater development in the Table Mountain Group (TMG) aquifer system. Water research commission report 1327/1/08

Di Capua G, Peppoloni S, Bobrowsky PT (2017) The Cape Town statement on geo-ethics. Ann Geophys 60, Fast Track 7. https://doi.org/10.4401/ag-7553

Hartnady CJH and Hay ER (2002) The use of structural geology and remote sensing in hydrogeological exploration of the Olifants and Doring River catchments. In: Pietersen K, Parsons R (eds) A synthesis of the hydrogeology of the Table Mountain Group—formation of a research strategy. Water research commission report TT158/01, pp 19–30

Hartnady CJH, Hay ER (2008) The assessment of water availability in the berg catchment (WMA 19) by means of water resource related models: groundwater model report, vol 3. Regional conceptual model. DWAF report no. P WMA 19/000/00/0408, 135pp

Hartnady CJH, Hay ER, Riemann K (2004) Table Mountain Group Aquifer feasibility study and pilot project: preliminary phase—hydrogeological report. TMG aquifer alliance/CMC administration, report no. 2, Oct 2004, 138pp

Hartnady CJH, Hay ER, Riemann KR, Mlisa A, Jackson C, Paige A Groenewald LG (2005) Deep artesian groundwater for oudtshoorn municipal supply: phase D—target generation and borehole/wellfield siting using structural geology and geophysical methods. Water research commission report 1254/1/05, 164pp

Hartnady CJH, Riemann K, Seyler H (2009) Sustainable groundwater management: a case study of the monitor, model, manage approach in Hermanus. Civ Eng 17(7):14–20

Humphreys AM, Govaerts R, Ficinski SZ, Lughadha EN, Vorontsova MS (2019) Global dataset shows geography and life form predict modern plant extinction and rediscovery. Nat Ecol Evol 3:1043–1047

Myers N, Mittermeier RA, Mittermeier CG, da Fonseca GAB, Kent J (2000) Biodiversity hotspots for conservation priorities. Nature 403:853–858

Rebelo AG, Holmes PM, Dorse C, Wood J (2011) Impacts of urbanization in a biodiversity hotspot: conservation challenges in Metropolitan Cape Town. S Afr J Bot 77(1):20–35

# Long-Term Planning During Emergency Response—A No Regrets Approach and Long-Term Vision for the Development of the Cape Flats Aquifer (Cape Town)

David McGibbon, Rui Hugman, Luke Towers, Kornelius Riemann, Rowena Hay, and Christopher Hartnady

**Abstract**

The Cape Flats Aquifer (CFA) is a primary coastal aquifer that underlies the City of Cape Town (South Africa), identified as a major source of water by hydrogeologists since the 1970s. Scientific specialists called for its protection so that it could be used for bulk water supply to the City. Unfortunately, the resource was largely ignored by the municipality, resulting in the degradation of the underlying aquifer over time to the point where groundwater from some portions is non-usable due to contamination. In 2015–2017, Cape Town experienced its worst drought since 1904, driving the municipality to look at alternative sources of water such as desalination, re-use of treated effluent and various groundwater schemes of which the CFA was one. Given the time-constraints of an emergency response project, long-term testing and study of the system to support design and implementation have been significantly reduced. A no regrets approach was therefore adopted by the team to reduce the chance of any negative impact such as saline intrusion, reduction in water levels that would negatively affect existing users that rely on groundwater for their livelihood and sensitive ecosystems. We aim to present the geoethical question of developing an emergency scheme with limited data on its sustainability and costing versus not developing it and risking a City of ~4 million people running out of water.

**Keywords**

Emergency response • No regrets • Groundwater scheme • Coastal aquifer • Urban environment

D. McGibbon (✉) · R. Hugman · L. Towers · K. Riemann · R. Hay · C. Hartnady
Umvoto Africa, Cape Town, South Africa
e-mail: david.m@umvoto.com

## 1 Introduction

Since the 1970s, hydrogeologists (e.g. Henzen 1973; Wessels and Greef 1980; Hay 1981; Roberts et al. 2006; DWAF 2008) have studied and identified the coastal primary Cape Flats Aquifer (CFA) as a major resource which underlies the City of Cape Town (CoCT) (DWS 2014). These authors called for its protection during the urban sprawl that accompanied considerable population growth in the City. Low cost surface water options and the implementation of a successful Water Conservation and Demand Management program curbed demand, resulting in the CFA not being developed as a bulk supply option.

Cape Town is in the Western Cape of South Africa (see Fig. 1). The CoCT and Western Cape as a whole experiences a Mediterranean climate of hot dry summers and cold wet winters but of late experienced its worst drought since 1904, with water storage capacity dropping down to 20% (CoCT 2017; see Fig. 2). The Premier of the Western Cape therefore declared Cape Town and other areas in the Western Cape as disaster areas.

As a result, the CoCT implemented the New Water Projects (NWP) to augment its potable water supply by between 100 and 500 mega litres (Ml) per day, through alternate sources such as desalination, re-use and various groundwater schemes as identified in the WCWSS Water Reconciliation Strategy (DWS 2016). This aimed to not only address their short-term requirements, but also provide enough flexibility to evaluate and implement their longer-term strategies as part of a holistic approach in solving the CoCT's water requirements with future variations in rainfall.

The CoCT aims to meet its immediate short-term water requirements through local scale groundwater abstraction and distribution from the Cape Flats Aquifer, while longer-term flexibility for broader scale distribution through the WCWSS, and implementation of such measures as Managed Aquifer Recharge (MAR) and bioremediation



M. Abrunhosa et al. (eds.), *Advances in Geoethics and Groundwater Management: Theory and Practice for a Sustainable Development*, Advances in Science, Technology & Innovation,
https://doi.org/10.1007/978-3-030-59320-9_91

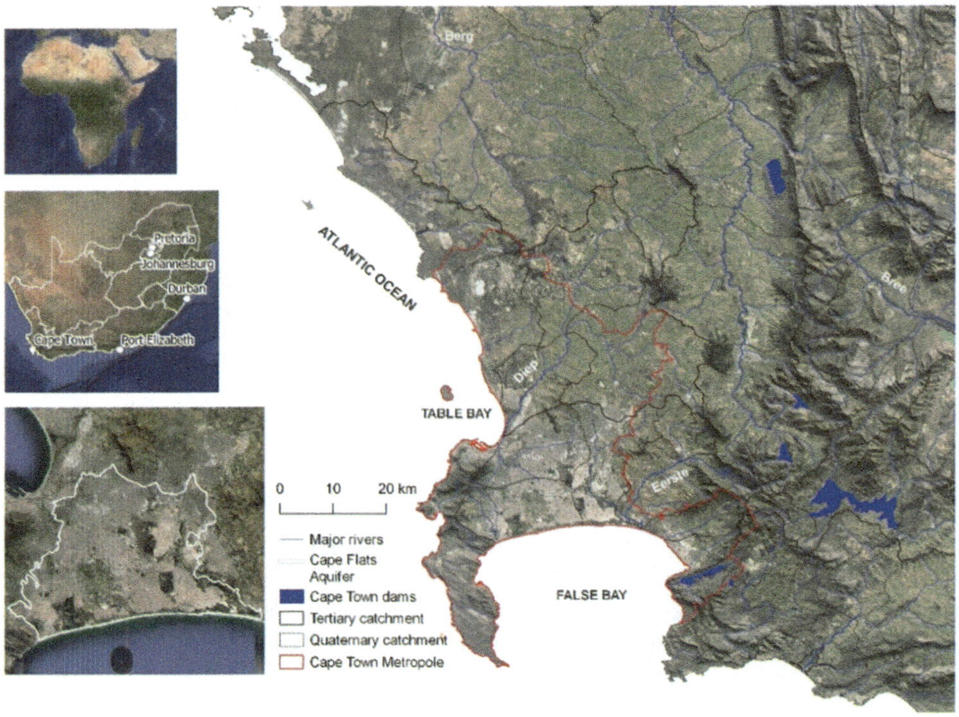


**Fig. 1** Location of the Cape Flats Aquifer within the City of Cape Town, South Africa

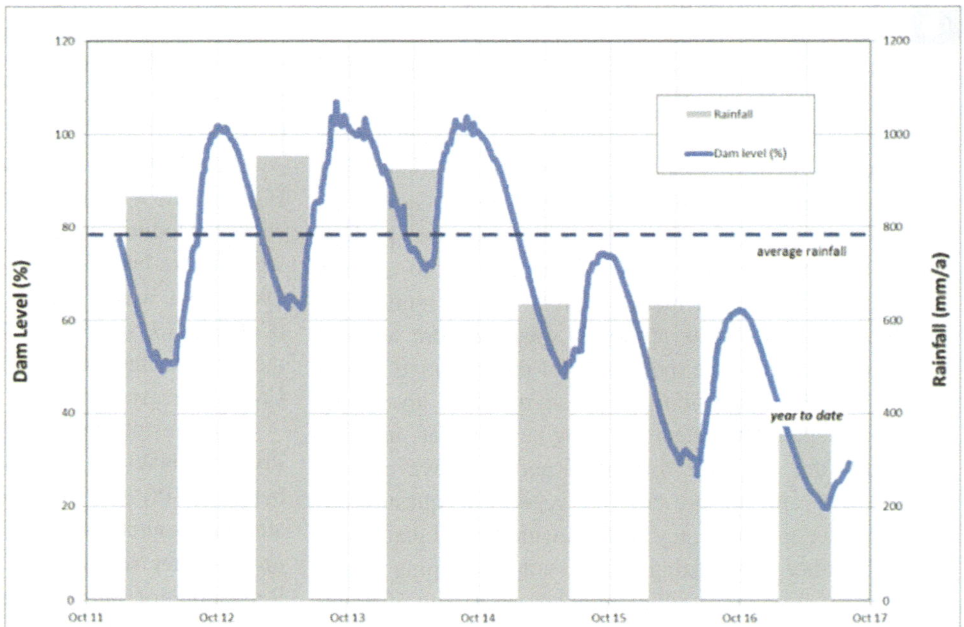


**Fig. 2** Seasonal rainfall trends with corresponding dam level fluctuations, reaching <20% in 2017 (Parson 2017)

processes will improve aquifer abstraction potential and water quality over the lifespan of the aquifer utilisation. The short-term response to the current drought experienced must ensure that the medium- to longer-term objectives of the CFA Management Strategy (DWS 2015) are not compromised, but are rather enhanced.

The CoCT was awarded a Water Use Licence under the Regulations of the National Water Act for the abstraction of groundwater from the Cape Flats Aquifer and recharge of water into the aquifer in December 2017 (see Table 1). Currently, only Phase 1 (average daily abstraction of 55 million litres) of the abstraction is authorised.

**Table 1** The authorised abstraction and recharge volumes for the Cape Flats Aquifer as stipulated in the water use licence

| Cape Flats Aquifer | Phase 1 | Phase 2 | Phase 3 |
|---|---|---|---|
| Abstraction | 20 million $m^3/a$ | 25 million $m^3/a$[a] | 30 million $m^3/a$[a] |
| Recharge | 12 million $m^3/a$ | 18 million $m^3/a$ | 25 million $m^3/a$ |

[a]Not authorised yet, would require an amendment to the licence

## 2 Project Implementation

The CFA Groundwater Scheme was implemented through a combination of exploration drilling, targeting appropriate City owned land to avoid time-consuming landowner agreements, and the simultaneous drilling of production boreholes in areas where high yields were realised during exploration.

In order to speed up exploration, airborne electromagnetic geophysical surveys were carried out over the Cape Flats in late 2017 to confirm the basic understanding of the aquifer from previous studies. During this time, drilling contractors descended on Cape Town, increasing drilling costs due to the high demand. These inflated prices were accepted because of the emergency, with drilling commencing in early 2018.

During the simultaneous expansion of the exploration and production drilling, tenders were put out for the construction of abstraction and recharge pipelines and the water treatment plants for the wellfields. The design of the pipelines and treatment plants was based on limited yield and quality data from existing literature and the few tested boreholes and the remainder was assumed based on this available data.

To date drilling and testing has continued with over 200 boreholes being drilled and tested in <2 years. Once drilled, boreholes are included in the ongoing monitoring of water quality and levels. The data gathered has refined the understanding of the underlying CFA and identified numerous hidden historic sources of contamination such as unlined wastewater treatment works (WWTW) sludge ponds, buried and burnt munition piles, "illegal dump sites", informal settlements and leaking canals.

This has resulted in the re-structuring of the projects implementation where certain wellfields have been put on hold and others expanded due to higher yields and better water quality. This has knock-on effects to the treatment plants design and costing, along with infrastructure lying dormant.

The City has undertaken to rehabilitate or facilitate the rehabilitation of contamination sources identified during the implementation of the CFA Groundwater Scheme.

## 3 Geoethical Considerations

The "Cape Town Statement on Geoethics" (Di Capua et al. 2017), prepared during the 35th International Geological Congress (Cape Town 2016), describes the purpose of geoethical thinking as: "to improve both the quality of professional work and the credibility of geoscientists, to foster excellence in geosciences, to assure sustainable benefits for communities, as well as to protect local and global environments; all with the aim of creating and maintaining the conditions for the healthy and prosperous development of future generations".

Taking the "Cape Town Statement on Geoethics" into consideration for the emergency implementation of the CFA Groundwater Scheme, the following question is raised: do you as a geoscientist undertake to develop an emergency scheme with limited data on its sustainability and ultimate costing versus not developing it in the eye of an emergency but rather risking a City of ~4 million people running out of water and the knock-on effects of that?

The decision was taken to implement the scheme with a no regrets approach to reduce the risk of any negative impacts to the natural environment, including seawater intrusion, sensitive ecosystems and existing users, such as the Philippi Horticultural Area (PHA) that relies solely on groundwater for irrigation. To achieve this, managed aquifer recharge (MAR) formed an essential component of the scheme where it aims to maintain water levels, increase productivity and improve water quality within the aquifer. Furthermore, a conservative approach was adopted for the operation of the scheme whereby operating water levels in abstraction boreholes near to the coast will be maintained above sea level and operating water levels in MAR injection boreholes will be kept well below ground level to reduce the risk of flooding.

By putting all of this in place and operating the scheme with a monitor, model and manage approach, the team felt confident enough that no major environmental catastrophes would take place and the City's water supply system would become more resilient. The downside to this decision is that as the project is re-structure based on gathered results, there is a risk that some of the costs become sunken and operating costs rise due to lower yields and poorer quality than originally expected.

However, this was felt to outweigh the cost of ~4 million people running out of water and the knock-on effects of this such as disease epidemics, rioting and ultimately people relocating from the City. Additionally, the project has resulted in unknown contamination sources been identified and rehabilitated which will improve ecological functioning of the aquifer, wetlands and the ocean that the aquifer discharges into.

**Acknowledgements** Our thanks to The City of Cape Town, Water and Sanitation Department, Bulk Water Branch.

## References

City of Cape Town (2017) Five-year integrated development plan 2017. https://www.csag.uct.ac.za/current-seasons-rainfall-in-cape-town/

Department of Water Affairs and Forestry (2008) The assessment of water availability in the Berg Catchment (WMA 19) by means of water resource related models: groundwater model report, vol 5. Cape Flats Aquifer model. Prepared by Umvoto Africa (Pty) Ltd in association with Ninham Shand (Pty) Ltd on behalf of the Directorate: National Water Resource Planning. DWAF report no. P WMA 19/000/00/0408

Department of Water and Sanitation (2014) Support to the continuation of the water reconciliation strategy for the western Cape Water Supply System—Cape Flats aquifer: situation assessment and potential for successful urban groundwater development and management. Prepared by Umvoto Africa (Pty) for the Department of Water Affairs, South Africa

Department of Water and Sanitation (2015) Support to the continuation of the water reconciliation strategy for the Western Cape Water Supply System—Cape Flats aquifer management strategy. Prepared by Umvoto Africa (Pty) for the Department of Water and Sanitation, South Africa

Department of Water and Sanitation (2016) Support to the continuation of the water reconciliation strategy for the western Cape Water Supply system: status report April 2016. Prepared by Umvoto Africa (Pty) in association with WorleyParsons RSA on behalf of Directorate: National Water Resource Planning

Di Capua G, Peppoloni S, Bobrowsky PT (2017) The Cape Town statement on geoethics. Ann Geophys 60, Fast Track 7. https://doi.org/10.4401/ag-7553

Hay ER (1981) A stratigraphic and sedimentological analysis of borehole data from a portion of the Cape Flats. Honours thesis

Henzen MR (1973) Die Herwinning, Opberging En Onttrekking Van Gesuiwerde Rioolwater In Die Kaapse Skiereiland. Report by the National Institute for Water Research, Council for Scientific and Industrial Research

Parsons R (2017) Op-Ed: Cape Town's water future is bleak—it's time to call in the groundwater experts. Daily Maverick newspaper. Retrieved 23 Nov 2017, from Daily Maverick: https://www.dailymaverick.co.za/article/2017-08-14-op-ed-cape-towns-water-future-is-bleak-its-time-to-call-in-the-groundwater-experts

Roberts DL, Botha GA, Maud RR, Pether J (2006) Coastal Cenozoic deposits. In: Johnson MR, Anhaeusser CR, Thomas RJ (eds) The geology of South Africa. Geological Society of South Africa, Johannesburg/Council for Geoscience, Pretoria, pp 605–628

Wessels WPJ, Greef GJ (1980) Ondersoek na die optimale benutting van Eersterivierwater deur opberging in sandafsettings of ander metodes. Department of Civil Engineering, University of Stellenbosch, Stellenbosch

# The Ethics of Groundwater Governance: Evaluating a Methodology in Philippi in Cape Town

Leanne Seeliger

**Abstract**

When South African policymakers compiled the National Water Act and started implementing Integrated Water Resources Management, they assumed that social cohesion and shared values existed in the divided communities that characterized the post-apartheid South Africa. Moreover, they naively anticipated that both governments and communities were able to engage effectively in transdisciplinary, multi-stakeholder platforms on water governance. Water governance is a normative concept that implies a shift from government being in control of water management to it being a shared responsibility between diverse stakeholders in a context. Value analysis can play a critical role in understanding the multitude of conflicts that arise among the different stakeholders in water governance. It can create platforms for meaningful engagement that could lead to the resolution of a conflict. This work evaluates the impact of an ethics methodology in South Africa, discussing how the value-driven process was able to achieve improved social cohesion within the divided community in the Philippi Horticultural Area (PHA). It compares this methodology with others arguing that its persistent focus on values and process, rather than content and outcome makes it more useful than other forms of participatory action research to hydrogeologists wanting to involve ethics in contested water governance issues.

**Keywords**

Ethics • Values • Water governance • Participatory action research • Integrated water resource management

L. Seeliger (✉)
Stellenbosch University Water Institute, Unit for Environmental Ethics, Stellenbosch, South Africa
e-mail: seeliger@sun.ac.za

## 1 Introduction

The purpose of an ethics methodology is to facilitate a process whereby stakeholders can jointly articulate their values and prioritize them to improve decisions around the governance of water in a context. The ethics methodology in the PHA centered around unlocking what the core values of stakeholders were through administering a questionnaire to individual stakeholders and then developing a set of communal values and a set of tangible indicators of those values. The social impact of this research process was profound and innovative, leading to the creation of several additional proposals that were subsequently supported by funders and government. This work seeks to explore how this ethics methodology compares with other participative action research methodologies and how it can assist practitioners to better understand and resolve water governance issues in groundwater conflicts.

## 2 Methods

The questionnaire asked a total of 34 individual stakeholders' factual questions about the case study to immerse the researcher in the contested water governance context of the PHA. It also asked the same stakeholders value-based questions about water governance in the PHA. The first workshop involved moving from the individual stakeholders' values to setting up a group problem identification, joint causes of the problems in the PHA, joint alternatives and joint solutions from the 34 people interviewed.

Workshop two involved identifying the joint values of the PHA stakeholders. These were identified as unity, sustainability and participation. The rules of engagement were then created to allow for respectful engagement. Moreover, it involved creating a draft set of proposed actions, in the immediate future, the medium term and eventually the long term.



M. Abrunhosa et al. (eds.), *Advances in Geoethics and Groundwater Management: Theory and Practice for a Sustainable Development*, Advances in Science, Technology & Innovation,
https://doi.org/10.1007/978-3-030-59320-9_92

Workshop three involved teasing out of a set of tangible indicators of success that would demonstrate that the values that had been agreed upon had been followed. Additional workshops followed where development proposals were proposed and fleshed out on the basis of the values given and indicators provided.

In the case of the PHA, a PHA Partnership Improvement Forum was established, a draft Constitution was drawn up and two proposals submitted for funding. A second questionnaire was then administered to evaluate the impact of the ethics process on improved water governance by asking the stakeholders some of the same factual questions again to see if they had altered their perspectives and secondly to ask them to evaluate the process and its impact on their values. The two key questions that were asked were: Has your interaction with other viewpoints in this research changed the way in which you understand water management in the PHA? Are you more or less positive about achieving improved water management in the Philippi Area?

## 3 Results

This research shows that part of the reason for the inability of people to articulate their core shared values is because stakeholders (retailers, farmers, business, ecologists, officials, consultants) have to work out what these principles mean for them in the context in which they find themselves. How does a sensitivity to the fact that water is a commodity essential for vegetable production equate with the human right for all human beings to have access to water even when they cannot pay for it. How does the necessity of maintaining the ecosystem health of the PHA mean for individual stakeholders in their daily lives and in their interaction with others. Who should be entitled to how much for how long? Are farmers' water rights just and fair?

The context in which water is shared is perceived differently by diverse stakeholders depending on their occupations, needs and histories. National policy and legislation is interpreted by officials, activists, ecologists, water scientists, farmers and informal settlement dwellers, on the ground, out of their socio-economic contexts. This research has demonstrated that there is a need to assist with the interpretation of these values at an individual and community level to implement water policy and legislation in an effective and just manner.

The chasm between policy implementation and community/local government uptake is created by two faulty assumptions: firstly, that various government departments do co-operate effectively to implement national water legislation and policy. Secondly, that communities have social cohesion and capacity to be able to work out how national policy and legislation on water should be interpreted to ensure that everyone's needs are respected and met in their various contexts (Republic of South Africa 1998).

The interviews with stakeholders revealed that both government departments and communities were not able to effectively engage in multi-stakeholder platforms on water management within contested contexts where diverse values were being prioritized and negotiated.

While the interviews with 34 stakeholders in the PHA, that included various government departments, activists and scientists living and operating in the PHA, showed that there was more convergence than divergence of values and concerns among all the stakeholders, they were not able to see this by themselves, and they were "stuck" in historical battles and largely swamped by the after effects of urbanization, crime and unemployment. Despite the extensive normative framework provided by the Integrated Water Resources Management, the National Water Act, the National Water Strategies or a Socio-Economic Study completed by the Western Cape Government before, the stakeholders were seldom able to internalize and apply these principles in a community context.

It is here where the ethics research methodology, which was piloted in the PHA, was useful. It developed a process that practitioners could use to assist local stakeholders to articulate their value frameworks and jointly debate them with other stakeholders, including government officials, who are tasked with implementing national policy and legislation norms and values.

The ethics methodology research process created a platform for engagement around values. It allowed stakeholders to air their concerns and grievances and engage with their values, with assistance to translate them into concrete development proposals on the ground. This occurred with the contextual principles of unity, sustainability and participation being identified in the second workshop and translated into tangible indictors in the third workshop. This later lead to the development of two further research/development proposals with the assistance of the researcher.

## 4 Discussion

An ethics methodology is likely to be most useful to water governance practitioners working in complex water governance issues where there are multiple stakeholders with longstanding, conflicting views on what the most appropriate course of action is. By including an ethics methodology in field work, geohydrologists, for example, will be able to ensure that the principles of the geoethical code of conduct (Peppoloni et al. 2019) are adhered to their daily practice.

The methodology engages stakeholders and creates social cohesion by identifying points of convergence and value alignment in conflicting water governance situations. It is a useful tool for water policy implementation, as it aims for consensus building. It unlocks the values that inform the decision-makers choices both as individual stakeholders and then as a group in the formation of a shared or consensual viewpoint among stakeholder in a water governance context.

The ethics methodology is a form of participatory action research (PAR) that involves the collection of data to take action and bring about change. It focuses on getting local people to participate in analyzing their own solutions. PAR methodologies in general create shared truth and joint social action by immersing itself in the contexts of divergent people's lives and creating a cyclic process of research, reflection, action to address specific issues (MacDonald 2012).

The ethics methodology in this research, however, differs from traditional PAR in that it specifically asked stakeholders to stay focused on values and the formation of joint values and the creation of indicators that would track the development of those values. It did not leave the process entirely up to stakeholders. The researcher was specifically focused on eliciting the values behind decisions and the values that would guide future decisions.

## 5 Concluding Remarks

The governance of aquifers is often by nature contested because of the large number of widely spread small stakeholders involved in the management of their own private boreholes and springs. Unlike surface water where the impacts of divergent users are visible, in aquifers there is often a time delay in the visible impact of unregulated water use (Custodio 2000).

Geohydrologists asked to consult in contested water governance contests, like in the Cape Flats Aquifer in the Philippi Horticultural Area, are often at a loss about how to go about engaging with communities on ethical issues around water resource management. The research methodology documented in this study not only pilots an example of what a hydrogeoethical process might look like, but it also demonstrates the success of it in improving water governance.

The social impact of the ethics research methodology piloted on the Cape Flats Aquifer was far-reaching in that it led to the formation of PHA People's Forum that united diverse water users in water governance in the PHA who previously were not engaging with one another. Emerging farmers and established commercial farmers as well as informal settlement dwellers and activists engaged actively and assisted the researcher with the procuring of additional funding to engage in a citizen science project that will encourage them to monitor their impact on water resources.

Collaborative water governance was initiated with multiple levels of government departments committing to work with the PHA People's Forum to clean the polluted channels on the Cape Flats Aquifer. The ethics methodology created the platform from which collaborative water governance could be built. This work argues this was partly due to the focus on values in the methodology and the decision to guide stakeholders to remain focused on this, rather than their various agendas.

**Acknowledgements** I acknowledge the support of the Water Research Commission in South Africa and the Stellenbosch University Water Institute that made this research possible.

## References

Custodio E (2000) Some relevant ethical issues in relation to freshwater resources and groundwater. Bol Geol Min Madrid 111(6):121–130

Macdonald C (2012) Understanding participatory action research: a qualitative research methodology option. Canad J Act Res 13 (2):34–50

Peppoloni S, Bilham N, Di Capua G (2019) Contemporary geoethics within geosciences. In: Martin B (ed) Exploring geoethics: ethical implications, societal contexts, and professional obligations of the geosciences. Palgrave Pivot, Cham, pp 25–70

Republic of South Africa (1998) National Water Act 36 of 1998. Available at www.gov.za. Last accessed 22 Feb 2020

# Decision-Making in Groundwater Management: Where Artificial Intelligence Can Really Lead Geoscientists?

Rodrigo Lilla Manzione and Mariana Matulovic

**Abstract**

In this modern and dynamic society, threatened by climate change, poverty, hungry and economical systems collapse, artificial intelligence (AI) emerged as a promise field to solve many actual problems. Although AI does not give absolute answers. The outputs of AI methods are subjective and in many situations depend on human-based decisions. It has a strong impact on decision-making processes and geoscientists are highly exposed to this question. Specifically, on groundwater, issues involving water quality and water quantity deserve special attention for monetary resources applications, urban supply, ecosystemical services should be balanced in order to avoid biased solutions. This paper aims to present some AI methods and discuss where it they can lead geoscientists with and without an ethical posture. A study case using monitoring water levels data is presented.

**Keywords**

Data analysis • Geosciences • Geoethics • Algorithmic responsibility

## 1 Introduction

One definition of artificial intelligence (AI) is: "*The theory and development of computer systems able to perform tasks normally requiring human intelligence, such as visual perception, speech recognition, decision-making and translation between languages*" (Oxford 2019). In this way, computers could be used to simulate certain important aspects of intelligence. In the 60s, artificial intelligence was oversold because the idea of general artificial intelligence problem solving was found to be much more difficult than originally suggested. Scientists had difficulty getting computers to solve problems that were routinely solved by human experts. Therefore, scientists instead started to investigate the development of artificial intelligence applications in micro-worlds, or very narrow topical areas, leading to the creation of the first useful artificial intelligence systems for select applications, e.g. games, disease diagnosis, spectrograph analysis (Jensen 2015).

Expert systems represent the expert's domain (i.e. subject matter) knowledge base as data and rules within the computer. The rules and data can be called upon when needed to solve problems. A completely different problem within the domain of the knowledge base can be solved without having to reprogram. In geosciences, the complexity of the systems in study makes extremely difficult to impose the same rules at different locations, making AI methods remaining in the shadows for many applications.

Rules specified by a human expert knowledge are established from the hypotheses to be tested and the rules (variables) determined by the expert. Note that the hypothesis represents the base of the decision tree. The expert identifies specific conditions that are associated with the target of the decision tree. The hypothesis, rules and conditions created by the expert are then passed to the inference engine, which processed the rules and conditions in conjunction with the required data producing a classification output. In general, geoscientists must comprehend the information and experiences and turn it into knowledge for it to be useful, considering that many human beings have trouble interpreting and understanding the information in images, books, articles, manuals and periodicals, although some others do not obtain much knowledge from field work. It would be much better if there were some way for the expert system to generate its own rules based on training data, leading us to machine learning.

R. L. Manzione (✉) · M. Matulovic
Biosystems Engineering Department, School of Sciences and Engineering (DEB/FCE), São Paulo State University (UNESP), Tupã, SP, Brazil
e-mail: lilla.manzione@unesp.br



M. Abrunhosa et al. (eds.), *Advances in Geoethics and Groundwater Management: Theory and Practice for a Sustainable Development*, Advances in Science, Technology & Innovation,
https://doi.org/10.1007/978-3-030-59320-9_93

The heart of an expert system is its knowledge base. The usual method of acquiring knowledge in a computer-usable format to build a knowledge base involves human domain experts and knowledge engineers. The human domain expert explicitly expresses his or her knowledge about a subject in a language that can be understood by the knowledge engineer. The knowledge engineer translates the domain knowledge into a computer-usable format and stores it in the knowledge base. Machine learning is defined as the science of computer modelling of learning processes. It enables a computer to acquire knowledge from existing data or theories using certain inference strategies such as induction or deduction. In machine learning, the process of inductive learning can be viewed as a heuristic search through a space of symbolic descriptions for plausible general descriptions, or concepts, that explain the input training data and are useful for predicting new data. Machine-learning decision trees can generate categorical output information while regression trees can predict continuous quantitative output. In these procedures are applied a new thinking in geosciences: let the geographical data themselves "have a stronger voice" rather than let statistics derived from the dataset dictate the analysis, as means and covariance matrices used in maximum likelihood classifications (Jensen 2015). Some other machine-learning algorithms make use of multiple decision trees at one time during the classification process. These so-called ensemble learning algorithms (e.g. random forests, bagging, boosting) have emerged as potentially more accurate and robust alternatives to conventional parametric (e.g. maximum likelihood) or other machine-learning algorithms (e.g. individual decision trees, neural networks).

The aim of this work was to analyse groundwater recharge data and related variables using an AI method and discuss how important are ethical choices in the decision-making processes in geosciences. For this purpose, we used fuzzy logic to select favourable areas for groundwater recharge from related variables sampled from nearby monitoring wells in a specific watershed, checking if the results could be extrapolated to other watersheds inside the study area.

## 2 Materials and Methods

### 2.1 Santa Barbara Ecological Station Groundwater Monitoring Network

The Santa Bárbara Ecological Station (EEcSB) is a conservation area located at Rodovia SP 261-km 58, coordinates 22° 48′ 59″ S and 49° 14′ 12″ W, in the municipality of Águas de Santa Bárbara/SP, Brazil (Fig. 1). The EEcSB has an area of 4,371 ha within the limits of the Santa Bárbara State Forest, of which 2,712 ha of native vegetation (cerrado, marshes and gallery forest) dividing the space with the reforestation with pine and eucalyptus trees. For water table monitoring, 55 wells were distributed in the Guarantã (14 wells), Bugre (13 wells), Santana (12 wells), Boi (9 wells) and Passarinho (5 wells) sub-basins were the levels are monitored on a monthly frequency (Fig. 2). From these, data was calculated the groundwater recharge for 2015 using water table fluctuation (WTF) method as described in Healy and Scanlon (2010) and the amplitude ($\Delta H$) of groundwater level. Additional data was collected in the surrounding areas of the monitoring wells and from satellite images, regarding to soil physics characteristics, relief and land use and management. Variables related with groundwater recharge were then selected to create an inference model for groundwater recharge. Hydraulic conductivity ($K$), slope, evapotranspiration (ET) and soil resistance to penetration (RP) variables were selected and reclassified as more and less favourable areas to groundwater recharge, following the established limits presented on Table 1. Details of this dataset can be found at Santarosa and Manzione (2018).

**Table 1** Fuzzy values for each input variables and the output value for groundwater recharge at each monitoring well of Boi watershed

| Well ID | slope | ET | $K$ | RP | $\Delta H$ | Fuzzy recharge (mm) | WTF recharge (mm) |
|---|---|---|---|---|---|---|---|
| Boi_D | 3 | 1 | 6 | 2 | 1 | 491.00 | 486.40 |
| Boi_1 | 1 | 1 | 6 | 2 | 1 | 295.00 | 284.60 |
| Boi_2 | 2 | 2 | 7 | 2 | 1 | 192.00 | 227.70 |
| Boi_3 | 1 | 2 | 7 | 3 | 1 | 85.70 | 94.40 |
| Boi_4 | 1 | 2 | 6 | 2 | 1 | 254.00 | 276.90 |
| Boi_5 | 2 | 2 | 6 | 2 | 1 | 222.00 | 215.50 |
| Boi_6 | 2 | 1 | 6 | 3 | 1 | 15.40 | 77.04 |
| Boi_7 | 2 | 1 | 7 | 1 | 1 | 192.00 | 190.08 |
| Boi_8 | 3 | 2 | 6 | 1 | 1 | 232.00 | 222.00 |

## 2.2 The Fuzzy System for Groundwater Recharge

Intelligent models developed for decision-making for the most diverse areas of knowledge carry with them a unique algorithmic responsibility, especially those based on concepts that work with inaccurate data, such as fuzzy set theory. Developed by Zadeh (1965), fuzzy sets and logic made possible an alternative (in terms of the standard classical sets) and formal treatment for a data class whose truth condition is defined over a numerical range corresponding to different degrees of truth.

The Mamdani's Fuzzy Inference System (MFIS) is the most commonly fuzzy methodology (Mamdani and Assilian 1975). Fuzzy sets theory has been applied to groundwater quality and quantity issues in recent studies. Mohamed et al. (2018) assessed groundwater quality unconsolidated sedimentary basin in China using fuzzy techniques. Duhalde et al. (2018) performed a fuzzy-based assessment of groundwater intrinsic vulnerability in Chile. Ghazavi et al. (2018) selected sites for artificial recharge wells in an urban area in Iran using fuzzy logic techniques. Nadiri et al. (2019) applied fuzzy logic to model groundwater level variations in East Azerbaijan province, Iran. The fuzzy logic used in this paper is characterized by being a system for representing and inferring inaccurate information for the purpose of making decisions through fuzzy set theory, based on the following steps: (i) Fuzzyfication: stage by which system variables are modelled by fuzzy sets, obtained from the variables, and converted into pertinence functions; (ii) Rule base: from the input data as well as the expert's expertise, the fuzzy controller syntactic core is elaborated; (iii) Fuzzy inferences: the rules are mathematically modulated through the techniques of fuzzy logic; and (iv) Defuzzification: reverse process to fuzzyfication, that is, it allows us to represent the fuzzy set by a real value. All calculation was performed using the Mathworks software tool MATLAB with the Fuzzy Logic Toolbox.

## 3 Results

The MFIS approach has shown to be a useful and practical tool to assess the intrinsic recharge of an aquifer under scarse data conditions. Table 1 presents the values for each variable at Boi watershed. Figure 3 shows the generalized fuzzy system established for groundwater recharge and the pertinence function associated with the output groundwater recharge.

The groundwater recharge was divided in six classes as very bad (from 0 to 80 mm/year), bad (from 80 to

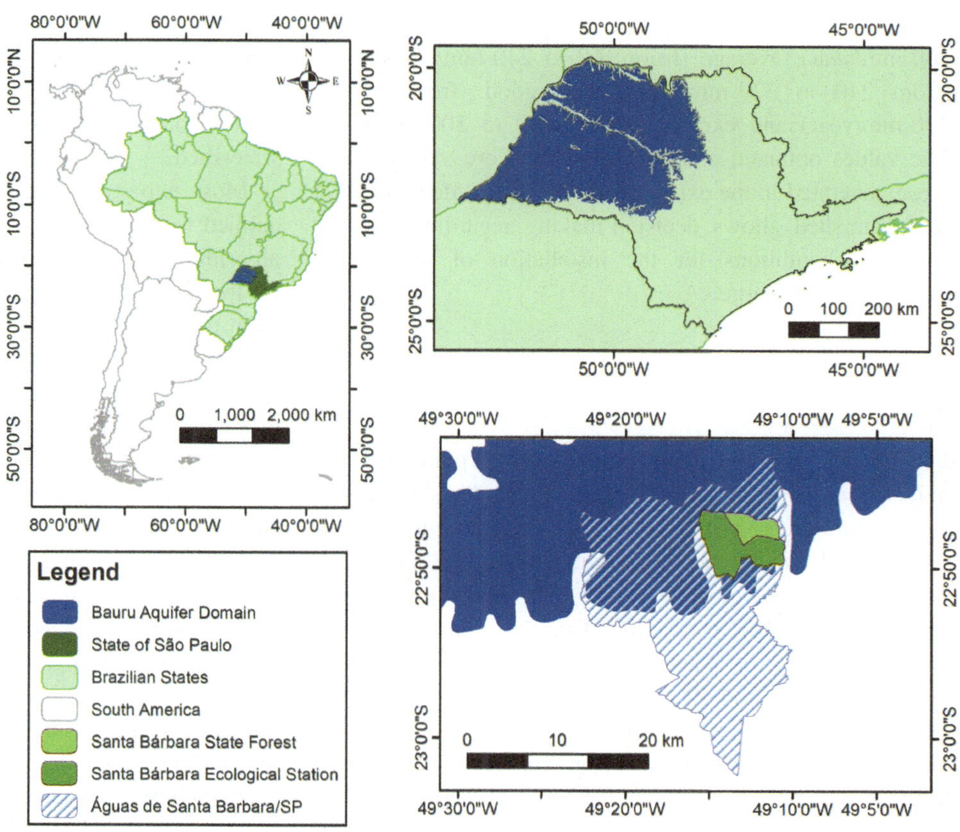


**Fig. 1** Location of the study area, Águas de Santa Barbara/SP-Brazil

Fig. 2 Details of the studied watersheds

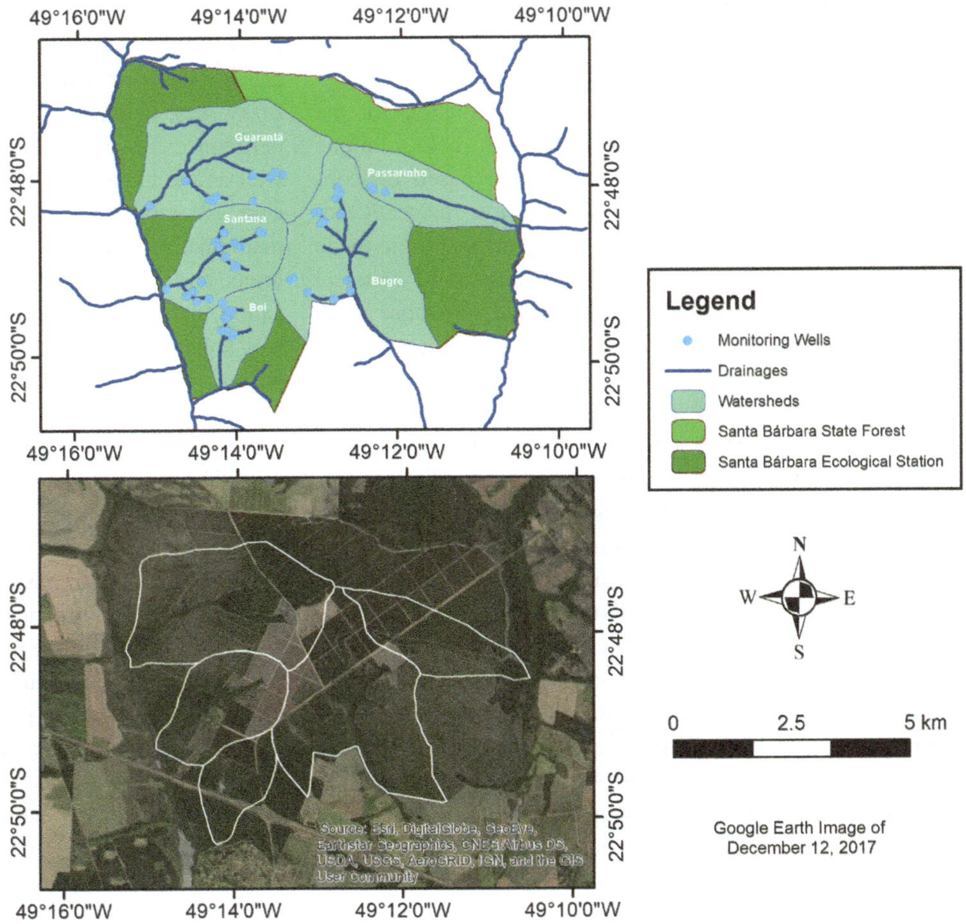


160 mm/year), average (from 160 to 240 mm/year), good (from 240 to 320 mm/year), very good (from 320 to 400 mm/year) and excellent (from 400 to 500 mm/year). The values obtained using WTF method are very close to those generated in the experiment. The model developed for Boi watershed allows decision-making regarding the best area and conditions for the installation of a well for groundwater recharge.

## 4 Discussion

Fuzzy logic has as one of its pillars the incorporation of the human way of thinking in a control and decision-making system, and the controller used in this study is an example of a model that contemplates field research and expert knowledge. However, this control only works for the data inherent to the Boi watershed. The same device, run on another dataset or identical dataset, but for different watershed inside the EEcSB shows different results than expected. The Pearson correlation coefficient calculated between WTF and MFIS predictions for Boi watershed was 0.99. On the other hand, using the same MFIS for the other four watersheds of EEcSB, the correlation coefficient was 0.72 for Passarinho watershed, 0.16 for Bugre watershed, −0.10 for Santana watershed and 0.16 for Guarantã watershed.

Most expert systems are shaped by rules that, when applied to a dataset, predict a result. In this case, we have algorithms whose protagonist is centred on the rules and not on the data itself. In reality, the models are not generated from patterns and groupings evidenced in the data analysis, but from guidelines programmed by an agent, whether ethical or not. The rules we use in this study are all built on the academic and life skills of an expert. But what would happen, if there are misconceptions in the analysis of data, considered relevant by this expert, for the purpose of this study? What if the expert acted in bad faith towards the researcher? And what guarantee do we have that the rules actually preserve the underlying theory? Accountability is characterized by the legal and ethical obligation, underlying an individual or organization, of accountability regarding their activities, the acceptability of their responsibility for them and the disclosure of results in a transparent and ethical manner (Diakopoulos 2016). In terms of algorithmic accountability, the same definition applies.

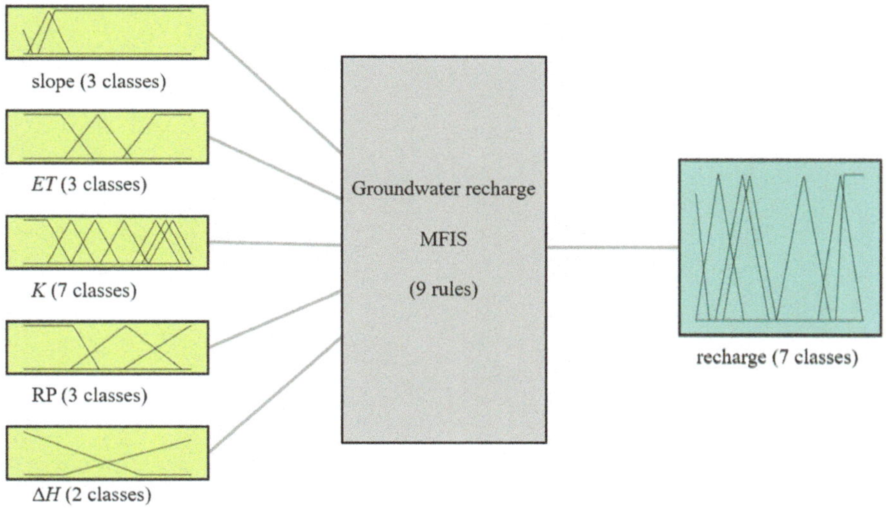


**Fig. 3** Generalized fuzzy system for groundwater recharge

Fox (1996) and Diakopoulos (2016) argue that although the essence of the algorithm power lies in its autonomy, human influences on them are still present (intensely) mainly because of the criteria determined for the selection of data, the choice of optimization functions, of categorical semantics, among others. By prioritizing one information over another, discriminatory trends may underlie the model. Algorithms can carry bias from the modelling agent or according to company operating policy. Geoscientists should be aware of their responsibilities when using AI methods and the implications of their choices. But as Morone and Peppoloni (2017) pointed out about ethical dilemmas in geosciences: we can ask, but can we answer? Following ethical codes (golden rules) and so on are key aspects that geoscientists should be aware for best serve our society.

## 5 Concluding Remarks

From the results obtained and the discussions presented, it is believed that ethical responsibility should be global and distributed, since each agent or organization chooses a path over others, according to the values intrinsic to each one of them. It needs to be global because of the breadth that moral issues can have because of the insertion of informational technologies in the conduct of human agents and distributed among all actors in the chain.

**Acknowledgements** This work was supported by the São Paulo State Foundation for Scientific Research (FAPESP—grants number 2014/04524-7 and 2016/09737-4) and the Brazilian National Council for Scientific and Technological Development (CNPq—grant number 421782/2016-1).

## References

Diakopoulos N (2016) Accountability in algorithmic decision making. Commun ACM 59(2):56–62

Duhalde DJ, Arumí JL, Oyarzún RA, Rivera DA (2018) Fuzzy-based assessment of groundwater intrinsic vulnerability of a volcanic aquifer in the Chilean Andean Valley. Environ Monit Assess 190:390

Fox J (1996) Expert systems and theories of knowledge. Artificial intelligence. Academic Press, Cambridge, pp 157–181

Ghazavi R, Babaei S, Erfanian M (2018) Recharge wells site selection for artificial groundwater recharge in an urban area using fuzzy logic technique. Water Resour Manage 32:3821

Healy R, Scanlon B (2010) Estimating groundwater recharge. Cambridge University Press, Cambridge

Jensen R (2015) Introductory digital image processing: a remote sensing perspective, 4th edn. Prentice Hall Press, Upper Saddle River

Mamdani EH, Assilian S (1975) An experiment in linguistic synthesis with a fuzzy logic controller. Int J Man Mach Stud 7(1):1–13

Marone E, Peppoloni S (2017) Ethical dilemmas in geosciences. We can ask, but, can we answer? Ann Geophys 60, Fast Track 7. https://doi.org/10.4401/AG-7445

Mohamed AK, Liu D, Mohamed MAA, Song K (2018) Groundwater quality assessment of the quaternary unconsolidated sedimentary basin near the Pi river using fuzzy evaluation technique. Appl Water Sci 8:65

Nadiri AA, Naderi K, Khatibi R, Gharekhani M (2019) Modelling groundwater level variations by learning from multiple models using fuzzy logic. Hydrolog Sci J 64(2):210–226

Oxford (2019) Definition of Artificial Intelligence. Oxford Dictionaries. Lexico, London

Santarosa LV, Manzione RL (2018) Soil variables as auxiliary information in spatial prediction of shallow water table levels for estimating recovered water volume. RBRH 23:e24

Zadeh LA (1965) Fuzzy sets. Inf Control 8(3):338–353

# The Geoethics of Using Geospatial Big Data in Water Governance

César de Oliveira Ferreira Silva, Mariana Matulovic, and Rodrigo Lilla Manzione

**Abstract**

Geoethics encourage us to reflect dialectically on the consequences, opportunities, risks and benefits of our actions when using geotechnologies. This paper presents insights into how geoethics can guide more conscious and transparent decisions in the use of geospatial data in water management. The concepts of microethics and macroethics are also presented in the context of geoethics. Water governance must provide water security in terms of quality and quantity for all citizens, ensuring that everyone receives water (equity) with transportation methods that avoid losses (efficiency), maintain quality (responsibility) with forms of monitoring and control that equalize the freedom (autonomy) and power (representativeness) of all agents involved. Following the principles of ensuring autonomy, equity, responsibility, efficiency and representativeness, the use of geospatial data must be made in order to achieve these objectives. For this, geoethics has a crucial role in providing guidelines to decision makers and society for guiding them in an inclusive, equitable and transparent way. The self (microethics) and societal (macroethics) responsibility and geoethical implication is distributed to all the components that constitute the complex system to which they are inserted, in different degrees of importance in decision-making.

**Keywords**

Geoethics • Guidelines • Risk • Allocation • Social

C. de O. F. Silva (✉)
School of Agriculture (FCA), São Paulo State University (UNESP), Botucatu, SP, Brazil
e-mail: cesaroliveira.f.silva@gmail.com

M. Matulovic · R. L. Manzione
Biosystems Engineering Department, School of Sciences and Engineering (DEB/FCE), São Paulo State University (UNESP), Tupã, SP, Brazil

## 1 Introduction

Geospatial data analysis is a quali-quantitative approach that focuses on the detection of patterns, their correlations and possible modelling in terms of geographical location. Geospatial Big Data are sets of patterns and correlations of data that go beyond human manipulation capacity, requiring highly complex computational systems and memory for its execution and analysis. Among Geospatial Technologies 4.0, as we call the technologies used to acquire, store, manipulate (model) and visualize this type of data, we highlight the Internet of things (IoT) and remote sensing. Limkar and Jha (2019) consider that most of the Big Data currently available is under the geospatial scope.

In the list of fundamental values set out in the Cape Town Statement on Geoethics (Di Capua et al. 2017), which require: honesty, integrity, transparency and reliability of the geoscientist, including strict adherence to scientific methods, it is clear that the concept of geoethics is not about a new discipline, it is a practical approach for promote new ethical basis for scientific studies in geosciences.

According to Peppoloni et al. (2019), geoscientists have grounded the concept of geoethics based on (i) concern with the individual responsibility of the self and profession in conducting actions in geosciences and (ii) analysing the context and making decisions in the social sphere on issues inherent to geoscience.

Ethics can be analysed in macro and microviews. When we talk about the microethics, the need for an awareness is based on necessary personal values (for personal and professional life). In the other hand, conducting consistent decision-making with usually established ethical conduct, such as the question of the programmer's algorithmic responsibility, is evidenced facing the demands of society and the researcher him/herself with the product of his research. Macroethics is centred on analysing the implications of decision-making in terms of society and planet Earth, in order to have an ethical responsibility included in



M. Abrunhosa et al. (eds.), *Advances in Geoethics and Groundwater Management: Theory and Practice for a Sustainable Development*, Advances in Science, Technology & Innovation,
https://doi.org/10.1007/978-3-030-59320-9_94

the entire research process, guaranteeing contributions to society (Mogk and Bruckner 2020).

Dealing with the challenge of provide water to different communities is a combination of management and development procedures to secure the safety of water resources. Identifying interconnected interactions add complexity to the water resources management. The concept of water governance aimed to analyse and assess the political, social, economic and legislative means to develop and manage water resources and services, delivered at different society levels (Rogers and Hall 2003). Water governance not only must address the natural dimension of water resources but also consider the anthropological aspect (Zolghadr-Asli et al. 2017).

Walker et al. (2015) suggest that environmental and water governance decision makers need to consider the social responses as well as the economic and environmental impacts of our decisions. These goals can be reached by including members of society (or at least their behaviour) in the policymaking process or making the policy adaptive considering monitoring and learning. In other words, policymaking must be able to be adapted in the complex reality that we are living.

The United Nations (UN) 2030 Agenda proposes 17 Sustainable Development Goals (SDGs) and 169 corresponding targets. SDG 6, or Sustainable Development Goal 6, consists of eight targets that aim to ensure the availability and sustainable management of water and sanitation for all.

This paper aimed to present insights into how geoethics can guide more conscious and transparent decisions in the use of geospatial data in water management.

## 2 Geoethics in Water Governance

An example of geoethical guidelines adopted by a community is compiled by the Agenda del Agua Cochabamba (AdA). According to Bellaubi and Bustamante (2018), this agenda sets the basic principles for water management under the scope of the autonomy, equity and responsibility. We use there three basic principles to start our discussion.

When it comes to autonomy, this concept differs from decentralization, delegation or devolution. Here we understand it as the recognition of the relationship between the communities and the water, which determines, within the interrelations, the laws to which it submits, considering its social, political, economic and survival reality as a specie. Mogk and Bruckner (2020) state that provide training is required to allow the community to recognize, prevent and mitigate ethical issues and dilemmas.

Water is a common living being. Therefore, equity refers to a fair relationship of all humans with water, sharing it in an equitable way of distributing the benefits and costs of this relationship under the solidarity principle. In a more general proposal, Peña (2011) list as equity issues in water management (1) nominal or practical absence of a normative framework, (2) asymmetry problems, (3) corruption and lack of transparency in procedures, (4) discrimination, and (5) problems of collective action and agency. As guidelines for equity implementation, Peña (2011) suggests (1) guarding basic needs and minimum requirements, (2) initial water allocation between different sectors and users, (3) distribution of benefits in geographical space, (4) re-allocation between different users and sectors.

When we talk about responsibility with water, the match point is to provide efficient and effective conditions for water use based on real needs and availability through service of multiple uses, recognizing the right of present and future generations. Jonas (1985) formulated an ethical principle based on responsibility to future generations, declaring as an imperative the responsibility for all human beings (present and future) as an end in itself.

Now we add two new principles, the efficiency and representativeness. Efficiency is needed because it is necessary to ensure that there are means to reach water to everyone. Although it may seem implicit to the concept of equity, it is not enough that there is water availability, but also the resources of engineering and administration should minimize losses in order to distribute water to citizens. The Target 6.4 from SDG 6 is explicit: by 2030, substantially increase water use efficiency across all sectors (ANA 2019). Representativeness is required in order to know which the requirements are to be met (to guarantee equity), being necessary to have means of communication for the abstractions existing in the models correspond to the reality of the communities served. The Target 6.B from SDG 6 highlights that: support and strengthen the participation of local communities in improving water and sanitation management (ANA 2019).

Water governance must provide water security in terms of quality and quantity for all citizens, making it necessary to ensure that everyone receives water (equity) with transportation methods that avoid losses (efficiency), maintain quality (responsibility) with forms of monitoring and control that balance the freedom (autonomy) and power (representativeness) of all agents involved. Therefore, these are the principles that guide the geoethical decisions inherent to water governance.

According to the Aristotelian conception of equity, we have to consider, understand, map and assume the hydrographic differences in the world to design actions in water governance that respect life in the context where it is inserted. Therefore, we conjecture that the ethical process related to decision-making in water governance should be based on the following scheme: (i) identification of the water problem/issue; (ii) fact checking; (iii) identification of factors effectively relevant to the process and resolution of the problem; (iv) development of a list of options for geotechnologies and processes for data modelling, considering the impartiality of the model in terms of the programmer's values and the subjectivity of the available methods; (v) testing the options and confronting the data through the analysis of a specialist; (vi) decision-making based on the aforementioned geoethical principles; (vii) act.

## 3 Geoethical Roadmap for Geospatial Data Assessment

Within the context of autonomy, equity, responsibility, efficiency and representativeness, geospatial data need to be analysed in order to avoid potential risks.

Crampton (1995) proposes an interesting four-stage sequence of ethical practice for the geographic information systems (GIS) environment, as following:

a. Ignoring ethics (or rather being unaware of ethical issues);
b. Considering ethics from an internal perspective only;
c. Considering ethics from both an internal and an external perspective; and
d. Establishing a dialectical relationship, which modifies both internal and external perspectives.

Crampton (1995) considers that GIS has achieved the second stage practice and the challenge is to make its users aware that the ethics of GIS is not suited to a rigid code of ethics. In other words, the geoethics of geospatial analysis is constructed by a series of principles that guide the researcher, in order to continuously improve his/her scientific conduct, as suggested in the Crampton's fourth stage. There is also a list of principles that can guide geoethical thinking with respect to spatial data, specifically in the context of water governance, based and adapted from UNICEF (2018):

- Negative political associations with visible data capture devices;
- Unintended or unknown surveillance;
- Privacy and safety risks related to transmission, sharing and storing of geospatial data;
- Lack of representativeness, robustness or usefulness of data;
- Discrimination can be consciously or unconsciously built into algorithms—without the final user's knowledge;
- Data stored on servers could be accessed or stolen by governments, militants or malicious parties; and
- Uncertain consent when using data from third-party providers, for example when provided by Unmanned Aerial Vehicle (UAVs).

The data limitations can be derived from space–time gaps, which generate zones with greater uncertainty, fusion of incompatible or outdated data sets. Thus, it is necessary for experts to be consulted to assess whether the available data are adequate for the purposes for which they are intended.

Walker et al. (2015) built a conceptual model of a coupled social-water system considering that each component influencing decisions that affect the other over time. These approaches include the use of (1) agent-based models in order to simulate the behaviour of individual or collective entities, (2) stakeholders who represent the social component interacting with computer models of the water systems within a decision support framework or within the framework, or (3) dynamic adaptive policies, including monitoring and adapting rather than implementing a determined policy. These approaches can be studied to understand the coupled social and natural components of water resource systems in simulation models looking for policymaking insights.

Liu et al. (2019) performed a water footprint allocation model with scenarios under different levels of equity and efficiency for different water destinations, considering industry, cultivated crops and livestock husbandry water footprints of the Yangtze River Economic Belt in China. The authors found difficulties in calculating the representativeness of cultivated crops due to lack of reliable data, only adjusting the crop coefficient data according to the growth stages of local crops in the studied period (2013), while the data for industry were more accurate.

The possibility of creating different and varied models is necessary to project possible scenarios in the face of uncertainties such as climate change and variations in social indicators. Yan et al. (2017) found that, by applying a model structure that combines robust decision-making with many objectives and biophysical modelling, the subjective design choices of researchers and water managers can potentially affect the capacity of the model structure and hinder the design. The development of robust allocation plans in a watershed that suffer from water scarcity requires efforts from the researchers and water managers to characterize well the future climate changes in the regions studied and the vulnerabilities of their communities.

Correlations obtained from the relationship between location, poverty and gender can result in geographical trends and predictive models that discriminate against certain people in relation to access to services and opportunities

(UNICEF 2018). The use of geospatial technologies and data should be reconsidered as well as any resulting decision-making whenever a possibility of discrimination is detected, and it is necessary to use other data sources to avoid this qualitative bias.

## 4 Concluding Remarks

Geoethics encourage us to take a proactive attitude, where the agent imposes him/herself to reflect dialectically on the consequences, opportunities, risks and benefits of his actions using geotechnologies in a rational way. The proposal for water governance must be water security in terms of quality and quantity for all citizens. For this, geoethics has a crucial role in providing decision makers and society with guidelines that guide them in an inclusive, equitable and transparent way, ensuring autonomy, equity, responsibility, efficiency and representativeness in the water governance. A geoethical thought starts on education (basic, undergraduate, graduate), running through professional guidance, displaying the role of schools, universities and professional association in this scope. The use of geospatial technologies and the resulting data need to be critically assessed through an analysis and ethical review before the implementation of programs, models or partnerships between systems and people. This oversight requires an explicit consideration of the possible negative consequences of adoption and a clear articulation of the specific contexts and conditions under which the benefits can be achieved, what is essential in terms of water accesses, distribution and sanitation. For a hidden resource such as groundwater, this becomes even more important, considering pollution and overexploitation risks.

The microethical and macroethical consequences are correlated and constitute a feedback system. Microethical decisions interfere in the universe of macroethics, which performs an ethical-conceptual rearrangement in its own conceptual framework that, through feedback, returns the same or different ethical consequences for the context of microethics and vice-versa. The self (microethics) and societal (macroethics) responsibility and geoethical implication is distributed to all the components in the geoethical decision-making process in (ground) water management.

## References


ANA (Brazilian National Water Agency) (2019) SDG 6 in Brazil: ANA's vision of the indicators. Brasília, National Water Agency

Bellaubi F, Bustamante R (2018) Towards a new paradigm in water management: cochabamba's water agenda from an ethical approach. Geosciences 8:177

Crampton J (1995) The ethics of GIS. Cartogr Geogr Informat Syst 22(1):84–89

Di Capua G, Peppoloni S, Bobrowsky P (2017) The Cape Town statement on geoethics. Ann Geophys 60: Fast Track 7. https://doi.org/10.4401/ag-7553

Jonas H (1985) The imperative of responsibility: in search of ethics for the technological age. University of Chicago Press, Chicago

Limkar SV, Jha RK (2019) A novel method for parallel indexing of real time geospatial big data generated by IoT devices. Future Gener Comput Syst 97:433–452

Liu G, Wang W, Li KW (2019) Water footprint allocation under equity and efficiency considerations: a case study of the Yangtze river economic belt in China. Int J Environ Res Pub Health 16:743

Mogk DW, Bruckner MZ (2020) Geoethics training in the earth and environmental sciences. Nat Rev Earth Environ 1:81–83

Peña H (2011) Social equity and integrated water resources management. TEC background paper 15. Global Water Partnership Technical Committee (TEC)

Peppoloni S, Bilham N, Di Capua G (2019) Contemporary geoethics within the geosciences. In: Bohle M (ed) Exploring geoethics—ethical implications, societal contexts, and professional obligations of the geosciences. Springer International Publishing, Cham

Rogers P, Hall AW (2003) Effective water governance. Global Water Partnership, Stockholm, Sweden

UNICEF (2018) Ethical considerations when using geospatial technologies for evidence generation. Discussion paper DP-2018–02, June 2018

Walker WE, Loucks DP, Carr G (2015) Social responses to water management decisions. Environ Proc 2:485–509

Yan D, Ludwig F, Qing HH, Werners SE (2017) Many-objective robust decision making for water allocation under climate change. Sci Total Environ 607–608:294–303

Zolghadr-Asli B, Bozorg-Haddad O, Chu X (2017) Strategic importance and safety of water resources. J Irrig Drain Eng 143(7): 02517001

# Groundwater Mismanagement: Impacts on Society Due to a Response Lacking Geoethics in Mexico

José Joel Carrillo-Rivera, Samira Ouysse, and Gonzalo Hatch Kuri

**Abstract**

Correct information dissemination on the best way the environment may be managed is seldom a practice reaching society. Public opinion is constructed by the media often neglecting scientific results. Furthermore, many enforced legal actions are based on words and concepts lacking an accepted sound definition. Regularly, words are used proposing kind of synonymous suggesting the integrity of the water supply is endangered of getting dry or having a saline-water inflow that would eventually make a city or agriculture water user to collapse. A lack of education or teaching on the actual functioning of groundwater is recurrently missing not only on the public domain but in the decision-making. Proper groundwater management is ignored in cases where sustainability should have been the core of the water policy. Environmental impacts are practically divorced from groundwater issues by well-established myths; developing strategies bring society away from proper land planning and social justice. The resulting application of terms as overexploitation, among others, is making Mexican small private groundwater users scared of using groundwater, abandoning agricultural practices that has resulted in increased immigration affecting them beyond the prevailing water market where groundwater efficient use is repeatedly invoked but is actual functioning is neglected.

**Keywords**

Overexploitation • Contamination • Water scarcity • Water rights • Groundwater functioning

J. J. Carrillo-Rivera (✉) · S. Ouysse
Institute of Geography, National Autonomous University of México, México City, México
e-mail: joeljcr@igg.unam.mx

G. Hatch Kuri
College of Geography, National Autonomous University of México, México City, México

## 1 Introduction

Correct information on the best way to manage the environment is seldom reaching society. Public opinion is constructed by the media often neglecting scientific facts, resulting in legal actions based on concepts lacking sound definitions. Regularly, used words propose synonymous suggesting water supply integrity is endangered of getting depleted or contaminated, conditions that would eventually make a city or general water user to collapse. A lack of education on actual groundwater functioning is recurrently missing in public domain and decision-making. Environmental impacts are practically divorced from groundwater issues by well-established myths; developed strategies bring society away from proper land planning resulting in increased emigration and loss of social justice. Consequently, the aim of this paper is to introduce to the injustice that the wrong use of terms as related to groundwater has been made on users in general and, particularly, in those having a water right or water concession in Mexico, and surely elsewhere.

## 2 General Setting and Methods

Mexico, as many other northern hemisphere countries, has a water law (CONAGUA 2016) that is meant to provide a legal reference to regulate water extraction, usage, distribution and control, as well as to preserve its quality and quantity to reach a sustainable development. That is CONAGUA, a federal administration office, is the administrative entity to manage water in Mexico, that is to gather and systematized information, to analyse, made diagnoses, programs and related actions in relation to water occurrence in quantity and quality as well as its extraction and usage. However, there is a lack of direct measures to determine the amounts of water involved as well as their quality distribution; that is, a reasonable knowledge and understanding of


M. Abrunhosa et al. (eds.), *Advances in Geoethics and Groundwater Management: Theory and Practice for a Sustainable Development*, Advances in Science, Technology & Innovation,
https://doi.org/10.1007/978-3-030-59320-9_95

water functioning (surface and groundwater) still waits for an acknowledgement. This lacking is tightly linked to several inconsistencies in the legal terms used. Terms, that continuously receive a synergy input of the media, where decision makers are not foreign to the misconception place on society by the terms currently used. Indeed, related terms to the presence of water as overexploitation, scarcity, demand or the ill-defined concept of availability are as commonly in usage as the well-defined but ill-applied concept of contamination. Further, there is an additional void in the obscure definition of water and resource, mainly when the meaning of commodity and common good, is concern.

The applied method in this work is related to a critical observation of facts connected to groundwater regulations and their application onto societal issues as well as to reported comments and experience of groundwater users in several parts of the world. The method of critical analysis existing in the 1970s appears to have lost grip on the present professional society; however, it seems that it was bought into a reviewed edition by the general public producing a social uprising in response to what the affected farmers define as cheating on them by the federal administration and its assisting in the dispossession of their water rights.

## 3 Results

### 3.1 About World-Wide Resolutions

The way information is disseminated on how is best to have a healthy environment appears to be planned as to provide the required setting to be reached by society. Concepts such as Right to Water and Sanitation as defined by the UN (2010) has been ill-used by some NGO's as Right to Water, where the concept of affordable water has been forgotten in their rhetoric, creating in the public an important misconception of free water for all. A second identified concept is that of Water Security (UN 2013) which is "The capacity of a population to safeguard sustainable access to adequate quantities of and acceptable quality water for sustaining livelihoods, human well-being, and socio-economic development, for ensuring protection against water-borne pollution and water-related disasters, and for preserving ecosystems in a climate of peace and political stability" which however, it is inappropriately used as the need to devise with the "required paraphernalia to have a safe water yield or constant water supply". The concept of the capacity of the population is basically ignored. From a worldwide perspective, such capacity has been virtually diminished since 1998 where related activities were intended to be reduced in China or by 2000 when UCL-GB groundwater M.Sc. and Ph.D. programs totally vanished, or research was eliminated by 2014 in Birmingham University. Related capacity has had a nil start in Latin America and North Africa, for example.

### 3.2 About Local Water Users

Public opinion is constructed by the media often neglecting scientific results and enforcing legal actions against groundwater users based on words and concepts lacking an accepted sound definition, both technically and legally. It has been found that words been brought into the water language proposed as kind of synonymous which suggest an entirely different concept to what is observed in the field or to public understanding. The idea is to pass a message that the integrity of the water supply is endangered of getting dry or having a saline-water inflow that would eventually make a city or agriculture water user to collapse. Integrity in which the users appear to be responsible where the concept of inefficient water management is never place on debate. That is, the obligation of the water administrator has automatically disappeared and the direct responsible becomes the deprived water users. In various localities in Mexico, the dispossession of groundwater, taking water from small agricultural users, has been accomplished basically using three words: overexploitation, contamination and scarcity. For example, farmers were truly convinced by the constant reiteration of these words in the rhetoric from CONAGUA authorities, as well as the media. In La Laguna, in northern Mexico, Hoogesteger (2018) reported that 12,000 out of 22,000 *ejidos* (agricultural communities) sold or rented their water licence to four private holdings. Recently, in the nearby locality of Saltillo, several small owners have openly complained that they have been cheated, as they discovered that the "new" owners have drilled and found water.

## 4 Discussion

The lack of clear, updated and strictly defined concepts has brought insecurity to groundwater users. The lack of proper education and teaching on the actual functioning of groundwater is recurrently missing in both the public domain and decision-making. Proper groundwater management is ignored in cases the sustainability of land and ecosystems should have been the core of any public and management policy to groundwater. Environmental impacts are practically neglected from groundwater issues by well-established myths; developing strategies bring society away from proper land planning and social justice. Current Mexican water law place an obligation to the responsible federal organism, CONAGUA for the proper management of both surface and groundwater. The resulting application of terms as overexploitation, scarcity, among others, is making

Mexican small private groundwater users scared of using groundwater, leaving agricultural practices that has resulted in increased emigration affecting them beyond the prevailing "water market" where groundwater efficient use is repeatedly invoked but is actual functioning is neglected. The concept of the capacity of the population (UN 2013) is basically ignored. From a worldwide perspective, such capacity has been virtually jeopardized since early this century where activities related to groundwater systemic functioning have been vanished from higher institutions and have not started in several parts of the word.

## 5 Concluding Remarks

Many enforced government legal actions related to groundwater are based on data, arguments, and concepts often used with a double, a sluggish, a wrong, or with a lack of meaning, all of which characterized of a desirable hydrogeoethical approach. Decision makers see groundwater limited by administrative boundaries with no connective dynamics, with limited ecosystem consideration. Use and misuse of related technical information are associated to international macroeconomic strategy neglecting the nation's interest. Economical decisions hide consequences for people under induced poverty of both: money and related knowledge.

## References

CONAGUA (2016) Ley de aguas nacionales (law of national waters). Last published version DOF, 24 Mar 2016, p 110

Hoogesteger J (2018) The ostrich politics of groundwater development and neoliberal regulation in Mexico. Water Altern 11(3):552–571

UN (2010) The human right to water and sanitation, resolution 64/292. The United Nations General Assembly, 28 July 2010

UN (2013) Water security and the global water agenda. A UN-water analytical brief. United Nations

# Science and Policy: How Ethical is Groundwater Management in Mexico (1948–2018)

Gonzalo Hatch-Kuri, José Joel Carrillo-Rivera, and Samuel Schmidt

**Abstract**

The management of groundwater in Mexico corresponds to the President of the Republic, who has three legal instruments for its ordering: forbidden decrees, regulated zones, and reserve zones. In 2013, President Peña to preserve groundwater decreed a ban affecting 332 administrative aquifers covering 99% of the national territory. In 2018, the same president decreed the temporary suspension of the previous decree to issue entitlements and renewing groundwater concessions, contravening the objective to protect and preserve groundwater, and contradicting international commitments, which can be considered unethical. Our essay uses an interdisciplinary approach (political geography, hydrogeology, and political science), to identify and analyze the components of groundwater rights management policy in Mexico and warns that its spatial manifestation should be the relationship between technical science, regulatory framework of water, and public policy. In the period, from 1948 to 2018, groundwater management has responded, fundamentally, to political and water rights management criteria, which favor political loyalty to the president and ignoring ethical aspects of environmental management and a general groundwater administration based on ethical criteria.

**Keywords**

Water power • Water rights • Groundwater • Presidential decrees • Political geography

## 1 Introduction

Governmental data estimate that groundwater supplies 60 million out of 130 million Mexicans and irrigates two million hectares; 75% of the water supplied in Mexico comes from the underground (CONAGUA 2019). Despite the strategic value of groundwater, it has not been possible to reform its management towards a democratic model that guarantees equitable access. Two of the main obstacles for democratization are the authoritarian nature of the political system and lack of sound scientific research about underground water.

Carrillo-Rivera et al. (2016) show the methodological limitations that characterize the evaluation of groundwater in Mexico, which is one of the main reasons for the growing number of legal, environmental, and social conflicts over this water. Schmidt and Hatch (2012) suggest that water, especially transborder underground water is a national security issue. Rivera and Candela (2018) and Megdal (2018) suggest that to guarantee a democratic and equity model in water rights, and its effective environmental conservation lies on the adequate institutional articulation between the agencies responsible for the scientific evaluation of water, government decentralization, transparency in decision-making, and participation of all those interested.

This research seeks to identify how the components that define groundwater policy in Mexico are articulated, the link between legal instruments that contain provisions for its use and distribution, the role of scientific-technical groundwater evaluation methodologies, the spatial dimension in the national territory and the political criteria, e.g., the effect of the political control of groundwater management, as it is an

G. Hatch-Kuri (✉)
Natural Sciencies Faculty, Universidad Autonoma de Queretaro, Carretera a Chichimequillas s/n, Ejido Bolaños, Santiago de Queretaro, 76140, Queretaro, Mexico
e-mail: ghatch@comunidad.unam.mx

J. J. Carrillo-Rivera
Institute of Geography, National Autonomous University of Mexico, Circuito de la Investigacion Cientifica, Ciudad Universitaria, Mexico City, 04510, Mexico

S. Schmidt
Teresa Lozano Long Institute of Latin American Studies, University of Texas At Austin, 2300 Red River Street D0800, Austin, 78712, TX, USA



M. Abrunhosa et al. (eds.), *Advances in Geoethics and Groundwater Management: Theory and Practice for a Sustainable Development*, Advances in Science, Technology & Innovation,
https://doi.org/10.1007/978-3-030-59320-9_96

authoritarian component and clientelist policies among the main users of water and its unethical consequences.

## 2 General Setting and Method

President Enrique Peña Nieto (2012–2018) made three decisions related to groundwater management. The first was the Presidential Decree of April 5, 2013; in which, practically, all groundwater extraction was banned in the entire national territory; drilling of new wells was prohibited, as well as obtaining or renewing water concessions. Entitlements are "Defined as the use, exploitation or utilization of the volume that is given in the concession, which must be respected by concessioners. Water entitlements are unbundled from property titles. Water entitlements can be traded, leased or transferred and last between 5 and 30 years with the expectation of periodic renewal,"[1] It is unclear the renewal process but the COTAs play a certain role. The second decree was on October 31, 2017, and the National Water Commission (CONAGUA) created the Technical Committee for Groundwater Management (COTEMA), a collegiate body responsible with analyzing and implementing alternative solutions to problems on groundwater management (pollution, groundwater depletion, reduction of base flow to rivers, soil subsidence, yield reduction of wells, seawater intrusion, etc.). The third one was the decree of March 23, 2018, reversing the provisions of the April 5th, 2013 decree, establishing a temporary suspension until December 31, 2018, in order to regulate, extend, and grant new water concessions and groundwater allocation. Between these two decrees, unlike the first ten decrees, groundwater provisions went unnoticed, almost entirely for public opinion.

The Peña administration failed to structurally reform the water legal framework, due among other things to the deregulatory policy (Hatch et al. 2017) geared to maintain and expand water supply to specially big private users; the 2018 decrees were the legal instrument used to renew and grant new concession and water entitlements without proper scientific-technical support. CONAGUA (2018) supports the notion that groundwater management is strictly a technical issue, but the decrees manifest an "invisible" political component: the absolute power of the president to decide the water rights policy and groundwater conservation policy in Mexico.

All groundwater decrees issued by the Federal Executive Power from 1948 to 2018 were reviewed, to determine management and the main classification (banned, reserve zone or regulated zone), their justification, and spatial display expressed in two concepts "administrative aquifer" and "irregular polygon" which have a potential unethical component.

Methodologically, we use an interdisciplinary approach that combines political geography, hydrogeology, and political science, and we look into the connection between the category of hydro-social cycle as it relates to the water–power relationship and Mexican authoritarianism and presidentialism as some of the main components of the political system, which in turn partially explains the water rights policy. One key element is the superficial and biased use of scientific-technical elements for decision-making, for example, the "determination of the average annual water availability" to grant water rights and the symbolic legislation to create rules in paper which are not translated into real life.

## 3 Results

1. Twelve presidents of the republic from 1948 to 2018 issued 105 provisions for legal groundwater ordering (see Table 1). Luis Echeverría Álvarez (1970–1976) issued the highest number of decrees (20), but, considering the spatial scope of the legal provisions, the 19 decrees issued by Adolfo Ruíz Cortines (1952–1958) had effects on 89 "irregular polygons."[2] Enrique Peña Nieto (2012–2018) decreed a ban on 332 administrative "aquifers" in Mexico (98% of the national territory). "The state territory is divided in 37 hydrological regions (accounting for 731 basins) which are grouped in 13 hydrological-administrative regions and the national level that oversees these regions" (https://www.oecd.org/mexico/Water-Resources-Allocation-Mexico.pdf).

2. Groundwater management seen through presidential decrees reveals the political decision to use two concepts to define the territorial scope of these provisions, and both of which lack scientific criteria. The first is the "irregular polygon," which refers to a geographical area in which the imposition of a banned reserve zone or a regulated zone have been imposed; it is determined by unclear criteria by CONAGUA. At the middle of the

[1]https://www.oecd.org/mexico/Water-Resources-Allocation-Mexico.pdf.

[2]"The area of interest for a ground-water-quality sample network commonly is not rectangular and could consist of one or more irregular polygons" (Alley 1993, p. 77).

**Table 1** Number of banned decrees imposed in Mexico 1948–2018

| President of the republic | Government period | Decrees | Irregular polygons |
|---|---|---|---|
| Miguel Alemán Valdés | 1946–1952 | 10 | 11 |
| Adolfo Ruiz Cortinés | 1952–1958 | 19 | 89 |
| Adolfo López Mateos | 1958–1964 | 17 | 42 |
| Gustavo Díaz Ordaz | 1964–1970 | 12 | 52 |
| Luis Echeverría Álvarez | 1970–1976 | 20 | 63 |
| José López Portillo y Pacheco | 1976–1982 | 17 | 43 |
| Miguel de la Madrid Hurtado | 1982–1988 | 8 | 19 |
| Enrique Peña Nieto | 2012–2018 | 2 | 332 |

twentieth century, the concept "Aquifer" was formalized, institutionalized, and characterized by administrative and political purposes, since its definition and spatial dimension considers only the surface of the geometry of the hydrogeological aquifer, without considering its depth and thickness. This way water administration fails to consider the determination of the systemic functioning of groundwater flows and their interdependent relationship with other environmental components.

3. The groundwater presidential decrees reveals a rhetoric that changed over time. From 1948 to 1972, the presidents justified the issue of banning decrees or reserve zones, to expropriate (public interest or public domain declaration) for works for artificial lighting, which included the infrastructure and previously acquired groundwater rights. In 1972, the first presidential decree that justifies the imposition of a ban to "conserve and protect the balance of water in aquifers" was recorded, although the concept aquifer was not yet defined by law; at present, it refers to a groundwater definition. This justification has a direct relationship with the agreements of various international environmental conferences, such as the Stockholm International Conference (1972), where it was agreed that the states will develop environmental protection policies, which include water.
4. The groundwater legal decrees for the decades 2000–2020, even though supposedly are based on environmental conservation of groundwater, they actually respond to water management rights policy, which responds to the logic of economic growth without scientific rigor, and it delivers water rights according to presidential whim, even though in many cases contradict the fundamentals of preserving the environment, such as the installation of industrial zones in regions, where aquifers were declared over exploited, thus, governing water management policy might lack ethical considerations in favor for political criteria, because it doesn't consider the rights of the original people, or the well-being of society at large.

## 4 Discussion

Mexican groundwater management based upon the absence of solid scientific-technical aspects has a close relationship with the application of regulatory criteria reflecting negligible updating of the scientific methodology for its correct assessment, according to international standards. These rules have facilitated that the president, in turn, can have political control and discretionary management in providing water to users as some users are more influential than others, which reinforces inequality. The lack of rigorous conceptual and scientific criteria for concepts such as "irregular polygon" and "aquifer" (administrative), also marginalize the objectives shared by Mexico's adherence to international policies, such as actions against climate change, which can also be unethical.

It is essential to conduct groundwater interdisciplinary studies, to define the functioning of the science-policy concatenation, since it has allowed to manipulate the destination of national water use in Mexico for some "big" users, privatizing the water de facto; the attempt having the greatest implications was during the Peña administration. These studies will help educated policy makers and society at large on the real conditions of groundwater (quality and quantity); at the same time, they will provide quality information to Basin Councils (Consejo de Cuenca) and to Groundwater Technical Committees (COTAS); nowadays, both of them depend on CONAGUA and have a marginal role on the formulation and implementation of groundwater policy in Mexico.

## 5 Concluding Remarks

Contrary to the usual argument that groundwater management is exclusively a technical issue; groundwater management in Mexico is based on political decisions geared to creating clientelist policies and loyalty among water users to support the authoritarian regime.

Water policy is part of the symbolic legislation that shows existing laws and regulations which are hardly applied, either because of political commitments or due to corruption. This contradicts the international commitments made by Mexico and affects water preservation and environmental conservation.

This authoritarian use of water policymaking is used by the government to justify the legal system and authoritarian decisions; and this in turn reinforces the unethical use of concepts that lack consistency and move away from a proper scientific application of modern hydrogeology, political geography, and democracy.

## References

Alley MW (1993) Regional groundwater quality. Van Nostrand Reinhold, New York

Carrillo-Rivera JJ, Peñuela L, Huízar R, Cardona A, Ortega M, Vallejo J, Hatch G (2016) Conflictos por el agua subterránea. In: Moncada-Maya O, López A (eds) Geografía de México. Una reflexión espacial contemporánea, Tomo I, Instituto de Geografía, UNAM, pp 151–166

CONAGUA (2018) Instala el director general de la CONAGUA el Comité Técnico de Manejo de Aguas Subterráneas. Press Release/Comunicado de Prensa No. 660–17, 31 Oct 2018

CONAGUA (2019) Estadísticas del agua en México. Edición 2019. Comisión Nacional del Agua. SEMARNAT, Ciudad de México

Hatch G, Schmidt S, Carrillo-Rivera JJ (2017) Elementos de análisis de la propuesta de Ley General de Aguas en México a partir del derecho humano al agua y sus repercusiones en el que hacer científico, docente y de la investigación. Rev El Colegio De San Luis 13(8):30–61

Megdal SB (2018) Invisible water: the importance of good groundwater governance and management. NPJ Clean Water 1(1):1–5

Rivera A, Candela L (2018) Fifteen-year experiences of the internationally shared aquifer resources management initiative (ISARM) of UNESCO at the global scale. J Hydrol Reg Stud 20:5–14

Schmidt S, Hatch G (2012) El agua en México: una cuestión de seguridad nacional. For Aff Latinoam 4(12):89–96

# What is the Way Forward to Protect the Ecological Values of Groundwater?

Elisabet V. Wehncke

**Abstract**

In recent years, groundwater systems have finally been seen as true ecosystems, and not only their enormous value as a resource has been appreciated, but also their relative susceptibility to contamination, which increases the need to learn more about its function and resilience capacity. However, there are several gaps in knowledge and a lot of uncertainty for what is important in decision-making in water and ecosystems connections and management programmes. The difficult accessibility has long challenged the evaluation and development of scientific theories and how to advance in the understanding of their functioning. In general, they are not easy to perceive; however, in recent years, it has been found that they can contain high diversity of living forms with particular adaptive characteristics that maintain water quality and particular ecological functions providing numerous services to humanity. Due to the increase of natural resources demands and the intensive and unplanned land use in the last years, which has led to environmental degradation and diverse social conflicts, studies on groundwater and ecosystems, or the ecology of groundwater, are gaining momentum. Thus, holistic perspectives are encouraged in order to properly understand these connections, their functional roles and visualize proper management perspectives under different scenarios of global changes.

**Keywords**

Biodiversity • Contamination • Groundwater • Geoethics • Human impacts

E. V. Wehncke (✉)
Centro de Investigación en Biodiversidad y Conservación, Universidad Autónoma del Estado de Morelos, Cuernavaca, México
e-mail: lizwehncke@gmail.com

## 1 Introduction

Globally, we have recently begun to struggle with the reality of recognizing the ecological values of groundwater in water planning and management (Tomlinson 2011; Goonan et al. 2015). There are several gaps in knowledge and a lot of uncertainty for what is important for the scientific integrity in decision-making in water and ecosystems management programmes. Given this high uncertainty, it must be recognized that current groundwater management for ecological purposes is entirely experimental. Therefore, it is essential to promote holistic perspectives and to adequately understand the connections between groundwater flows and ecosystems, their functional roles and to visualize adequate management proposals under different scenarios of global changes, if we plan to conciliate the development needs of humanity and the maintenance of ecosystem services in a sustainable way. Thus, there is a need of the following:

- Carefully planned adaptive monitoring and evaluation are crucial to inform decision-making (Tuinhof et al. 2002–2006).
- Integrated assessment of ecological quality and groundwater parameters is necessary, because it is not at all evident that environmental policies converge towards common objectives such as guidelines that protect the quality of drinking water that does not necessarily contribute to the protection and maintenance of groundwater biodiversity and ecosystems (Michel et al. 2009).
- The training of skills in methodologies and techniques is mandatory for the scientific determination of ecological water requirements in order to establish a robust and accurate water provision programme.
- Progress in water governance should consider protecting the full spectrum of values and benefits of groundwater.
- Improve the understanding of the interlinkages between human actions and groundwater impacts (e.g. diffuse pollution by nitrates, see Aldaya et al. (2020).



M. Abrunhosa et al. (eds.), *Advances in Geoethics and Groundwater Management: Theory and Practice for a Sustainable Development*, Advances in Science, Technology & Innovation,
https://doi.org/10.1007/978-3-030-59320-9_98

## 2 Ecological Values of Groundwater

The daily life of the human being depends on many services given by countless ecosystems on the planet (Daily et al. 1997; MEA 2005). Groundwater provides ecosystem services with immense social impact and great economic value; however, the protection and management of these services require a quantitative and qualitative knowledge of the processes at different spatial and temporal scales and an understanding of their resilience to the various anthropogenic impacts. The concept of ecosystem services points to a tendency of understanding beyond purely hydrogeological processes, to focus on the functions and processes that are connected to them. An approach leading to an understanding of ecological principles will allow predicting the responses of ecosystems as a whole and their services in the face of disturbances such as pollutants spills, climate changes, alterations to land use, among others. For this purpose, it is necessary to develop simple predictive tools so that decision makers and environmental managers can adequately protect ecosystems and generate instruments for risk assessment (WLE 2015).

The presence of groundwater by itself is a support service and every aquatic and terrestrial ecosystem depends on its availability in good quantity and quality. Regulating services include water purification, biodegradation of pollutants and the elimination of pathogenic microorganisms and viruses, which in turn contributes to the control of diseases. Cultural services include large bodies of water in caves and hot springs that are tourist attractions and of great spiritual importance to many indigenous communities. The conceptual framework of ecosystem services has been a very powerful weapon to raise awareness about their importance to humanity; however, the next step that is the protection and the adequate and sustainable management of the ecosystems, their organisms, and the functions they represent still awaits an efficient implementation.

## 3 How Much Water Does the Environment Need?

To integrate groundwater, it is necessary to consider water regimes necessary to maintain the ecological values of groundwater dependent ecosystems at low levels of risk (Tomlinson 2011). This implies:

- to identify ecological values (such as biodiversity and ecological processes),
- to understand the dynamics of water regime that maintains these values, and
- to evaluate the risks that these values may have due to disturbances to the groundwater regime and those due to changes and alterations in the surrounding environment (Carrillo-Rivera et al. 2007).

It is also necessary to know the management unit, the tributaries involved, their potential flow, and their chemical quality; a similar knowledge should be required to define the dynamics of subterranean flow systems involved. Both surface and groundwater face similar challenges in their management such as the confused effects of multiple factors that can alter the aquatic ecosystem, the uncertainty in the magnitude, scale and variability of the responses and the lack of integration in conceptual models of hydrological systems.

## 4 What is a Sustainable Extraction of Groundwater?

With surface water a sustainable extraction would be that extraction pattern that reflects the natural hydrological regime composed of the magnitude, frequency, duration, periodicity, and rate of change (Hamstead 2009; Poff et al. 1997). Groundwater still lacks a more complete understanding to develop an expression of a sustainable extraction, since it is still very unclear that how much water requires their dependent ecosystems. In groundwater studies, the hydrogeological concept of *safe yield* refers to the amount of groundwater that can be extracted without exhausting reserves (Kalf and Woolley 2005). This rather physical concept does not take into account the concept of ecological value supported by groundwater or the interconnection of ecosystems, the actual size of the reserve and the hydraulic interconnection between the various surface watersheds involved in the short and long terms, and apparently, does not consider any of the principles of environmental water provision. Thus, the global operational rule seems to be that water for the environment is considered after consumptive demands which have been satisfied.

The management of surface water is inconceivable beyond the management of groundwater. Methods for determining *environmental flows* are fairly well developed for surface water (Dyson et al. 2003; Tharme 2003; Acreman and Dunbar 2004; Poff and Zimmerman 2010), although there is still a lack of knowledge about their effectiveness, and flow-ecology relationships in general to the scale necessary to achieve objectives of water planning and management (National Water Commission 2011). However, the recognition of the role played by groundwater in issues of environmental flows is generally inadequate and considered an additional and not at all priority. The complexities of surface water dynamics that have not yet been resolved in terms of environmental flow are multiplied when considering groundwater. Groundwater flows are not always unidirectional; there are fissures, ducts and pores, where water can pass quickly or very slowly with gradients of variable magnitude and direction such as in seasonal climates. In addition to the temporal variability, there is a spatial distribution in patches in flow rates (Stanford and Ward 1993; Boulton et al. 2010), which direct vertical and lateral redox gradients, and a chemical and nutrient zoning are generated impacting microbial processes that provide ecosystem services. Key attributes of groundwater that can be recognized are: levels, flow, pressure and water quality. Residence times are also key in biogeochemical functions (Boano et al. 2010).

## 5 Main Groundwater Dependent Ecosystems

A first step has been to conceptualize the presence of groundwater in the landscape, and many questions add for wanting to understand more fully how ecosystems function as a whole. In some countries, these ecosystems have been recategorized; this new classification includes aquifers and caverns, and ecosystems that depend on the surface expression of groundwater as different discharge zones. One way to classify ecosystems related to groundwater is by its geomorphological environment (aquatic, terrestrial, coastal, etc.) and the associated groundwater flow system (deep or shallow).

Focusing on understanding how groundwater flows work would give us a better understanding of the requirements of ecosystems and biota in general, which can be variable and in accordance with the flow system involved. The development of more detailed studies on these issues may reveal connectivity between biodiversity and groundwater along hydro-ecological gradients and provide a model to gain a fuller understanding of the role of groundwater and the ecological value of these systems.

## 6 What Do We Know About the Connection Between Wetlands and Groundwater?

- Connection potential: Many wetlands are hydrologically and ecologically linked to adjacent masses of groundwater, but the degree of interaction can vary greatly. The degree of wetlands dependence on groundwater can be complete or very limited, for example, only in conditions of extreme dryness; and some may lack complete connection to groundwater. In turn, some aquifers depend almost entirely on wetlands for recharge or may not receive any recharge from wetlands.
- Hydrological links: Knowing the way in which water enters and leaves the wetland (the so-called water transfer mechanisms) and quantifying their frequency is a prerequisite to assess the consequences for a wetland of any type of external hydrological impact. In addition to geographical (horizontal or planiform) analyses of wetlands, by mapping open water bodies and zoning vegetation, understanding the interactions with groundwater requires a three-dimensional geological view, that is, to look at soil vertical sections that lie beneath the wetland.
- Uncertainty: Water balance cannot provide a definitive determination and prediction of wetland responses due to hydrological impacts, such as excessive groundwater extraction, because we are dealing with natural systems. It also does not provide information on the rhythm and frequency of hydrological events. To define these hydrological properties detailed modelling is necessary. Additionally, it is advisable that uncertainty (or level of confidence) should be calculated for any water transfer mechanism using several different methods.
- Modelling: Sometimes complex models do not guarantee the understanding of phenomena. After a correct conceptual understanding of water transfer mechanisms resorting to modelling implies: (i) generate quantitative information about the processes that drive water transfer mechanisms, (ii) understand temporal and spatial variability of processes and (iii) predict what will happen in different possible climatic or water management situations (Secretariat of the Ramsar Convention 2010).

## 7 Concluding Remarks: Why Do We Need an Integrated View of Groundwater?

Groundwater is an ecosystem in its own right and plays an integrating role in supporting other types of aquatic, terrestrial and coastal ecosystems, and the landscapes associated with them, in both humid and arid regions. Therefore, it is a key factor difficult to isolate from other key ecosystems due to the multiple processes, connections and functions that they all perform in an integrated manner.

The lack of control in the exploitation and protection of groundwater has had negative impacts, mainly on aquatic flora and fauna due to changes in flow systems. In some aquifers under intensive exploitation (especially in arid or highly populated regions), the ecological function of groundwater has largely been lost as a result of the depletion of groundwater levels. In other cases, it is threatened by the deterioration of groundwater quality caused by diffuse contamination (mainly due to excess nutrients and pesticides).

Maintaining the connections between ecosystems and groundwater as healthy as possible are one of the key aspects to achieve the sustainability of these complex systems. Various sources of contamination on these connections through runoff from agricultural land, soil erosion, deforestation, urban development, industries, mining and the introduction of exotic species threaten such sustainability. There are various ethical approaches to regulate all these sources of environmental impact. Identifying the way in which these should be considered and applied is not only part of the political will but also a moral responsibility of all human beings if we want to articulate sustainable mechanisms of governance of our present and future natural resources. Joint efforts are required at various levels of organization, politics and society to make visible and more compatible the use of groundwater and environmental conservation.

**Acknowledgements** I am grateful to Dr. José Joel Carrillo Rivera, who is responsible for my growing interest in the connections between groundwater and ecosystems. I would also like to thank to some anonymous members of the IAH Mexican chapter, for also contributing with good advices from the hydrogeology to the ecology field. Finally, I thank the Biodiversity and Conservation Research Centre of the Autonomous University of Morelos, for giving me the necessary space to spread this knowledge.

## References

Acreman MC, Dunbar MJ (2004) Defining environmental river flow requirements? a review. Hydrol Earth Syst Sci Discuss 8:861–876

Aldaya MM, Rodriguez CI, Fernandez-Poulussen A, Beriain MJ, Llamas R (2020) Grey water footprint as an indicator for diffuse nitrogen pollution: the case of Navarra, Spain. Sci Total Environ 698:134338

Boano F, Demaria A, Revelli R, Ridolfi L (2010) Biogeochemical zonation due to intrameander hyporheic flow. Water Resour Res 46(2): 1–13

Boulton AJ, Datry T, Kasahara T, Mutz M, Stanford JA (2010) Ecology and management of the hyporheic zone: stream–groundwater interactions of running waters and their floodplains. J North Am Benthol Soc 29:26–40

Carrillo-Rivera JJ, Cardona A, Huizar-Alvarez R, Graniel E (2007) Response of the interaction between groundwater and other components of the environment in Mexico. Environ Geol 55:303–319

Daily GC, Alexander S, Ehrlich PR, Goulder L, Lubchenco J, Matson PA, Mooney H, Postel S, Schneider SH, Tilman D (1997) Ecosystem services: benefits supplied to human societies by natural ecosystems. Issues Ecol 2:116

Dyson M, Bergkamp G, Scanlon J (2003) Flow: the essentials of environmental flows. IUCN, Gland, Switzerland, Cambridge, UK

Goonan P, Jenkins C, Hill R, Kleinig T (2015) Subsurface groundwater ecosystems: a briefing report on the current knowledge, monitoring considerations and future plans for South Australia. Environment Protection Authority, Adelaide. https://www.epa.sa.gov.au

Hamstead M (2009) Improving environmental sustainability in water planning. Waterlines report. National Water Commission, Canberra

Kalf FRP, Woolley DR (2005) Applicability and methodology of determining sustainable yield in groundwater systems. Hydrogeol J 13:295–312

Millennium Ecosystem Assessment (MEA) (2005) Ecosystems and human well being: wetland and water. Synthesis World Resources Institute, Washington, DC

Michel G, Malard F, Deharveng L, Di Lorenzo T, Sket B, De Broyer C (2009) Reserve selection for conserving groundwater biodiversity. Freshw Biol 54:861–876

National Water Commission (2011) The national water initiative-securing Australia's water future: 2011 assessment. NWC, Canberra

Poff NL, Zimmerman JKH (2010) Ecological responses to altered flow regimes: a literature review to inform the science and management of environmental flows. Freshw Biol 55:194–205

Poff NL, Allan JD, Bain MB, Karr JR, Prestegaard KL, Richter BD, Sparks RE, Stromberg JC (1997) The natural flow regime. Bioscience 47:769–784

Secretariat of the Ramsar Convention (2010) Manejo de Aguas Subterráneas: Lineamientos para el manejo de las aguas subterráneas a fin de mantener las características ecológicas de los humedales Manuales Ramsar para el uso racional de los humedales, vol 11, 4th edn. Secretaría de la Convención de Ramsar, Gland, Suiza

Stanford J, Ward J (1993) An ecosystem perspective of alluvial rivers—connectivity and the hyporheic corridor. J North Am Benthol Soc 12:48–60

Tharme RE (2003) A global perspective on environmental flow assessment: emerging trends in the development and application of environmental flow methodologies for rivers. River Res Appl 19:397–441

Tomlinson M (2011) Ecological water requirements of groundwater systems: a knowledge and policy review Waterlines report. National Water Commission, Canberra, Australia

Tuinhof A, Foster S, Kemper K, Garduño H, Nanni M (2002–2006) Requerimientos de monitoreo del agua subterránea: para manejar la respuesta de los acuíferos y las amenazas a la calidad del agua Serie de Notas Informativas Nr 9, GW-MATE Banco Mundial. EEUU, Washington, DC

Water, Land and Ecosystems (WLE) CRP (2015) Groundwater and ecosystem services: a framework for managing smallholder groundwater-dependent agrarian socio-ecologies—applying an ecosystem services and resilience approach. International Water Management Institute (IWMI)-CGIAR Research Program on Water, Land and Ecosystems (WLE)

# Ethical Dilemmas Behind Groundwater Sampling, Laboratory Testing Processes and Data Analysis Outcomes

Samira Ouysse and José Joel Carrillo-Rivera

**Abstract**

In modern hydrogeology, it is essential to interpret information following an integrative approach which considers actual groundwater conditions rather than biasing results to meet certain predetermined concept or models. In the last decade, the usage of statistical methods (i.e. multivariate analysis) to characterize and monitor groundwater quality has become an unethical myth that has been largely used; this brings to question the margin of uncertainty, reliability and degree of risk of having results which are not explicitly representing the real field conditions. An ethical dilemma rises here as usually, and most of these statistical analyses are based on estimates and inferences of a selected population omitting and standardizing the samples that might present a degree of "anomaly" which may reflect end members. In hydrogeological sciences where exact boundaries between different components are unclear, data cannot be treated based on inductive logic since all parameters are connected and data analysis must be conducted following a multidisciplinary characterization where all information including anomalies are reported and thoroughly examined.

**Keywords**

Groundwater • Multivariate analyses • Outcome • Flow systems

## 1 Introduction

Groundwater is an important freshwater source and a significant moderator of hydrologic processes at the land surface and in the atmosphere. Despite research on regional groundwater dynamics dating back to more than a century (Dupuit 1863; Forchheimer 1886; King 1899), there are still open questions regarding the controls of regional groundwater flow and its associated environmental manifestations.

Since the 1960s, a fundamental shift in the understanding of groundwater hydrology was initiated, using analytical solutions of idealized rolling terrain with a homogenous subsurface. Tóth (1963) demonstrated connections between topography and groundwater fluxes. He identified three types of groundwater flow that can occur simultaneously: local flow is recharged and discharged at adjacent local maxima and minima, respectively; intermediate flow has multiple topographic highs and lows between discharge and recharge areas, and regional flow is recharged at the highest water divide and discharged at the lowest part of the basin. He further demonstrated that the relative importance of local flow systems increases with topographic relief. Subsequent work showed that topography contributes to groundwater movement across many spatial scales; steeper topography can be associated with deeper water table depths, more regional groundwater flow and increased groundwater imports and exports to surface water bodies (Tóth 1963; Marklund and Worman 2007; Schaller and Fan 2009). This has emphasized the dimensionality role of groundwater been a geologic agent and helped in the understanding of a diverse range of physical, chemical and ecological phenomena. In this framework, the methodology adapted is based on groundwater flow system in which the importance of the study scale is fundamental, as the understanding of groundwater functioning needs to be achieved at a regional scale. This methodology is not limited to the surface watershed but adapts regional boundary conditions which are the topography, the depth to the basement rock and the regional groundwater divide.

In the last decades, hydrological studies have taken a direction where the application of multivariate statistical methods (i.e. clustering and factorial analyses) has been widely used for hydrogeochemical characterization of groundwater quality. This kind of study is not presenting an

S. Ouysse (✉) · J. J. Carrillo-Rivera
Institute of Geography, National Autonomous University of Mexico, Mexico City, Mexico
e-mail: samira@igg.unam.mx



M. Abrunhosa et al. (eds.), *Advances in Geoethics and Groundwater Management: Theory and Practice for a Sustainable Development*, Advances in Science, Technology & Innovation,
https://doi.org/10.1007/978-3-030-59320-9_99

alternative or continuity of the modern hydrogeology context, rather constitutes a large challenge regarding the degree of accuracy of the outcomes, especially when these methods have been made much easier to use with the development of inexpensive, fast computers and powerful analytical software. Subsequently, various investigations are carried out lacking elementary and basic background in statistics, which conducts to misleading conclusions. Groundwater studies are multivariate and multi-dimensional in nature and relying solely on finding patterns, and clustering reduces the information associated with dimensionality and limits the general understanding of the involved thermo-physical and chemical processes. In addition to data standardization which considers that all variables have the same initial weight or orients results towards variables that have more variance, parametric statistics remove outliers from the dataset, which evokes the question of reliability of the interpretation which may affect flow systems components and indexes (end members). Subsequently, the results of statistical methods can be reasonable but not significant, and interpretations are largely subjective lacking an insight and control in the natural and technical processes that could affect groundwater from its sampling phase to the step of analyses and conceptualization.

## 2 Materials and Methods

To achieve a better groundwater understanding that might lead to a suitable management, the strictness in water sampling method presents a basic phase towards more rational subsequent interpretation processes. Commonly, the international groundwater sampling guidelines are not strictly followed, and sometimes ignored due to the limited knowledge, technical skills and research infrastructure. Hydro-physico-chemical analyses when done in certified laboratories are supposed to provide information with relatively high accuracy. Yet, the obtained results need to be controlled and verified when processing the data; however, the existing control measures might be irrelevant when results come from uncertified laboratories as their equipment facilities, performance, precision, standard and accredited personal cannot be ensured. In the groundwater flow system method, a procedure is to follow as to control the study shortcoming and support the conclusion. The use of geochemistry and isotope methods in hydrogeology usually involves a well-planned sampling and analytical program Clark (2015).

### (a) **Sampling phase**

During groundwater sampling, the process must be strictly conducted according to international guidelines. Geochemical and isotope sampling requires measurement of several field parameters that are sensitive to change during sampling, transport and storage Clark (2015). The important parameters to be measured in the field are: pH, temperature, alkalinity, redox potential (Eh), electrical conductivity (EC) and dissolved oxygen (DO). Samples should be preserved accordingly avoiding the expulsion of $CO_2$. All geochemical analyses of anions and cations must be made in a certified laboratory. To assure data quality, at least one duplicate set per sampling event per analysis is required.

### (b) **Data filtering**

The first step of verification process is the cation/anion balance equation to check the sampling and laboratory analyses. This procedure will help to filter data and eliminate or resample cases with pronounced errors. Since it is important to understand groundwater functioning under natural conditions, the second step will concern the elimination of contaminated waters; i.e. groundwater presenting a value of nitrates >10 mg/L ($NO_3$) can be good indicator for modern contaminated waters or mixing; however, this presents a limiting factor in achieving good understanding of groundwater origin and associated natural physico-chemical processes.

### (c) **Graphical and numerical analyses**

Alike statistical studies which reduce the given information of original data, in groundwater flow system concept, pristine data are key factors in classifying and understanding groundwater evolution. The objective is to use dimensionality and cross-check all variables as to reach a meaningful and scientifically based conclusion. This is achieved using major ions graphical display, correlation, modelling of geochemical reactions, residence time tracer's verification, groundwater flow conceptualization and modelling where further morphological, geological and biotic variables are involved. Correlation is highly used to detect relationships between variables, hidden patterns and end members in the data sets; however, it is important to indicate that correlation does not necessary imply causation; thus, the multi-dimensional scheme of groundwater flow methodology helps to analyse data with a critical view and assess whether the chemical and isotopic results are sound and in agreement with other environmental components.

## 3 Results

Groundwater is a complex system which is in a simultaneous interaction with the environmental components and is controlled by topography, geological features and climate;

nonlinear interactions between these drivers within large real-world systems are not well understood and are difficult to characterize given sparse groundwater observations in time and space. Studies that are often depending on statistical bias without considering a fully integrated physical and conceptual groundwater flow functioning are profoundly misrepresenting what modern hydrogeology has demonstrated. Groundwater flow system methodology limitations and uncertainties are likely small, since the results are in perpetual validation with updated and higher-resolution data set. Thus, it is judicious to adapt this integrated approach in order to advance our understanding of groundwater in large heterogeneous domains that are difficult to evaluate with limited analytical approaches or observations.

## 4 Concluding Remarks

Groundwater functioning and its associated environmental problems are multidimensional; thus, in modern hydrogeology, it is essential to interpret information following an integrative approach that discusses in-depth the actual groundwater conditions. In the last few decades, the use of statistical explorative methods to reduce dimensionality has increased in the study of groundwater quality; this indiscriminate use of these methods has conducted to subjective unverified results where data sets had lost their interpretability. In groundwater studies, dimensionality reduction and data compression need to be evaluated and constrained by modern hydrogeology methods in order to have better control and understanding of the underlying natural processes in their large extent.

## References

Clark I (2015) Groundwater geochemistry and isotopes. CRC Press, Taylor and Francis group

Dupuit J (1863) Études théoriques et pratiques sur le mouvement des eaux dans les canaux découverts et à travers les terrains perméables. Dunod, Paris

Forchheimer P (1886) Über die ergiebigkeit von Brunnen-Anlagen un Sickershlitzen. Z Arch Ing Ver Hannover 32:539–564

King FH (1899) Principals and conditions of the movement of ground water. U.S. Geological Survey, Washington, DC, pp 89–91

Marklund L, Worman A (2007) The impact of hydraulic conductivity on topography driven groundwater flow. Publ Inst Geophys Polish Acad Sci E-7(401):159–167

Schaller MF, Fan Y (2009) River basins as groundwater exporters and importers: implications for water cycle and climate modelling. J Geophys Res 114:D04103

Tóth J (1963) A theoretical analysis of groundwater flow in small drainage basins. J Geophys Res 68(16):4795–4812

# Groundwater: Geological, Legal, Social, and Ethical Challenges of a Unique Natural Resource

**Introductory Note**

Patrícia Ferraz de Matos (ISC, University of Lisbon, Lisbon, Portugal)

This part aims to reflect on the challenges that groundwater, as a unique natural resource, raises, both in the present and in the past, and in different geographical contexts (Europe, Latin America, North Africa, and Asia). The contributions include aspects related to climate change, sustainability, and efficiency, in rural and urban contexts. It is intended to have technical perspectives from hydrogeology and groundwater, but also takes into account aspects provided in the law that protect, or not, groundwater, and to reflect on the effects that these subjects have on people's lives, seeking, whenever possible, a social perspective, since the reflection on ethics should include this knowledge and the learnings acquired with the ethnographic methodology.

The part has two keynote lectures, one on the protection and management of groundwater and another on the groundwater-fed plot-and-berm agroecosystems. It also has several chapters on subjects that include ancestral groundwater techniques, governance and social participation, management of aquifer systems, and impressive examples of music inspired in groundwater and the hydrological cycle.

The purpose of the part is to encompass the knowledge of hydrogeosciences, and specifically about groundwater science and engineering, with the ethical issues raised, seeking to contribute with new approaches.

**Highlights**

- Ancestral groundwater techniques and ethics;
- Governance and social participation;
- Sustainable management of aquifer systems;
- Impressive examples of the music inspired in groundwater and the hydrological cycle;
- Interlinkage of the knowledge of geology, and groundwater, with ethical issues.

# Protection and Management of Groundwater: An Invisible Vital Resource

Carlos Almeida


## Abstract

In many regions, groundwater is the main source for water supply, irrigation, and industry. Groundwater reservoirs, the aquifers, can contain important reserves of groundwater mitigating the effects of drought and other phenomena related to climate change. Furthermore, groundwater is less vulnerable to many types of contamination, although when contaminated its restoration could be more problematic. Groundwater resources are renewable but not unlimited, often subject to overexploitation which has negative consequences: economic, environmental, subsidence, etc. Definitions of criteria to be used as a guide to sustainable management have been proposed by several authors. However, the enforcement of those criteria can be challenging because they have to reconcile opposed interests. Aquifer modeling and monitoring are useful tools for guidance in aquifer management. This one can be modified according to the results of the actual exploitation (adaptive management). The most widespread type of contamination results from agriculture. This problem is difficult to deal with because population growth implies more food production. Population growth and the crescent tendency to migration from the country to big cities pose big stress on water resources resulting in widespread contamination due to poor sanitation conditions.



## Keywords

Groundwater • Sustainability • Overexploitation • Contamination • Monitoring



C. Almeida (✉)
Department of Geology, University of Lisbon, Lisbon, Portugal
e-mail: calmeida96@gmail.com




M. Abrunhosa et al. (eds.), *Advances in Geoethics and Groundwater Management: Theory and Practice for a Sustainable Development*, Advances in Science, Technology & Innovation,
https://doi.org/10.1007/978-3-030-59320-9_100

## 1 Introduction

Due to the transversal character of geoethics and groundwater, there will be presented a review of some key concepts aimed at a wide audience that is to some extent involved in groundwater management: scientists, engineers, and technicians working in water management agencies, economists, social scientists, and stakeholders.

Groundwater is by far the most important source of liquid freshwater. Groundwater accounts for about 96.2% of all freshwater in liquid form, whereas superficial water amounts to only 0.3%. Unlike many other resources (ores, fossil fuels, etc.), water resources are renewable, with just a few exceptions. However, this does not mean that they are not limited, and its exploitation must take into account its sustainability. According to FAO, groundwater supplies 50% of the world population and represents 43% of irrigation water. About 2.500 billion people depend exclusively on groundwater. About 70% of the population of China depends on groundwater. Furthermore, groundwater reservoirs, the aquifers, contain a volume of water that exceeds by far the water contained in superficial reservoirs. Just to give an example, the biggest aquifer in the world, the Alter do Chão Aquifer (Brazil), contains more than 3.5 times the volume of water of Lake Baikal, the world's biggest superficial reservoir of freshwater: 86.000 $km^3$ versus 23.600 $km^3$. As a consequence of the great volume of resources stored in aquifers, they are more resilient to climate fluctuations, e.g., droughts, acting as a cushion that permits easier water management: Shah (2009).

Alley et al. (1999) define "*ground-water sustainability as development and use of ground water in a manner that can be maintained for an indefinite time without causing unacceptable environmental, economic, or social consequences.*" They recognize that this definition is largely subjective and can lead to a wide set of criteria. Several concepts are related to sustainability, like *safe yield*, *available groundwater resources*, *overexploitation*, etc. According to the

DIRECTIVE 2000/60/EC, "*available groundwater resources means the long-term annual average rate of overall recharge of the body of groundwater less the long-term annual rate of flow required to achieve the ecological quality objectives for associated surface waters specified under Article 4, to avoid any significant diminution in the ecological status of such waters and to avoid any significant damage to associated terrestrial ecosystems*". It is important to stress that to ensure an ethical development of groundwater resources, we have to try not only to satisfy the current needs for various purposes but also to preserve the resource for future generations. Furthermore, avoiding adverse effects such as land subsidence, seawater intrusion, degradation of wetlands and water quality, negative impacts on superficial water bodies and rivers should also be considered (Liu et al. 2010).

## 2 Evaluating Sustainable Groundwater

To quantify the groundwater that can be exploited sustainably, several concepts were developed; all of them are more or less dependent on the groundwater budget of the aquifer being exploited. The belief that the groundwater budget of an aquifer in his natural state, i.e., before development, could provide guides to sustainability is profoundly wrong and is known as the groundwater myth (Bredehoeft et al. 1982). It is wrong because it does not consider the character dynamic of an aquifer system which reacts to stresses imposed on it. When development starts, the abstractions can induce more recharge (induced recharge) and reduce natural discharges. In consequence, the functioning of an aquifer in natural conditions is different when the development starts. So, the mean recharge in those conditions cannot be used to estimate the safe yield. The sum of induced recharge and the quantity of reduced discharge is known as capture (Lohman 1972). So, how much discharge could be reduced and how the impacts on recharge are acceptable are options of the management.

To stress the notion that the groundwater myth was a wrong concept, Bredehoeft et al. (1982) made a strong and provocative statement: After all, the virgin recharge, i.e., before development, was not at all significant. The groundwater myth was revisited in other papers: Bredehoeft (1997, 2002) and was criticized by Zhou (2009). Zhou, starting from the formulation proposed by Bredehoeft et al. (1982), presents a modified equation where the recharge is explicitly present.

According to Bredehoeft et al. (1982), in natural conditions, the balance equation is

$$R_0 = D_0 \tag{1}$$

where $R_0$ is the natural recharge and $D_0$ is the natural discharge. When the development starts, Eq. (1) should be modified as follows:

$$(R_0 + \Delta R_0) - (D_0 - \Delta D_0) - P = \mathrm{d}V/\mathrm{d}t \tag{2}$$

where $\Delta R_0$ is the increased recharge induced by pumping, $\Delta D_0$ is the decreased discharge, $P$ is the abstraction rate, and $\mathrm{d}V/\mathrm{d}t$ is the time variation of the volume of water stored in the aquifer. In the case of sustainable development, $\mathrm{d}V/\mathrm{d}t$ should be null and the equation reads:

$$(R_0 + \Delta R_0) - (D_0 - \Delta D_0) = P_s \tag{3}$$

where $P_s$ represents the sustainable abstraction. Inserting (1) in (3), one gets $(\Delta R_0 + \Delta D_0) = P_s$ which means that the sustainable abstraction is originated by the sum of the induced recharge and the decrease of discharge (capture). Zhou (2009) introduced the variable *residual discharge*: $D_r = D_0 - \Delta D_0$ and introducing this new variable in Eq. (3), the balance equation becomes:

$$P_s = R_0 + \Delta D_0 - D_R \tag{4}$$

In this new equation, the natural recharge is explicitly present, a formulation that seems much more intuitive. Now, the sustainable abstraction equals the natural recharge plus the induced recharge, minus the residual discharge. The evaluation of the variables of the balance equation is not an easy task (Healy et al. 2007). Perhaps the most problematic term in the Eq. (4) is the evaluation of the recharge. Many numbers of methods for its evaluation have been proposed, but the degree of uncertainty is discouraging. Mathematical modeling and monitoring are valuable tools for the management of water resources.

## 3 Agriculture, Groundwater Overexploitation and Impacts on the Quality

### 3.1 Agriculture and Groundwater Overexploitation

Groundwater sustainability and overexploitation are complex concepts as it was extensively discussed by several authors, e.g., Custodio (2002) and Llamas (2004).

Agriculture is the most important user of groundwater. The big change occurs when the so-called green revolution has started. The green revolution permitted a massive production of foods, even in arid or semi-arid regions by changing from rainfed farming to irrigated farming. Furthermore, in the green revolution, the production is enhanced by the use of fertilizers, better seeds and pest control. This change leads to frequent overexploitation of aquifers with negative consequences, and in the most important food producers, China, India, and the USA, the situation is already critical requiring urgent measures. India is the most important water user in the world with an estimated 230 $km^3$ by year, more than a quarter of the total global (FAO 2010), corresponding to 60% of irrigated area and 85% of rural water supplies (FAO 2010).

The consequences of overexploitation are multiple: environmental like drying wetlands, decreasing spring discharges and river baseflow, economic, due to the lowering of water level, and subsidence of the terrains, increasing their susceptibly to floods, and ruining buildings and other infrastructures, changes in water quality, etc. Overexploitation is an important issue as it can compromise food security.

The import/export of agricultural products represents a virtual transfer of groundwater. This *Virtual Groundwater Transfer* (VGT) is often overlooked (Fig. 1).

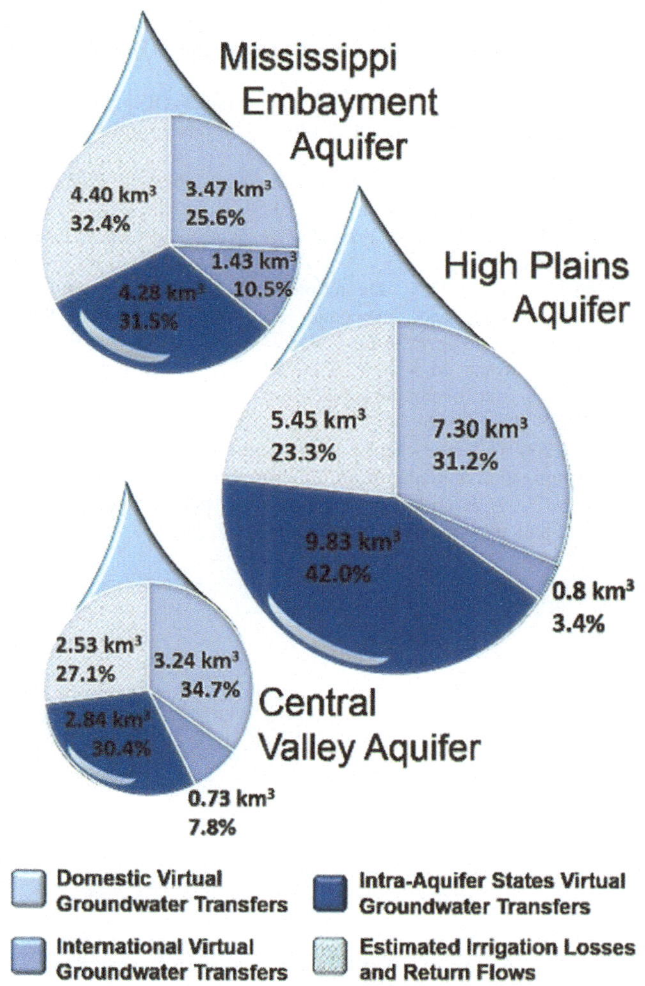


**Fig. 1** Volumes of groundwater withdrawals for agriculture, indicated by the size of each circle, in three of the main aquifers of U.S. VGT which are also indicated (Marston et al. 2015)

When overexploitation occurs, the abstraction exploits the reserves of the aquifer, eventually until its complete exhaustion (groundwater mining). This situation is not necessarily a bad option because it will permit the development of a region and provide enough time to look for different solutions (Llamas 2004). A famous example is the exploitation of the Nubian Sandstone Aquifer System the world's largest fossil water aquifer, containing an estimated 150.000 $km^3$ of water. In Spain, there are several cases of water mining, as was pointed out by Custodio (2000, 2002). By 2025, the number of people living in water-stressed countries is projected to climb to 3 billion—more than a sixfold increase (Cosgrove and Rijsberman 2000).

Diagnosing overexploitation could be made by observing some of the aforementioned effects. Monitoring water levels is one of the most cheap and important tools. However, the analysis of the evolution of water levels should be made carefully. As it was already referred, aquifers react sluggishly, and a transient trend should not be confounded with a persistent one. The time to adapt to a new situation depends on aquifer dimensions and hydraulic properties.

## 3.2 Agriculture and Impacts on Water Quality

The most widespread issue with groundwater quality is the increase in nitrates concentration, mainly because of the application of fertilizers. The solution for this trend is problematic because one has to reconcile the crescent pressure for more foods and to maintain or decrease the nitrate levels to acceptable values.

The definition of nitrate vulnerable zones and the acceptance of good agricultural practices have permitted some mitigation. However, the high concentration of nitrates in many areas is making the water unusable for human consumption or requiring proper treatment.

In general, groundwater is less vulnerable than superficial waters to other compounds used in agriculture, like phosphates, potassium, pesticides, etc. Potassium and phosphates have low mobility as they are prone to be adsorbed, precipitated, or coprecipitated. The modern pesticides have a short half-life and have high affinity to adsorption by soils, mainly if they are rich in organic matter, which facilitates its decay.

## 3.3 Karst Aquifers

Karst aquifers are among the most productive aquifers in the world and are of extreme importance in some regions, e.g., Mediterranean Basin, the USA, and China.

Karst aquifers present some special characteristics like high recharge rates, sometimes more than 50% of the precipitation, a scarce capacity of regulation, i.e., great part of the recharge has short residence time, the presence of big springs that can be used for supply and irrigation and high vulnerability. Due to these special characteristics, they are not often overexploited, because they can store big reserves that are not easily accessed. As a result of more intensive exploitation in some regions or due to natural fluctuations of the recharge, it is relatively frequent the occurrence of collapses, which sometimes can be very damaging for people and properties.

To face the vulnerability of karst aquifers, the recharge zones should be subject to special measures of protection.

The Edwards Aquifer (Texas) is an example of an important karst aquifer as it is one of the most prolific artesian aquifers in the world. It supplies about two million people and is the principal source for agriculture and industry in the region. The Floridan aquifer system is also one of the most productive aquifers and supplies drinking water for nearly 10 million people.

## 4 Concluding Remarks

Groundwater is a vital resource, being one of the most important sources of freshwater for human consumption, agriculture, and industry. Although groundwater is a renewable resource, it is not unlimited, and its management should be made in a sustainable way to avoid overexploitation. The management of groundwater resources is a difficult challenge for all people involved in it because the increase in population causes an increase of consumption and an increase in food production, which implies more consumption for irrigation and threats on water quality (increase in the concentration of nitrates, pesticides, and other constituents). The import/export of foods represents a virtual transfer of water, sometimes from regions with scarcity to regions with groundwater surplus. The effect of climate change on groundwater balance is another factor that should be considered.

## References


Alley WM, Reilly TE, Franke OL (1999) Sustainability of ground-water resources. US Geological Survey Circular 1186, 79 pp

Bredehoeft JD (1997) Safe yield and the water budget myth. Ground Water 35(6):929

Bredehoeft JD (2002) The water budget myth revisited: why hydrogeologists model. Ground Water 40(4):340–345

Bredehoeft JD, Papadopulus SS, Cooper HH (1982) Groundwater: the water-budget myth. Sci Basis Water-Resour Manage Stud Geophys 51–57

Cosgrove WJ, Rijsberman FR (2000) World water vision: making water every-body's business. Earthscan, London, UK

Custodio E (2000) The complex concept of groundwater overexploitation. Papeles del Proyecto Aguas Subterráneas. No. A1. Fundación Marcelino Botín. Santander, 58 pp

Custodio E (2002) Aquifer overexploitation: what does it mean? Hydrogeol J 10:254–277

FAO (2010) Deep wells and prudence: towards pragmatic action for addressing ground-water overexploitation in India. The International Bank for Reconstruction and Development/The World Bank, 120 pp

Healy RW, Winter TC, LaBaugh JW, Franke OL (2007) Wa-ter budgets: foundations for effective water-resources and environmental management

Liu J, Zhang M, Zheng C (2010) Role of ethics in groundwater management. Ground Water 48(1):1

Llamas MR (2004) Use of groundwater. Water and Ethics, Essay 7. UNESCO, 34 pp

Lohman SW (1972) Definitions of selected ground-water terms-revisions and conceptual refinements. Geological Survey Water-Supply Paper, 26 pp

Marston L, Konar M, Cai X, Troy TJ (2015) Virtual groundwater transfers from overexploited aquifers in the United States. PNAS 112(28):8561–8566

Shah T (2009) Climate change and groundwater: India's opportunities for mitigation and adaptation. Environ Res Lett 4:035005, 13 pp

Zhou Y (2009) A critical review of groundwater budget myth, safe yield and sustainability. J Hydrol 370:207–213

# Groundwater-Fed Plot-and-Berm Agroecosystems in Aeolian Sand in the Mediterranean Basin

Joel Roskin and Itamar Taxel

**Abstract**

Overcoming liabilities of loose sand such as scarce nutrients and low water retention remains an agricultural challenge. "plot-and-berm" (P&B) agroecosystems, situated in dune fields, consist of sunken agricultural plots between 1 and 5 m high berms, overcome these constraints. The plots situated over a 1–3 m deep and perched groundwater table ease and control access for crop roots and/or human water extraction and may make this agrotechnology resilient to short-term drought. Refuse and organic material enrich the sandy soil in plots. The agroecosystems require significant resources for construction and maintenance. The earliest agroecosystems are Early Islamic (nineth-early twelfth centuries AD) ones along Mediterranean coastal dune fields of Israel. Arabic literature reviews have not found descriptions of P&Bs, but this effort may be an original type of *mawāt* (Arabic: "dead") unowned wastelands reclamation, an important issue in Islamic economic history. Partially active and similar P&B agroecosystems along the southern Mediterranean region in Northern Sinai/Gaza Strip, Algerian Sahara, and Iberia yield fruit and vegetable crops and date back to the Middle Ages. Their concept may have originated from the Early Islamic ones. Being in stressed environments, P&B agroecosystems can be re-established into groundwater-efficient community-farming econiches that can nurture geoethical relations between science and local populations.

**Keywords**

Groundwater • Agroecosystems • Early Islamic period • Mediterranean • Water harvesting • Geoethics

## 1 Introduction

Ancient water harvesting and agricultural systems in marginal zones have intrigued generations of hydrogeologists, archaeologists, agronomists, and geomorphologists (Evenari et al. 1982; Beckers et al. 2013; Avni 2018). One of the reasons driving the study of ancient water harvesting is the effect of climate change and anthropogenic activity on the world's dwindling and contaminated water resources and the degrading quality and often overexploited global reserves of suitable soil for agriculture.

Ancient water harvesting developed in various fashions and in different climate zones of the Mediterranean basin and western Asia. People usually harvested surface/overland waters for agriculture, but rarely shallow groundwater. The distance between water sources and suitable areas for agricultural cultivation presented engineering and maintenance challenges (Beckers et al. 2013). Water loss end route to the fields via conduits lime channels, pipes and tunnels was imperative for farmers.

Sporadic research has shown that mainly since the past two millennia, high water tables covered by sand sheets have been utilized for agricultural purposes in discrete locations in the Americas, Asia, North Africa and southwest Europe (Roskin and Taxel 2018 and references within). Populations of these regions developed unique methods to overcome and harness the challenges of inert sand properties such as scarce nutrients

J. Roskin (✉)
Analytical Laboratory, The Artifacts Treatment, Conservation and Laboratories Dept., Israel Antiquities Authority, Jerusalem, Israel
e-mail: yoelr@post.bgu.ac.il

J. Roskin
Geomorphology and Portable Luminescence Laboratory, Leon Recanati Institute for Maritime Studies, University of Haifa, Haifa, Israel

I. Taxel
Archaeological Research Department, Israel Antiquities Authority, Jerusalem, Israel



M. Abrunhosa et al. (eds.), *Advances in Geoethics and Groundwater Management: Theory and Practice for a Sustainable Development*, Advances in Science, Technology & Innovation,
https://doi.org/10.1007/978-3-030-59320-9_101

and low water retention (Tsoar and Zohar 1985). However, the reasons and knowhow for undertaking such agro-engineering efforts and the resulting productivity are unclear.

In the Mediterranean basin, the Roman and Byzantine periods are well known for significant technological advancements in overland water harvesting for agriculture (Evenari et al. 1982; Butzer et al. 1985; Lucke et al. 2019). Recent studies have demonstrated that Early Islamic-era populations, as part of their expansions into Mediterranean zones furthered such efforts, by expanding existing methods of shallow groundwater exploitation such as the qanat irrigation system (Avni 2018).

The well-known qanat system extracts shallow groundwater from upper reaches of alluvial fan aquifers and transports the water downhill by tunnels to farmlands. The system originated in ancient Persia in the 1st millennium BCE and was later adopted by the Muslims who were responsible for its introduction via the Levant along the southern Mediterranean region to Iberia during the Umayyad period (Avni 2018). However, qanats are not the only Early Islamic agrotechnological innovation.

Plot-and-berm (P&B) agroecosystems (Taxel et al. 2018) comprise a notable and relatively unrecognized phenomenon of agricultural groundwater harvesting within loose sand. Remains of Early Islamic—to early Crusader-period (tenth-twelfth centuries CE) agroecosystems of Israel's Mediterranean coastal dune fields point to sophisticated agricultural utilization of a shallow groundwater table in agricultural hinterlands (Fig. 1a–c). The climate here is Mediterranean, with cool, rainy winters and hot rainless summers. Mean annual precipitation, 400–600 mm from south to north, is understood to annually recharge the underlying phreatic coastal aquifer that as a major water source has undergone significant exploitation (Schwarz et al. 2016). This paper reviews Early Islamic and active groundwater-fed P&B agroecosystems in the Mediterranean basin that is hypothesized to have stemmed from the former and their potent to be preserved and redeveloped for groundwater-efficient community farming.

## 2 Methodology

This study relies on geological, geomorphological and archaeological surveys complemented by field and laboratory work on Early Islamic P&B agroecosystems in Israel (details in Shtienberg et al. 2017; Taxel et al. 2018). Spatial coverage of the Mediterranean P&B agroecosystems systems is based on Google Earth Pro, regional contacts, peer-reviewed papers, reports and traveller accounts (in English and French) and a review of medieval Islamic agricultural manuals (*kutub al-filāḥa*) in Arabic.

## 3 Results

### 3.1 Early Islamic P&B Agroecosystems in Israel

Plot-and-berm agroecosystems have been studied in the coastal dune fields of Israel, south of *Caesarea Maritima* (Porath 1975; Roskin et al. 2015; Shtienberg et al. 2017), between Tel Yavneh (Early Islamic Yubnā) and its harbour town Yavneh-Yam (which in Early Islamic times functioned as a military stronghold called Māḥūz Yubnā; Taxel et al. 2018), and near Ziqim several km south of Tel Ashkelon (Fig. 1a). These chessboard-shaped agroecosystems consist of human-made rectangular depressions (plots) tens to several hundred m wide, confined by sand berms with varying degrees of preservation that have a spatial crisscross pattern (Fig. 1b). The berms, usually 1–5 m high, have a 1.5–5 m wide flat crest, and a slope angle range of 20°–30°. They are topped by a scatter of often weathered artifacts (notably pottery sherds, but also glass, stone and metal objects) usually 5–20 cm deep, mixed with grey sand/sandy loam. This coating/mix protects the berms from erosion until today.

The Caesarea and Ziqim agroecosystems, situated in coastal lowlands by stream mouths, border the backbeach (Fig. 1b, c). The Yavneh agroecosystem, several km from the beach, lays in an interdune trough, between coast-parallel aeolianite ridges. Modern depth ranges of seasonal groundwater, perched upon clay loam, is 1.5–2.5 m (Roskin et al. 2015; Taxel et al. 2018). Hydrophilic vegetation is absent in the plots.

A 5–50 cm thick grey sand/sandy loam unit in the plots is interpreted as an anthrosol (Fig. 1c). This anthrosol, often close to the surface, has well-preserved upper and lower boundaries within the encompassing "clean" sand. These anthrosols consist of several per cent of soil nutrients, charcoal bits, ash, artifacts, and organic refuse mixed with the sand substrate for soil enrichment. The location of the Caesarea and Yavneh P&B agroecosystems and the artifactual contents of their anthrosols indicate that the refuse originated from their nearby town dumps (Shtienberg et al. 2017; Taxel et al. 2018).

Optically stimulated luminescence (OSL) dating based on the underlying sand beneath the anthrosol at Yavneh showed that the age of construction or the last natural deposition of aeolian sand prior to P&B construction was $1.1 \pm 0.04$ ka (Taxel et al. 2018). The luminescence age of the overlaying anthrosol sand of $0.88 \pm 0.04$ ka (1096–1176 CE) was strikingly similar to OSL ages from the Caesarea agroecosystem (Roskin et al. 2015; Shtienberg et al. 2017), suggesting a similarity in abandonment time and possibly similar circumstances. Pottery found within the anthropogenic refuse of the P&Bs was dated to the Early Islamic

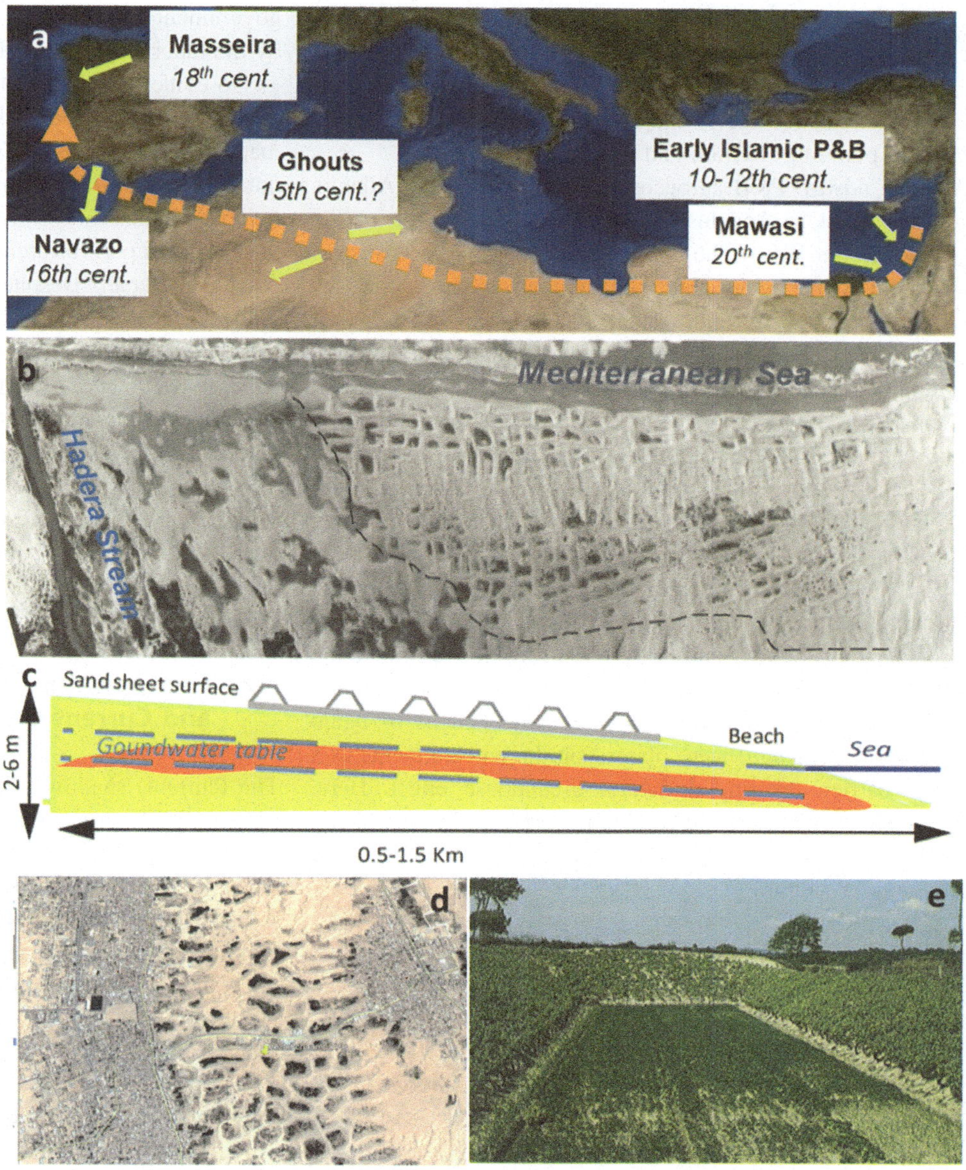


**Fig. 1** **a** Google Earth Pro image of P&B agroecosystems, current but partial understanding of their establishment times and proposed tempo-spatial dissemination route (orange dashed line). **b** Aerial photograph (1946) of the Early Islamic Caesarea P&B agroecosystem. The (dark) western plots near the beach (upper part of image) seem to be flooded. **c** Scheme of the Caesarea P&B agroecosystem (anthrosols in grey) and slightly fluctuating perched groundwater table (light blue dashed line) (modified from Taxel et al., 2018). **d** Google Earth image of Oued city, NE Algeria (33° 17.364′ N/6°54.996′ E) where ~9500 active Ghouts are undergoing sewage contamination and flooding and also groundwater level drop. **e** Active Portuguese Masseira P&B agroecosystem (details in Matos 1973). Note crops on plot-and-berms. Image from: http://saberdageografia.blogspot.com/2012/04/os-campos-masseira.html; accessed by April 2020

period (not before the ninth century CE) to sometime after the Crusader conquest (Porath 1975). OSL ages of the anthrosols that was younger than the surface artifacts showed a stratigraphic age inversion, suggesting that the artifacts are not in situ and were placed there. The spatial distribution of P&B agroecosystems suggests that during Early Islamic times this previously unrecognized type of agrarian activity was well established along the Mediterranean coast of Israel.

## 3.2 Mediterranean Distribution of Partially Active P&B Agroecosystems

Among the challenges of analysing ancient water harvesting systems is the lack of modern agricultural analogues (Evenari et al. 1982). Intermittently along the southern continuation of the Israeli coast in the southern Gaza Strip and northeastern Sinai Peninsula stands the *mawasi* plot-and-berm agroecosystem. The *mawasi* is known since the twentieth century (Tsoar and Zohar 1985) (Fig. 1a).

West of the Nile, only in Algeria and Iberia, active P&B agroecosystems are identified. The *ghout* agroecosystem in the NW and SE Algerian Sahara sand sheets and dunes is comprised of patches of thousands of man-made sunken plots between sand berms (Fig. 1a, d). The origins of the *ghout* method, believed to have first appeared in the fifteenth century (Remini and Kechad 2011), are unknown. Date palms are cultivated in *ghout* plots, slightly above a high groundwater table while additional crops are grown at higher levels. This means that the crops that require no external water source or irrigation are highly dependent on climate stability, i.e. the corresponding seasonal and periodic changes in the groundwater table and its water quality. The hydrology of the *ghout* agroecosystems is currently degrading due to infiltration of urban sewage or/and decrease in groundwater levels due to pumping (Remini and Kechad 2011).

The *navazo* P&B agroecosystems are along the Atlantic Ocean coast in Andalusia, especially at the mouth of the Guadalquivir River. The origin of the *navazo* agroecosystem and the source of its name are unresolved. Spanish documents since the sixteenth century mention *navazos* though it may be associated with the ninth-fourteenth-century Islamic domination in southern Spain. Today, these agroecosystem remains boost biodiversity and are utilized for community horticulture Spain (Sánchez and Cuellar 2016).

The *masseira* (or *gamela*) is an interdune agroecosystem found along the northern Atlantic coast of NW Portugal (Matos 1973), between Esposende and Vila do Conde (Fig. 1a, e). Its names allude to the shape of a wooden trough used to knead dough. Here, until the nineteenth century, abandoned sandy plots were developed into fertile farmlands. The berms (called *valados*) were traditionally stabilized with grapevines. Occasionally, hedges of ngaio (*Myoporum laetum*) were planted on the embankments as a heat barrier, creating a natural greenhouse effect. Here, seaweed provides for an excellent fertilizer for the sandy soils since ancient times (Sánchez and Cuellar 2016 and references within).

## 4 Discussion

### 4.1 P&B Agroecosystems Origins and Tempo-spatial Distribution Pattern

Early Islamic agroecosystems of the Israeli coast are very similar to the later and active agroecosystems in Gaza, northern Sinai, Algeria and Iberia though each agroecosystem has slightly different characteristics (Fig. 1). The P&B agroecosystems require considerable resources for construction and maintenance, which raises intriguing questions concerning their construction motives and relevant agrotechnological expertise.

Early Islamic agroecosystems funded on adjacent hinterlands of coastal towns of ancient Israel may be a form of *mawāt* (Arabic: "dead") land reclamation, which was an important issue in Islamic economic history (Salasal 1998; Debs 2010). *Mawāt*, which we learn about in Early Islamic juristic documents, refers to unowned wastelands. Early Islamic jurists prescribed rules for *mawāt* vivification and acquisition and their typical cultivation using irrigation systems such as the qanat. It is unclear whether the presumably *mawāt* lands in the sand of the Israeli coastal plain were official land grants (Arabic: *qaṭā'i'*) provided by the government to individuals or were independently taken over by the latter (regardless of social/political rank).

No evidence of P&B agroecosystems has been found in Islamic texts from Iberia (Kirchner 2019) nor have historical maps (e.g., Jacotin 1826; Conder and Kitchener 1880) pointed out this unique landform in Israel. Islamic texts partially reviewed in this study have not referred to such methods, and there are no reports on archaeological remains of such systems outside of Israel. Despite the leading logic that the P&B innovation was developed by a centralized administration, the lack of documentation may indicate that the agroecosystems were not initiated by an administration but rather were sometimes a local/regional "bottom-up" initiative of "people without history" (Erickson 2006). The active agroecosystems may be an inherited or revived manifestation of the possibly original Early Islamic effort in coastal Israel that similar to the qanat dissemination route (Avni 2018), spread across Muslim-controlled North Africa to Iberia (Fig. 1a).

### 4.2 Groundwater Character, Utilization and Current Geoethical Significance

The (current) shallow groundwater depths reported for the various P&B agroecosystem plots reflect agricultural effectivity. For modern agriculture in groundwater-fed sand, a water table depth of 1.5–2.0 m is optimum (Kahlown and Ashraf 2005). This depth range is similar to those of the mentioned agroecosystems. Modern agriculture water allowance needs adjustments to avoid over irrigation and in-efficient use of water but the traditional groundwater-fed plot-agriculture does not generate water losses.

Coastal P&B agroecosystems rely on sufficient rainfall infiltration via aeolian sand that collects along with existing groundwater on a shallow impermeable substrate or salt-water surface near the shoreline. The part of the precipitation that infiltrates through the sand is slowly drained underground (Kutiel et al. 2016) flows towards the sea but remains fresh (Goldberg et al. 1970). Water of "*thamila*" 1–4 m deep hand-dug pits in the sand by Bedouins along the Sinai coast for potable consumption (Hellström 1953) may have been used for hand irrigation of plots during the summer. Here annual precipitation is only 100–200 mm, illustrating the resilience of this agrotechnology.

Today, *mawasi* farmers conduct pumping, flooding and mechanized change of the plot surface elevation (Tsoar and Zohar 1985). When the water table drops, the top layer of the plot sand is bulldozed onto the barrier berms. In inland cases like Algeria and Yavneh, a synclinal like geological structure channels groundwater to topographical/structural lows (Remini and Kechad 2011; Taxel et al. 2018). These systems may preserve shallow groundwater availability along short droughts.

*Navazos* differ in relation to their distance from the coast. *Navazos* near the coast are tidal *navazos* and largely rely on tidal forces for irrigation. The rising tide shifts the freshwater–saltwater interface inland, causing a rise in the water table. The design of the *masseira* accounts for the level of the water table in the summer and includes a drainage system for dealing with excess water in the rainy season. Folklore claims that *masseira* farmers possess a unique art to access the aquifer and for levelling the (plot) surface to grow vegetables (Sánchez and Cuellar 2016 and references within).

Early Islamic agroecosystems in the Mediterranean climate zones of Israel were probably never constrained by short-term drought. Their current shallow perched aquifers have potential to be re-established for small "plot"-scale/community organic horticulture (like the active agroecosystems). This proposed effort is anticipated to enliven sustainable and productive local cultural heritage and develop hands-on geoethical bonding to the vitality of local and shallow groundwater resources (Bobrowsky et al. 2017).

## 5 Concluding Remarks

Plot-and-Berm agroecosystems reflect an innovative and resilient adaptation in marginal lands to the challenges and opportunities of aeolian erosion, inert soil and shallow groundwater availability and variability, independent of natural or artificial overland flow, water collection and distribution. It seems that agroecosystems demanded considerable resources for construction, maintenance and management. The long-term resilience of these agroecosystems to natural erosion has been unable to spare substantial expanses of these systems from destruction and groundwater degradation by modern construction. This situation is attributable to the lack of academic, government and public familiarity (Sánchez and Cuellar 2016) with these systems that are unrecognized as cultural heritage sites.

Further study of P&B agroecosystems and their physical and juridical reliance on groundwater requires particular urgency. These scientific efforts can manifest cultural development of groundwater-efficient community horticulture in such agroecosystems that will enliven local heritage, implement green niches in industrialized/urban coastal zones and nurture geoethical relations between science and local populations.

**Acknowledgements** N. Amitai-Preiss analysed Arabic texts. The Katif centre supplied valuable archive material on the *mawasi*. H. Elbelrhiti provided information on traditional Moroccan irrigation systems. Ruben Sánchez Cáceres is thanked for information on Iberian P&B agroecosystems. This study was partly supported by Gerda Henkel Foundation grant AZ 42/F/19.

## References

Avni G (2018) Early Islamic irrigated farmsteads and the spread of Qanats in Eurasia. Wat Hist 10(4):313–338

Beckers B, Berking J, Schütt B (2013) Ancient water harvesting methods in the drylands of the Mediterranean and western Asia. eTopoi J Ancient Stud 2:145–164

Bobrowsky P, Cronin VS, Di Capua G, Kieffer SW, Peppoloni S (2017) The emerging field of geoethics. In: Gundersen LC (ed) Scientific integrity and ethics with applications to the geosciences. Special Publication American Geophysical Union, vol 73. Wiley, Inc, pp 175–212

Butzer KW, Mateu JF, Butzer EK, Kraus P (1985) Irrigation agrosystems in eastern Spain: Roman or Islamic origins? Ann Assoc Amer Geogr 75(4):479–509

Conder CR, Kitchener HH (1880) Map of Western Palestine: from surveys conducted for the Committee of the Palestine Exploration Fund during the years 1872-1877, London

Debs RA (2010) Islamic law and civil code: the law of property in Egypt. Columbia University Press, New York

Erickson CL (2006) Intensification, political economy, and the farming community. in defense of a bottom-up perspective of the past. In: Marcus J, Stanish C (eds) Agricultural strategies. UCLA, Los Angeles, pp 233–265

Evenari M, Shanan L, Tadmor N (1982) The Negev: the challenge of a desert. Harvard University Press, Harvard

Goldberg D, Gornat B, Ben-Asher J (1970) Background material for the agricultural development of the Al-Arish Region and Nahal Sinai. In: The Society for Scientific Publications, Bet Dagan, Israel Bulletin 136. Volcani Center, Bet Dagan (in Hebrew)

Hellström B (1953) The ground water supply of North-Eastern Sinai. Geogr Annal 35(2):61–74

Jacotin P (1826) Mémoire sur la construction de la Carte de l'Égypte. In: Panckoucke CLF (ed) Description de l'Egypte ou recueil des observations et des recherches qui ont été faites en Égypte pendant l'expédition de l'armée française, vol VIII. Carte Topographique de l'Égypte, Paris, pp 1–102

Kahlown MA, Ashraf M (2005) Effect of shallow groundwater table on crop water requirements and crop yields. Agric Wat Manag 76 (1):24–35

Kirchner H (2019) Water management and irrigation in Medieval Mediterranean societies: an overview. In: Abate M (ed) Convivencia and Medieval Spain. Mediterranean Perspectives. Palgrave Macmillan, Cham pp 65–98

Kutiel PB, Katz O, Ziso-Cohen V, Divinsky I, Katra I (2016) Water availability in sand dunes and its implications for the distribution of *Artemisia monosperma*. Cat 137:144–151

Lucke B, Roskin J, Vanselow KA, Bruins HJ, Abu-Jaber N, Deckers K, Lindauer S, Porat N, Reimer PJ, Bäumler R, Erickson-Gini T, Kouki P (2019) Character, rates, and environmental significance of Holocene dust accumulation in archaeological hilltop ruins in the Southern Levant. Geoscience 9(4):190

Matos MF (1973) A vida rural na Apúlia. Finisterra, Rev Portg Geogr Lisboa 8(15):67–103. https://doi.org/10.18055/Finis2393

Porath Y (1975) The gardens of Caesarea. Qadmoniot 30–31:90–93 (in Hebrew)

Remini B, Kechad R (2011) Impact of the water table razing on the date degradation of El Oued palm plantation (Angeria): mechanisms and solutions. Geog Tech 1:48–56

Roskin J, Sivan D, Shtienberg G, Roskin E, Porat N, Bookman R (2015) Natural and human controls of the Holocene evolution of the beach, aeolian sand and dunes of Caesarea (Israel). Aeol Res 19:65–85

Roskin J, Taxel I (2018) Plot-and-berm agroecosystems in aeolian sand sheets: a case of global/regional cultural connectivity? In: EGU General Assembly Conference Abstracts, vol 20, p 128

Salasal SMMS (1998) The concept of land ownership: Islamic perspective. Bull Geoinformasi 2(2):285–304

Sánchez R, Cuellar M (2016) Coastal interdune agroecosystems in the Mediterranean: a case study of the Andalusian Navazo. Agroecol Sustain Food Syst 40(9):895–921

Schwarz J, Bear J, Dagan G (2016) Groundwater development in Israel. Ground Water 54(1):143–148

Shtienberg G, Shalev S, Sivan D, Shahack-Gross R, Dix J, Waldmann N, Roskin J, Yasur-Landau A, Bookman R (2017) Anthropogenic overprints on natural coastal aeolian sediments: a case study from the periphery of ancient Caesarea, Israel. Anthrop 19:22–34

Taxel I, Sivan D, Bookman R, Roskin J (2018) An early Islamic inter-settlement agroecosystem in the coastal sand of the Yavneh Dunefield, Israel. J Field Archaeol 43(7):551–569

Tsoar H, Zohar Y (1985) Desert dune sand and its potential for modern agricultural development. In: Gradus Y (ed) Desert development. man, and technology in Sparselands, D. Reidel, Dordrecht, pp 184–200

# Music Inspired in Groundwater and Other Components of the Hydrological Cycle

Luís Ribeiro

**Abstract**

Water, as a symbol of fertility, death and rebirth, appears associated with many classical music works. Rivers, seas, lakes and glaciers are the leitmotivs of many composers who used them in different musical forms: symphonies, suites, symphonic poems and operas. The river, as a watercourse in continuous flow, is described in various stages of its life (Smetana: Vlatva), in its exuberance (J. Strauss: Blue Danube), as a privileged setting (Handel: Water Music) or as a celebration (Telemann: Wassermusik), or as a journey of fantastic trips (Wagner: Siegfried's voyage along the Rhine) or as a generator of life (Schubert: The trout). In the sea, the water appears metamorphosed, in the movement, in the sound and in the colour that constitute the elements of the waves (Debussy: La Mer), represented as a terrific force (Verdi: Otello), as a scenario of storms (Vivaldi: La Tempesta del Mare), or pacificator (Mendelssohn: Calm Sea and Happy Travel) and as a stage for ancestral legends (Wagner: Der Fliegende Holländer) or as theme of several narratives: (Rimsky-Korsakov: Shererazade; Britten: Peter Grimes; Zemlinsky: Die Seejungfrau) or pure poetry (Vaughan Williams: Symphony The Sea). Also, in the lakes, it is represented as a heady element (Liadov: The Enchanted Lake) or as a spell (Tchaikowsky: The Swan Lake). The water appears invisible (Denisov: Légendes des Eaux Souterraines), static, in magnificent glacial landscapes (Vaughan Williams: Antarctic Symphony) or as a scenario of a battle (Prokofiev: Alexander Nevsky) or pretext for beautiful architectures (Respighi: The Fountains of Rome). Rain inspires storms (Sibelius: The Tempest) and bonanzas (Beethoven: Pastoral Symphony) or as daydreams (Takemitsu: I Hear the Water Dreaming).

L. Ribeiro (✉)
CERIS, Instituto Superior Técnico, University of Lisbon, Lisbon, Portugal
e-mail: luis.ribeiro@tecnico.ulisboa.pt

**Keywords**

Classic music • Water • Debussy • Glass • Takemitsu • Smetana • Denisov

## 1 Introduction

Bachelard (1942) used to say that « ... *pour certaines âmes, l'eau tient vraiment la mort dans sa substance. Elle communique une rêverie où l'horreur est lente et tranquille* » .

Water is ubiquitous in most of the founding myths of civilizations. Its strong symbolic dimension has always inspired painters, poets and musicians.

Nature is full of music. There is the music of the water falling in cascades, there is the music of the water discharging in the springs, there is the music of the rain, there is the music of the rivers, and there is the music of the sea. Because it is natural, despite providing us with multiple sensory images for free, it sometimes goes unnoticed.

However, many composers dared to deepen its mystery, its rhythm, its strength, its charm, offering us unforgettable masterpieces. Real or imaginary aquatic landscapes have been both a reason for the construction of arguments of many operas and dances, as well as for the construction of scenarios where celebrations and festivities take place, as well as for the composition of impressionist musical poems through the apprehension filtered by the senses.

Nature as a source of musical inspiration is a recent invention. It will be with the romantics, realists and impressionists that the happiest musical works in which water is the protagonist will be created.

Orchestral instruments, taken in isolation, lend themselves poorly to representing water. Once combined, the result is however excellent. The same is true with the voice. Only the piano has a special propensity to express the magic of water. The crystalline notes of the piano fit very well with drops of water. There is transparency in the sounds emitted



M. Abrunhosa et al. (eds.), *Advances in Geoethics and Groundwater Management: theory and practice for a sustainable development*, Advances in Science, Technology & Innovation,
https://doi.org/10.1007/978-3-030-59320-9_102

by the piano. Notice how the musical form "barcarola" is neither more nor less than an extension of water music. The rhythm of this musical form immediately suggests the balancing of the boat, the sweet and monotonous movement of the oar. Fauré was one of the composers who was able to offer us this sensation.

The impressionists were able to capture the luminosity, colour and rhythm of the water better than anyone, as is the case with Debussy in "Reflets dans L´Eau" or Ravel in "Jeux d´Eaux". However, romantics seem to have been the first to introduce the theme of water to music. "The Water Games at Franz Liszt's Villa d'Est" is a paradigmatic example in this context.

Water is always on movement. From the time the earth was formed, it has been endlessly circulating through the hydrologic cycle. Groundwater is an important part of this continuous cycle. The hydrological cycle is a dynamic system. It is a closed system, meaning that nothing can be lost, and it can only be relocated to another part of the system. To start the cycle, water evaporates from the oceans and condenses as clouds that eventually float across the landscape and deliver their moisture in liquid (rain) or solid (snow, ice pellets) form. Some of this water runs off the landscape back to rivers that eventually flow back to the oceans, completing that part of the cycle.

Water that enters the subsurface as infiltration similarly flows back to rivers through small- to large-scale flow systems and discharges through their bases to form "baseflow", or it may recharge to aquifers and transiently stored rapidly as groundwater.

The water cycle can be considered a ballet staged on Earth. This essay proposes a brief journey through this cycle through musical examples of their main hydrological components: sea, clouds, rain, rivers and groundwater (Ribeiro 2017).

## 2 Sea

In 1905, Claude Debussy composes "La Mer"—three symphonic sketches for orchestra. Ignoring the classical rules of composition, Debussy created a new world based on a combination of textures and sounds using the entire instrument palette of the orchestra for this purpose.

To translate the impressions, Debussy uses sounds to imitate the voice of the sea, rhythms to suggest the constant movement of the waves and harmonies and timbres to translate nuances and reflections of the water.

The composer manages to capture in his work the luminosity and colours of the sea and the sky, the enigmas of nature, the eternal ebb and flow of the waters. Debussy himself defined his work as a maritime landscape without figures (Fig. 1).

**Fig. 1** Claude Debussy (1862–1918)

As the critic, Camille Mauclair wrote at the time (Ribeiro 2017): "light is used in impressionist painting in the same way that a theme is symphonically developed in impressionist musical composition". In this respect, Claude Monet's landscapes are in fact symphonies of luminous waves just as Debussy's works are a kaleidoscope of sounds of different tones.

The work is structured in three movements: the first movement, entitled: "De l'aube to midi sur la mer" (From dawn to noon over the sea) mysteriously begins as an echo in a shell—The immensity of the sea. There is the hiss of the winds and the ripple of the waves. These become more tumultuous. Debussy seems to have been inspired by Turner's seascapes that he had seen in an exhibition in 1902. The second movement entitled "Jeux de Vagues" (Game of the Waves) acting as a scherzo is a capricious dance of the elements. The third and final movement, "Dialogue du vent et de la mer" (Dialogue of the Wind and the Sea) depicts a wild and threatening sea—The mystery of a mermaid's voice. An increasingly insistent and voluptuous cry. If Turner's painting influenced the writing of the first movement, for the third it is the paintings of the monumental waves of the Japanese painter Hokusai that served as inspiration.

## 3 Clouds

The ethereal character of the clouds, their diaphanous, intangible, immaterial nature is the raw material for the elaboration of various musical compositions.

Koyaanisqatsi (Life Out of Balance) is an experimental film produced and directed by Geoffrey Reggio between 1975 and 1982. The title is a word in Hopi dialect that means Life in disaggregation, Life in turmoil. The film is an apocalyptic view of the conflict between two different worlds, urban life and associated technology and nature. The soundtrack was composed by Philip Glass (Fig. 2) and consists of six movements. In the third part of this authentic visual symphony entitled "Cloudscape"; Glass describes the movement of clouds and their shadows across New York cityscapes.

**Fig. 2** Philip Glass (1937–)

**Fig. 4** Bedřich Smetana (1824–1884)

## 4 Rain

The fascination of the Japanese composer Toru Takemitzu (Fig. 3) for the theme of water in all its manifestations is very evident in the large number of opus in which water is the protagonist, such as "Toward the sea" and "I hear the water dreaming".

Takemitsu stated that thinking in a musical way meant for him to think in a liquid way and that he wanted the musical changes to be as gradual as the tides.

This is the case of the cycle called "Waterscape", with a set of pieces whose common motive is rain: "Rain Coming", "Garden Rain", "Rain Tree" and "Rain Spell". Abstract musical pieces that, through symbolism, allow the composer to move freely establishing a network of musical meanings. The composer himself gives us the meaning of his music: «*My intention is that my compositions capture the different metamorphoses of the sea, the different shades with which water circulates in the Universe*».

## 5 River

Smetana (Fig. 4) was an indefatigable defender of the music of Czech countries. "Ma Vlast" (My Fatherland) is a great fresco that includes six symphonic poems, each evoking an episode in Czech history that pays tribute to the country of Bohemia.

**Fig. 3** Toru Takemitsu (1930–1996)

The second symphonic painting, entitled "Vltava", describes the different phases of the life of this famous river, which rises in the mountains of Šumava and flows into the Elbe, in Melnik north of Prague. The song describes its birth from two sources, one hot and one cold, the way these water lines come together to form the river, its passage through forests and plains through an enchanting landscape where castles rise over rocks, medieval and ruins of ancient civilizations, where charming parties and happy festivals take place. Then, the Vltava river turns into a whirlwind in the rapids of Saint John, launching itself into a great torrent near Prague, passes by the hill of Vyšehrad and finally disappears on the Elbe.

## 6 Groundwater

The "Légendes des Eaux Souterraines" (Legends of Subterranean Waters) were composed by Edison Denisov in 1989 on texts by the French poet Yves Bergeret. It is a choral work sung a capella by 12 soloist voices. The poet wrote the poems after a trip to Bosnia, in 1978 where he had the opportunity to visit a huge karst cave, in the midst of which a large watercourse passed, before resurfacing in a

**Fig. 5** Edison Denisov (1929–1996)

spring. The poet gazed in awe at this magnificent spectacle of dark and secret places that served as inspiration for the making of the poem. Denisov's composition has an epic dimension, translating this magical environment, in a network of musical textures, in a counterpoint, hieratic, slow and very unconventional writing (Fig. 5).

## 7 Concluding Remarks

Bachelard (1942) used to say: «*L'eau est une substance pleine de réminiscences et rêveries divinatrices*».

Music as a universal language has the power to convey the "magic" nature of water, to transport us to a spiritual dimension like no other art form can. We can navigate suggested by a melody through visual impressions.

Music stimulates the imagination by allowing you to dream with infinite images, to travel through psychic states where the combination of references is unlimited. The various states of mind can be represented simultaneously by the appearance of clouds, the flow of rivers, the coming of the waves of the sea, the rain that falls and translated into musical movements: *the andante, the adagio, the allegro, the vivace, the presto*, performed either by the piano, violin or cello, by trios and quartets, or by large orchestral or choral ensembles, ballets, operas and other scenic actions. Often, this symbiosis is perfect and when that happens the masterpiece appears.

Being the elements of nature, and especially water, a source of inspiration in the history of music, there is a sensation in this journey through the musical production of different eras that music can also be an inspiration and a means of reconnecting man with nature.

## References

Bachelard G (1942) L'eau et les rêves: essai sur l'imagination de la matière. Libraire José Corti, Paris

Ribeiro L (2017) O ciclo musical da água. Instituto Superior Técnico, Lisboa

# Revisiting Ancestral Groundwater Techniques as Nature Based Solutions for Managing Water

Luís Ribeiro

**Abstract**

To achieve water sustainability and a more efficient use of water we should base on the ancestral water and territory management knowledge and grained in the culture of the people, This work is inspired in Nature Based Solutions (NBS) for managing water availability, particularly groundwater and aquifer-related NBS that hold major unrealized potential for alleviating adverse impacts of progressive climate change, namely to increase water security/drought resilience. In some cases, more ecosystem-friendly forms of water storage, such as natural wetlands, improvements in soil moisture and more efficient recharge of groundwater, could be more sustainable and cost-effective than traditional grey infrastructure such as dams. The core of this study is centred in the pre-Inca and Inca civilizations and how these communities have developed ingenious NBS solutions to adapt to extreme climate scenarios such as prolonged droughts, managing water resources in a holistic way and how they understand clearly the global water cycle in all the components specially groundwater. The work is divided in three interlinked phases: to sow water, by implementing ancestral aquifer recharge solutions, to retain water by improve hydraulic efficiency in terms of infiltration and drainage and to collect water by improve the performance of extraction in the subterranean aqueducts in arid regions.

**Keywords**

Groundwater • Sustainability • Nature-based solutions • Recharge • Climate change

## 1 Introduction

For some decades, the world has shown its concern for a better use of water, ending with the paradigm that it is an unlimited natural resource and almost worthless. The management of water resources requires new solutions to counteract the growing challenges of water security arising from population growth and climate change. Today more than ever we must work with nature, rather than against it. The challenge we must all face is to meet this demand in a way that does not exacerbate negative impacts on ecosystems. These trends pose broader challenges, stemming from the increased risk of floods and droughts, which in turn has an impact on our ability to adapt to climate change. The great challenge is to take full advantage of nature's potential to contribute to the achievement of the three main objectives of water management—increase the availability of water resources, improve their quality, and reduce water-related risks.

The key global problem nowadays persists: the need to achieve a sustainable balance between water availability and demand especially in these times of the threats of the effects of climate change on the water resources of the planet.

The water sustainability can be advantageously achieved based on culturally anchored local knowledge and practices, by revaluating the ancestral wisdom in the management of water and territory how they have been able to overcome water problems throughout their histories, and on that basis build processes for a more efficient use of water (Vargas 2006; Yapa 2013).

This article is inspired in Nature Based Solutions (NBS) for managing water availability. The option of building more reservoirs is increasingly limited by silting, decrease of available runoff, environmental concerns, and restrictions. In many cases, more ecosystem-friendly forms of water storage, such as natural wetlands, improvements in soil moisture and more efficient recharge of groundwater, could be more sustainable and cost-effective than traditional grey infrastructure such as dams (WWAP 2018).

L. Ribeiro (✉)
CERIS, Instituto Superior Técnico, University of Lisbon, Lisbon, Portugal
e-mail: luis.ribeiro@tecnico.ulisboa.pt



M. Abrunhosa et al. (eds.), *Advances in Geoethics and Groundwater Management: Theory and Practice for a Sustainable Development*, Advances in Science, Technology & Innovation,
https://doi.org/10.1007/978-3-030-59320-9_103

As the largest liquid freshwater reservoir on earth, groundwater has both a huge environmental and economic value, and will be an essential resource for adaptation to climate change and reduction of socio-economic vulnerability, particularly in regions where freshwater availability is highly variable and frequently limited.

Groundwater and aquifer related NBS hold major unrealized potential for alleviating adverse impacts of both floods and droughts in the same region/basin and impacts of progressive climate change overall.

Groundwater is less vulnerable to impacts of climate change, such as increasing temperatures (WWAP 2018). A related aspect is the role of improved soil management (a NBS) for managing infiltration, and therefore both runoff and groundwater recharge, as well as soil moisture retention, which is a particularly important factor regarding water security for crop production.

Aquifer-centric NBS, such as large-scale MAR interventions, may be applied in certain physiographic conditions to alleviate the risks of both floods and droughts in the same river basin in order to (a) increase water security/drought resilience; (b) increase food security, agricultural production, employment and farmer income; and (c) increased dry-season base flows to rivers and wetlands. Water storage is far from being realized. For example, trees can increase or decrease groundwater recharge according to their type, density, location, size, and age (Borg et al. 1988; Ilstedt et al. 2016).

The core of this contribution is centred in the pre-Inca and Inca civilizations and how these communities have developed ingenious NBS solutions to adapt to extreme climate scenarios such as prolonged droughts, managing water resources in a holistic way and how they understood clearly the global water cycle in all the components specially groundwater.

Why Peru? Among the most vulnerable areas to the catastrophic effects of climate change is Peru (Vuille et al. 2007). These effects include severe droughts, which affect everything from Peru's agricultural output to the resilience of the surrounding lands. Another prominent Peru effect is glacial melt. Peru has nearly 70% of the world's tropical glaciers. Both tropical glaciers and arctic glaciers are in peril, as rising air temperatures are causing them to lose ice more quickly than they acquire new snow. In 8 years (from 1989 to 1997), 447 km$^2$ of glaciers (21.85%) were lost corresponding to 12.3 billion cubic meters of water reserves (Vuille et al. 2007).

In the past, the ancestral civilizations of Peru have developed extraordinary solutions to adapt to adverse climate scenarios such as very prolonged droughts or the periodically El Niño events (Yapa 2013). Hence, Peru is a living laboratory where we can use the wisdom of the ancestor cultures to solve the present and the future of water management not only in this country, but also the problems that arise in another areas of Planet where the impact of climate change is also critical such as the Mediterranean region.

This article is divided in three interlinked phases: (1) To sow water by implementing ancestral aquifer recharge solutions, (2) To retain water by improve hydraulic efficiency in terms of infiltration and drainage and (3) To collect water by improving the performance of extraction in the subterranean aqueducts in arid regions.

## 2 To Sow Water

In Peru, the pre-Inca populations observed that the melting of the snows disappeared among the fissures of the Rocky Mountains feeding an aquifer or connecting directly to springs some kilometres further. The long fissures in the rocks that feed the springs in Peru are known as *amunas* (Fig. 1) The *amunas* constitute a complex hydro-geo-cultural system of artificial recharge of the aquifers, which begins with the rainy season and lasts in active state throughout the agricultural cycle. They are part of the inheritance of the pre-Hispanic times, reproduced and managed by three Huarochiran communities, in order to increase the volume of water from the springs that are used in their different economic and social activities, and mainly to productive purposes. The *amunas* currently in operation are located within the Huarochiran communities: San Andrés de Tupicocha, Santiago de Tuna and La Merced de Chaute (Apaza et al. 2006).

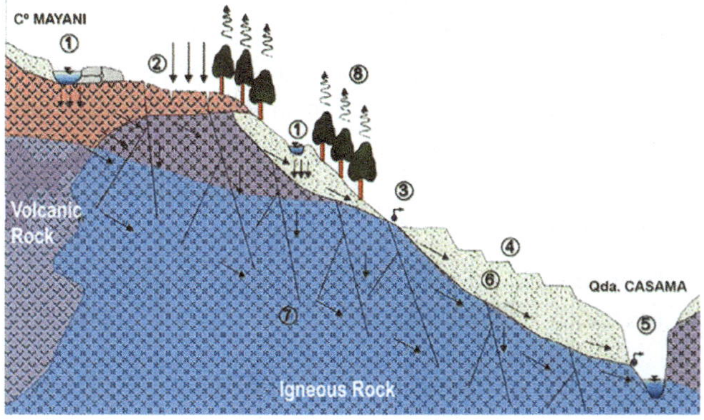


**Fig. 1** Conceptual scheme of an *amuna*

## 3 To Retain Water

Ancestral farmers have learned to survive in harsh climate, cultivating many different products at as many different ecological niches as they can gain access to. This practice does not reduce the risk of loss in each plot but avoids the total loss of food and seeds and provides them with a cushion to survive until the next harvest. Wherever possible, they also attempt to modify their farming landscape, by preparing terraces, for example. Terracing can be defined as man-made alteration of sloping land topography, with the object of making better use of natural resources such as soil, water, and climate. Regarding climate, terracing enables the creation and improvement of microclimates, protecting the environment from advective transferal of energy, reducing temporal changes, conserving the components of greenhouse effects in the atmosphere, creating an environment of turbulence or air mixture that reduces the loss of preserved energy, and favour the development of crops. In addition, a terrace reduces soil erosion to a minimum; helps infiltrate the rain or irrigation water; and maintains the moisture longer in soil (Fig. 2).

The importance and usefulness of terraces may have decreased in modern times; but even in locations where irrigated terraces have been dismantled, irrigation is still in use, since natural precipitation is insufficient for agriculture.

In the South American Andes, farmers have used terraces, known as *andenes* (Fig. 2), for over a thousand years to farm potatoes, maize, and other native crops. Terraced farming was developed by the Wari culture and other peoples of the south-central Andes. It is estimated that its construction started approximately 3000 years BC, developing along with the expansion of corn cultivation. The terraces purpose was to stop erosion, expand the agricultural frontier, retain moisture, and form microclimates.

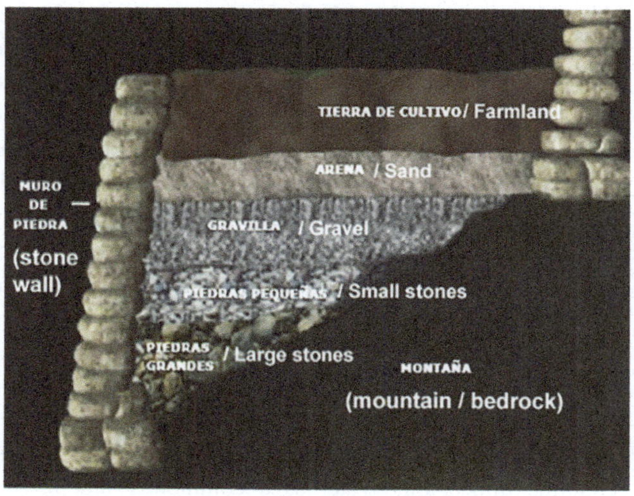


**Fig. 2** Cross section of an *andene*

**Fig. 3** Andenes of Tipón (Peru)

In Peru, there is an approximate area of one million hectares of terraces, of which approximately 10% is in permanent use, 20% in temporary or stationary use and 70% abandoned or destroyed (Masson 1986). One of the highest expressions is Tipon (Fig. 3) located 21 km from Cuzco consisted in 12 agricultural terraces with areas ranging from 1200 to 4500 m$^2$ (Wright 2006).

## 4 To Collect Water

Qanats have played a vital role in groundwater extraction since ancient times. These subterranean aqueducts run across the desert like the body's veins, bringing life and prosperity to the people who used to live off the water flowing down them. This technology is the focal point of the genesis of civilization in some parts of the world. The harsh environment drove people to invent the technology of the qanat and the know-how revolving around it. Due to their distinctive features, qanats discharge aquifer water continuously (Fig. 4), so that users (farmers, settlers, nomads, etc.) can perfectly adapt themselves to water fluctuations affected by droughts and wet years (Yazdi and Khaneiki 2010).

In the south coast of Peru, in the inhospitable desert of Nazca, there are more than 50 subterranean galleries, called *puquios* (Fig. 5) that connect the point of a surface with the groundwater and discharges into a small reservoir (*cocha*). Spaced along the gallery there are open holes commonly called eyes that connected to the surface of the earth with the gallery to provide a means of access to the galleries for cleaning (Schreiber and Rojas 2006). These structures have emerged in response to a long period of drought in the region around 560 AD (Thompson et al. 1985).

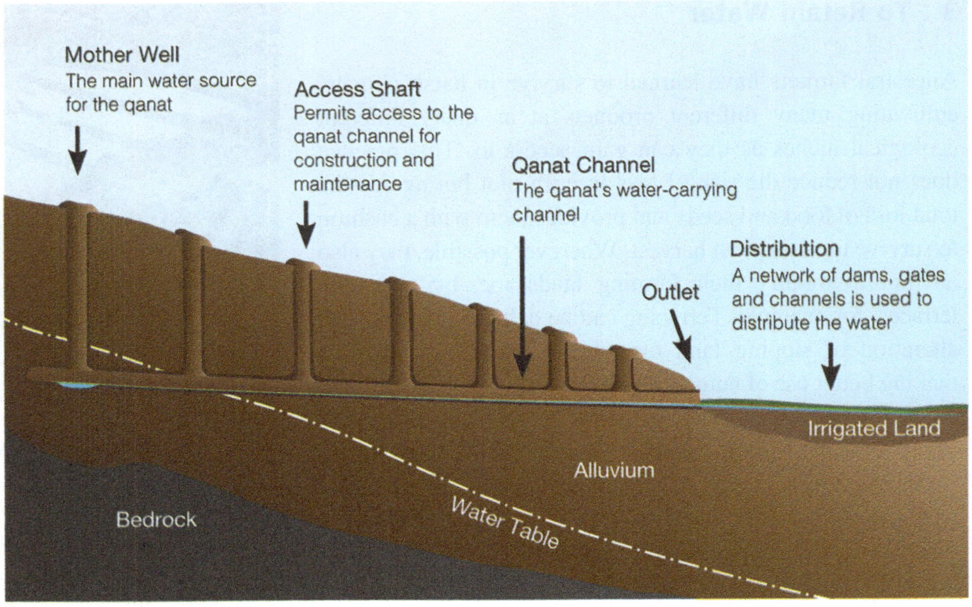


**Fig. 4** Cross-section of a qanat

**Fig. 5** Aerial view of a *puquio*

## 5 Concluding Remarks

The focus of ethics is not on water, seen in isolation, but on the water cycle and how the cycle connects the earth and the atmosphere. To live in an ethically responsible society, we need to modify water planning and management in order to achieve fairer access to drinking water, as well as providing effective social responses to public health concerns without jeopardizing ecosystems.

All these ancestral techniques are a clear example of symbiosis between human being and his environment. For indigenous peoples of Amerindian civilizations water is sacred. They celebrate rituals of requesting rains, they thank for the fruits, give offerings to the different deities that provoke the rains. These ceremonies are practiced at the top of the hills, at the source of springs, at the wells and in the riverbeds.

For indigenous societies, rivers, lakes, and meteorological phenomena that cause rain, snow or hail are part of a cosmogonic and spiritual universe, which is lived and reformulated on a daily basis. Water is not so much a "resource" as it is the core of the entire network of planetary life.

Indigenous societies do not give to water an economic valuation. Unlike those who see in the water an element that is applied for productive purposes, especially for industry and agribusiness or for urban and tourist development, indigenous people consider that water is the origin of life. Nowadays there is a growing need to recognize the cosmogonic and spiritual representations of indigenous societies around water and their possible contributions to a more balanced vision for their use and conservation. In this sense. Although the western and indigenous conceptions of water resources can cause conflicts, they can be a factor of complementarity and cooperation that feed strategies for a sustainable development. There are several lessons to be learned in contemporary times that highlighted on the causes of the environmental disaster and the loss of common goods essential for life, such as water. In particular, the hegemony and cult of "modern" technology as a unique solution for water management and its dissociation of cultural dimensions and ecological contexts. As well as the fragmentation of the socio-cultural matrix of water, which undermines the holistic and integral vision, and privileges the privatization of a common good in a scenario of globalization and free market, in order to submit the cultural dimension to the economic dimension: Water as a commodity and private good. The pre-Colombian communities developed ingenious adaptation measures to meet their needs. Today, the problem remains: the need to achieve a sustainable balance between availability and demand for water. In these times of the threats of climate change impacts on water resources of the planet, in which one example is the accelerated reduction of the volume of water contained in the glaciers of the Andes, it would be very useful to revisit these measures adaptation implemented by our ancestors and apply today in a new scientific and technological context.

## References


Apaza D, Hurtado RA, Calderón A (2006) Las Amunas de Huarochirí: recarga de los acuíferos de los Andes. Ediciones GSAAC, Lima

Borg H, Stoneman GL, Ward CG (1988) The effect of logging and regeneration on groundwater, streamflow and stream salinity in the southern forest of Western Australia. J Hydrol 99(3–4):253–270

Ilstedt U, Bargués Tobella A, Bazié HR, Bayala J, Verbeeten E, Nyberg G, Sanou J, Benegas L, Murdiyarso D, Laudon H, Sheil D, Malmer A (2016) Intermediate tree cover can maximize groundwater recharge in the seasonally dry tropics. Scient Rep 6(21930). https://doi.org/10.1038/srep21930

Masson L (1986) Rehabilitación de andenes en la comunidad de San Pedro de Casta, Lima. In De la Torre C, Burga M (eds) Andenes y Camellones en el Perú Andino: Historia, Presente y Futuro, Consejo Nacional de Ciencia y Tecnología (CONCYTEC), Lima, pp 207–216

Schreiber K, Rojas JL (2006) Aguas en el desierto: los puquios de Nasca. Fondo Editorial de la PUCP, Lima, Peru

Thompson LG, Mosley-Thompson E, Bolzan JF, Koci BR (1985) A 1500-year record of tropical precipitation in ice cores from the Quelccaya Ice Cap, Peru. Science 229:971–973

Vargas R (2006) La Cultura del agua: Lecciones de la América Indígena, Programa Hidrológico Internacional de la UNESCO para América Latina y el Caribe PHI VI/Serie Agua y Cultura/N° 1

Vuille M, Francou B, Wagnon P, Juen I, Kaser G, Mark BG, Bradley M (2007) Climate change and tropical Andean glaciers: past, present and future. Earth-Sci Rev 89:79–96

Wright KR (2006) Tipon, masterpiece of hydraulic engineering Inca empire. ASCE press, NY

WWAP (United Nations World Water Assessment Programme)/ UN-Water (2018) The united nations world water development report, nature-based solutions for water. UNESCO, Paris

Yapa K (2013) Prácticas ancestrales de crianza de agua. Una guía de campo: estrategias para adaptarnos a la escasez de agua. United Nations Development Program (UNDP) – Ecuador, Riobamba, Ecuador

Yazdi SSA, Khaneiki ML (2010) Veins of Desert: a review on Qanat/Falaj/Karez. IWRMO. River Publishers, Gistrup, Denmark

# Ancestral Techniques of Water Sowing and Harvesting in Ibero-America: Examples of Hydrogeoethical Systems

Sergio Martos-Rosillo, Alfredo Durán, Milka Castro, Jorge Julián Vélez, Gricelda Herrera, José María Martín-Civantos, Luciano Mateos, Juan José Durán, Jorge Jódar, Carlos Gutiérrez, Rosa María Hermoza, and Fluquer Peña

**Abstract**

Water sowing and harvesting (WS&H) consists of a series of ancestral procedures by which humans collect and infiltrate (sow) rainwater and runoff underground, so as to recover (harvest) it downgradient at some later time. This management of the water has made it possible for various regions of Ibero-America that is, Latin America plus the Iberian Peninsula to overcome dramatic cultural and climatic changes over the centuries. The principles governing WS&H coincide with those pursued under the present paradigm of Integrated Water Resource Management. Moreover, WS&H implies a better use of water and enhanced conservation of the environment and patrimony, as well as recognition of rural communities as vital custodians of the land and of its relevant cultural aspects. The main WS&H systems that serve Ibero-American countries are described here, emphasizing the principles underlying this means of water management as exemplary of hydrogeoethical systems.

**Keywords**

Recharge • Ibero-America • Integrated Water Resources Management • Resilience • Sustainability

## 1 Introduction

Water sowing and harvesting (WS&H) a term adopted from the Spanish language, *Siembra y Cosecha del Agua* (SyCA), is a process used by man that involves gathering and infiltrating (sowing) rainwater, surface runoff, and hypodermic or groundwater to retrieve (harvest) it at a later date and/or place. The concept evokes that of the Managed Aquifer Recharge of, understood as intentional recharge, for scientific/technical reasons, of an aquifer´s water in order to recover it later on or to generate environmental benefits. The concept behind WS&H, as developed in Ibero-America, is held to be of great utility and relevance in that it emphasizes ecological knowledge of the environs where our ancestors, over many centuries observing nature, transmitted this knowledge from generation to generation up to the present. The infiltration of water in an aquifer for its posterior retrieval at another site, either by means of wells and

S. Martos-Rosillo (✉)
Geological Survey of Spain, Granada, Spain
e-mail: s.martos@igme.es

A. Durán
Universidad Mayor de San Simón, Cochabamba, Bolivia,

M. Castro
Facultad de Derecho, University of Chile, Santiago de Chile, Chile

J. J. Vélez
Universidad Nacional de Colombia Sede Manizales, Manizales, Colombia

G. Herrera
Universidad Estatal Península de Santa Elena, Santa Elena, Ecuador

J. M. Martín-Civantos
Universidad de Granada, Granada, Spain

L. Mateos
Consejo Superior de Investigaciones Científicas, Córdoba, Spain

J. J. Durán · J. Jódar
Geological Survey of Spain, Madrid, Spain

C. Gutiérrez
Instituto Mexicano de Tecnología del Agua, Jiutepec, Morelos, Mexico

R. M. Hermoza
Universidad Nacional Agraria de La Molina, Lima, Peru

F. Peña
Superintendencia Nacional de Servcios de Saneamiento, Lima, Peru


M. Abrunhosa et al. (eds.), *Advances in Geoethics and Groundwater Management: Theory and Practice for a Sustainable Development*, Advances in Science, Technology & Innovation,
https://doi.org/10.1007/978-3-030-59320-9_104

draining galleries, or through an increase in the flow of springs, rivers or streams, is a procedure that calls for sound hydrogeological knowledge of the setting. Different WS&H techniques had been applied in the Andes Cordillera even before the arrival of the Spaniards (Yapa 2016; Ochoa-Tocachi et al. 2019), and in Sierra Nevada, in southern Spain, at least since the late Middle Ages (Martos-Rosillo et al. 2019). In fact, when the first Spaniards began exploring America, they encountered impressive hydraulic and agrarian infrastructures that evidenced an advanced utilization of the land and water. The newly arrived Spaniards did not understand the workings of the system, despite the fact that in some of the mountains of Spain itself, for instance around Sierra Nevada, similar techniques for water management were in use.

The objectives of this paper are threefold: (i) to explain the concept of WS&H, (ii) to divulge the main places and techniques of WS&H within the countries of Ibero-America, and (iii) present the main hydrogeoethical principles underlying WS&H as living yet ancient systems for water management.

## 2 Materials and Methods

Data retrieval on WS&H systems in Ibero-America was facilitated by the formation of a research network, financed by the Ibero-American Program of Science and Technology for Development (*Programa Iberoamericano de Ciencia y Tecnología para el Desarrollo*, CYTED). This network, denominated "Sowing and Harvesting Water in Natural Protected Areas", is made up of a group of 66 researchers and technicians from seven Ibero-American countries: Bolivia, Chile, Colombia, Ecuador, Mexico, Peru, and Spain. The network is open and diverse, including specialists in anthropology, archeology, agronomy, hydrology, hydrogeology, and law. It has been active since the beginning of 2019, and during this time the persons integrating it have carried out a search and synthesis of information surrounding these ancestral water management systems.

The information retrieved has been integrated into a database that is still being elaborated. Nevertheless, this information has allowed to classify and georeference the main systems of WS&H known so far. It is important to stress that most of these WS&H systems have not been investigated yet from a hydrological or hydrogeological point of view. This is precisely one of the aims of the WS&H network: to divulge their existence and workings, in terms of scientific knowledge, to planners, politicians, and society in general, and expound the benefits of recovering and replicating these systems in viable and adequate places, for improved water management.

## 3 Results

### 3.1 Main WS&H Systems in Ibero-America

Table 1 lists the places and the WS&H systems identified through the retrieval efforts undertaken by the WS&H network that is the places involving operations of water sowing and harvesting in Ibero-America.

One of the most widely employed WS&H systems in Ibero-America is that of *qochas*, also known as *albarradas* (and locally, as *atajados, jagüeyes* or *pataquis*). They are natural depressions in the terrain that allow for the temporary or permanent retention of rainwater or runoff. Their storage capacity can be increased by building a dyke of land. The *qochas* are not impermeable, so that besides serving as surface water reservoirs, the water recharges aquifers and increases the volume and the period of flow of nearby streams or *arroyos* as well as that of springs situated downstream.

The *tapes*, as they are called in Ecuador, consist of small walls built in the main channels of some streams and rivers of discontinuous flow, trapping the water during rainy periods. They favor infiltration of the water, which can then be tapped downstream through wells or draining galleries.

One of the most widely used WS&H systems in Peru is that of *zanjas de infiltración*, or infiltrate ditches. This system imitates the ancient *cuchacuchas*, very abundant in the dry Andean high plains or *altiplanos* (over 4000 m a.s.l.) of Ayacucho (Peru). The *cuchacuchas* are small, circular, permeable ponds dug out of the earth, maintained by the shepherds of the region. They are two to 12 m in diameter and have depths between 0.3 and 0.6 m (Yapa 2016). In addition to helping generate fodder for the herds, they recharge springs that appear downstream. In zones where this type of WS&H is developed, there may be hundreds of *cuchacuchas* per hectare. The more modern infiltration ditches are excavations in the land in the form of channels with a rectangular or trapezoidal section, built following altitudinal curves, to intercept the surface runoff generated by rains to increase recharge while preventing erosion. They are usually executed with a length of 5 and 0.5 m of width and of depth. In the lower part of the ditches, it is common to sow local vegetable species, which take advantage of part of the infiltrated water for their growth, thereby contributing to decreased surface runoff and defending the soil from erosion.

The pre-Hispanic *amunas* of Peru, also known as *mamanteos*, (Ochoa-Tocachi et al. 2019) and the Medieval *acequias de careo* of Sierra Nevada (southern Spain) (Martos-Rosillo et al. 2019) consist of channels dug in the earth and designed to derive surplus water generated in the rainy season or during the spring thaw. The water derived

**Table 1** Sites identified as featuring WS&H in Ibero-America

| Site | Country | WS&H technique | Ancestral versus Replicate |
|---|---|---|---|
| Bofedales de Parinacota, Caquena, Guallatire, Sorasorani | Chile | *Bofedal* | A |
| Chiapa, Poroma, Macaya | Chile | *Acequias* | A |
| Manglar Alto, Santa Elena | Ecuador | *Albarradas, tapes* | A/R |
| Catacocha, Palta, Loja | Ecuador | *Qochas* | A |
| Granada and Almería. S. Nevada | Spain | *Acequias de careo* | A |
| La Valduerna (León) | Spain | *Zayas* | A |
| Quispillacta (Ayacucho) | Peru | *Qochas* | R |
| Santo Dominigo de Capillas, (Huaytara, Huancavelica) | Peru | *Zanjas, qochas* | R |
| Chaclla (Huarochiri, Lima) | Peru | *Zanjas, amunas* | R |
| Huacrahuacho (Cuzco) | Peru | *Zanjas, qochas* | R |
| Ayacucho | Peru | *Cuchacuchas* | A |
| Pillao y Matao, San Jerónimo, Salkantay (Cuzco) | Peru | *Zanjas, amunas* | R |
| Chichilla (Arequipa), Antacollana (Cuzco) | Peru | *Qochas* | R |
| Tupicocha, Santiago de Tuna, Chaute, Huamantanga | Peru | *Amunas* | A |

from the head of the streams and the high springs circulates through these channels, causing their infiltration over different stretches of their extension, particularly in zones where the terrain is known to be quite permeable. Once the water infiltrates into the higher parts of the hillsides of a valley, it goes on to circulate slowly through thin superficial aquifers surrounded by the hard rock that tends to crop out in large mountains. The water surges at some point later downgradient, through springs, through drainage galleries, or directly through the riverbed of the rivers, helping maintain the continuity of the flow.

The so-called *andenerías* and irrigation terraces were not included as systems of WS&H in this work; although they may provide returns of water that contribute substantially to the recharge of aquifers. It is believed that this type of action lies beyond the objectives of analysis set forth here. Included, however, are the *bofedales* in which man intervenes to artificially increase the land surface and the amount of water that feeds it by creating *acequias* and *camellones* (the latter term being used in Chile, Peru, and Bolivia) for irrigation. As seen in Table 1, the places where the WS&H infrastructure is original are highlighted. That is, the places where structures are conserved from long ago, but perhaps rehabilitated and recovered, are distinguished from the zones where replicate forms based on the ancestral procedures of recharge are either in use or under construction. The SyCA is of special interest to increase water security in the supply to the rural population, for irrigation and for the maintenance of ecosystems in protected natural areas. For this reason, new areas of SyCA are being replicated.

## 3.2 WS&H as a Hydrogeoethical System

"Hydrogeoethics" contemplates the relations between man and groundwater, embracing an ethical standpoint regarding human conduct as well as ethical aspects of hydrogeological behavior. Several hydrogeoethical issues are closely related with WS&H, yet the foremost ones are environmental sustainability and human safety. The use of groundwater must be sustainable, hence entailing respect of the environment in both the short and the long terms. As upheld here, WS&H represents a form of water resource management that allows for the conditions of life (mankind and nature) to be improved, reducing the poverty of populations where it is put into practice. In this sense, it should be stressed that the regions where WS&H is practiced derive substantial hydrological, environmental, and social benefits. The underground infiltration of water reduces losses by evaporation, as it occurs in reservoirs and pools and regulates the flow of rivers in a natural manner, by increasing the groundwater component of their hydrograms (Jódar et al. 2017, 2018; Barberá et al. 2018, Martos-Rosillo et al. 2019). This is beneficial both for users of the upper part of the riverbed and for those accessing the lower part, which would have a greater flow rate in its streams during dry or drought periods (Ochoa-Tocachi et al. 2019). Furthermore, WS&H reduces the loss of fertile soil through erosion and the effects of flooding along plains, by fomenting the natural growth of vegetation on hillsides and shorelines. Water filtered through the ground may present improved quality for local populations, with greater contents in mineral salts (Barberá et al. 2018). In addition to acquiring a better quality for irrigation and for urban supply, its microbiological contamination is significantly reduced. At the same time, biodiversity is enhanced: A wide variety of floral species of considerably different water requirements can coexist, because their needs are covered thanks to the WS&H, along with other species adapted to dry conditions.

Meanwhile, WS&H stands as a collective effort, with the participation of both men and women in the tasks of conservation, constituting systems of communal governance. Their maintenance favors conservation of the patrimony as well as social cohesion. The principle underlying WS&H coincides with those pursued under the current paradigm of the Integrated Water Resources Management. These are the coordinated use and management of water, of the land, and of associated resources, with an aim to achieve a just balance between social and economic well-being and the needs of ecosystems, in space and in time (GWP 2000). For this reason, we may state that the modern paradigm of management of hydric resources, the IWRM, has in fact been carried out in some remote areas of Ibero-America for over one thousand years (Martos-Rosillo et al. 2019). Its antiquity comes as sound proof of its sustainability, as well of its resilient capacity, thanks to the forms of communal governance with local ecological knowledge that created these systems and kept them working during centuries.

## 4 Concluding Remarks

Certain principles of hydrogeoethics, such as the sustainable use of groundwater and of human safety, are found to be upheld by the water management practices promoted by the ancestral systems of WS&H, long in use in a number of Ibero-American countries. Overall, the WS&H process implies a sustainable and efficient use of water and land. It is founded on the use of green infrastructures, implementing nature-based solutions for resource management. This is possible because it takes advantage of the ecological and ancestral knowledge of rural communities that live in harmony with nature. Prospection and investigation of ancient WS&H systems, with a scientific focus like the one applied here, are still in early stages of development. One of the many tasks to be undertaken by the new WS&H network is to present to the scientific community related to hydrology and hydrogeology an inventory or map of these ancestral water management systems. It is hoped that such efforts will encourage further research and eventually serve to recover or replicate these sustainable and resilient IWRM systems in places where they may prove adequate.

**Acknowledgements** The authors thank the "*Programa Iberoamericano de Ciencia y Tecnología para el Desarrollo* (CYTED)" (Ibero-Americam Program of Science and Technology for Development) for the financial support of the network "Siembra y Cosecha del Agua en Áreas Naturales Protegidas" (419RT0577) (WS&H in Natural Protected Areas).

## References

Barberá JA, Jódar J, Custodio E, González-Ramón A, Jiménez-Gavilán P, Vadillo I, Pedrera A, Martos-Rosillo S (2018) Groundwater dynamics in a hydrologically-modified alpine watershed from an ancient managed recharge system (Sierra Nevada National Park, Southern Spain): Insights from hydrogeochemical and isotopic information. Sci Total Environ 640:874–893

Global Water Partnership (2000) Integrated Water Resources Management, technical Advisory Committee (TAC). Background Paper nº 9. Stockholm, Sweden

Jódar J, Cabrera JA, Martos-Rosillo S, Ruiz-Costán A, González-Ramón A, Lambán LJ, Custodio E (2017) Groundwater discharge in Mediterranean high-mountain watersheds: a valuable resource. The case of the Bérchules river in Sierra Nevada (Southern Spain). Sci Total Envron 593–594: 760–772

Jódar J, Carpintero E, Martos-Rosillo S, Ruiz-Constán A, Marín-Lechado C, Cabrera-Arrabal JA, González-Dugo MP (2018) Combination of lumped hydrological and remote-sensing models to evaluate water resources in a semi-arid high altitude ungauged watershed of Sierra Nevada (Southern Spain). Sci Total Environ 625:285–300

Martos-Rosillo S, Ruiz-Constán A, González-Ramón A, Mediavilla R, Martín-Civantos JM, Martínez-Moreno FJ, Jódar J, Marín-Lechado C, Medialdea A, Galindo-Zaldívar J, Pedrera A, Durán JJ (2019) The oldest managed aquifer recharge system in Europe: New insights from the Espino recharge channel (Sierra Nevada, southern Spain). J Hydrol 578:124147

Ochoa-Tocachi BF, Bardales JD, Antiporta J, Pérez K, Acosta L, Mao F, Zulkafli Z, Gil-Ríos J, Angulo O, Grainger S, Gammie G, De Bièvre B, Buytaert W (2019) Potential contributions of pre-Inca infiltration infrastructure to Andean water security. Nat Sustain 2:584–593. https://doi.org/10.1038/s41893-019-0307-1

Yapa KA (2016) Nurturing ancestral ground water recharging in the Americas. In: 7th RWSN Forum "Water for Everyone", Abidjan, Ivory Coast

# Groundwater Vulnerability Mapping and Ancestral Systems of Water Galleries (Porto Urban Area, NW Portugal): A Design on Nature-Based Solutions

Liliana Freitas, Maria José Afonso, Nicole Devy-Vareta, Alcides J. S. C. Pereira, José Martins Carvalho, and Helder I. Chaminé

**Abstract**

Groundwater resources are crucial to the settlement of populations, and their quantity and quality are essential to the development of urban areas. In fact, nature-based solutions for water were considered in many places using ancestral systems of water galleries and springs to supply urban areas. That design based in natural solutions had proven during centuries to be much less demanding and resourceful. A multidisciplinary approach was applied in Porto urban area (NW Portugal), to assess the urban groundwater supply and ancestral network of water galleries and springs. The infiltration potential index in urban areas (IPI-Urban) is dependent on several parameters (e.g. lithology, structure, weathering grade, morphotectonics, land use, drainage, slope, rainfall, anthropogenic and urban hydraulic features, like the water supply, the sewer and the stormwater networks) which can be overlapped and cross-linked in a GIS environment. Moreover, several vulnerability indexes (DRASTIC, GODS, DRASTIC-Fm, SINTACS and SI) were outlined within a combined approach, and an evaluation of urban recharge was performed. All these permitted to improve the hydrogeological conceptual model for Porto urban area. Therefore, those old underground structures could be a positive asset as socio-economic, environmental and heritage drivers if are used nature-based solutions and good geoethical practices.

**Keywords**

Urban groundwater • Vulnerability • Hydrogeomorphology • Nature-based solutions • Geoethics

## 1 Introduction

The impressive words of Luís Ribeiro are topical: "*Currently, due to the fast technological progress — but not always beneficial to people's lives many galleries, were abandoned although since the oldest times, the populations have been supplied with water captured in several galleries evenly distributed in the mountains, which never failed.*" (Ribeiro 2019; p. 225). In fact, Nature-based solutions (NBS) are inspired and supported by nature and use, or mimic, natural processes to contribute to the improved management of water (Kabisch et al. 2017; WWAP/UN-Water 2018, 2020). That is the key approach to achieve in any project aiming at a sustainable design that is compatible with nature, environment and society (McHarg 1992).

The *social ethics* related to the moral and behaviour issues, alongside with aesthetics and values, are rooted in the classic philosophers. The concept of *land ethics* stated by Leopold (1949) triggered in a consciousness about the fragile balance of the environment in planet Earth. Later, several authors highlighted the urgent needs to have water ethics and eco-responsibility in exploitation of water resources, groundwater uses and management (e.g. Wittfogel 1956; Carson 1962; Llamas 1975, 2004; Leopold 1990; Custodio 2000, 2002; Datta 2005; Llamas et al. 2009; Chaminé and Carvalho 2015; Chaminé et al. 2016;

L. Freitas · M. J. Afonso · J. M. Carvalho · H. I. Chaminé
Laboratory of Cartography and Applied Geology (LABCARGA), Department of Geotechnical Engineering, School of Engineering (ISEP), Polytechnic of Porto, Porto, Portugal

L. Freitas (✉) · A. J. S. C. Pereira
CITEUC, Department of Earth Sciences, Faculty of Sciences and Technology, University of Coimbra, Coimbra, Portugal
e-mail: lfsfr@isep.ipp.pt

M. J. Afonso · J. M. Carvalho · H. I. Chaminé
Centre GeoBioTec UA, Aveiro, Portugal

N. Devy-Vareta
CEGOT, Department of Geography, Faculty of Arts, University of Porto, Porto, Portugal



M. Abrunhosa et al. (eds.), *Advances in Geoethics and Groundwater Management: Theory and Practice for a Sustainable Development*, Advances in Science, Technology & Innovation,
https://doi.org/10.1007/978-3-030-59320-9_105

Abrunhosa et al. 2018; Chaminé and Gómez-Gesteira 2019). Recently, emerge the *geoethics* concept embracing the ethics, culture, sociology and scientific integrity in earth sciences practice and research (e.g. Wyss and Peppoloni 2015; Peppoloni and Di Capua 2017; Bobrowsky et al. 2017; and references therein).

New challenges under the framework of Sustainable Development Goals emerge related to the interlinkage of the assessment, abstracting and modelling of the urban water cycle (Guppy et al. 2018; Freitas et al. 2019b). Urbanisation increases the number of users; consequently, the urban water resources are contaminated and polluted. In addition, the changes in climatological patterns aggravate these effects in many regions in terms of human, sanitation, economic and environmental insecurity dimensions (Tortajada and Biswas 2017; Tortajada 2020).

In Northern and Central Portugal, urban water systems were absent or seriously degraded, in quantity and quality, by flawed sanitation infrastructures and hygiene practices only set-up in the late nineteenth century and early twentieth century (Ribeiro 2019). A good example of that reality is the Porto urban area on water supply, sewage and sanitation issues, particularly in the period between 1860 and 1920 (e.g. Silva 2004; Chaminé et al. 2010, 2014; Freitas et al. 2014; Afonso et al. 2018) that illustrates the importance of this ancient underground water galleries in urban areas.

## 2 Porto Urban Area: Historical and Hydrogeological Setting

The Porto urban area, located in NW Portugal, is one of the oldest on the European continent (Costa-Lobo 1991). The first settlements date from the *Castro* culture, before the expansion of the Romans, who settled in this place in the first century BC on the north bank of the Douro River (Azevedo 1960). The Romans named the village *Portus*, which became a very important administrative and commercial centre. The designation *Portus Cale*, which later became *Portucale* and, in its non-Latin form, Portugal, was the first name given to the towns of Porto and Gaia (Rebello da Costa 1788; Azevedo 1960). After Suevians and Visigoths era and until the middle of the ninth century, the region was under the control of the Muslims, who were ousted from the Iberian Peninsula definitively in 1492. Porto has become an important city since the twelfth century, having expanded along the stepped granitic slopes of the Douro River. The patrimonial attributes of its old neighbourhoods led the historical downtown of Porto to be recognised by UNESCO as World Heritage Site in 1996.

Despite its small area (41.3 km$^2$), the city of Porto is one of the Portuguese cities with the highest population density, with 237.559 inhabitants ( INE – Instituto Nacional de Estatística 2011). The development of the city was accompanied by the evolution of its urban groundwater systems. For six centuries, the water supply to the city of Porto was carried out through small water courses, springs, water mines, or wells. The water was then provided to the population from public fountains, which were fed by the numerous springs existing inside and outside the city limits. The spring water ran through extensive underground galleries and aqueducts to supply the fountains. It is known that the network of galleries has at least 64 mapped water galleries, some of them with 3.3 km length (COBA 2003; Chaminé et al. 2010). Some springs and fountains, currently, are not found in the original places, because they were either destroyed or shifted; those that remain in place are mostly deactivated (Fig. 1).

Figure 2a illustrates the regional hydrogeological units (HGU) background in Porto urban area. It is a heterogeneous area, mainly formed by fissured media (granitoids and metasedimentary rocks) and porous media (alluvia and fluvial or marine deposits) (Chaminé et al. 2010). Most of the area is constituted by granites, namely two-mica, medium to fine-grained facies (e.g. Carríngton da Costa 1938; Almeida 2006).

The Porto granite facies arises weathered to different grades, from fresh rock to residual soil in short distances, showing highly variable conditions, resulting in arenisation and kaolinisation that may reach depths above 30 m. These residual soils play a major role in the local aquifer recharge, and their permeability is low to moderate ($K < 1$ m/day). Saprolite masses, tightly related to the Porto granite, comprise primarily kaolinitic silty sands, with a very low permeability ($K < 0.1$ m/day). Moreover, alluvial deposits consist mainly of sands and silty clays with small thicknesses (<6 m), representing shallow aquifers with lesser significance, with a low to moderate permeability ($K < 2$ m/day). Moreover, the hydrogeochemistry of Porto groundwater shows that these waters are of neutral character (pH 6–7), have a moderate mineralisation and are nitrate enriched (median $NO_3$ > 50 mg/L), and their chemical signatures are mostly Cl–Na to Cl–$SO_4$–Na. Nowadays, most of the groundwater in Porto urban area has mainly an industrial use, although private uses are common (Afonso et al. 2007, 2010, 2016, 2019).

The Asprela catchment is located in NE of Porto urban area and constitutes a stream of the Leça River (Matosinhos municipality). According to the land use map (Fig. 2b), it is dominated by an urban fabric setting, occupying 52.2% of the Asprela catchment. Industrial, commercial and transport units are the second most representative class (32.4%). The rest of the area (15.3%) is distributed between green urban areas, forest and natural/seminatural areas and agricultural areas.

The Asprela catchment is in the highest altitude region of this urban area. In geomorphological terms, flattened areas

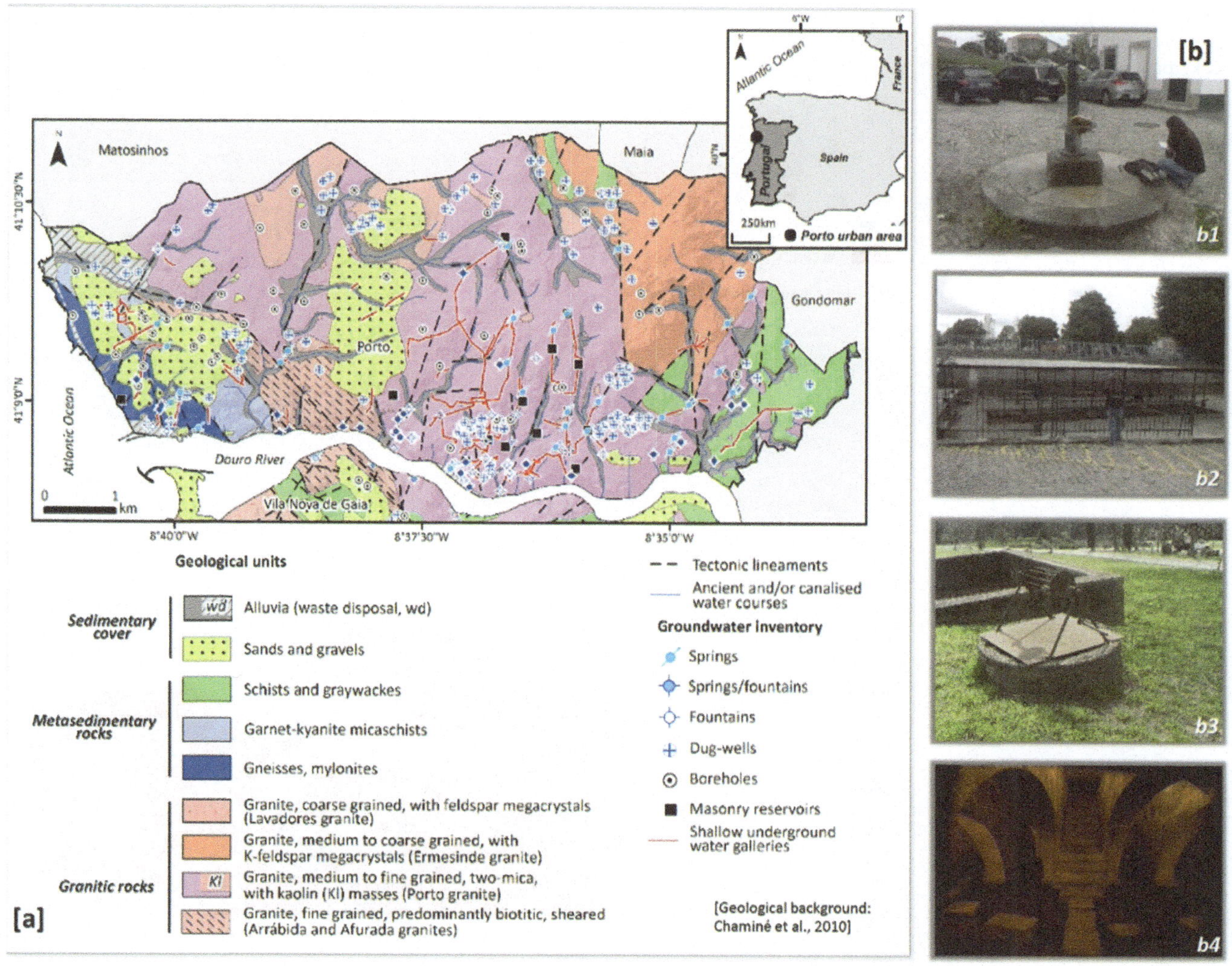


**Fig. 1** Geological framework and groundwater inventory from the early twentieth century to present (over 410 water sites) from Porto urban area (adapted from Chaminé et al. 2010, 2014); b1 spring/fountain; b2 washing place; b3 dug-well; b4 Arca D'Água masonry reservoir

with altitudes in the range 80–125 m are predominant. In the southern part of this hydrographic basin, flattened areas occur with altitudes in the range 130–160 m. According to the regional hydrogeological units, most of the Asprela catchment (73.1%) is occupied by HGU6 and HGU5. The HGU2 (alluvia) and the HGU4 (micaschist, schist and metagreywacke) occupies, respectively, 18.2% and 8.8%. On the Asprela catchment, 55 potential contamination activities have been inventoried, and these sources are concentrated in the North and South areas, being dispersed in the central area of the basin (Fig. 2b). They are categorised according to their origin (Vrba and Zaporozec 1994; Zaporozec 2004).

The details about the methodology approach followed in this study could be found in the works of Freitas (2014, 2019a, b). This approach couples GIS-based mapping and Analytical Hierarchical Process (AHP) techniques. In addition, this methodology encompasses also remote sensing, hydrogeology and urban hydraulic studies, accompanied by a detailed hydrogeological fieldwork and an inventory of groundwater potential contamination activities.

## 3 Results and Discussion

A vulnerability assessment of groundwater to contamination has been performed, using several vulnerability indexes. Figure 2 presents the GODS, DRASTIC-Fm, SINTACS and SI vulnerability maps and a matrix that relates hydrogeological units and vulnerability indexes. The location of potential contamination activities and the type of contamination that they can release to water resources are essential to any understanding of their impact.

The potential contamination load (Foster et al. 2002) can be classified into four classes: (i) low; (ii) moderate; (iii) moderate to high; (iv) high. In the Asprela catchment,

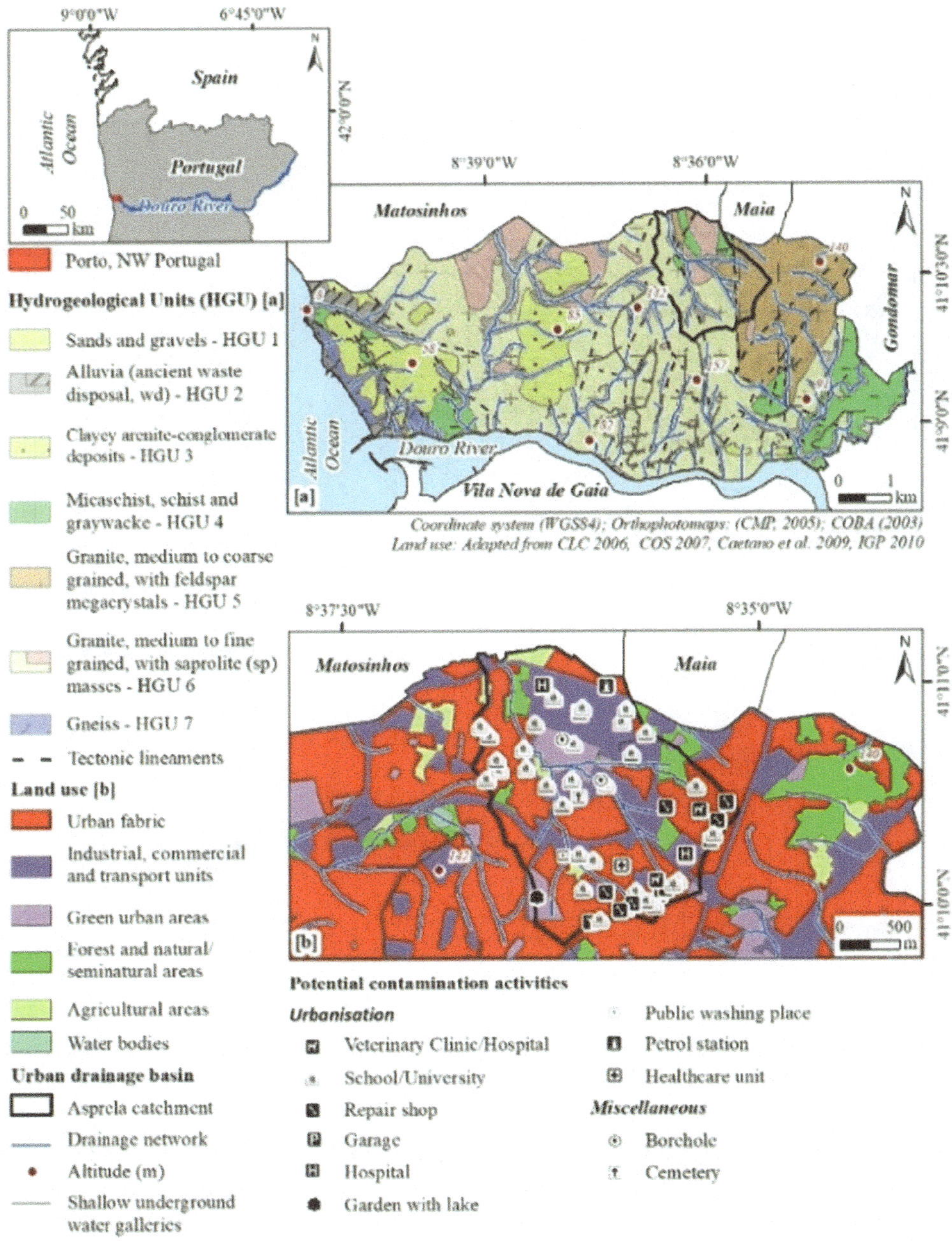


**Fig. 2** Porto urban area setting (NW Portugal): hydrogeological background (**a**) (adapted from Afonso et al. 2007, 2016); land use map and potential contamination activities in Asprela catchment (**b**)

72.7% of the potential contamination activities have a moderate potential for contamination (e.g. school/university). There are activities with a moderate to high contamination potential (e.g. petrol station), corresponding to 23.6%. Only 3.7% of potential contamination activities have a low potential for contamination (e.g. public washing place). Therefore, most of the contamination sources have a moderate to high contamination potential (96.4%), which may contribute to locally increase the vulnerability to contamination, especially in the areas where alluvia (HGU2) and granitoid rocks (HGU5 and HGU6) occur, since these units have a low to moderate permeability.

The lithological heterogeneity justifies the permeability differences and consequently the results in the vulnerability indexes. Figure 3a shows the GODS index that classifies HGU2 with high vulnerability, clearly distinctive from the other hydrogeological units, where vulnerability varies between negligible (HGU4 and HGU6—saprolite), low (HGU5) and low to moderate (HGU6). In this urban hydrogeological context, the application of DRASTIC-Fm, SINTACS and SI was fundamental. This fact is justified by the addition of parameters that characterises the area according to its geostructure (Fm parameter) and urban conditions (land use) particularities. According to these

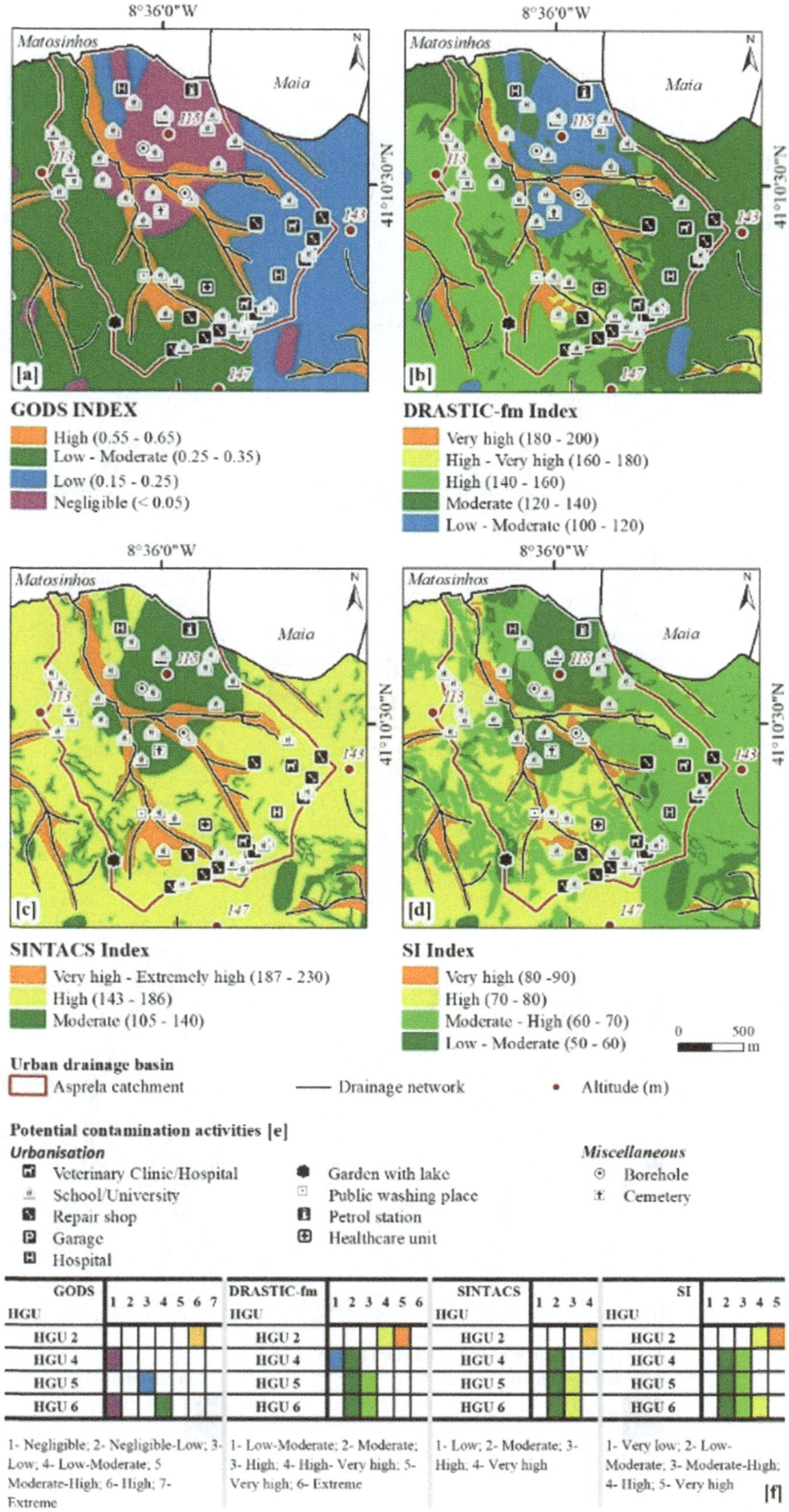


**Fig. 3** Urban groundwater vulnerability assessment in the Asprela catchment (GODS (**a**), DRASTIC-Fm (**b**), SINTACS (**c**) and SI (**d**) indexes); potential contamination activities inventory [e] and matrix of hydrogeological units *vs.* groundwater vulnerability indexes [f]

indexes, HGU2 presents a high to very high vulnerability to contamination. The metasedimentary rocks (HGU4) are classified as having a low to moderate vulnerability and the granitic rocks (HGU5 and HGU6) as having a moderate to high vulnerability.

Figure 4 presents the urban infiltration potential index (IPI-Urban) and the urban groundwater recharge for summer and winter scenarios (details in Freitas et al. 2019a, 2019b). In the Asprela catchment, the IPI-Urban ranges from very low to very high.

The value for urban groundwater recharge was achieved by multiplying IPI-Urban by the mean precipitation. For the summer scenario, the average annual precipitation of the dry months was calculated (165.1 mm). Moreover, for the winter scenario, the average annual precipitation of the wet months was calculated (1071.7 mm). Regional

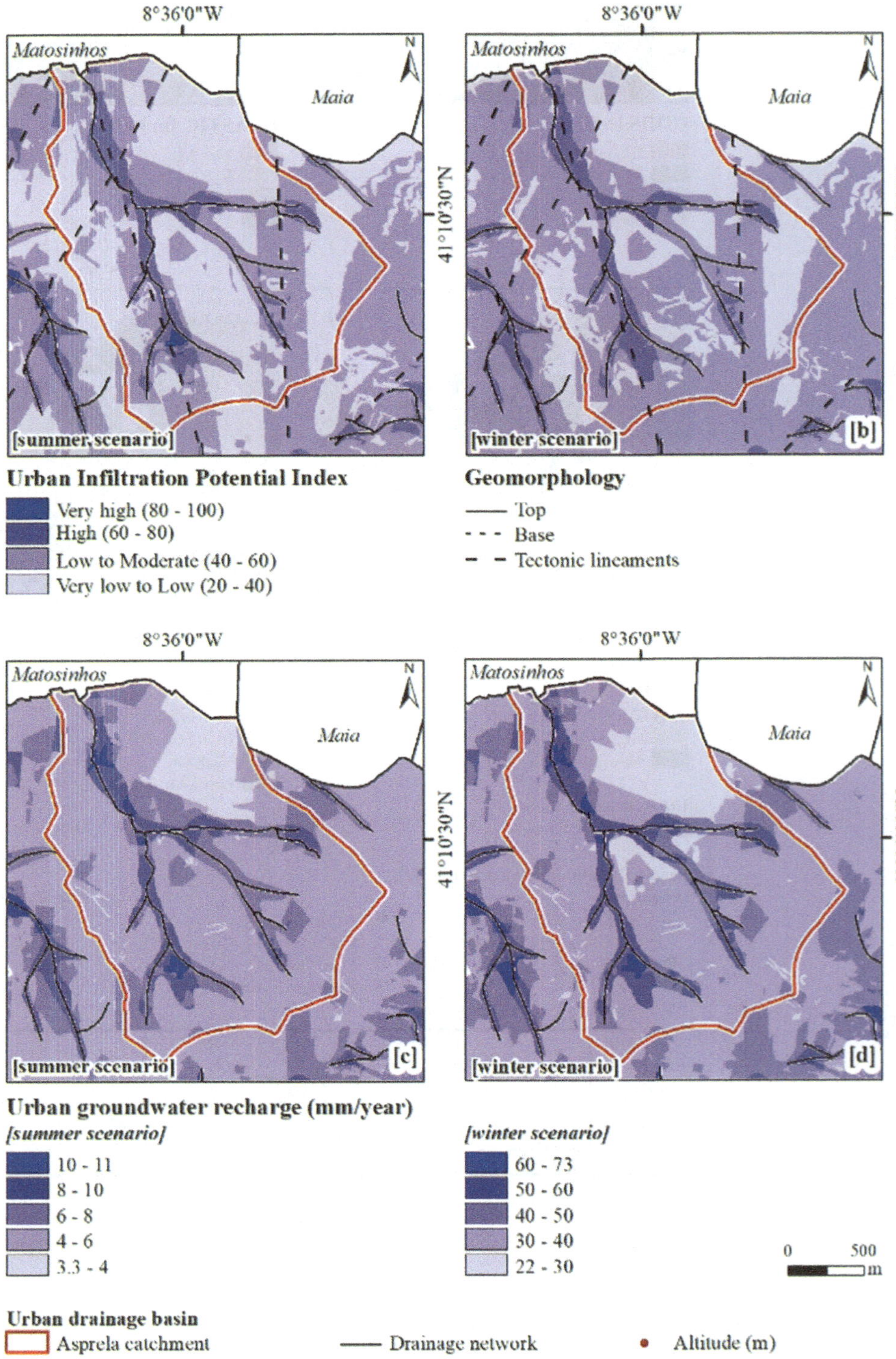


**Fig. 4** Urban groundwater mapping in the Asprela catchment: urban infiltration potential index (IPI-Urban) for summer (**a**) and winter scenarios (**b**); Urban groundwater recharge for summer (**c**) and winter scenarios (**d**)

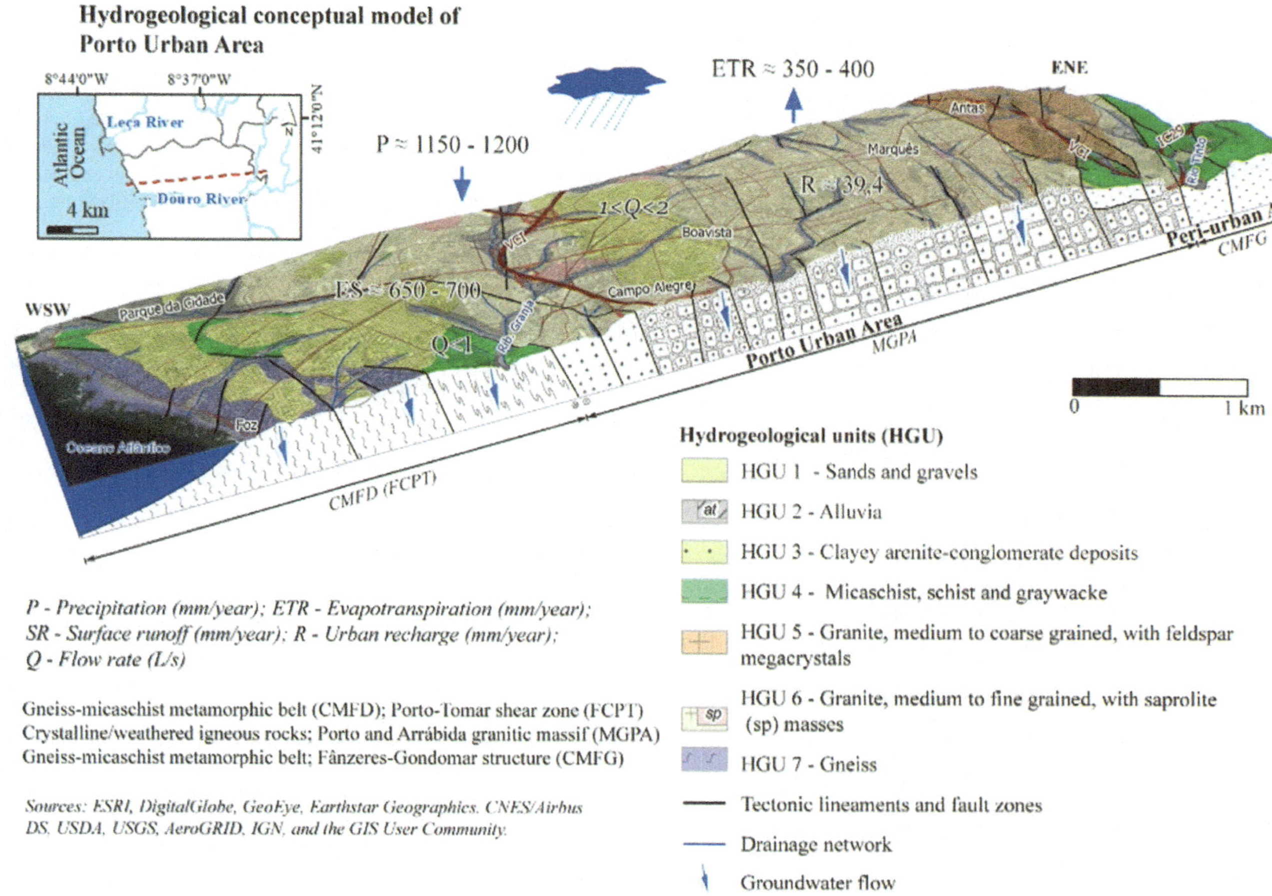


**Fig. 5** Hydrogeological conceptual model of Porto urban area

hydrogeological studies (Afonso et al. 2007, 2019) suggest the value of 8% to be the initial urban recharge rate. Therefore, in the summer scenario, the recharge varies between 3.3 and 10.9 mm/year, which corresponds to an average of 5.7 mm/year. In the winter scenario, recharge is in the range 2.2–65.9 mm/year, with an average value of 37.5 mm/year.

The development of a hydrogeological conceptual model is the first step in the entire process of modelling of Earth systems. A multicriteria methodology supported by GIS mapping allows the contribution of geosciences to the study of smart cities, using a holistic approach (Chaminé et al. 2014, 2016). Figure 5 shows a hydrogeological conceptual model for Porto urban area, where climatology, geology and hydrogeology contribute for the assessment of the urban groundwater resources.

## 4 Concluding Remarks

Conceptual hydrogeological models can contribute to a balanced decision-making in terms of water resources management in urban areas to a better design on nature-based solutions. It is important to highlight a better planning of hydrogeological investigations, the identification of potential contamination/pollution areas, the reduction of the cost of advanced studies and the definition of protection areas for natural discharge (springs) in historical areas. It is also important to be aware to the use of ancient urban groundwater systems, namely to evaluate the groundwater supply that might be available for urban non-potable practices, such as street cleaning, irrigation of parks and lawns and firefighting. These models are an appropriate tool, combined with mapping, to communicate with the different stakeholders involved in the urban integrated groundwater management, if used ethically, eco-responsible, rigorously and with sound hydrological sense.

**Acknowledgements** This study was carried out partially under the framework of the LABCARGA|ISEP re-equipment programme (IPP-ISEP|PAD'2007/08), GeoBioTec|UA (UID/GEO/04035/2020), CITEUC (UID/Multi/00611/2020). The research was also funded by a doctoral scholarship to L. Freitas (SFRH/BD/117927/2016) by FCT (Portuguese Foundation for Science and Technology) and, currently, by Project ReNATURE – Valorization of the Natural Endogenous Resources of the Centro Region (Centro 2020, Centro-01-0145-FEDER-000007, Portugal). This work is dedicated to the colleague Luís Ribeiro (1955–2020), a skilled professor of quantitative hydrogeology and mining engineer, with a unique outlook of the interlinkage of water, environment, Earth systems, arts and culture.

## References


Abrunhosa MJ, Peppoloni S, Chambel A (2018) Introducing geoethics in groundwater science, technology, and management. In: Abstracts Book, Session T1.1 Socioeconomic Aspects of Groundwater Resources Management, 45th IAH Congress, Daejeon, Korea, pp 26

Afonso MJ, Chaminé HI, Carvalho JM, Marques JM, Gomes A, Araújo MA, Fonseca PE, Teixeira J, Marques da Silva MA, Rocha F (2007) Urban groundwater resources: a case study of Porto City in north-west Portugal. In: Howard KWF (ed) Urban aquifer management, urban groundwater: meeting the challenge. Selected papers on hydrogeology, IAH Special Publications. Taylor & Francis CRC Press, SP8, pp 278–294

Afonso MJ, Chaminé HI, Marques JM, Carreira PM, Guimarães L, Guilhermino L, Gomes A, Fonseca PE, Pires A, Rocha F (2010) Environmental issues in urban groundwater systems: a multidisciplinary study of the Paranhos and Salgueiros spring waters, Porto (NW Portugal). Environ Eart Sci 61(2):379–392

Afonso MJ, Freitas L, Chaminé HI (2019) Groundwater recharge in urban areas (Porto, NW Portugal): the role of GIS hydrogeology mapping. Sust Wat Resour Manag 5:203–216

Afonso MJ, Freitas L, Devy-Vareta N, Teixeira J, Fontes G, Chaminé HI (2018) Cartografia subterrânea dos antigos mananciais de Paranhos e Salgueiros para o abastecimento de água à cidade do Porto (NW de Portugal): Contexto e marcos históricos. In: Blanco-Chao R, Castillo-Rodríguez F, Costa-Casáis M, García JH, Valcárcel-Díaz M (eds) Xeomorfoloxía e Paisaxes Xeográficas: Catro décadas de investigación e Ensino, Homenaxe a Augusto Pérez Alberti. Universidade de Santiago de Compostela Publicacións, Santiago de Compostela, Série Homenaxes, pp 261–292

Afonso MJ, Freitas L, Pereira AJSC, Neves LJPF, Guimarães L, Guilhermino L, Mayer B, Rocha F, Marques JM, Chaminé HI (2016) Environmental groundwater vulnerability assessment in urban water mines (Porto, NW Portugal). Water MDPI 8(11):499

Almeida A (2006) Geology and urban landscape: the granite in Oporto, NW Portugal. Eur Geol J 21:4–8

Azevedo RS (1960) O Porto desde a proto-história à época do Infante D. Publicações do Centro de Estudos Humanísticos, Porto, Henrique

Bobrowsky P, Cronin VS, Di Capua G, Kieffer SW, Peppoloni S (2017) The emerging field of geoethics. In: Gundersen LC (ed) Scientific integrity and ethics with applications to the geosciences, Special Publication American Geophysical Union, vol 73. Wiley, pp 175–212

Carríngton da Costa J (1938) O Pôrto: geografia-geologia. In: Bastos C (org.), Nova Monografia do Porto, Companhia Portuguesa do Porto, Porto, pp 3–32

Carson R (1962) Silent spring. Houghton Mifflin Company, Boston

Chaminé HI, Afonso MJ, Freitas L (2014) From historical hydrogeological inventory through GIS mapping to problem solving in urban groundwater systems. Eur Geol J 38:33–39

Chaminé HI, Afonso MJ, Robalo PM, Rodrigues P, Cortez C, Monteiro Santos FA, Plancha JP, Fonseca PE, Gomes A, Devy-Vareta N, Marques JM, Lopes ME, Fontes G, Pires A, Rocha F (2010) Urban speleology applied to groundwater and geo-engineering studies: underground topographic surveying of the ancient Arca D'Água galleries catchworks (Porto, NW Portugal). Int J Speleol 39(1):1–14

Chaminé HI, Carvalho JM (2015) Meeting sustainability and water resources: a dialogue about groundwater science and hydrogeology practice. Environ Eart Sci 73(6):2531–2542

Chaminé HI, Gómez-Gesteira M (2019) Sustainable resource management: water practice issues. Sustain Water Resour Manag 5:3–9

Chaminé HI, Teixeira J, Freitas L, Pires A, Silva RS, Pinho T, Monteiro R, Costa AL, Abreu T, Trigo JF (2016) From engineering geosciences mapping towards sustainable urban planning. Eur Geol J 41:16–25

COBA – Consultores de Engenharia e Ambiente, SA (2003) Notícia explicativa da Carta Geotécnica do Porto. 2ª edição, COBA/FCUP/CMP, Porto

Costa-Lobo M (1991) Oporto: city profile. Cities 8:38–43

Custodio E (2000) Some relevant ethical issues in relation to freshwater resources and groundwater. Bol Geol Min, Madrid 111(6):121–130

Custodio E (2002) Overexploitation, what does it mean? Hydrogeol J 10(2):254–277

Datta PS (2005) Groundwater ethics for its sustainability. Curr Sci 89 (5):1–6

Foster SSD, Hirata R, Gomes D, D'Elia M, Paris M (2002) Groundwater quality protection: a guide for water utilities, municipal authorities and environment agencies. The World Bank, Washington, DC

Freitas L, Afonso MJ, Devy-Vareta N, Marques JM, Gomes A, Chaminé HI (2014) Coupling hydrotoponymy and GIS cartography: a case study of hydro-historical issues in urban groundwater systems, Porto, NW Portugal. Geogr Res 52(2):182–197

Freitas L, Afonso MJ, Pereira AJSC, Delerue-Matos C, Chaminé HI (2019a) Assessment of sustainability of groundwater in urban areas (Porto, NW Portugal): a GIS mapping approach to evaluate vulnerability, infiltration and recharge. Environ Eart Sci 78:140

Freitas L, Chaminé HI, Pereira AJSC (2019b) Coupling groundwater GIS mapping and geovisualisation techniques in urban hydrogeomorphology: focus on methodology. SN Appl Sci 1:490

Guppy L, Uyttendaele P, Villholth KG, Smakhtin V (2018) Groundwater and sustainable development goals: analysis of interlinkages. UNU-INWEH Report Series, Issue 04, United Nations, University Institute for Water, Environment and Health, Hamilton, ON, Canada

INE – Instituto Nacional de Estatística (2011) Informação estatística sobre a população portuguesa: Concelho de Penafiel. http://www.ine.pt/. Last accessed 27 Jan 2020

Kabisch N, Korn H, Stadler J, Bonn A (2017) Nature-based solutions to climate change adaptation in urban areas: Linkages between science, policy and practice, Springer Open, Cham, Switzerland

Leopold A (1949) A sand county almanac: and sketches here and there. Oxford University Press, Oxford

Leopold LB (1990) Ethos, equity and the water resources. Environ Sci Policy Sustain 32(2):16–41

Llamas MR (1975) Non economic motivations in ground water use: hydroschizophrenia. Groundwater 13(3):296–300

Llamas MR (2004) Water and ethics: use of groundwater. UNESCO series on water and ethics, essay 7. UNESCO, New York

Llamas MR, Martínez-Cortina L, Mukherji A (2009) Water ethics. 3rd marcelino botín foundation water workshop. CRC Press, Boca Raton

McHarg IL (1992) Design with nature Wiley series in sustainable design, 25th edn. Wiley, New York

Peppoloni S, Di Capua G (2017) Geoethics: ethical, social and cultural implications in geosciences. Annals of Geophysics, 60, Fast Track 7. https://doi.org/10.4401/ag-7473

Rebello da Costa A (1788) Descripção topografica, e historica da Cidade do Porto. Que contém a sua origem, situação e antiguidades: a magnificencia dos seus templos, mosteiros, hospitaes, ruas, praças, edificios e fontes. Officina de Antonio Alvarez Ribeiro, Porto

Ribeiro L (2019) Groundwater public supply based on an ancestral system of galleries. In: Gómez-Hernández JJ, Andreo-Navarro B (eds) Proceedings 46th annual congress IAH, groundwater management and governance coping with uncertainty, Malaga, Spain, p 225

Silva AF (2004) Problems of agency and technological issues in the municipalisation of water supply systems, Portugal (1850–1930). In: Proceedings of the seventh international conference on urban history: european city in comparative perspective, Athens, Greece, pp 1–35

Tortajada C (2020) Contributions of recycled wastewater to clean water and sanitation Sustainable Development Goals. Clean Water 3:22

Tortajada C, Biswas AK (2017) Water as a human right. Int J Water Res Develop 33:509–511

Vrba J, Zaporozec A (1994) Guidebook on mapping groundwater vulnerability. International Contributes to Hydrogeology, vol 16. Verlag Heinz, Hannover

Wittfogel KA (1956) The hydraulic civilizations. In: Thomas WL (ed) Man's role in changing the face of the earth. The University of Chicago Press, Illinois, pp 152–164

WWAP (United Nations World Water Assessment Programme)/ UN-Water (2018) The United Nations World Water Development Report: Nature-Based Solutions for Water. UNESCO, Paris

WWAP (United Nations World Water Assessment Programme)/ UN-Water (2020) The United Nations World Water Development Report: Water and Climate Change. UNESCO, Paris

Wyss M, Peppoloni S (2015) Geoethics: ethical challenges and case studies in earth sciences. Elsevier, Amsterdam

Zaporozec A (2004) Groundwater contamination inventory: a methodological guide with a model legend for groundwater contamination inventory and risk maps. UNESCO, Paris

# Rethinking the Role of Science in Society? Groundwater Science, Critical Reflections and Learnt Lessons

Sofia Bento and M. Teresa Condesso de Melo

**Abstract**

The need to rethink the role of groundwater science in risk assessment and society results of a joint reflection between a sociologist and a hydrogeologist and arises from a research collaboration in a long-term highly contaminated area. The objective is to bridge social and natural sciences in the reflection of ethical questions and their implications for technical reporting, scientific production and societal impacts. The authors aim to understand the various social, economic and political components of long-term hydrogeological investigations and to test the application of ethnographic methodology in ethical issues. The investigation addresses three experiences that pose ethical challenges: the access and use of public domain monitoring data; the dissemination of scientific research data and its controversial character; and the lessons and necessary connections between scientific results, policy regulation and the design of water governance programs. Ethical issues should be addressed by the scientific community as are technical results discussed and scrutinized in conferences and scientific papers. In the framework of hydrogeological investigations in contaminated areas, ethics should help scientists to move from the present situation (based on science) to what should be in the future (based on environmental, human health and societal protection goals).

**Keywords**

Groundwater • Contamination • Social participation • Geoethics • Governance

## 1 Introduction

Research is a quest of knowledge through methodical process of systematic observation and reflection. However, doing research can pose ethical dilemmas and integrity challenges. As other sciences, groundwater science generates ethical issues that are rarely investigated from the perspective of the hydrogeological community.

In social sciences, Merton (1938, 1942) is an important reference for understanding ethics and science. In his research, he demonstrated how norms that govern the researchers' behaviour constitute the ethos for science. The role of the norms was considered to regulate the researchers' practices. These norms, not codified, are instead interiorized by the scientific community and obviously linked to the values existing in each institution. Merton has highlighted four main ethical norms in the scientific activity: universalism, (intellectual) communism, disinterestedness and organized scepticism that together constitute what he called the 'ethos of science', i.e. the institutional logics that scientific communities use for distinguishing appropriate and inappropriate values, beliefs, attitudes and behaviours (Huutoniemi 2015).

Ethical challenges exist in many different aspects of science, including mentoring, education, collaborative research, data sharing, peer review, technical writing and publication. In fact, we know that education and research activity depend a great deal on cooperation based on shared expectations and understandings (Shamoo and Resnik 2015) and those incorrect behaviours exist as well. Ethical issues are also present when researchers make discovers or work for other institutions that need their expertise. Researchers have social

S. Bento (✉)
CSG-SOCIUS, School of Economics & Management (ISEG), University of Lisbon, Lisbon, Portugal
e-mail: sbento@iseg.ulisboa.pt

M. T. C. de Melo
CERIS, Instituto Superior Técnico (IST), Universidade de Lisboa, Lisbon, Portugal



M. Abrunhosa et al. (eds.), *Advances in Geoethics and Groundwater Management: Theory and Practice for a Sustainable Development*, Advances in Science, Technology & Innovation,
https://doi.org/10.1007/978-3-030-59320-9_106

responsibilities and can be held publicly accountable (Davis 1995), and this pushes them for more ethical requirements. Simultaneously, more and more ethical guidelines are recommended for researchers and for submission of international funding for research (Carvallo 2019).

Groundwater researchers with their specific scientific knowledge may have an important contribution as consultants for companies with significant economic interests or as experts for the governmental departments with water management and monitoring public responsibilities. These types of collaborations are even more frequent in countries like Portugal where the number of groundwater professionals is limited and the market is often too small to justify a clear separation between public and private sectors of activity despite the fact that these sectors operate with different understandings, strategies and goals. The duplication role of the groundwater researcher may rise several geoethical questions to be explored later.

Groundwater research is not just theoretical but an imminently practical science that requires contacts in the field where researchers collect data and interact with the locals. It is also a truly interdisciplinary activity where community is mobilized for consulting for institutions. Can we seriously and responsibly face the black box of this scientific activity in the groundwater field? How can the scientific community be more prepared to deal with these ethical dimensions? It is important to boost this particular community to imagine the limits and the frontiers of its scientific production for the collective good and for improving quality for living communities in specific contexts.

## 2 Settings of Reflection

### 2.1 The Chemical Industry and the Study Area

The setting of this exercise is a highly contaminated area that has been severely impacted by chemical industry activity in the last decades. The chemical industry uses a wide variety of raw materials (from air, water to minerals and oil) to produce an enormous range of chemical products that influence practically every moment of human lives and socio-ecological systems. Some of these products are final goods for the public consumers but most are intermediate goods to be used by industries or by chemical industry itself. Many of these products are potentially hazardous during their manufacture, transport and use, and they may affect living organisms, human health and the environment.

The birth of the chemical industry in Estarreja (NW Portugal) dates back to the 1940s when Amoniaco Português (now, part of Bondalti) and SAPEC (Belgium) installed their units in Estarreja to produce fertilizers for agriculture. Nowadays, the Chemical Complex is constituted by different companies (Air Liquide, Cires, Bondalti and Dow Portugal) with a significant supply chain integration and accounting for 10% of the total Portuguese chemical industry. The location of the industrial Chemical Complex was indeed a strategic decision that prioritized: (1) the access to raw materials; (2) reliable water and energy supplies; and (3) good transportation system (road, rail and harbour facilities), disregarding the environmental suitability of the areas and the environmental footprint of the chemical industry.

In Estarreja, the Chemical Complex is built in a very vulnerable area in terms of natural environment, surface and groundwater bodies characteristics and biodiversity aspects, where the industrial human activities have been always a severe constraint to natural environmental sustainability. The area is bounded to the west by the Atlantic Ocean and the 'Ria de Aveiro', a shallow coastal lagoon with both marine and estuarine waters that has been classified as a Special Protection Area (SPA) under the EC Directive on the Conservation of Wild Birds. It forms a gently sloping coastal plain in the lower end of Vouga River Basin with extensive mudflats, sandbanks and salt marshes developed in its intertidal zone and drained by a complex network of water ditches and hundreds of wells. From the hydrogeological point of view, the Aveiro Quaternary groundwater body forms a shallow aquifer system that occurs locally as a surficial layer or as layers of variable thickness that are very vulnerable to soil and groundwater contamination. In the context of the River Basin Management Plans (APA 2016), the Aveiro groundwater body is defined as in poor chemical status due to diffuse contamination from agriculture and the Estarreja Chemical Complex is inventoried as a point source of contamination and is described as an environmental liability of the chemical industry that impacts soil, surface and groundwater bodies.

However, it is important to emphasize, that in the last 20 years, the modernization processes in the Estarreja Chemical Complex to improve efficiency and reduce environmental footprint required several environmental impact assessment studies that resulted in many mitigation measures of the contamination and the set-up of detailed soil and groundwater monitoring programs. All the monitoring data is regularly reported to the Portuguese Environmental Agency and is publicly available.

### 2.2 Methodology

The method used in this investigation is based on a reflexive exercise undertaken by the authors taking advantage of their scientific background on sociology and hydrogeology. It is anchored in qualitative approaches in social sciences; it was chosen because it emphasizes personal and singular experiences lived by the hydrological researcher in a context (scientific and professional) that is also the fieldwork of the sociologist as researcher. Autobiography tells a story and gives the reader a distance from the events described to think them in a more abstract way. Autoethnography is a method that combines the characteristics of autobiography (significant moments of the life of the person) with ethnography (study of the relational practices of a culture), using participant observation, interviews, analysis of artefacts and ways of being (Ellis et al. 2011). Both autobiography and ethnography seek to give a picture to the phenomenon studied through the voices of the elements of the community and through the social dynamics observed (Creswell 2014). In this specific case, the sociologist had long interviews with the hydrogeologist; usually, they are called comprehensive conversation (Mason 2012). The autobiographical interviews were taped, transcribed and analysed through content analysis of the textual discourse. The method used included also reading of documents and collaborative analysis of situations. The results present a short summary of some of the most important some learnt lessons based on the interviews.

## 3 Results

A quick search in the Science Direct webpage (accessed April 2020) associates to the keyword Estarreja to more than 181 research papers, 9 review articles and 5 book chapters. If we narrow this research to the keywords Estarreja, soil and water, a total number of 95 contributions is achieved.

All these research papers produced a huge amount of data that is difficult to integrate because of the different methodologies, objectives and time spans. Many of these papers refer to the environmental or health impacts of air, soil, water and groundwater contamination related to the environmental liabilities of the Estarreja Chemical Complex. However, these scientific achievements remain most of the times within the scientific community as the dissemination of scientific data is not an easy nor a neutral task, especially when political, social and ethical issues are at stake.

The main ethical question does not involve the building of fraudulent data in the scientific research. In fact, the cases of fraud exist but they consist of what Merton (1942) defended as 'abnormal' situations. So, this will not be our focus. However, nowadays, more or less with the globalization and the incentive for funding research in projects, several studies underline the complexity of scientific activity and the existence of strong dilemma for researchers and a grey environment for them whenever they have to choose to whom disseminate the data, or what implications the results have on local populations.

In this way, the different monitoring programmes carried out by the companies within the Estarreja Chemical Complex provide unique and valuable data on soil and groundwater quality of the study area. That data is reported to APA and despite it is of public domain, it is very hardly accessible to the water end-user community.

In the following sections, there is a first critical reflection about the importance of ethics on the way groundwater researchers and consultants plan, do and disseminate their work and how it limits the reach of science in terms of environmental, human health and societal protection goals.

### 3.1 From Field Work to Data Interpretation

For hydrogeologists, the planning of a field campaign for collecting groundwater samples is essential to guarantee that the point samples collected in springs, piezometers, wells and boreholes are representative of the water quality in the aquifer. The spatial distribution of the sampling points needs also to be carefully chosen because it will affect the results and will bias the interpretations. The sampling network must be spatially representative of the study area, and contaminated areas should be selected together with non-contaminated ones so that is possible to characterize the groundwater natural background and infer the contamination processes.

Another crucial step is the sampling process. It is essential that the international protocols for groundwater sampling are followed and that the samples are properly filtrated, preserved and transported to the laboratory or otherwise the analytical results are not representative of the in situ groundwater quality. It is important to understand if for the characterization of the contamination a flowing sample is needed, or a slow purge sample or a depth specific sample because the three methods will provide different results.

The uncertainty for any methodology used to interpret the data should be defined and described. And if risk assessment for human health is considered, it is relevant that the water consuming habits are well documented for the local populations and compared to a non-target region.

### 3.2 From Analytical Data to End-Users

For sampling groundwater, the researchers and sometimes the consultants need to access private wells and boreholes. Analytical results should be provided and explained to the well owners; the project research is not always compatible with this practice. If the results may have adverse health effects, researchers should communicate to the local environmental and health authorities, discuss with them the potential sources and help them to mitigate the contamination. These should include an open dialogue with the companies from the Chemical Complex, but many researchers skip this process because they are afraid of facing limitations to their research in the future.

### 3.3 Consultancy and Research Work

In Portugal, the community of hydrogeologists is very small (<30) and this situation leads to very close relations between the academy researchers and the private sector where the researchers may be consultants. For both scientists and consultants, there are several ethical issues at stake. The most important is for the scientist her/himself which means that she/he will be doing and reporting the technical work the best way she/he knows. Another situation is that although the researcher respects the public or private company that she/he is working for, the researcher should never accept pressures on the interpretation of the results and dissemination of the data. Reporting data in contaminated areas can be absolutely biased if researchers just emphasize, for example, the positive achievements of the monitoring programmes or the other way round. The last situation identified is that researcher should never ignore that surrounding the Chemical Complex. There are villages, farmers that use groundwater for irrigation and sometimes for domestic uses, vulnerable ecosystems and that the main objective is to protect both. On the other hand, it is important to define the limit between a job as a consultant and as a researcher because the data to analyse crosses both borders. If the researcher is limiting her/himself in data analysis, then it is recommended from the ethical point of view, to leave one of the roles, as scientist or as consultant.

Moreover, the APA and its regional offices are very short staffed in terms of technical professionals and this is particularly limiting in the understanding of a hidden resource such as groundwater that requires specific expertise.

## 4 Concluding Remarks

The paper seeks to explore qualitative approaches for the study of ethical issues in groundwater sciences. The results show that ethics are crucial to help groundwater researchers to move from the case study investigation (based on science) to what should be in the future (based on environmental, human health and societal protection goals).

The analysed examples are related to the access and use of both scientific and technical data and are based on personal experiences that were considered relevant for ethical issues. The purpose here is more than to enlighten their capacity to express responsible stories about science, in particular groundwater, and to function as ethical recommendations. The main objective is that those stories or situations serve as cases for reflection questioning science in the way it relates with society when potential risks are at stake and when authorities are not able to open the black box.

In this sense, the cases used surpass an anecdotal perspective; instead, they seek to describe more attentively the conditions and the context of the both scientific production and technical reporting, and they wish to demonstrate the relevance of an ethnographical perspective for the study of ethics in geological and groundwater studies.

**Acknowledgements** The authors would like to dedicate the paper to the memory of Luís Ribeiro.

## References

APA (2016) Avaliação Ambiental Estratégica. Plano de Gestão de Região Hidrográfica do Vouga, Mondego e Lis (RH4). Plano de Gestão dos Riscos de Inundações (RH4). Relatório Ambiental Final. https://www.apambiente.pt/_zdata/Politicas/Agua/PlaneamentoeGestao/AAE_PGRH_PGRI/2016-2021/RelatorioAmbiental_RH4A.pdf

Beck U (1992) Risk society: towards a new modernity. Sage Publications, London

Carvallo S (2019) L'éthique de la recherché entre réglementation et réflexivité. Revue d'Anthropologie 13(2):299–326

Creswell JW (2014) Research design: qualitative, quantitative and mixed methods approaches. Sage Publications, California

Davis M (1995) A preface to accountability in the professions. Account Res 4:2, 81–90

Ellis C, Adam TE, Bochner AP (2011) Autoethnography: an overview. Qualitative Health Research 12(1): Art. 10

Huutoniemi K (2015) Peer review: organized skepticism. International Encyclopedia of the Social & Behavioral Sciences, 2nd edn. , pp 685–689

Mason J (2012) Qualitative researching. London: Sage Publications. ISBN 0 7619 7427 X

Merton R (1938) Science, Technology and Society in seventeenth century. Fertig, England, New York

Merton R (1942) Science and technology in a democratic order. J Legal Polit Sociol 1:15–26

Shamoo AE, Resnik D (2015) Responsible conduct of research. Oxford University Press, New York

# The International Agreement of the Guarani Aquifer System: A Transboundary Aquifer

Luciana Cordeiro de Souza Fernandes

**Abstract**

The hydrogeology associated with Environmental Law is essential in the discussion of transboundary aquifer management. The "Guarani Aquifer Agreement", although not yet in force, presents the instruments for shared management of the Guarani Aquifer System (SAG), located in parts of the territories of Argentina, Brazil, Paraguay and Uruguay countries that form MERCOSUR. It is recognized as the largest aquifer on the planet in superficial extension, crossing political borders, river basins, and hydrogeological provinces, as well as it supplies large part of the population of these countries. This agreement is one of the few treaties signed in a precautionary and preventive context; both are fundamental principles of Environmental Law. It is noteworthy that this agreement does not come from a prior existence of conflicts, as has already happened in several regions of the planet when dealing with transboundary water resources. The Guarani Aquifer System Environmental Protection and Sustainable Development Project (PSAG) was an important instrument that brought together both the researchers and the political power of the four countries to identify the characteristics and potentialities of the SAG, so that cooperation existing during the project and the results obtained favored the creation of this Agreement.

**Keywords**

Groundwater • International environmental law • Geo-law • Hydrology law • Groundwater management

L. C. de Souza Fernandes (✉)
Faculdade de Ciências Aplicadas, Universidade Estadual de Campinas, São Paulo, Brazil
e-mail: lucord@unicamp.br

## 1 Introduction

Globally, invisible groundwater resources are under increasing pressure due to human activities and climate change. Our response to this pressure is often not adequate, also due to limited awareness of the importance of groundwater resources (IGRAC 2015).

A recent study of Hirata et al. (2019) indicates that the estimated annual groundwater abstraction worldwide from 2010 onwards exceeds 1,000,000 $Mm^3$, which set it in the position of substance with the highest level of underground extraction. Groundwater plays a key role in many countries, being present in population supply, irrigation, and industry.

According to the IGRAC (2019), considering this worldwide exploited volume, 67% of all groundwater is used for irrigation (food production), 22% is used for domestic purposes (drinking water and sanitation) and 11% is used for industry.

However, innumerous aquifers that promote global water sustainability are not restricted to a country's territory, i.e., political boundaries are not limiting. There are now 592 identified transboundary aquifers, including 226 transboundary "groundwater bodies" as defined in the European Union Water Framework Directive, EU WFD, underlying almost every nation (IGRAC 2015).

## 2 Materials and Methods

This paper aims to present the importance of technical studies of hydrogeology to support the elaboration of international agreements in a preventive way, using as an example International Agreement of the Guarani Aquifer System (IA-GAS).

The analysis of transboundary management of aquifers, as in the GAS case, was carried out through bibliographic research, with the survey of international technical and legal documents from various programs and entities, such as the


M. Abrunhosa et al. (eds.), *Advances in Geoethics and Groundwater Management: Theory and Practice for a Sustainable Development*, Advances in Science, Technology & Innovation,
https://doi.org/10.1007/978-3-030-59320-9_107

International Hydrology Program (IHP-UNESCO), International Law Association (ILA), the International Association of Hydrogeologists (IAH), the International Groundwater Resources Assessment Center (IGRAC), the Internationally Shared Aquifer Resources Management (ISARM), the Global Water Partnership (GWP). Regarding the legal aspects, special attention has been devoted to Resolution 63/124 of the United Nations General Assembly (UNGA) on the Law of Transboundary Aquifers, and were referenced in the IA-GAS.

## 3 Results

The IA-GAS, which has as signatory countries Argentina, Brazil, Paraguay and Uruguay, is one of the few treaties signed in a precautionary and preventive context; both are fundamental principles of Environmental Law. It is noteworthy that this agreement does not come from a prior existence of conflicts, as has already happened in several regions of the planet when dealing with transboundary water resources.

In this sense, the Guarani Aquifer System Environmental Protection and Sustainable Development Project (PSAG) was an important instrument that brought together both the researchers and the political power of the four countries to identify the characteristics and potentialities of the GAS, so that cooperation existing during the project and the results obtained favored the creation of this Agreement.

In legal terms, this Agreement was based on UNGA Resolution 63/124, approved in 2009 and which deals with the law of Transboundary Aquifers, as well as UNGA Resolution 1803, of December 14, 1962, which refers to permanent sovereignty over natural resources.

This Agreement was signed in the San Juan city, Argentine Republic, on October 2, 2010, between the four countries, stating that the GAS is a transboundary water resource that integrate the sovereign territorial domain of the Argentine Republic, Federative Republic of Brazil, Republic of Paraguay and the Oriental Republic of Uruguay, which are the sole holders of this resource and hereafter will be referred to as "States Parties" or simply "Parties". Among other objectives, this agreement aim to expand the levels of cooperation for greater scientific knowledge about the GAS and responsible management of your water resources.

In this regard, it is noteworthy that States Parties undertake to exchange technical information on studies, activities, and works that contemplate the sustainable use of water resources of the GAS, giving effectiveness to the principle of prevention. Similarly, the obligation of the Parties to inform each other of all activities and works they intend to undertake is established, so that the exchange of environmental information between the Parties will be the key to the continuity of the success of GAS shared management for present and future generations.

It also provides for the establishment of a Commission composed of the Parties, which will coordinate the fulfillment of the principles and objectives of this Agreement, with its own regulation. In addition, that in dispute situations, they shall seek to negotiate directly between the Parties in an amicable manner, and when the dispute cannot be resolved, they may resort to arbitration proceedings. It is a document that demonstrates a deep knowledge of the technical characteristics of the aquifer, since PSAG offers this support, combining technical studies with legislation.

However, despite the importance of this Agreement, its structure and cooperation of the States Parties, it is still in no force, because Paraguay has recently ratified, and only on November 26, 2020, will it enter into force (MRE 2020).

## 4 Discussion

With one-third of humanity totally dependent on groundwater for their daily needs and 98% of the Planet Earth's easily accessible freshwater found in aquifers, there is an urgent need for improved governance to halt over exploitation and degradation of this vital "hidden" groundwater resource (IGRAC 2015).

Transboundary aquifer or transboundary aquifer system means, respectively, an aquifer or aquifer system, part of which are situated in different States. The international aspect of a transboundary aquifer makes its management more complex than in a case of an aquifer located within the State borders. An informed and sustainable management of commonly shared aquifer asks for adequate knowledge of its characteristics, present state, and trends. In order to acquire this knowledge, regular monitoring and assessment of the transboundary aquifer need to be performed. The document describes a methodology for multidisciplinary assessment of transboundary aquifers and gives the guidelines for its implementation (IGRAC 2019). Subsequent paragraphs, however, are indented (here insert the second paragraph).

The GAS is transboundary aquifer, considered the largest aquifer in the world in territorial extension, located in South America, has a total area of 1,087,879 km$^2$, according to surveys carried out in the SAG by PSAG. Under part of the territories of four countries, as shown in Fig. 1, its occurrence covers an area of 735,918 km$^2$ (61.65%) in Brazil; 36,170 km$^2$ in Uruguay (3.32%); 228,255 km$^2$ (20.98%) in Argentina, and 87,536 km$^2$ (8.05%) in Paraguay (OEA 2009). Due to its hydrogeological and hydrochemical properties, it is considered as the main groundwater reserve in this region.

The extensive area underlain by the SAG has a present population of about 15 million (although including large

**Fig. 1** Spatial location of the Guarani Aquifer System ("Aquífero Guarani") in South America (after ECOA)

cities in its proximity this figure increases to about 90 million), a mainly sub-tropical climate, and abundant (but often polluted) surface-water resources which experience occasional droughts. Thus, the need for reliable potable water-supply sources and industrial supplies (of low-treatment cost) is likely to grow significantly, especially in some climate-change scenarios (which imply increased water demand due to ambient temperature increase and more frequent and intense surface-water droughts) (Foster et al. 2009).

According to recent data, the use of SAG waters has become increasingly extensive, with a substantial increase in its exploitation in the four countries, notably in Brazil, being the only source of supply in some regions. The current resource exploitation exceeds 1.0 $km^3/a$, being 93.6% in Brazil (of which about 80% are in São Paulo State), 2.8% in Uruguay, 2.3% in Paraguay, and 1.3% in Argentina. Approximately, 80% of the total is used for public water supply, 15% for industrial processes, and 5% by geothermal spas (Foster et al. 2009). Although this water extraction is essential, it should be considered that it is still a small fraction of its full capacity (Sindico et al. 2018).

Considering the legal aspect, since 2002, the codification of a law on transboundary aquifers has been the subject of the Commission on International Law-United Nations (ILC-UN), aiming to create an appropriate legal regime for aquifer management, given its importance as a source of fresh water. In a rather short period of 6 years, the ILC was able to adopt a set of 19 draft articles at its session in 2008. It sent the draft articles together with the commentaries there to the 63rd session (2008) of the UN General Assembly (Stephan 2009). This articles set aims to regulate the relationship between persons (physical and legal) and between the people and the state administration on water resources; it includes all legal provisions on development, use, protection, and management of groundwater resources, which may be either scattered in various enactments or integrated into a comprehensive water law (Nani et al. 2004).

Finally and without exhausting the subject, it is worth emphasizing that the technical data are essential for groundwater management, protection of recharge areas, and appropriate land use in these areas, as well as regulating the exploitation according to aquifer characteristics. However, its effective protection requires legislation that incorporates the technical aspects of governance tools and instruments, notably for transboundary aquifers, as provided for in legal regulations. Especially for transboundary aquifers, international agreements and treaties should be established between countries that share groundwater sources.

## 5 Concluding Remarks

The cooperative development between technicians and legal specialists who worked in the PSAG, based on ethics and commitment, can be seen as a model for the elaboration of projects and international legal diplomas for the management of other transboundary aquifers, to be prepared pre-emptively to conflicts, combining geology and law and striving for cooperation between countries.

Although the International Agreement of GAS is not in force, we find that its preparation is due to the linkage of the technical data obtained from the PSAG to the UNGA Resolution 63/124, the Law of Transboundary Aquifers. This gives rise to a new branch of law, which goes beyond Environmental Law and International Law, wholly dependent on geology we may call Geo-Law, since for the effectiveness of legal protection geology becomes protagonist.

**Acknowledgements** The author was supported by the "FAPESP Projects. 2013/10689-6 and 2017/19973-0".

## References

ECOA: https://ecoa.org.br/aquifero-guarani-como-se-formou-esse-imenso-oceano-que-esta-sob-nossos-pes/. Last accessed 14 Nov 2019

Foster S, Hirata R, Vidal A, Schmidt G, Garduño H (2009) The Guarani aquifer initiative—towards realistic groundwater management in a transboundary context. Sustainable Groundwater Management Lessons from Practice. Case Profile Collection Number 9. World Bank, Washington D.C., USA, pp 1–28

Hirata R, Suhogusoff AV, Marcelini SS, Lillar PC, Marcellini L (2019) A revolução silenciosa das águas subterrâneas no Brasil: uma análise da importância do recurso e os riscos pela falta de saneamento. Trata Brasil, Brasília, DF, Brazil

IGRAC (2015) Transboundary aquifers of the world map. IGRAC Publications, Delft https://www.un-igrac.org/tbamap. Last accessed 14 Nov 2019

IGRAC (2019) Guidelines for multidisciplinary TBA assessment (draft version). IGRAC Publications, Delft. https://www.un-igrac.org/news/draft-guidelines-multidisciplinary-tba-assessment-now-available-online. Last accessed 14 Nov 2019

MRE (2020) Ministerio de Relaciones Exteriores. https://www.mre.gov.py/index.php/noticias-de-embajadas-y-consulados/el-26-de-noviembre-entra-en-vigor-el-acuerdo-sobre-el-acuifero-guarani. Last accessed 12 Nov 2020

Nani M, Foster S, Dumars C, Garduño H, Kemper K, Tuinhof A (2004) Groundwater Legislation & Regulatory Provision from customary rules to integrated catchment planning. Sustainable Groundwater Management: Concepts and Tools GWMATE Briefing Note Series. Note 4. World Bank, Washington D.C., pp 1–6

ONU. Resolution adopted by the General Assembly 63/124. The law of transboundary aquifers. https://www.internationalwaterlaw.org/documents/intldocs/UNGA_Resolution_on_Law_of_Transboundary_Aquifers.pdf. Last accessed 12 Nov 2019

OEA – Organização dos Estados Americanos (2009) Aquífero Guarani: programa estratégico de ação/Acuífero Guaraní: programa estatégico de acción. Organização dos Estados Americanos. https://www.mma.gov.br/informma/item/8617-aquifero-guarani. Last accessed 12 Nov 2019

Sindico F, Hirata R, Manganelli A (2018) The Guarani aquifer system: from a beacon of hope to a question mark in the governance of transboundary aquifers. J Hydrol Reg Stud 20:40–49

Stephan RM (2009) Transboundary aquifers: managing a vital resource. UNESCO, Paris

# Role of Groundwater as a Climate Change Adaptation Strategy in Dry Zone Farming Systems, Sri Lanka

M. M. G. S. Dilini, S. Pathmarajah, and E. R. N. Gunawardena

**Abstract**

Water-related uncertainties are a troubling concern for farming communities in the dry zone Sri Lanka. Farmers have to manage available water or seek alternative ways to fulfil the water requirements needed to maintain their livelihoods. Within this context, a comparative study was conducted in Horivila-Palugaswewa and Sivalakulama dry zone farming systems in order to discover the role of certain climate change adaptation strategies to cope under climate-related vulnerability. Key informant interviews and questionnaire surveys facilitated the data collection process. The data was analysed through the development of an Adaptation Strategy Index and the application of Cronbach's Alpha Reliability Test. The findings show that the use of alternative water sources in terms of agro-wells as the first-ranked adaptation strategy in both study sites. This finding was further confirmed with Cronbach's Alpha Reliability Test as the same strategy has taken the highest reliability in the test results also. Although number of adaptation strategies have been included in the Adaptation Strategy Index development, overall, the farmers tend to prioritize the use of alternative water sources (agro-wells) while neglecting methods of managing the available water sources. With a growing reliance on water from agro-wells, there is a clear need to focus towards groundwater management to ensure that over extraction does not become a major issue. This is particularly important as agro-wells have an increasingly enormous role in supporting farming communities in both study sites.

M. M. G. S. Dilini
Postgraduate Institute of Agriculture, University of Peradeniya, Peradeniya, Sri Lanka

S. Pathmarajah · E. R. N. Gunawardena (✉)
Department of Agricultural Engineering, Faculty of Agriculture, University of Peradeniya, Peradeniya, Sri Lanka
e-mail: nimalgun@pdn.ac.lk



M. Abrunhosa et al. (eds.), *Advances in Geoethics and Groundwater Management: Theory and Practice for a Sustainable Development*, Advances in Science, Technology & Innovation,
https://doi.org/10.1007/978-3-030-59320-9_108



**Keywords**

Adaptation • Agro-wells • Alternative • Strategies • Water

## 1 Introduction

The natural geography of the dry zone Sri Lanka has resulted in a water-scarcity scenario in the region, exposing vulnerability on agro based farming communities. *Yala* and *Maha* are the two major growing seasons in the area. *Yala* season is the minor growing season which receives First Inter Monsoon and South West Monsoon. As the South West Monsoon rains are not very effective in the dry zone, *Yala* season rains are confined to two months, showing a longer dry period which extends from May to September. *Maha* season is the major cultivation season which receives high rainfall with North East Monsoon over the December to February period. Dating back to ancient times, Sri Lankan farmers have developed tanks (reservoirs) to store water during *Maha* season and utilize them in times of water scarcity. However, this water is increasingly becoming not enough under current climatic uncertainties, population increases, and the continued expansion of cultivation area.

Given this concern, many dry zone farmers have adopted different coping mechanisms. What is still uncertain is whether these farmers have achieved the anticipated benefits in a sustainable manner. Therefore, it is increasingly necessary to understand which strategies have been effective in the context of increasing water scarcity. Identifying which aspects of particular adaptation strategies have been successful and those that have failed can provide needed insights to address future climate change and water scarcity concerns.

## 2 Methodology

This study was carried out as a comparative study in Horivila-Palugaswewa and Sivalakulama cascades which belongs to Malwathu oya drainage basin in the dry zone Sri Lanka. Firstly, all native and modern climate change adaptation strategies were listed in the study sites based on the Key Informant Interviews (KIIs) with experienced farmers, Farmer Organization (FO) leaders, Agricultural Research Production Assistants, and Agriculture Instructors. The Adaptation Strategy Index (ASI) was developed in order to identify the most effective climate change adaptation strategies in these farming systems as per farmers' point of view. The study needed to assess different adaptation strategies by using the four-point rating scale to rate the importance of each strategy. For this purpose, questionnaire surveys were conducted with a purposive sampling group of 50 farmers from the Horivila-Palugaswewa cascade and 50 farmers from the Sivalakulama cascade who are involved in the climate change adaptation strategies. The Adaptation Strategy Index, outlined in Index 1, has been utilized in other studies by Uddin et al. (2014) and Herath and Thirumarpan (2017).

$$ASI = (ASn \times 0) + (ASl \times 1) + (ASm \times 2) + (ASh \times 3)$$

ASI = Adaptation Strategy Index
ASn = Frequency of farmers rating adaptation strategy as having no importance
ASl = Frequency of farmers rating adaptation strategy as having low importance
ASm = Frequency of farmers rating adaptation strategy as having moderate importance
ASh = Frequency of farmers rating adaptation strategy as having high importance

**Index 1** Adaptation Strategy Index

Cronbach's Alpha Reliability Test was adopted on the assumption that farmer's attitude or intention should reflect the real means of listed strategies to their agricultural livelihood. The test was undertaken with the use of SPSS 16.0 software. In this study, farmers' intentions were measured concerning familiarity with the strategy, helpfulness of the strategy in farming in cases of climate change volatilities during recent five-year time period, and personal interest towards the strategy. Responses for each question were recorded for each strategy using *Likert* scale. Alpha values were calculated using SPSS 'Reliability Analysis'. After performing Cronbach's Alpha Reliability Test, the study shortlisted the adaptation strategies only accepting the strategies which shows score of $\geq .7$. A rule of thumb to interpret alpha as in Table 1.

Only the accepted strategies with Cronbach's Alpha Reliability Test were considered in developing the ASI for the current study. Final rankings have been made relative to the ASI.

## 3 Results and Discussion

For the Horivila-Palugaswewa cascade, the use of alternative water sources (agro-wells) was found to be the first-ranked climate change adaptation strategy as per its ASI value 226. As identified through the Divisional Secretariat Resource Profile (2018), altogether there are 89 agro-wells in the Horivila-Palugaswewa cascade and 87 of them are in usable condition provisioning water for 332 ac. of area cultivation during *Yala* season. With an ASI value 220, the use of alternative water sources (agro-wells) was found to be the first-ranked climate change adaptation strategy in the Sivalakulama cascade as well. According to the data obtained through the Galenbindunuwewa Divisional Secretariat Resource Profile (2017), there are 329 agro-wells in the Sivalakulama cascade. As per the Agricultural Instructors, number of agro-wells likely to exceed 400 according to the latest count. Farmers are highly depending on these water sources for *Yala* season cultivation where tank cascade water cannot meet the required water needs, especially given the current remarkable water scarcity scenario in the Sivalakulama cascade. Therefore, alternative water sources (agro-wells) can be recognized as a huge relief for this farming community. It is noteworthy that the highest value of Cronbach's Alpha Reliability Test for the Horivila-Palugaswewa cascade is .888 and .885 for the Sivalakulama cascade. Both values clarifying familiarity, helpfulness of this strategy to farming in cases of climate change volatilities during the recent five-year time period, and farmers' personal interest towards use of agro-wells.

As the results in Table 2 denote, shifts to shorter cycle paddy varieties can be identified as the second-ranked adaptation strategy in the Horivila-Palugaswewa cascade. As Agrarian Service Centre *Kanna* (seasonal) meeting records (2018) reveal, BG 300, BG 358, and BG 359 (3½ months) are the paddy varieties most commonly cultivated under the water scarcity scenarios during *Yala* season. Connection with market chains was identified as the second-ranked adaptation strategy in the Sivalakulama cascade. Sivalakulama cascade farmers have market chains with 'Prima' and 'Samaposha'. Connections with market chains are helpful for these rural farmers as they can receive quality seeds, fertilizer, farming advice, and especially buy back guarantees for their crops. As the Sivalakulama cascade farmers are more oriented towards maize farming, they may feel that they are more secure under this support. The calculated ASI

**Table 1** A rule of thumb to interpret Cronbach's alpha

| Cronbach's Alpha | Internal Consistency |
|---|---|
| $\propto \geq 0.9$ | Excellent |
| $0.9 > \propto \geq 0.8$ | Good |
| $0.8 > \propto \geq 0.7$ | Acceptable |
| $0.7 > \propto \geq 0.6$ | Questionable |
| $0.6 > \propto \geq 0.5$ | Poor |
| $0.5 > \propto$ | Unacceptable |

*Source* Stephanie (2014)

**Table 2** Cronbach's alpha values and ASI values calculated for listed climate change adaptation strategies in Horivila-Palugaswewa and Sivalakulama cascades

| Strategy | Cronbach's alpha | | ASI | |
|---|---|---|---|---|
| | Horivila-Palugaswewa cascade | Sivalakulama cascade | Horivila-Palugaswewa cascade | Sivalakulama cascade |
| 1. Native weather forecasting system | 0.685 | 0.595 | – | – |
| 2. Communication for updating on rainfalls based on data from Department of Meteorology | 0.771 | 0.756 | 186 | 168 |
| Crop cultivation plans guided by weather forecasting | 0.797 | 0.794 | 206 | 154 |
| 3. Utilize rainwater during *Maha* season to carry out farming under direct rainfall | 0.719 | 0.716 | 148 | 144 |
| 4. Shift to shorter cycle paddy varieties | 0.815 | 0.840 | 210 | 152 |
| 5. Sharing paddy plots among head and tail farmers of the irrigation network (*Bethma*) | 0.727 | 0.715 | 160 | 122 |
| 6. Dry sowing of paddy (*Kekulama*) | 0.809 | 0.832 | 200 | 158 |
| 7. Parachute method for crop establishment | 0.748 | 0.766 | 136 | 124 |
| 8. Restriction on cultivation water uses during water scarcity period (*Mura Kramaya*) | 0.765 | 0.747 | 196 | 126 |
| 9. Adjusting timing of irrigation | 0.754 | 0.726 | 136 | 124 |
| 10. Changing the timing of activities (sowing, planting) | 0.787 | 0.815 | 170 | 122 |
| 11. Changing harvesting time | 0.730 | 0.764 | 178 | 120 |
| 12. Reusing drainage water for cultivation | 0.807 | 0.816 | 202 | 164 |
| 13. Conserving soil moisture using green mulch | 0.642 | 0.703 | – | 84 |
| 14. OFC cultivation under smaller tanks | 0.709 | 0.779 | 144 | 152 |
| 15. Use of alternative water sources (agro-wells) | 0.888 | 0.885 | 226 | 220 |
| 16. Organic fertilization for other crops | 0.591 | 0.558 | – | – |
| 17. Crop diversification | 0.820 | 0.803 | 164 | 166 |
| 18. Selection of tolerant OFC varieties | 0.827 | 0.829 | 204 | 150 |
| 19. Shifting cultivation | 0.758 | 0.770 | 124 | 140 |
| 20. Fruits and vegetable cultivation under 'Good Agricultural Practice' programme | 0.743 | 0.747 | 140 | 120 |
| 21. Support through loan facilities for cultivation | 0.821 | 0.873 | 128 | 178 |
| 22. Agricultural insurance schemes | 0.758 | 0.751 | 134 | 148 |
| 23. Crop damage compensations | 0.709 | 0.701 | 94 | 146 |
| 24. Access to agriculture advisory and extension services | 0.795 | 0.792 | 116 | 128 |
| 25. Connection with market chains | 0.861 | 0.876 | 118 | 210 |

*Note* ASI was not calculated for rejected strategies which did not meet the condition of acceptance at $\geq .7$ with Cronbach's Alpha Reliability Test value is same as it denotes 210 for second prioritized strategy in both cascades.

Crop cultivation plans guided by weather forecasting indicate an ASI value of 206 making it the third-ranked adaptation strategy in the Horivila-Palugaswewa cascade. Agrarian Service Centres disseminate weather information to the FO and farmers based on the report received by them. Accordingly, if there is normal rainfall, farmers are instructed to cultivate paddy. If there is below normal rainfall, farmers are instructed to cultivate OFC (Other Field Crop). Support through loan facilities for cultivation could be recognized as the third-ranked adaptation strategy in the Sivalakulama cascade with an ASI value of 178. Some of the reasons that lead to the prioritization of this strategy may be linked with the second-rank strategy of the Sivalakulama cascade, as market chains provide support through loans. Beyond this, agriculture-related facilitations with government bank loans and agro bank loans can be recognized in this context.

It is notable, apart from the three prioritized strategies in the Horivila-Palugaswewa cascade, there are few strategies that denote an ASI value more than 200. Furthermore, it is worth noting that even the third strategy of the Sivalakulama cascade has obtained low value. A reason for this could be that all these prioritizations are dependent on what the farmers have really experienced.

Crop damage compensations have been ranked as least important strategy in the Horivila-Palugaswewa study site. Crop damages occur in this study site due to water scarcity, water excessiveness, and wild animal invasions. As per the farmers' point of view, though there are estimations, farmers who always deal with the claiming process receive money and others do not. A lack of awareness regarding claiming processes may also be another aspect on this issue. Conserve soil moisture using green mulch was identified as the least important adaptation strategy in the Sivalakulama cascade.

## 4 Concluding Remarks

Although both cascade belongs to same geographical context, farmers in respective cascades have prioritized adaptation strategies as they receive benefits from those. Given that water scarcity is the main overarching concern for dry zone farming systems, farmers have prioritized the use of alternative water sources (agro-wells). Not only with the ASI, Cronbach's Alpha Reliability Test also configured the same result. Though these are tank-based farming systems, farmers are most reliant on agro-wells in the water scarcity scenarios at present. Unregulated amounts of agro-wells in a small area can impact the groundwater table. Therefore, there is an urgent need to pay attention to the groundwater management by managing tank water, water management at field level, building up farmers' awareness, and promoting alternative livelihood practices.

**Acknowledgements** This work was carried out with the aid of a grant from the International Development Research Centre, Ottawa, Canada. Their financial support is greatly appreciated.

## References

Agrarian Service Centre Records (2018) Agrarian Service Centre, Palugaswewa, Anuradhapura District, Sri Lanka

Herath HML, Thirumarpan K (2017) Climate Change Induced Adaptation Strategies by Paddy Farmers: Special Emphasis on Socio Economic Insights. J Agric Sci Sri Lanka 12(2):124–137

Resource Profile (2017) DS Office, Galenbindunuwewa, Anuradhapura District, Sri Lanka

Resource Profile (2018) DS Office, Palugaswewa, Anuradhapura District, Sri Lanka

Stephanie (2014) Cronbach's Alpha: Simple Definition, Use and Interpretation. https://www.statisticshowto.datasciencecentral.com/cronbachs-alpha-spss/. Last accessed 28 Oct 2019

Uddin MN, Bokelmann W, Entsminger JS (2014) Factors affecting farmers' adaptation strategies to environmental degradation and climate change effects: a farm level study in Bangladesh. Climate MDPI 2(4):223–241

# Safeguarding of Groundwater Abstractions by Enforcement of Source Protection Zones

Jane Dottridge, Aidan Foley, and Nick Walters

**Abstract**

Groundwater provides 80% of the drinking water in southern and eastern England. To protect water quality, the environmental regulator defined groundwater Source Protection Zones, to show the risk of contamination based on travel times to the abstraction. Application of this policy requires the regulator to oppose developments with bulk storage of hazardous substances within the inner, 50-day zone. It appears that the policy is not consistently applied; thus, allowing recent developments close to strategically important groundwater abstractions. The ongoing dialogue on the real levels of risk, effectiveness of engineered mitigation, and threats to security of public water supplies requires hydrogeologists to use their geoscience knowledge, apply the principles of hydrogeoethics, and influence awareness of society. The issues are illustrated with an example of a proposed petrol filling station inside the inner source protection zone and only 200 m from a public water supply borehole. The borehole abstracts groundwater from the Cretaceous Chalk aquifer, with karstic features and potential preferential flow pathways requiring a large source protection zone. Although the environmental regulator allowed reliance on engineering solutions, the water company's objection was upheld by the planning system because of unacceptable long-term risk to strategic public water supplies.

**Keywords**

Groundwater • Abstraction • Protection • Policy • Enforcement

J. Dottridge (✉) · A. Foley · N. Walters
Mott MacDonald, Cambridge, UK
e-mail: jane.dottridge@mottmac.com

## 1 Introduction

Groundwater provides a third of the drinking water in England, but the proportion rises to 80% in some densely populated areas of southern and eastern England. Environmental regulations to manage groundwater resources have been in force since abstraction licensing was introduced in 1963. This licensing system balances the quantity of water available for abstraction with the needs of the environment, and is strictly enforced by the regulator, the Environment Agency.

In addition, the quality of drinking water for public supply is protected by groundwater Source Protection Zones (SPZs), which show the level of risk to a groundwater source of contamination from any polluting activities. Application of this policy thus requires that the Environment Agency will normally oppose new developments close to public water supplies which involve large-scale storage of hazardous substances, cemeteries, liquid effluent disposal, and some infrastructure schemes. However, this policy is not consistently applied, thus allowing some developments close to strategically important groundwater abstractions. This results in potential threats to security of public water supplies, which is illustrated by a case study of a proposed petrol filling station within 200 m of an abstraction borehole.

## 2 Source Protection Zones

The Environment Agency has a statutory duty to monitor and protect the quality of groundwater in England, and to conserve its use for water resources. The Approach to Groundwater Protection (Environment Agency 2018) contains position statements which provide information about the Environment Agency's approach to managing and protecting groundwater. They detail how the Environment Agency delivers government policy for groundwater and adopts a risk-based approach where legislation allows.



M. Abrunhosa et al. (eds.), *Advances in Geoethics and Groundwater Management: Theory and Practice for a Sustainable Development*, Advances in Science, Technology & Innovation,
https://doi.org/10.1007/978-3-030-59320-9_109

To protect the quality of drinking water, the Environment Agency has defined Source Protection Zones (SPZs) around public water supply abstractions (Fig. 1). The policy states that the zones are used for pollution prevention and monitoring. The zones are based on travel times to the source in the saturated aquifer, assuming a porous medium: inner zone SPZ1—50-days, outer zone SPZ2—300 days, and SPZ3—total groundwater catchment. Where karstic features exist increasing heterogeneity of the aquifer, the equivalent porous medium approach used to define the SPZs is not appropriate, and a much larger SPZ is required to protect the source (Pochon et al. 2008) and this is acknowledged in the guidance on definition of SPZs (Environment Agency 2019).

The Environment Agency uses a risk-based approach to regulate activities that could impact groundwater resources. The position statement for underground storage and associated pipework is that they will normally object to new and increased underground storage of hazardous substances in SPZ1. This includes hydrocarbons and other chemicals. As the environmental regulator, they adopt the precautionary principle with respect to protecting groundwater due to:

- Difficulties associated with observing and remediating leaks from underground storage and transmission facilities;
- Previous history of pollution from such facilities.

## 3 Cretaceous Chalk Aquifer

The Cretaceous Chalk, a soft white limestone, is the major aquifer of southern and eastern England, supplying up to 80% of the drinking water in this densely populated region. Although not generally considered to be a karstic aquifer, the Chalk has karstic features (Price 1987; Mylroie and Carew 1988), including (Fig. 2):

- Solution-enlarged fissures, characterized by rapid groundwater travel times and high borehole yields;
- Solution features, for example, swallow holes, that focus groundwater recharge.

Consequently, the published source protection zones, based on assumption of a porous medium, do not consider preferential pathways and may underestimate the extent of SPZ1 and SPZ2 (Environment Agency 2019).

## 4 Case Study

The construction of a large petrol station was proposed, with three 75.000 l underground storage tanks, in SPZ1 and only 200 m from an abstraction borehole in eastern England used for public water supply. The abstraction supplies groundwater from the Chalk aquifer to 60.000 people and forms a strategic resource for the city of Norwich. This borehole was constructed in 2008 to replace an existing abstraction which was deemed by the Environment Agency to have unacceptable environmental impacts on nearby protected wetlands. The water company required five years and several million pounds to find and develop the replacement site, which was subsequently licensed by the Environment Agency.

During construction, borehole logging identified 'A very large 'tunnel-like' fissure at 29.8 m, with medium aperture fissures present at 26.4, 29.2, 31.6 and 32.1 m and numerous small fissures between 25.2 and 38.3 m with others at 44.1, 67.3 and 68.0 m' (Anglian Water Services 2010). This is evidence of karstic features and preferential fast flow horizons, indicating that the SPZs based on Darcian flow may be too small.

The petrol station required planning consent from the local authority (UK government 2019), with the Environment Agency as a statutory consultee. The applicant proposed tertiary containment for the tanks and associated pipework, with both construction and operation following current UK good practice (APEA and Energy Institute 2018). Although the Environment Agency initially objected to the development, it did not maintain its regulatory position and allowed reliance on engineering solutions for groundwater protection. After the water company strongly objected on the grounds of unacceptable risk to the groundwater abstraction, the elected councillors refused planning consent.

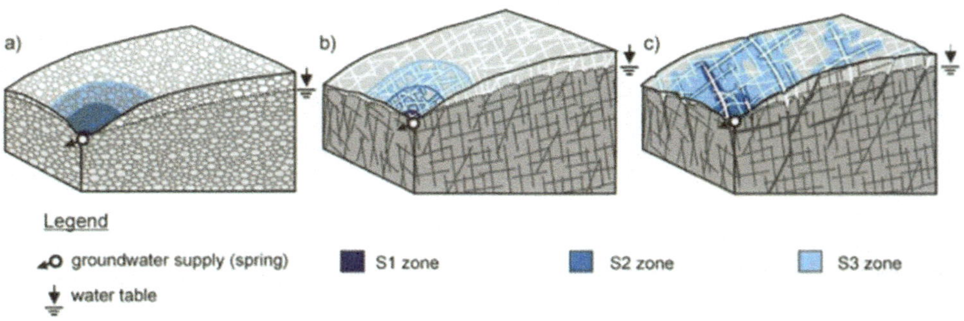


**Fig. 1** Extent of source protection zones in aquifers with increasing heterogeneity (Pochon et al. 2008)

**Fig. 2** Karstic features in Chalk, showing fissures and solution enlarged bedding planes

The applicant appealed; thus, the decision was referred to the national Planning Inspectorate for a formal hearing with representations from all parties. The key issue was whether the proposed engineering solutions were adequate to prevent leaks and ensure groundwater protection throughout the operational lifetime of the petrol station and its subsequent decommissioning.

The inspector refused the appeal, commenting that 'the implications of a leak would be very serious with the consequences being of a high magnitude to the public water supply. Although the probability of a leak could be reduced to an acceptable extent through a tertiary containment system, there was insufficient information to demonstrate that the installation would be adequately managed and monitored to maintain the low probability of a leak over the life term of the development. In the circumstances, it is sensible to take a precautionary approach.'

## 5 Concluding Remarks

The case study illustrates several issues of hydrogeoethics because the case required the hydrogeologists advising the water company, developer and regulator to apply their geoscience knowledge to protect the aquifer by managing the anthropogenic risks and potential pollution impacts of the proposed development. It also provided an opportunity to fulfil a social role and responsibilities by communicating the technical issues and potential risks clearly, in order to influence the awareness of wider society, including local residents, councils, and the planning inspector.

This case raised some strategic questions about approaches to risk, the value of groundwater, and understanding of the need for groundwater protection. It shows that the water supply companies need to be vigilant to protect their assets and customers, especially in areas where finding an alternative supply would be extremely impractical. One possible interpretation is that water supplies are sometimes considered of secondary importance to other elements of infrastructure, even in water-stressed regions.

It also revealed that the environmental regulator does not consistently implement its own guidance; thus, there is a risk that public water supplies could become contaminated through ineffective or inadequate regulation. Considering the fractured nature of many aquifers, it is apparent that Source Protection Zones are often too small, providing insufficient protection for the source, because they do not consider karstic features and preferential fast pathways for contaminant transport. This is important for the Chalk aquifer, which is not generally considered to be truly karstic although karstic features are widespread across its outcrop in southern and eastern England.

Reliance solely on engineering to mitigate all risks is not a precautionary approach in the longer term, when considering operational risks, including adequate training of staff, maintenance and monitoring, the need for prompt action in case of any leaks or spills and potential future deterioration of the structures. Fortunately, in the UK, the national planning policy framework provided another opportunity to ensure groundwater protection, as it considers wider public interests, taking a balanced view of the risks, sustainability, social implications, and benefits of developments.

**Acknowledgements** Thanks to Anglian Water Services for the permission to use the case study.

## References

Anglian Water Services (2010) Postwick Construction and Test Pump Report (unpublished report)

Association for Petroleum and Explosives Administration (APEA) and Energy Institute, London (2018) Design, construction, modification, maintenance and decommissioning of filling stations, 4th edn. Energy Institute, London

Environment Agency (2018) Groundwater protection position statements. https://www.gov.uk/government/publications/groundwater-protection-position-statements. Last accessed 15 Nov 2019

Environment Agency (2019) Manual for the production of Groundwater Source Protection Zones. https://assets.publishing.service.gov.uk/government/uploads/system/uploads/attachment_data/file/822402/Manual-for-the-production-of-Groundwater-Source-Protection-Zones.pdf. Last accessed 6 Dec 2019

Mylroie JE, Carew JL (1988) Solution conduits as indicators of Late Quaternary sea level position. Quat Sci Rev 7:55–64

Pochon A, Tripet JP, Kozel R, Meylan B, Sinreich M, Zwahlen F (2008) Groundwater protection in fractured media: a vulnerability-based approach for delineating protection zones in Switzerland. Hydrogeol J 16:1267–1281

Price M (1987) Fluid flow in the Chalk of England. In: Goff JC, Williams BPJ (eds) Fluid flow in sedimentary basins and aquifers, vol 34. Geological Society London Special Publications, pp 141–156

UK government (2019) National Planning Policy Framework. https://www.gov.uk/guidance/national-planning-policy-framework. Last accessed 15 Nov 2019

# Sustainability and Management of the Menzel Habib Aquifer System, Southeastern Tunisia

Oussama Dhaoui, Isabel Margarida Horta Ribeiro Antunes, and Belgacem Agoubi

**Abstract**

In arid and semi-arid areas, the water quantity and quality are a great problem. Salinization is the major threat in the region of Menzel Habib (north-western Gabès, southeastern Tunisia). The region is a large basin which is essentially represented by sandy-clay formations and bordered by cretaceous reliefs. Geochemical and statistical approaches are reported in the Menzel Habib aquifer system to examine groundwater salinization processes and factors controlling its mineralization. Geochemical studies were developed in 25 groundwater samples from the shallow aquifer to identify the origin of groundwater salinization. Groundwater geochemistry shows a high correlation between salinity and Na, Cl, Ca, Mg and $SO_4$. These elements are mainly associated to the evaporitic Triassic by dissolution of halite, anhydrite and gypsum which occur on the area and are related to the tectonic context of the region. Additionally, bivariate diagram between Na and Cl, and Ca and $SO_4$ have also provided a comprehensive understanding of other salinization processes that are involving in Menzel Habib shallow aquifer such as cation exchange and reverse cation exchange.

**Keywords**

Salinization • Geochemistry • Menzel Habib • Tunisia

O. Dhaoui (✉) · B. Agoubi
Higher Institute of Water Sciences and Techniques of Gabès, Gabès, Tunisia
e-mail: dhaoui.oussama2013@gmail.com

I. M. H. R. Antunes
ICT, Department of Earth Sciences, University of Minho, Braga, Portugal

## 1 Introduction

Most of the scenarios of the future water resources are predicting water scarcity, with a decrease in precipitation and limitation in groundwater recharge for the next five decades (Doll and Fiedler 2005). Thus, salinization of groundwater aquifers is one of the principal causes of degradation of water resources worldwide. This phenomenon is a major drawback for water use for irrigation and drinking water supply in arid and semi-arid regions, such as Menzel Habib. Menzel Habib aquifer is in the northwest of Gabès, southeastern Tunisia, between latitudes 34° and 34° 20′ N and longitudes 9° 15′ and 9° 58′ E. The average annual precipitation and temperature are 150 mm and 20 °C, respectively. The low recharge and the agricultural activities are an evidence to suggest that the quality of groundwater is threatened by the salinization risk and water quality problems superpose to a general quantitative limitation of freshwater availability. Previous studies have revealed that groundwater salinization could have many origins occurring alone or concomitantly, with dominant mechanisms, described as natural mechanisms, such as seawater intrusion (Kharroubi et al. 2012; Ben Hamouda et al. 2013), dissolution of evaporites (Rosenthal et al. 2007) and infiltration of saline surface water; but also anthropogenic sources (e.g., agricultural, industrial and domestic activities).

The origin groundwater salinization could be investigated using different approaches: from geochemical analysis to isotopic tracers and using statistical approaches. Indeed, the main subject of this study is to investigate the sources of groundwater salinization in Menzel Habib shallow aquifer using geochemical tools.



M. Abrunhosa et al. (eds.), *Advances in Geoethics and Groundwater Management: Theory and Practice for a Sustainable Development*, Advances in Science, Technology & Innovation,
https://doi.org/10.1007/978-3-030-59320-9_110

## 2 Geological and Hydrogeological Settings

Menzel Habib area is represented by a large syncline basin with plio-quaternary filling which is essentially represented by sandy-clay and sandy-loam formations and bordered by cretaceous reliefs. Menzel Habib area is complex and was affected by different series of geological structures (Fig. 1). The Triassic lithologies have been identified as diapiric extrusions in Hadifa Mountain and are related to the tectonic context of the region (Ouled-Ghrib and Slimane 1994).

The aquifer system of Menzel Habib is composed of three layers. The reservoir formations are distributed between quaternary deposits and limestones of the upper cretaceous (Cenomanian-Turonian and Senonian). The quaternary layer is logged in a sandy-loam formation, containing a stack of aquifer levels dependent on rain inputs that are too limited in this region, the depth of water varies from 7 to 60 m. The Senonian is logged in marl levels with limestone passages, and this aquifer is exploited in the region of Menzel Habib by some public boreholes partly meeting the needs of agricultural activities. The Cenomanian-Turonian layer contained in limestone and marl-limestone levels, and it is characterized by a strong karstification and contains a water of high salinity exceeding 19 g/L preventing its exploitation for human supply and agriculture.

## 3 Results

### 3.1 Geochemical Data

In this study, 25 water samples were taken from the shallow aquifer of Menzel Habib. Electrical conductivity (EC) and pH had been measured in the field using a C933 multi-parameter analyzer. All samples were filtered with a 0.45 µm Millipore filter paper. Bicarbonate ($HCO_3-$) was determined through the titration method with hydrochloric acid. Major cations (Na+, K+, Ca2+, Mg2+) and major anions (F-, Cl-, Br-, SO42-) were measured by an ionic liquid chromatography Metrohm 850 Professional IC in the Integrated Laboratory of Water Sciences in the Higher Institute of Water Sciences and Techniques of Gabès (Tunisia). The pH values vary from 6.95 to 7.91 with an average of 7.63 and a standard deviation of 0.168 indicating a slight alkalinity of groundwater samples. Electrical conductivity ranges from

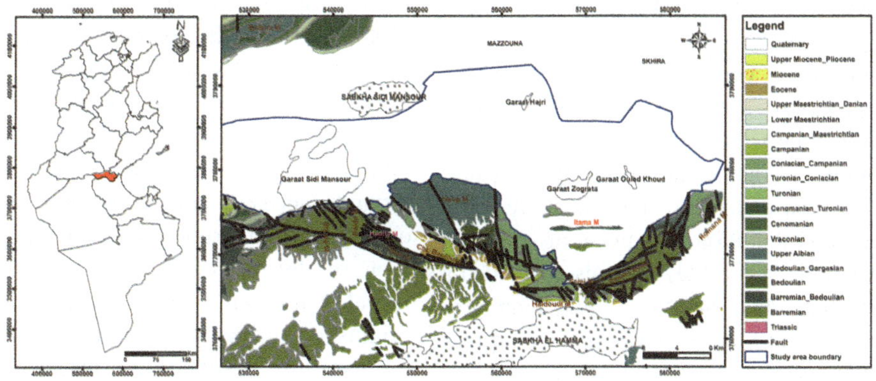


**Fig. 1** Geographical and geological setting of the study area

**Table 1** Statistical characteristics of the hydrochemical parameters

| Parameters | N | Minimum | Maximum | Mean | Std. deviation |
|---|---|---|---|---|---|
| $HCO_3$ | 25 | 66.625 | 188.651 | 137.219 | 29.562 |
| F | 25 | 0.802 | 1.254 | 0.955 | 0.133 |
| Cl | 25 | 489.628 | 3264.997 | 1415.123 | 771.375 |
| Br | 25 | 0.751 | 1.326 | 0.984 | 0.146 |
| $SO_4$ | 25 | 878.991 | 2876.339 | 1506.06 | 571.089 |
| Na | 25 | 318.591 | 2064.597 | 946.286 | 456.452 |
| K | 25 | 18.267 | 53.009 | 28.937 | 8.588 |
| Ca | 25 | 176.264 | 894.697 | 366.624 | 190.468 |
| Mg | 25 | 109.864 | 341.567 | 169.969 | 58.319 |
| EC (25 °C) | 25 | 3.060 | 14.980 | 7.836 | 2.978 |
| Ph | 25 | 6.950 | 7.910 | 7.630 | 0.168 |
| TDS | 25 | 1.598 | 8.706 | 4.332 | 1.774 |

3.06 to 14.98 ms $cm^{-1}$ with an average of 7.84 and a standard deviation of 2.98. Total dissolved solids (TDS) is heterogeneous and varies strongly from 1.598 to 8.706 g/L with an average and standard deviation of 4.33 g/L and 1.77 g/L, respectively.

The major ion composition is represented in Table 1 and shows that sodium is the major cation and sulfate is the dominant anion in groundwater composition. Piper diagram (Fig. 2) shows three types of water: Na-Cl, Na-$SO_4$ and mixed type.

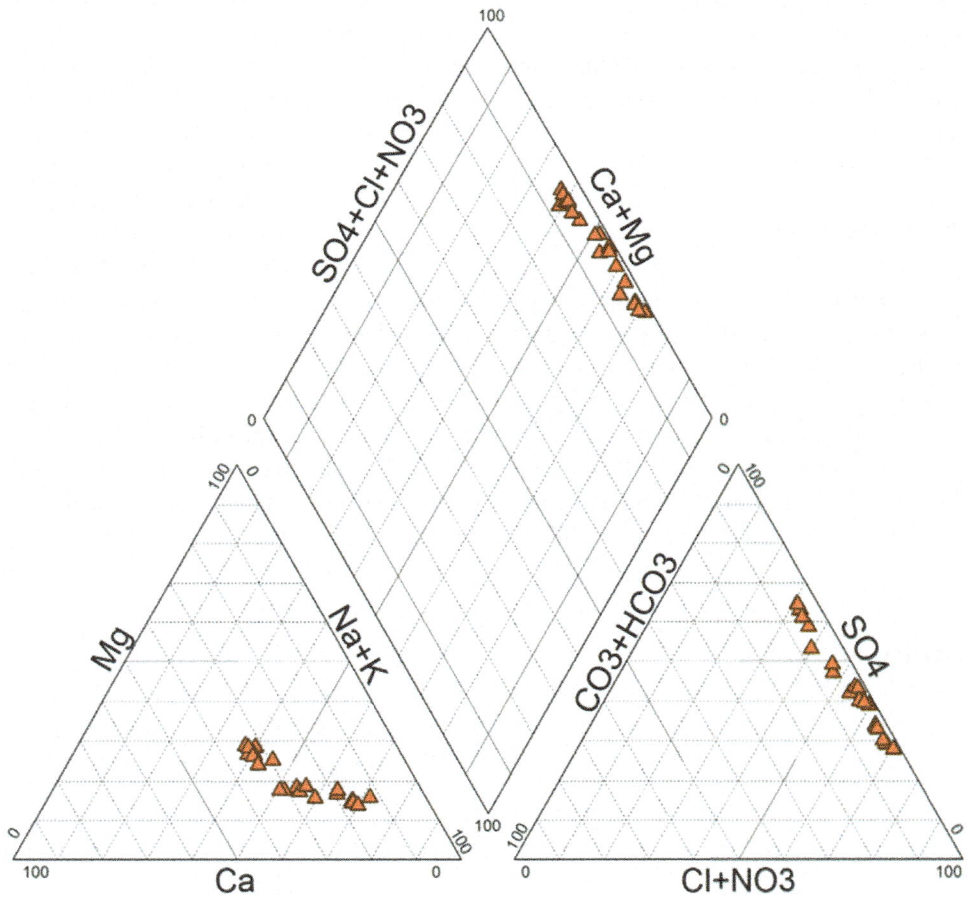


**Fig. 2** Projection of water samples in Piper diagram

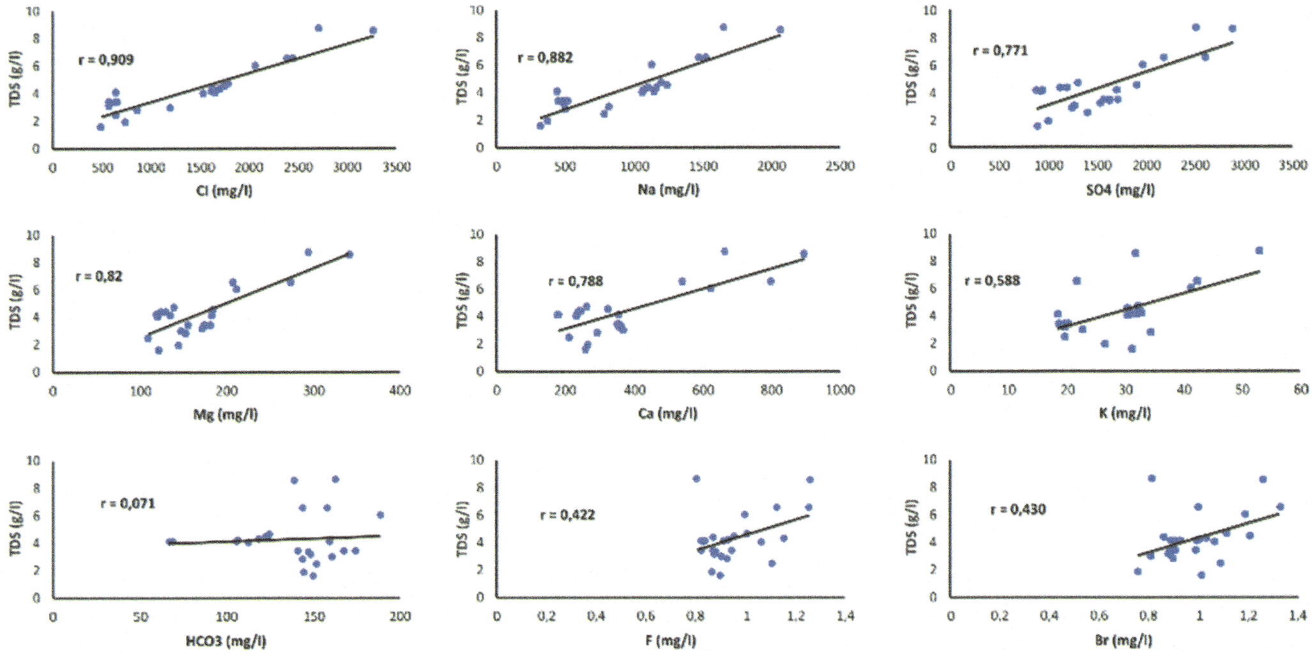


**Fig. 3** Bivariate hydrochemical relations between TDS and chemical elements

## 3.2 Spatial Correlations

To examine spatial relations between TDS and chemical elements of groundwater, bivariate diagrams were plotted. As a result, they show a positive high correlation ($r > 0.7$ at 5% significance level) between TDS and Cl, $SO_4$, Na, Mg and Ca (Fig. 3).

## 4 Discussion

The bivariate diagrams are an important tool to identify the origin of the water salinization of an aquifer. The high correlation between TDS and Cl, $SO_4$, Na, Mg and Ca may be explained by the dissolution of evaporites such as halite, gypsum and anhydrite (Agoubi et al. 2013). This process can be justified by the dissolution of halite and gypsum/anhydrite (Fig. 4). The diagram of sodium versus chloride could be identified on the distinction of the different mechanisms for acquiring salinity in arid and semi-arid regions. The positive correlation ($r = 0.977$) suggests that sodium and chloride are derived from the same origin which is the dissolution of halite that occurs in the aquifer. These elements could have its origin on the Triassic diapir of Hadifa or the evaporation dominance that characterizes the arid and semi-arid areas. However, some of groundwater samples have a relative excess or deficit of sodium content relative to the halite dissolution line, explained by the intervention of another phenomenon controlling their salinization. The deficit of sodium content which characterizes some groundwater samples indicate the presence of reverse cation exchange mechanism, which contributes to Na adsorption on clay minerals and simultaneous release of Ca. Some other water samples have an enriched Na against Cl and a deficit on Ca against $SO_4$. This relation could be explained by another mechanism that can also be involved in the contamination of Menzel Habib aquifer, which is the cation exchange phenomenon, with a release of sodium and binding of calcium.

**Fig. 4** Correlation between Na-Cl, Ca-$SO_4$

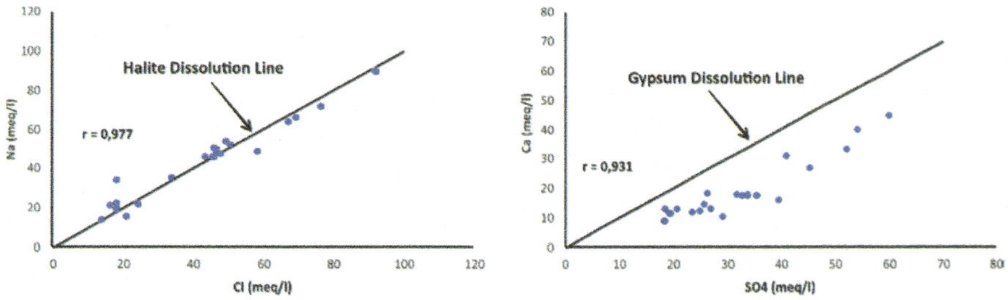

## 5 Concluding Remarks

The results of geochemical and statistical approaches have shown that the salinity of Menzel Habib shallow aquifer is due to different mechanisms, such as dissolution of evaporites, cation exchange, reverse cation exchange and dominance of evaporation that can threat the shallow groundwater. To resolve this problem, a sustainable management of aquifer is required as a main issue in the area. Indeed, the construction of additional groundwater exploitation should not be allowed in the threatened parts of the study area, and the delimitation of groundwater protection zones should be respected. The irrigation with saltwater must be avoided with the consequence of soil salinization and therefore, an increase of water salinity by infiltration. The simultaneous pumping of adjacent wells must be avoided with a daily temporal control on wells pumping. Therefore, artificial recharge can also be applied for dilution and thus decreasing of salinity.

**Acknowledgements** The authors gratefully acknowledge the contributions of the technical staff at the Laboratory Applied Hydrosciences Research Unit of Higher Institute of Water Sciences and Techniques of Gabès (Tunisia) for their help during laboratory analysis. This work is co-funded by the European Union through the European Regional Development Fund, based on COMPETE2020, project ICT (UID/GEO/04683/2013) with reference POCI- 01-0145-FEDER-007690 by FCT.

## References

Agoubi B, Kharroubi A, Ta Abichou, Abida H (2013) Hydrochemical and geoelectrical investigation of Marine Jeffara Aquifer, southeastern Tunisia. Appl Water Sci 15:415–430

Ben Hamouda MF, Carreira P, Marques JM, Eggenkamp H (2013) Geochemical and isotopic investigations to study the origin of mineralization of the coastal aquifer of Sousse, Tunisia. Proc Earth Planet Sci 7:61–64

Doll P, Fiedler M (2005) Global-scale estimation of diffuse groundwater recharge. Hydrol Earth Syst Sci 12:863–885

Kharroubi A, Tlahigue F, Agoubi B, Azri C, Bouri S (2012) Hydrochemical and statistical studies of the groundwater salinization in Mediterranean arid zones: case of the Jerba coastal aquifer in southeast Tunisia. Environ Earth Sci 67(7):2089–2100

Ouled-Ghrib A, Slimane MF (1994) Nouvelles données géologiques sur l'Atlas méridional de la Tunisie: mise en évidence du Trias dans la chaîne de Gafsa. Not Serv Géol Tunisie 60:5–10

Rosenthal E, Zilberbrand M, Livshitz Y (2007) The hydrochemical evolution of brackish groundwater in central and northern Sinai (Egypt) and in the western Negev (Israel). J Hydrol 337(3):294–314

The manufacturer's authorised representative in the EU is Springer
Nature Customer Service Centre GmbH, Europaplatz 3, 69115 Heidelberg,
Germany. If you have any concerns regarding our products, please
contact ProductSafety@springernature.com

Printed and bound by CPI Group (UK) Ltd, Croydon, CR0 4YY
30/09/2026
02225271-0006